Physical Audio Signal Processing

for Virtual Musical Instruments and Audio Effects

Julius O. Smith III

Center for Computer Research in Music and Acoustics (CCRMA)
Department of Music, Stanford University, Stanford, California 94305 USA

PHYSICAL AUDIO SIGNAL PROCESSING

FOR VIRTUAL MUSICAL INSTRUMENTS AND AUDIO EFFECTS

JULIUS O. SMITH III

W3K Publishing (http://books.w3k.org)

ISBN 978-0-9745607-2-4

Printed in the United States of America

Printed from PostScript™ files generated by LaTeX 2ε and dvips using standard 10-point book format, and the packages vmargin, fancyheadings, graphicx, subfigure, makeidx, latexsym, html, multicol, and boxedminipage

Contents Overview

Appendices

Contents

Preface

This book was developed for my course[1] entitled "Signal Processing Models in Musical Acoustics," which I have given at the Center for Computer Research in Music and Acoustics (CCRMA) every year since 1984. The course was created primarily as a research preparation and dissemination vehicle intended for graduate students in computer music and engineering interested in efficient computational modeling of musical instruments. Ideally, in addition to a first course in digital signal processing [451, 449], the student will also have studied elementary physics, including waves, and a prior first course in acoustics is desirable. The Web version[2] of this book contains hypertext links to more elementary material, thus rendering it more self contained.

The driving goal behind the research and course leading to this book is the development of "virtual musical instruments" and audio effects in the form of efficient algorithms suitable for real-time execution on general purpose computers or embedded processors. As a result, the emphasis is on "signal processing models of physical models" of musical instruments and audio effects. The starting point is typically a mathematical model of a musical instrument from the field of musical acoustics, or a circuit description of an audio effect, and the final algorithms are expressed as computational forms from the field of signal processing. In the realm of computational physics, such algorithms might be called "real-time finite-difference/solution-propagation schemes".

In one sense, this book is about how to avoid the computational expense associated with using general purpose differential equation solvers, such as most finite difference schemes, applied in a "brute force" way. In other respects, it is about the art of homing in on the "essential ingredients" of an acoustic instrument and taking advantage of "data reduction" inherent in human hearing in order to minimize computational expense. In the early days of computer music, it was not uncommon to run "acoustic

[1] `http://ccrma.stanford.edu/~jos/#courses`
[2] `http://ccrma.stanford.edu/~jos/pasp/`

compilers" orders of magnitude slower than real time to compute sound. Nowadays, computers are so fast that physical modeling synthesis can be (and is) integrated in software synthesizers running on inexpensive personal computers without special synthesizer hardware. However, to obtain the best results on a given machine, it is still necessary to simplify computational complexity relative to more general numerical simulation techniques.

As indicated in the foregoing, the material of this book is multidisciplinary, building on results from physics, musical acoustics, psychoacoustics, signal processing, control engineering, computer music, and computer science. Such diversity is typical of applied research.

Organization

The chapters are generally organized as a series of "theory" and "application" chapters, working up from delay effects through virtual musical instruments. The purpose of mixing theory and application is to put useful techniques to work soon after they are covered, instead of spending forever in preliminaries before getting to musically interesting applications. Thus, for example, acoustic modeling with delay is followed by artificial reverberation, delay-line interpolation is followed by time-varying delay effects, and so on.

The style of the chapters is relatively concise and bottom-line oriented, with more detailed coverage deferred to the appendices when reasonable. In the Web version, many technical terms are linked to associated tutorials (and this work is ongoing). All software examples in the text are freely available, and perhaps most easily obtained via copy/paste from the Web version.

For class use, the design is approximately one chapter per week, spanning a quarter. There is a significant rise in difficulty level when "lumped models" are reached, presumably due to the use of complex impedances in the Laplace and/or z domains. Extra time should be allowed to practice problems such as a point-mass colliding with an ideal string (§9.3.1). For a semester course, one could include more material on digitizing differential equations, such as in the appendices regarding finite difference schemes and/or wave digital filters. Alternatively, one could expand the final chapter entitled Virtual Musical Instruments to include more case studies.

Book Series Overview

This book is the third in my music signal processing series,[3] after [451] and [449]. The books can be loosely summarized by the following "design goals":

1. **Mathematics of the Discrete Fourier Transform**
   All about the DFT formula and its constituents, with frequent references to audio applications.

2. **Introduction to Digital Filters**
   A gentle introduction to the analysis and implementation of digital filters, with particular emphasis on audio applications.

3. **Physical Audio Signal Processing**
   Efficient computational physical models for delay effects and virtual acoustic musical instruments.

4. **Spectral Audio Signal Processing**
   Analysis, processing, and synthesis of audio signals in terms of spectral representations computed using the Fast Fourier Transform (FFT).

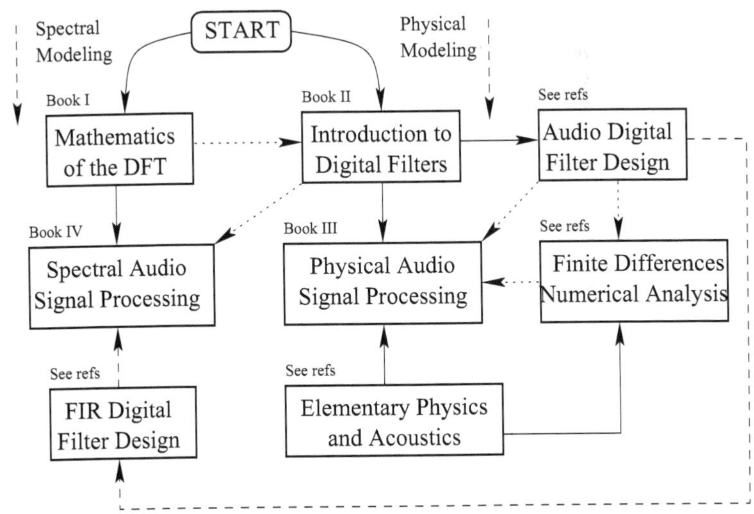

Figure 1: Schematic of interdependencies in the music signal processing book series, along with some closely related topics.

[3]http://ccrma.stanford.edu/~jos/pubs.html

Figure 1 illustrates the dependencies. A solid line indicates a strong dependence, while a dotted line indicates a weaker (optional) dependence. The student is expected to pick up elementary physics [180] and programming skills [443, 86] elsewhere. In all books, the main chapters contain approximately what is covered in class, while the appendices provide both elementary background material and additional advanced topics.

Acknowledgments

Thanks to my current and former graduate students Edgar Berdahl, Juan Pablo Caceres, Patty Huang, Reeve Ingle, Tatsuki Kashitani, Minjong Kim, John Nolting, Gary Scavone, and Matt Wright for useful feedback regarding the contents of this book. Matt Wright, in particular, contributed by far the largest number of errata notes and suggested improvements. Thanks also to netizens Arnold Bomans, Ayoub Banoushi, Niels Farver, Todd Lehman, Alastair Moore, Niels Moseley, and Bernhard Neugebauer for reporting errata they noted in the earlier Web versions. The exercises in this book were developed with much help from my former teaching assistants, Patty Huang and Pamornpol (Tak) Jinachitra. Last but certainly not least, thanks to my wife Carol and son Harrison for allowing me to bear down on this book with relatively obsessed focus at "finishing time".

Errata

Like any relatively large work, this book is sure to have some remaining typos and other errors. The Web page `http://books.w3k.org/pasp1p1/` will list known errata, clarifications, or other information pertaining to this book.

The author and publisher make no warranties, express or implied, regarding the contents of this book.

Chapter 1

Introduction to Physical Signal Models

This book is about techniques for building *real-time computational physical models* of musical instruments and audio effects. So, why would anyone want to do this, and what exactly is a "computational physical model"?

There are several reasons one might prefer a computational model in place of its real-world counterpart:

- A *virtual musical instrument* (or audio effect) is typically much less expensive than the corresponding real instrument (effect). Consider, for example, the relative expense of a piano versus its simulation in software. (We discuss real-time piano modeling in §9.4.)
- Different instruments can share common controllers (such as keyboards, wheels, pedals, etc.). Any number of virtual instruments and/or effects can be quickly loaded as "presets".
- Sound quality ("signal to noise ratio") can far exceed what is possible with a recording. This is because we can use any number of bits per sample, rendering the "noise floor" completely inaudible at all times.
- Software implementations are exactly repeatable. They never need to be "tuned" or "calibrated" like real-world devices.
- It is useful to be able to "archive" and periodically revive rare or obsolete devices in virtual form.
- The future evolution of virtual devices is less constrained than that of real devices.

But How Does It Sound?

The challenge of physical modeling, however, is that the model must really sound like the instrument or effect it is imitating. In some cases, particularly with wind instruments, this can be difficult to achieve. Thus, while noise need never be a problem, it can be difficult to generate exactly the right timbre throughout the control space. For this reason, many expressive sound synthesizers employ *both* modeling structure and measured data. One name for this is *structured sampling* (discussed further below).

What is a Model?

Model-building is a fundamental human activity. For our purposes, a *model* can be defined as any form of computation that *predicts* the behavior of a physical object or phenomenon based on its initial state and any "input" forces. Our first successful models occurred in our heads [185]. By mentally predicting future effects of seasonal climate change, for example, we were able to plan for winter food supplies. Current theories of human evolution postulate that increased brain size came about as an adaptation to larger scale climate changes, including centuries of drought. Mental simulation was later extended by *mathematical simulation*, enabling, for example, accurate prediction of astronomical events and physical phenomena at all scales. A *mathematical physical theory*, such as Newton's laws of motion, gives us a mathematical model that readily predicts the outcome of any physical experiment in the real world for which the theory is applicable. More recently, our brains have been supplemented even further by *computer simulation*, enabling formerly intractable problems to be simulated as well, thereby further extending our ability to predict. For example, it is now possible to predict even the weather some days into the future with a useful degree of accuracy.

The Basic Science Loop

The basic "science loop" consists of the following steps:

1. Observations
2. Model construction
3. Comparison of simulated observations to real-world observations
4. Model refinement
5. Go to step 3

A successful model will accurately predict the outcome of any experiment given the same input conditions and initial state.

Models for Music and Audio

For virtual musical instruments and audio effects, the model *replaces* the real thing. Therefore, the model's *prediction* from its inputs (controls and possible input sounds), is simply the desired output sound. Since a sound is a special case of a *signal*, we can say that all models of interest in this book are *signal models*.

Also in the musical case, we often desire that our models run in *real time*. That is, we may supply model inputs from a live human performer, and the task of the model becomes rendering high-quality sound based on performance gestures. Real-time computational forms are ubiquitous in the field of *digital signal processing*, and in large measure, this book is about signal processing methods most applicable to real-time computational physical models.

Physical models are desirable when *many dimensions of expressive control* are needed. This is generally best appreciated by *performing musicians*. Audience members only hear the final music played, so they are not in a position to judge the quality of multidimensional control, as skilled musicians utilize only a narrow control subspace in any given performance.[1]

1.1 Overview of Model Types

There are many kinds of signal models to consider for musical instruments and audio effects. This book emphasizes *physical signal models*, that is, signal models derived from mathematical physical descriptions of acoustic instruments or electronic effects. To set a more general context, some of the more successful non-physical signal models are mentioned below. After that, we will consider only physics-based models in this book. The next volume [457] in the music signal processing series discusses *spectral signal models*, which are models based on the encoding of sound by the ear (the "receiver") as opposed to the mechanics of sound production (the "source"). In all cases, a good signal model must have the property that it generates the desired sound in response to intuitive controls by the performing musician.

[1]This distinction is also important when evaluating real-world musical instruments. It is better to ask a skilled performer to comment on the quality of an instrument than a listener. A good musician can make almost any instrument sound good, while appreciating its defects and working around them in performance.

1.2 Signal Models

To simplify terminology, the term "signal model" will henceforth mean "non-physical signal model".

1.2.1 Recordings (Samples)

Perhaps the simplest possible signal "model" is a *recording* of the desired sound indexed by the controller state used to produce it. In other words, for each possible input condition (such as a key-press on a keyboard, pedal press, etc.) we record the sound produced in the desired acoustic space (which itself can be a recording parameter). Such a procedure is called instrument *sampling* [303]. Sampling is of course highly laborious, but it is actually the current common practice for electronic musical instruments, such as "sampled pianos".[2]

A problem with sample-based synthesizers, aside from the enormous recording effort required, is that the dimensions of expressive performance are invariably limited. While impressive results have been obtained for struck-string instruments such as the piano, continuously-controlled instruments such as bowed-string, wind, and brass instruments are reduced to highly oversimplified shadows of themselves. This can be appreciated by considering that the only non-pedal control parameters for the piano are key-number and key-velocity, while for bowed-string and wind instruments, the player manages multiple continuous dimensions of control. Skilled performers do not wish to give up these dimensions.

Another source of sonic richness routinely given up by sampling synthesis is the interaction between the performer and the instrument. For example, in a long-sustaining electric guitar performance, there is significant interaction between a ringing string and its subsequent re-excitation. Such effects are also audible (though subtle) on a piano when restriking a ringing string (with the sustain pedal down).

While sample-based sound synthesis can be frustrating (or at least constraining) for the performing musician, its best feature is the high quality of the ultimate sound to the listener. The sound quality is limited only by the quality of the original recordings and subsequent signal processing.

A major advantage of physical models, especially relative to sample-based signal models, is that the *internal state is automatically maintained.* That is, sample-based models should in principle index each acoustic[3]

---

[2]For example, the *Synthogy Ivory* (a $349 software product in 2006), ships as 40 Gigabytes on ten DVDs (three sampled pianos). Every key is sampled, with 4–10 "velocity layers", separate recordings with the soft pedal down, and separate "release" recordings, for multiple striking velocities. (Source: *Electronic Musician* magazine, October 2006)

[3]In the case of audio effects, "acoustic" recordings are normally replaced by "elec-

recording by not only the input state, but also by the internal state of the instrument (which is prohibitive and rarely done). Physical models, in contrast, propagate some kind of simulation of the internal state, so that realistic interactions between the external inputs and the internal state are provided "for free".

1.2.2 Structured Sampling

Structured sampling refers to the use of a combination of sampling and model-based methods. Instead of sampling the acoustic pressure wave, as in any typical audio recording, we sample more fundamental physical quantities such as an *impulse response* [449] that can be used to provide the desired level of both audio quality and model flexibility.

For example, in "commuted waveguide synthesis" (§8.7), the body resonator of a stringed instrument is efficiently modeled by its impulse response.

Another example is measuring the *frequency response* of a vibrating string so that a digital filter can be fit to that instead of being designed from first principles.

An advantage of sampling more fundamental characteristic signals such as impulse-responses is that they are often largely *invariant* with respect to controller state. This yields a far smaller memory footprint relative to brute force sampling of the acoustic pressure wave as a function of controller state.

There is an approximate continuum between sampling and physical modeling. That is, there is a wide range of possible hybrids between computational physical modeling and interpolation/manipulation of recorded samples. More computing power generally enables more accurate modeling and less memory usage.

1.2.3 Spectral Models

As discussed in [457], *spectral* models are inspired by the mechanics of *hearing*. Typically they are based on the *Short-Time Fourier Transform* (STFT), but there are also signal models, such as *Linear Predictive Coding* (LPC) whose success derives from how well they match spectral characteristics of hearing [457]. Additionally, *Frequency-Modulation* (FM) synthesis is typically developed by tweaking FM parameters to match short-term audio spectra [82]. Other well known signal models rooted in the spectral point of view include the phase vocoder, *additive synthesis*, and so-called

tronic" recordings. The same applies to the sampling of vintage electronic instruments, such as the Fender Rhodes electric piano, etc.

spectral modeling synthesis [457].

1.2.4 Virtual Analog

Analog synthesizers, such as the modular Moog and ARP synthesizers, typically used elementary waveforms such as sawtooth, pulse train, and the like. A variety of signal models has been developed for generating such waveforms digitally without incurring the aliasing associated with simply sampling the ideal waveforms. A special issue of the IEEE Transactions on Audio, Speech, and Language Processing is scheduled for May 2010 in this area. Specific papers accepted for publication include [517, 324]. The same issue also includes papers on more model-based approaches in which the original analog circuit is digitized using general methods [555], some of which will be discussed here in later chapters. Waveform signal models will not be discussed further in this book, so see the references for further guidance in this area [517, 324, 323, 344, 505, 506, 478, 504, 479, 67, 78].

1.3 Physical Models

We now turn to the main subject of this book, *physical models of musical instruments and audio effects.* In contrast to the non-physical signal models mentioned above, we will consider a signal model to be a *physical signal model* when there is an explicit representation of the relevant physical *state* of the sound source. For example, a string physical model must offer the possibility of exciting the string at any point along its length.

We begin with a review of physical models in general, followed by an overview of computational subtypes, with some indication of their relative merits, and what is and is not addressed in this book.

1.3.1 All We Need is Newton

Since there are no relativistic or quantum effects to worry about in musical instruments or audio effects (at least not yet), good old Newtonian mechanics will suffice for our purposes. Newton's three laws of motion can be summarized by the classic equation[4]

$$f = m\,a \qquad \text{(Force = Mass times Acceleration).}$$

However, the apparent mathematical simplicity of Newton's basic laws of motion does not mean that our task will always be easy. Models based on Newton's laws can quickly become prohibitively complex. We will usually

[4]See Appendix B for further discussion of Newton's laws of motion.

need many further simplifications that preserve both sound quality and expressivity of control.

1.3.2 Formulations

Below are various physical-model representations we will consider:

- Ordinary Differential Equations (ODE)
- Partial Differential Equations (PDE)
- Difference Equations (DE)
- Finite Difference Schemes (FDS)
- (Physical) State Space Models
- Transfer Functions (between physical signals)
- Modal Representations (Parallel Second-Order Filter Sections)
- Equivalent Circuits
- Impedance Networks
- Wave Digital Filters (WDF)
- Digital Waveguide (DW) Networks

We will briefly introduce these topics below, and the remaining chapters and appendices will develop further details.

ODEs and PDEs are purely mathematical descriptions (being differential equations), but they can be readily "digitized" to obtain computational physical models.[5] *Difference equations* are simply digitized differential equations. That is, digitizing ODEs and PDEs produces DEs. A DE may also be called a *finite difference scheme*. A discrete-time *state-space* model is a special formulation of a DE in which a vector of *state variables* is defined and propagated in a systematic way (as a vector first-order finite-difference scheme). A linear difference equation with constant coefficients—the Linear, Time-Invariant (LTI) case—can be reduced to a collection of *transfer functions*, one for each pairing of input and output signals (or a single *transfer function matrix* can relate a vector of input signal z transforms to a vector of output signal z transforms). An LTI state-space model can be *diagonalized* to produce a so-called *modal representation*, yielding a computational model consisting of a parallel bank of second-order digital filters. *Impedance networks* and their associated *equivalent circuits* are at the foundations of electrical engineering, and analog circuits have been used extensively to model linear systems and provide many useful functions. They are also useful intermediate representations

[5] *Digitizing* a system generally means converting it from continuous-time to discrete-time form. For an ODE, for example, this typically involves algebraically replacing the time differential dt in the ODE by a practical sampling interval T, as will be discussed below and in §7.3.1.

for developing computational physical models in audio. *Wave Digital Filters* (WDF) were introduced as a means of digitizing analog circuits element by element, while preserving the "topology" of the original analog circuit (a very useful property when parameters are time varying as they often are in audio effects). *Digital waveguide networks* can be viewed as highly efficient computational forms for propagating solutions to PDEs allowing wave propagation. They can also be used to "compress" the computation associated with a sum of quasi harmonically tuned second-order resonators.

All of the above techniques are discussed to varying extents in this book. The following sections provide a bit more introduction before plunging into the chapters that follow.

1.3.3 ODEs

Ordinary Differential Equations (ODEs) typically result directly from Newton's laws of motion, restated here as follows:

$$f(t) = m\,\ddot{x}(t) \qquad \text{(Force = Mass times Acceleration)}$$

This is a second-order ODE relating the force $f(t)$ on a mass m at time t to the second time-derivative of its position $x(t)$, *i.e.*, $a(t) \triangleq \ddot{x}(t) \triangleq d^2x(t)/dt^2$. A physical diagram is shown in Fig. 1.1. From this ODE we can see that a constant applied force $f(t) = f_0 > 0$ results in a constant acceleration $a(t)$, a linearly increasing velocity $v(t) = \int a(t)\,dt$, and quadratically increasing position $x(t) = \int v(t)\,dt$. The initial position $x(0)$ and velocity $v(0)$ of the mass comprise the *initial state* of mass, and serve as the *boundary conditions* for the ODE. The boundary conditions must be known in order to determine the two constants of integration needed when computing $x(t)$ for $t > 0$.

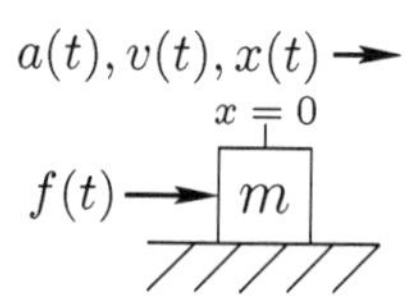

Figure 1.1: Physical diagram of an external force driving a mass sliding on a frictionless surface.

If the applied force $f(t)$ is due to a spring with spring-constant k, then we may write the ODE as

$$k\,x(t) + m\,\ddot{x}(t) = 0 \qquad \text{(Spring Force + Mass Inertial Force = 0).}$$

This case is diagrammed in Fig. 1.2.

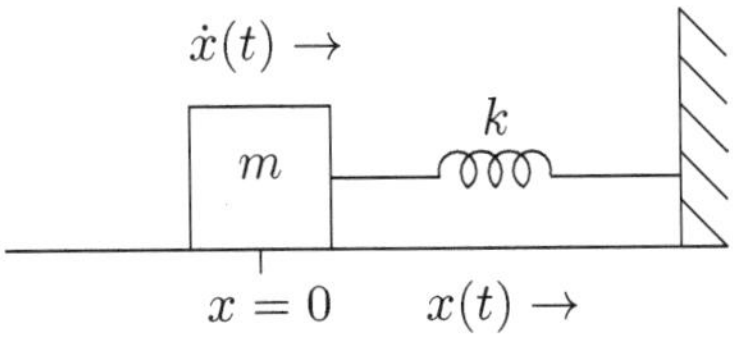

Figure 1.2: Mass-spring-wall diagram.

If the mass is sliding with *friction*, then a simple ODE model is given by

$$k\,x(t) + \mu\,\dot{x}(t) + m\,\ddot{x}(t) = 0 \qquad \text{(Spring + Friction + Inertial Forces = 0).}$$

as depicted in Fig. 1.3.

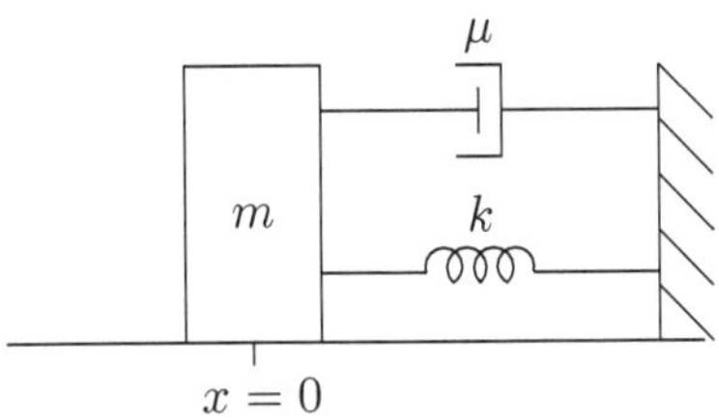

Figure 1.3: Mass-spring-dashpot-wall diagram.

We will use such ODEs to model mass, spring, and dashpot[6] elements in Chapter 7.

1.3.4 PDEs

A *partial* differential equation (PDE) extends ODEs by adding one or more independent variables (usually spatial variables). For example, the wave equation for the ideal vibrating string adds one spatial dimension x (along the axis of the string) and may be written as follows:

$$K\,y''(x,t) = \epsilon\,\ddot{y}(t) \qquad \text{(Restoring Force = Inertial Force),} \tag{1.1}$$

[6]A *dashpot* is the idealized mechanical equivalent of a *resistor* in electrical circuit theory. Its compression velocity is proportional to applied force, *i.e.*, $f_\mu = \mu\dot{x}$. Dashpots are often used to model forces due to *friction* and are typically valid over a restricted frequency range. Masses and springs are mechanical equivalents of electrical inductors and capacitors, respectively. More about these elements will be discussed below in this chapter and later in this book.

where $y(x,t)$ denotes the *transverse* displacement of the string at position x along the string and time t, and $y'(x,t) \triangleq \partial y(x,t)/\partial x$ denotes the *partial derivative* of y with respect to x.[7] The physical parameters in this case are string tension K and string mass-density ϵ. This PDE is the starting point for both digital waveguide models (Chapter 6) and finite difference schemes (§C.2.1).

1.3.5 Difference Equations (Finite Difference Schemes)

There are many methods for converting ODEs and PDEs to difference equations [53, 55, 482]. As will be discussed in §7.3, a very simple, order-preserving method is to replace each derivative with a *finite difference*:

$$\dot{x}(t) \triangleq \frac{d}{dt}x(t) \triangleq \lim_{\delta\to 0}\frac{x(t)-x(t-\delta)}{\delta} \approx \frac{x(nT)-x[(n-1)T]}{T} \tag{1.2}$$

for sufficiently small T (the sampling interval). This is formally known as the *backward difference* operation for approximating differentiation. We will discuss a variety of such methods in §7.3 and Appendix D.

As a simple example, consider a mass m driven along a frictionless surface by a driving force $f(t)$, as in Fig. 1.1, and suppose we wish to know the resulting velocity of the mass $v(t)$, assuming it starts out with position and velocity 0 at time 0 (*i.e.*, $x(0) = v(0) = 0$). Then, from Newton's $f = ma$ relation, the ODE is

$$f(t) = m\,\dot{v}(t),$$

and the difference equation resulting from the backward-difference substitution is

$$f(nT) = m\frac{v(nT)-v[(n-1)T]}{T}, \quad n = 0,1,2,\ldots.$$

Solving for $v(nT)$ yields the following *finite difference scheme:*

$$v(nT) = v[(n-1)T] + \frac{T}{m}f(nT), \quad n = 0,1,2,\ldots \tag{1.3}$$

with $v(-T) \triangleq 0$. In general, the driving force f could depend on the current state of the system (*e.g.*, if a spring and/or dashpot were introduced). In such a case, Eq. (1.3) may not be *computable.* (A *delay-free loop* could appear in the signal flow diagram.) Also, a finite force at time t cannot produce an instantaneous velocity at time t, so Eq. (1.3) is not "physical" in

[7] Transverse displacement is displacement in a plane orthogonal to the string axis x.

that sense, since $v(nT)$ depends on $f(nT)$. To address both of these issues, we can instead use the *forward difference* approximation to the derivative:

$$\dot{x}(t) = \lim_{\delta\to 0} \frac{x(t+\delta)-x(t)}{\delta} \approx \frac{x[(n+1)T]-x(nT)}{T} \qquad (1.4)$$

As $T \to 0$, the forward and backward difference operators approach the same limit (since $x(t)$ is presumed continuous). Using this we obtain what is called an *explicit finite difference scheme:*

$$v[(n+1)T] = v(nT) + \frac{T}{m} f(nT), \quad n = 0, 1, 2, \dots \qquad (1.5)$$

with $v(0) \triangleq 0$.

A finite difference scheme is said to be *explicit* when it can be computed forward in time in terms of quantities from previous time steps, as in this example. Thus, an explicit finite difference scheme can be implemented in real time as a *causal digital filter.*

There are also *implicit* finite-difference schemes which may correspond to *non-causal* digital filters [449]. Implicit schemes are generally solved using iterative and/or matrix-inverse methods, and they are typically used *offline* (not in real time) [556].

There is also an interesting class of explicit schemes called *semi-implicit finite-difference schemes* which are obtained from an implicit scheme by imposing a fixed upper limit on the number of iterations in, say, Newton's method for iterative solution [556]. Thus, any implicit scheme that can be quickly solved by iterative methods can be converted to an explicit scheme for real-time usage. One technique for improving the iterative convergence rate is to work at a very high sampling rate, and initialize the iteration for each sample at the solution for the previous sample [556].

In this book, we will be concerned almost exclusively with *explicit* linear finite-difference schemes, *i.e.*, *causal digital filter models* of one sort or another. That is, the main thrust is to obtain as much "physical modeling power" as possible from ordinary digital filters and delay lines. We will also be able to easily add memoryless nonlinearities where needed (such as implemented by table look-ups and short polynomial evaluations) as a direct result of the physical meaning of the signal samples.

1.3.6 State Space Models

Equations of motion for any physical system may be conveniently formulated in terms of the *state* of the system [330]:

$$\underline{\dot{x}}(t) = f_t[\underline{x}(t), \underline{u}(t)] \qquad (1.6)$$

Here, $\underline{x}(t)$ denotes the *state* of the system at time t, $\underline{u}(t)$ is a vector of *external inputs* (typically forces), and the general vector function f_t specifies how the current state $\underline{x}(t)$ and inputs $\underline{u}(t)$ cause a change in the state at time t by affecting its time derivative $\dot{\underline{x}}(t)$. Note that the function f_t may itself be time varying in general. The model of Eq. (1.6) is extremely general for causal physical systems. Even the functionality of the human brain is well cast in such a form.

Equation (1.6) is diagrammed in Fig. 1.4.

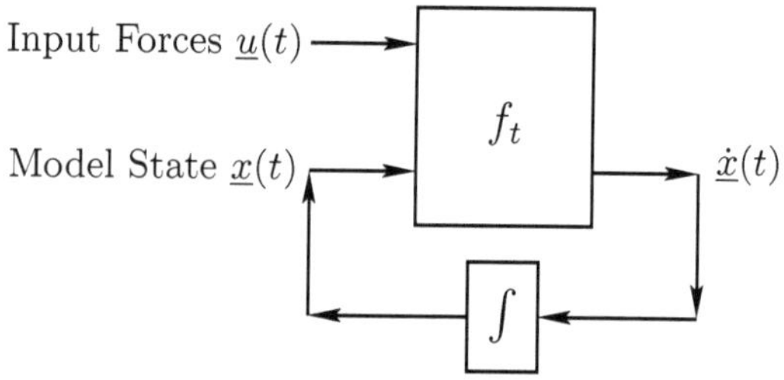

Figure 1.4: Continuous-time state-space model $\dot{\underline{x}}(t) = f_t[\underline{x}(t), \underline{u}(t)]$.

The key property of the state vector $\underline{x}(t)$ in this formulation is that it *completely determines the system at time t*, so that future states depend only on the current state and on any inputs at time t and beyond.[8] In particular, all past states and the entire input history are "summarized" by the current state $\underline{x}(t)$. Thus, $\underline{x}(t)$ must include all "memory" of the system.

Forming Outputs

Any system *output* is some function of the state, and possibly the input (directly):

$$\underline{y}(t) \triangleq o_t[\underline{x}(t), \underline{u}(t)]$$

The general case of output extraction is shown in Fig. 1.5.

The output signal (vector) is most typically a *linear combination* of

[8] It is interesting to note that in the development of *quantum mechanics* in the early 1900s, it was necessary to replace deterministic Newtonian dynamics with a *probabilistic* model. In quantum mechanics, probability distributions may follow deterministic trajectories as in Newtonian mechanics (see, for example, Schrödinger's equation), but they are only probability distributions, so there is no deterministic "clock works" at the smallest physical scales. We are fortunate to be able to use Newtonian dynamics with such great accuracy. Our difficulty will be complexity, not randomness. Paradoxically, the typical way to deal with overwhelming complexity is to model it as random (*e.g.*, filtered white noise).

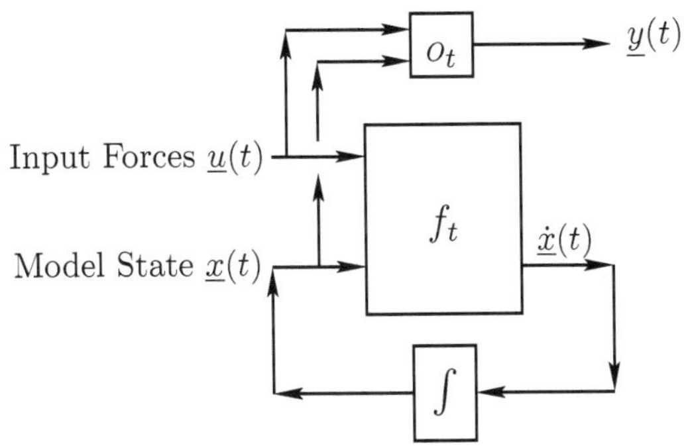

Figure 1.5: Continuous-time state-space model with output vector $\underline{y}(t) = o_t[\underline{x}(t), \underline{u}(t)]$.

state variables and possibly the current input:

$$\underline{y}(t) \triangleq C\underline{x}(t) + D\underline{u}(t)$$

where C and D are constant matrices of linear-combination coefficients.

State-Space Model of a Force-Driven Mass

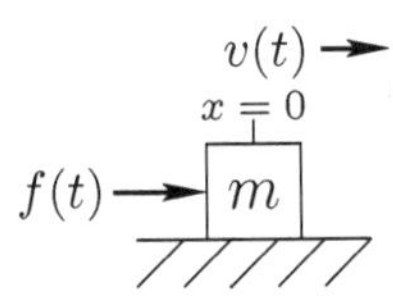

Figure 1.6: Ideal mass m on frictionless surface driven by force $f(t)$.

For the simple example of a mass m driven by external force f along the x axis, we have the system of Fig. 1.6. We should choose the state variable to be velocity $v = \dot{x}$ so that Newton's $f = ma$ yields

$$\dot{v} = \frac{1}{m} f.$$

This is a first-order system (no vector needed). We'll look at a simple vector example below in §1.3.7.

Numerical Integration of General State-Space Models

An approximate discrete-time numerical solution of Eq. (1.6) is provided by

$$\underline{x}(t_n + T_n) = \underline{x}(t_n) + T_n\, f[\underline{x}(t_n), \underline{u}(t_n)], \quad n = 0, 1, 2, \ldots. \tag{1.7}$$

Let

$$g_{t_n}[\underline{x}(t_n), \underline{u}(t_n)] \triangleq \underline{x}(t_n) + T_n\, f_{t_n}[\underline{x}(t_n), \underline{u}(t_n)].$$

Then we can diagram the time-update as in Fig. 1.7. In this form, it is clear that g_{t_n} *predicts* the next state $\underline{x}(t_n + T_n)$ as a function of the current state $\underline{x}(t_n)$ and current input $\underline{u}(t_n)$. In the field of computer science, computations having this form are often called *finite state machines* (or simply *state machines*), as they compute the next state given the current state and external inputs.

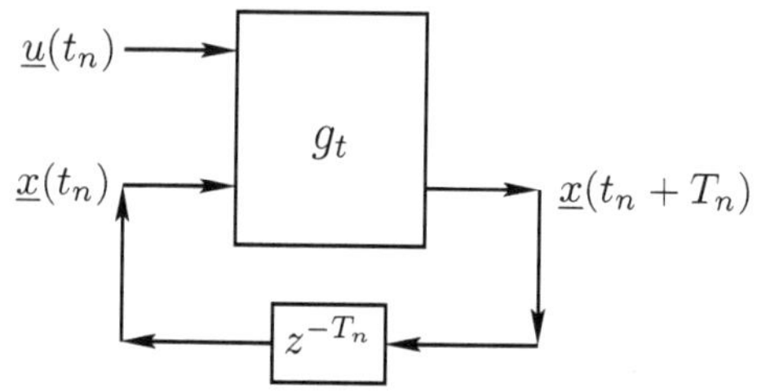

Figure 1.7: Discrete-time state-space model viewed as a state predictor, or finite state machine.

This is a simple example of *numerical integration* for solving an ODE, where in this case the ODE is given by Eq. (1.6) (a very general, potentially nonlinear, vector ODE). Note that the initial state $\underline{x}(t_0)$ is required to start Eq. (1.7) at time zero; the initial state thus provides boundary conditions for the ODE at time zero. The time sampling interval T_n may be fixed for all time as $T_n = T$ (as it normally is in linear, time-invariant digital signal processing systems), or it may vary adaptively according to how fast the system is changing (as is often needed for nonlinear and/or time-varying systems). Further discussion of nonlinear ODE solvers is taken up in §7.4, but for most of this book, linear, time-invariant systems will be emphasized.

Note that for handling *switching states* (such as op-amp comparators and the like), the discrete-time state-space formulation of Eq. (1.7) is more conveniently applicable than the continuous-time formulation in Eq. (1.6).

State Definition

In view of the above discussion, it is perhaps plausible that the *state* $\underline{x}(t) = [x_1(t), \ldots, x_N(t)]^T$ of a physical system at time t can be defined as a collection of *state variables* $x_i(t)$, wherein each state variable $x_i(t)$ is a *physical amplitude* (pressure, velocity, position, ...) corresponding to a *degree of freedom* of the system. We define a *degree of freedom* as a single dimension of *energy storage*. The net result is that it is possible to compute the stored energy in any degree of freedom (the system's "memory") from its corresponding state-variable amplitude.

For example, an *ideal mass* m can store only *kinetic energy* $E_m = \frac{1}{2}m\,v^2$, where $v = \dot{x}$ denotes the mass's velocity along the x axis. Therefore, *velocity* is the natural choice of state variable for an ideal point-mass. Coincidentally, we reached this conclusion independently above by writing $f = ma$ in state-space form $\dot{v} = (1/m)f$. Note that a point mass that can move freely in 3D space has three degrees of freedom and therefore needs three state variables (v_x, v_y, v_z) in its physical model. In typical models from musical acoustics (*e.g.*, for the piano hammer), masses are allowed only one degree of freedom, corresponding to being constrained to move along a 1D line, like an ideal spring. We'll study the ideal mass further in §7.1.2.

Another state-variable example is provided by an *ideal spring* described by Hooke's law $f = kx$ (§B.1.3), where k denotes the spring constant, and x denotes the spring displacement from rest. Springs thus contribute a force proportional to displacement in Newtonian ODEs. Such a spring can only store the physical *work* (force times distance), expended to displace, it in the form of *potential energy* $E_k = \frac{1}{2}k\,x^2$. More about ideal springs will be discussed in §7.1.3. Thus, *spring displacement* is the most natural choice of state variable for a spring.

In so-called RLC electrical circuits (consisting of resistors R_i, inductors L_i, and capacitors C_i), the state variables are typically defined as all of the capacitor voltages (or charges) and inductor currents. We will discuss RLC electrical circuits further below.

There is no state variable for each resistor current in an RLC circuit because a resistor dissipates energy but does not store it—it has no "memory" like capacitors and inductors. The state (current I, say) of a resistor R is determined by the voltage V across it, according to Ohm's law $V = IR$, and that voltage is supplied by the capacitors, inductors, and voltage-sources, etc., to which it is connected. Analogous remarks apply to the *dashpot*, which is the mechanical analog of the resistor—we do not assign state variables to dashpots. (If we do, such as by mistake, then we will obtain state variables that are linearly dependent on other state variables, and the order of the system appears to be larger than it really is. This does not normally cause problems, and there are many numerical ways to later "prune" the state down to its proper order.)

Masses, springs, dashpots, inductors, capacitors, and resistors are examples of so-called *lumped* elements. Perhaps the simplest *distributed* element is the continuous ideal delay line. Because it carries a *continuum* of independent amplitudes, the order (number of state variables) is *infinity* for a continuous delay line of any length! However, in practice, we often work with *sampled, bandlimited* systems, and in this domain, delay lines have a finite number of state variables (one for each delay element). Networks of lumped elements yield finite-order state-space models, while even one distributed element jumps the order to infinity until it is bandlimited and sampled.

In summary, a state variable may be defined as a physical amplitude for some energy-storing degree of freedom. In models of mechanical systems, a state variable is needed for each ideal spring and point mass (times the number of dimensions in which it can move). For RLC electric circuits, a state variable is needed for each capacitor and inductor. If there are any switches, their state is also needed in the state vector (*e.g.*, as boolean variables). In discrete-time systems such as digital filters, each unit-sample delay element contributes one (continuous) state variable to the model.

1.3.7 Linear State Space Models

As introduced in Book II [449, Appendix G], in the linear, time-invariant case, a discrete-time *state-space model* looks like a vector first-order finite-difference model:

$$\begin{aligned} \underline{y}(n) &= C\underline{x}(n) + D\underline{u}(n) \\ \underline{x}(n+1) &= A\underline{x}(n) + B\underline{u}(n) \end{aligned} \tag{1.8}$$

where $\underline{x}(n)$ is the length N *state vector* at discrete time n, $\underline{u}(n)$ is in general a $p \times 1$ vector of inputs, and $\underline{y}(n)$ the $q \times 1$ output vector. A is the $N \times N$ *state transition matrix*, and it determines the *dynamics* of the system (its *poles*, or *modal resonant frequencies and damping*).

The state-space representation is especially powerful for *multi-input, multi-output* (MIMO) linear systems, and also for *time-varying* linear systems (in which case any or all of the matrices in Eq. (1.8) may have time subscripts n) [220].

To cast the previous force-driven mass example in state-space form, we may first observe that the state of the mass is specified by its velocity $v(t)$ and position $x(t) = \int v(t)\,dt$, or $x[(n+1)T] = x(nT) + T\,v(nT)$.[9] Thus, to Eq. (1.5) we may add the explicit difference equation

$$x[(n+1)T] \;=\; x(nT) + T\,v(nT) \;=\; x(nT) + T\,v[(n-1)T] + \frac{T^2}{m} f[(n-1)T]$$

[9] As discussed in §1.3.6 on page 14, only velocity $v(t)$ is needed for the state variable of a mass, since a mass moving in one dimension has only one degree of freedom (with energy $mv^2/2$). However, since it is physically reasonable to expect both velocity and position to be needed for the state of a point mass, let's "play that out" and see how it goes. This also gives us a simple *vector* example to study.

which, in canonical state-space form, becomes (letting $x_n \triangleq x(nT)$, etc., for notational simplicity)

$$\left[\begin{array}{c} x_{n+1} \\ v_{n+1} \end{array}\right] = \left[\begin{array}{cc} 1 & T \\ 0 & 1 \end{array}\right] \left[\begin{array}{c} x_n \\ v_n \end{array}\right] + \left[\begin{array}{c} 0 \\ T/m \end{array}\right] f_n, \quad n = 0, 1, 2, \ldots, \tag{1.9}$$

with $x_0 = v_0 = 0$ being a typical initial state.

General features of this example are that the entire physical state of the system is collected together into a single vector, and the elements of the (A, B, C, D) matrices include physical parameters (and the sampling interval, in the discrete-time case). The parameters may also vary with time (time-varying systems), or be functions of the state (nonlinear systems).

The general procedure for building a state-space model is to label all the state variables and collect them into a vector $\underline{x}$, and then work out the state-transition matrix A, input gains B, output gains C, and any direct coefficient D. A state variable $\underline{x}_i(n)$ is needed for each lumped energy-storage element (mass, spring, capacitor, inductor), and one for each sample of delay in sampled distributed systems. After that, various equivalent (but numerically preferable) forms can be generated by means of *similarity transformations* [449, pp. 360–374]. We will make sparing use of state-space models in this book, because they can be linear-algebra intensive, and therefore rarely used in practical real-time signal processing systems for music and audio effects. However, the state-space framework is an important general-purpose tool that should be kept in mind [220], and there is extensive support for state-space models in the matlab ("matrix laboratory") language and its libraries. We will use it mainly as an analytical tool from time to time.

As noted earlier, a point mass only requires a first-order model:

$$v_{n+1} = v_n + \frac{T}{m} f_n$$

Position x_n can always be computed from initial position x_0 and velocity v_n by numerical integration:

$$x_{n+1} = x_n + T\, v_n$$

Impulse Response of State Space Models

As derived in Book II [449, Appendix G], the impulse response of the state-space model can be summarized as

$$\mathbf{h}(n) = \begin{cases} D, & n = 0 \\ CA^{n-1}B, & n > 0 \end{cases} \tag{1.10}$$

Thus, the nth "sample" of the impulse response is given by $CA^{n-1}B$ for $n \geq 0$. Each such "sample" is a $p \times q$ matrix, in general.

In our force-driven-mass example, we have $p = q = 1$, $B = [0, T/m]^T$, and $D = 0$. For a position output we have $C = [1, 0]$ while for a velocity output we would set $C = [0, 1]$. Choosing $C = \mathbf{I}$ simply feeds the whole state vector to the output, which allows us to look at both simultaneously:

$$\begin{aligned}\mathbf{h}(n+1) &= \left[\begin{array}{cc} 1 & 0 \\ 0 & 1 \end{array}\right]\left[\begin{array}{cc} 1 & T \\ 0 & 1 \end{array}\right]^n\left[\begin{array}{c} 0 \\ T/m \end{array}\right] \\ &= \left[\begin{array}{cc} 1 & nT \\ 0 & 1 \end{array}\right]\left[\begin{array}{c} 0 \\ T/m \end{array}\right] \\ &= \frac{T}{m}\left[\begin{array}{c} nT \\ 1 \end{array}\right]\end{aligned}$$

Thus, when the input force is a *unit pulse*, which corresponds physically to imparting momentum T at time 0 (because the time-integral of force is momentum and the physical area under a unit sample is the sampling interval T), we see that the velocity after time 0 is a constant $v_n = T/m$, or $m v_n = T$, as expected from conservation of momentum. If the velocity is constant, then the position must grow linearly, as we see that it does: $x_{n+1} = n(T^2/m)$. The finite difference approximation to the time-derivative of $x(t)$ now gives $(x_{n+1} - x_n)/T = T/m = v_n$, for $n \geq 0$, which is consistent.

Zero-Input Response of State Space Models

The response of a state-space model Eq. (1.8) to *initial conditions*, *i.e.*, its *initial state* $\underline{x}(0)$, is given by

$$\underline{y}_x(n) \;=\; CA^{n-1}\underline{x}(0), \quad n = 0, 1, 2, \ldots,$$

and the *complete response* of a linear system is given by the sum of its *forced response* (such as the impulse response) and its *initial-condition response.*

In our force-driven mass example, with the external force set to zero, we have, from Eq. (1.9) or Eq. (1.11),

$$\left[\begin{array}{c} x_{n+1} \\ v_{n+1} \end{array}\right] = \left[\begin{array}{cc} 1 & T \\ 0 & 1 \end{array}\right]\left[\begin{array}{c} x_n \\ v_n \end{array}\right] = \left[\begin{array}{cc} 1 & T \\ 0 & 1 \end{array}\right]^{n-1}\left[\begin{array}{c} x_0 \\ v_0 \end{array}\right] = \left[\begin{array}{c} x_0 + v_0 nT \\ v_0 \end{array}\right].$$

Thus, any initial velocity v_0 remains unchanged, as physically expected. The initial position x_0 remains unchanged if the initial velocity is zero. A nonzero initial velocity results in a linearly growing position, as physically expected. This response to initial conditions can be added to any forced response by superposition. The forced response may be computed as the convolution of the input driving force f_n with the impulse response Eq. (1.11).

1.3.8 Transfer Functions

As developed in Book II [449], a discrete-time *transfer function* is the z transform of the *impulse response* of a linear, time-invariant (LTI) system. In a physical modeling context, we must specify the input and output signals we mean for each transfer function to be associated with the LTI model. For example, if the system is a simple mass sliding on a surface, the input signal could be an external applied force, and the output could be the velocity of the mass in the direction of the applied force. In systems containing many masses and other elements, there are many possible different input and output signals. It is worth emphasizing that a system can be reduced to a set of transfer functions only in the LTI case, or when the physical system is at least *nearly linear* and only *slowly* time-varying (compared with its impulse-response duration).

As we saw in the previous section, the state-space formulation nicely organizes all possible input and output signals in a linear system. Specifically, for inputs, each input signal is multiplied by a "B vector" (the corresponding column of the B matrix) and added to the state vector; that is, each input signal may be arbitrarily scaled and added to any state variable. Similarly, each state variable may be arbitrarily scaled and added to each output signal via the row of the C matrix corresponding to that output signal.

Using the closed-form sum of a matrix geometric series (again as detailed in Book II), we may easily calculate the transfer function of the state-space model of Eq. (1.8) above as the z transform of the impulse response given in Eq. (1.10) above:

$$H(z) \;=\; D + C\,(zI - A)^{-1}\,B \tag{1.11}$$

Note that if there are p inputs and q outputs, $H(z)$ is a $p \times q$ *transfer-function matrix* (or "matrix transfer function").

In the force-driven-mass example of the previous section, defining the input signal as the driving force f_n and the output signal as the mass velocity v_n, we have $p = q = 1$, $B = [0, T/m]^T$, $C = [0, 1]$, and $D = 0$, so that the force-to-velocity transfer function is given by

$$\begin{aligned}
H(z) &= D + C\,(zI - A)^{-1}\,B \\
&= \begin{bmatrix} 0 & 1 \end{bmatrix} \begin{bmatrix} z-1 & -T \\ 0 & z-1 \end{bmatrix}^{-1} \begin{bmatrix} 0 \\ T/m \end{bmatrix} \\
&= \frac{1}{(z-1)^2} \begin{bmatrix} 0 & 1 \end{bmatrix} \begin{bmatrix} z-1 & T \\ 0 & z-1 \end{bmatrix} \begin{bmatrix} 0 \\ T/m \end{bmatrix} \\
&= \frac{T}{m}\frac{z-1}{(z-1)^2} = \boxed{\frac{T}{m}\frac{z^{-1}}{1-z^{-1}}}.
\end{aligned}$$

Thus, the force-to-velocity transfer function is a one-pole filter with its pole at $z = 1$ (an integrator). The unit-sample delay in the numerator guards against delay-free loops when this element (a mass) is combined with other elements to build larger filter structures.

Similarly, the force-to-position transfer function is a two-pole filter:

$$\begin{aligned}
H(z) &= \frac{1}{(z-1)^2} \begin{bmatrix} 1 & 0 \end{bmatrix} \begin{bmatrix} z-1 & T \\ 0 & z-1 \end{bmatrix} \begin{bmatrix} 0 \\ T/m \end{bmatrix} \\
&= \frac{T^2/m}{(z-1)^2} = \frac{T^2}{m}\frac{z^{-2}}{1-2z^{-1}+z^{-2}}
\end{aligned}$$

Now we have two poles on the unit circle at $z = 1$, and the impulse response of this filter is a ramp, as already discovered from the previous impulse-response calculation.

Once we have transfer-function coefficients, we can realize any of a large number of digital filter types, as detailed in Book II [449, Chapter 9].

1.3.9 Modal Representation

One of the filter structures introduced in Book II [449, p. 209] was the *parallel second-order filter bank*, which may be computed from the general transfer function (a ratio of polynomials in z) by means of the *Partial Fraction Expansion* (PFE) [449, p. 129]:

$$H(z) \triangleq \frac{B(z)}{A(z)} = \sum_{i=1}^{N} \frac{r_i}{1 - p_i z^{-1}} \tag{1.12}$$

where

$$\begin{aligned}
B(z) &= b_0 + b_1 z^{-1} + b_2 z^{-2} + \cdots + b_M z^{-M} \\
A(z) &= 1 + a_1 z^{-1} + a_2 z^{-2} + \cdots + a_N z^{-N}, \quad M < N
\end{aligned}$$

The PFE Eq. (1.12) expands the (strictly proper[10]) transfer function as a parallel bank of (complex) *first-order* resonators. When the polynomial coefficients b_i and a_i are real, complex poles p_i and residues r_i occur in conjugate pairs, and these can be combined to form second-order sections [449, p. 131]:

$$\begin{aligned} H_i(z) &= \frac{r_i}{1-p_i z^{-1}} + \frac{\overline{r_i}}{1-\overline{p_i} z^{-1}} = \frac{r_i - r_i\overline{p_i}z^{-1} + \overline{r_i} - \overline{r_i}p_i z^{-1}}{(1-p_i z^{-1})(1-\overline{p_i}z^{-1})} \\ &= \frac{2\text{re}\left\{r_i\right\} - 2\text{re}\left\{r_i\overline{p_i}\right\}z^{-1}}{1-2\text{re}\left\{p_i\right\}z^{-1} + \left|p_i\right|^2 z^{-2}} = 2G_i\frac{\cos(\phi_i) - \cos(\phi_i-\theta_i)z^{-1}}{1-2R_i\cos(\theta_i)z^{-1} + R_i^2 z^{-2}}. \end{aligned}$$

where $p_i \triangleq R_i e^{j\theta_i}$ and $r_i \triangleq G_i e^{j\phi_i}$. Thus, every transfer function $H(z)$ with real coefficients can be realized as a parallel bank of real first- and/or second-order digital filter sections, as well as a parallel FIR branch when $M \geq N$.

As we will develop in §8.5, *modal synthesis* employs a "source-filter" synthesis model consisting of some driving signal into a parallel filter bank in which each filter section implements the transfer function of some *resonant mode* in the physical system. Normally each section is second-order, but it is sometimes convenient to use larger-order sections; for example, fourth-order sections have been used to model piano partials in order to have beating and two-stage-decay effects built into each partial individually [30, 29].

For example, if the physical system were a row of *tuning forks* (which are designed to have only one significant resonant frequency), each tuning fork would be represented by a single (real) second-order filter section in the sum. In a modal vibrating string model, each second-order filter implements one "ringing partial overtone" in response to an excitation such as a finger-pluck or piano-hammer-strike.

State Space to Modal Synthesis

The partial fraction expansion works well to create a modal-synthesis system from a transfer function. However, this approach can yield inefficient realizations when the system has multiple inputs and outputs, because in that case, each element of the transfer-function *matrix* must be separately expanded by the PFE. (The poles are the same for each element, unless they are canceled by zeros, so it is really only the residue calculations that must be carried out for each element.)

If the second-order filter sections are realized in direct-form-II or transposed-direct-form-I (or more generally in any form for which the poles effectively precede the zeros), then the poles can be *shared* among all the

[10] See [449, p. 133] regarding the cases $M \geq N$ for which a "long division" is first performed to obtain an FIR part in parallel with a strictly proper part.

outputs for each input, since the poles section of the filter from that input to each output sees the same input signal as all others, resulting in the same filter state. Similarly, the recursive portion can be shared across all inputs for each output when the filter sections have poles implemented after the zeros in series; one can imagine "pushing" the identical two-pole filters through the summer used to form the output signal. In summary, when the number of inputs exceeds the number of outputs, the poles are more efficiently implemented before the zeros and shared across all outputs for each input, and vice versa. This paragraph can be summarized symbolically by the following matrix equation:

$$\begin{bmatrix} y_1 \\ y_2 \end{bmatrix} = \frac{1}{A} \begin{bmatrix} B_1 & B_2 \\ B_3 & B_4 \end{bmatrix} \begin{bmatrix} u_1 \\ u_2 \end{bmatrix} = \begin{bmatrix} B_1 & B_2 \\ B_3 & B_4 \end{bmatrix} \left\{ \frac{1}{A} \begin{bmatrix} u_1 \\ u_2 \end{bmatrix} \right\}$$

What may not be obvious when working with transfer functions alone is that it is possible to share the poles across all of the inputs *and* outputs! The answer? Just *diagonalize* a state-space model by means of a *similarity transformation* [449, p. 360]. This will be discussed a bit further in §8.5. In a diagonalized state-space model, the A matrix is diagonal.[11] The B matrix provides routing and scaling for all the input signals driving the modes. The C matrix forms the appropriate linear combination of modes for each output signal. If the original state-space model is a physical model, then the transformed system gives a parallel filter bank that is excited from the inputs and observed at the outputs in a physically correct way.

Force-Driven-Mass Diagonalization Example

To diagonalize our force-driven mass example, we may begin with its state-space model Eq. (1.9) on page 17:

$$\begin{bmatrix} x_{n+1} \\ v_{n+1} \end{bmatrix} = \begin{bmatrix} 1 & T \\ 0 & 1 \end{bmatrix} \begin{bmatrix} x_n \\ v_n \end{bmatrix} + \begin{bmatrix} 0 \\ T/m \end{bmatrix} f_n, \quad n = 0, 1, 2, \dots$$

which is in the general state-space form $\underline{x}(n+1) = A\,\underline{x}(n) + B\,\underline{u}(n)$ as needed (Eq. (1.8) on page 16). We can see that A is already a Jordan block of order 2 [449, p. 368]. (We can change the T to 1 by scaling the physical units of $x_2(n)$.) Thus, the system is already as diagonal as it's going to get. We have a repeated pole at $z = 1$, and they are effectively in series (instead of parallel), thus giving a "defective" A matrix [449, p. 136].

[11] More precisely, A is diagonal when the poles are *distinct*. A repeated pole can result in a block of A having 1s along its superdiagonal.

Typical State-Space Diagonalization Procedure

As discussed in [449, p. 362] and exemplified in §C.17.6 on page 627, to diagonalize a system, we must find the eigenvectors of A by solving

$$A\underline{e}_i = \lambda_i \underline{e}_i$$

for $\underline{e}_i$, $i = 1, 2$, where λ_i is simply the ith pole (eigenvalue of A). The N eigenvectors $\underline{e}_i$ are collected into a *similarity transformation matrix:*

$$E = \left[\begin{array}{cccc} \underline{e}_1 & \underline{e}_2 & \cdots & \underline{e}_N \end{array}\right]$$

If there are coupled repeated poles, the corresponding missing eigenvectors can be replaced by generalized eigenvectors.[12] The E matrix is then used to diagonalize the system by means of a simple *change of coordinates:*

$$\underline{x}(n) \stackrel{\Delta}{=} E\,\underline{\tilde{x}}(n)$$

The new diagonalized system is then

$$\begin{aligned} \underline{\tilde{x}}(n+1) &= \tilde{A}\,\underline{\tilde{x}}(n) + \tilde{B}\,\underline{u}(n) \\ \underline{y}(n) &= \tilde{C}\,\underline{\tilde{x}}(n) + \tilde{D}\,\underline{u}(n), \end{aligned} \tag{1.13}$$

where

$$\begin{aligned} \tilde{A} &= E^{-1}AE \\ \tilde{B} &= E^{-1}B \\ \tilde{C} &= CE \\ \tilde{D} &= D. \end{aligned} \tag{1.14}$$

The transformed system describes the same system as in Eq. (1.8) relative to new state-variable coordinates $\underline{\tilde{x}}(n)$. For example, it can be checked that the transfer-function matrix is unchanged.

Efficiency of Diagonalized State-Space Models

Note that a general Nth-order state-space model Eq. (1.8) requires around N^2 multiply-adds to update for each time step (assuming the number of inputs and outputs is small compared with the number of state variables, in which case the $A\underline{x}(n)$ computation dominates). After diagonalization by a similarity transform, the time update is only order N, just like any other efficient digital filter realization. Thus, a diagonalized state-space model

[12] A *generalized eigenvector* $\mathbf{p}$ of matrix A corresponding to eigenvalue λ having multiplicity k is a nonzero solution of $(A - \lambda I)^k \mathbf{p} = 0$.

(modal representation) is a strong contender for applications in which it is desirable to have independent control of resonant modes.

Another advantage of the modal expansion is that frequency-dependent characteristics of hearing can be brought to bear. Low-frequency resonances can easily be modeled more carefully and in more detail than very high-frequency resonances which tend to be heard only "statistically" by the ear. For example, rows of high-frequency modes can be collapsed into more efficient digital waveguide loops (§8.5) by retuning them to the nearest harmonic mode series.

1.3.10 Equivalent Circuits

The concepts of "circuits" and "ports" from classical circuit/network theory [35] are very useful for *partitioning* complex systems into self-contained sections having well-defined (small) interfaces. For example, it is typical in analog electric circuit design to drive a high-input-impedance stage from a low-output-impedance stage (a so-called "voltage transfer" connection). This large impedance ratio allows us to neglect "loading effects" so that the circuit sections (stages) can be analyzed separately.

The name "analog circuit" refers to the fact that electrical capacitors (denoted C) are analogous to physical springs, inductors (L) are analogous to physical masses, and resistors (R) are analogous to "dashpots" (which are idealized physical devices for which compression velocity is proportional to applied force—much like a shock-absorber ("damper") in an automobile suspension). These are all called *lumped elements* to distinguish them from *distributed parameters* such as the capacitance and inductance per unit length in an electrical transmission line. Lumped elements are described by ODEs while distributed-parameter systems are described by PDEs. Thus, RLC analog circuits can be constructed as *equivalent circuits* for lumped dashpot-mass-spring systems. These equivalent circuits can then be digitized by *finite difference* or *wave digital* methods. PDEs describing distributed-parameter systems can be digitized via finite difference methods as well, or, when wave propagation is the dominant effect, *digital waveguide* methods.

As discussed in Chapter 7 (§7.2), the *equivalent circuit* for a force-driven mass is shown in Fig. F.10. The mass m is represented by an *inductor* $L = m$. The driving force $f(t)$ is supplied via a *voltage source*, and the mass velocity $v(t)$ is the *loop current*.

As also discussed in Chapter 7 (§7.2), if two physical elements are connected in such a way that they share a *common velocity*, then they are said to be formally connected *in series*. The "series" nature of the connection becomes more clear when the *equivalent circuit* is considered.

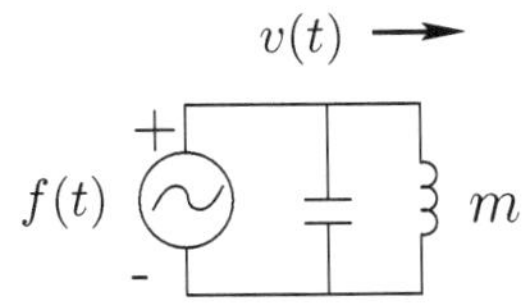

Figure 1.8: Electrical equivalent circuit of the force-driven mass in Fig. 1.6 on page 13.

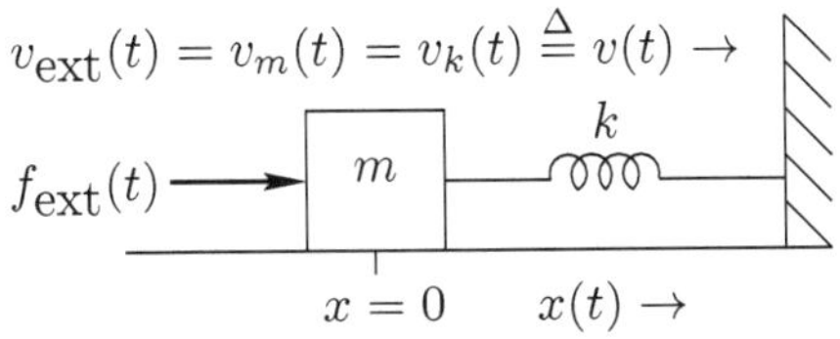

Figure 1.9: A mass and spring connected in series.

For example, Fig. 1.9 shows a mass connected to one end of a spring, with the other end of the spring attached to a rigid wall. The driving force $f_{\text{ext}}(t)$ is applied to the mass m on the left so that a positive force results in a positive mass displacement $x_m(t)$ and positive spring displacement (compression) $x_k(t)$. Since the mass and spring displacements are physically the same, we can define $x_m = x_k \triangleq x$. Their velocities are similarly equal so that $v_m = v_k \triangleq v$. The equivalent circuit has their electrical analogs connected in series, as shown in Fig. 1.10. The common mass and spring velocity $v(t)$ appear as a single current running through the inductor (mass) and capacitor (spring).

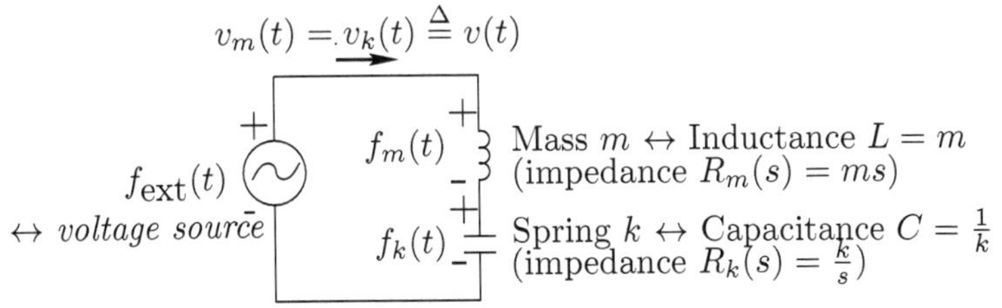

Figure 1.10: Electrical equivalent circuit of the series mass-spring driven by an external force diagrammed in Fig. 1.9.

By Kirchoff's loop law for circuit analysis, the sum of all voltages around a loop equals zero.[13] Thus, following the direction for current v in

[13]The other law routinely used in circuit analysis is that the *sum of all currents*

Fig. 1.10, we have $f_m + f_k - f_{\text{ext}} = 0$ (where the minus sign for $f_{\text{ext}} = 0$ occurs because the current enters its minus sign), or

$$f_{\text{ext}}(t) = f_m(t) + f_k(t).$$

Thus, the equivalent circuit agrees with our direct physical analysis that the applied force $f_{\text{ext}}(t)$ is equal at all times to the sum of the mass inertial force $f_m(t) = m\ddot{x}(t)$ and spring force $f_k(t) = kx(t)$, *i.e.*,

$$f_{\text{ext}}(t) = m\ddot{x}(t) + kx(t).$$

1.3.11 Impedance Networks

The concept of impedance is central in classical electrical engineering. The simplest case is *Ohm's Law* for a resistor R:

$$V(t) = R\,I(t)$$

where $V(t)$ denotes the voltage across the resistor at time t, and $I(t)$ is the current through the resistor. For the corresponding mechanical element, the *dashpot*, Ohm's law becomes

$$f(t) = \mu\,v(t)$$

where $f(t)$ is the force across the dashpot at time t, and $v(t)$ is its compression velocity. The dashpot value μ is thus a *mechanical resistance.*[14]

Thanks to the *Laplace transform* [449][15] (or *Fourier transform* [451]), the concept of impedance easily extends to masses and springs as well. We need only allow impedances to be *frequency-dependent.* For example, the Laplace transform of Newton's $f = ma$ yields, using the *differentiation theorem* for Laplace transforms [449],

$$F(s) = m\,A(s) = m\,sV(s) = m\,s^2X(s)$$

where $F(s)$ denotes the Laplace transform of $f(t)$ ("$F(s) = \mathcal{L}_s\{f\}$"), and similarly for displacement, velocity, and acceleration Laplace transforms. (It is assumed that all initial conditions are zero, *i.e.*, $f(0) = x(0) = v(0) = a(0) = 0$.) The *mass impedance* is therefore

$$R_m(s) \triangleq \frac{F(s)}{V(s)} = ms.$$

entering a circuit node (connection of wires) is zero. This kind of analysis will be revisited in §9.3.1.

[14]Note that models of damping in practical physical systems are rarely completely independent of frequency, as is the ideal dashpot.

[15]For a short online introduction to Laplace transforms, see, *e.g.*, `http://ccrma.stanford.edu/~jos/filters/Laplace_Transform_Analysis.html`.

Specializing the Laplace transform to the Fourier transform by setting $s = j\omega$ gives

$$R_m(j\omega) = jm\omega.$$

Similarly, the impedance of a *spring* having spring-constant k is given by

$$\begin{aligned} R_k(s) &= \frac{k}{s} \\ R_k(j\omega) &= \frac{k}{j\omega}. \end{aligned}$$

The important benefit of this frequency-domain formulation of impedance is that it allows every interconnection of masses, springs, and dashpots (every RLC equivalent circuit) to be treated as a simple *resistor network*, parametrized by frequency.

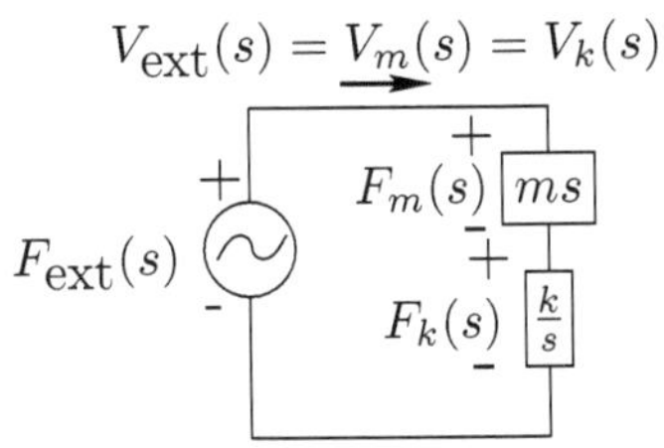

Figure 1.11: Impedance diagram for the force-driven, series arrangement of mass and spring shown in Fig. 1.9.

As an example, Fig. 1.11 gives the impedance diagram corresponding to the equivalent circuit in Fig. 1.10. Viewing the circuit as a (frequency-dependent) resistor network, it is easy to write down, say, the Laplace transform of the force across the spring using the *voltage divider* formula:

$$F_k(s) = F_{ext}(s)\frac{R_k(s)}{R_m(s)+R_k(s)} = F_{ext}(s)\frac{k/m}{s^2+k/m}$$

These sorts of equivalent-circuit and impedance-network models of mechanical systems, and their digitization to digital-filter form, are discussed further in Chapter 7.

1.3.12 Wave Digital Filters

The idea of wave digital filters is to digitize RLC circuits (and certain more general systems) as follows:

1. Determine the ODEs describing the system (PDEs also workable).

2. Express all physical quantities (such as force and velocity) in terms of *traveling-wave components.* The traveling wave components are called *wave variables.* For example, the force $f(n)$ on a mass is decomposed as $f(n) = f^+(n) + f^-(n)$, where $f^+(n)$ is regarded as a traveling wave propagating *toward* the mass, while $f^-(n)$ is seen as the traveling component propagating *away from* the mass. A "traveling wave" view of force mediation (at the speed of light) is actually much closer to underlying physical reality than any instantaneous model.

3. Next, digitize the resulting traveling-wave system using the *bilinear transform* (§7.3.2,[449, p. 386]). The bilinear transform is equivalent in the time domain to the *trapezoidal rule for numerical integration* (§7.3.2).

4. Connect N elementary units together by means of *N-port scattering junctions.* There are two basic types of scattering junction, one for parallel, and one for series connection. The theory of scattering junctions is introduced in the *digital waveguide* context (§C.8).

A more detailed introduction to WDFs is provided in Appendix F. In particular, the force-driven mass is considered in §F.3.4, and its wave digital model is shown in Fig. 1.12.

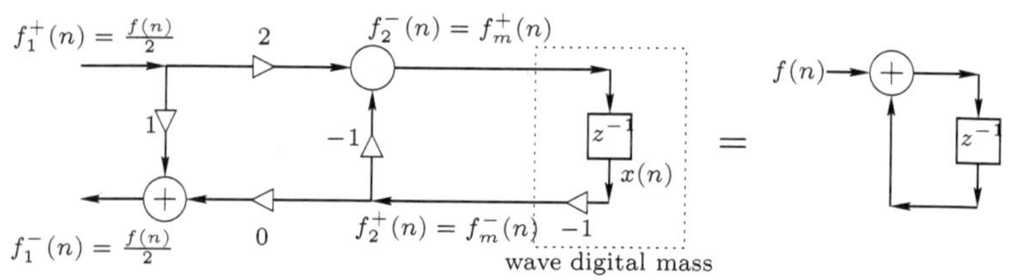

Figure 1.12: Wave digital mass driven by external force $f(n)$.

We will not make much use of WDFs in this book, preferring instead more prosaic finite-difference models for simplicity. However, we will utilize closely related concepts in the digital waveguide modeling context (Chapter 6).

1.3.13 Digital Waveguide Modeling Elements

As mentioned above, digital waveguide models are built out of digital delay-lines and filters (and nonlinear elements), and they can be understood as propagating and filtering sampled traveling-wave solutions to the wave equation (PDE), such as for air, strings, rods, and the like [433, 437]. It is noteworthy that strings, woodwinds, and brasses comprise three of the four principal sections of a classical orchestra (all but percussion). The digital waveguide modeling approach has also been extended to propagation in 2D, 3D, and beyond [519, 396, 523, 400]. They are not finite-difference models, but paradoxically they are equivalent under certain conditions (Appendix E). A summary of historical aspects appears in §A.9.

As mentioned at Eq. (1.1) on page 9, the ideal wave equation comes directly from Newton's laws of motion ($f = ma$). For example, in the case of vibrating strings, the wave equation is derived from first principles (in Chapter 6, and more completely in Appendix C) to be

$$\begin{aligned} Ky'' &= \epsilon\ddot{y} \\ \text{(Restoring Force Density} &= \text{Mass Density times Acceleration)}, \end{aligned}$$

where

$$\begin{aligned} K &\stackrel{\Delta}{=} \text{string tension} \\ \epsilon &\stackrel{\Delta}{=} \text{linear mass density.} \end{aligned}$$

Defining $c = \sqrt{K/\epsilon}$, we obtain the usual form of the PDE known as the *ideal 1D wave equation*.

$$y'' = \frac{1}{c^2}\ddot{y} \tag{1.15}$$

where $y(t, x)$ is the string displacement at time t and position x. (We omit the time and position arguments (t, x) when they are the same for all signal terms in the equation.) For example, y can be the transverse displacement of an ideal stretched string or the longitudinal displacement (or pressure, velocity, etc.) in an air column. The independent variables are time t and the distance x along the string or air-column axis. The partial-derivative notation is more completely written out as

$$\begin{aligned} \ddot{y} &\stackrel{\Delta}{=} \frac{\partial^2}{\partial t^2}y(t,x) \\ y'' &\stackrel{\Delta}{=} \frac{\partial^2}{\partial x^2}y(t,x). \end{aligned}$$

As has been known since d'Alembert [100], the 1D wave equation is obeyed by arbitrary *traveling waves* at speed c:

$$y(t, x) = y_r(t - x/c) + y_l(t + x/c)$$

To show this, just plug $y_r(t-x/c)$ or $y_l(t+x/c)$ (or any linear combination of them) into the wave equation Eq. (1.15). Thus, $c = \sqrt{K/\epsilon}$ is the traveling-wave propagation speed expressed in terms of the string tension K and mass density ϵ.

In digital waveguide modeling, the traveling-waves are *sampled:*

$$\begin{aligned} y(nT, mX) &= y_r(nT - mX/c) + y_l(nT + mX/c) \qquad \text{(set } X = cT\text{)} \\ &= y_r(nT - mT) + y_l(nT + mT) \\ &\triangleq y^+(n-m) + y^-(n+m) \end{aligned}$$

where T denotes the time sampling interval in seconds, $X = cT$ denotes the spatial sampling interval in meters, and y^+ and y^- are defined for notational convenience.

An ideal string (or air column) can thus be simulated using a *bidirectional delay line*, as shown in Fig. 1.13 for the case of an N-sample section of ideal string or air column. The "R" label denotes its *wave impedance* (§6.1.5) which is needed when connecting digital waveguides to each other and to other kinds of computational physical models (such as finite difference schemes). While propagation speed on an ideal string is $c = \sqrt{K/\epsilon}$, we will derive (§C.7.3) that the wave impedance is $R = \sqrt{K\epsilon}$.

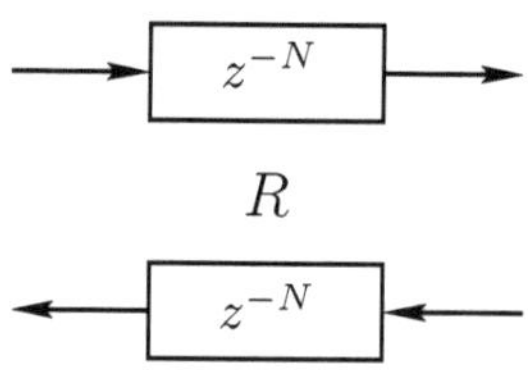

Figure 1.13: A digital waveguide—a sampled traveling-wave simulation for waves in ideal strings or acoustic tubes having wave impedance R.

Figure 1.14 (from Chapter 6, §6.3), illustrates a simple digital waveguide model for rigidly terminated vibrating strings (more specifically, one polarization-plane of transverse vibration). The traveling-wave components are taken to be *displacement* samples, but the diagram for velocity-wave and acceleration-wave simulation are identical (inverting reflection at each rigid termination). The output signal $y(nT, \xi)$ is formed by summing traveling-wave components at the desired "virtual pickup" location (position $x = \xi$ in this example). To drive the string at a particular point, one simply takes the *transpose* [449] of the output sum, *i.e.*, the input excitation is summed equally into the left- and right-going delay-lines at the same x position (details will be discussed near Fig. 6.14 on page 226).

In Chapter 9 (example applications), we will discuss digital waveguide models for single-reed instruments such as the clarinet (Fig. 1.15), and bowed-string instruments (Fig. 1.16) such as the violin.

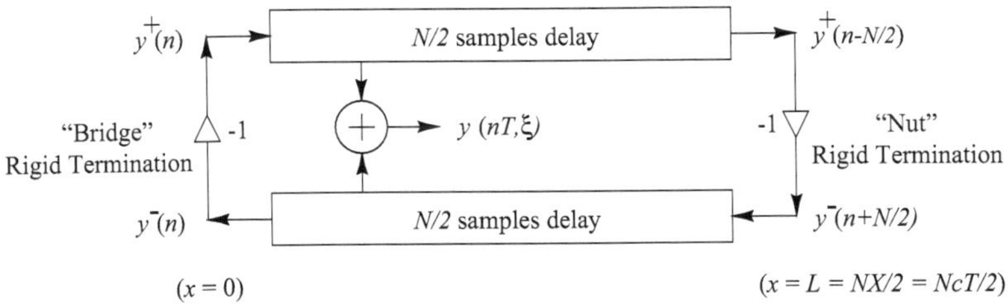

Figure 1.14: Digital waveguide model of a rigidly terminated ideal string, with a displacement output indicated at position $x = \xi$. Rigid terminations reflect traveling displacement, velocity, or acceleration waves with a sign inversion. Slope or force waves reflect with no sign inversion.

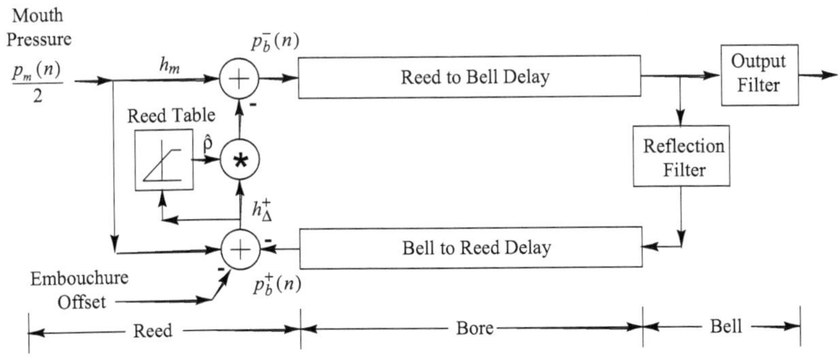

Figure 1.15: Digital waveguide model of a single-reed, cylindrical-bore woodwind, such as a clarinet (copy of Fig. 9.39 on page 404).

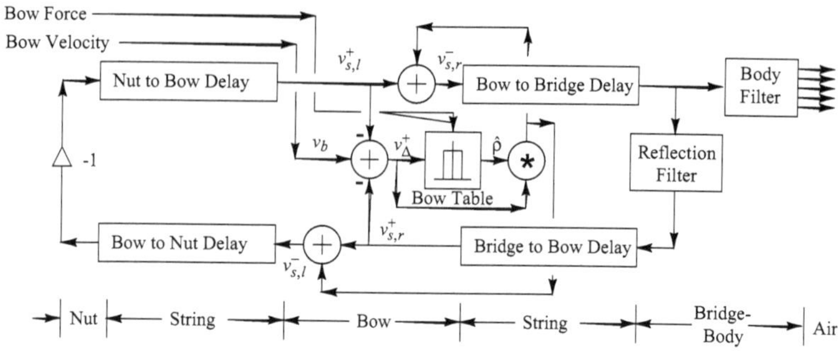

Figure 1.16: Digital waveguide model for a bowed-string instrument, such as a violin (copy of Fig. 9.52 on page 429).

1.3.14 General Modeling Procedure

While each situation tends to have special opportunities, the following procedure generally works well:

1. Formulate a *state-space model.*

2. If it is nonlinear, use *numerical time-integration*:

    - Explicit (causal finite difference scheme)
    - Implicit (iteratively solved each time step)
    - Semi-Implicit (truncated iterations of Implicit)

    There are many other descriptions of nonlinear systems to consider, some of which will be noted as we proceed.

3. In the linear case, *diagonalize* the state-space model to obtain the *modal representation.*

    - Implement isolated modes as second-order filters ("biquads").
    - Implement *quasi-harmonic* mode series as *digital waveguides.*

1.4 Our Plan

A logical organization for this book would be along the lines above, starting with Newton's laws, differential equations, digitization methods, and so on through the various methods. However, we proceed instead as follows:

1. Delay-line and digital-waveguide acoustic modeling
    (a) Time-invariant case (acoustic multipath, artificial reverberation)
    (b) Time-varying case (phasing, flanging, chorus, Leslie, vibrating strings and air columns)
2. Impedance modeling and equivalent circuits
3. Finite differences and transfer functions
4. Application examples
5. Appendices on C++ software, history, Newtonian physics, digital waveguide theory details, more about finite difference schemes, and wave digital filters

The reason for this ordering is that it starts out really simple, with no calculus required, and plenty of signal processing and practical applications along the way. It is easily possible to fill a first course without reaching impedance models (which require calculus and frequency-domain facility for both continuous- and discrete-time systems). The appendices provide both supporting and relatively advanced material.

1.4.1 A Simplified Starting Theory

For delay-line and simple digital waveguide modeling, we can accept it as an *experimental fact* that "traveling waves happen". That is, traveling waves are simply observed to propagate at speed c through air and stretched strings, and they are also observed to obey the superposition principle (traveling waves pass through each other as if they were ghosts). Lossy and dispersive propagation can similarly be observed to correspond to a fixed linear filtering per unit length of propagation medium. When the propagation path-length is doubled, the filter transfer function is squared.

The simple paradigm of using a "filtered delay line" as a computational modeling element for a single "ray" of acoustic propagation is a quite general building block that can be pushed far without getting into the full theory.

However, to take digital waveguide modeling to the next level, and to include lumped modeling elements such as masses and springs, we ultimately get to impedance concepts and simple differential equations built around Newton's $f = ma$. This gives us a "deeper theory" in which sound speed can be predicted from basic physical properties, as we saw above. We also then know how distributed and lumped elements should interact, such as when a mass (computed as a first-order digital filter) "strikes" a vibrating string (computed as a digital waveguide).

Elementary Physical Modeling Problems

1. Derive Eq. (1.11) on page 19:

$$H(z) = D + C(zI - A)^{-1} B$$

2. In Newton's famous equation $f(t) = m\,a(t)$, suppose $f(t)$ is regarded as an "input signal" and the position $x(t)$ of the mass m is regarded as an "output signal". Find the s-plane *transfer function* $H(s) \triangleq X(s)/F(s)$. State any assumptions that are necessary.

3. Referring to Fig. 1.9 on page 25,

   (a) Write down the ODE that describes the system using the sign conventions in Fig. 1.10 on page 25.

   (b) Write down the ODE that results if the signs on the "force source" (voltage source) in Fig. 1.10 are reversed.
   In the following, assume the signs Fig. 1.10 as they are given.

   (c) Find the s-plane transfer function relating the driving-force input signal $f_{\text{ext}}(t)$ to the mass-velocity output signal $v_m(t)$.

   (d) Find the poles of the system.

   (e) Where are the zeros?

   (f) Write down the equations describing a linear, time-invariant state-space model, choosing the spring compression $x(t)$ as state-variable 1 and the mass velocity $v_m(t)$ as state-variable 2.

   (g) Under what conditions does this system become identical to the force-driving mass system in Fig. 1.6? Explain how this makes physical sense.

   (h) Digitize the ODE, converting it to an explicit difference equation that computes a sample of velocity $v_m(nT)$ as a function of previous velocity and driving-force samples.

   (i) Find the z-plane transfer function $H(z) = V(z)/F(z)$ of the difference equation.

   (j) Find the poles of the digitized system.

   (k) Find the zeros of the digitized system.

   (l) Discuss the quality of the digitization in terms of how accurately the resonance frequency and damping characteristics are carried over from continuous to discrete time.

Synthesis Tool Kit (STK) Orientation

1. Follow the instructions in

   `https://ccrma.stanford.edu/~jos/stkintro/`

   regarding getting started with the Synthesis Tool Kit (STK) for Music 420 at CCRMA.

2. Read the STK documentation at `https://ccrma.stanford.edu/software/stk/`

3. Look over the library modules, especially anything named `Delay*`.

4. Compile and run the `sineosc` example.

5. Familiarize yourself with the STK to the extent you have time. The demo script `projects/demo/Physical` is especially relevant.

Lab Problem

Write an STK "effect" that implements your difference equation derived above using the BiQuad two-pole filter class. The input to its `tick()` method should be a force sample, and the output of `tick()` is a sample of mass velocity. The spring constant k and mass m should be set at instantiation time. Write a main program that provides a force impulse input $f(n) = \delta(n)$, and computes the first 10 samples of the impulse response. Turn in a printout and a plot of the output for the parameters $k = 100\pi^2$, $m = 1$, $T = 0.1$.

Chapter 2

Acoustic Modeling with Digital Delay

Delay effects, such as *phasing*, *flanging*, chorus, and artificial reverberation, as well as digital waveguide models (Chapter 6), are built using delay lines, simple digital filter sections, and sometimes nonlinear elements and modulation. We will focus on these elements in the simpler context of delay effects before using them for sound synthesis.

2.1 Delay Lines

The *delay line* is an elementary functional unit which models *acoustic propagation delay*. It is a fundamental building block of both delay-effects processors and digital-waveguide synthesis models. The function of a delay line is to introduce a time delay between its input and output, as shown in Fig. 2.1.

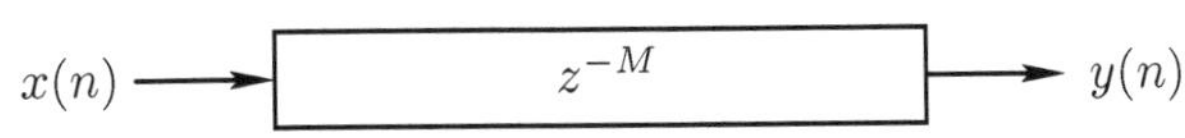

Figure 2.1: The M-sample delay line.

Let the input signal be denoted $x(n)$, $n = 0, 1, 2, \ldots$, and let the delay-line length be M samples. Then the output signal $y(n)$ is specified by the relation

$$y(n) = x(n - M), \quad n = 0, 1, 2, \ldots \tag{2.1}$$

where $x(n) \triangleq 0$ for $n < 0$.

Before the digital era, delay lines were expensive and imprecise in "analog" form. For example, "spring reverberators" (common in guitar amplifiers) use metal springs as analog delay lines; while adequate for that purpose, they are highly dispersive and prone to noise pick-up. Large delays require prohibitively long springs or coils in analog implementations. In the digital domain, on the other hand, delay by N samples is trivially implemented, and non-integer delays can be implemented using interpolation techniques, as discussed later in §4.1.

2.1.1 A Software Delay Line

In software, a delay line is often implemented using a *circular buffer*. Let `D` denote an array of length M. Then we can implement the M-sample delay line in the `C` programming language as shown in Fig. 2.2.

```
     /* delayline.c */
     static double D[M];  // initialized to zero
     static long ptr=0;   // read-write offset

     double delayline(double x)
     {
       double y = D[ptr];      // read operation
       D[ptr++] = x;           // write operation
       if (ptr >= M) { ptr -= M; } // wrap ptr
//     ptr %= M; // modulo-operator syntax
       return y;
     }
```

Figure 2.2: The M-sample delay line.

Delay lines of this type are typically used in *digital reverberators* and other acoustic simulators involving *fixed* propagation delays. Later, in Chapter 5, we will consider *time-varying* delay lengths.

2.2 Acoustic Wave Propagation Simulation

Delay lines can be used to simulate *acoustic wave propagation*. We start with the simplest case of a pure *traveling wave*, followed by the more general case of *spherical waves*. We then look at the details of a simple *acoustic echo* simulation using a delay line to model the difference in time-of-arrival between the direct and reflected signals.

2.2.1 Traveling Waves

In acoustic wave propagation, pure delays can be used to simulate *traveling waves*. A traveling wave is any kind of wave which propagates in a single direction with negligible change in shape. An important class of traveling waves is "plane waves" in air which create "standing waves" in rectangular enclosures such as "shoebox" shaped concert halls. Also, far away from any acoustic source (where "far" is defined as "many wavelengths"), the direct sound emanating from any source can be well approximated as a plane wave, and thus as a traveling wave.

Another case in which plane waves dominate is the *cylindrical bore*, such as the bore of a clarinet or the straight tube segments of a trumpet. Additionally, the *vocal tract* is generally simulated using plane waves, though in this instance there is a higher degree of approximation error.

Transverse and *longitudinal* waves in a vibrating string, such as on a guitar, are also nearly perfect traveling waves, and they can be simulated to a very high degree of perceptual accuracy by approximating them as ideal, while implementing slight losses and dispersion once per period (*i.e.*, at one particular point along the "virtual string").

In a conical bore, we find sections of *spherical waves* taking the place of plane waves. However, they still "travel" like plane waves, and we can still use a delay line to simulate their propagation. The same applies to spherical waves created by a "point source." Spherical waves will be considered on page 41.

2.2.2 Damped Traveling Waves

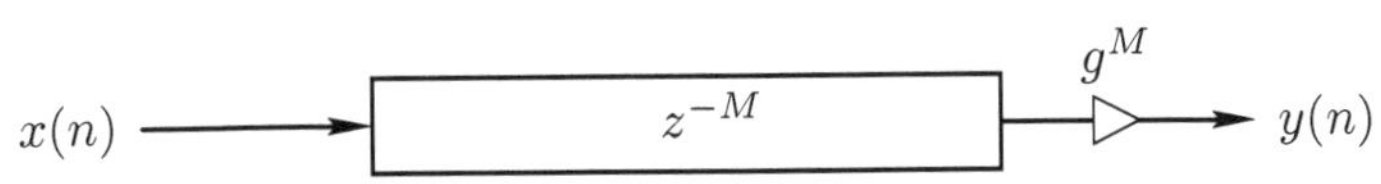

Figure 2.3: Damped traveling-wave simulator. In addition to propagation delay by M samples, there is attenuation by $g^M < 1$.

The delay line shown in Fig. 2.1 on page 37 can be used to simulate any traveling wave, where the traveling wave must propagate in one direction with a fixed waveshape. If a traveling wave *attenuates* as it propagates, with the same attenuation factor at each frequency, the attenuation can be simulated by a simple *scaling* of the delay line output (or input), as shown in Fig. 2.3. This is perhaps the simplest example of the important principle of *lumping distributed losses* at discrete points. That is, it is not necessary to implement a small attenuation g for each time-step of wave propagation; the same result is obtained at the delay-line output if propagation is "lossless" within the delay line, and the total cumulative attenuation g^M is applied at the output. The input-output simulation is exact, while the signal samples inside the delay line are simulated with a slight gain error. If the internal signals are needed later, they can be tapped out using correcting gains. For example, the signal half way along the delay line can be tapped using a coefficient of $g^{M/2}$ in order to make it an exact second output. In summary, computational efficiency can often be greatly increased at no cost to accuracy by lumping losses only at the outputs and points of interaction with other simulations.

Modeling traveling-wave attenuation by a scale factor is only exact physically when all frequency components decay at the same rate. For accurate acoustic modeling, it is usually necessary to replace the constant scale factor g by a *digital filter* $G(z)$ which implements *frequency-dependent attenuation*, as depicted in Fig. 2.4. In principle, a linear time-invariant (LTI) filter can provide an independent attenuation factor at each frequency. Section 2.3 addresses this case in more detail. Frequency-dependent damping substitution will be used in artificial reverberation design in §3.7.4.

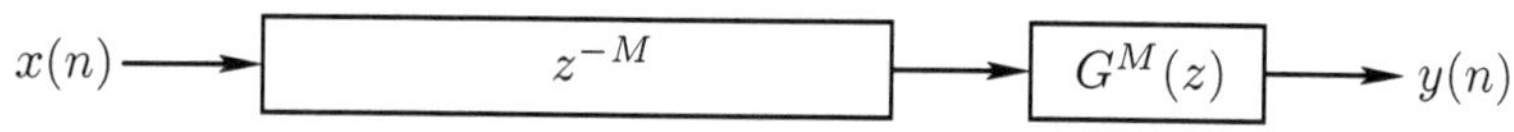

Figure 2.4: Damped traveling-wave simulator. In principle, the digital filter $G(z)$ provides an arbitrary attenuation corresponding to one sample of wave propagation at each frequency. Physically, we must have $|G(e^{j\omega T})| \leq 1$ for all ω.

2.2.3 Dispersive Traveling Waves

In many acoustic systems, such as *piano strings* (§9.4.1,§C.6), wave propagation is also significantly *dispersive*. A wave-propagation medium is said to be dispersive if the speed of wave propagation is not the same at all frequencies. As a result, a propagating wave shape will "disperse" (change shape) as its various frequency components travel at different speeds. Dispersive propagation in one direction can be simulated using a delay line in series with a *nonlinear phase* filter, as indicated in Fig. 2.5. If there is no damping, the filter $A(z)$ must be *all-pass* [449], *i.e.*, $|A(e^{j\omega T})| = 1$ for all $\omega T \in [-\pi, \pi]$.

2.2.4 Converting Propagation Distance to Delay Length

We may regard the delay-line memory itself as the fixed "air" which propagates sound samples at a fixed speed c ($c = 345$ meters per second at 22

Figure 2.5: Dispersive traveling-wave simulator. In principle, the digital filter $A(z)$ provides an arbitrary nonnegative *delay* corresponding to one sample of wave propagation at each frequency.

degrees Celsius and 1 atmosphere). The input signal $x(n)$ can be associated with a sound source, and the output signal $y(n)$ (see Fig. 2.1 on page 37) can be associated with the listening point. If the listening point is d meters away from the source, then the delay line length M needs to be

$$M = \frac{d}{cT} \quad \text{samples,}$$

where T denotes the discrete-time sampling interval. In other words, the number of samples delay is the propagation distance d divided by cT, the distance sound propagates in one sampling interval. In practice, rounding $M = d/cT$ to the nearest integer causes no audible difference, unless the echo time is so short that the system is not really perceived as an echo (we'll learn about "comb filters" in §2.6 below).

2.2.5 Spherical Waves from a Point Source

Acoustic theory tells us that a *point source* produces a *spherical wave* in an ideal isotropic (uniform) medium such as air. Furthermore, the sound from any radiating surface can be computed as the sum of spherical wave contributions from each point on the surface (including any relevant reflections). The *Huygens-Fresnel principle* explains wave propagation itself as the superposition of spherical waves generated at each point along a wavefront (see, *e.g.*, [349, p. 175]). Thus, all linear acoustic wave propagation can be seen as a superposition of spherical traveling waves.

To a good first approximation, wave energy is *conserved* as it propagates through the air. In a spherical pressure wave of radius r, the energy of the wavefront is spread out over the spherical surface area $4\pi r^2$. Therefore, the energy per unit area of an expanding spherical pressure wave decreases as $1/r^2$. This is called *spherical spreading loss*. It is also an example of an *inverse square law* which is found repeatedly in the physics of conserved quantities in three-dimensional space. Since energy is proportional to amplitude squared, an inverse square law for energy translates to a $1/r$ decay law for amplitude.

The sound-pressure amplitude of a traveling wave is proportional to the square-root of its energy per unit area. Therefore, in a spherical traveling

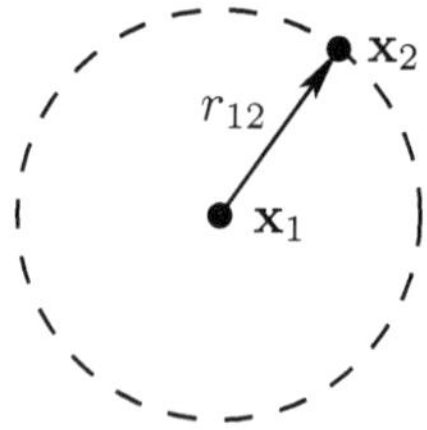

Figure 2.6: Geometry of wave propagation from a point source $\mathbf{x}_1$ to a listening point $\mathbf{x}_2$.

wave, acoustic amplitude is proportional to $1/r$, where r is the radius of the sphere. In terms of Cartesian coordinates, the amplitude $p(\mathbf{x}_2)$ at the point $\mathbf{x}_2 = (x_2, y_2, z_2)$ due to a point source located at $\mathbf{x}_1 = (x_1, y_1, z_1)$ is given by

$$p(\mathbf{x}_2) = \frac{p_1}{r_{12}}$$

where p_1 is defined as the pressure amplitude one radial unit from the point source located at $\mathbf{x} = \mathbf{x}_1$ (*i.e.*, $p_1 = p(\mathbf{x}_1 + \mathbf{e})$ where $\|\mathbf{e}\| = 1$), and r_{12} denotes the distance from the point $\mathbf{x}_1$ to $\mathbf{x}_2$:

$$r_{12} \triangleq \|\mathbf{x}_2 - \mathbf{x}_1\| = \sqrt{(x_2 - x_1)^2 + (y_2 - y_1)^2 + (z_2 - z_1)^2}$$

This geometry is depicted for the 2D case in Fig. 2.6.

In summary, every point of a radiating sound source emits spherical traveling waves in all directions which decay as $1/r$, where r is the distance from the source. The amplitude-decay by $1/r$ can be considered a consequence of energy conservation for propagating waves. (The energy spreads out over the surface of an expanding sphere.) We often visualize such waves as "rays" emanating from the source, and we can simulate them as a delay line along with a $1/r$ scaling coefficient (see Fig. 2.7). In contrast, since plane waves propagate with no decay at all, each "ray" can be considered lossless, and the simulation involves only a delay line with no scale factor, as shown in Fig. 2.1 on page 37.

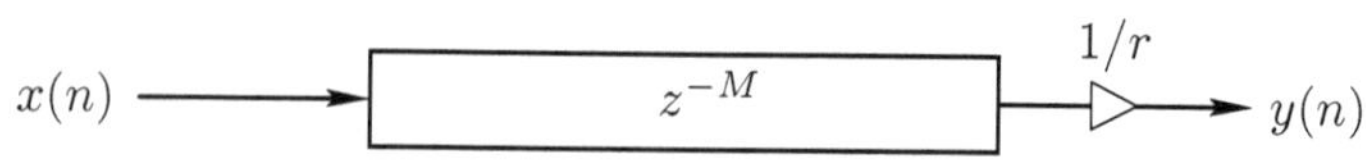

Figure 2.7: Point-to-point spherical wave simulator. In addition to propagation delay, there is attenuation by $g = 1/r$.

2.2.6 Reflection of Spherical or Plane Waves

When a spreading spherical wave reaches a wall or other obstacle, it is either reflected or scattered. A wavefront is *reflected* when it impinges on a surface which is flat over at least a few wavelengths in each direction.[1] Reflected wavefronts can be easily mapped using *ray tracing*, *i.e.*, the reflected ray leaves at an angle to the surface equal to the angle of incidence ("law of reflection"). Wavefront reflection is also called *specular reflection*, especially when considering light waves.

A wave is *scattered* when it encounters a surface which has variations on the scale of the spatial wavelength. A scattering reflection is also called a *diffuse reflection*. As a special case, objects smaller than a wavelength yield a diffuse reflection which approaches a spherical wave as the object approaches zero volume. More generally, each point of a scatterer can be seen as emitting a new spherically spreading wavefront in response to the incoming wave—a decomposition known as Huygen's principle, as mentioned in the previous section. The same process happens in reflection, but the hemispheres emitted by each point of the flat reflecting surface combine to form a more organized wavefront which is the same as the incident wave but traveling in a new direction.

The distinction between specular and diffuse reflections is dependent on frequency. Since sound travels approximately 1 foot per millisecond, a cube 1 foot on each side will "specularly reflect" directed "beams" of sound energy above 1 KHz, and will "diffuse" or scatter sound energy below 1 KHz. A good concert hall, for example, will have plenty of diffusion. As a general rule, reverberation should be diffuse in order to avoid "standing waves" (isolated energetic modes). In other words, in reverberation, we wish to spread the sound energy uniformly in both time and space, and we do not want any specific spatial or temporal patterns in the reverberation.

2.2.7 An Acoustic Echo Simulator

An acoustic *echo* is one of the simplest acoustic modeling problems. Echoes occur when a sound arrives via more than one acoustic propagation path, as shown in Fig. 2.8. We may hear a discrete echo, for example, if we clap our hands standing in front of a large flat wall outdoors, such as the side of a building. To be perceived as an echo, however, the reflection must arrive well after the direct signal (or previous echo).

A common cause of echoes is "multipath" wave propagation, as diagrammed in Fig. 2.8. The acoustic source is denoted by 'S', the listener by

[1] To determine whether a surface is effectively flat, it may first be smoothed so that variations less than a wavelength in size are ignored. That is, waves do not "see" variations on a scale much less than a wavelength.

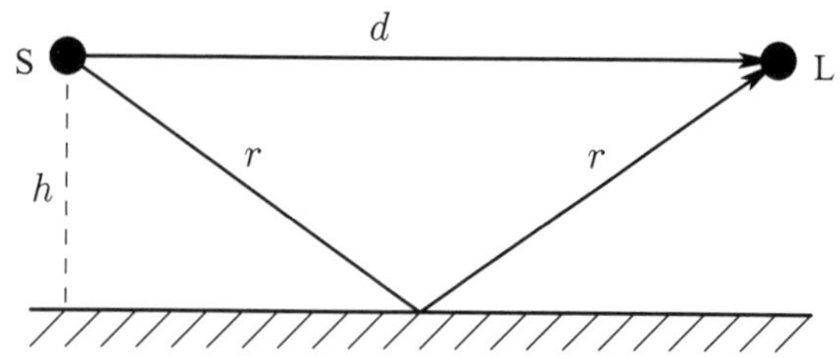

Figure 2.8: Geometry of an acoustic echo caused by "multipath" propagation. A direct signal and a "floor bounce" are received from the source **S** at the listening point **L**.

'L', and they are at the same height h meters from a reflecting surface. The direct path is d meters long, while the length of the single reflection is $2r$ meters. These quantities are of course related by the Pythagorean theorem:

$$r^2 = h^2 + \left(\frac{d}{2}\right)^2$$

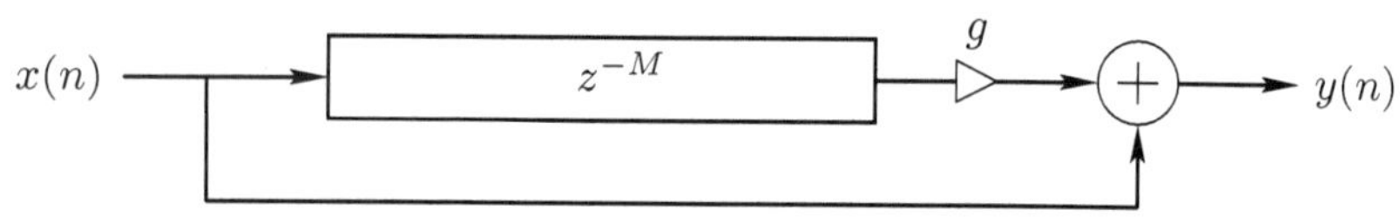

Figure 2.9: Acoustic echo simulation using a delay line, gain, and summer.

Figure 2.9 illustrates an echo simulator for the case of a direct signal and single echo, as shown in Fig. 2.8. It is common practice to pull out and discard any *common delay* which affects all signals equally, since such a delay does not affect timbre; thus, the direct signal delay is not implemented at all in Fig. 2.9. Similarly, it is not necessary to implement the *attenuation* of the direct signal due to propagation, since it is the *relative amplitude* of the direct signal and its echoes which affect timbre.

From the geometry in Fig. 2.8, we see that the delay-line length in Fig. 2.9 should be

$$M = \frac{2r - d}{cT}$$

and

$$g = \frac{1/2r}{1/d} = \frac{d}{2r},$$

where c is the speed of sound and T denotes the sampling interval. We may eliminate r using the substitution $r = \sqrt{h^2 + (d/2)^2}$, leaving only two independent variables: the height h of the source above the reflecting surface, and the distance d between the source and listener:

$$g = \frac{1}{\sqrt{1+(2h/d)^2}}$$

2.2.8 Program for Acoustic Echo Simulation

The following main program (Fig. 2.10) simulates a simple acoustic echo using the `delayline` function in Fig. 2.2. It reads a sound file and writes a sound file containing a single, discrete echo at the specified delay. For simplicity, utilities from the free Synthesis Tool Kit (STK) (Version `4.2.x`) are used for sound input/output [86].[2]

In summary, a delay line simulates the *time delay* associated with wave propagation in a particular direction. Attenuation (*e.g.*, by $1/r$) associated with ray propagation can of course be simulated by multiplying the delay-line output by some constant g.

2.3 Lossy Acoustic Propagation

Attenuation of waves by spherical spreading, as described in §2.2.5 above, is not the only source of amplitude decay in a traveling wave. In air, there is always significant additional loss caused by *air absorption*. Air absorption varies with frequency, with high frequencies usually being more attenuated than low frequencies, as discussed in §B.7.15 on page 513. Wave propagation in *vibrating strings* undergoes an analogous absorption loss, as does the propagation of nearly every other kind of wave in the physical world. To simulate such propagation losses, we can use a delay line in series with a nondispersive filter, as illustrated in §2.2.2 above. In practice, the desired attenuation at each frequency becomes the desired magnitude frequency-response of the filter in Fig. 2.4 on page 40, and filter-design software (typically matlab) is used to compute the filter coefficients to approximate the desired frequency response in some optimal way. The phase response may be linear, minimum, or left unconstrained when damping-filter dispersion is not considered harmful. There is typically a frequency-dependent weighting on the approximation error corresponding to audio perceptual importance (*e.g.*, the weighting $1/f$ is a simple example that increases accuracy at low

[2]In the code of Fig. 2.10, the first executable line changes the global STK sampling-rate from 44100 Hz (the default) to that of the input soundfile. If this is not done, there is a hidden sampling-rate conversion using linear interpolation when the input file is read, if its sampling rate is different from the global STK sampling-rate. This default rate conversion could alternatively be canceled by saying `input.setRate(1.0);` after the input file is opened.

```
/* Acoustic echo simulator, main C++ program.
   Compatible with STK version 4.2.1.
   Usage: main inputsoundfile
   Writes main.wav as output soundfile
 */

#include "FileWvIn.h"  /* STK soundfile input support */
#include "FileWvOut.h" /* STK soundfile output support */

static const int M = 20000; /* echo delay in samples */
static const int g = 0.8;   /* relative gain factor */

#include "delayline.c" /* defined previously */

int main(int argc, char *argv[])
{
  long i;
  Stk::setSampleRate(FileRead(argv[1]).fileRate());
  FileWvIn input(argv[1]);  /* read input soundfile */
  FileWvOut output("main"); /* creates main.wav */
  long nsamps = input.getSize();
  for (i=0;i<nsamps+M;i++)   {
    StkFloat insamp = input.tick();
    output.tick(insamp + g * delayline(insamp));
  }
}
```

Figure 2.10: Program in C++ for implementing the echo simulator of Fig. 2.9 using the free, open-source Synthesis Tool Kit (STK).

frequencies). Some filter-design methods are summarized in §8.6 on page 290.

2.3.1 Exponentially Decaying Traveling Waves

Let $g(r, \omega)$ denote the decay factor associated with propagation of a plane wave over distance r at frequency ω rad/sec. For an ideal plane wave, there is no "spreading loss" (attenuation by $1/r$). Under uniform conditions, the amount of attenuation (in dB) is proportional to the distance traveled; in other words, the attenuation factors for two successive segments of a propagation path are *multiplicative*:

$$g(r_1 + r_2, \omega) = g(r_1, \omega)g(r_2, \omega)$$

This property implies that g is an *exponential* function of distance r.[3]

Frequency-independent air absorption is easily modeled in an acoustic simulation by making the substitution

$$z^{-1} \leftarrow gz^{-1}$$

in the transfer function of the simulating delay line, where g denotes the attenuation associated with propagation during one sampling period (T seconds). Thus, to simulate absorption corresponding to an M-sample delay, the difference equation Eq. (2.1) on page 37 becomes

$$y(n) = g^M x(n - M),$$

as depicted in Fig. 2.9.

2.3.2 Frequency-Dependent Air-Absorption Filtering

More generally, *frequency-dependent* air absorption can be modeled using the substitution

$$z^{-1} \leftarrow G(z)z^{-1}$$

where $G(z)$ denotes the *filtering per sample* in the propagation medium. Since air absorption cannot amplify a wave at any frequency, we have

---

[3]It is easy to see that the exponential function $f(x) = e^x$ has the property $f(a+b) = f(a)f(b)$. To show that all differentiable functions with this property are exponentials, one can look at the definition of the derivative of $f(x)$ with respect to x:

$$f'(x) \triangleq \lim_{\Delta \to 0} \frac{f(x+\Delta) - f(x)}{\Delta} = \lim_{\Delta \to 0} \frac{f(x)f(\Delta) - f(x)}{\Delta} = f(x) \cdot \lim_{\Delta \to 0} \frac{f(\Delta) - 1}{\Delta}.$$

Since $f(a + 0) = f(a)f(0)$, we must have $f(0) = 1$. Therefore, the last limit above converges to some constant $\alpha = f'(0)$, and $f'(x) = \alpha f(x)$. In this way, it is shown that f satisfies a differential equation whose solution space consists of exponential functions of the form $f(x) = ae^{\alpha x}$, and to obtain $f(0) = 1$, we must have $a = 1$.

$\left|G(e^{j\omega T})\right| \leq 1$. A lossy delay line for plane-wave simulation is thus described by

$$Y(z) = G^M(z)z^{-M}X(z)$$

in the frequency domain, and

$$y(n) = \underbrace{g * g * \ldots * g}_{M \text{ times}} * x(n-M)$$

in the time domain, where '$*$' denotes convolution, and $g(n)$ is the impulse response of the per-sample loss filter $G(z)$. The effect of $G(z)$ on the poles of the system is discussed in §3.7.4 on page 114.

For spherical waves, the loss due to spherical spreading is of the form

$$Y(z) \propto \frac{G^M(z)z^{-M}}{r}X(z)$$

where r is the distance from X to Y. We see that the spherical spreading loss factor is "hyperbolic" in the propagation distance r, while air absorption is exponential in r.

2.3.3 Dispersive Traveling Waves

In addition to frequency-dependent attenuation, LTI filters can provide a *frequency-dependent delay*. This can be used to simulate *dispersive wave propagation*, as introduced in §2.2.3 on page 40.

2.3.4 Summary

Up to now, we have been concerned with the simulation of *traveling waves* in *linear, time-invariant (LTI) media*. The main example considered was wave propagation in air, but waves on vibrating strings behave analogously. We saw that the point-to-point propagation of a traveling plane wave in an LTI medium can be simulated simply using only a *delay line* and an *LTI filter*. The delay line simulates propagation delay, while the filter further simulates (1) an independent attenuation factor at each frequency by means of its amplitude response (*e.g.*, to simulate air absorption), and (2) a frequency-dependent propagation speed using its phase response (to simulate dispersion). If there is additionally spherical spreading loss, the amplitude may be further attenuated by $1/r$, where r is the distance from the source. For more details about the acoustics of plane waves and spherical waves, see, *e.g.*, [318, 349]. Appendix B contains a bit more about elementary acoustics,

So far we have considered only traveling waves going in one direction. The next simplest case is 1D acoustic systems, such as vibrating strings and acoustic tubes, in which traveling waves may propagate in *two* directions. Such systems are simulated using a pair of delay lines called a *digital waveguide*.

2.4 Digital Waveguides

A (lossless) *digital waveguide* is defined as a *bidirectional delay line* at some wave impedance R [430, 433]. Figure 2.11 illustrates one digital waveguide.

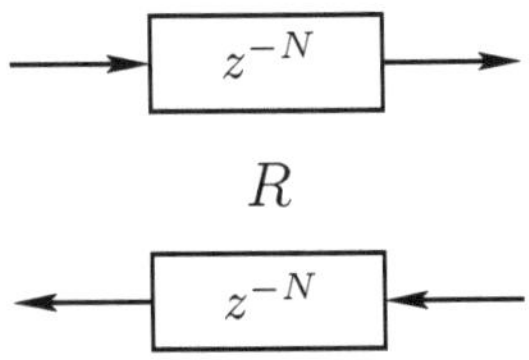

Figure 2.11: A digital waveguide N samples long at wave-impedance R.

As before, each delay line contains a sampled acoustic traveling wave. However, since we now have a *bidirectional* delay line, we have *two* traveling waves, one to the "left" and one to the "right", say. It has been known since 1747 [100] that the vibration of an ideal string can be described as the sum of two traveling waves going in opposite directions. (See Appendix C for a mathematical derivation of this important fact.) Thus, while a single delay line can model an acoustic plane wave, a *bidirectional* delay line (a digital waveguide) can model any one-dimensional linear acoustic system such as a violin string, clarinet bore, flute pipe, trumpet-valve pipe, or the like. Of course, in real acoustic strings and bores, the 1D waveguides exhibit some loss and dispersion[4] so that we will need some *filtering* in the waveguide to obtain an accurate physical model of such systems. The *wave impedance* R (derived in Chapter 6) is needed for connecting digital waveguides to other physical simulations (such as another digital waveguide or finite-difference model).

2.4.1 Physical Outputs

Physical variables (force, pressure, velocity, ...) are obtained by *summing* traveling-wave components, as shown in Fig. 2.12, and more elaborated in Fig. 2.13.

[4] *Dispersion* occurs in a traveling wave whenever the propagation speed is different at two or more frequencies.

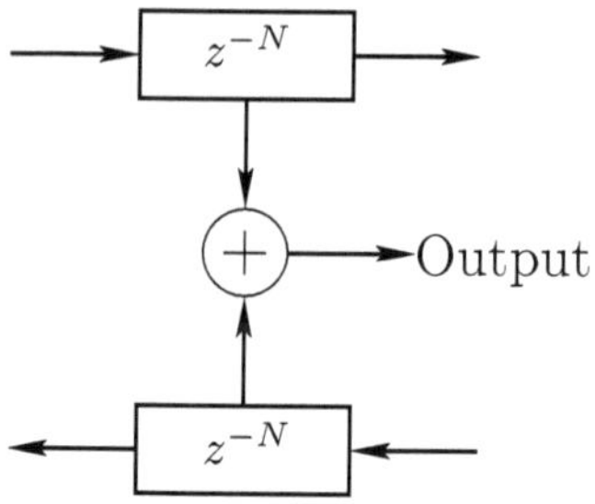

Figure 2.12: Extracting a physical signal from a digital waveguide using delay-line *taps*.

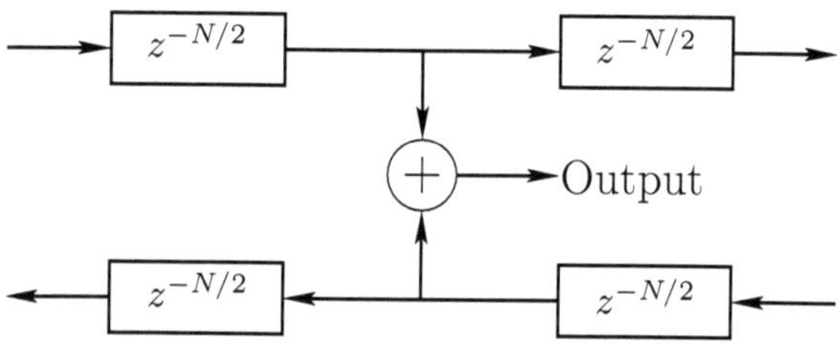

Figure 2.13: More detailed diagram of Fig. 2.12.

It is important to understand that the two traveling waves in a digital waveguide are now *components* of a more general acoustic vibration. The physical wave vibration is obtained by *summing* the left- and right-going traveling waves. A traveling wave by itself in one of the delay lines is no longer regarded as "physical" unless the signal in the opposite-going delay line is zero. Traveling waves are efficient for simulation, but they are not easily estimated from real-world measurements [477], except when the traveling-wave component in one direction can be arranged to be zero.

Note that traveling-wave components are not necessarily *unique*. For example, we can add a constant to the right-going wave and subtract the same constant from the left-going wave without altering the (physical) sum [263]. However, as derived in Appendix C (§C.3.6), 1D traveling-wave components are uniquely specified by *two* linearly independent physical variables along the waveguide, such as position and velocity (vibrating strings) or pressure and velocity (acoustic tubes).

2.4.2 Physical Inputs

A digital waveguide *input signal* corresponds to a *disturbance* of the 1D propagation medium. For example, a vibrating string is *plucked* or *bowed* by such an external disturbance. The result of the disturbance is wave propagation to the left and right of the input point. By physical symmetry, the amplitude of the left- and right-going propagating disturbances will normally be equal.[5] If the disturbance *superimposes* with the waves already passing through at that point (an idealized case), then it is purely an *additive input*, as shown in Fig. 2.14.

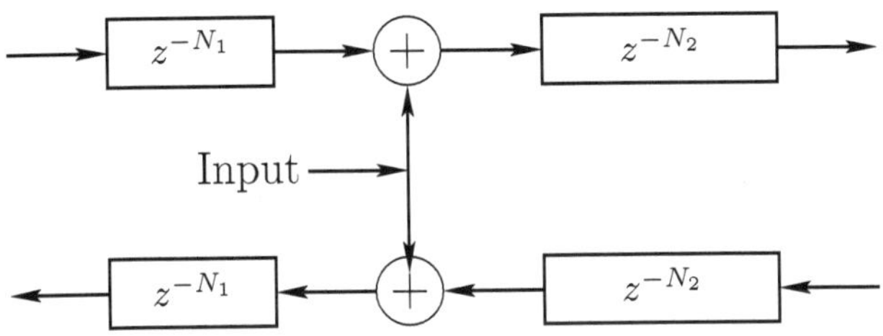

Figure 2.14: Summing a signal into a digital waveguide corresponding to a superimposing disturbance at one point. The original state is unaffected, *i.e.*, the input signal enters the waveguide in *superposition* with whatever is already going on.

Note that the superimposing input of Fig. 2.14 is the graph-theoretic *transpose* of the ideal output shown in Fig. 2.13. In other words, the superimposing input injects by means of two *transposed taps*. Transposed taps are discussed further in §2.5.2 below.

In practical reality, physical driving inputs do not merely superimpose with the current state of the driven system. Instead, there is normally some amount of *interaction* with the current system state (when it is nonzero), as discussed further in the next section. Note that there are similarly no ideal outputs as depicted in Fig. 2.13. Real physical ouputs must present some kind of *load* on the system (energy must be extracted). Superimposing inputs and non-loading outputs are ideals that are often approximated in real-world systems. Of course, in the virtual world, they are no problem at all—in fact, they are usually easier to implement, and more efficient.

[5]The signs may differ, however. For example, imagine *twisting* a stretched string about a point along its length in the plane of transverse vibration. In that case, the transverse-displacement pulses propagating away from the twist have opposite sign. This opposite-sign effect happens also for *slope waves* emanating from a pluck disturbance, as we'll see in Chapter 6. (A positive pluck corresponds to a positive slope on its left and a negative slope on its right.)

2.4.3 Interacting Physical Input

Figure 2.15 shows the general case of an input signal that interacts with the state of the system at one point along the waveguide. Since the interaction is physical, it only depends on the "incoming state" (traveling-wave components) and the driving input signal.

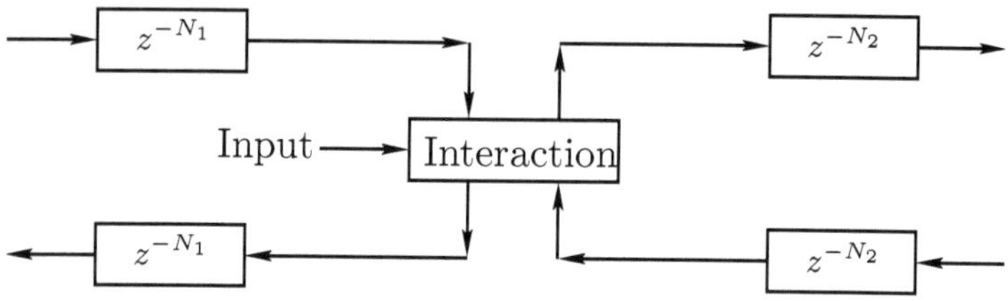

Figure 2.15: Driving a digital waveguide with physical *interaction* between the driving input signal and the current state of the waveguide.

A less general but commonly encountered case is shown in Fig. 2.16. This case requires the "outgoing disturbance" to be distributed equally to the left and right, and it sums with the incoming waves to produce the outgoing waves.

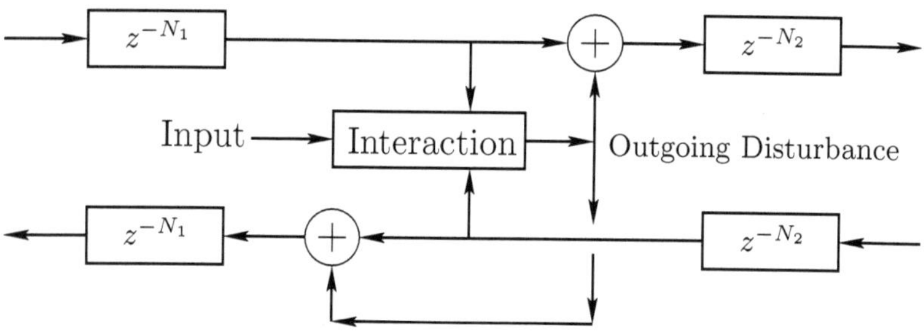

Figure 2.16: Symmetric outgoing disturbance in superposition with incoming waves.

Figure 2.17 shows a further reduction in generality—also commonly encountered—in which the interaction depends only on the *amplitude* of the simulated physical variable (such as string velocity or displacement). The incoming amplitude is formed as the sum of the incoming traveling-wave components. We will encounter examples of this nature in later chapters (such as Chapter 9). It provides realistic models of physical excitations such as a guitar plectra, violin bows, and woodwind reeds.

If an output signal is desired at this precise point, it can be computed as

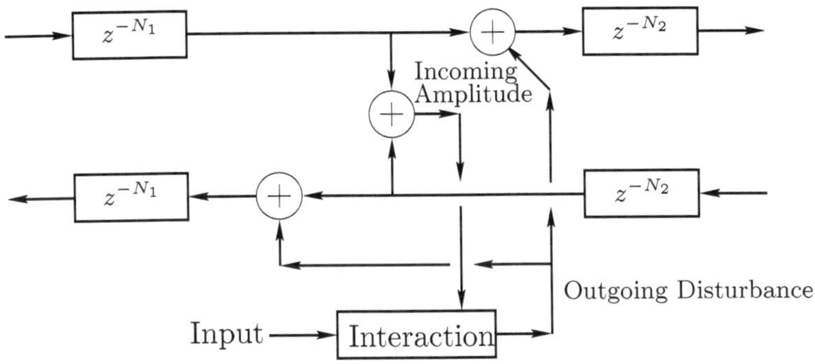

Figure 2.17: Same as Fig. 2.15 for the case in which the interaction depends only upon incoming *amplitude*.

the incoming amplitude plus twice the outgoing disturbance signal (equivalent to summing the inputs of the two outgoing delay lines).

Note that the above examples all involve waveguide excitation at a single spatial point. While this can give a sufficiently good approximation to physical reality in many applications, one should also consider excitations that are spread out over multiple spatial samples (even just two).

We will develop the topic of digital waveguide modeling more systematically in Chapter 6 and Appendix C, among other places in this book. This section is intended only as a high-level preview and overview. For the next several chapters, we will restrict attention to normal signal processing structures in which signals may have physical units (such as acoustic pressure), and delay lines hold sampled acoustic waves propagating in one direction, but successive processing blocks do not "load each other down" or connect "bidirectionally" (as every truly physical interaction must, by Newton's third law[6]). Thus, when one processing block feeds a signal to a next block, an "ideal output" drives an "ideal input". This is typical in digital signal processing: Loading effects and return waves[7] are neglected.[8]

[6]Newton's third law states that "every action produces an equal and opposite reaction." Therefore, every "physical" signal connection must be *bidirectional*, in order to model both the applied force and the equal reaction force, or "loading effect". Bidirectional signal connections are typically called *ports*. A "port" (one physical connection point) in electrical circuit theory consists of two variables, typically voltage and current. In acoustics, corresponding port variables would be pressure and velocity. As discussed later (*e.g.*, for Wave Digital Filters in Appendix F), a simple 2x2 matrix linear transformation converts "force and flow" port variables, such as these, to "traveling-wave component" port variables.

[7]Loading effects can be viewed as the result of instantaneous return waves. This equivalence is seen clearly, for example, in Wave Digital Filters (Appendix F).

[8]This follows years of standard practice in analog electric circuit design using *voltage*

2.5 Tapped Delay Line (TDL)

A *tapped delay line* (TDL) is a delay line with at least one "tap". A delay-line *tap* extracts a signal output from somewhere within the delay line, optionally scales it, and usually sums with other taps for form an output signal. A tap may be *interpolating* or *non-interpolating.* A non-interpolating tap extracts the signal at some fixed integer delay relative to the input. Thus, a tap implements a shorter delay line within a larger one, as shown in Fig. 2.18.

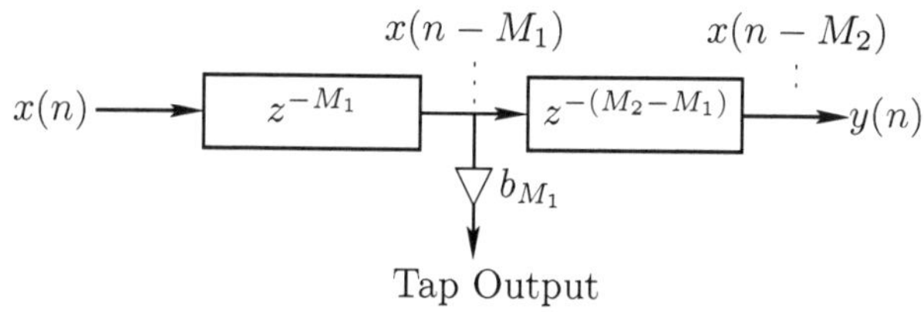

Figure 2.18: A delay line with one *tap*.

Tapped delay lines efficiently simulate *multiple echoes* from the same source signal. As a result, they are extensively used in the field of *artificial reverberation.*

2.5.1 Example Tapped Delay Line

An example of a TDL with two internal taps is shown in Fig. 2.19. The total delay line length is M_3 samples, and the internal taps are located at delays of M_1 and M_2 samples, respectively. The output signal is a linear combination of the input signal $x(n)$, the delay-line output $x(n-M_3)$, and the two tap signals $x(n-M_1)$ and $x(n-M_2)$.

The difference equation of the TDL in Fig. 2.19 is, by inspection,

$$y(n) = b_0 x(n) + b_{M_1} x(n-M_1) + b_{M_2} x(n-M_2) + b_{M_3} x(n-M_3)$$

corresponding to the transfer function

$$H(z) = b_0 + b_{M_1} z^{-M_1} + b_{M_2} z^{-M_2} + b_{M_3} z^{-M_3}$$

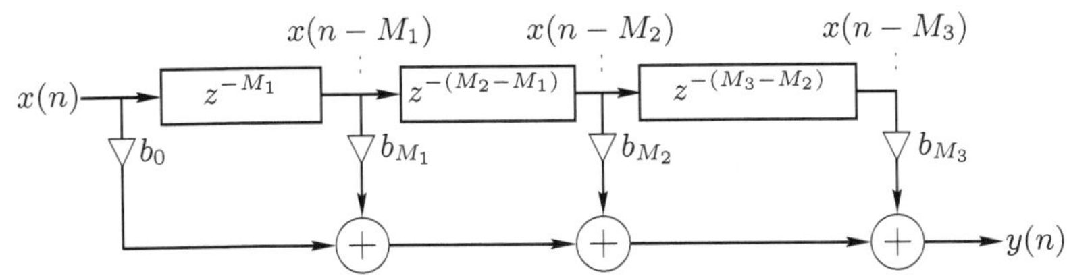

Figure 2.19: Tapped Delay Line (TDL).

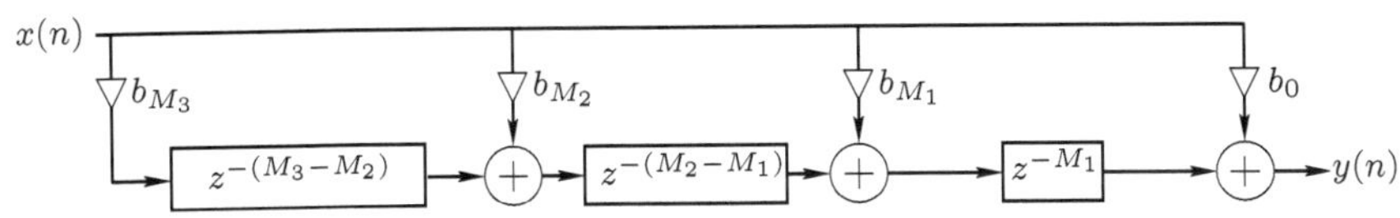

Figure 2.20: *Transposed* Tapped Delay Line (TTDL).

2.5.2 Transposed Tapped Delay Line

In many applications, the *transpose* of a tapped delay line is desired, as shown in Fig. 2.20, which is the transpose of the tapped delay line shown in Fig. 2.19. A transposed TDL is obtained from a normal TDL by formal *transposition* of the system diagram. The transposition operation is also called *flow-graph reversal* [333, pp. 153–155]. A flow-graph is transposed by reversing all signal paths, which necessitates signal branchpoints becoming sums, and sums becoming branchpoints. For single-input, single-output systems, the transfer function is the same, but the input and output are interchanged. This "flow-graph reversal theorem" derives from *Mason's gain formula* for signal flow graphs. Transposition is used to convert direct-forms I and II of a digital filter to direct-forms III and IV, respectively [333].

2.5.3 TDL for Parallel Processing

When multiplies and additions can be performed in parallel, the computational complexity of a tapped delay line is $\mathcal{O}(1)$ multiplies and $\mathcal{O}(\lg(K))$ additions, where K is the number of taps. This computational complexity is achieved by arranging the additions into a *binary tree*, as shown in Fig. 2.21 for the case $K = 4$.

---

transfer interconnections. For "voltage transfer" from one circuit-stage to another with minimized loading effects, output impedances are made as low as practical, while input impedances are set very high. For example, in analog synthesizers built during the 1970s, input impedances were typically around 100 times output impedances (*e.g.*, 100K and 1K Ohms, respectively in a typical op-amp circuit) [204].

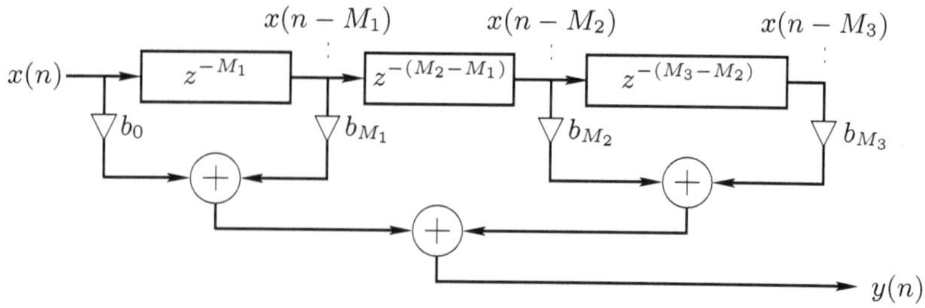

Figure 2.21: An example Tapped Delay Line (TDL), with additions organized into a *binary tree* for maximized parallel computation.

2.5.4 General Causal FIR Filters

The most general case—a TDL having a tap after *every* delay element—is the general causal *Finite Impulse Response (FIR)* filter, shown in Fig. 2.22. It is restricted to be *causal* because the output $y(n)$ may not depend on *"future"* inputs $x(n+1)$, $x(n+2)$, etc. The FIR filter is also called a *transversal filter*. FIR filters are described in greater detail in [449].

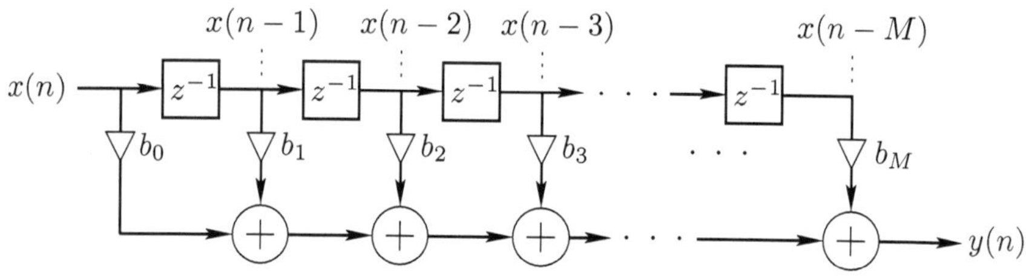

Figure 2.22: The general, causal, finite-impulse-response (FIR) digital filter.

The *difference equation* for the Mth-order FIR filter in Fig. 2.22 is, by inspection,

$$y(n) = b_0 x(n) + b_1 x(n-1) + b_2 x(n-2) + b_3 x(n-3) + \cdots + b_M x(n-M)$$

and the *transfer function* is

$$H(z) = b_0 + b_1 z^{-1} + b_2 z^{-2} + b_3 z^{-3} + \cdots + b_M z^{-M} = \sum_{m=0}^{M} b_m z^{-m} \triangleq B(z).$$

The STK class for implementing arbitrary direct-form FIR filters is called Fir[9]. (There is also a class for IIR filters named Iir.) In Matlab and Octave, the built-in function `filter` is normally used.

[9] `http://ccrma.stanford.edu/software/stk/classes.html`

2.6 Comb Filters

Comb filters are basic building blocks for digital audio effects. The acoustic echo simulation in Fig. 2.9 on page 44 is one instance of a comb filter. This section presents the two basic comb-filter types, *feedforward* and *feedback*, and gives a frequency-response analysis.

2.6.1 Feedforward Comb Filters

The *feedforward comb filter* is shown in Fig. 2.23. The direct signal "feeds forward" around the delay line. The output is a linear combination of the direct and delayed signal.

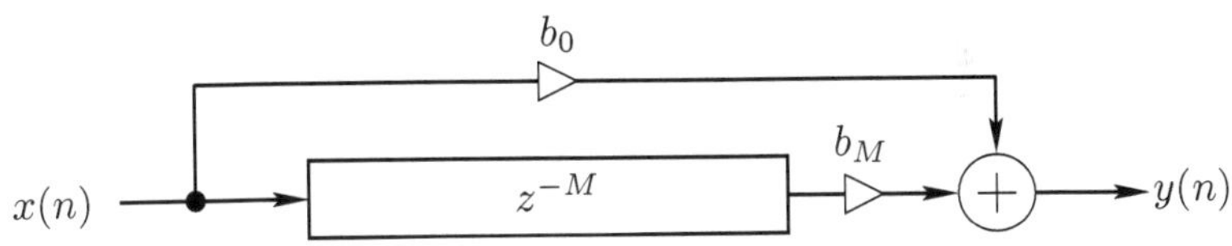

Figure 2.23: The feedforward comb filter.

The "difference equation" [449] for the feedforward comb filter is

$$y(n) = b_0 x(n) + b_M x(n - M). \tag{2.2}$$

We see that the feedforward comb filter is a particular type of FIR filter. It is also a special case of a TDL.

Note that the feedforward comb filter can implement the echo simulator of Fig. 2.9 by setting $b_0 = 1$ and $b_M = g$. Thus, it is is a *computational physical model* of a single discrete echo. This is one of the simplest examples of acoustic modeling using signal processing elements. The feedforward comb filter models the superposition of a "direct signal" $b_0 x(n)$ plus an attenuated, delayed signal $b_M x(n-M)$, where the attenuation (by $|b_M| < 1$) is due to "air absorption" and/or spherical spreading losses, and the delay is due to acoustic propagation over the distance cMT meters, where T is the sampling period in seconds, and c is sound speed. In cases where the simulated propagation delay needs to be more accurate than the nearest integer number of samples M, some kind of *delay-line interpolation* needs to be used (the subject of §4.1). Similarly, when air absorption needs to be simulated more accurately, the constant attenuation factor b_M can be replaced by a linear, time-invariant filter $G(z)$ giving a different attenuation at every frequency. Due to the physics of air absorption, $G(z)$ is generally lowpass in character [349, p. 560], [47, 318].

2.6.2 Feedback Comb Filters

The *feedback comb filter* uses feedback instead of a feedforward signal, as shown in Fig. 2.24 (drawn in "direct form 2" [449]).

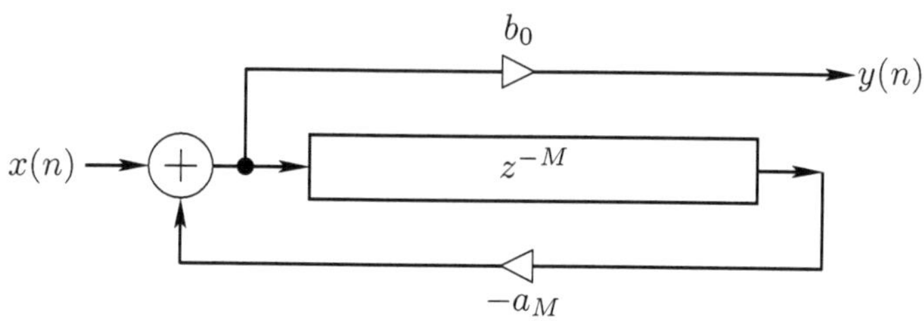

Figure 2.24: The feedback comb filter.

A difference equation describing the feedback comb filter can be written in "direct form 1" [449] as[10]

$$y(n) = b_0 x(n) - a_M y(n - M).$$

The feedback comb filter is a special case of an Infinite Impulse Response (IIR) ("recursive") digital filter, since there is *feedback* from the delayed output to the input [449]. The feedback comb filter can be regarded as a computational physical model of a *series* of echoes, exponentially decaying and uniformly spaced in time. For example, the special case

$$y(n) = x(n) + g\,y(n - M)$$

is a computational model of an ideal plane wave bouncing back and forth between two parallel walls; in such a model, g represents the total round-trip attenuation (two wall-to-wall traversals, including two reflections).

For *stability*, the feedback coefficient a_M must be less than 1 in magnitude, *i.e.*, $|a_M| < 1$. Otherwise, if $|a_M| > 1$, each echo will be louder than the previous echo, producing a never-ending, growing series of echoes.

Sometimes the output signal is taken from the end of the delay line instead of the beginning, in which case the difference equation becomes

$$y(n) = b_M x(n - M) - a_M y(n - M).$$

This choice of output merely delays the output signal by M samples.

[10] Note that Fig. 2.24 is given in direct form 2, so its describing equations are $v(n) = x(n) - a_M v(n - M)$; $y(n) = b_0 v(n)$. To make the figure correspond exactly to the direct-form-1 difference equation, simply move b_0 from the output arrow to the input arrow. That is, commute the scaling by b_0 and the feedback loop so that b_0 is first.

2.6.3 Feedforward Comb Filter Amplitude Response

Comb filters get their name from the "comb-like" appearance of their amplitude response (gain versus frequency), as shown in Figures 2.25, 2.26, and 2.27. For a review of frequency-domain analysis of digital filters, see, *e.g.*, [449].

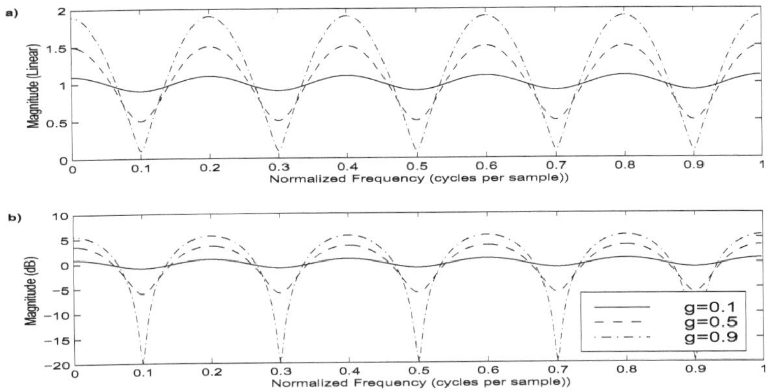

Figure 2.25: Amplitude responses of the *feed forward* comb-filter $H(z) = x(n) + gx(n-M)$ (diagrammed in Fig. 2.23 on page 57) with $M = 5$ and $g = 0.1$, 0.5, and 0.9. a) Linear amplitude scale. b) Decibel scale. The frequency axis goes from 0 to the *sampling rate* (instead of only *half* the sampling rate, which is more typical for real filters) in order to display the fact that the number of notches is exactly $M = 5$ (as opposed to "2.5").

The transfer function of the feedforward comb filter Eq. (2.2) is

$$H(z) = b_0 + b_M\, z^{-M}, \tag{2.3}$$

so that the amplitude response (gain versus frequency) is

$$G(\omega) \triangleq \left|H(e^{j\omega})\right| = \left|b_0 + b_M e^{-j\omega M}\right|, \quad -\pi \le \omega \le \pi. \tag{2.4}$$

This is plotted in Fig. 2.25 for $M = 5$, $b_0 = 1$, and $b_M = 0.1$, 0.5, and 0.9. When $b_0 = b_M = 1$, we get the simplified result

$$G(\omega) = \left|1 + e^{-j\omega M}\right| = \left|e^{-j\omega M/2}\right| \left|e^{j\omega M/2} + e^{-j\omega M/2}\right| = 2\left|\cos\left(\omega \frac{M}{2}\right)\right|.$$

In this case, we obtain M *nulls*, which are points (frequencies) of zero gain in the amplitude response. Note that in *flangers*, these nulls are *moved* slowly over time by modulating the delay length M. Doing this smoothly requires interpolated delay lines (see Chapter 4 and Chapter 5).

2.6.4 Feedback Comb Filter Amplitude Response

Figure 2.26 shows a family of *feedback*-comb-filter amplitude responses, obtained using a selection of feedback coefficients.

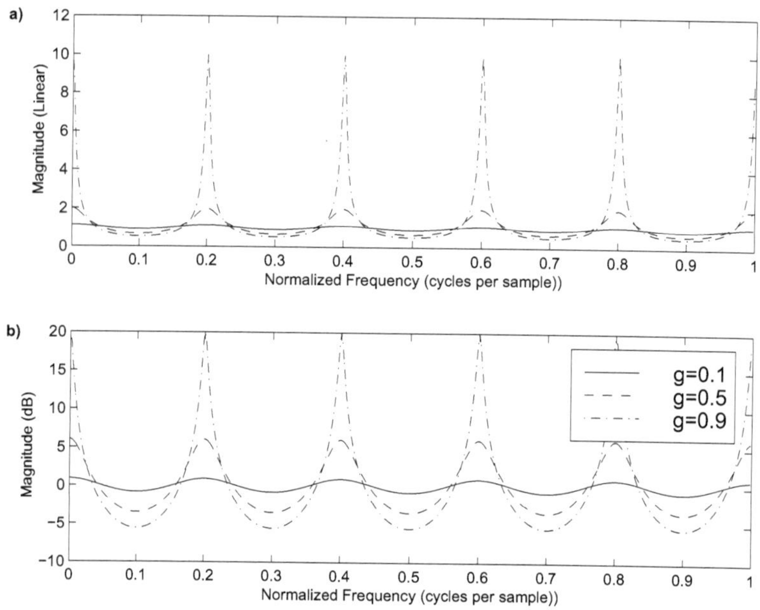

Figure 2.26: Amplitude response of the *feedback* comb-filter $H(z) = 1/(1-gz^{-M})$ (Fig. 2.24 with $b_0 = 1$ and $-a_M = g$) with $M = 5$ and $g = 0.1, 0.5,$ and 0.9. a) Linear amplitude scale. b) Decibel scale.

Figure 2.27 shows a similar family obtained using *negated* feedback coefficients; the opposite sign of the feedback exchanges the peaks and valleys in the amplitude response.

As introduced in §2.6.2 above, a class of feedback comb filters can be defined as any difference equation of the form

$$y(n) = x(n) + g\,y(n-M).$$

Taking the z transform of both sides and solving for $H(z) \triangleq Y(z)/X(z)$, the transfer function of the feedback comb filter is found to be

$$H(z) = \frac{1}{1 - g\,z^{-M}}, \tag{2.5}$$

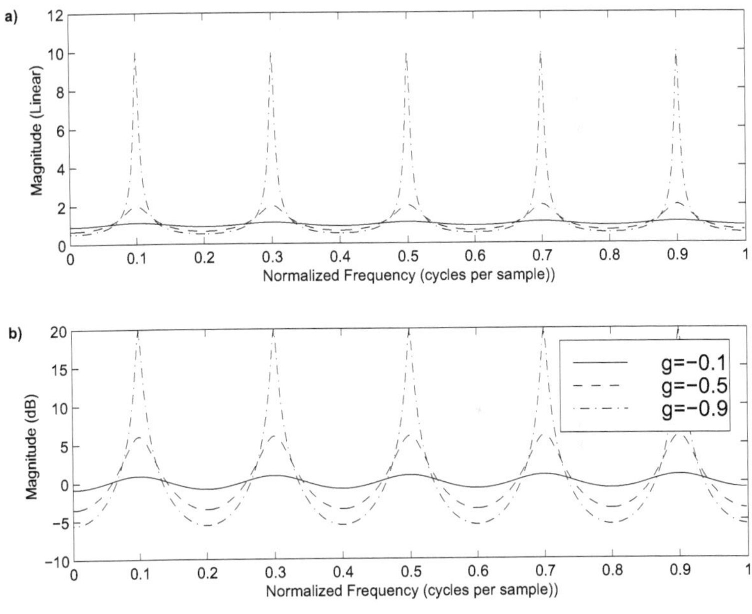

Figure 2.27: Amplitude response of the *phase-inverted* feedback comb-filter, *i.e.*, as in Fig. 2.26 with negated $g = -0.1$, -0.5, and -0.9. a) Linear amplitude scale. b) Decibel scale.

so that the amplitude response is

$$G(\omega) \stackrel{\Delta}{=} \left|H(e^{j\omega})\right| = \frac{1}{|1 - ge^{-j\omega M}|}, \quad -\pi \leq \omega \leq \pi.$$

This is plotted in Fig. 2.26 for $M = 5$ and $g = 0.1$, 0.5, and 0.9. Figure 2.27 shows the same case but with the feedback *sign-inverted.*

For $g = 1$, the feedback-comb amplitude response reduces to

$$G(\omega) = \frac{1}{2\left|\sin(\omega M/2)\right|},$$

and for $g = -1$ to

$$G(\omega) = \frac{1}{2\left|\cos(\omega M/2)\right|},$$

which exactly inverts the amplitude response of the feedforward comb filter with gain $g = 1$ (Eq. (2.4)).

Note that $g > 0$ produces resonant peaks at

$$\omega_k = 2\pi\frac{k}{M}, \quad k = 0, 1, 2, \ldots, M-1,$$

while for $g < 0$, the peaks occur midway between these values.

2.6.5 Filtered-Feedback Comb Filters

The *filtered-feedback comb filter* (FFBCF) uses filtered feedback instead of just a feedback gain.

Denoting the feedback-filter transfer function by $H_l(z)$, the transfer function of the filtered-feedback comb filter can be written as

$$H(z) = \frac{b_0}{1 - H_l(z)z^{-M}}.$$

Note that when $H_l(z)$ is a causal filter, the FFBCF can be considered mathematically a special case of the general allpole transfer function in which the first $M-1$ denominator coefficients are constrained to be zero:

$$H(z) = \frac{b_0}{1 + 0 + \cdots + 0 + a_M z^{-M} + a_{M+1} z^{-M+1} + \cdots}$$

It is this "sparseness" of the filter coefficients that makes the FFBCF more computationally efficient than other, more general-purpose, IIR filter structures.

In §2.6.2 above, we mentioned the physical interpretation of a feedback-comb-filter as simulating a plane-wave bouncing back and forth between two walls. Inserting a lowpass filter in the feedback loop further simulates frequency dependent *losses* incurred during a propagation round-trip, as naturally occurs in real rooms.

The main physical sources of plane-wave attenuation are *air absorption* (§B.7.15 on page 513) and the *coefficient of absorption* at each wall [349]. Additional "losses" for plane waves in real rooms occur due to *scattering*. (The plane wave hits something other than a wall and reflects off in many different directions.) A particular scatterer used in concert halls is *textured wall surfaces*. In ray-tracing simulations, reflections from such walls are typically modeled as having a *specular* and *diffuse* component. Generally speaking, wavelengths that are large compared with the "grain size" of the wall texture reflect specularly (with some attenuation due to any wall motion), while wavelengths on the order of or smaller than the texture grain size are scattered in various directions, contributing to the diffuse component of reflection.

The filtered-feedback comb filter has many applications in computer music. It was evidently first suggested for artificial reverberation by Schroeder [412, p. 223], and first implemented by Moorer [314]. (Reverberation applications are discussed further in §3.6 on page 101.) In the physical interpretation [428, 207] of the Karplus-Strong algorithm [236, 233], the FFBCF can be regarded as a transfer-function physical-model of a vibrating string. In digital waveguide modeling of string and wind instruments, FFBCFs are typically derived routinely as a computationally optimized equivalent forms based on some initial waveguide model developed in terms of bidirectional delay-lines ("digital waveguides") (see §6.10.1 on page 227 for an example).

For *stability*, the amplitude-response of the feedback-filter $H_l(z)$ must be less than 1 in magnitude at all frequencies, *i.e.*, $\left|H_l(e^{j\omega T})\right| < 1$, $\forall \omega T \in [-\pi, \pi)$.

2.6.6 Equivalence of Parallel Combs to TDLs

It is easy to show that the TDL of Fig. 2.19 is equivalent to a *parallel combination* of three feedforward comb filters, each as in Fig. 2.23. To see this, we simply add the three comb-filter transfer functions of Eq. (2.3) and equate coefficients:

$$\begin{aligned} H(z) &= \left(1+g_1 z^{-M_1}\right) + \left(1+g_2 z^{-M_2}\right) + \left(1+g_3 z^{-M_3}\right) \\ &= 3 + g_1 z^{-M_1} + g_2 z^{-M_2} + g_3 z^{-M_3} \end{aligned}$$

which implies

$$b_0 = 3,\ b_{M_1} = g_1,\ b_{M_2} = g_2,\ b_{M_3} = g_3.$$

We see that parallel comb filters require *more delay memory* ($M_1 + M_2 + M_3$ elements) than the corresponding TDL, which only requires $\max(M_1, M_2, M_3)$ elements.

2.6.7 Equivalence of Series Combs to TDLs

It is also straightforward to show that a *series combination* of feedforward comb filters produces a sparsely tapped delay line as well. Considering the case of two sections, we have

$$\begin{aligned} H(z) &= \left(1 + g_1 z^{-M_1}\right)\left(1 + g_2 z^{-M_2}\right) \\ &= 1 + g_1 z^{-M_1} + g_2 z^{-M_2} + g_1 g_2 z^{-(M_1+M_2)} \end{aligned}$$

which yields

$$b_0 = 1,\ b_{M_1} = g_1,\ b_{M_2} = g_2,\ M_3 = M_1 + M_2,\ b_{M_3} = g_1 g_2.$$

Thus, the TDL of Fig. 2.19 is equivalent also to the *series* combination of *two* feedforward comb filters. Note that the same TDL structure results irrespective of the series ordering of the component comb filters.

2.6.8 Time Varying Comb Filters

Comb filters can be changed slowly over time to produce the following digital audio "effects", among others:

- Phasing
- Flanging
- Chorus
- Leslie

Since all of these effects involve modulating *delay length* over time, and since time-varying delay lines typically require *interpolation*, these applications will be discussed after Chapter 5 which covers variable delay lines. For now, we will pursue what can be accomplished using *fixed* (time-invariant) delay lines. Perhaps the most important application is *artificial reverberation*, addressed in Chapter 3.

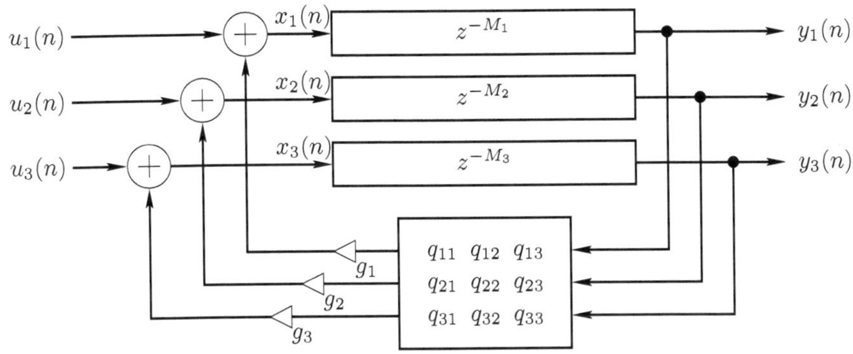

Figure 2.28: Order 3 MIMO Feedback Delay Network (FDN).

2.7 Feedback Delay Networks (FDN)

The FDN can be seen as a *vector feedback comb filter*,[11] obtained by replacing the delay line with a diagonal delay matrix (defined in Eq. (2.10) below), and replacing the feedback gain g by the product of a diagonal matrix $\mathbf{\Gamma}$ times an orthogonal matrix $\mathbf{Q}$, as shown in Fig. 2.28 for $N = 3$. The time-update for this FDN can be written as

$$\begin{bmatrix} x_1(n) \\ x_2(n) \\ x_3(n) \end{bmatrix} = \begin{bmatrix} g_1 & 0 & 0 \\ 0 & g_2 & 0 \\ 0 & 0 & g_3 \end{bmatrix} \begin{bmatrix} q_{11} & q_{12} & q_{13} \\ q_{21} & q_{22} & q_{23} \\ q_{31} & q_{32} & q_{33} \end{bmatrix} \begin{bmatrix} x_1(n-M_1) \\ x_2(n-M_2) \\ x_3(n-M_3) \end{bmatrix} + \begin{bmatrix} u_1(n) \\ u_2(n) \\ u_3(n) \end{bmatrix} \tag{2.6}$$

with the outputs given by

$$\begin{bmatrix} y_1(n) \\ y_2(n) \\ y_3(n) \end{bmatrix} = \begin{bmatrix} x_1(n-M_1) \\ x_2(n-M_2) \\ x_3(n-M_3) \end{bmatrix}, \tag{2.7}$$

or, in frequency-domain vector notation,

$$\mathbf{X}(z) = \mathbf{\Gamma}\mathbf{Q}\mathbf{D}(z)\mathbf{X}(z) + \mathbf{U}(z) \tag{2.8}$$

$$\mathbf{Y}(z) = \mathbf{D}(z)\mathbf{X}(z) \tag{2.9}$$

[11]Readers of the main chapters of this book are not required to know linear algebra, but one does not have to learn much to follow the discussion in this section. See, for example, [451, available online] for the basic matrix operations. An excellent linear algebra text is [329]. See also the lecture videos by Strang by searching for "linear algebra" at `http://ocw.mit.edu/`.

where

$$\mathbf{D}(z) \triangleq \begin{bmatrix} z^{-M_1} & 0 & 0 \\ 0 & z^{-M_2} & 0 \\ 0 & 0 & z^{-M_3} \end{bmatrix}. \tag{2.10}$$

2.7.1 FDN and State Space Descriptions

When $M_1 = M_2 = M_3 = 1$ in Eq. (2.10), the FDN (Fig. 2.28) reduces to a normal *state-space model* (§1.3.7, p. 16),

$$\begin{aligned} \mathbf{x}(n+1) &= \mathbf{A}\,\mathbf{x}(n) + \mathbf{u}_+(n) \\ \mathbf{y}_+(n) &= \mathbf{x}(n) \end{aligned} \tag{2.11}$$

The matrix $\mathbf{A} = \mathbf{\Gamma Q}$ is the *state transition matrix*. The vector $\mathbf{x}(n) = [x_1(n), x_2(n), x_3(n)]^T$ holds the *state variables* that determine the state of the system at time n. The *order* of a state-space system is equal to the number of state variables, *i.e.*, the dimensionality of $\mathbf{x}(n)$. The input and output signals have been trivially redefined as

$$\begin{aligned} \mathbf{u}_+(n) &\triangleq \mathbf{u}(n+1) \\ \mathbf{y}_+(n) &\triangleq \mathbf{y}(n+1) \end{aligned}$$

to follow normal convention for state-space form.

Thus, an FDN can be viewed as a generalized state-space model for a class of Nth-order linear systems—"generalized" in the sense that unit delays are replaced by arbitrary delays. This correspondence is valuable for analysis because tools for state-space analysis are well known and included in many software libraries such as with matlab.

2.7.2 Single-Input, Single-Output (SISO) FDN

When there is only one input signal $u(n)$, the input vector $\mathbf{u}(n)$ in Fig. 2.28 can be defined as the scalar input $u(n)$ times a vector of gains:

$$\mathbf{u}(n) = \mathbf{B}u(n)$$

where $\mathbf{B}$ is an $N \times 1$ matrix. Similarly, a single output can be created by taking an arbitrary linear combination of the N components of $\mathbf{y}(n)$. An example single-input, single-output (SISO) FDN for $N = 3$ is shown in Fig. 2.29.

Note that when $M_1 = M_2 = M_3 = 1$, this system is capable of realizing *any* transfer function of the form

$$H(z) = \frac{\beta_1 z^{-1} + \beta_2 z^{-2} + \beta_3 z^{-3}}{1 + a_1 z^{-1} + a_2 z^{-2} + a_3 z^{-3}}.$$

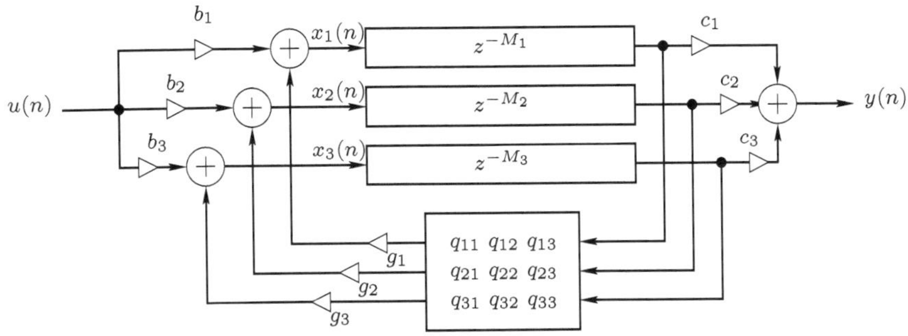

Figure 2.29: Order 3 SISO Feedback Delay Network (FDN).

By elementary state-space analysis [449, Appendix E], the transfer function can be written in terms of the FDN system parameters as

$$H(z) = \mathbf{C}^T(z\mathbf{I} - \mathbf{A})^{-1}\mathbf{B}$$

where $\mathbf{I}$ denotes the 3×3 identity matrix. This is easy to show by taking the z transform of the impulse response of the system.

The more general case shown in Fig. 2.29 can be handled in one of two ways: (1) the matrices $\mathbf{A}, \mathbf{B}, \mathbf{C}$ can be *augmented* to order $N = M_1 + M_2 + M_3$ such that the three delay lines are replaced by N unit-sample delays, or (2) ordinary state-space analysis may be *generalized* to non-unit delays, yielding

$$H(z) = \mathbf{C}^T\mathbf{D}(z)\left[\mathbf{I} - \mathbf{A}\mathbf{D}(z)\right]^{-1}\mathbf{B}$$

where $\mathbf{C}^T$ denotes the matrix transpose of $\mathbf{C}$, and

$$\mathbf{D}(z) \isdef \left[\begin{array}{ccc} z^{-M_1} & 0 & 0 \\ 0 & z^{-M_2} & 0 \\ 0 & 0 & z^{-M_3} \end{array}\right].$$

In FDN reverberation applications, $\mathbf{A} = \mathbf{\Gamma}\mathbf{Q}$, where $\mathbf{Q}$ is an orthogonal matrix, for reasons addressed below, and $\mathbf{\Gamma}$ is a diagonal matrix of lowpass filters, each having gain bounded by 1. In certain applications, the subset of orthogonal matrices known as *circulant matrices* have advantages [385].

2.7.3 FDN Stability

Stability of the FDN is assured when some *norm* [451] of the state vector $\mathbf{x}(n)$ decreases over time when the input signal is zero [220, "Lyapunov stability theory"]. That is, a sufficient condition for FDN stability is

$$\|\,\mathbf{x}(n+1)\,\| < \|\,\mathbf{x}(n)\,\|\,, \tag{2.12}$$

for all $n \geq 0$, where $\| \mathbf{x}(n) \|$ denotes the norm of $\mathbf{x}(n)$, and

$$\mathbf{x}(n+1) = \mathbf{A} \begin{bmatrix} x_1(n-M_1) \\ x_2(n-M_2) \\ x_3(n-M_3) \end{bmatrix}.$$

Using the augmented state-space analysis mentioned above, the inequality of Eq. (2.12) holds under the L^2 norm [451] whenever the feedback matrix $\mathbf{A}$ in Eq. (2.6) satisfies [474]

$$\| \mathbf{Ax} \|_2 < \| \mathbf{x} \|_2 \tag{2.13}$$

for all $\mathbf{x}$, where $\| \cdot \|_2$ denotes the L^2 *norm*, defined by

$$\| \mathbf{x} \|_2 \triangleq \sqrt{x_1^2 + x_2^2 + \cdots + x_N^2}.$$

In other words, stability is guaranteed when the feedback matrix decreases the L^2 norm of its input vector.

The *matrix norm* corresponding to any vector norm $\| \cdot \|$ may be defined for the matrix $\mathbf{A}$ as

$$\| \mathbf{A} \| \triangleq \max_{\mathbf{x} \neq \mathbf{0}} \frac{\| \mathbf{Ax} \|}{\| \mathbf{x} \|}$$

where $\| \mathbf{x} \|$ denotes the norm of the vector $\mathbf{x}$. In other words, the matrix norm "induced" by a vector norm is given by the maximum of $\| \mathbf{Ax} \|$ over all unit-length vectors $\mathbf{x}$ in the space. When the vector norm is the L^2 norm, the induced matrix norm is often called the *spectral norm*. Thus, Eq. (2.13) can be restated as

$$\| \mathbf{A} \|_2 < 1 \tag{2.14}$$

where $\| \mathbf{A} \|_2$ denotes the spectral norm of $\mathbf{A}$.

It can be shown [167] that the spectral norm of a matrix $\mathbf{A}$ is given by the largest singular value of $\mathbf{A}$ ("$\| \mathbf{A} \|_2 = \sigma_1(\mathbf{A})$"), and that this is equal to the square-root of the largest eigenvalue of $\mathbf{AA}^T$, where $\mathbf{A}^T$ denotes the matrix transpose of the real matrix $\mathbf{A}$.[12]

Since every *orthogonal matrix* $\mathbf{Q}$ has spectral norm 1,[13] a wide variety of stable feedback matrices can be parametrized as

$$\mathbf{A} = \mathbf{\Gamma Q}$$

---

[12]If $\mathbf{A}$ is complex, transposition must include complex conjugation, *i.e.*, $\mathbf{A}^* \triangleq \overline{\mathbf{A}^T}$. The conjugate-transpose $\mathbf{A}^*$ is called the *Hermitian transpose* of $\mathbf{A}$.

[13]By definition, $\mathbf{QQ}^T = \mathbf{I}$, so that all singular values of an orthogonal matrix are equal to 1.

where $\mathbf{Q}$ is any orthogonal matrix, and $\mathbf{\Gamma}$ is a diagonal matrix having entries less than 1 in magnitude:

$$\mathbf{\Gamma} = \left[\begin{array}{cccc} g_1 & 0 & \dots & 0 \\ 0 & g_2 & \dots & 0 \\ \vdots & \vdots & \ddots & \vdots \\ 0 & 0 & \dots & g_N \end{array}\right], \quad |g_i| < 1.$$

An alternative stability proof may be based on showing that an FDN is a special case of a passive digital waveguide network (derived in §C.15 on page 612). This analysis reveals that the FDN is lossless if and only if the feedback matrix $\mathbf{A}$ has unit-modulus eigenvalues and linearly independent eigenvectors.

2.8 Allpass Filters

The *allpass filter* is an important building block for digital audio signal processing systems. It is called "allpass" because all frequencies are "passed" in the same sense as in "lowpass", "highpass", and "bandpass" filters. In other words, the amplitude response of an allpass filter is 1 at each frequency, while the phase response (which determines the delay versus frequency) can be arbitrary.

In practice, a filter is often said to be allpass if the amplitude response is any nonzero constant. However, in this book, the term "allpass" refers to *unity gain* at each frequency.

In this section, we will first make an allpass filter by cascading a feedback comb-filter with a feedforward comb-filter. This structure, known as the *Schroeder allpass comb filter*, or simply the *Schroeder allpass* section, is used extensively in the fields of artificial reverberation and digital audio effects. Next we will look at creating allpass filters by *nesting* them; allpass filters are nested by replacing delay elements (which are allpass filters themselves) with arbitrary allpass filters. Finally, we will consider the general case, and characterize the set of all single-input, single-output allpass filters. The general case, including multi-input, multi-output (MIMO) allpass filters, is treated in [449, Appendix D].

2.8.1 Allpass from Two Combs

An *allpass filter* can be defined as any filter having a gain of 1 at all frequencies (but typically different delays at different frequencies).

It is well known that the series combination of a feedforward and feedback comb filter (having equal delays) creates an allpass filter when the feedforward coefficient is the negative of the feedback coefficient.

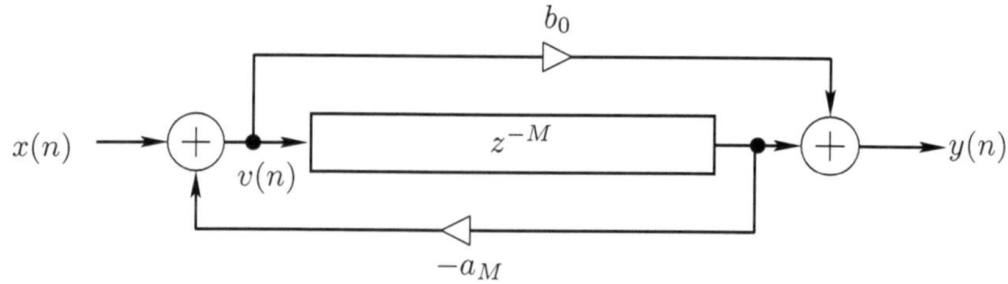

Figure 2.30: A combined feedback/feedforward comb filter which gives an allpass filter when $b_0 = a_M$.

Figure 2.30 shows a combination feedforward/feedback comb filter structure which shares the same delay line.[14] By inspection of Fig. 2.30, the difference equation is

$$\begin{aligned} v(n) &= x(n) - a_M v(n-M) \\ y(n) &= b_0 v(n) + v(n-M). \end{aligned}$$

This can be recognized as a digital filter in direct form II [449]. Thus, the system of Fig. 2.30 can be interpreted as the series combination of a feedback comb filter (Fig. 2.24 on page 58) taking $x(n)$ to $v(n)$ followed by a feedforward comb filter (Fig. 2.23) taking $v(n)$ to $y(n)$. By the commutativity of LTI systems, we can interchange the order to get

$$\begin{aligned} v(n) &= b_0 x(n) + x(n-M) \\ y(n) &= v(n) - a_M y(n-M). \end{aligned}$$

Substituting the right-hand side of the first equation above for $v(n)$ in the second equation yields more simply

$$y(n) = b_0 x(n) + x(n-M) - a_M y(n-M). \tag{2.15}$$

This can be recognized as direct form I [449], which requires $2M$ delays instead of M; however, unlike direct-form II, direct-form I cannot suffer from "internal" overflow—overflow can happen only at the output.

The coefficient symbols b_0 and a_M here have been chosen to correspond to standard notation for the *transfer function*

$$H(z) = \frac{b_0 + z^{-M}}{1 + a_M z^{-M}}.$$

[14] This structure may be quickly derived by forming a series combination of a feedback comb filter followed by a feedforward comb filter, and noticing that the two delay lines contain the same numbers at all times. Therefore, the two delay lines can be replaced by a single shared delay line.

The frequency response is obtained by setting $z = e^{j\omega T}$, where ω denotes radian frequency and T denotes the sampling period in seconds [449]. For an allpass filter, the frequency magnitude must be the same for all $\omega \in [-\pi/T, \pi/T]$.

An allpass filter is obtained when $b_0 = \overline{a_M}$, or, in the case of real coefficients, when $b_0 = a_M$. To see this, let $a \triangleq a_M = \overline{b_0}$. Then we have

$$\left|H(e^{j\omega T})\right| = \left|\frac{\overline{a} + e^{-j\omega MT}}{1 + ae^{-j\omega MT}}\right| = \left|\frac{\overline{a} + e^{-j\omega MT}}{e^{j\omega MT} + a}\right| = \left|\frac{\overline{a + e^{j\omega MT}}}{a + e^{j\omega MT}}\right| = 1.$$

2.8.2 Nested Allpass Filters

An interesting property of allpass filters is that they can be *nested* [412, 152, 153]. That is, if $H_1(z)$ and $H_2(z)$ denote unity-gain allpass transfer functions, then both $H_1(H_2(z))$ and $H_2(H_1(z))$ are allpass filters. A proof can be based on the observation that, since $|H_i(e^{j\omega})| = 1$, $H_i(z)$ can be viewed as a conformal map [326] which maps the unit circle in the z plane to itself; therefore, the set of all such maps is closed under functional composition. Nested allpass filters were proposed for use in artificial reverberation by Schroeder [412, p. 222].

An important class of nested allpass filters is obtained by nesting first-order allpass filters of the form

$$S_i(z) = \frac{k_i + z^{-1}}{1 + k_i z^{-1}}.$$

The nesting begins with $H_1(z) \triangleq S_1(z)$, and $H_2(z)$ is obtained by replacing z^{-1} in $H_1(z)$ by $z^{-1}S_2(z)$ to get

$$H_2(z) \triangleq S_1\left(\left[z^{-1}S_2(z)\right]^{-1}\right) \triangleq \frac{k_1 + z^{-1}S_2(z)}{1 + k_1 z^{-1}S_2(z)}.$$

Figure 2.31a depicts the first-order allpass $S_1(z)$ in direct form II. Figure 2.31b shows the same filter redrawn as a *two-multiplier lattice filter* section [297]. In the lattice form, it is clear that replacing z^{-1} by $z^{-1}S_2(z)$ just extends the lattice to the right, as shown in Fig. 2.32 on page 73.

The equivalence of nested allpass filters to lattice filters has computational significance since it is well known that the two-multiply lattice sections can be replaced by one-multiply lattice sections [297, 314].

In summary, *nested first-order allpass filters are equivalent to lattice filters made of two-multiply lattice sections.* In §C.8.4, a one-multiply section is derived which is not only less expensive to implement in hardware, but it additionally has a direct interpretation as a physical model.

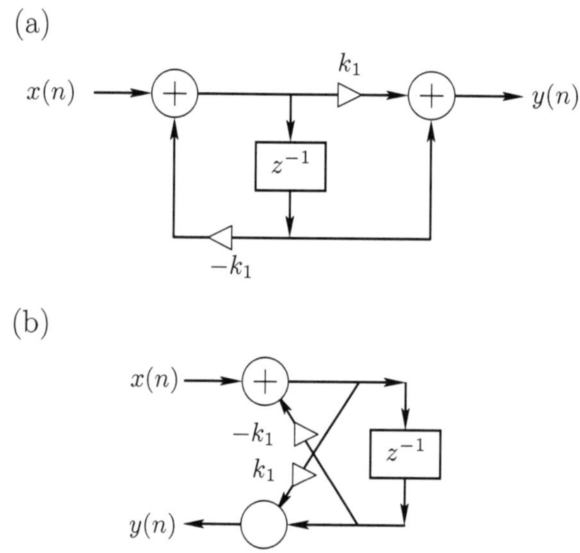

Figure 2.31: First-order allpass filter: (a) Direct form II. (b) Two-multiply lattice section. (b) is just (a) folded over.

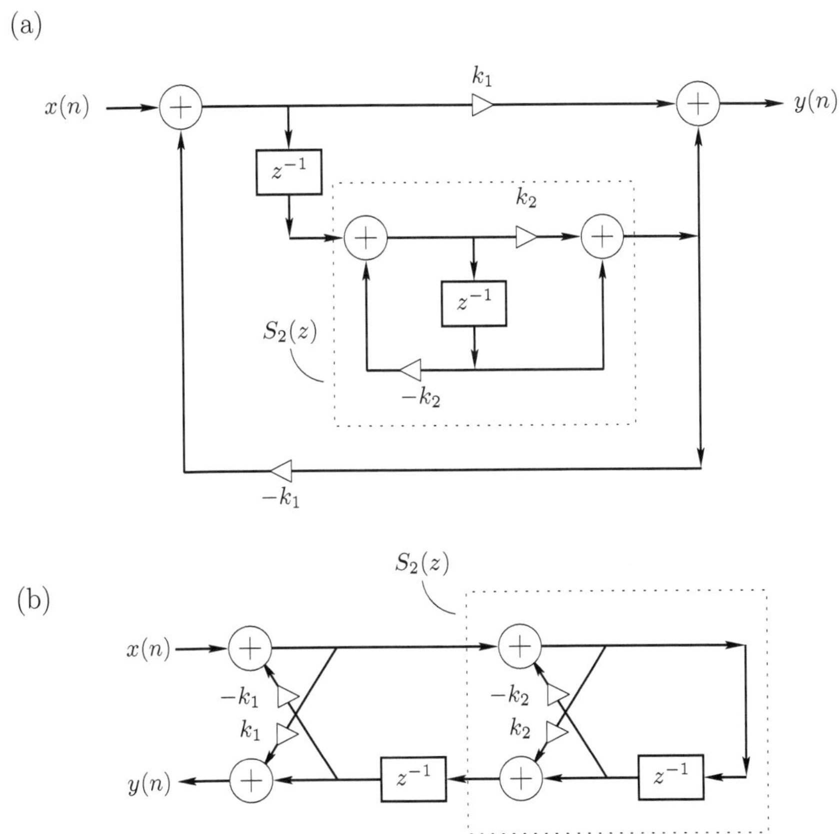

Figure 2.32: Second-order allpass filter: (a) Nested direct-form II. (b) Consecutive two-multiply lattice sections.

2.8.3 More General Allpass Filters

We have so far seen two types of allpass filters:

- The series combination of feedback and feedforward comb-filters is allpass when their delay lines are the same length and their feedback and feedforward coefficents are the same. An example is shown in Fig. 2.30.

- Any delay element in an allpass filter can be replaced by an allpass filter to obtain a new (typically higher order) allpass filter. The special case of nested first-order allpass filters yielded the lattice digital filter structure of Fig. 2.32 on page 73.

We now discuss allpass filters more generally in the SISO case. (See Appendix D of [449] for the MIMO case.)

Definition: A linear, time-invariant filter $H(z)$ is said to be *lossless* if it *preserves signal energy* for every input signal. That is, if the input signal is $x(n)$, and the output signal is $y(n) = (h * x)(n)$, we must have

$$\sum_{n=-\infty}^{\infty} |y(n)|^2 = \sum_{n=-\infty}^{\infty} |x(n)|^2 .$$

In terms of the L^2 signal norm $\| \cdot \|_2$, this can be expressed more succinctly as

$$\| y \|_2^2 = \| x \|_2^2 .$$

Notice that only stable filters can be lossless since, otherwise, $\| y \|$ is generally infinite, even when $\| x \|$ is finite. We further assume all filters are *causal*[15] for simplicity. It is straightforward to show the following:

It can be shown [449, Appendix C] that stable, linear, time-invariant (LTI) filter transfer function $H(z)$ is lossless if and only if

$$\left|H(e^{j\omega})\right| = 1, \quad \forall \omega.$$

That is, the frequency response must have magnitude 1 everywhere over the unit circle in the complex z plane.

Thus, "lossless" and "unity-gain allpass" are synonymous. For an allpass filter with gain g at each frequency, the energy gain of the filter is g^2 for every input signal x. Since we can describe such a filter as an allpass times a constant gain, the term "allpass" will refer here to the case $g = 1$.

---

[15] A filter is said to be causal if its impulse response $h(n)$ is zero for $n < 0$.

2.8.4 Example Allpass Filters

- The simplest allpass filter is a unit-modulus gain

$$H(z) = e^{j\phi}$$

where ϕ can be any phase value. In the real case ϕ can only be 0 or π, in which case $H(z) = \pm 1$.

- A lossless FIR filter can consist only of a single nonzero tap:

$$H(z) = e^{j\phi} z^{-K}$$

for some fixed integer K, where ϕ is again some constant phase, constrained to be 0 or π in the real-filter case. Since we are considering only causal filters here, $K \geq 0$. As a special case of this example, a unit delay $H(z) = z^{-1}$ is a simple FIR allpass filter.

- The transfer function of every finite-order, causal, lossless IIR digital filter (recursive allpass filter) can be written as

$$H(z) = e^{j\phi} z^{-K} \frac{\tilde{A}(z)}{A(z)} \tag{2.16}$$

where $K \geq 0$, $A(z) = 1 + a_1 z^{-1} + a_2 z^{-2} + \cdots + a_N z^{-N}$, and $\tilde{A}(z) \triangleq z^{-N}\overline{A}(z^{-1})$. The polynomial $\tilde{A}(z)$ can be obtained by reversing the order of the coefficients in $A(z)$ and conjugating them. (The factor z^{-N} serves to restore negative powers of z and hence causality.)

In summary, every SISO allpass filter can be expressed as the product of a unit-modulus gain factor, a pure delay, and an IIR transfer function in which the numerator is the "flip" of the denominator, as in Eq. (2.16).

2.8.5 Gerzon Nested MIMO Allpass

An interesting generalization of the single-input, single-output Schroeder allpass filter (defined in §2.8.1 on page 69) was proposed by Gerzon [157] for use in artificial reverberation systems.

The starting point can be the first-order allpass of Fig. 2.31a on page 72, or the allpass made from two comb-filters depicted in Fig. 2.30 on page 70.[16] In either case,

- all signal paths are converted from scalars to *vectors* of dimension N,

[16] Gerzon's starting point was Schroeder's allpass which differs slightly from Eq. (2.15),

- the delay element (or delay line) is replaced by an arbitrary $N \times N$ *unitary matrix frequency response*.[17]

Let $\underline{x}(n)$ denote the $N \times 1$ input vector with components $x_i(n), i = 1, \ldots, N$, and let $\underline{X}(z) = [X_1(z), \ldots, X_N(z)]$ denote the corresponding vector of z transforms. Denote the $N \times 1$ output vector by $\underline{y}(n)$. The resulting vector difference equation becomes, in the frequency domain (cf. Eq. (2.15))

$$\underline{Y}(z) = \overline{g}\underline{X}(z) + \mathbf{U}(z)\underline{X}(z) - g\mathbf{U}(z)\underline{Y}(z)$$

which leads to the matrix transfer function

$$\mathbf{H}(z) = [\mathbf{I} + g\mathbf{U}(z)]^{-1}[\overline{g}\mathbf{I} + \mathbf{U}(z)]$$

where $\mathbf{I}$ denotes the $N \times N$ identity matrix, and $\mathbf{U}(z)$ denotes any *paraunitary matrix transfer function* [501], [449, Appendix C].

Note that to avoid implementing $\mathbf{U}(z)$ twice, $\mathbf{H}(z)$ should be realized in vector direct-form II, *viz.*,

$$\begin{aligned}
\underline{v}_d(n) &= \mathbf{U}(d)\underline{v}(n) = \mathcal{Z}_n^{-1}\left\{\mathbf{U}(z)\underline{v}(z)\right\} \\
\underline{v}(n) &= \underline{x}(n) - g\underline{v}_d(n) \\
\underline{y}(n) &= \underline{v}(n) + \overline{g}\underline{v}_d(n)
\end{aligned}$$

where d denotes the *unit-delay operator* ($d^k x(n) \triangleq x(n-k)$).

To avoid a delay-free loop, the paraunitary matrix must include at least one pure delay in every row, *i.e.*, $\mathbf{U}(z) = z^{-1}\mathbf{U}'(z)$ where $\mathbf{U}'(z)$ is paraunitary and causal.

In [157], Gerzon suggested using $\mathbf{U}(z)$ of the form

$$\mathbf{U}(z) = \mathbf{D}(z)\mathbf{Q}$$

---

having difference equation [412]

$$\begin{aligned}
v(n) &= x(n) + gv(n-M) \\
y(n) &= -gx(n) + (1-g^2)v(n-M),
\end{aligned}$$

with $|g| < 1$ required for stability. This structure can be derived from Eq. (2.15) on page 70 by moving the feedforward path to the other side of the input summer and deriving the new gain for $v(n-M)$. Gerzon's idea was to replace the M-sample delay by a multi-input, multi-output (MIMO) allpass-filter matrix; Gerzon's "unitary frequency-response matrices" correspond to *paraunitary matrix transfer functions* [449, Appendix D].

[17] A (possibly complex) matrix $\mathbf{U}$ is said to be *unitary* if $\mathbf{U}^*\mathbf{U} = \mathbf{I}$, where $W^* \triangleq \overline{W^T}$ is the *Hermitian transpose* of W. A real unitary matrix is said to be *orthogonal*, which is the case most commonly used in practice.

where $\mathbf{Q}$ is a simple $N \times N$ *orthogonal* matrix, and

$$\mathbf{D}(z) = \begin{bmatrix} z^{-m_1} & 0 & 0 & \dots & 0 \\ 0 & z^{-m_2} & 0 & \dots & 0 \\ 0 & 0 & z^{-m_3} & \dots & 0 \\ \vdots & \vdots & \vdots & \ddots & \vdots \\ 0 & 0 & 0 & \dots & z^{-m_N} \end{bmatrix} \tag{2.17}$$

is a diagonal matrix of pure delays, with the lengths m_i chosen to be mutually prime (as suggested by Schroeder [417] for a series combination of Schroeder allpass sections). This structure is very close to the that of typical feedback delay networks (FDN), but unlike FDNs, which are "vector feedback comb filters," the vectorized Schroeder allpass is a true multi-input, multi-output (MIMO) *allpass* filter.

Gerzon further suggested replacing the feedback and feedforward gains $\pm g$ by digital filters $\pm G(z)$ having an amplitude response bounded by 1. In principle, this allows the network to be arbitrarily different at each frequency.

Gerzon's vector Schroeder allpass is used in the IRCAM Spatialisateur [218].

2.9 Allpass Digital Waveguide Networks

We now describe the class of multi-input, multi-output (MIMO) allpass filters which can be made using *closed waveguide networks*. We will see that feedback delay networks can be obtained as a special case.

2.9.1 Signal Scattering

The digital waveguide was introduced in §2.4 on page 49. A basic fact from acoustics is that traveling waves only happen in a *uniform medium*. For a medium to be uniform, its *wave impedance*[18] must be *constant*. When a traveling wave encounters a *change* in the wave impedance, it will *reflect*, at least partially. If the reflection is not total, it will also partially *transmit* into the new impedance. This is called *scattering* of the traveling wave.

Let R_1 denote the constant impedance in some waveguide, such as a stretched steel string or acoustic bore. Then signal scattering is caused by a change in wave impedance from R_1 to R_2. We can depict the partial reflection and transmission as shown in Fig. 2.33.

The computation of reflection and transmission in both directions, as shown in Fig. 2.33 is called a *scattering junction*.

[18] See §7.1 on page 252 for definitions and discussion regarding *impedance*.

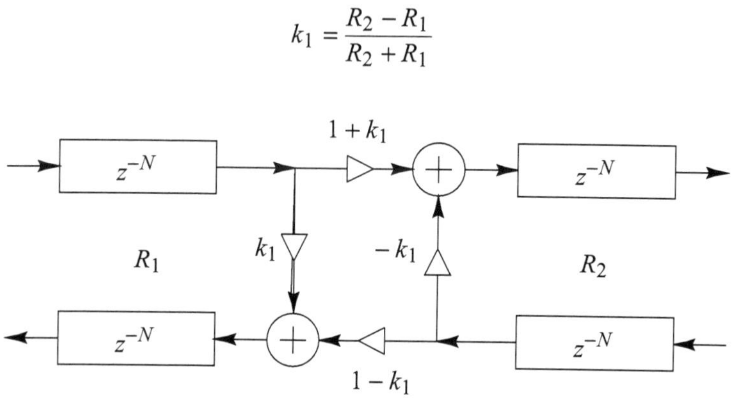

Figure 2.33: Signal scattering at a junction of different wave impedances R_1 & R_2.

As derived in Appendix C, for force or pressure waves, the *reflection coefficient* k_1 is given by

$$k_1 = \frac{R_2 - R_1}{R_2 + R_1} \tag{2.18}$$

That is, the coefficient of reflection for a traveling pressure wave leaving impedance R_1 and entering impedance R_2 is given by the *impedance step over the impedance sum*. The *reflection coefficient* k_1 fully characterizes the scattering junction.

For *velocity* traveling waves, the reflection coefficient is just the negative of that for force/pressure waves, or $-k_1$ (see Appendix C).

Signal scattering is *lossless*, *i.e.*, wave energy is neither created nor destroyed. An implication of this is that the *transmission coefficient* for a traveling pressure wave leaving impedance R_1 and entering impedance R_2 is given by

$$t_1 = 1 + k_1.$$

For velocity waves, the transmission coefficient is $t_1 = 1 - k_1$, which is perhaps more intuitive.

2.9.2 Digital Waveguide Networks

A *Digital Waveguide Network* (DWN) consists of any number of digital waveguides interconnected by scattering junctions. For example, when two digital waveguides are connected together at their endpoints, we obtain a two-port scattering junction as shown in Fig. 2.33. When three or more waveguides are connected at a point, we obtain a *multiport scattering junction*, as discussed in §C.8 on page 560. In other words, a digital waveguide network is formed whenever digital waveguides having arbitrary wave impedances are interconnected. Since DWNs are lossless, they provide a systematic means of building a very large class of MIMO allpass filters.

Consider the following question:

> *Under what conditions may I feed a signal from one point inside a given allpass filter to some other point (adding them) without altering signal energy at any frequency?*

In other words, how do we add *feedback paths* anywhere and everywhere, thereby maximizing the richness of the recursive feedback structure, while maintaining an overall allpass structure?

The *digital waveguide* approach to allpass design [430] answers this question by maintaining a *physical interpretation* for all delay elements in the system. Allpass filters are made out of *lossless digital waveguides* arranged in *closed, energy conserving networks.* See Appendix C for further discussion.

2.10 Exercises

Comb Filter Problems

1. Referring to §2.6.3 on page 59, analyze and plot the frequency response (both amplitude and phase responses) for the feedforward comb filter
$$y(n) = x(n) - x(n-10)$$

2. Given the difference equation for the general feedback comb filter
$$y(n) = x(n) + gy(n-M)$$
Derive the transfer function $H(z) = Y(z)/X(z)$.

3. Find the poles of the recursive comb filter
$$y(n) = x(n) + y(n-8)$$

4. Find the poles of the recursive comb filter
$$y(n) = x(n) + 0.5y(n-8)$$

5. Plot the poles in the previous two problems on the same plot (you may want to refer to functions `pzmap` and `tf` in Matlab).

6. Let $H_1(z)$ denote the transfer function of the filter in problem 3, and let $H_2(z)$ denote that for problem 4. We can express $H_2(z)$ as a conformal mapping of $H_1(z)$, viz., $H_2(z) = H_1(\alpha z)$. Find α.

7. Find an approximate formula for the poles of the recursive comb filter

$$y(n) = x(n) + \alpha y(n-7) + (1-2\alpha)y(n-8) + \alpha y(n-9)$$

under the assumption $G(z) = G(z/|z|)$. Your result should be exact when $\alpha = 0$. [Hint: Analyze this as $Y(z) = X(z) + G(z)z^{-8}Y(z)$ and approximate $G(z)$ near the unit circle by $G(z/|z|)$.]

8. Plot the exact and approximate poles in the previous problem on the same plot for $\alpha = 0.25$. Discuss where the approximation is good and where it is bad and why.

9. Express the impulse response of

$$y[n] = x[n] - y[n-3] - 0.25y[n-6]$$

as a sum of N exponentially decaying sinusoids. List explicitly the decay time-constant for each sinusoid.

Acoustic Echo Simulation Problems

1. Assuming sound propagates at $c = 343$ meters per second (at 20 degrees Celsius), find the delay-line length M which can be used to simulate sound propagation delay over a distance of approximately 10 meters (rounded to the nearest sample), where the simulation sampling rate is 44.1 kHz.

2. Suppose that pressure-wave amplitude was observed to decay by a factor of 2 in the geometry of the previous problem. In other words, sound propagation over a distance of 10 meters is associated with 6.02 dB of loss at all frequencies. What is the loss associated with propagation over one spatial sample?

3. Suppose the high E string of an electric guitar vibrates at 330 Hz and is measured to be 25 inches long from nut to bridge,

    (a) Find the speed of sound for transverse traveling-waves on the string

    (b) Find the spatial sampling interval at 44.1 kHz.

4. Sketch a delay-line simulation of an acoustic point-source at $\mathbf{x} = (1,1,0)$ meters, a listener at $\mathbf{x} = (2,0,0)$, a reflecting wall at $\mathbf{x} = (x,2,z)$, and a second reflecting wall at $\mathbf{x} = (0,y,z)$. Assume the wall reflection coefficient is 1, and that air absorption is 0.1 dB per meter at all frequencies.

   (a) Find all acoustic path lengths.

   (b) Find all acoustic path gains.

   (c) Assuming sound speed is $c = 343$ meters per second, and the sampling rate is 44.1 kHz, find all delay-line lengths.

   (d) Find the transfer function from source to listener.

   (e) Using Matlab, plot the echogram (impulse response) and amplitude response (magnitude frequency response).

5. Sketch a delay-line simulation of an acoustic *plane-wave source* at $\mathbf{x} = (x,1,z)$ meters, a listener at $\mathbf{x} = (0,1,0)$, a reflecting wall at $\mathbf{x} = (x,2,z)$, and a second reflecting wall at $\mathbf{x} = (x,0,z)$. Assume air absorption is 0.01 dB per meter and the wall reflection coefficient is 0.9.

   (a) Find all acoustic path lengths.

   (b) Find all acoustic path gains.

   (c) Assuming sound speed is $c = 343$ meters per second, and the sampling rate is 44.1 kHz, find all delay-line lengths.

   (d) Find the transfer function from source to listener.

   (e) Using Matlab, plot the echogram and amplitude response.

Allpass Problems

1. **Lattice/Ladder Allpass Filters**
   It is shown in §2.8.2 on page 71 that the nested allpass filter is equivalent to the 2-multiplier lattice filter shown in Fig. 2.34. This problem explores the relationship of this structure to the *Kelly Lochbaum ladder comb filter* depicted in Fig. 2.35.

   (a) Find the impulse response of the two-multiply filter in Fig. 2.34.

   (b) Find the impulse response of the four-multiply filter in Fig. 2.35.

   (c) Find the transfer function of the two-multiply filter in Fig. 2.34.

   (d) Find the transfer function of the four-multiply filter in Fig. 2.35.

   (e) Find a one-multiply implementation of the four-multiply filter in Fig. 2.35, and draw the implementation.

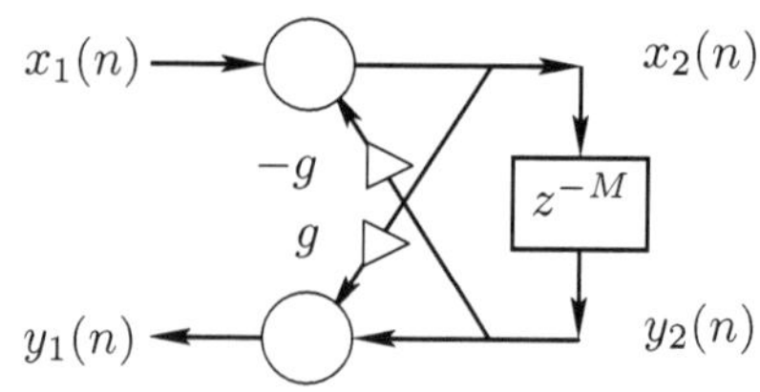

Figure 2.34: Two-multiply lattice filter.

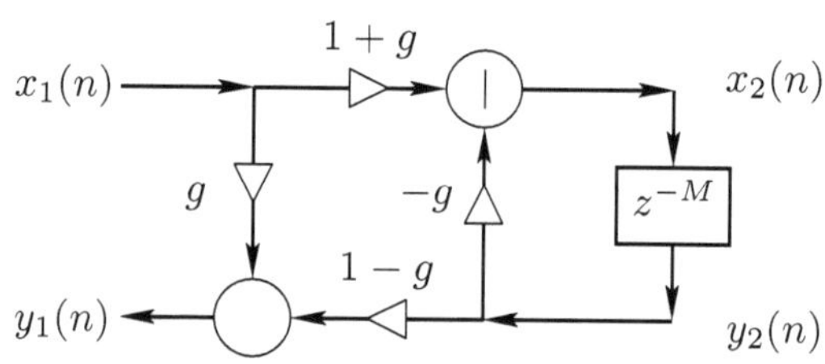

Figure 2.35: Four-multiply ladder filter.

Allpass Comb Laboratory Exercise

1. Write an STK module called `Rev.cpp` which implements the two-multiply lattice filter shown in Fig. 2.34. The input arguments should be the two coefficients and the delay-line length. Write a small main program to test your module with an impulse input, and verify that you get the same numbers calculated by hand in problem 1a above.

2. Write an STK module called `RevLP.cpp` which extends `Rev.cpp` to include the one-pole lowpass filter $y(n) = (1-a)x(n) + ay(n-1)$ *in the feedback loop only*. Thus, there should be one new argument a which can be set to 0 to obtain the former case (a useful intermediate test).

FDN Problems

1. Using Eq. (2.11) on page 66 as a definition of a state-space description,[19] find the state space description for the single-input, single-

[19]Or see [449, Appendix E]:
`http://ccrma.stanford.edu/~jos/filters/State_Space_Models.html`.

output (SISO) Feedback Delay Network (FDN) described by

$$\begin{bmatrix} x_1(n) \\ x_2(n) \\ x_3(n) \end{bmatrix} = \begin{bmatrix} g_{11} & g_{12} & g_{13} \\ g_{21} & g_{22} & g_{23} \\ g_{31} & g_{32} & g_{33} \end{bmatrix} \begin{bmatrix} x_1(n-1) \\ x_2(n-2) \\ x_3(n-3) \end{bmatrix} + \begin{bmatrix} 1 \\ 1 \\ 1 \end{bmatrix} u(n)$$

$$y(n) = x_1(n) + x_2(n) + x_3(n)$$

2. In §2.7.3 on page 67, it was noted that FDN stability is guaranteed when the spectral norm of the feedback matrix $\mathbf{A}$ is less than 1, where the spectral norm is given by the largest singular value of $\mathbf{A}$ (or square-root of the largest eigenvalue of $\mathbf{A}\mathbf{A}^T$ when $\mathbf{A}$ is real). Using the example

$$\mathbf{A} = \begin{bmatrix} \lambda & c \\ 0 & \lambda \end{bmatrix},$$

show that stability is *not* assured when it is only given that all *eigenvalues* of $\mathbf{A}$ have magnitude less than 1.

3. Show that stability is assured when it is given that (1) all eigenvalues of $\mathbf{A}$ have magnitude less than 1, and (2) $\mathbf{A}$ is a Hermitian matrix ($\mathbf{A}^* = \mathbf{A}$).

Chapter 3

Artificial Reverberation

This chapter summarizes basic results in systems for *artificial reverberation.* Such systems make extensive use of tapped delay lines, comb filters, and allpass filters (described in the previous chapter).

3.1 The Reverberation Problem

Consider the requirements for acoustically simulating a concert hall or other listening space. Suppose we only need the response at one or more discrete *listening points* in space ("ears") due to one or more discrete *point sources* of acoustic energy.

First, as discussed in §2.2, the *direct signal* propagating from a sound source to a listener's ear can be simulated using a single delay line in series with an attenuation scaling or lowpass filter. Second, each sound ray arriving at the listening point via one or more *reflections* can be simulated using a delay-line and some scale factor (or filter). Two rays create a feed-forward comb filter, like the one in Fig. 2.9 on page 44. More generally, a *tapped delay line* (Fig. 2.19, p. 55) can simulate many reflections. Each tap brings out one echo at the appropriate delay and gain, and each tap can be independently filtered to simulate air absorption and lossy reflections. In principle, tapped delay lines can accurately simulate *any* reverberant environment, because reverberation really does consist of many paths of acoustic propagation from each source to each listening point. As we will see, the only limitations of a tapped delay line are (1) it is *expensive* computationally relative to other techniques, (2) it handles only one "point to point" transfer function, *i.e.*, from one point-source to one ear,[1] and (3)

[1] One tapped delay line can also model one source to multiple ears, or vice versa (Fig. 5.6, p. 195).

it should be changed when the source, listener, or anything in the room moves.

3.1.1 Exact Reverb via Transfer-Function Modeling

Figure 3.1 depicts the general reverberation scenario for three sources and one listener (two ears). In general, the filters should also include filtering by the pinnae of the ears, so that each echo can be perceived as coming from the correct angle of arrival in 3D space; in other words, at least some reverberant reflections should be *spatialized* so that they appear to come from their natural directions in 3D space [248]. Again, the filters change if anything changes in the listening space, including source or listener position. The artificial reverberation problem is then to implement some approximation of the system in Fig. 3.1.

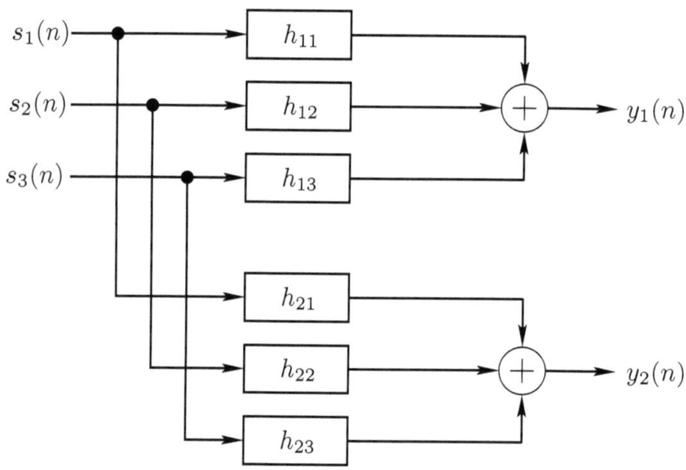

Figure 3.1: General reverberation simulation for three sources and one listener (two ears). Each filter h_{ij} can be implemented as a tapped delay line FIR filter.

In the frequency domain, it is convenient to express the input-output relationship in terms of the *transfer-function matrix*:

$$\begin{bmatrix} Y_1(z) \\ Y_2(z) \end{bmatrix} = \begin{bmatrix} H_{11}(z) & H_{12}(z) & H_{13}(z) \\ H_{21}(z) & H_{22}(z) & H_{23}(z) \end{bmatrix} \begin{bmatrix} S_1(z) \\ S_2(z) \\ S_3(z) \end{bmatrix}$$

Denoting the impulse response of the filter from source j to ear i by

$h_{ij}(n)$, the two output signals in Fig. 3.1 are computed by six convolutions:

$$y_i(n) = \sum_{j=1}^{3} s_j * h_{ij}(n) = \sum_{j=1}^{3} \sum_{m=0}^{M_{ij}} s_j(m) h_{ij}(n-m), \quad i = 1, 2,$$

where M_{ij} denotes the order of FIR filter h_{ij}. Since many of the filter coefficients $h_{ij}(n)$ are zero (at least for small n), it is more efficient to implement them as tapped delay lines (§2.5) so that the inner sum becomes *sparse*. For greater accuracy, each tap may include a lowpass filter which models *air absorption* [314] and/or spherical spreading loss (see §2.3 on page 45). For large n, the impulse responses are not sparse, and we must either implement very expensive FIR filters, or approximate the tail of the impulse response using less expensive IIR filters; this subject—"late reverberation" approximation—is taken up in §3.4.

3.1.2 Complexity of Exact Reverberation

For music, a typical reverberation time[2] is on the order of one second. Suppose we choose exactly one second for the reverberation time. At an audio sampling rate of 50 kHz, each filter in Fig. 3.1 requires 50,000 multiplies and additions per sample, or 2.5 *billion* multiply-adds per second. Handling three sources and two listening points (ears), we reach 30 billion operations per second for the reverberator. This computational load would require at least 10 Pentium CPUs clocked at 3 Gigahertz, assuming they were doing nothing else, and assuming both a multiply and addition can be initiated each clock cycle, with no wait-states caused by the three required memory accesses (input, output, and filter coefficient). While these numbers can be improved using FFT convolution instead of direct convolution (at the price of introducing a throughput delay which can be a problem for real-time systems), it remains the case that exact implementation of all relevant point-to-point transfer functions in a reverberant space is very expensive computationally.

While a tapped delay line FIR filter can provide an accurate model for any point-to-point transfer function in a reverberant environment, it is rarely used for this purpose in practice because of the extremely high computational expense. While there are specialized commercial products that implement reverberation via direct convolution of the input signal with the impulse response, the great majority of artificial reverberation systems use other methods to synthesize the late reverb more economically.

[2] *Reverberation time* is typically defined as t_{60}, the time, in seconds, to decay by 60 dB.

3.1.3 Possibility of a Physical Reverb Model

One disadvantage of the point-to-point transfer function model depicted in Fig. 3.1 is that some or all of the filters must change when anything moves. If instead we had a computational model of the whole acoustic space, sources and listeners could be moved as desired without affecting the underlying room simulation. Furthermore, we could use "virtual dummy heads" as listeners, complete with pinnae filters, so that all of the 3D directional aspects of reverberation could be captured in two extracted signals for the ears. Thus, there are compelling reasons to consider a full 3D model of a desired acoustic listening space.

Let us briefly estimate the computational requirements of a "brute force" acoustic simulation of a room. It is generally accepted that audio signals require a 20 kHz bandwidth. Since sound travels at about a foot per millisecond (see §B.7.14 on page 513 for a more precise value), a 20 kHz sinusoid has a wavelength on the order of 1/20 feet, or about half an inch. Since, by elementary sampling theory, we must sample faster than twice the highest frequency present in the signal, we need "grid points" in our simulation separated by a quarter inch or less. At this grid density, simulating an ordinary 12'x12'x8' room in a home requires more than 100 million grid points. Using finite-difference (Appendix D) or waveguide-mesh techniques (§C.14,Appendix E) [519, 396], the average grid point can be implemented as a multiply-free computation; however, since it has waves coming and going in six spatial directions, it requires on the order of 10 additions per sample. Thus, running such a room simulator at an audio sampling rate of 50 kHz requires on the order of 50 billion additions per second, which is comparable to the three-source, two-ear simulation of Fig. 3.1. However, scaling up to a 100'x50'x20' concert hall requires more than 5 quadrillion operations per second. We may conclude, therefore, that a fine-grained physical model of a complete concert hall over the audio band is prohibitively expensive.

The remainder of this chapter will be concerned with ways of reducing computational complexity without sacrificing too much perceptual quality.

3.2 Perceptual Aspects of Reverberation

Artificial reverberation is an unusually interesting signal processing problem because, as discussed in the previous sections, the "obvious" methods based on physical modeling or input-output modeling are too expensive computationally for most applications. This leads to the question of what are the perceptually important aspects of reverberation, and how can these be provided by efficient computational structures.

3.2.1 Perception of Echo Density and Mode Density

The reverberation problem can be greatly simplified without sacrificing perceptual quality. For example, it can be shown[3] that for typical rooms, the echo density increases as t^2, where t is time. Therefore, beyond some time, the echo density is so great that it can be modeled as some uniformly sampled stochastic process without loss of perceptual fidelity. In particular, there is no need to explicitly compute multiple echoes per sample of sound. For smoothly decaying late reverb (the desired kind), an appropriate random process sampled at the audio sampling rate will sound equivalent perceptually.

Similarly, it can be shown[4] that the number of resonant modes in any given frequency band increases as frequency squared, so that above some frequency, the modes are so dense that they are perceptually equivalent to a *random frequency response* generated according to some statistics. In particular, there is no need to explicitly implement resonances so densely packed that the ear cannot hear them all.

---

[3] A simple proof for rectangular rooms can be based on considering a single spherical wave produced from a point source in the room. Imagine tesselating all of 3D space with copies of the original room (minus the source), all in the same orientation. As the spherical wave expands, it intersects a number of rooms that is roughly proportional to its radius squared. Since the radius is proportional to time for an expanding spherical wavefront ($r = ct$), the number of rooms containing a wave section grows as t^2. By adding all the rooms together (flipping every other one along x and y), we obtain the acoustic field in the original reverberant room at time t, for the case of lossless wall reflections of pressure waves. (This is the dual of the usual *image method* for computing the impulse response of a room [11].) Since each wave-section traverses each point exactly once in each room image (disregarding reflections, which are accounted for by different room images), the number of echoes at any point in the room during the time it takes a plane wave to traverse the room is very close to the number of wave sections in the room at the beginning of that time interval. Incidentally, this construction gives, in the limit as $r \rightarrow \infty$, a derivation and vivid visualization of the *diffuse field*—the superposition of equal-amplitude traveling plane-waves arriving from all directions and having uniformally distributed random arrival phase [349, 47].

[4] As described in virtually any acoustics textbook, the resonant modes of a rectangular room are given by (see, *e.g.*, [349, pp. 284–286])

$$k^2(l, m, n) = k_x^2(l) + k_y^2(m) + k_z^2(n) \tag{3.1}$$

where $k_x(l) = l\pi/L_x$ denotes the lth harmonic frequency (radian/meter) of the fundamental standing wave in the x direction (L_x being the length of the room along x), and similarly for the y and z directions. Thus, the mode frequencies of a room can be enumerated on a uniform 3D Cartesian grid indexed by (l, m, n). The grid spacings along x, y, and z are taken to be π/L_x, π/L_y, and π/L_z, respectively. From Eq. (3.1), the spatial frequency k of mode (l, m, n) is given by the *distance* from the origin of the grid to the point indexed by (l, m, n). Therefore, the number of room modes having spatial frequency between k and $k + \Delta$ is equal to the number of grid points lying in the spherical annulus between radius k and $k + \Delta$. Since the grid is uniform, this number grows as k^2.

In summary, we see that, based on limits of perception, the *impulse response* of a reverberant room can be divided into two segments. The first segment, called the *early reflections*, consists of the relatively sparse first echoes in the impulse response. The remainder, called the *late reverberation*, is so densely populated with echoes that it is best to characterize the response *statistically* in some way. Section 3.3 discusses methods for simulating early reflections in the reverberation impulse response.

Similarly, the *frequency response* of a reverberant room can be divided into two segments. The low-frequency interval consists of a relatively sparse distribution of resonant modes, while at higher frequencies the modes are packed so densely that they are best characterized statistically as a random frequency response with certain (regular) statistical properties. Section 3.4 describes methods for synthesizing hiqh quality late reverberation.

3.2.2 Perceptual Metrics for Ideal Reverberation

Some desirable controls for an artificial reverberator include [218]

- $t_{60}(f)$ = desired reverberation time at each frequency f
- $G^2(f)$ = signal power gain at each frequency
- $C(f)$ = "clarity" = ratio of impulse-response energy in early reflections to that in the late reverb
- $\rho(f)$ = *interaural correlation coefficient* at left and right ears

The time to decay 60 dB (t_{60}) is a classical objective parameter used as a measure of perceived reverberation time. Classically, t_{60} was measured for the whole response. More recently [216], it has become more common to design for a given t_{60} at more than one frequency, *e.g.*, one for low frequencies, another for high frequencies, and interpolated values at intermediate frequencies. Perceptual studies indicate that reverberation time should be independently adjustable in at least three frequency bands [217].

Energy Decay Curve

For measuring and defining reverberation time t_{60}, Schroeder introduced the so-called *energy decay curve (EDC)* which is the *tail integral* of the squared impulse response at time t:

$$\text{EDC}(t) \triangleq \int_t^{\infty} h^2(\tau)d\tau$$

Thus, EDC(t) is the total amount of signal energy remaining in the reverberator impulse response at time t. The EDC decays more smoothly than the impulse response itself, and so it is more useful than ordinary amplitude envelopes for estimating t_{60}.

Energy Decay Relief

The *energy decay relief (EDR)* is a time-frequency distribution which generalizes the EDC to multiple frequency bands [215]:

$$\text{EDR}(t_n, f_k) \triangleq \sum_{m=n}^{M} |H(m,k)|^2$$

where $H(m,k)$ denotes bin k of the short-time Fourier transform (STFT) at time-frame m [12, 451], and M denotes the total number of time frames. The FFT within the STFT is typically used with a window, such as a Hann window of length 30 or 40 ms.

Thus, $\text{EDR}(t_n, f_k)$ is the total amount of signal energy remaining in the reverberator's impulse response at time $t_n = nT$ in a frequency band centered about $f_k = kf_s/N$ Hz, where N denotes the FFT length.

The EDR of a violin-body impulse response is shown in Fig. 3.2. For better correspondence with audio perception, the frequency axis is warped to the Bark frequency scale [460], and energy is summed within each Bark band (one critical band of hearing equals one Bark). A violin body can be regarded as a very small reverberant room, with correspondingly "magnified" spectral structure relative to reverberant rooms.

The EDR of the Boston Symphony Hall is displayed in [153, p. 96].

The EDR is used to measure partial overtone dampings from recordings of a vibrating string in §6.11.5 on page 236.

3.3 Early Reflections

The "early reflections" portion of the impulse response of a reverberant environment is often taken to be the first 100ms or so [314]. However, for greater accuracy, it should be extended to the time at which the reverberation reaches its asymptotic statistical behavior.[5]

[5]One definition of when *late reverberation* begins is when it begins to look *Gaussian*, since a random sum of plane waves uniformly distributed over all angles of arrival yields a Gaussian distributed pressure field. (The fact that late reverberation in a rectangular room always approaches the *diffuse field*—a superposition of plane waves traveling in all directions can be seen from the "image-method dual" argument in §3.2.1.) One way to test for Gaussianness is to form a *histogram* of impulse response sample values over a finite window (say 10ms, or roughly 500 samples), and compare the normalized histogram $h(p)$ to a standard Gaussian bell curve $G(p)$ computed using the measured sample variance. A threshold can be placed on $||\,h - G\,||$, or some number of higher order sample

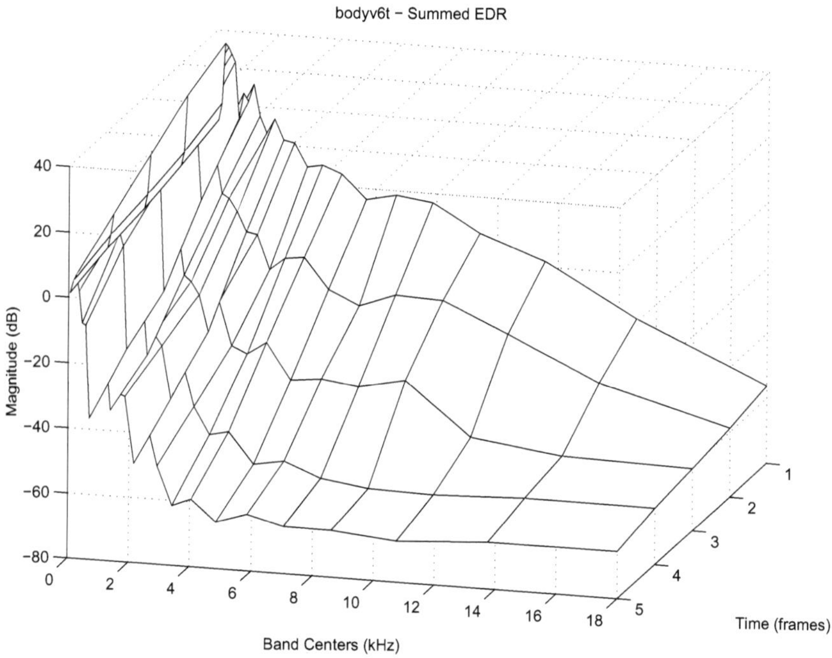

Figure 3.2: Energy Decay Relief of a violin-body impulse response (from [203]).

Since the early reflections are relatively sparse and span a relatively short time, they are often implemented using *tapped delay lines* (TDL).[6]

If the computation is affordable, it is best to *spatialize* the early reflections [248] so that they come from the right directions in 3D space. It is known [61] that the early reflections have a strong influence on *spatial impression*, *i.e.*, the listener's perception of the listening-space shape.

Figure 3.3 shows a general schematic of a reverberator with separate implementations of early and late reverberation. The taps on the TDL may include lowpass filtering for simulation of air absorption.

While the late reverb logically begins when the early reflections end, as implemented in Fig. 3.3, it may be more cost-effective in practice to feed the "late reverb" unit from an earlier tap (or set of taps) from the TDL, thus overlapping them somewhat. This can help when the late reverberator needs time to build up to full density.

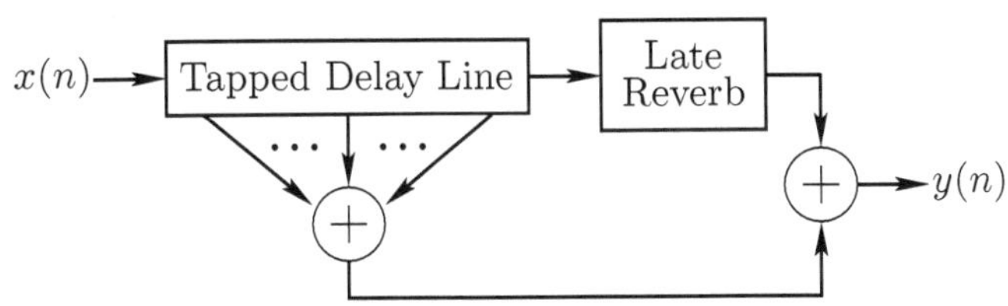

Figure 3.3: Division of a reverberator into early and late sections.

It is often the case that early reflections can be worked into the late-reverberation simulation. For example, usually there are long delay lines in which the input signal can be summed at various points, thereby implementing a transposed tapped delay line (see §2.5.2).

Good concert halls are observed to have to have stereo-recorded impulse responses that quickly "lateralize", with a smooth decay and overall duration of approximately 1.9 seconds [48].

moments can be compared. A simpler test is to determine when roughly 30% of the samples in a frame have magnitude larger than one standard deviation ($|x(n)| > \sigma_x$) [3], since the probability of this occurring in truly Gaussian sequences is $1-\text{erf}(1/\sqrt{2}) = 0.3173\ldots$, where $\text{erf}(x)$ denotes the "error function" (integral from $-x$ to x of the Gaussian probability density function with normalized variance $\sigma^2 = 1/2$). For a zero-mean signal $x(n)$, the sample standard deviation σ_x is given by the root-mean-square (RMS level). Another desired property of "late reverb" is that the remaining energy in the impulse response (Schroeder's *Energy Decay Curve* (EDC) [413]), has begun what appears to be an exponential decay. This can be ascertained by fitting an exponential to the EDC using, *e.g.*, Prony's method.

[6]This idea was apparently first suggested by Schroeder in 1970 [415] and evidently first implemented by Moorer [314].

3.4 Late Reverberation Approximations

3.4.1 Desired Qualities in Late Reverberation

From a perceptual standpoint, the main qualities desired of a good late-reverberation impulse response are

1. a smooth (but not too smooth) decay, and
2. a smooth (but not too regular) frequency response.

Providing an exponential decay is no problem since both stable linear systems and natural reverberation decay exponentially. The harder issue is making it *smooth*, that is, free of "flutter," "beating," or other unnatural irregularities. In general, smooth decay results when the echo density is sufficiently high. Note, however, that some fluctuation in the short-term energy is needed to sound natural [58, 104], corresponding to that of a decaying *noise* process [314].[7]

A smooth frequency response exhibits no large, isolated gaps or boosts. It is generally provided when the mode density is sufficiently large in the frequency domain, with the modes being spread out uniformly, as opposed to piling up in the same place or separated to form gaps. On the other hand, the modes should not be too regularly spaced, since this can produce audible periodicity in the time-domain impulse response.

An interesting experiment by Moorer [314] was to try exponentially decaying white noise as a late reverberation impulse response. This signal satisfies both smoothness criteria (time domain and frequency domain), and it sounds quite natural. However, since natural reverberation decays faster at high frequencies, it is better to say that the ideal late reverberation impulse response is exponentially decaying "colored" noise, with the high-frequency energy decaying faster than the low-frequency energy.

Schroeder's rule of thumb for echo density in the late reverb is 1000 echoes per second or more [417]. However, for impulsive sounds, 10,000 echoes per second or more may be necessary for a smooth response [217, 153].

3.4.2 Schroeder Allpass Sections

Manfred Schroeder's original papers on the use of allpass filters for artificial reverberation [417, 412, 152, 153] started a lively thread of research which continues to the present. For many years thereafter, digital reverberation

[7] In computer music, an old trick for making a synthesized tone sound reverberated is to randomly modulate its amplitude envelope, and to append a low-level exponentially decaying tail to the amplitude envelope (also modulated). This can produce a very convincing illusion.

algorithms were designed along the lines suggested by Schroeder using delay lines, comb filters, and allpass filters—elements described in Chapter 2. There was even special-purpose hardware developed to implement these structures efficiently in real time [393]. Today, these elements continue to serve as the basis for commercial devices for artificial reverberation and related effects [104]. They are also still typically used in software for artificial reverberation [86]. We will see some examples starting in §3.5 below.

Schroeder's suggested use of allpass filters was especially brilliant because there is nothing in nature to suggest them. Instead, he recognized the conceptual and practical utility of separating the coloration of reverberation from its duration and density aspects. While Schroeder's 1961 paper is entitled "Colorless Artificial Reverberation," there is no such thing as colorless (exactly allpass) reverberation in the real world. However, it makes sense as an idealization of natural reverb. Colorless reverberation is an idealization only possible in the "virtual world".

In Schroeder's original work, and in much work which followed, allpass filters are arranged in *series*, as shown in Fig. 3.4.

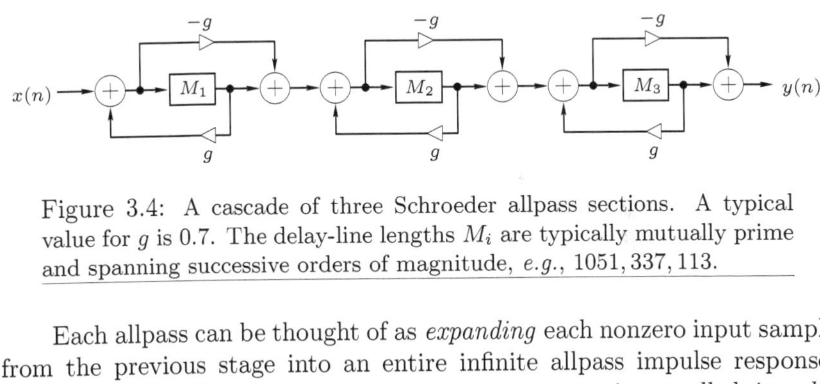

Figure 3.4: A cascade of three Schroeder allpass sections. A typical value for g is 0.7. The delay-line lengths M_i are typically mutually prime and spanning successive orders of magnitude, *e.g.*, $1051, 337, 113$.

Each allpass can be thought of as *expanding* each nonzero input sample from the previous stage into an entire infinite allpass impulse response. For this reason, Schroeder allpass sections are sometimes called *impulse expanders* [484] or *impulse diffusers.* While not a physical model of diffuse reflection, single reflections are expanded into many reflections, which is qualitatively what is desired.

Another interesting interpretation of a Schroeder allpass section is as a *digital waveguide model* of the *driving-point impedance* of an ideal string (or cylindrical acoustic tube) which is reflectively terminated at a real impedance. That is, if the input signal is regarded as samples of a physical *velocity*, then the output signal is proportional to samples of the corresponding *force* (or pressure) at the same physical point. The delay line contains traveling-wave samples; the first half corresponds to traveling waves heading toward the far end of the string (or tube), while the second half holds traveling-wave samples coming back the other way toward the driving point.

For the mathematical details of this interpretation, see Appendix C.

3.4.3 Nested Allpass Filters

Another common method for increasing the density of an allpass impulse response is to *nest* two or more allpass filters, as described in §2.8.2 and shown in Fig. 2.32 on page 73. In general, a nested allpass filter is created when one or more of its delay elements is replaced by another allpass filter. As we saw in §2.8.2, first-order nested allpass filters are equivalent to *lattice filters*. This equivalence implies that any order N transfer function (any N poles and zeros) may be obtained from a linear combination of the delay elements of nested first-order allpass filters, since this is a known property of the lattice filter [297].

In general, any delay-element or delay-line inside a stable allpass-filter can be replaced with any stable allpass-filter, and the result will be a stable allpass.

3.5 Schroeder Reverberators

The subject of *artificial reverberation* was initiated in the early 1960s by Manfred Schroeder [417, 412]. Early Schroeder reverberators consisted of the following elements [412]:

- A series connection of several *allpass filters*
- A parallel bank of *feedback comb filters*
- A *mixing matrix*

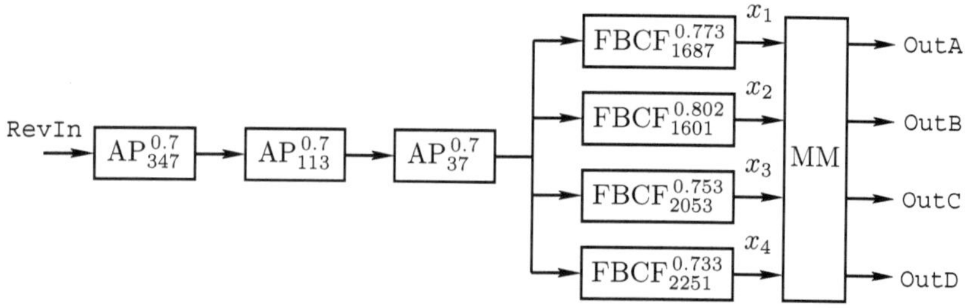

Figure 3.5: A Schroeder reverberator we will call JCREV developed by Prof. John Chowning, founding director of CCRMA (drawn from a 1972 MUS10 software listing, where MUS10 was an "acoustic compiler" language descended from Music V [305]).

An example is shown in Fig. 3.5, where, in that figure,

$$\text{AP}_N^g \triangleq \frac{-g + z^{-N}}{1 - gz^{-N}} \tag{3.2}$$

denotes a *Schroeder allpass* section with delay length N samples and coefficients g (see Fig. 2.30 on page 70 and associated discussion),

$$\text{FBCF}_N^g \triangleq \frac{1}{1 - g\, z^{-N}} \tag{3.3}$$

denotes a *feedback comb* filter with delay length N and coefficient g (diagrammed in Fig. 2.24 on page 58), and MM in Fig. 3.5 denotes the *mixing matrix*

$$\texttt{MM} = \begin{bmatrix} 1 & 1 & 1 & 1 \\ -1 & -1 & -1 & -1 \\ -1 & 1 & -1 & 1 \\ 1 & -1 & 1 & -1 \end{bmatrix}$$

which can be efficiently implemented using four adders and two negations:

$$\begin{aligned} \texttt{OutA} &= s_1 + s_2 \\ \texttt{OutB} &= -\texttt{OutA} \\ \texttt{OutC} &= -\texttt{OutD} \\ \texttt{OutD} &= s_1 - s_2 \end{aligned} \qquad \begin{aligned} s_1 &= x_1 + x_3 \\ s_2 &= x_2 + x_4 \end{aligned}$$

As discussed above in §3.4.2, the allpass filters provide "colorless" high-density echoes in the late impulse response of the reverberator [417]. These allpass filters may also be referred to as *diffusers*. While allpass filters are "colorless" in theory, perceptually, their impulse responses are only colorless when they are extremely short (less than 10 ms or so). Longer allpass impulse responses sound similar to feedback comb-filters. For steady-state tones, however, such as sinusoids, the allpass property gives the same gain at every frequency, unlike comb filters.

Schroeder [412, p. 221] suggests a progression of allpass delay-line lengths close to

$$M_i T \approx \frac{100 \text{ ms}}{3^i}, \quad i = 0, 1, 2, 3, 4,$$

and chosen to be mutually prime (no common factors). The 100 ms value was chosen so that when $g = 0.708$ in Eq. (3.2), the time to decay 60 dB (t_{60}) would be 2 seconds. Thus, for $i = 0$, $t_{60} \approx 2$, and each successive allpass has an impulse-response duration that is about a third of the previous one. Using 5 series allpasses in this way yields an impulse-response echo density of about 810 per second, which is close to the desired thousand per second [412, p. 221].

The parallel comb-filter bank is intended to give a psychoacoustically appropriate fluctuation in the reverberator frequency response. As discussed in Chapter 2 (§2.6.2), a feedback comb filter can simulate a pair of parallel walls, so one could choose the delay-line length in each comb filter to be the number of samples it takes for a plane wave to propagate from one wall to the opposite wall and back. However, in his original paper [412], Schroeder describes a more psychoacoustically motivated approach:

> "There are about 15 large response peaks in every 100 cps [Hz] interval for a room with 1 sec reverberation time. Thus, one might hope that if an artificial reverberator has a comparable number of response peaks it might sound just as good as a real room. We have been able to confirm this expectation by subjective evaluations of the responses of reverberators consisting of several comb filters ... connected in parallel. For a delay of 0.04 sec, the number of response peaks per 100 cps [Hz] is 4. Thus, between 3 and 4 comb filters in parallel ... with incommensurate delays, are required to approximate the number of peaks in the frequency reponse of a room having a reverberation time of T[60] = 1 sec. Also, the open loop gain of the comb filters should not exceed about 0.85 or -1.4 dB to keep the response fluctuations from being excessive."

Thus, one may choose the comb-filter delay-line lengths more or less arbitrarily, and then use enough of them in parallel (with mutually prime delay-line lengths) to achieve a perceptually adequate fluctuation density in the frequency-response magnitude. In [412], four such delays are chosen between 30 and 45 ms, and the corresponding feedback coefficients g_i are set to give the desired overall decay time.

The delay lengths shown in Fig. 3.7 were optimized by ear by John Chowning (and perhaps others at CCRMA) for an audio sampling rate of $f_s = 25$ kHz.

Finally, for multichannel listening, Schroeder suggested [412] a *mixing matrix* at the reverberator output. The goal of the mixing matrix is to bring out any number of uncorrelated audio channels of reverberation (for any number of output speakers) [153, p. 111–112].

3.5.1 Example Schroeder Reverberators

Additional example Schroeder Reverberators, drawn from CCRMA software listings, are shown in Figures 3.6 and 3.7. The notation used in the figures is explained above in Equations (3.2–3.3).

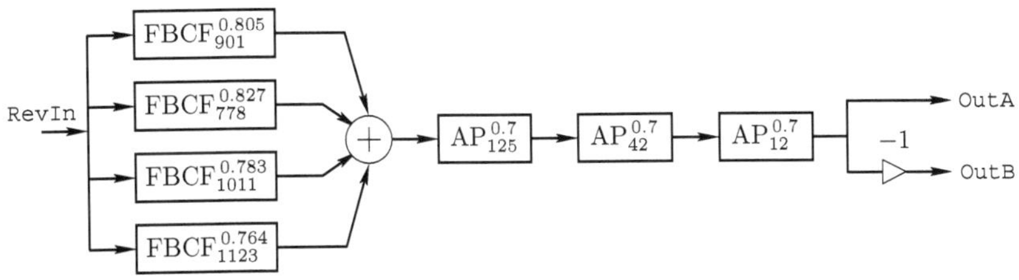

Figure 3.6: Schroeder reverberator **SATREV** by Prof. John Chowning at CCRMA (drawn from a 1971 MUS10 software listing).

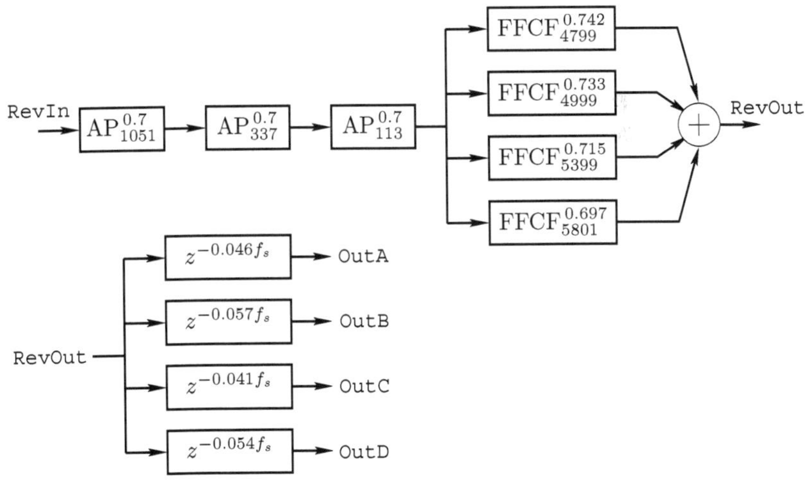

Figure 3.7: A later Schroeder reverberator for the Samson Box [393] at CCRMA, based on (and still called) JCREV.

A software musical instrument using one of these reverberators simply adds its output, suitably scaled, to the real-time variable `RevIn` (the global reverberator input sample at the current time). In both examples, we again see three Schroeder allpass filters in series (Schroeder suggested five, as disussed above).

The rather small reverberator of Fig. 3.6 is thought to have been used in John Chowning's often-heard FM-brass canon sound examples.[8] Like the more computationally expensive four-channel-audio example in Fig. 3.5, it is designed for a 25 kHz sampling rate. Its "mixing matrix" is simply a negation of the right stereo channel.

In addition to the allpass chain in Figures 3.6 and 3.7, there is a parallel bank of four feedback comb filters. Since all of the filters are linear and time-invariant [449], the series allpass chain can go either before or after the parallel comb-filter bank. Unlike the allpass filters, the comb filters have an irregular magnitude frequency response, and they can be considered a simulation of four specific echo sequences. The delay lengths in these comb filters may be used to adjust the illusion of "room size", although if they are shortened, there should be more of them in parallel, according to Schroeder's quote above.

In Fig. 3.7, the reverbator output signal `RevOut` is fed to the four audio output channels via four delay lines. These delay lengths are specified relative to the sampling rate f_s.[9] Output delay lines can substitute for or supplement a mixing matrix as a means for *decorrelating* the reverberation output channels (to minimize reverberation *imaging* between speakers). In this particular case, however, the delays are evidently not optimized for decorrelation. Also, for purposes of decorrelation, the shortest delay can be subtracted from the other three and its corresponding delay-line eliminated.

A Schroeder reverberator along the above lines may be found in the Synthesis Tool Kit (STK) [86]. See files `JCRev.cpp` and `JCRev.h`.

The reverberators shown in Figures 3.5 and 3.6 above are included in the Faust distribution. See functions `jcrev`, `satrev`, and `reverb_demo` in the `effect.lib` library. While these Schroeder reverberators are quite small by today's standards, they are well tuned for their size. More commonly used today is `freeverb` (included in the examples directory of the Faust distribution as `freeverb.dsp`), discussed next.

---

[8] `https://ccrma.stanford.edu/~jos/wav/FM_BrassCanon2.wav`

[9] As an alternative to the output delay values shown in Fig. 3.7, the values $[0.013, 0.011, 0.019, 0.017]f_s$ have been used.

3.6 Freeverb

A more recently developed Schroeder reverberator is "Freeverb" — a public domain `C++` program by "Jezar at Dreampoint" used extensively in the free-software world. It uses four Schroeder allpasses in series and eight parallel Schroeder-Moorer filtered-feedback comb-filters (§2.6.5, p. 62) for each audio channel, and is said to be especially well tuned.

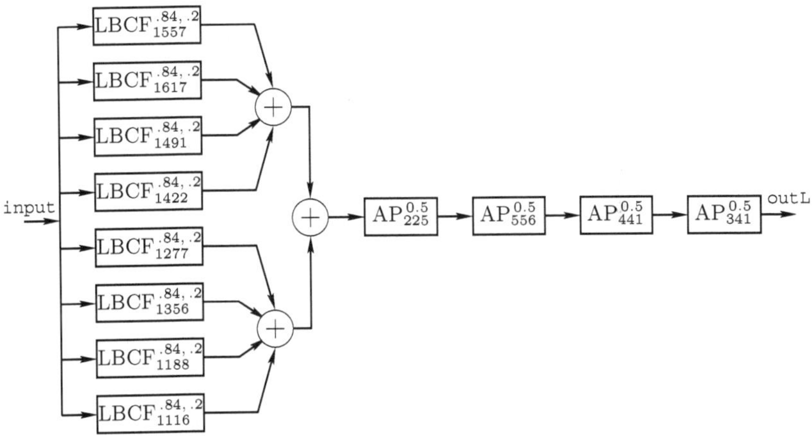

Figure 3.8: Freeverb block diagram (left stereo channel). The Schroeder-Moorer lowpass-feedback-comb-filter, denoted $\text{LBCF}_N^{f,\,d}$ in the figure, is defined in §3.6.2 on page 102 below. The use of three summers instead of one is for drawing convenience.

Figure 3.8 shows the default signal-processing settings for the Freeverb left stereo channel. Processing for the right channel is obtained by adding an integer to each of the twelve delay-line lengths. This integer is called `stereospread`, and its default value is 23. (See the file `tuning.h` for all constants and default values used by Freeverb.) Different software distributions may include slightly different default values in `tuning.h`. The values in Fig. 3.8 were found in the `ladspa-cmt-plugins` package ("Computer Music Toolkit") which is included in the Planet CCRMA distribution, and which is based on the June 2000 version of Freeverb, as of this writing. There are at least six more instances of Freeverb in the Planet CCRMA distribution alone.[10]

[10] There is a Freeverb `pd` "external" in both the `pd-extended` and `faust-pd` packages, and it appears in the `muse` sequencer, `kdemultimedia`, `clm`, and `faust` language packages, all included with Planet CCRMA. Other versions may be found by a Web search for "Freeverb source", but I have not found any version that isn't essentially the same June 2000 version by Jezar at Dreampoint.

3.6.1 Freeverb Main Loop

The C++ code for the main processing loop of Freeverb is shown in Fig. 3.9. Notice that it sums the two stereo input channels to create a mono signal that is fed to the reverberator, which then computes a stereo output signal.

From the code in Fig. 3.9, we see that the left and right reverberator output channels `outL` and `outR` are combined with the left and right input channels `inputL` and `inputR` as follows:

$$\begin{bmatrix} \texttt{outputL} \\ \texttt{outputR} \end{bmatrix} = \texttt{dry} \begin{bmatrix} \texttt{inputL} \\ \texttt{inputR} \end{bmatrix} + \begin{bmatrix} \texttt{wet1} & \texttt{wet2} \\ \texttt{wet2} & \texttt{wet1} \end{bmatrix} \begin{bmatrix} \texttt{outL} \\ \texttt{outR} \end{bmatrix}$$

The `dry` parameter (initially set to 0 in `tuning.h`, and typically changed using a graphical user interface (GUI)), determines how much of the original (stereo) input signal is mixed to the output, and `wet1` and `wet2` determine how much of the reverberated signal is mixed to the output, and also how much "stereo separation" occurs in the reverberation. In particular, if `wet1` = `wet2`, there is no stereo separation since the same reverberator output signal is sent to the left and right output channels. (Note that there is no decorrelating delay-line like we saw at the output of `JCRev` in Fig. 3.7 on page 99.) Setting `wet2` = 0 yields maximally different left and right reverberation signals, thus maximizing "stereo separation" in the reverb. The default values, determined by `tuning.h` and `revmodel.cpp`, are `wet1` = 1 and `wet2` = 0.

3.6.2 Lowpass-Feedback Comb Filter

Inspection of `comb.h` in the Freeverb source shows that Freeverb's "comb" filter is more specifically a *lowpass-feedback-comb filter* (LBCF[11]—§2.6.5 on page 62). It is constructed using a delay line whose output is lowpass-filtered and summed with the delay-line's input. The particular lowpass used in Freeverb is a *unity-gain one-pole lowpass* having the transfer function

$$H(z) = \frac{1-d}{1-d\,z^{-1}}.$$

When $d = 0$, the LBCF reduces to the feedback comb filter (FBCF) of §2.6.2 on page 58 in which the feedback was not filtered. The overall LBCF transfer function is then

$$\text{LBCF}_N^{f,\,d} \triangleq \frac{1}{1 - f\frac{1-d}{1-d\,z^{-1}}\,z^{-N}}.$$

[11] "Lowpass-Back-Comb-Filter"—to keep all the comb-filter acronyms down to four characters: FFCF, FBCF, LBCF. Author's perogative!

```
void revmodel::processreplace(float *inputL, float *inputR,
  float *outputL, float *outputR, long numsamples, int skip)
{
  float outL,outR,input;
  int i;

  while(numsamples-- > 0)
    {
      outL = outR = 0;
      input = (*inputL + *inputR) * gain;

      // Accumulate comb filters in parallel
      for(i=0; i<numcombs; i++) {
        outL += combL[i].process(input);
        outR += combR[i].process(input);
      }

      // Feed through allpasses in series
      for(i=0; i<numallpasses; i++) {
        outL = allpassL[i].process(outL);
        outR = allpassR[i].process(outR);
      }

      // Calculate output REPLACING anything already there
      *outputL = outL*wet1 + outR*wet2 + *inputL*dry;
      *outputR = outR*wet1 + outL*wet2 + *inputR*dry;

      // Increment sample pointers, allowing for interleave
      // (if any)
      inputL += skip; // For stereo buffers, skip = 2
      inputR += skip;
      outputL += skip;
      outputR += skip;
    }
}
```

Figure 3.9: Freeverb processing function.

This structure was introduced for artificial reverberation by Schroeder [412] and Moorer [314].

In Freeverb's `comb` section (`comb.h` and `comb.cpp`), the "damping" d is set initially to

$$d = \texttt{damp = initialdamp * scaledamp} = 0.5 \cdot 0.4 = 0.2\ .$$

The lowpass scale-factor f is called **`feedback`** in the source, and it is set initially to

$$\begin{aligned} f &= \texttt{roomsize = initialroom * scaleroom + offsetroom} \\ &= 0.5 \cdot 0.28 + 0.7 = 0.84\ . \end{aligned}$$

Increasing the **`roomsize`** parameter (typically brought out to a GUI slider) increases f and hence the reverberation time. Since $f < 1$ is required for dc stability, the **`roomsize`** must be less than 1.0714, and so the GUI slider max is typically 1 ($f = 0.98$).

The **`feedback`** variable f mainly determines reverberation time at low-frequencies at which the feedback lowpass has negligible effect. The feedback lowpass causes the reverberation time to decrease with frequency, which is natural. At very high frequencies—those for which the lowpass gain times f is much less than 0.5—the reverberation time becomes dominated by the diffusion allpass filters (which have a fixed feedback coefficient of $g = 0.5$). Thus, in Freeverb, the "room size" parameter can be interpreted as setting the low-frequency T60 (time to decay 60 dB), while the "damping" parameter controls how rapidly T60 shortens as a function of increasing frequency. A lower-limit on T60 is given by the four diffusion allpass filters.

In terms of the physical interpretation of the filtered-feedback comb-filter discussed in §2.6.5 on page 62, Freeverb's **`roomsize`** parameter can be interpreted as the square-root of the low-frequency reflection-coefficient of each wall. That is, when a planewave bounces back and forth between two walls, the attenuation coefficient is **`roomsize`** after one round trip (two wall reflections). Therefore, a better name in this interpretation would be **`liveness`** or **`reflectivity`**. Since the round-trip delay is given in samples by the delay-line length, changing the roomsize requires changing the delay-line lengths in this interpretation.

3.6.3 Freeverb Allpass Approximation

In Eq. (3.2) on page 97 we defined the allpass notation AP_N^g by

$$\text{AP}_N^g \triangleq \frac{-g + z^{-N}}{1 - gz^{-N}}$$

A look at `allpass.h` reveals that Freeverb implements

$$\text{AP}_N^g \approx \frac{-1+(1+g)z^{-N}}{1-gz^{-N}}.$$

As a result, each of the four Freeverb "allpass" sections is really a feedback comb-filter FBCF_N^g in series with a feedforward comb-filter $\text{FFCF}_N^{-1,1+g}$, where (cf. §2.6)

$$\begin{aligned}\text{FBCF}_N^g &\stackrel{\Delta}{=} \frac{1}{1-g\,z^{-N}}\\ \text{FFCF}_N^{-1,1+g} &\stackrel{\Delta}{=} -1+(1+g)z^{-N}.\end{aligned}$$

A true allpass is obtained only for $g=(\sqrt{5}-1)/2 \approx 0.618$ (reciprocal of the "golden ratio"). The default value used in Freeverb (see `revmodel.cpp`) is 0.5. A detailed discussion of feedforward and feedback comb filters appears in §2.6 on page 57, and corresponding Schroeder allpass filters are described in §2.8 on page 69.

3.6.4 Conclusions

Since Freeverb is a Schroeder-Moorer reverberator, and such reverbs have been around since the 1970s, its relatively recent success underscores the value of careful parameter tuning (typically by ear, but automatic optimizations are possible).

The idea of synthesizing right-channel processing from left-channel processing to obtain "stereo spreading" by enlarging all the delay lines by a fixed amount appears to have been introduced by Freeverb.

3.7 FDN Reverberation

Feedback Delay Networks (FDN) were introduced earlier in §2.7. An example is shown in Fig. 2.29 on page 67. After a brief historical summary, this section will cover some practical considerations for the use of FDNs as reverberators.

3.7.1 History of FDNs for Artificial Reverberation

Feedback delay networks were first suggested for artificial reverberation by Gerzon [156], who proposed an "orthogonal matrix feedback reverberation unit". He noted that individual feedback comb filters yielded poor quality, but that several such filters could sound good when cross-coupled. An "orthogonal matrix feedback" around a parallel bank of delay lines was suggested as a means of obtaining maximally rich cross-coupling. He was especially concerned with good stereo spreading of the reverberation at a time when most artificial reverberators sought merely to decorrelate the reverberation in each output channel.

Later, and apparently independently, Stautner and Puckette [474] suggested a specific four-channel FDN reverberator and gave general stability conditions for the FDN. They proposed the feedback matrix

$$\mathbf{A} = g\frac{1}{\sqrt{2}}\begin{bmatrix} 0 & 1 & 1 & 0 \\ -1 & 0 & 0 & -1 \\ 1 & 0 & 0 & -1 \\ 0 & 1 & -1 & 0 \end{bmatrix}$$

which can be understood as a permutation (with one row sign inversion) of a 4×4 Hadamard matrix (see §3.7.2).

More recently, Jot [217, 216] developed a systematic FDN design methodology allowing largely independent setting of reverberation time in different frequency bands. Using Jot's methodology, FDN reverberators can be polished to a high degree of quality, and they are presently considered to be among the best choices for high-quality artificial reverberation.

Jot's early work was concerned only with single-input, single-output (SISO) reverberators. Later work [218] with Jullien and others at IRCAM was concerned also with *spatializing* the reverberation.

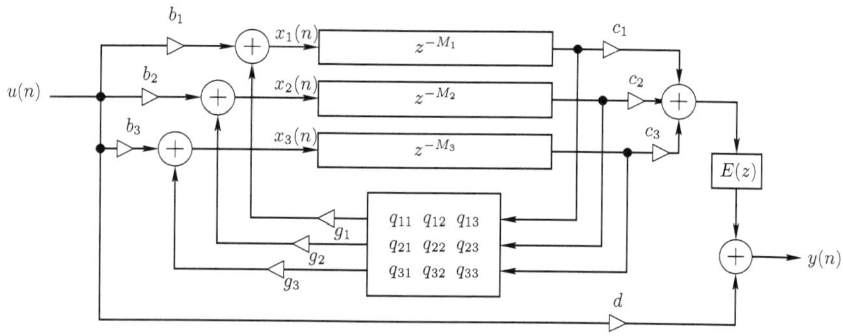

Figure 3.10: Feedback Delay Network (FDN) structure proposed for artificial reverberation by Jot.

An example FDN reverberator using three delay lines is shown in Fig. 3.10. It can be seen as an FDN (introduced in §2.7 on page 65), plus an additional low-order filter $E(z)$ applied to the non-direct signal. This filter is called a "tonal correction" filter by Jot, and it serves to equalize modal energy irrespective of the reverberation time in each band. In other words, if the decay time is made very short in some band, $E(z)$ will have a large gain in that band so that the total energy in the band's impulse-response is unchanged. This is another example of *orthogonalization* of reverberation parameters: In this case, adjustments in reverberation time, in any frequency band, do not alter total signal energy in the impulse response in that band.

3.7.2 Choice of Lossless Feedback Matrix

As mentioned in §3.4, an "ideal" late reverberation impulse response should resemble exponentially decaying noise [314]. It is therefore useful when designing a reverberator to start with an infinite reverberation time (the "lossless case") and work on making the reverberator a good "noise generator". Such a starting point is ofen referred to as a *lossless prototype* [153, 430]. Once smooth noise is heard in the impulse response of the lossless prototype, one can then work on obtaining the desired reverberation time in each frequency band (as will be discussed in §3.7.4 below).

In reverberators based on feedback delay networks (FDNs), the smoothness of the "perceptually white noise" generated by the impulse response of the lossless prototype is strongly affected by the choice of FDN feedback matrix as well as the (ideally mutually prime) delay-line lengths in the FDN (discussed further in §3.7.3 below). Following are some of the better known feedback-matrix choices.

Hadamard Matrix

A second-order *Hadamard matrix* may be defined by

$$\mathbf{H}_2 \triangleq \frac{1}{\sqrt{2}} \left[\begin{array}{rr} 1 & 1 \\ -1 & 1 \end{array} \right],$$

with higher order Hadamard matrices defined by recursive embedding, *e.g.*,

$$\mathbf{H}_4 \triangleq \frac{1}{\sqrt{2}} \left[\begin{array}{rr} \mathbf{H}_2 & \mathbf{H}_2 \\ -\mathbf{H}_2 & \mathbf{H}_2 \end{array} \right]. = \frac{1}{2} \left[\begin{array}{rrrr} 1 & 1 & 1 & 1 \\ -1 & 1 & -1 & 1 \\ -1 & -1 & 1 & 1 \\ 1 & -1 & -1 & 1 \end{array} \right].$$

When n is a power of 4, the Hadamard matrix $\mathbf{H}_n$ of that order requires no multiplies in fixed-point arithmetic. An $n \times n$ Hadamard matrix has the maximum possible determinant of any $n \times n$ complex matrix containing elements which are bounded by 1 in magnitude. This can be seen as an optimal *mixing and scattering* property of the matrix.

As of version 0.9.30, Faust's `math.lib`[12] contains a function called `hadamard(n)` for generating an $n \times n$ Hadamard matrix, where n must be a power of 2. A Hadamard feedback matrix is used in the programming example `reverb_designer.dsp` (a configurable FDN reverberator) distributed with Faust.

A Hadamard feedback matrix is said to be used in the IRCAM Spatialisateur [218].

Householder Feedback Matrix

One choice of lossless feedback matrix $\mathbf{A}_N$ for FDNs, especially nice in the 4×4 case, is a specific *Householder reflection* proposed by Jot [217]:

$$\mathbf{A}_N = \mathbf{I}_N - \frac{2}{N} \underline{u}_N \underline{u}_N^T \tag{3.4}$$

where $\underline{u}_N^T = [1, 1, \ldots, 1]$ can be interpreted as the specific vector about which the input vector is reflected in N-dimensional space (followed by a sign inversion). More generally, the identity matrix $\mathbf{I}_N$ can be replaced by any $N \times N$ permutation matrix [153, p. 126].

It is interesting to note that when N is a power of 2, no multiplies are required [430]. For other N, only one multiply is required (by $2/N$).

Another interesting property of the Householder reflection $\mathbf{A}_N$ given by Eq. (3.4) (and its permuted forms) is that an $N \times N$ matrix-times-vector operation may be carried out with only $2N-1$ additions (by first forming $\underline{u}_N^T$ times the input vector, applying the scale factor $2/N$, and subtracting the result from the input vector). This is the same computation as physical *wave scattering* at a junction of identical waveguides (§C.8 on page 560).

An example implementation of a Householder FDN for $N=3$ is shown in Fig. 3.11. As observed by Jot [153, p. 216], this computation is equivalent to N parallel feedback comb filters with one new feedback path from the output to the input through a gain of $-2/N$.

A nice feature of the Householder feedback matrix A_N is that for $N \neq 2$, *all entries in the matrix are nonzero.* This means every delay line feeds back to every other delay line, thereby helping to maximize echo density as soon as possible.

Furthermore, for $N=4$, all matrix entries have the *same magnitude*:

$$\mathbf{A}_4 = \frac{1}{2} \left[\begin{array}{rrrr} 1 & -1 & -1 & -1 \\ -1 & 1 & -1 & -1 \\ -1 & -1 & 1 & -1 \\ -1 & -1 & -1 & 1 \end{array} \right].$$

[12]Details about Faust may be found at `http://faust.grame.fr/`.

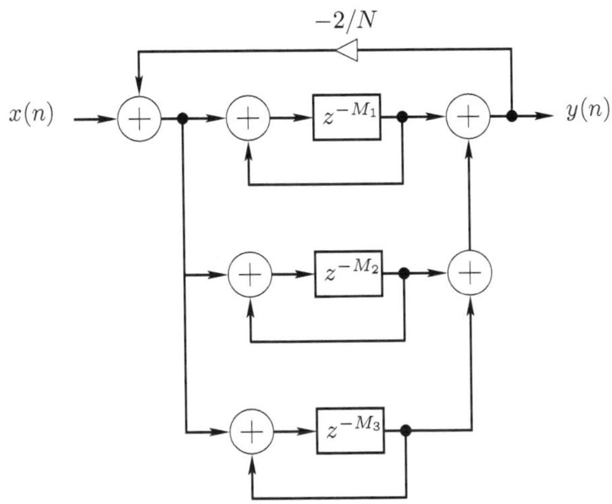

Figure 3.11: FDN using a Householder-reflection feedback matrix.

Only the $N = 4$ case is "balanced" in this way. For larger N, the diagonal becomes larger than the off-diagonal elements, and as N becomes very large, the FDN approaches a bank of *decoupled* parallel comb filters.

Due to the elegant balance of the $N = 4$ Householder feedback matrix, Jot [216] proposes an $N = 16$ FDN based on an embedding of $N = 4$ feedback matrices:

$$\mathbf{A}_{16} = \frac{1}{2}\left[\begin{array}{rrrr} \mathbf{A}_4 & -\mathbf{A}_4 & -\mathbf{A}_4 & -\mathbf{A}_4 \\ -\mathbf{A}_4 & \mathbf{A}_4 & -\mathbf{A}_4 & -\mathbf{A}_4 \\ -\mathbf{A}_4 & -\mathbf{A}_4 & \mathbf{A}_4 & -\mathbf{A}_4 \\ -\mathbf{A}_4 & -\mathbf{A}_4 & -\mathbf{A}_4 & \mathbf{A}_4 \end{array}\right]$$

Another method is to replace each of the four delay lines in an FDN(4) by a Gerzon vector allpass (see §2.8.5 on page 75) which is 4×4 and contains four delay lines.

Householder Reflections

For completeness, this section derives the Householder reflection matrix from geometric considerations [451]. Let $\mathbf{P}_{\underline{u}}$ denote the *projection matrix* which orthogonally projects vectors onto $\underline{u}$, *i.e.*,

$$\mathbf{P}_{\underline{u}} = \frac{\underline{u}\,\underline{u}^T}{\underline{u}^T\underline{u}} = \frac{\underline{u}\,\underline{u}^T}{\|\,\underline{u}\,\|^2}$$

and

$$\mathbf{P}_{\underline{u}}\,\underline{x} = \underline{u}\,\frac{\langle \underline{u}, \underline{x} \rangle}{\|\,\underline{u}\,\|^2}$$

specifically projects $\underline{x}$ onto $\underline{u}$. Since the projection is orthogonal, we have

$$\langle \underline{x} - \mathbf{P}_{\underline{u}}\underline{x}, \underline{u} \rangle = \langle (\mathbf{I} - \mathbf{P}_{\underline{u}})\underline{x}, \underline{u} \rangle = \underline{0}.$$

We may interpret $(\mathbf{I} - \mathbf{P}_{\underline{u}})\underline{x}$ as the *difference vector* between $\underline{x}$ and $\mathbf{P}_{\underline{u}}\underline{x}$, its orthogonal projection onto $\underline{u}$, since

$$(\mathbf{I} - \mathbf{P}_{\underline{u}})\underline{x} + \mathbf{P}_{\underline{u}}\underline{x} = \underline{x}$$

and we have $(\mathbf{I} - \mathbf{P}_{\underline{u}})\underline{x} \perp \mathbf{P}_{\underline{u}}\underline{x}$ by definition of the orthogonal projection. Consequently, the projection onto $\underline{u}$ *minus* this difference vector gives a *reflection* of the vector $\underline{x}$ about $\underline{u}$:

$$\underline{y} = \mathbf{P}_{\underline{u}}\underline{x} - (\mathbf{I} - \mathbf{P}_{\underline{u}})\underline{x} = (2\mathbf{P}_{\underline{u}} - \mathbf{I})\underline{x}$$

Thus, $\underline{y}$ is obtained by *reflecting* $\underline{x}$ about $\underline{u}$—a so-called *Householder reflection.*

Most General Lossless Feedback Matrices

As shown in §C.15.3 on page 615, an FDN feedback matrix $\mathbf{A}_N$ is lossless if and only if its eigenvalues have modulus 1 and its N eigenvectors are linearly independent.

A *unitary matrix* Q is any (complex) matrix that is inverted by its own (conjugate) transpose:

$$Q^{-1} = Q^H,$$

where Q^H denotes the *Hermitian conjugate* (*i.e.*, the complex-conjugate transpose) of Q. When Q is real (as opposed to complex), we may simply call it an *orthogonal matrix*, and we write $Q^{-1} = Q^T$, where T denotes matrix transposition.

All unitary (and orthogonal) matrices have unit-modulus eigenvalues and linearly independent eigenvectors. As a result, when used as a feedback matrix in an FDN, the resulting FDN will be lossless (until the delay-line damping filters are inserted, as discussed in §3.7.4 below).

Triangular Feedback Matrices

An interesting class of feedback matrices, also explored by Jot [216], is that of *triangular matrices*. A basic fact from linear algebra is that triangular matrices (either lower or upper triangular) have all of their eigenvalues along the diagonal.[13] For example, the matrix

$$\mathbf{A}_3 = \left[\begin{array}{ccc} \lambda_1 & 0 & 0 \\ a & \lambda_2 & 0 \\ b & c & \lambda_3 \end{array}\right]$$

is lower triangular, and its eigenvalues are $(\lambda_1, \lambda_2, \lambda_3)$ for all values of a, b, and c.

It is important to note that not all triangular matrices are lossless. For example, consider

$$\mathbf{A}_2 = \left[\begin{array}{cc} 1 & 0 \\ 1 & 1 \end{array}\right]$$

It has two eigenvalues equal to 1, which looks lossless, but a quick calculation shows that there is only one eigenvector, $[0, 1]^T$. This happens because this matrix is a Jordan block of order 2 corresponding to the repeated eigenvalue $\lambda = 1$. A direct computation shows that

$$\mathbf{A}_2^n = \left[\begin{array}{cc} 1 & 0 \\ n & 1 \end{array}\right]$$

which is clearly not lossless.

One way to avoid "coupled repeated poles" of this nature is to use non-repeating eigenvalues. Another is to convert $\mathbf{A}$ to Jordan canonical form by means of a similarity transformation, zero any off-diagonal elements, and transform back [329].

3.7.3 Choice of Delay Lengths

Following Schroeder's original insight, the delay line lengths in an FDN (M_i in Fig. 3.10) are typically chosen to be *mutually prime*. That is, their prime factorizations contain no common factors. This rule maximizes the number of samples that the lossless reverberator prototype must be run before the impulse response repeats.

The delay lengths M_i should be chosen to ensure a *sufficiently high mode density* in all frequency bands. An insufficient mode density can be heard as "ringing tones" or an uneven amplitude modulation in the late reverberation impulse response.

[13]This is easy to show by performing an expansion by minors to calculate $\det(z\mathbf{I} - \mathbf{A})$, always choosing to expand along the top row (for lower triangular) or first column (for upper triangular).

Mean Free Path

A rough guide to the average delay-line length is the "mean free path" in the desired reverberant environment. The *mean free path* is defined as the average distance a ray of sound travels before it encounters an obstacle and reflects. An approximate value for the mean free path, due to Sabine, an early pioneer of statistical room acoustics, is

$$\bar{d} = 4\frac{V}{S} \qquad \text{(mean free path)}$$

where V is the total volume of the room, and S is total surface area enclosing the room. This approximation requires the *diffuse field* assumption, *i.e.*, that plane waves are traveling randomly in all directions [349, 47] (see §3.2.1, p. 89 for a simple construction). Normally, late reverberation satisfies this assumption well, away from open doors and windows, provided the room is not too "dead". Regarding each delay line as a mean-free-path delay, the average can be set to the mean free path by equating

$$\frac{\bar{d}}{cT} = \frac{1}{N}\sum_{i=1}^{N} M_i$$

where c denotes sound speed and T denotes the sampling period. This number should be treated as a lower bound because in real rooms reflections are often *diffuse*, especially at high frequencies. In a diffuse reflection, a single incident plane wave reflects in many directions at once.

Mode Density Requirement

A guide for the *sum* of the delay-line lengths is the desired *mode density.* The sum of delay-line lengths M_i in a lossless FDN is simply the *order* of the system M:

$$M \triangleq \sum_{i=1}^{N} M_i \qquad \text{(FDN order)}$$

The order increases slightly when lowpass filters are introduced after the delay lines to achieve a specific reverberation time at low and high frequencies (as described in the next subsection).

Since the order of a system equals the number of poles, we have that M is the number of poles on the unit circle in the lossless prototype. If the modes were uniformly distributed, the mode density would be $M/f_s = MT$ modes per Hz. Schroeder [417] suggests that, for a reverberation time of 1 second, a mode density of 0.15 modes per Hz is adequate. Since the mode widths are inversely proportional to reverberation time, the mode density for a reverberation time of 2 seconds should be 0.3 modes per Hz, etc. In summary, for a sufficient mode density in the frequency domain, Schroeder's formula is

$$M \geq 0.15 t_{60} f_s$$

For a sampling rate of 50 kHz and a reverberation time (t_{60}) equal to 1 second, we obtain $M \geq 7500$.

Prime Power Delay-Line Lengths

When the delay-line lengths need to be varied in real time, or interactively in a GUI, it is convenient to choose each delay-line length $\hat{M}_i$ as an integer power of a distinct prime number p_i [458]:

$$\hat{M}_i \;\stackrel{\Delta}{=}\; p_i^{m_i}$$

where we call $m_i \geq 1$ the "multiplicity" of the prime p_i. With this choice, the delay-line lengths are always coprime (no factors in common other than 1), and yet we can lengthen or shorten each delay line individually (by factors of p_i) without affecting the mutually prime property.

Suppose we are initially given desired delay-line lengths M_i arranged in ascending order so that

$$M_1 < M_2 < \cdots < M_N.$$

Then good prime-power approximations $\hat{M}_i$ can be expected using the prime numbers in their natural order:

$$p_i \in \{2, 3, 5, 7, 11, 13, 17, 19, 23, 29, 31, 37, 41, 43, 47, 53, \ldots\}$$

Since $\hat{M}_i = p_i^{m_i} \;\Rightarrow\; \log(\hat{M}_i) = m_i \log(p_i)$ (for any logarithmic base), an optimal (in some sense) choice of prime multiplicity m_i is

$$m_i \;\stackrel{\Delta}{=}\; \text{round}\left[\frac{\log(M_i)}{\log(p_i)}\right] \;\stackrel{\Delta}{=}\; \left\lfloor 0.5 + \frac{\log(M_i)}{\log(p_i)} \right\rfloor.$$

where M_i is the desired length in samples. That is, m_i can be simply obtained by *rounding* $\log(M_i)/\log(p_i)$ to the nearest integer (max 1). The prime-power delay-line length approximation is then of course

$$\hat{M}_i \;\stackrel{\Delta}{=}\; p_i^{m_i},$$

and the multiplicative approximation error is bounded by $p_i^{\pm 1/2}$ (when $M_i \geq \sqrt{p_i}$).

This prime-power length scheme is used to keep 16 delay lines both variable and mutually prime in Faust's `reverb_designer.dsp` programming example (via the function `prime_power_delays` in `effect.lib`).

3.7.4 Achieving Desired Reverberation Times

A lossless prototype reverberator, as in Fig. 3.10 on page 106 when $g_i = 1$, has all of its poles on the unit circle in the z plane, and its reverberation time is infinity. To set the reverberation time to a desired value, we need to move the poles slightly inside the unit circle. Furthermore, due to air absorption (§2.3,§B.7.15), we want the high-frequency poles to be more damped than the low-frequency poles [314]. As discussed in §2.3 on page 45, this type of transformation can be obtained using the substitution

$$z^{-1} \leftarrow G(z)z^{-1}, \tag{3.5}$$

where $G(z)$ denotes the *filtering per sample* in the propagation medium (a lowpass filter with gain not exceeding 1 at all frequencies).[14] Thus, to set the FDN reverberation time to $t_{60}(\omega) \triangleq n_{60}(\omega)T$ at frequency ω, we want propagation through n_{60} samples to result in attenuation by 60 dB, *i.e.*,

$$\left[G(e^{j\omega T})\right]^{n_{60}(\omega)} = 0.001. \tag{3.6}$$

Solving for G, the propagation attenuation per-sample, gives

$$\begin{aligned} G(e^{j\omega T}) &= (0.001)^{\frac{1}{n_{60}(\omega)}} = 10^{-3/n_{60}} = \left(e^{\ln(10)}\right)^{-3/n_{60}} = e^{-3\ln(10)/n_{60}} \\ &= e^{-T/\tau(\omega)} \end{aligned} \tag{3.7}$$

The last form comes from $t_{60} = 3\ln(10)\tau \approx 6.91\tau$, where τ denotes the time constant of decay (time to decay by $1/e$) [451], *i.e.*,

$$e^{-t_{60}/\tau} = 0.001 \quad \Leftrightarrow \quad t_{60} = -3\ln(10)\tau. \tag{3.8}$$

Series expanding $e^{-T/\tau(\omega)}$ and assuming $n_{60}(\omega) \gg 7$ samples ($\tau(\omega) \gg T$ seconds) provides the practically useful approximation

$$e^{-T/\tau(\omega)} = 1 - \frac{T}{\tau(\omega)} + \frac{1}{2!}\left[\frac{T}{\tau(\omega)}\right]^2 - \cdots \approx 1 - \frac{3\ln(10)}{n_{60}} \approx 1 - \frac{6.91}{n_{60}}.$$

Conformal Map Interpretation of Damping Substitution

The relation $G(e^{j\omega T}) \approx e^{-T/\tau(\omega)}$ [Eq. (3.7)] can be written down directly from $z^{-1} \leftarrow G(z)z^{-1}$ [Eq. (3.5)] by interpreting Eq. (3.5) as an approximate *conformal map* [326] which takes each pole $p_k = e^{j\omega_k T}$, say, from the unit

[14] This substitution was first applied to complete reverberators by Jot [216, 217]. The idea is closely related to the approximate pole analysis in [314, p. 17], [428, pp. 170–172], and [207], and somewhat also to the allpass conformal maps used in [315, 460].

circle to the point $p'_k \approx G(e^{j\omega_k T})e^{-j\omega_k T}$. Thus, the new pole radius is approximately $|p'_k| \approx |G(e^{j\omega_k T})|$, where the approximation is valid when $G(z)$ is approximately constant between the new pole location and the unit circle. To see this, consider the partial fraction expansion [449] of a proper Nth-order lossless transfer function $H(z)$ mapped to $H'(z) \isdef H[z/G(z)]$:

$$H'(z) = \sum_{k=1}^{N} \frac{r_k}{1 - p_k G(z) z^{-1}} = \sum_{k=1}^{N} r_k \left[1 + p_k G(z) z^{-1} + p_k^2 G^2(z) z^{-2} + \cdots\right],$$

where $p_k = \exp(j\omega_k T)$ denotes the kth original pole on the unit circle. Then $H'(z)$ has a pole at $z'_k = p_k G(z'_k)$, which must be solved iteratively for z'_k, in general, since $G(z)$ can be a complicated function of z. However, if $G(z'_k) \approx G(p_k)$, which is typically true when damping digital waveguides for music applications, then $z'_k \approx p_k G(p_k) = G[\exp(j\omega_k T)]\exp(j\omega_k T)$. In other words, we can think of the pole p_k as moving from $\exp(j\omega_k T)$ to near $G[\exp(j\omega_k T)]\exp(j\omega_k T)$, provided it doesn't move too far compared with the near-constant behavior of $G(z)$. Another way to say it is that we need $G(z)$ to be approximately the same at the new pole location and its initial location on the unit circle in the lossless prototype.

Happily, while we may not know precisely where our poles have moved as a result of introducing the per-sample damping filter $G(z)$, the relation $G^{n_{60}(\omega)}(e^{j\omega T}) = 0.001$ [Eq. (3.6)] remains exact at every frequency by construction, as it is based only on the physical interpretation of each unit delay as a propagation delay for a plane wave across one sampling interval T, during which (zero-phase) filtering by $G(z)$ is assumed (§2.3). More generally, we can design minimum-phase filters for which $|G^{n_{60}(\omega)}(e^{j\omega T})| = 0.001$, and neglect the resulting phase dispersion.

In summary, we see that replacing z^{-1} by $G(z)z^{-1}$ everywhere in the FDN lossless prototype (or any lossless LTI system for that matter) serves to move its poles away from the unit circle in the z plane onto some contour inside the unit circle that provides the desired decay time at each frequency.

A general design guideline for artificial reverberation applications [217] is that *all pole radii in the reverberator should vary smoothly with frequency.* This translates to $G(z)$ having a *smooth frequency response.* To see why this is desired, consider momentarily the frequency-independent case in which we desire the same reverberation time at all frequencies (Fig. 3.10 with real $g_i \leq 1$, as drawn). In this case, it is ideal for all of the poles to have this decay time. Otherwise, the late decay of the impulse response will be dominated by the poles having the largest magnitude, and it will be "thinner" than it was at the beginning of the response when all poles were contributing to the output. Only when all poles have the same magnitude will the late response maintain the same modal density throughout the decay.

Damping Filters for Reverberation Delay Lines

In an FDN, such as the one shown in Fig. 3.10, delays z^{-1} appear in long delay-line chains z^{-M_i}. Therefore, the filter needed at the output (or input) of the ith delay line is $G^{M_i}(z)$ (replace g_i by $G^{M_i}(z)$ in Fig. 3.10).[15] However, because $G(e^{j\omega T})$ is so close to 1 in magnitude, and because it varies so weakly across the frequency axis, we can design a much lower-order filter $H_i(z) \approx G^{M_i}(z)$ that provides the desired attenuation versus frequency to within psychoacoustic resolution. In fact, perfectly nice reverberators can be designed in which $H_i(z)$ is merely *first order* for each i [314, 217].

3.7.5 Delay-Line Damping Filter Design

Let $t_{60}(\omega)$ denote the desired reverberation time at radian frequency ω, and let $H_i(z)$ denote the transfer function of the lowpass filter to be placed in series with the ith delay line which is M_i samples long. The problem we consider now is how to design these filters to yield the desired reverberation time. We will specify an *ideal* amplitude response for $H_i(z)$ based on the desired reverberation time at each frequency, and then use conventional filter-design methods to obtain a low-order approximation to this ideal specification.

In accordance with Eq. (3.6) on page 114, the lowpass filter $H_i(z)$ in series with a length M_i delay line should approximate

$$H_i(z) = G^{M_i}(z)$$

which implies

$$\left|H_i(e^{j\omega T})\right|^{\frac{t_{60}(\omega)}{M_i T}} = 0.001.$$

Taking $20\log_{10}$ of both sides gives

$$20\log_{10}\left|H_i(e^{j\omega T})\right| = -60\frac{M_i T}{t_{60}(\omega)}. \tag{3.9}$$

This is the same formula derived by Jot [217] using a somewhat different approach.

Now that we have specified the ideal delay-line filter $H_i(e^{j\omega T})$ in terms of its amplitude response in dB, any number of filter-design methods can be used to find a low-order $H_i(z)$ which provides a good approximation to satisfying Eq. (3.9). Examples include the functions `invfreqz` and `stmcb` in Matlab. Since the variation in reverberation time is typically very smooth with respect to ω, the filters $H_i(z)$ can be very low order.

---

[15]Note that, as developed here, the delay-line filters $G_i(z)$ should be located either before or after its associated delay line, while Fig. 3.10 associates it with the feedback matrix. The difference is typically not audible.

First-Order Delay-Filter Design

The first-order case is very simple while enabling separate control of low-frequency and high-frequency reverberation times. For simplicity, let's specify $t_{60}(0)$ and $t_{60}(\pi/T)$, denoting the desired decay-time at dc ($\omega = 0$) and half the sampling rate ($\omega = \pi/T$). Then we have determined the coefficients of a one-pole filter:

$$H_i(z) = \frac{g_i}{1 - p_i z^{-1}}$$

The dc gain of this filter is $H_i(1) = g_i/(1-p_i)$, while the gain at $\omega = \pi/T$ is $H_i(-1) = g_i/(1+p_i)$. From Eq. (3.9) (or Eq. (3.8) on page 114), we obtain two equations in two unknowns:

$$\begin{aligned} \frac{g_i}{1-p_i} &= 10^{-3M_iT/t_{60}(0)} = e^{-M_iT/\tau(0)} \triangleq R_0^{M_i} \\ \frac{g_i}{1+p_i} &= 10^{-3M_iT/t_{60}(\pi/T)} = e^{-M_iT/\tau(\pi/T)} \triangleq R_\pi^{M_i} \end{aligned}$$

where $D_i \triangleq M_i T$ denotes the ith delay-line length in seconds. These two equations are readily solved to yield

$$\begin{aligned} p_i &= \frac{R_0^{M_i} - R_\pi^{M_i}}{R_0^{M_i} + R_\pi^{M_i}} \\ g_i &= \frac{2R_0^{M_i} R_\pi^{M_i}}{R_0^{M_i} + R_\pi^{M_i}} \end{aligned}$$

The truncated series approximation

$$R_\omega^{M_i} \triangleq e^{-\frac{M_iT}{\tau(\omega)}} \approx 1 - \frac{M_iT}{\tau(\omega)} \approx 1 - \frac{6.91\,M_iT}{t_{60}(\omega)} \triangleq 1 - \frac{6.91\,M_i}{n_{60}(\omega)}$$

has been found to work well in practical FDN reverberators.

Orthogonalized First-Order Delay-Filter Design

In [217], first-order delay-line filters of the form

$$H_i(z) = g_i' \frac{1 - p_i}{1 - p_i z^{-1}}$$

are proposed. Clearly $g_i = g_i' \cdot (1 - p_i)$. This form has the advantage that the dc gain is always $H_i(1) = g_i'$ for all (stable) values of p_i. Therefore, we can set g_i' to give a desired reverberation time at dc, and not have to change it when p_i is varied to modulate the high-frequency decay rate. As in the previous section, from Eq. (3.9), we obtain

$$g_i' = 10^{-3M_iT/t_{60}(0)}.$$

A calculation given in [217] arrives at

$$p_i = \frac{\ln(10)}{4}\log_{10}(g_i)\left(1 - \frac{1}{\alpha^2}\right)$$

where

$$\alpha \triangleq \frac{t_{60}(\pi/T)}{t_{60}(0)} \tag{3.10}$$

denotes the ratio of reverberation time at half the sampling rate divided by the reverberation time at dc.[16]

Multiband Delay-Filter Design

In §3.7.5, we derived first-order FDN delay-line filters which can independently set the reverberation time at dc and at half the sampling rate. However, perceptual studies indicate that reverberation time should be independently adjustable in at least three frequency bands [217]. To provide this degree of control (and more), one can implement a multiband delay-line filter using a general-purpose *filter bank* [370, 501]. The output, say, of each delay line is split into K bands, where $K \geq 3$ is recommended, and then, from Eq. (3.6) on page 114, the gain in the kth band for a length M_i delay-line can be set to

$$G^{M_i}(e^{j\omega_k T}) = 10^{-\frac{3M_i}{n_{60}(\omega_k)}} = e^{-\frac{3\,\ln(10)\,M_i}{n_{60}(\omega_k)}} \approx 1 - \frac{3\,\ln(10)\,M_i}{n_{60}(\omega_k)} \approx 1 - \frac{6.91\,M_i}{n_{60}(\omega_k)}$$

to produce the desired decay-time in that band, where $n_{60}(\omega) = t_{60}(\omega)/T$ denotes the desired 60-dB decay time in samples. Faust implementations of FDN reverberator along these lines are described in §3.7.9 below.

3.7.6 Spectral Coloration Equalizer

In the previous section, a "graphical equalizer" was used to set the reverberator decay time independently in each spectral band slice. While this gives much control over decay time, there is no control over the initial spectral gain in each band. Therefore, another good place for a graphical equalizer is at the reverberator input or output. Such an equalizer allows control of the

[16] In a Faust implementation, this form was found to be more numerically sensitive than that of the previous section when changing the t_{60}s in real time (see §3.7.9).

initial spectral coloration of the reverberator. In the example of Fig. 3.10 on page 106, a spectral coloration equalizer is most efficiently applied to the input signal $u(n)$, before entering the FDN (but after splitting off the direct signal to be scaled by d and added to the output), or the output of $E(z)$ in Fig. 3.10.

3.7.7 Tonal Correction Filter

Let $h_k(n)$ denote the component of the impulse response arising from the kth pole of the system. Then the energy associated with that pole is

$$\mathcal{E}_k = \sum_{n=0}^{\infty} |h_k(n)|^2 .$$

All other factors being equal, if the decay time of the mode is shortened by half, it follows that the total energy contributed by that mode to the impulse response is also reduced by half. To compensate for this effect, Jot introduced a *tonal correction filter* $E(z)$ to be placed in series with the FDN, as shown in Fig. 3.10 on page 106.

In the case of the first-order delay-line filters discussed in §3.7.5, good tonal correction is given by the following one-zero filter:

$$E(z) = \frac{1 - bz^{-1}}{1 - b}$$

where

$$b = \frac{1 - \alpha}{1 + \alpha}$$

and α is defined in Eq. (3.10).

3.7.8 FDNs as Digital Waveguide Networks

As discussed in §C.15 on page 612, the FDN using a Householder-reflection feedback matrix $\mathbf{A}_N = \mathbf{I}_N - (2/N)\underline{u}\underline{u}^T$ is equivalent to a network of N digital waveguides intersecting at a single scattering junction [464, 465, 385]. The wave impedance in the ith waveguide is simply $\underline{u}[i]$, the ith element of the axis-of-reflection vector $\underline{u}$. The choice $\underline{u}^T = [1, 1, \ldots, 1]$ corresponds to all of the waveguides having the same impedance (the "isotrophic junction" case).

3.7.9 FDN Reverberators in Faust

The Faust example `reverb_designer.dsp` brings up a 16×16 FDN reverberator in which the signal out of each delay line is split into five bands so that $t_{60}(\omega_k)$ can be controlled independently in each band. The 16 delay-line lengths are distributed exponentially between a minimum and maximum length set by two min/max-length sliders, but rounded to the nearest integer-power of a distinct prime, as introduced above in §3.7.3, p. 113). The FDN reverberator is implemented in Faust's `effect.lib`. The band-splitting is carried out by the `filterbank` function in Faust's `filter.lib`.

The Faust function `filterbank(order,freqs)` implements a filter bank having the needed properties using Butterworth lowpass/highpass band-splitting arranged in a dyadic tree (normally a good choice for audio filter banks). That is, the whole spectrum is split at the highest crossover frequency, the lowpass region is then split into two bands at the next crossover frequency down, and so on, splitting the lowpass band at each stage in the dyadic tree [455, 501]. The number of poles in each Butterworth lowpass/highpass filter is specified by `order`, and `freqs` contains a list of desired crossover frequencies separating the bands. A certain amount of dispersion is also introduced, since the filter bank is causal and delay-equalized (so that the bands may be summed without phase cancellation artifacts at the band edges). Also note that the lower bands are effectively produced by higher order filters than the upper bands. When the reverberation time is longer than the dispersion delay, the dispersion should not be audible as such, although it can affect the "sound" of the reverberation. In general, however, artificial reverberators normally benefit from additional allpass dispersion.

Figure 3.12 shows the block diagram of a 4×4 FDN reverberator made from Faust's `reverb_designer.dsp` by changing 16 to 4. Figure 3.13 shows the Faust block diagram of the associated 4×4 Hamard matrix multiplication. As it shows, multiplication by a Hadamard matrix can be implemented (ignoring the normalizing scale factor) as a series of block sums and differences (often called *butterflies* or *shufflers*) in which the block size decreases by a factor of 2 each stage. Figures for the remaining components of the reverberator may be perused via the shell command `faust2firefox reverb_designer.dsp` followed by clicking on the blocks in the browser.

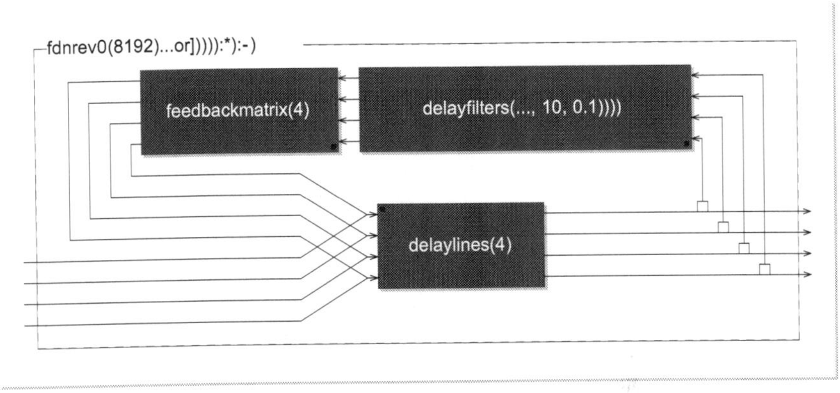

Figure 3.12: FDN reverberator implemented in the Faust example **reverb_designer.dsp**, but scaled down from order 16 to order 4.

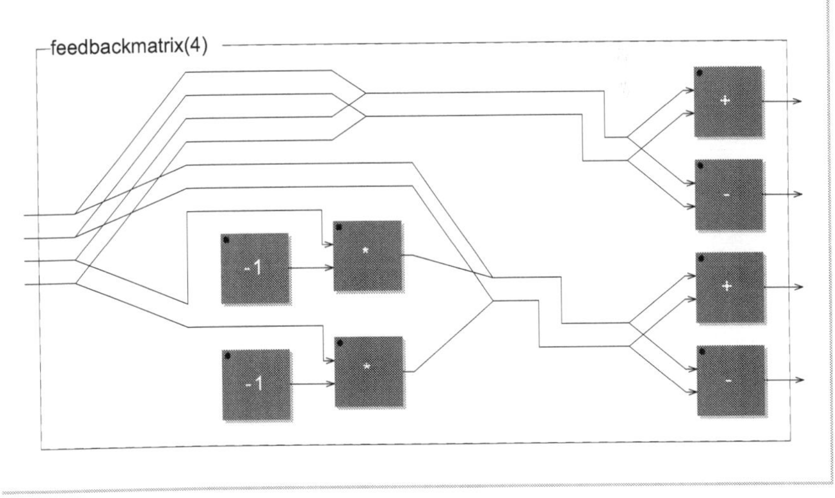

Figure 3.13: Hadamard processing used in the Faust FDN reverberator.

3.7.10 Zita-Rev1

A FOSS[17] reverberator that combines elements of Schroeder (§3.5) and FDN reverberators (§3.7) is **zita-rev1**,[18] written in C++ for Linux systems by Fons Adriaensen. A Faust version of the **zita-rev1** stereo-mode functionality is **zita_rev1** in Faust's **effect.lib**. A high-level block diagram appears in Fig. 3.14.

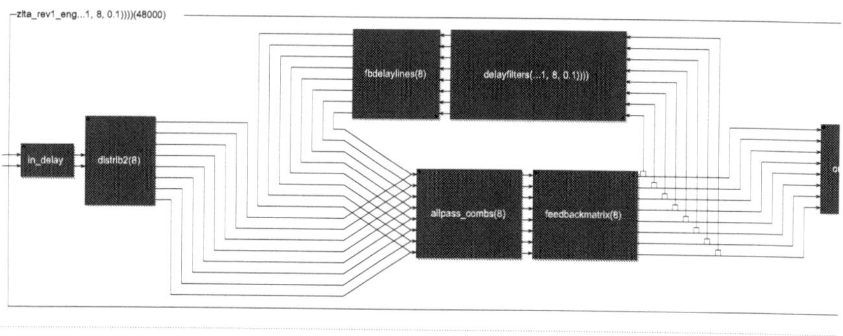

Figure 3.14: High-level block diagram of **zita_rev1_engine()** in Faust's **effect.lib** generated by the shell script **faust2firefox**.

The main high-level addition relative to an 8th-order FDN reverberator is the block labeled **allpass_combs** in Fig. 3.14. This block inserts a Schroeder allpass comb filter (Fig. 2.30, p. 70) in series with each delay line. In **zita-rev1** (as of this writing), the allpass-comb feedforward/feedback coefficients are all set to ± 0.6. The delay-line lengths and other details are readily found in the freely available source code (or by browsing the Faust-generated block diagram).

Zita-Rev1 Delay-Line Filters

In **zita-rev1**, the damping filter for each delay line consists of a low-shelf filter $H_l(z)$ [449],[19] in series with a unique first-order lowpass filter $H_h(z)$ that sets the high-frequency t_{60} to be half that of the middle-band at a particular frequency f_h (specified as "HF Damping" in the GUI). Since the filter H_h is constrained to be a lowpass, $t_{60}(f) < t_{60}(f_h)$ for $f > f_h$, *i.e.*, the decay time gets shorter at higher frequencies.

Viewing the resulting damping filter $H_d(z) = H_l(z)H_h(z)$ as a three-band filter bank (§3.7.5), let g_0 and g_m denote the desired band gains at

[17] FOSS $\triangleq$ Free Open Source Software

[18] `http://www.kokkinizita.net/linuxaudio/`

[19] `https://ccrma.stanford.edu/~jos/filters/Low_High_Shelving_Filters.html`

dc and "middle frequencies", respectively.[20] Then the low shelf may be set for a desired dc-gain of g_0/g_m, and its input (or output) signal multiplied by g_m to obtain the resulting filter

$$H_l(z) = g_m + (g_0 - g_m)\frac{1-p_l}{2}\frac{1+z^{-1}}{1-p_l z^{-1}},$$

where p_l denotes the (real) first-order lowpass pole, given by [449]

$$p_l \triangleq \frac{1-\pi f_1 T}{1+\pi f_1 T},$$

where f_1 specifies (in Hz) the crossover point between "low" and "middle" frequencies, and T denotes the sampling interval as usual.

The lowpass filter $H_h(z)$ is also first order, and to provide half the middle-band t_{60} at the beginning of the "high" band, the lowpass should "break" to a gain of g_m at the "HF Damping" frequency f_h specified in the GUI. A unity-dc-gain one-pole lowpass has the form [449]

$$H_h(z) = \frac{1-p_h}{1-p_h z^{-1}},$$

where the pole p_h must be found to give a gain of g_M at frequency f_h:

$$\left|H_h\left(e^{j2\pi f_h T}\right)\right| = \left|\frac{1-p_h}{1-p_h e^{-j2\pi f_h T}}\right| = g_M$$

Squaring and normalizing yields a quadratic equation of the form $p_h^2 + b\,p_h + 1 = 0$. Solving for p_h using the quadratic formula yields

$$p_h = -\frac{b}{2} - \sqrt{\left(\frac{b}{2}\right)^2 - 1},$$

where

$$-\frac{b}{2} = \frac{1-g_M^2\cos(2\pi f_h T)}{1-g_M^2} > 1,$$

and the unstable solution $-b/2 + \sqrt{(b/2)^2 - 1} > 1$ is discarded. To ensure $|g_M| < 1$, the GUI must limit the middle-band t_{60} to finite values. (The upper limit is presently 8 seconds for both low and middle frequencies.)

---

[20] This section describes the Faust version `zita_rev1`, which may not be exactly the same as the C++ version `zita-rev1`, although each control parameter range is the same, and the filter-orders are the same, so at least near-equivalance is expected.

3.7.11 Further Extensions

Schroeder's original structures for artificial reverberation were comb filters and allpass filters made from two comb filters. Since then, they have been upgraded to include specific early reflections and per-sample air-absorption filtering (Moorer, Schroeder), precisely specified frequency dependent reverberation time (Jot), and a nearly independent factorization of "coloration" and "duration" aspects (Jot). The evolution from comb filters to feedback delay networks (Gerzon, Stautner, Puckette, Jot) can be seen as a means for obtaining greater richness of feedback, so that the diffuseness of the impulse response is greater than what is possible with parallel and/or series comb filters. In fact, an FDN can be seen as a richly cross-coupled bank of feedback comb filters whenever the diagonal of the feedback matrix is nonzero. The question then becomes what aspects of artificial reverberation have not yet been fully addressed?

Spatialization of Reverberant Reflections

While we did not go into the subject here, the early reflections should be *spatialized* by including a head-related transfer function (HRTF) on each tap of the early-reflection delay line [248].[21]

Some kind of spatialization may be needed also for the late reverberation. A true diffuse field (§3.2.1, p. 89) consists of a sum of plane waves traveling in *all* directions in 3D space. Since we do not know how to achieve this effect using current systems for reverberation, the typical goal is to simply extract *uncorrelated* outputs from the reverberation network and feed them to the various output channels, as discussed in §3.5 on page 96. However, this is not ideal, since the resulting sound field consists of wavefronts arriving from each of the speakers, and it is possible for the reverberation to sound like it is emanating from discrete speaker locations. It may be that spatialization of some kind can better fool the ear into believing the late reverberation is coming from all directions.

Distribution of Mode Frequencies

Another way in which current reverberation systems are "artificial" is the unnaturally uniform distribution of resonant modes with respect to frequency. Because Schroeder, FDN, and waveguide reverbs are all essentially a collection of N delay lines with feedback around them, the modes tend to be distributed as the superposition of the resonant modes of N feedback comb filters. Since a feedback comb filter has a nearly harmonic set of

[21] Of course, correct angles-of-arrival for early reflections are more straightforward to implement using an array of loudspeakers enclosing the listener.

modes (see §2.6.2 on page 58), aggregates of comb filters tend to provide a *uniform* modal density in the frequency domain. In real reverberant spaces, the mode density increases as frequency squared, so it should be verified that the uniform modes used in a reverberator are perceptually equivalent to the increasingly dense modes in nature. Another aspect of perception to consider is that frequency-domain perception of resonances actually *decreases* with frequency. To summarize, in nature the modes get denser with frequency, while in perception they are less resolved, and in current reverberation systems they stay more or less uniform with frequency; perhaps a uniform distribution is a good compromise between nature and perception?

At low frequencies, however, resonant modes are accurately perceived in reverberation as boosts, resonances, and cuts. They are analogous to early reflections in the time domain, and we could call them the "early resonances." It is interesting that no system for artificial reverberation except waveguide mesh reverberation (of which the author is aware) explicitly attempts precise shaping of the low-frequency amplitude response of a desired reverberant space, at least not directly. The low-frequency response is shaped indirectly by the choice of early reflections, and the use of parallel comb-filter banks in Schroeder reverberators serves also to shape the low-frequency response significantly. However, it would be possible to add filters for shaping more carefully the low-frequency response. Perhaps a reason for this omission is that hall designers work very hard to *eliminate* any explicit resonances or antiresonances in the response of a room. If uneven resonance at low frequencies is always considered a defect, then designing for a maximally uniform mode distribution, as has been discussed for the high-frequency modes, would be ideal also at low frequencies. Quite the opposite situation exists when designing "small-box reverberators" to simulate musical instrument resonators [428, 203]; there, the low-frequency modes impart a characteristic timbre on the low-frequency resonance of the instrument (see Fig. 3.2 on page 92).

Digital Waveguide Reverberators

It was mentioned in §3.7.8 above that FDNs can be formulated as special cases of Digital Waveguide Networks (DWN) (see Appendix C for a fuller development of DWNs). Specifically, an FDN is obtained from a DWN consisting of a single scattering junction (§C.15 on page 612). It follows that the DWN paradigm provides a more generalized framework in which to pursue further improvements of reverberation architecture. For example, when multiple FDNs are embedded within a single DWN, it becomes possible to richly cross-couple them in an energy-controlled manner in order to create richer recursive structures than either alone. General DWNs were proposed for artificial reverberation in [430, 433].

The Digital Waveguide Mesh for Reverberation

A special case of digital waveguide networks known as the *digital waveguide mesh* has also been proposed for use in artificial reverberation systems [396, 519].

As discussed in §2.4 on page 49, a digital waveguide (bidirectional delay line) can be considered a computational acoustic model for traveling waves in opposite directions. A *mesh* of such waveguides in 2D or 3D can simulate waves traveling in *any* direction in the space. As an analogy, consider a tennis racket in which a rectilinear mesh of strings forms a pseudo-membrane.

A major advantage of the waveguide mesh for reverberation applications is that wavefronts are explicitly simulated in all directions, as in real reverberant spaces. Therefore, a true diffuse field can be developed in the late reverberation. Also, the echo density grows with time and the mode density grows with frequency in a natural manner for the 2D and 3D mesh. Finally, the low-frequency modes of the reverberant space can be simulated very precisely (for better or worse).

The computational cost of a waveguide mesh is made tractable relative to more conventional finite-difference simulations by (1) the use of multiply-free scattering junctions and (2) very coarse meshes. Use of a coarse mesh means that the "physical modeling" aspects of the mesh are only valid at low frequencies. As practical matter, this works out well because the ear cannot hear mode tuning errors at high frequencies. There is no error in the mode dampings in a lossless reverberator prototype, because the waveguide mesh is lossless by construction. Therefore, the only errors relative to an ideal simulation of a lossless membrane or space are (1) mode tuning error, and (2) finite band width (cut off at half the sampling rate). The tuning error can be understood as due to *dispersion* of the traveling waves in certain directions [519, 399]. Much progress has been made on the problem of correcting this dispersion error in various mesh geometries (rectilinear, triangular, tetrahedral, etc.) [522, 398, 399].

See §C.14 on page 605 for an introduction to the digital waveguide mesh and a few of its properties.

Time Varying Reverberators

In real rooms, thermal convention currents cause the propagation path delays to vary over time [58]. Therefore, for greater physical accuracy, the delay lines within a digital reverberator should *vary* over time. From a more practical perspective, time variation helps to break up and obscure unwanted repetition in the late reverberation impulse response [430, 104].

3.8 Exercises

EDC/EDR Laboratory Exercise

1. Plot the Energy Decay Curve (EDC) and Energy Decay Relief (EDR) of the Jot reverberator you built in the previous assignment.

2. Using the EDR near dc and near half the sampling rate, measure the decay times at these two frequencies, and compare them to values desired at design time.

Reverberation Laboratory Exercises

1. Write an STK module called `FDNJot4.cpp` which implements an order 4 SISO feedback delay network with $B^T = [1,1,1,1]$, $C^T = [1,1,1,1]$, and feedback matrix $A = A_4$, where

   $$A_N \triangleq g\left(I_N - \frac{2}{N}u_N u_N^T\right)$$

   where I_N denotes the $N \times N$ identity matrix, and $u_N = [1,\ldots,1]^T$. (It may be helpful to refer to the third-order example depicted in Fig. 2.29 on page 67.) When $g = 1$, A_N is a so-called *Householder reflection* about the vector u_N in N-space. Note that A_N is orthogonal ($A_N^T A_N = I_N$) when $g = \pm 1$ (as is any real rotation matrix). [Hint: It is more efficient to implement the Householder reflection as written above than to calculate and implement the resulting four-by-four matrix.] The input arguments should be g and the four delay-line lengths $M_1,\ldots,M_4$. Write a small main program to test `FDNJot4` with an impulse input. Turn in your commented program listings and a print-out of your test results for the case $g = 1$ and M_i set to the first four odd primes $(3,5,7,11)$. (include at least 50 output samples).

   Find a set of four mutually prime delay lengths which make the impulse response sound maximally smooth and dense (again for $g = 1$). [Hint: start with Schroeder's suggested lengths given in §3.5 on page 96.]

2. Write an STK module called `FDNJot4LP.cpp` in which each of the four delay lines is cascaded with one-pole lowpass filter

   $$H_i(z) = g_i \frac{1 - a_i}{1 - a_i z^{-1}}, \quad i = 1,2,3,4$$

where g_i and a_i are automatically computed using Jot's formulas. The inputs to Jot's formulas are the desired reverberation time at dc and at half the sampling rate.

3. Write an STK module called `FDNJot16LP.cpp` which combines four instances of `FDNJot4LP.cpp`. It is important that combined system remain lossless when the lowpass filters are disabled, so that the lossless prototype can be checked for its quality as a "white noise" generator. [Hint: Consider creating A_{16} from A_4 using the same type of embedding used to create H_{16} from H_4 in the case of Hadamard matrices. Alternative hint: Consider adding a 4×4 feedforward matrix to each FDN to make a Gerzon vector allpass.]

   The input arguments should be the desired reverberation time at dc, the desired reverberation time at half the sampling rate, and the 16 delay-line lengths. Alternatively, you may want to synthesize the 16 delay-line lengths according to an algorithm of your choosing, but there should be at least one parameter such as "room size" that affects the choice of delay length. Optionally, all reverberation parameters can be arranged as a set of "presets" which are invoked by name, such as "plate," "small room," "cathedral," etc.

   Experiment with the the delay lengths and try to tune them for best results (i.e., sounding closest to a white noise output when running in lossless mode—smooth in both time and frequency).

4. Write a small main program to test `FDNJot16LP` with an impulse input.

5. With your "best" reverberator set in lossless mode, save its impulse response and amplitude response to disk. Turn in (1) an "echogram" plot for the first half second of the impulse response, and (2) an amplitude response on a dB scale. Finally, turn in also your commented program listings together with a discussion of how you determined your delay-line lengths, any design decisions, etc.

State-Space Formulation Problems

State-space models are described in the second book of this series [449]. The following problems provide exercises pertaining to casting delay effects in state-space form.

1. Consider the MIMO state space description

$$\begin{aligned} \mathbf{x}(n+1) &= A\mathbf{x}(n) + B\mathbf{u}(n) \\ \mathbf{y}(n) &= C^T\mathbf{x}(n) + D\mathbf{u}(n) \end{aligned}$$

(a) Find the impulse response ("Markov parameters") in closed form by direct calculation of the output given an impulse input.

(b) Find the transfer function by taking the z transform of the impulse response.

2. Find a state space description (A, B, C, D) for the following two-input, two-output digital filter

$$\begin{aligned} y_1(n) &= u_1(n) + 2u_2(n) + u_2(n-1) + y_2(n-1) \\ y_2(n) &= u_1(n) + u_1(n-1) + y_1(n-1) + 0.5y_2(n-1) \end{aligned}$$

where $u_i(n)$ denotes the ith input signal at time n, and $y_i(n)$ is the ith output signal at time n.

3. Find the *poles* of the system in the previous problem. Is the system stable?

4. Find a state space description (A, B, C, D) for the *series* (cascade) combination of two Schroeder allpass filters, with transfer functions

$$H_i(z) = \frac{g_i + z^{-M_i}}{1 + g_i z^{-M_i}}$$

for $i = 1, 2$, where $M_1 = 2$ and $M_2 = 3$.

5. Find a state space description (A, B, C, D) for the *parallel* combination of the two Schroeder allpass filters in the previous problem.

6. Repeat the previous two problems with $M_1 = 3$ and $M_2 = 5$.

7. From the previous problem, what can you say about the relationship between reverberators made using Schroeder allpass sections and those made using feedback delay networks? [Hint: You can focus on the feedback matrix matrix. Try to accurately characterize the class of A matrices corresponding to (1) cascade and (2) parallel Schroeder reverberators.]

8. Find a state space description (A, B, C, D) for the single-input, single-output (SISO) Feedback Delay Network (FDN) described by

$$\begin{bmatrix} x_1(n) \\ x_2(n) \end{bmatrix} = \begin{bmatrix} 1 & 0 \\ 1 & -1 \end{bmatrix} \begin{bmatrix} x_1(n-3) \\ x_2(n-5) \end{bmatrix} + \begin{bmatrix} 1 \\ 1 \end{bmatrix} u(n)$$

$$y(n) = x_1(n) + x_2(n)$$

and find all system poles. [Hint: A should be an 8×8 matrix.]

9. Consider the following state-space-like recursion:

$$\begin{aligned} x_1(n+1) &= x_1(n) + s \cdot x_2(n) \\ x_2(n+1) &= -s \cdot x_1(n+1) + x_2(n) \end{aligned}$$

   (a) Find the state transition matrix A for this system, assuming the state vector is defined as $[x_1(n), x_2(n)]^T$.
   (b) Find the eigenvalues of A.
   (c) What range of s, if any, results in the state variables oscillating sinusoidally in response to nonzero initial conditions $[x_1(0), x_2(0)]^T$? Find a formula for the oscillation frequency as a function of s in this range.
   (d) Find the eigenvectors of A.
   (e) Suppose the state vector at time n is written in the form

$$\begin{bmatrix} x_1(n) \\ x_2(n) \end{bmatrix} = \begin{bmatrix} A_1 \cos(\omega_1 nT + \phi_1) \\ A_2 \cos(\omega_2 nT + \phi_2) \end{bmatrix}$$

   where the sinusoidal parameters are functions of s and the initial conditions. In particular

      i. Find the amplitudes of oscillation A_1 and A_2 as function of s, given the initial conditions $x_1(0) = 1$, and $x_2(0) = 0$.
      ii. Find the oscillation phases ϕ_1 and ϕ_2 as functions of s, given $x_1(0) = 1$, and $x_2(0) = 0$.

   (f) If $x_1(n)$ and $x_2(n)$ are used as Cartesian coordinates for drawing in a plane, what geometrical figures can be drawn using various values of s? For what values of s, if any, does a circle result?

10. In §C.17 on page 619, the digital waveguide oscillator is found to have two "natural" outputs given by the delay-element signals. One is

$$y_1(n) = \cos(\omega nT).$$

Find the other signal $y_2(n) = x_2(n)$. [Hint: It should be 90 degrees out of phase with respect to $y_1(n) = x_1(n)$.]

Chapter 4

Delay-Line and Signal Interpolation

It is often necessary for a delay line to vary in length. Consider, for example, simulating a sound ray as in Fig. 2.8 on page 44 when either the source or listener is moving. In this case, separate read and write pointers are normally used (as opposed to a shared read-write pointer in Fig. 2.2 on page 38). Additionally, for good quality audio, it is usually necessary to *interpolate* the delay-line length rather than "jumping" between integer numbers of samples. This is typically accomplished using an *interpolating read*, but *interpolating writes* are also used (*e.g.*, for true Doppler simulation, as described in §5.9).

4.1 Delay-Line Interpolation

As mentioned above, when an audio delay line needs to vary smoothly over time, some form of *interpolation* between samples is usually required to avoid "zipper noise" in the output signal as the delay length changes. There is a hefty literature on "fractional delay" in discrete-time systems, and the survey in [267] is highly recommended.

This section will describe the most commonly used cases. *Linear interpolation* is perhaps most commonly used because it is very straightforward and inexpensive, and because it sounds very good when the signal bandwidth is small compared with half the sampling rate. For a delay line in a nearly lossless feedback loop, such as in a vibrating string simulation, *allpass interpolation* is sometimes a better choice since it costs the same as linear interpolation in the first-order case and has no gain distortion.

(Feedback loops can be very sensitive to gain distortions.) Finally, in later sections, some higher-order interpolation methods are described.

4.1.1 Linear Interpolation

Linear interpolation works by effectively drawing a straight line between two neighboring samples and returning the appropriate point along that line.

More specifically, let η be a number between 0 and 1 which represents how far we want to interpolate a signal y between time n and time $n+1$. Then we can define the linearly interpolated value $\hat{y}(n+\eta)$ as follows:

$$\hat{y}(n+\eta) = (1-\eta) \cdot y(n) + \eta \cdot y(n+1) \tag{4.1}$$

For $\eta = 0$, we get exactly $\hat{y}(n) = y(n)$, and for $\eta = 1$, we get exactly $\hat{y}(n+1) = y(n+1)$. In between, the interpolation error $|\hat{y}(n+\eta) - y(n+\eta)|$ is nonzero, except when $y(t)$ happens to be a linear function between $y(n)$ and $y(n+1)$.

One-Multiply Linear Interpolation

Note that by factoring out η, we can obtain a *one-multiply* form,

$$\hat{y}(n+\eta) = y(n) + \eta \cdot [y(n+1) - y(n)] .$$

Thus, the computational complexity of linear interpolation is one multiply and two additions per sample of output.

Fractional Delay Filtering by Linear Interpolation

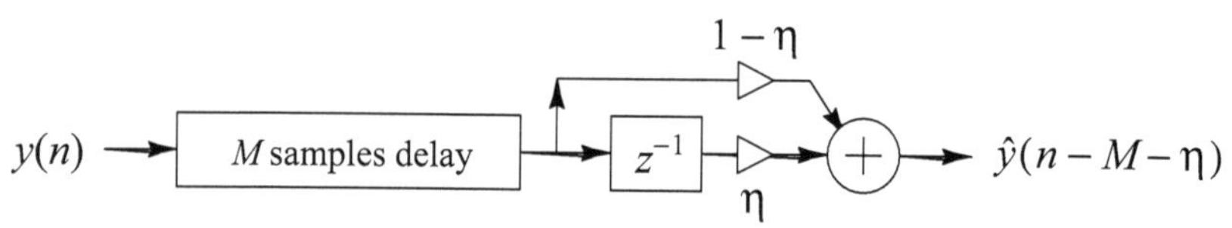

Figure 4.1: Linearly interpolated delay line.

A linearly interpolated delay line is depicted in Fig. 4.1. In contrast to Eq. (4.1), we interpolate linearly between times $n-M$ and $n-M-1$, and η is called the *fractional delay* in samples. The first-order (linear-interpolating) filter following the delay line in Fig. 4.1 may be called a *fractional delay filter* [267]. Equation (4.1), on the other hand, expresses the more general case of an *interpolated table lookup*, where $y(n)$ is regarded as a table of samples and $\hat{y}(n+\eta)$ is regarded as an interpolated table-lookup based on the samples stored at indices n and $n+1$.

The difference between a fractional delay filter and an interpolated table lookup is that table-lookups can "jump around," while fractional delay filters receive a *sequential* stream of input samples and produce a corresponding sequential stream of interpolated output values. As a result of this sequential access, fractional delay filters may be *recursive* IIR digital filters (provided the desired delay does not change too rapidly over time). In contrast, "random-access" interpolated table lookups are typically implemented using weighted linear combinations, making them equivalent to nonrecursive FIR filters in the sequential case.[1]

The `C++` class implementing a linearly interpolated delay line in the Synthesis Tool Kit (`STK`) is called `DelayL`.

The frequency response of linear interpolation for fixed fractional delay (η fixed in Fig. 4.1) is shown in Fig. 4.2. From inspection of Fig. 4.1, we see that linear interpolation is a one-zero FIR filter. When used to provide a fixed fractional delay, the filter is linear and time-invariant (LTI). When the fractional delay η changes over time, it is a linear time-varying filter.

Linear interpolation sounds best when the signal is *oversampled.* Since natural audio spectra tend to be relatively concentrated at low frequencies, linear interpolation tends to sound very good at high sampling rates.

When interpolation occurs inside a *feedback loop*, such as in digital waveguide models for vibrating strings (see Chapter 6), errors in the amplitude response can be highly audible (particularly when the loop gain is close to 1, as it is for steel strings, for example). In these cases, it is possible to eliminate amplitude error (at some cost in delay error) by using an *allpass filter* for delay-line interpolation.

4.1.2 First-Order Allpass Interpolation

A delay line interpolated by a first-order allpass filter is drawn in Fig. 4.3.

Intuitively, ramping the coefficients of the allpass gradually "grows" or "hides" one sample of delay. This tells us how to handle resets when crossing sample boundaries.

---

[1] A causal IIR filter can be used for random-access table lookup by simply starting the filter at a point sufficiently far away in the table and "running it" until it reaches steady-state at the desired interpolation point. While this requires at least as many multiply-adds as an FIR filter spanning the same duration, the number of IIR filter coefficients can be much less than the number of FIR filter coefficients required. IIR interpolators are therefore possibly interesting for certain memory-starved VLSI applications.

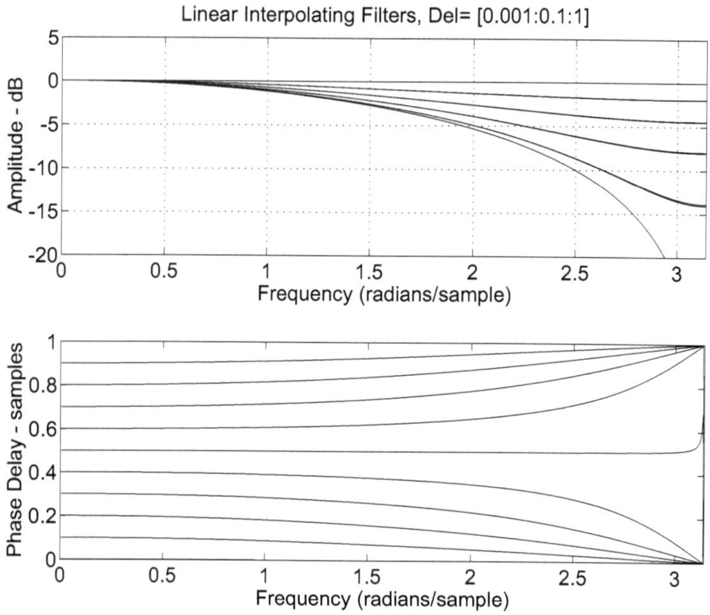

Figure 4.2: Linear interpolation frequency responses for delays between 0 and 1. Note the higher accuracy at low frequencies, reaching zero error at dc for all fractional delays.

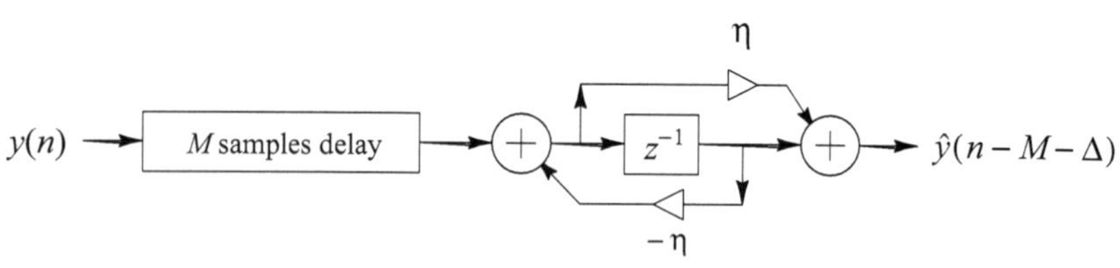

Figure 4.3: Allpass-interpolated delay line.

The difference equation is

$$\begin{aligned}\hat{x}(n-\Delta) \stackrel{\Delta}{=} y(n) &= \eta \cdot x(n) + x(n-1) - \eta \cdot y(n-1) \\ &= \eta \cdot [x(n) - y(n-1)] + x(n-1).\end{aligned}$$

Thus, like linear interpolation, first-order allpass interpolation requires only one multiply and two adds per sample of output.

The transfer function is

$$H(z) = \frac{\eta + z^{-1}}{1 + \eta z^{-1}}. \tag{4.2}$$

At low frequencies ($z \to 1$), the delay becomes

$$\Delta \approx \frac{1-\eta}{1+\eta} \tag{4.3}$$

Figure 4.4 shows the *phase delay* of the first-order digital allpass filter for a variety of desired delays at dc. Since the amplitude response of any allpass is 1 at all frequencies, there is no need to plot it.

The first-order allpass interpolator is generally controlled by setting its dc delay to the desired delay. Thus, for a given desired delay Δ, the allpass coefficient is (from Eq. (4.3))

$$\eta \approx \frac{1-\Delta}{1+\Delta}$$

From Eq. (4.2), we see that the allpass filter's pole is at $z = -\eta$, and its zero is at $z = -1/\eta$. A pole-zero diagram for $\Delta = 0.1$ is given in Fig. 4.5. Thus, zero delay is provided by means of a *pole-zero cancellation*! Due to inevitable round-off errors, pole-zero cancellations are to be avoided in practice. For this reason and others (discussed below), allpass interpolation is best used to provide a delay range lying wholly above zero, *e.g.*,

$$\Delta \in [0.1, 1.1] \longleftrightarrow \eta \in [-0.05, 0.82]$$

Note that, unlike linear interpolation, allpass interpolation is not suitable for "random access" interpolation in which interpolated values may be requested at any arbitrary time in isolation. This is because the allpass is *recursive* so that it must run for enough samples to reach steady state. However, when the impulse response is reasonably short, as it is for delays near one sample, it can in fact be used in "random access mode" by giving it enough samples with which to work.

The `STK` class implementing allpass-interpolated delay is `DelayA`.

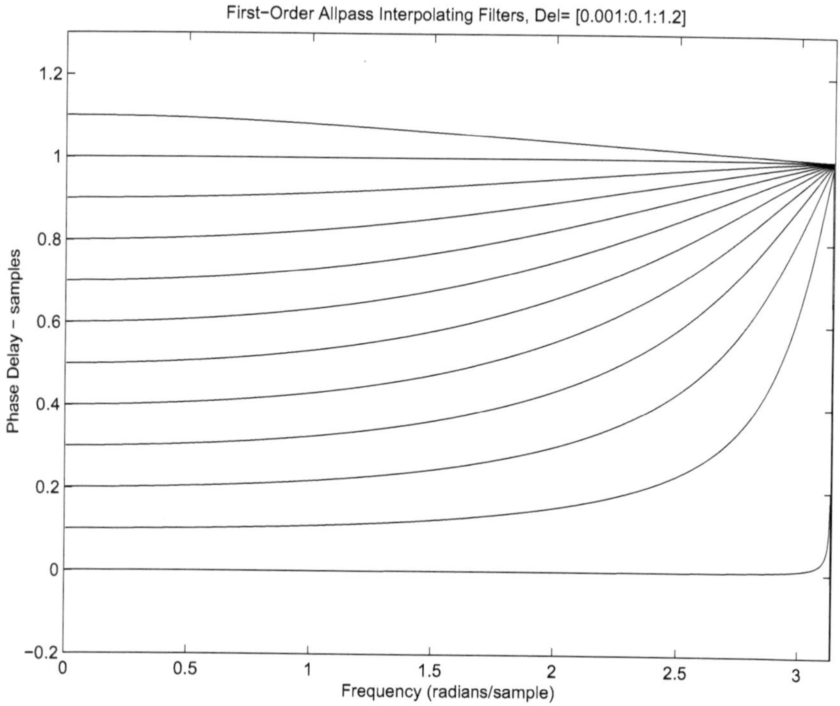

Figure 4.4: Allpass-interpolation phase delay for a variety of desired delays (exact at dc).

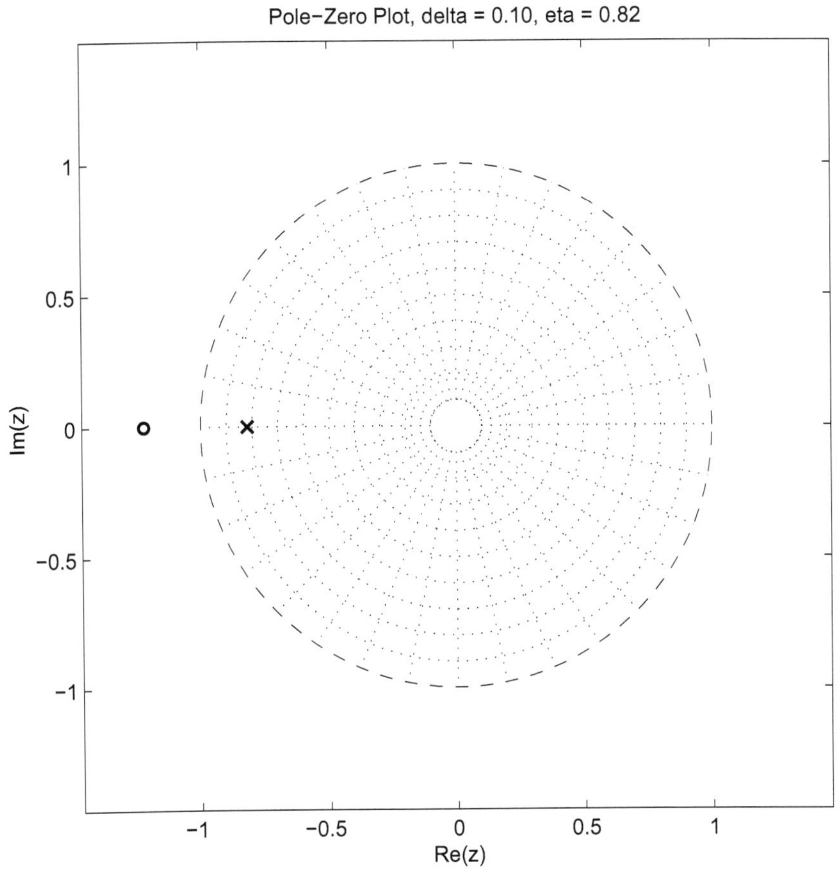

Figure 4.5: Allpass-interpolator pole-zero diagram for $\Delta = 0.1$.

Minimizing First-Order Allpass Transient Response

In addition to approaching a pole-zero cancellation at $z = -1$, another undesirable artifact appears as $\Delta \to 0$: The *transient response* also becomes long when the pole at $z = -\eta$ gets close to the unit circle.

A plot of the impulse response for $\Delta = 0.1$ is shown in Fig. 4.6. We see a lot of "ringing" near half the sampling rate. We actually should expect this from the *nonlinear-phase distortion* which is clearly evident near half the sampling rate in Fig. 4.4. We can interpret this phenomenon as the signal components near half the sampling rate being delayed by different amounts than other frequencies, therefore "sliding out of alignment" with them.

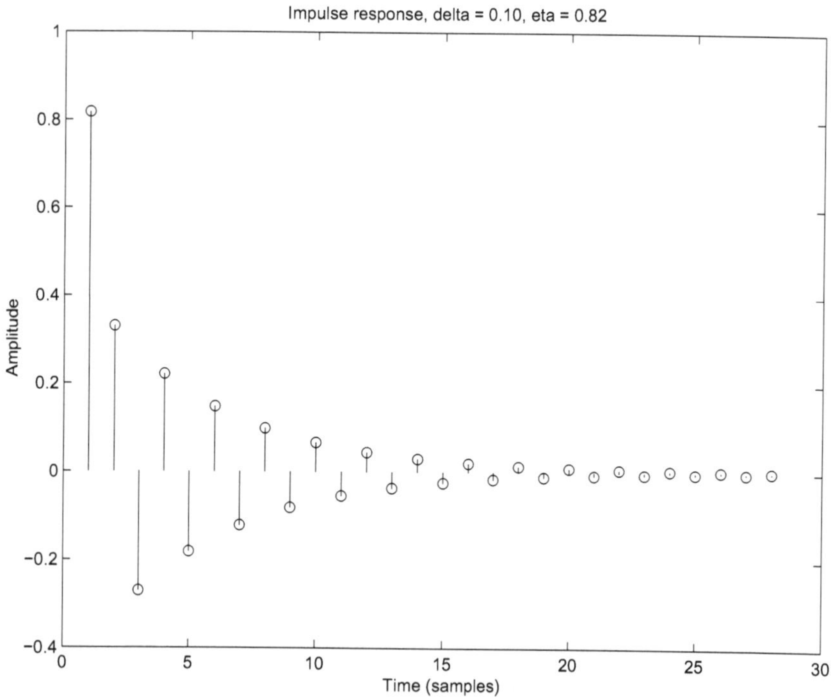

Figure 4.6: Impulse response of the first-order allpass interpolator for $\Delta = 0.1$.

For audio applications, we would like to keep the impulse-response duration short enough to sound "instantaneous." That is, we do not wish to have audible "ringing" in the time domain near $f_s/2$. For high quality sampling rates, such as larger than $f_s = 40$ kHz, there is no issue of direct audibility, since the ringing is above the range of human hearing. However, it is often convenient, especially for research prototyping, to work at lower sampling rates where $f_s/2$ is audible. Also, many commercial products use such sampling rates to save costs.

Since the time constant of decay, in samples, of the impulse response of a pole of radius R is approximately

$$\frac{\tau}{T} \approx \frac{1}{1-R},$$

and since a 60-dB decay occurs in about 7 time constants ("t_{60}") [451, p. 38], we can limit the pole of the allpass filter to achieve any prescribed specification on maximum impulse-response duration.

For example, suppose 100 ms is chosen as the maximum t_{60} allowed at a sampling rate of $f_s = 10,000$. Then applying the above constraints yields $\eta \leq 0.993$, corresponding to the allowed delay range $[0.00351, 1.00351]$.

4.1.3 Linear Interpolation as Resampling

Convolution Interpretation

Linearly interpolated fractional delay is equivalent to filtering and resampling a weighted impulse train (the input signal samples) with a continuous-time filter having the simple triangular impulse response

$$h_l(t) = \begin{cases} 1 - |t/T|, & |t| \leq T, \\ 0, & \text{otherwise.} \end{cases} \tag{4.4}$$

Convolution of the weighted impulse train with $h_l(t)$ produces a continuous-time linearly interpolated signal

$$x(t) = \sum_{n=-\infty}^{\infty} x(nT) h_l(t - nT). \tag{4.5}$$

This continuous result can then be resampled at the desired fractional delay.

In discrete time processing, the operation Eq. (4.5) can be approximated arbitrarily closely by digital *upsampling* by a large integer factor M, delaying by L samples (an integer), then finally downsampling by M, as depicted in Fig. 4.7 [96]. The integers L and M are chosen so that $\eta \approx L/M$, where η the desired fractional delay.

The convolution interpretation of linear interpolation, Lagrange interpolation, and others, is discussed in [407].

Frequency Response of Linear Interpolation

Since linear interpolation can be expressed as a convolution of the samples with a triangular pulse, we can derive the *frequency response* of linear

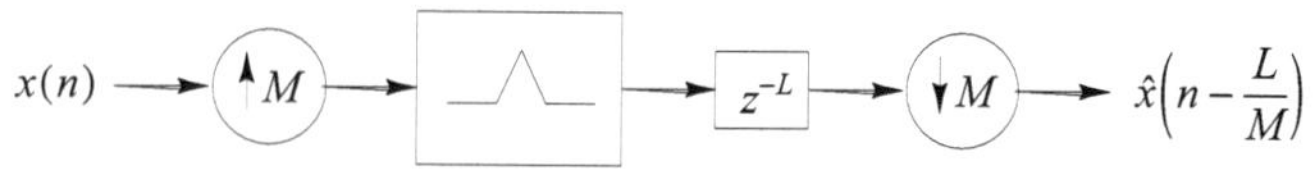

Figure 4.7: Linear interpolation as a convolution.

interpolation. Figure 4.7 indicates that the triangular pulse $h_l(t)$ serves as an *anti-aliasing lowpass filter* for the subsequent downsampling by M. Therefore, it should ideally "cut off" all frequencies higher than $0.5f_s/M$.

Triangular Pulse as Convolution of Two Rectangular Pulses

The 2-sample wide triangular pulse $h_l(t)$ (Eq. (4.4)) can be expressed as a convolution of the one-sample rectangular pulse with itself.

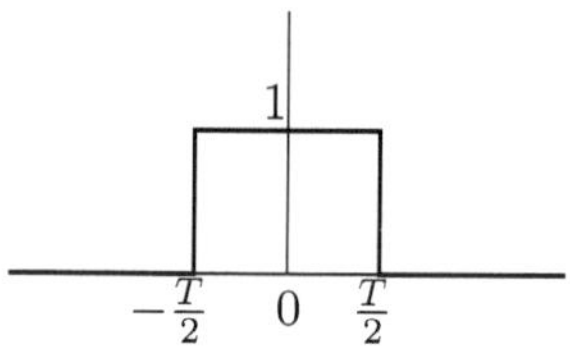

Figure 4.8: The width T rectangular pulse.

The one-sample *rectangular pulse* is shown in Fig. 4.8 and may be defined analytically as

$$p_T(t) \triangleq u\left(t+\frac{T}{2}\right) - u\left(t-\frac{T}{2}\right),$$

where $u(t)$ is the *Heaviside unit step function*:

$$u(t) \triangleq \begin{cases} 1, & t \geq 0 \\ 0, & t < 0 \end{cases}.$$

Convolving $p_T(t)$ with itself produces the two-sample triangular pulse $h_l(t)$:

$$h_l(t) = (p_T * p_T)(t) \triangleq \int_{-\infty}^{\infty} p_T(\tau)p_T(t-\tau)d\tau$$

While the result can be verified algebraically by substituting $u(t+T/2) - u(t-T/2)$ for $p_T(t)$, it seen more quickly via graphical convolution[2].

[2]`http://ccrma.stanford.edu/~jos/mdft/Graphical_Convolution.html`

Linear Interpolation Frequency Response

Since linear interpolation is a convolution of the samples with a triangular pulse $h_l(t) = (p_T * p_T)(t)$ (from Eq. (4.5)), the frequency response of the interpolation is given by the Fourier transform $H_l(f)$, which yields a sinc^2 function. This frequency response applies to linear interpolation from discrete time to continuous time. If the output of the interpolator is also sampled, this can be modeled by sampling the continuous-time interpolation result in Eq. (4.5), thereby *aliasing* the sinc^2 frequency response, as shown in Fig. 4.9.

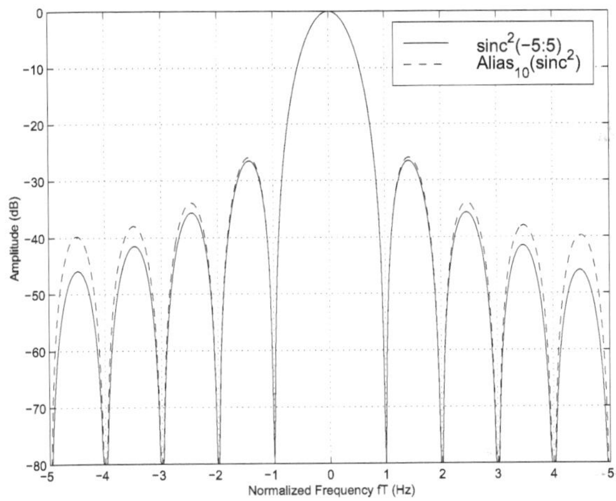

Figure 4.9: (a) $\text{sinc}^2(fT) = \mathcal{F}_f^2\{p_T/T\}$.
(b) $\text{A}\text{LIAS}_{10}\{\text{sinc}^2(fT)\} \propto \text{DTFT}\{\text{S}\text{AMPLE}_{T/10}\{p_T * p_T\}\}$.

In slightly more detail, from $h_l(t) = (p_T * p_T)(t)$, and $\mathcal{F}_f\{p_T\} = T\text{sinc}(fT)$, we have

$$H_l(f) \triangleq \mathcal{F}_f\{h_l\} = \mathcal{F}_f\{p_T * p_T\} = \mathcal{F}_f^2\{p_T\} = T^2\text{sinc}^2(fT)$$

where we used the convolution theorem for Fourier transforms, and the fact that $P_T(f) = T\text{sinc}(fT) = \sin(\pi fT)/(\pi f)$.

The Fourier transform of $h_l(nT/L)$ is the same function aliased on a block of size $f_s = L/T$ Hz. Both H_l and its alias are plotted in Fig. 4.9. The example in this figure pertains to an output sampling rate which is $L = 10$ times that of the input signal. In other words, the input signal is upsampled by a factor of $L = 10$ using linear interpolation. The "main lobe" of the interpolation frequency response $H_l(f)$ contains the original signal bandwidth; note how it is attenuated near half the original sampling rate ($fT = 1$ in Fig. 4.9). The "sidelobes" of the frequency response contain attenuated *copies* of the original signal bandwidth (see the DFT stretch theorem), and thus constitute *spectral imaging distortion* in the final output (sometimes also referred to as a kind of "aliasing," but, for clarity, that term will not be used for imaging distortion in this book). We see that the frequency response of linear interpolation is less than ideal in two ways:

- The spectrum is "rolled" off near half the sampling rate. In fact, it is nowhere flat within the "passband" (-1 to 1 in Fig. 4.9).
- Spectral imaging distortion is suppressed by only 26 dB (the level of the first sidelobe in Fig. 4.9.

These qualitative remarks apply to all upsampling factors $L \geq 2$ using linear interpolation. The case $L = 1$ is considered in the next section.

Special Cases

In the limiting case of $L = 1$, the input and output sampling rates are equal, and all sidelobes of the frequency response $H_l(f)$ (partially shown in Fig. 4.9) alias into the main lobe.

If the output is sampled at the same exact time instants as the input signal, the input and output are identical. In terms of the aliasing picture of the previous section, the frequency response aliases to a perfect flat response over $fT \in [-1, 1]$, with all spectral images combining coherently under the flat gain. It is important in this reconstruction that, while the frequency response of the underlying continuous interpolating filter is aliased by sampling, the signal spectrum is only imaged—not aliased; this is true for all positive *integers* L and M in Fig. 4.7 on page 140.

More typically, when linear interpolation is used to provide *fractional delay*, identity is not obtained. Referring again to Fig. 4.7, with L considered to be so large that it is effectively infinite, fractional-delay by $\tau < 1$ can be modeled as convolving the samples $x(n)$ with $h_l(t - \tau)$ followed by sampling at $t = nT$. In this case, a linear phase term has been introduced in the interpolator frequency response, giving,

$$H_\tau(f) \triangleq e^{-j\tau 2\pi f} H_l(f)$$

so that now downsampling to the original sampling rate does *not* yield a perfectly flat amplitude response for $fT \in [-1, 1]$ (when τ is non-integer). Moreover, the *phase response is nonlinear* as well; a sampled symmetric triangular pulse is only linear phase when the samples fall symmetrically about the midpoint. Some example frequency-responses for various delays τ are graphed in Fig. 4.2 on page 134.

4.1.4 Large Delay Changes

When implementing large delay length changes (by many samples), a useful implementation is to *cross-fade* from the initial delay line configuration to the new configuration:

- Computational requirements are doubled during the cross-fade.
- The cross-fade should occur over a time interval long enough to yield a smooth result.
- The new delay interpolation filter, if any, may be initialized in advance of the cross-fade, for maximum smoothness. Thus, if the transient response of the interpolation filter is N samples, the new delay-line + interpolation filter can be "warmed up" (executed) for N time steps before beginning the cross-fade. If the cross-fade time is long compared with the interpolation filter duration, "pre-warming" is not necessary.
- This is not a true "morph" from one delay length to another since we do not pass through the intermediate delay lengths. However, it avoids a potentially undesirable Doppler effect.
- A single delay line can be *shared* such that the cross-fade occurs from one *read-pointer* (plus associated filtering) to another.

4.2 Lagrange Interpolation

Lagrange interpolation is a well known, classical technique for interpolation [193]. It is also called Waring-Lagrange interpolation, since Waring actually published it 16 years before Lagrange [309, p. 323]. More generically, the term *polynomial interpolation* normally refers to Lagrange interpolation. In the first-order case, it reduces to *linear interpolation*.

Given a set of $N+1$ known samples $f(x_k)$, $k = 0, 1, 2, \ldots, N$, the problem is to find the *unique* order N polynomial $y(x)$ which *interpolates* the samples.[3] The solution can be expressed as a linear combination of elementary Nth order polynomials:

$$y(x) = \sum_{k=0}^{N} l_k(x) f(x_k) \tag{4.6}$$

where

$$l_k(x) \triangleq \frac{(x - x_0)\cdots(x - x_{k-1})(x - x_{k+1})\cdots(x - x_N)}{(x_k - x_0)\cdots(x_k - x_{k-1})(x_k - x_{k+1})\cdots(x_k - x_N)}.$$

From the numerator of the above definition, we see that $l_k(x)$ is an order N polynomial having zeros at all of the samples except the kth. The denominator is simply the constant which normalizes to give $l_k(x_k) = 1$. Thus, we have

$$l_k(x_j) = \delta_{kj} \triangleq \begin{cases} 1, & j = k, \\ 0, & j \neq k. \end{cases}$$

The polynomial l_k can be interpreted as the kth *basis polynomial* for constructing a polynomial interpolation of order N over the $N+1$ sample points x_k. It is an order N polynomial having zeros at all of the samples except the kth, where it is 1. An example of a set of eight basis functions l_k for randomly selected interpolation points x_k is shown in Fig. 4.10.

[3] Classically, an interpolation $\hat{f}(x)$ of discrete data points $f(x_k)$ always required $\hat{f}(x_k) = f(x_k)$; that is, the interpolating function must always pass through the given data points. More recently, however, particularly in the field of computer graphics, interpolation schemes such as *Bezier splines* have been defined which do not always pass through the known points.

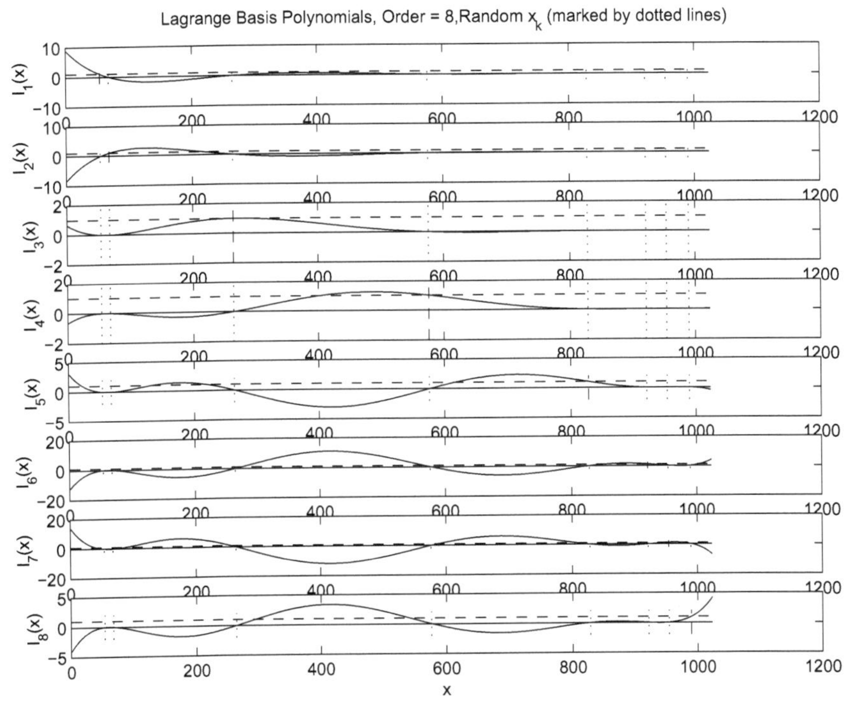

Figure 4.10: Example Lagrange basis functions in the eighth-order case for randomly selected interpolation points (marked by dotted lines). The unit-amplitude points are marked by dashed lines.

4.2.1 Interpolation of Uniformly Spaced Samples

In the uniformly sampled case ($x_k = kT$ for some sampling interval T), a Lagrange interpolator can be viewed as a Finite Impulse Response (FIR) filter [449]. Such filters are often called *fractional delay filters* [267], since they are filters providing a non-integer time delay, in general. Let $h(n)$ denote the *impulse response* of such a fractional-delay filter. That is, assume the interpolation at point x is given by

$$\begin{aligned} y(x) &= h(0)\, f(x_N) + h(1)\, f(x_{N-1}) + \cdots h(N)\, f(x_0) \\ &= h(0)\, y(N) + h(1)\, y(N-1) + \cdots h(N)\, y(0). \end{aligned}$$

where we have set $T = 1$ for simplicity, and used the fact that $y(x_k) = f(x_k)$ for $k = 0, 1, \ldots, N$ in the case of "true interpolators" that pass through the given samples exactly. For best results, $y(x)$ should be evaluated in a one-sample range centered about $x = N/2$. For delays outside the central one-sample range, the coefficients can be shifted to translate the desired delay into that range.

4.2.2 Fractional Delay Filters

In *fractional-delay filtering* applications, the interpolator typically slides forward through time to produce a time series of interpolated values, thereby implementing a non-integer signal delay:

$$\hat{y}\left(n - \frac{N}{2} - \eta\right) = h(0)\, y(n) + h(1)\, y(n-1) + \cdots h(N)\, y(0)$$

where $\eta \in [-1/2, 1/2]$ spans the central one-sample range of the interpolator. Equivalently, the interpolator may be viewed as an FIR filter having a linear phase response corresponding to a delay of $N/2 + \eta$ samples. Such filters are often used in series with a *delay line* in order to implement an *interpolated delay line* (§4.1) that effectively provides a continuously variable delay for discrete-time signals.

The *frequency response* [449] of the fractional-delay FIR filter $h(n)$ is

$$H(e^{j\omega}) = \sum_{n=0}^{N} h(n) e^{-j\omega n}.$$

For an ideal fractional-delay filter, the frequency response should be equal to that of an ideal delay

$$H^*(e^{j\omega}) = e^{-j\omega\Delta}$$

where $\Delta \triangleq N/2 + \eta$ denotes the total desired delay of the filter. Thus, the ideal desired frequency response is a linear phase term corresponding to a delay of Δ samples.

4.2.3 Lagrange Interpolation Optimality

As derived in §4.2.14 on page 160, Lagrange fractional-delay filters are *maximally flat* in the frequency domain at dc. That is,

$$\left.\frac{d^m E(e^{j\omega})}{d\omega^m}\right|_{\omega=0} = 0, \quad m = 0, 1, 2, \ldots, N,$$

where $E(e^{j\omega})$ is the interpolation error expressed in the frequency domain:

$$E(e^{j\omega}) \;\stackrel{\Delta}{=}\; H^*(e^{j\omega}) - H(e^{j\omega}),$$

where H and H^* are defined in §4.2.2 above. This is the same optimality criterion used for the power response of (recursive) *Butterworth filters* in classical analog filter design [343, 449]. It can also be formulated in terms of "Pade approximation" [373, 374]. To summarize, the basic idea of maximally flat filter design is to match exactly as many leading terms as possible in the Taylor series expansion of the desired frequency response. Equivalently, we zero the maximum number of leading terms in the Taylor expansion of the frequency-response error.

Figure 4.11 compares Lagrange and optimal Chebyshev fractional-delay filter frequency responses. Optimality in the *Chebyshev sense* means minimizing the worst-case error over a given frequency band (in this case, $|\omega| \in [0, 0.8\pi]$). While Chebyshev optimality is often the most desirable choice, we do not have closed-form formulas for such solutions, so they must be laboriously pre-calculated, tabulated, and interpolated to produce variable-delay filtering [358].

4.2.4 Explicit Lagrange Coefficient Formulas

Given a desired fractional delay of Δ samples, the Lagrange fraction-delay impulse response can be written in closed form as

$$h_\Delta(n) = \prod_{\substack{k=0\\k\neq n}} \frac{\Delta - k}{n - k}, \quad n = 0, 1, 2, \ldots, N. \tag{4.7}$$

The following table gives specific examples for orders 1, 2, and 3:

h_Δ Order	$h_\Delta(0)$	$h_\Delta(1)$	$h_\Delta(2)$	$h_\Delta(3)$
$N=1$	$1-\Delta$	Δ	0	0
$N=2$	$\frac{(\Delta-1)(\Delta-2)}{2}$	$-\Delta(\Delta-2)$	$\frac{\Delta(\Delta-1)}{2}$	0
$N=3$	$-\frac{(\Delta-1)(\Delta-2)(\Delta-3)}{6}$	$\frac{\Delta(\Delta-2)(\Delta-3)}{2}$	$-\frac{\Delta(\Delta-1)(\Delta-3)}{2}$	$\frac{\Delta(\Delta-1)(\Delta-2)}{6}$

Note that, for $N = 1$, Lagrange interpolation reduces to *linear interpolation*, *i.e.*, the interpolator impulse response is $h = [1 - \Delta, \Delta]$. Also, remember

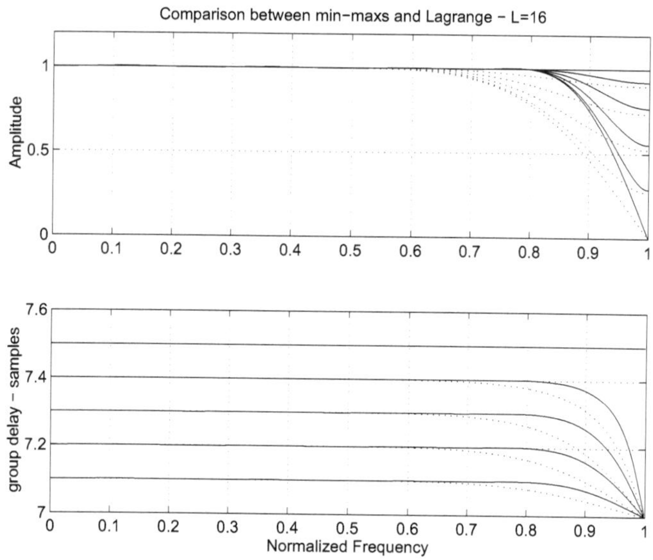

Figure 4.11: Comparison of Lagrange and Optimal Chebyshev Fractional-Delay Filter Frequency Responses

that, for order N, the desired delay should be in a one-sample range centered about $\Delta = N/2$.

4.2.5 Lagrange Interpolation Coefficient Symmetry

As shown in [503, §3.3.3], directly substituting into Eq. (4.7) derives the following *coefficient symmetry* property for the interpolation coefficients (impulse response) of a Lagrange fractional delay filter:

$$h_\Delta(n) \;=\; h_{N-\Delta}(N-n), \quad n = 0, 1, \ldots, N, \tag{4.8}$$

where N is the order of the interpolator. Thus, the interpolation coefficients for delay $N - \Delta$ are the "flip" (time reverse) of the coefficients for delay Δ.

4.2.6 Matlab Code for Lagrange Interpolation

A simple matlab function for computing the coefficients of a Lagrange fractional-delay FIR filter is as follows:

```
function h = lagrange( N, delay )
```

```
n = 0:N;
h = ones(1,N+1);
for k = 0:N
    index = find(n ~= k);
    h(index) = h(index) *  (delay-k)./ (n(index)-k);
end
```

4.2.7 Maxima Code for Lagrange Interpolation

The maxima program is free and open-source, like Octave for matlab:[4]

```
(%i1) lagrange(N, n) :=
         product(if equal(k,n) then 1
                 else (D-k)/(n-k), k, 0, N);

(%o1) lagrange(N, n) := product(if equal(k, n) then 1

                                             D - k
                                        else -----, k, 0, N)
                                             n - k
```

Usage examples in maxima:

```
(%i2) lagrange(1,0);
(%o2)                             1 - D

(%i3) lagrange(1,1);
(%o3)                               D

(%i4) lagrange(4,0);
                         (1 - D) (D - 4) (D - 3) (D - 2)
(%o4)                  - -------------------------------
                                       24
(%i5) ratsimp(lagrange(4,0));
                          4      3       2
                         D  - 10 D  + 35 D  - 50 D + 24
(%o5)                   ------------------------------
                                      24

(%i6) expand(lagrange(4,0));
                          4      3       2
                         D    5 D    35 D    25 D
```

[4]Say "help;" within maxima and/or "man maxima" in a shell to get started. Emacs users should say "M-x maxima".

```
(%o6)                        -- - ---- + ----- - ---- + 1
                             24    12     24       12

(%i7) expand(lagrange(4,0)), numer;
                             4                      3
(%o7) 0.041666666666667 D  - 0.41666666666667 D
                             2
    + 1.458333333333333 D  - 2.083333333333333 D + 1.0
```

4.2.8 Faust Code for Lagrange Interpolation

The Faust programming language for signal processing [453, 450] includes support for Lagrange fractional-delay filtering, up to order five, in the library file `filter.lib`. For example, the fourth-order case is listed below:

```
// fourth-order (quartic) case, delay d in [1.5,2.5]
fdelay4(n,d,x) = delay(n,id,x)   * fdm1*fdm2*fdm3*fdm4/24
               + delay(n,id+1,x) * (0-fd*fdm2*fdm3*fdm4)/6
               + delay(n,id+2,x) * fd*fdm1*fdm3*fdm4/4
               + delay(n,id+3,x) * (0-fd*fdm1*fdm2*fdm4)/6
               + delay(n,id+4,x) * fd*fdm1*fdm2*fdm3/24
with {
  o = 1.49999;
  dmo = d - o; // assumed nonnegative
  id = int(dmo);
  fd = o + frac(dmo);
  fdm1 = fd-1;
  fdm2 = fd-2;
  fdm3 = fd-3;
  fdm4 = fd-4;
};
```

An example calling program is shown in Fig. 4.12.

```
// tlagrange.dsp - test Lagrange interpolation in Faust

import("filter.lib");

N = 16; % Allocated delay-line length

% Compare various orders:
D = 5.4;
process = 1-1' <: fdelay1(N,D),
                  fdelay2(N,D),
                  fdelay3(N,D),
                  fdelay4(N,D),
                  fdelay5(N,D);
// To see results:
// [in a shell]:
//   faust2octave tlagrange.dsp
// [at the Octave command prompt]:
//   plot(db(fft(faustout,1024)(1:512,:)));

// Alternate example for testing a range of 4th-order cases
// (change name to "process" and rename "process" above):
process2  = 1-1' <: fdelay4(N, 1.5),
                    fdelay4(N, 1.6),
                    fdelay4(N, 1.7),
                    fdelay4(N, 1.8),
                    fdelay4(N, 1.9),
                    fdelay4(N, 2.0),
                    fdelay4(N, 2.1),
                    fdelay4(N, 2.2),
                    fdelay4(N, 2.3),
                    fdelay4(N, 2.4),
                    fdelay4(N, 2.499),
                    fdelay4(N, 2.5);
```

Figure 4.12: Faust program `tlagrange.dsp` used to generate Figures 4.13 through 4.16.

4.2.9 Lagrange Frequency Response Examples

The following examples were generated using Faust code similar to that in Fig. 4.12 and the `faust2octave` command distributed with Faust.

Orders 1 to 5 on a fractional delay of 0.4 samples

Figure 4.13 shows the amplitude responses of Lagrange interpolation, orders 1 through 5, for the case of implementing an interpolated delay line of length 5.4 samples. In all cases the interpolator follows a delay line of appropriate length so that the interpolator coefficients operate over their central one-sample interval. Figure 4.14 shows the corresponding phase delays. As discussed in §4.2.10, the amplitude response of every odd-order case is constrained to be zero at half the sampling rate when the delay is half-way between integers, which this example is near. As a result, the curves for the two even-order interpolators lie above the three odd-order interpolators at high frequencies in Fig. 4.13. It is also interesting to note that the 4th-order interpolator, while showing a wider "pass band," exhibits more attenuation near half the sampling rate than the 2nd-order interpolator.

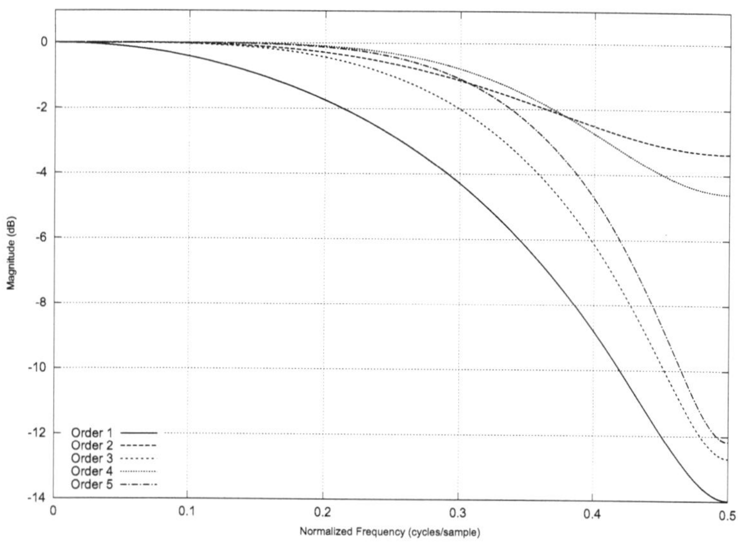

Figure 4.13: Amplitude responses, Lagrange interpolation, orders 1 to 5, for an interpolated delay of 5.4 samples. From the bottom-right corner along the right edge, the curves represent orders 1,3,5,4,2.

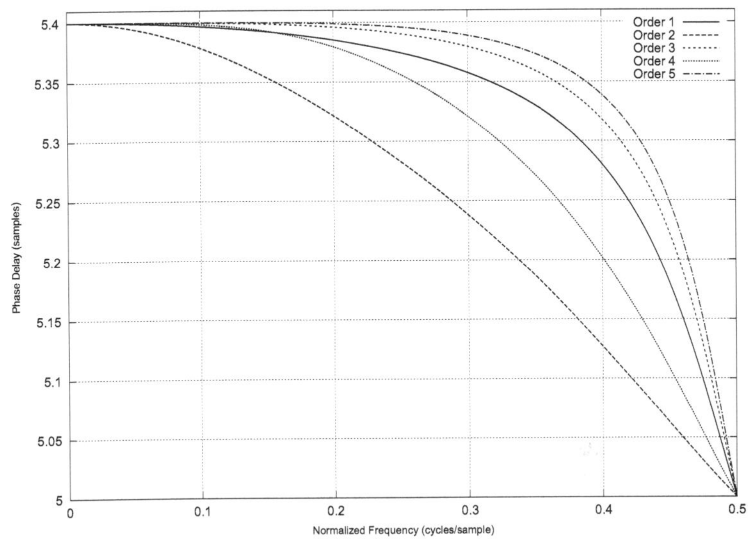

Figure 4.14: Phase delays, Lagrange interpolation, orders 1 to 5, for an interpolated delay of 5.4 samples. From bottom to top, the curves represent orders 2,4,1,3,5.

In the phase-delay plots of Fig. 4.14, all cases are exact at frequency zero. At half the sampling rate they all give 5 samples of delay.

Note that all three odd-order phase delay curves look generally better in Fig. 4.14 than both of the even-order phase delays. Recall from Fig. 4.13 that the two even-order amplitude responses outperformed all three odd-order cases. This illustrates a basic trade-off between gain accuracy and delay accuracy. The even-order interpolators show generally less attenuation at high frequencies (because they are not constrained to approach a gain of zero at half the sampling rate for a half-sample delay), but they pay for that with a relatively inferior phase-delay performance at high frequencies.

Order 4 over a range of fractional delays

Figures 4.15 and 4.16 show amplitude response and phase delay, respectively, for 4th-order Lagrange interpolation evaluated over a range of requested delays from 1.5 to 2.5 samples in increments of 0.1 samples. The amplitude response is ideal (flat at 0 dB for all frequencies) when the requested delay is 2 samples (as it is for any integer delay), while there is maximum high-frequency attenuation when the fractional delay is half a sample. In general, the closer the requested delay is to an integer, the flatter the amplitude response of the Lagrange interpolator.

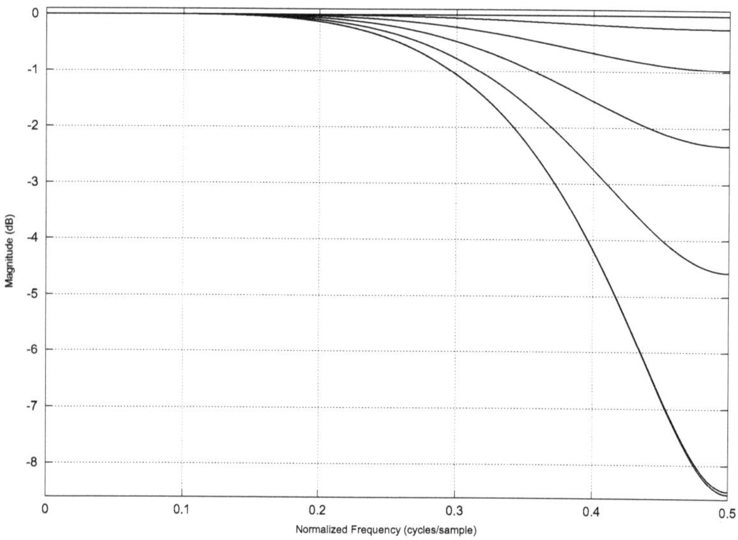

Figure 4.15: Amplitude responses, Lagrange interpolation, order 4, for the range of requested delays [1.5 : 0.1 : 2.5], with 2.499 thrown in as well (see next plot for why). From bottom to top, ignoring the almost invisible split in the bottom curve, the curves represent requested delays $1.5, 1.6, 1.7, 1.8, 1.9$, and 2.0. Then, because the curve for requested delay $2+\eta$ is the same as the curve for delay $2-\eta$, for $|\eta| < 1/2$, the same curves, from top to bottom, represent requested delays $2.0, 2.1, 2.2, 2.3, 2.4$ and 2.5 (which is nearly indistinguishable from 2.499).

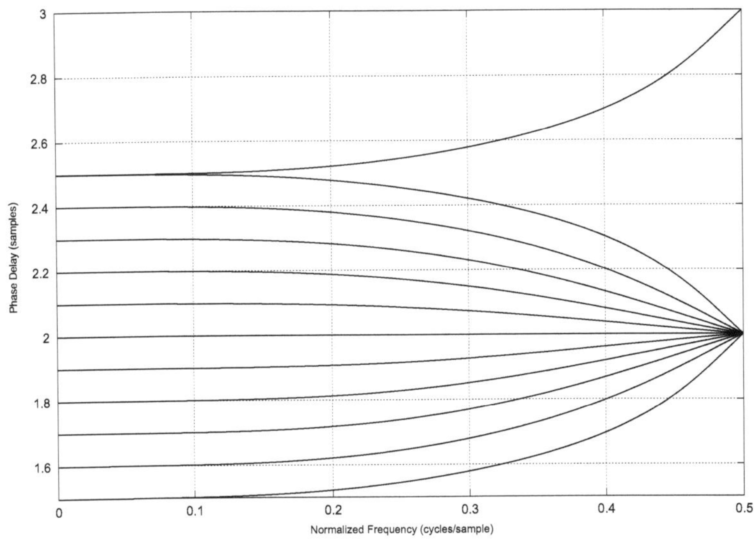

Figure 4.16: Phase delays, Lagrange interpolation, order 4, for the range of requested delays [1.5 : 0.1 : 2.5], and additionally 2.499.

Note in Fig. 4.16 how the phase-delay jumps discontinuously, as a function of delay, when approaching the desired delay of 2.5 samples from below: The top curve in Fig. 4.16 corresponds to a requested delay of 2.5 samples, while the next curve below corresponds to 2.499 samples. The two curves roughly coincide at low frequencies (being exact at dc), but diverge to separate integer limits at half the sampling rate. Thus, the "capture range" of the integer 2 at half the sampling rate is numerically suggested to be the half-open interval $[1.5, 2.5)$.

Order 5 over a range of fractional delays

Figures 4.17 and 4.18 show amplitude response and phase delay, respectively, for 5th-order Lagrange interpolation, evaluated over a range of requested delays between 2 and 3 samples in steps of 0.1 samples. Note that the vertical scale in Fig. 4.17 spans 100 dB while that in Fig. 4.15 needed less than 9 dB, again due to the constrained zero at half the sampling rate for odd-order interpolators at the half-sample point.

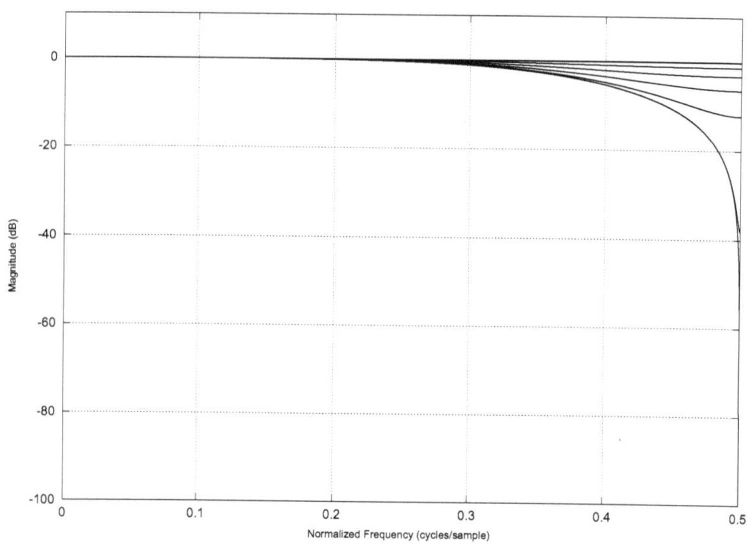

Figure 4.17: Amplitude responses, Lagrange interpolation, order 5, for the range of requested delays [2.0 : 0.1 : 3.0], with 2.495 and 2.505 included as well (see next plot for why).

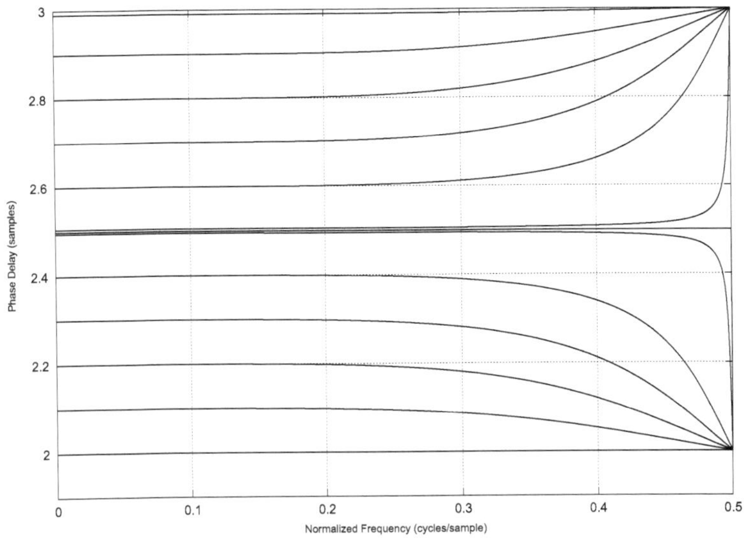

Figure 4.18: Phase delays, Lagrange interpolation, order 5, for the range of requested delays [2.0 : 0.1 : 3.0], with 2.495 and 2.505 included as well.

Notice in Fig. 4.18 how suddenly the phase-delay curves near 2.5 samples delay jump to an integer number of samples as a function of frequency near half the sample rate. The curve for 2.495 samples swings down to 2 samples delay, while the curve for 2.505 samples goes up to 3 samples delay at half the sample rate. Since the gain is zero at half the sample rate when the requested delay is 2.5 samples, the phase delay may be considered to be exactly 2.5 samples at all frequencies in that special case.

4.2.10 Avoiding Discontinuities When Changing Delay

We have seen examples (*e.g.*, Figures 4.16 and 4.18) of the general fact that every Lagrange interpolator provides an integer delay at frequency $\omega T = \pi$, except when the interpolator gain is zero at $\omega T = \pi$. This is true for any interpolator implemented as a real FIR filter, *i.e.*, as a linear combination of signal samples using real coefficients.[5] Therefore, to avoid a relatively large discontinuity in phase delay (at high frequencies) when varying the

[5]To see this, consider that a unit-amplitude sinusoid at half the sampling rate $f_s/2$ is either $(-1)^n$ or $-(-1)^n$. Therefore, the frequency response $H(\omega)$ of every real filter at $f_s/2$ is a real number $H(\pi f_s) = g$ that is either positive or negative (or zero). When $g > 0$, the filter's phase delay must be an even integer number of samples. When $g < 0$, the filter delay is constrained to be some odd number of samples. Taken together, the possible delays at $f_s/2$ for real filters are the integers. The particular integer is obtained by finding the limit as $f \to f_s/2$. The case $g = 0$ can be associated with a

delay over time, the requested interpolation delay should stay within a half-sample range of some fixed integer, irrespective of interpolation order. This provides that the requested delay stays within the "capture zone" of a single integer at half the sampling rate. Of course, if the delay varies by more than one sample, there is no way to avoid the high-frequency discontinuity in the phase delay using Lagrange interpolation.

Even-order Lagrange interpolators have an integer at the midpoint of their central one-sample range, so they spontaneously offer a one-sample variable delay free of high-frequency discontinuities.

Odd-order Lagrange interpolators, on the other hand, must be shifted by 1/2 sample in either direction in order to be centered about an integer delay. This can result in stability problems if the interpolator is used in a feedback loop, because the interpolation gain can exceed 1 at some frequency when venturing outside the central one-sample range (see §4.2.11 below).

In summary, discontinuity-free interpolation ranges include

$$\begin{aligned} \frac{N}{2} - \frac{1}{2} < D < \frac{N}{2} + \frac{1}{2} \qquad & (N \text{ even}), \\ \frac{N+1}{2} - \frac{1}{2} < D < \frac{N+1}{2} + \frac{1}{2} \qquad & (N \text{ odd}). \end{aligned}$$

Wider delay ranges, and delay ranges not centered about an integer delay, will include a phase discontinuity in the delay response (as a function of delay) which is largest at frequency $\omega T = \pi$, as seen in Figures 4.16 and 4.18.

4.2.11 Lagrange Frequency Response Magnitude Bound

The amplitude response of fractional delay filters based on Lagrange interpolation is observed to be bounded by 1 when the desired delay Δ lies within half a sample of the midpoint of the coefficient span [503, p. 92], as was the case in all preceeding examples above. Moreover, *even*-order interpolators are observed to have this boundedness property over a *two*-sample range centered on the coefficient-span midpoint [503, §3.3.6]. These assertions are easily proved for orders 1 and 2. For higher orders, a general proof appears not to be known, and the conjecture is based on numerical examples. Unfortunately, it has been observed that the gain of some odd-order Lagrange interpolators do exceed 1 at some frequencies when used outside of their central one-sample range [503, §3.3.6].

non-integer limit, as suggested by the line for 2.5 samples delay in Fig. 4.18 above, and as easily shown analytically for order 1 (simple linear interpolation with coefficients 0.5, 0.5, corresponding to a desired delay of 0.5 samples) [449].

4.2.12 Even-Order Lagrange Interpolation Summary

We may summarize some characteristics of even-order Lagrange fractional delay filtering as follows:

- Two-sample bounded-by-1 delay-range instead of only one-sample
- No gain zero at half the sampling rate for the middle delay
- No phase-delay discontinuity when crossing the middle delay
- Optimal (central) delay range is centered about an integer

To stay within the central, one-sample delay range for even-order Lagrange interpolators, the delay range should be

$$\Delta \in \left(\frac{N}{2} - \frac{1}{2}, \frac{N}{2} + \frac{1}{2}\right),$$

where N is the order of the interpolator.

4.2.13 Odd-Order Lagrange Interpolation Summary

In contrast to even-order Lagrange interpolation, the odd-order case has the following properties (in fractional delay filtering applications):

- Improved phase-delay accuracy at the expense of decreased amplitude-response accuracy (low-order examples in Fig. 4.14 on page 153)

- Optimal (centered) delay range lies between two integers

The usual centered delay range is

$$\Delta \in \left(\frac{N}{2} - \frac{1}{2}, \frac{N}{2} + \frac{1}{2}\right),$$

which is between integers, and in this range, the amplitude response is observed to be bounded by 1.

To avoid a discontinuous phase-delay jump at high frequencies when crossing the middle delay, the delay range can be shifted to

$$\Delta \in \left(\frac{N \pm 1}{2} - \frac{1}{2}, \frac{N \pm 1}{2} + \frac{1}{2}\right),$$

but then the gain may exceed 1 at some frequencies.

4.2.14 Proof of Maximum Flatness at DC

The maximumally flat fractional-delay FIR filter is obtained by equating to zero all $N+1$ leading terms in the Taylor (Maclaurin) expansion of the frequency-response error at dc:

$$\begin{aligned} 0 &= \left.\frac{d^k}{d\omega^k}E(e^{j\omega})\right|_{\omega=0} = \left.\frac{d^k}{d\omega^k}e^{-j\omega\Delta} - \sum_{n=0}^{N} h(n)e^{-j\omega n}\right|_{\omega=0} \\ &= (-j\Delta)^k - \sum_{n=0}^{N}(-jn)^k h(n) \end{aligned}$$

$$\Rightarrow \boxed{\sum_{n=0}^{N} n^k h(n) = \Delta^k,\ k = 0, 1, \ldots, N}$$

This is a linear system of equations of the form $V\underline{h} = \underline{\Delta}$, where V is a Vandermonde matrix. The solution can be written as a ratio of Vandermonde determinants using Cramer's rule [329]. As shown by Cauchy (1812), the determinant of a Vandermonde matrix $[p_i^{j-1}]$, $i, j = 1, \ldots, N$ can be expressed in closed form as

$$\begin{aligned} \left|\left[p_i^{j-1}\right]\right| &= \prod_{j>i}(p_j - p_i) \\ &= (p_2 - p_1)(p_3 - p_1)\cdots(p_N - p_1)\cdots \\ &\quad (p_3 - p_2)(p_4 - p_2)\cdots(p_N - p_2)\cdots \\ &\quad (p_{N-1} - p_{N-2})(p_N - p_{N-2})\cdots \\ &\quad (p_N - p_{N-1}). \end{aligned}$$

Making this substitution in the solution obtained by Cramer's rule yields that the impulse response of the order N, maximally flat, fractional-delay FIR filter may be written in closed form as

$$h(n) = \prod_{\substack{k=0 \\ k\neq n}}^{N} \frac{D-k}{n-k}, \quad n = 0, 1, \ldots N,$$

which is the formula for Lagrange-interpolation coefficients (Eq. (4.6) on page 144) adapted to this problem (in which abscissae are equally spaced on the integers from 0 to $N-1$).

Further details regarding the theory of Lagrange interpolation can be found (online[6]) in [503, Ch. 3, Pt. 2, pp. 82–84].

[6] `http://www.acoustics.hut.fi/~vpv/publications/vesan_vaitos/ch3_pt2_lagrange.pdf`

4.2.15 Variable Filter Parametrizations

In practical applications of Lagrange Fractional-Delay Filtering (LFDF), it is typically necessary to compute the FIR interpolation coefficients $h_\Delta(n)$ as a function of the desired delay Δ, which is usually time varying. Thus, LFDF is a special case of FIR *variable filtering* in which the FIR coefficients must be time-varying functions of a single delay parameter $\Delta(t)$.

Table Look-Up

A general approach to variable filtering is to tabulate the filter coefficients as a function of the desired variables. In the case of fractional delay filters, the impulse response $h_\Delta(n)$ is tabulated as a function of delay $\Delta = N/2+\eta$, $\eta \in [-1/2, 1/2)$, $n = 0, 1, 2, \ldots, N$, where N is the interpolation-filter order. For each n, Δ may be sampled sufficiently densely so that linear interpolation will give a sufficiently accurate "interpolated table look-up" of $h_\Delta(n)$ for each n and (continuous) Δ. This method is commonly used in closely related problem of sampling-rate conversion [463].

Polynomials in the Delay

A more parametric approach is to formulate each filter coefficient $h_\Delta(n)$ as a *polynomial* in the desired delay Δ:

$$h_\Delta(n) \triangleq \sum_{m=0}^{M} c_n(m)\Delta^m, \quad n = 0, 1, 2, \ldots, N \tag{4.9}$$

Taking the z transform of this expression leads to the interesting and useful *Farrow structure* for variable FIR filters [134].

Farrow Structure

Taking the z transform of Eq. (4.9) yields

$$\begin{aligned} h_\Delta(n) &\triangleq \sum_{m=0}^{M} c_n(m)\Delta^m, \quad n = 0,1,2,\ldots,N \\ \Longleftrightarrow \quad H_\Delta(z) &\triangleq \sum_{n=0}^{N} h_\Delta(n) z^{-n} \\ &= \sum_{n=0}^{N}\left[\sum_{m=0}^{M} c_n(m)\Delta^m\right] z^{-n} \\ &= \sum_{m=0}^{M}\left[\sum_{n=0}^{N} c_n(m) z^{-n}\right]\Delta^m \\ &\triangleq \sum_{m=0}^{M} C_m(z)\Delta^m \end{aligned} \tag{4.10}$$

Since $H_\Delta(z)$ is an Nth-order FIR filter, at least one of the $C_m(z)$ must be Nth order, so that we need $M \geq N$. A typical choice is $M = N$.

Such a parametrization of a variable filter as a polynomial in *fixed* filters $C_m(z)$ is called a *Farrow structure* [134, 503]. When the polynomial Eq. (4.10) is evaluated using *Horner's rule*,[7] the efficient structure of Fig. 4.19 is obtained. Derivations of Farrow-structure coefficients for Lagrange fractional-delay filtering are introduced in [503, §3.3.7].

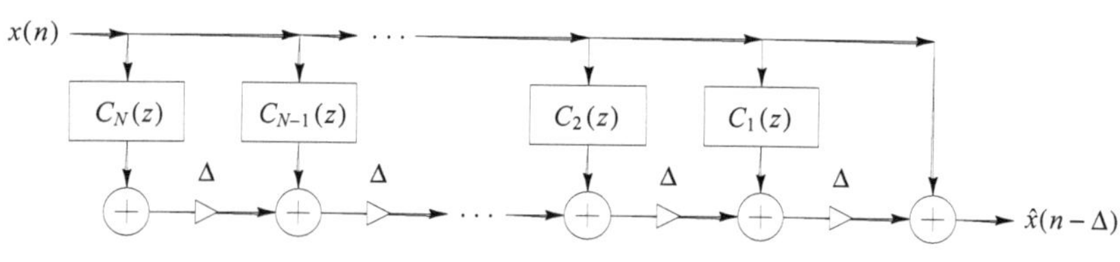

Figure 4.19: Farrow structure for implementing parametrized filters as a fixed-filter polynomial in the varying parameter.

As we will see in the next section, Lagrange interpolation can be implemented exactly by the Farrow structure when $M = N$. For $M < N$, approximations that do not satisfy the exact interpolation property can be computed [148].

[7] Evaluation of a polynomial $p(x) = p_0 + p_1 x + p_2 x^2 + \cdots + p_N x^N$ by *Horner's rule* can be expressed as $p(x) = p_0 + x \cdot (p_1 + x \cdot (p_2 + \cdots + x \cdot p_N)\cdots)$.

Farrow Structure Coefficients

Beginning with a restatement of Eq. (4.9),

$$h_\Delta(n) \triangleq \sum_{m=0}^{M} c_n(m)\Delta^m, \quad n = 0, 1, 2, \ldots, N,$$

we can express each FIR coefficient $h_\Delta(n)$ as a vector expression:

$$h_\Delta(n) = \underbrace{\left[\begin{array}{ccccc} 1 & \Delta & \Delta^2 & \cdots & \Delta^M \end{array}\right]}_{\underline{V}_\Delta^T} \left[\begin{array}{c} C_n(0) \\ C_n(1) \\ \vdots \\ C_n(M) \end{array}\right]$$

Making a row-vector out of the FIR coefficients gives

$$\underbrace{\left[\begin{array}{cccc} h_\Delta(0) & h_\Delta(1) & \cdots & h_\Delta(N) \end{array}\right]}_{\underline{h}_\Delta} = \underline{V}_\Delta^T \underbrace{\left[\begin{array}{cccc} C_0(0) & C_1(0) & \cdots & C_N(0) \\ C_0(1) & C_1(1) & \cdots & C_N(1) \\ \vdots & \vdots & & \vdots \\ C_0(M) & C_1(M) & \cdots & C_N(M) \end{array}\right]}_{\mathbf{C}}$$

or

$$\underline{h}_\Delta = \underline{V}_\Delta^T \mathbf{C}.$$

We may now choose a set of parameter values $\underline{\Delta}^T = [\Delta_0, \Delta_1, \ldots, \Delta_L]$ over which an optimum approximation is desired, yielding the matrix equation

$$\mathbf{H}_{\underline{\Delta}} = \mathbf{V}_{\underline{\Delta}}\mathbf{C}, \tag{4.11}$$

where

$$\mathbf{H}_{\underline{\Delta}} \triangleq \left[\begin{array}{c} \underline{h}_{\Delta_0}^T \\ \vdots \\ \underline{h}_{\Delta_L}^T \end{array}\right] \quad \text{and} \quad \mathbf{V}_{\underline{\Delta}} \triangleq \left[\begin{array}{c} \underline{V}_{\Delta_0}^T \\ \vdots \\ \underline{V}_{\Delta_L}^T \end{array}\right].$$

Equation (4.11) may be solved for the polynomial-coefficient matrix $\mathbf{C}$ by usual least-squares methods. For example, in the unweighted case, with $L \geq M$, we have

$$\boxed{\mathbf{C} = \left(\mathbf{V}_{\underline{\Delta}}^T \mathbf{V}_{\underline{\Delta}}\right)^{-1} \mathbf{V}_{\underline{\Delta}}^T \mathbf{H}_{\underline{\Delta}}.}$$

Note that this formulation is valid for finding the Farrow coefficients of any Nth-order variable FIR filter parametrized by a single variable Δ. Lagrange interpolation is a special case corresponding to a particular choice of $\mathbf{H}_{\underline{\Delta}}$.

Differentiator Filter Bank

Since, in the time domain, a Taylor series expansion of $x(n-\Delta)$ about time n gives

$$\begin{aligned} x(n-\Delta) &= x(n) - \Delta\, x'(n) + \frac{\Delta^2}{2!}x''(n) + \cdots + \frac{(-\Delta)^k}{k!}x^{(k)}(n) + \cdots \\ \longleftrightarrow \quad & X(z)\left[1 - \Delta\, D(z) + \frac{\Delta^2}{2!}D^2(z) + \cdots + \frac{(-\Delta)^k}{k!}D^k(z) + \cdots\right] \end{aligned}$$

where $D(z)$ denotes the transfer function of the *ideal differentiator*, we see that the mth filter in Eq. (4.10) should approach

$$C_m(z) = \frac{(-1)^m}{m!}D^m(z), \tag{4.12}$$

in the limit, as the number of terms M goes to infinity. In other terms, the coefficient $C_m(z)$ of Δ^m in the polynomial expansion Eq. (4.10) must become proportional to the *mth-order differentiator* as the polynomial order increases. For any finite N, we expect $C_m(z)$ to be close to some scaling of the mth-order differentiator. Choosing $C_m(z)$ as in Eq. (4.12) for finite N gives a *truncated Taylor series approximation* of the ideal delay operator in the time domain [184, p. 1748]. Such an approximation is "maximally smooth" in the time domain, in the sense that the first N derivatives of the interpolation error are zero at $x(n)$.[8] The approximation *error* in the time domain can be said to be *maximally flat*.

Farrow structures such as Fig. 4.19 may be used to implement any *one-parameter* filter variation in terms of several *constant* filters. The same basic idea of polynomial expansion has been applied also to *time-varying* filters ($\Delta \leftarrow t$).

4.2.16 Recent Developments in Lagrange Interpolation

Franck (2008) [148] provides a nice overview of the various structures being used for Lagrange interpolation, along with their computational complexities and depths (relevant to parallel processing). He moreover proposes a novel computation having linear complexity and log depth that is especially well suited for parallel processing architectures.

4.2.17 Relation of Lagrange to Sinc Interpolation

For an *infinite* number of *equally spaced* samples, with spacing $x_{k+1} - x_k = \Delta$, the Lagrangian basis polynomials converge to shifts of the *sinc function*,

[8] Of course, all derivatives of $x(n)$ must be calculated from the underlying continuous signal $x_c(t)$ represented by the samples $x(n)$ and then evaluated at $t = nT$.

i.e.,

$$l_k(x) = \text{sinc}\left(\frac{x - k\Delta}{\Delta}\right), \quad k = \dots, -2, -1, 0, 1, 2, \dots$$

where

$$\text{sinc}(x) \triangleq \frac{\sin(\pi x)}{\pi x}$$

A simple argument is based on the fact that any analytic function is determined by its zeros and its value at one point. Since $\text{sinc}(x)$ is zero on all the integers except 0, and since $\text{sinc}(0) = 1$, it must coincide with the infinite-order Lagrangian basis polynomial for the sample at $x = 0$ which also has its zeros on the nonzero integers and equals 1 at $x = 0$.

The equivalence of sinc interpolation to Lagrange interpolation was apparently first published by the mathematician Borel in 1899, and has been rediscovered many times since [309, p. 325].

A direct proof can be based on the equivalance between Lagrange interpolation and windowed-sinc interpolation using a "scaled binomial window" [262, 503]. That is, for a fractional sample delay of D samples, multiply the shifted-by-D, sampled, sinc function

$$h_s(n) = \text{sinc}(n - D) = \frac{\sin[\pi(n - D)]}{\pi(n - D)}$$

by a binomial window

$$w(n) = \binom{N}{n}, \quad n = 0, 1, 2, \dots N$$

and normalize by [503]

$$C(D) = (-1)^N \frac{\pi(N+1)}{\sin(\pi D)} \binom{D}{N+1},$$

which scales the interpolating filter to have a unit L_2 norm, to obtain the Nth-order Lagrange interpolating filter

$$h_D(n) = C(D)w(n)h_s(n), \quad n = 0, 1, 2, \dots, N$$

Since the binomial window converges to the Gaussian window as $N \to \infty$, and since the window gets wider and wider, approaching a unit constant in the limit, the convergence of Lagrange to sinc interpolation can be seen.

A more recent alternate proof appears in [558].

4.3 Thiran Allpass Interpolators

Given a desired delay $\Delta = N + \delta$ samples, an order N allpass filter

$$H(z) = \frac{z^{-N}A\left(z^{-1}\right)}{A(z)} = \frac{a_N + a_{N-1}z^{-1} + \cdots + a_1 z^{-(N-1)} + z^{-N}}{1 + a_1 z^{-1} + \cdots + a_{N-1}z^{-(N-1)} + a_N z^{-N}}$$

can be designed having *maximally flat group delay* equal to Δ at dc using the formula

$$a_k = (-1)^k \left(\begin{array}{c} N \\ k \end{array} \right) \prod_{n=0}^{N} \frac{\Delta - N + n}{\Delta - N + k + n}, \quad k = 0, 1, 2, \ldots, N$$

where

$$\left(\begin{array}{c} N \\ k \end{array} \right) = \frac{N!}{k!(N-k)!}$$

denotes the kth *binomial coefficient.* Note, incidentally, that a lowpass filter having maximally flat group-delay at dc is called a *Bessel filter* [362, pp. 228–230].

- $a_0 = 1$ without further scaling
- For sufficiently large Δ, stability is guaranteed
  Rule of thumb: $\Delta \approx$ order
- It can be shown that the mean group delay of any stable Nth-order allpass filter is N samples [449].[9]
- Only known closed-form case for allpass interpolators of arbitrary order
- Effective for delay-line interpolation needed for *tuning* since pitch perception is most acute at low frequencies.

[9] To see this, note that the integral of the group delay around the unit circle is simply minus the *winding number* of the phase response (the number of zeros minus the number of poles inside the unit circle):

$$\int_0^{2\pi} \Theta'(\omega) d\omega = \Theta(2\pi) - \Theta(0) = -2\pi N$$

By the graphical method for phase response calculation [449, Chapter 8], each zero contributes 2π radians to $\Theta(2\pi) - \Theta(0)$, and each pole contributes -2π radians. Thus, the average group delay around the unit circle is

$$\frac{1}{2\pi} \int_0^{2\pi} D(\omega) d\omega \triangleq -\frac{1}{2\pi} \int_0^{2\pi} \Theta'(\omega) d\omega = N$$

- Since Thiran allpass filters have maximally flat group-delay at dc, like Lagrange FIR interpolation filters, they can be considered the recursive extension of Lagrange interpolation.

4.3.1 Thiran Allpass Interpolation in Matlab

```
function [A,B] = thiran(D,N)
%  [A,B] = thiran(D,N)
%  returns the order N Thiran allpass interpolation filter
%  for delay D (samples).

A = zeros(1,N+1);
for k=0:N
    Ak = 1;
    for n=0:N
        Ak = Ak * (D-N+n)/(D-N+k+n);
    end
    A(k+1) = (-1)^k * nchoosek(N,k) * Ak;
end

B = A(N+1:-1:1);
```

4.3.2 Group Delays of Thiran Allpass Interpolators

Figure 4.20 shows a family of group-delay curves for Thiran allpass interpolators, for orders 1, 2, 3, 5, 10, and 20. The desired group delay was equal to the order plus 0.3 samples (which is in the "easy zone" for an allpass interpolator).

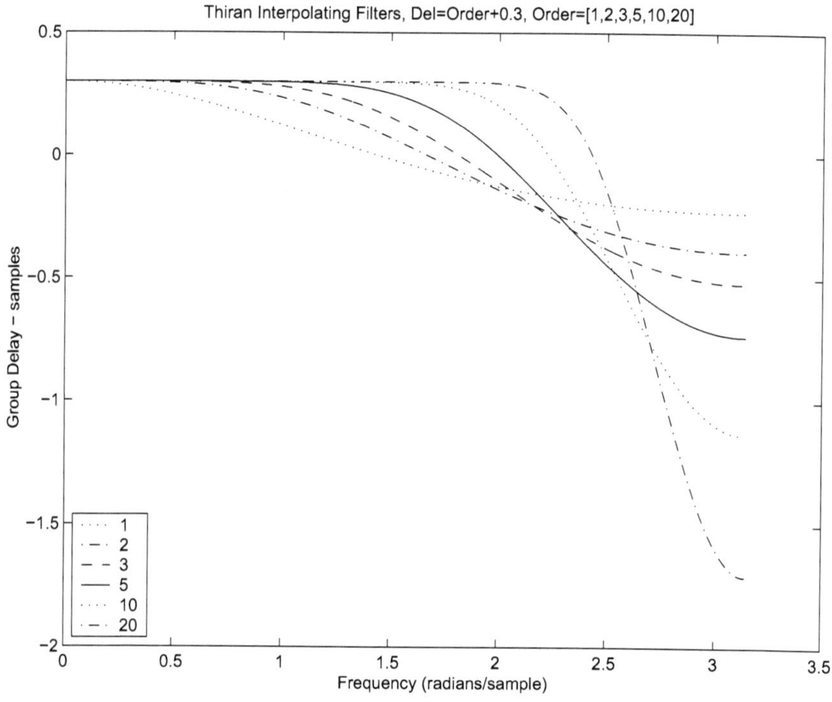

Thiran Interpolating Filters, Del=Order+0.3, Order=[1,2,3,5,10,20]
Group Delay - samples
Frequency (radians/sample)
0.5
0
-0.5
-1
-1.5
-2
0
0.5
1
1.5
2
2.5
3
3.5
1
2
3
5
10
20

Figure 4.20:

4.4 Windowed Sinc Interpolation

Bandlimited interpolation of discrete-time signals is a basic tool having extensive application in digital signal processing.[10] In general, the problem is to correctly compute signal values at arbitrary continuous times from a set of discrete-time samples of the signal amplitude. In other words, we must be able to interpolate the signal between samples. Since the original signal is always assumed to be *bandlimited* to half the sampling rate, (otherwise aliasing distortion would occur upon sampling), Shannon's sampling theorem tells us the signal can be exactly and uniquely reconstructed for all time from its samples by bandlimited interpolation.

Considerable research has been devoted to the problem of interpolating discrete points. A comprehensive survey of "fractional delay filter design" is provided in [267]. A comparison between classical (*e.g.*, Lagrange) and bandlimited interpolation is given in [407]. The book **Multirate Digital Signal Processing** [97] provides a comprehensive summary and review of classical signal processing techniques for sampling-rate conversion. In these techniques, the signal is first interpolated by an integer factor L and then decimated by an integer factor M. This provides sampling-rate conversion by any rational factor L/M. The conversion requires a digital lowpass filter whose cutoff frequency depends on $\max\{L, M\}$. While sufficiently general, this formulation is less convenient when it is desired to resample the signal at arbitrary times or change the sampling-rate conversion factor smoothly over time.

In this section, a public-domain resampling algorithm is described which will evaluate a signal at any time specifiable by a fixed-point number. In addition, one lowpass filter is used regardless of the sampling-rate conversion factor. The algorithm effectively implements the "analog interpretation" of rate conversion, as discussed in [97], in which a certain lowpass-filter impulse response must be available as a continuous function. Continuity of the impulse response is simulated by linearly interpolating between samples of the impulse response stored in a table. Due to the relatively low cost of memory, the method is quite practical for hardware implementation.

In the next section, the basic theory is presented, followed by sections on implementation details and practical considerations.

---

[10]The material in this section was adapted from `https://ccrma.stanford.edu/~jos/resample/`, which is an updated online version of [463].

4.4.1 Theory of Ideal Bandlimited Interpolation

We review briefly the "analog interpretation" of sampling rate conversion [97] on which the present method is based. Suppose we have samples $x(nT)$ of a continuous absolutely integrable signal $x(t)$, where t is time in seconds (real), n ranges over the integers, and T is the sampling period. We assume $x(t)$ is bandlimited to $\pm f_s/2$, where $f_s = 1/T$ is the sampling rate. If $X(\omega)$ denotes the Fourier transform of $x(t)$, *i.e.*, $X(\omega) = \int_{-\infty}^{\infty} x(t)e^{-j\omega t}dt$, then we assume $X(\omega) = 0$ for $|\omega| \geq \pi f_s$. Consequently, Shannon's sampling theorem gives us that $x(t)$ can be uniquely reconstructed from the samples $x(nT)$ via

$$\hat{x}(t) \triangleq \sum_{n=-\infty}^{\infty} x(nT)h_s(t-nT) \equiv x(t), \tag{4.13}$$

where

$$h_s(t) \triangleq \text{sinc}(f_s t) \triangleq \frac{\sin(\pi f_s t)}{\pi f_s t}.$$

To resample $x(t)$ at a new sampling rate $f_s' = 1/T'$, we need only evaluate Eq. (4.13) at integer multiples of T'.

When the new sampling rate f_s' is less than the original rate f_s, the lowpass cutoff must be placed below half the new lower sampling rate. Thus, in the case of an ideal lowpass, $h_s(t) = \min\{1, f_s'/f_s\}\text{sinc}(\min\{f_s, f_s'\}t)$, where the scale factor maintains unity gain in the passband.

A plot of the sinc function $\text{sinc}(t) \triangleq \sin(\pi t)/(\pi t)$ to the left and right of the origin $t = 0$ is shown in Fig. 4.21. Note that peak is at amplitude 1, and zero-crossings occur at all nonzero integers. The sinc function can be seen as a hyperbolically weighted sine function with its zero at the origin canceled out. The name *sinc function* derives from its classical name as the *sine cardinal* (or *cardinal sine*) function.

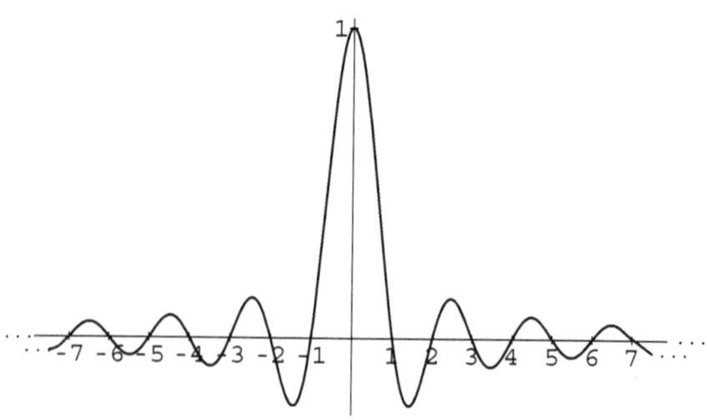

Figure 4.21: The sinc function plotted for seven zero-crossings to the left and right.

If "$*$" denotes the convolution operation for digital signals, then the summation in Eq. (4.13) can be written as $(x * h_s)(t)$.

Equation Eq. (4.13) can be interpreted as a superpositon of shifted and scaled sinc functions h_s. A sinc function instance is translated to each signal sample and scaled by that sample, and the instances are all added together. Note that zero-crossings of $\text{sinc}(z)$ occur at all integers except $z = 0$. That means at time $t = nT$, (*i.e.*, on a sample instant), the only contribution to the sum is the single sample $x(nT)$. All other samples contribute sinc functions which have a zero-crossing at time $t = nT$. Thus, the interpolation goes precisely through the existing samples, as it should.

A plot indicating how sinc functions sum together to reconstruct bandlimited signals is shown in Fig. 4.22. The figure shows a superposition of five sinc functions, each at unit amplitude, and displaced by one-sample intervals. These sinc functions would be used to reconstruct the bandlimited interpolation of the discrete-time signal $x = [\ldots, 0, 1, 1, 1, 1, 1, 0, \ldots]$. Note that at each sampling instant $t = nT$, the solid line passes exactly through the tip of the sinc function for that sample; this is just a restatement of the fact that the interpolation passes through the existing samples. Since the nonzero samples of the digital signal are all 1, we might expect the interpolated signal to be very close to 1 over the nonzero interval; however, this is far from being the case. The deviation from unity between samples can be thought of as "overshoot" or "ringing" of the lowpass filter which cuts off at half the sampling rate, or it can be considered a "Gibbs phenomenon" associated with bandlimiting.

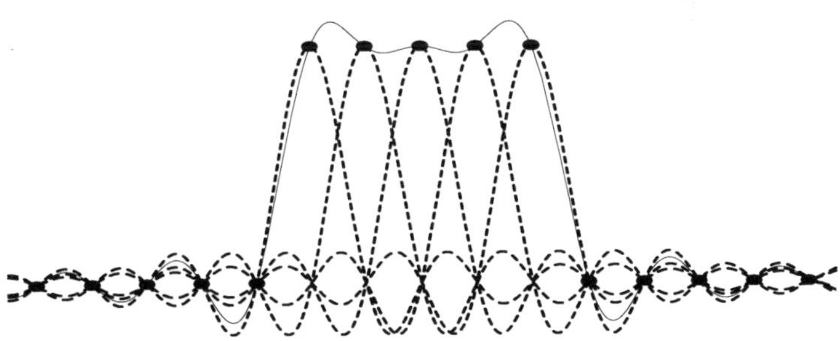

Figure 4.22: Bandlimited reconstruction of the signal $x(t)$ from its samples $x = [\ldots, 0, 1, 1, 1, 1, 1, 0, \ldots]$. The dots show the signal samples, the dashed lines show the component sinc functions, and the solid line shows the unique bandlimited reconstruction from the samples obtained by summing the component sinc functions.

A second interpretation of Eq. (4.13) is as follows: to obtain the interpolation at time t, shift the signal samples under *one* sinc function so that time t in the signal is translated under the peak of the sinc function, then create the output as a linear combination of signal samples where the coefficient of each signal sample is given by the value of the sinc function at the location of each sample. That this interpretation is equivalent to the first can be seen as a result of the fact that convolution is commutative; in the first interpretation, all signal samples are used to form a linear combination of shifted sinc functions, while in the second interpretation, samples from one sinc function are used to form a linear combination of samples of the shifted input signal. The practical bandlimited interpolation algorithm presented below is based on the second interpretation.

4.4.2 From Theory to Practice

The summation in Eq. (4.13) cannot be implemented in practice because the "ideal lowpass filter" impulse response $h_s(t)$ actually extends from minus infinity to infinity. It is necessary in practice to *window* the ideal impulse response so as to make it finite. This is the basis of the *window method* for digital filter design [115, 362]. While many other filter design techniques exist, the window method is simple and robust, especially for very long impulse responses. In the case of the algorithm presented below, the filter impulse response is very long because it is heavily oversampled. Another approach is to design optimal decimated "sub-phases" of the filter impulse response, which are then interpolated to provide the "continuous" impulse response needed for the algorithm [358].

Figure 4.23 shows the frequency response of the ideal lowpass filter. This is just the Fourier transform of $h_s(t)$.

If we truncate $h_s(t)$ at the fifth zero-crossing to the left and the right of the origin, we obtain the frequency response shown in Fig. 4.24. Note that the stopband exhibits only slightly more than 20 dB rejection.

If we instead use the Kaiser window [221, 438] to taper $h_s(t)$ to zero by the fifth zero-crossing to the left and the right of the origin, we obtain the frequency response shown in Fig. 4.25. Note that now the stopband starts out close to -80 dB. The Kaiser window has a single parameter which can be used to modify the stop-band attenuation, trading it against the transition width from pass-band to stop-band.

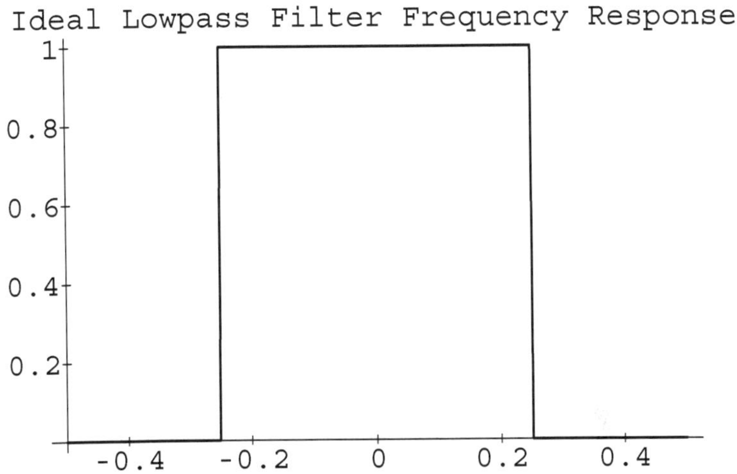

Figure 4.23: Frequency response of the ideal lowpass filter.

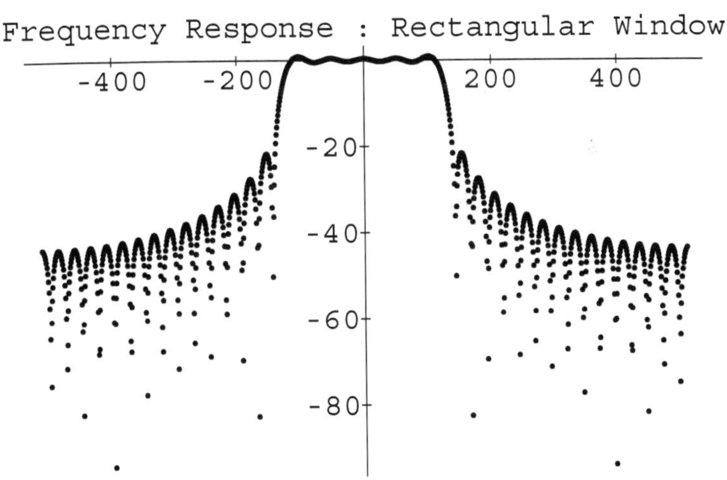

Figure 4.24: Frequency response of the ideal lowpass filter after rectangularly windowing the ideal (sinc) impulse response at the fifth zero crossing to the left and right of the time origin. The vertical axis is in units of decibels (dB), and the horizontal axis is labeled in units of spectral samples between plus and minus half the sampling rate.

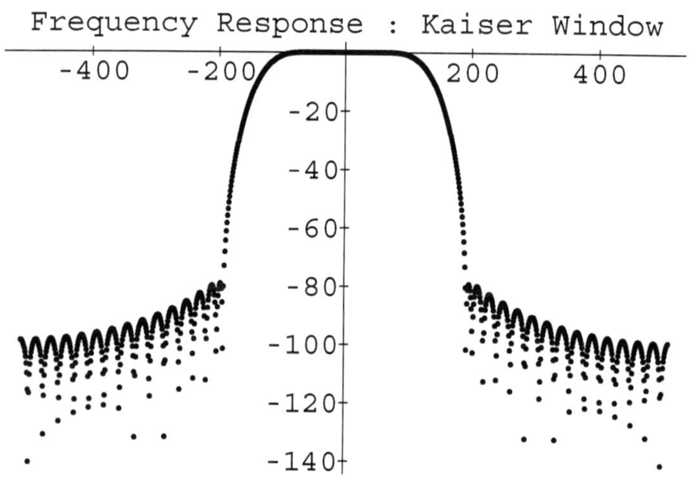

Figure 4.25: Frequency response of the ideal lowpass filter Kaiser windowed at the fifth zero crossing to the left and right.

4.4.3 Implementation

The implementation below provides signal evaluation at an arbitrary time, where time is specified as an unsigned binary fixed-point number in units of the input sampling period (assumed constant).

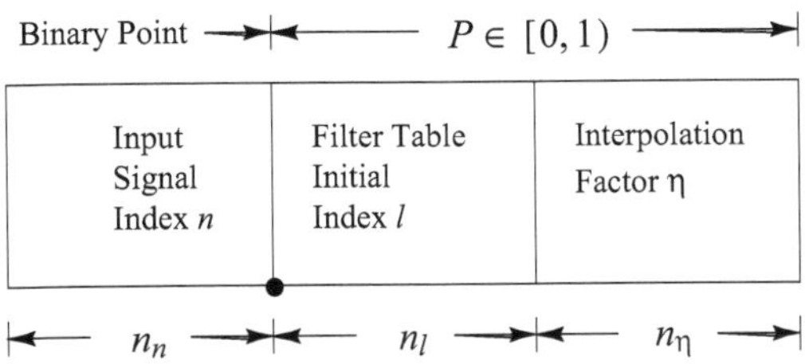

Figure 4.26: Time register format.

Figure 4.26 shows the time register t, and Figure 4.27 shows an example configuration of the input signal and lowpass filter at a given time. The time register is divided into three fields: The leftmost field gives the number n of samples into the input signal buffer, the middle field is an initial index l into the filter coefficient table $h(l)$, and the rightmost field is interpreted as a number ϵ between 0 and 1 for doing linear interpolation between samples l and $l+1$ (initially) of the filter table. The concatenation of l and ϵ are called $P \in [0, 1)$ which is interpreted as the position of the current time between samples n and $n+1$ of the input signal.

Let the three fields have n_n, n_l, and n_η bits, respectively. Then the input signal buffer contains $N = 2^{n_n}$ samples, and the filter table contains $L = 2_l^n$ "samples per zero-crossing." (The term "zero-crossing" is precise only for the case of the ideal lowpass; to cover practical cases we generalize "zero-crossing" to mean a multiple of time $t_c = 0.5/f_c$, where f_c is the lowpass cutoff frequency.) For example, to use the ideal lowpass filter, the table would contain $h(l) = \text{sinc}(l/L)$.

Our implementation stores only the "right wing" of a symmetric finite-impulse-response (FIR) filter (designed by the window method based on a Kaiser window [362]). Specifically, if $h(l)$, $l \in [-LN_z, LN_z]$, denotes a length $2LN_z + 1$ symmetric finite impulse response, then the *right wing* of h is defined as the set of samples $h(l)$ for $l \in [0, LN_z]$. By symmetry, the *left wing* can be reconstructed as $h(-l) = h(l)$, $l = 1, 2, \ldots, LN_z$.

Our implementation also stores a table of *differences* $\hbar(l) = h(l+1) - h(l)$ between successive FIR sample values in order to speed up the linear interpolation. The length of each table is $\hat{N} = LN_z + 1$, including the endpoint definition $\hbar(\hat{N}) = 0$.

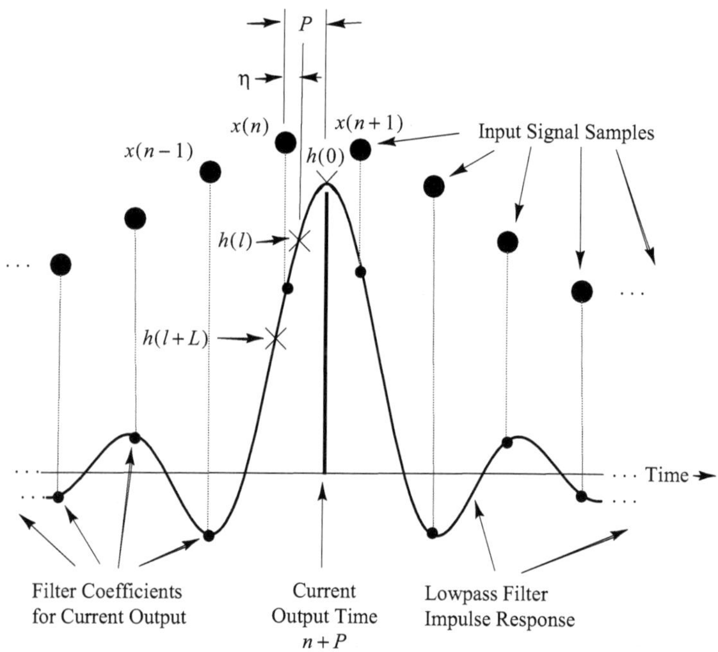

Figure 4.27: Waveforms and parameters in the interpolator.

Consider a sampling-rate conversion by the factor $\rho = f'_s/f_s$. For each output sample, the basic interpolation Eq. (4.13) is performed. The filter table is traversed twice—first to apply the left wing of the FIR filter, and second to apply the right wing. After each output sample is computed, the time register is incremented by $2^{n_l+n_\eta}/\rho$ (*i.e.*, time is incremented by $1/\rho$ in fixed-point format). Suppose the time register t has just been updated, and an interpolated output $y(t)$ is desired. For $\rho \geq 1$, output is computed via

$$\begin{aligned} v &\leftarrow \sum_{i=0}^{h\ \mathrm{end}} x(n-i)\left[h(l+iL)+\epsilon\bar{h}(l+iL)\right] \\ P &\leftarrow 1-P \\ y(t) &\leftarrow v+\sum_{i=0}^{h\ \mathrm{end}} x(n+1+i)\left[h(l+iL)+\epsilon\bar{h}(l+iL)\right], \end{aligned}$$

where $x(n)$ is the current input sample, and $\epsilon \in [0,1)$ is the interpolation factor. When $\rho < 1$, the initial P is replaced by $P' = \rho P$, $1-P$ becomes $\rho - P' = \rho(1-P)$, and the step-size through the filter table is reduced to ρL instead of L; this lowers the filter cutoff to avoid aliasing. Note that ϵ is fixed throughout the computation of an output sample when $\rho \geq 1$ but changes when $\rho < 1$.

When $\rho < 1$, more input samples are required to reach the end of the filter table, thus preserving the filtering quality. The number of multiply-adds per second is approximately $(2N_z+1)\max\{f_s, f'_s\}$. Thus the higher sampling rate determines the work rate. Note that for $\rho < 1$ there must be $\lceil N_z f_s/f'_s \rceil$ extra input samples available before the initial conversion time and after the final conversion time in the input buffer. As $\rho \to 0$, the required extra input data becomes infinite, and some limit must be chosen, thus setting a minimum supported ρ. For $\rho \geq 1$, only N_z extra input samples are required on the left and right of the data to be resampled, and the upper bound for ρ is determined only by the fixed-point number format, *viz.*, $\rho\mathrm{max} = 2^{n_l+n_\eta}$.

As shown below, if n_c denotes the word-length of the stored impulse-response samples, then one may choose $n_l = n_c/2$, and $n_\eta = n_c/2$ to obtain $n_c - 1$ effective bits of precision in the interpolated impulse response.

Note that rational conversion factors of the form $\rho = L/m$, where $L = 2^n_l$ and m is an arbitrary positive integer, do not use the linear interpolation feature (because $\epsilon \equiv 0$). In this case our method reduces to the normal type of bandlimited interpolator [97]. With the availability of interpolated lookup, however, the range of conversion factors is boosted to the order of $2^{n_l+n_\eta}/m$. *E.g.*, for $\rho \approx 1$, $n_l = 9, n_\eta = 8$, this is about 5.1 decimal digits of accuracy in the conversion factor ρ. Without interpolation, the number of significant figures in ρ is only about 2.7.

The number N_z of zero-crossings stored in the table is an independent design parameter. For a given quality specification in terms of aliasing rejection, a trade-off exists between N_z and sacrificed bandwidth. The lost bandwidth is due to the so-called "transition band" of the lowpass filter [362]. In general, for a given stop-band specification (such as "80 dB attenuation"), lowpass filters need approximately twice as many multiply-adds per sample for each halving of the transition band width.

As a practical design example, we use $N_z = 13$ in a system designed for high audio quality at 20% oversampling. Thus, the effective FIR filter is 27 zero crossings long. The sampling rate in this case would be 50 kHz.[11] In the most straightforward filter design, the lowpass filter pass-band would stop and the transition-band would begin at 20 kHz, and the stop-band would begin (and end) at 25 kHz. As a further refinement, which reduces the filter design requirements, the transition band is really designed to extend from 20 kHz to 30 kHz, so that the half of it between 25 and 30 kHz aliases on top of the half between 20 and 25 kHz, thereby approximately halving the filter length required. Since the entire transition band lies above the range of human hearing, aliasing within it is not audible.

Using 512 samples per zero-crossing in the filter table for the above example (which is what we use at CCRMA, and which is somewhat over designed) implies desiging a length $27 \times 512 = 13824$ FIR filter having a cut-off frequency near $\pi/512$. It turns out that optimal Chebyshev design procedures such as the Remez multiple exchange algorithm used in the Parks-McLellan software [362] can only handle filter lengths up to a couple hundred or so. It is therefore necessary to use an FIR filter design method which works well at such very high orders, and the window method employed here is one such method.

It is worth noting that a given percentage increase in the original sampling rate ("oversampling") gives a larger percentage savings in filter computation time, for a given quality specification, because the added bandwidth is a larger percentage of the filter transition bandwidth than it is of the original sampling rate. For example, given a cut-off frequency of 20 kHz, (ideal for audio work), the transition band available with a sampling rate of 44 kHz is about 2 kHz, while a 48 kHz sampling rate provides a 4 kHz transition band. Thus, a 10% increase in sampling rate *halves* the work per sample in the digital lowpass filter.

---

[11] We arbitrarily define the 20% guard band as a percentage of half the sampling rate actually used, not as 20% of the desired 20 kHz bandwidth which would call for a 48 kHz sampling rate.

Choice of Table Size and Word Lengths

It is desirable that the stored filter impulse response be sampled sufficiently densely so that interpolating linearly between samples does not introduce error greater than the quantization error. It is shown in [463] that this condition is satisfied when the filter impulse-response table contains at least $L = 2^{1+n_c/2}$ entries per "zero-crossing", where n_c is the number of bits allocated to each table entry. (A later, sharper, error bound gives that $L = 2^{n_c/2}$ is sufficient.) It is additionally shown in [463] that the number of bits in the interpolation between impulse-response samples should be near $n_c/2$ or more. With these choices, the linear interpolation error and the error due to quantized interpolation factors are each about equal to the coefficient quantization error. A signal resampler designed according to these rules will typically be limited primarily by the lowpass filter *design*, rather than by quantization effects.

4.4.4 Summary of Windowed Sinc Interpolation

The digital resampling method described in this section is convenient for bandlimited interpolation of discrete-time signals at arbitrary times and/or for arbitrary changes in sampling rate. The method is well suited for software or hardware implementation, and widely used free software[12] is available.

4.5 Delay-Line Interpolation Summary

The most commonly used closed-form methods for delay-line interpolation may be summarized by the following table:

Order

	1	N	Large N	∞
FIR	Linear	Lagrange	Windowed Sinc	
IIR	Allpass_1	Thiran		Sinc

In the first column, we have linear and first-order allpass interpolation, as discussed above in sections §4.1.1 and §4.1.2, respectively. These are the least-expensive methods computationally, and they find wide use, especially in audio applications with large sampling rates. While linear and first-order allpass interpolation cost the same, only linear interpolation offers "random access mode". That is, since the interpolated signal is a (finite) linear combination of known samples, the signal can be evaluated at any

[12] `http://ccrma.stanford.edu/~jos/resample/`

arbitrary time within the range of known samples — the interpolation can "jump around" as desired. On the other hand, allpass interpolation has no gain error, so it may preferred inside a feedback loop to provide slowly varying fractional delay filtering.

In the second column — Nth-order interpolation — we list Lagrange for the FIR case (top) and Thiran for the IIR case (bottom), and these are introduced in §4.2 and §4.3, respectively. Lagrange and Thiran interpolators, properly implemented, enjoy the following advantages:

- Gain bounded by 1 at all frequencies
- Coefficients known in closed form as a function of desired delay
- Maximally flat at low frequencies:
  - *Lagrange:* maximally flat *frequency response* at dc
  - *Thiran:* maximally flat *group delay* at dc

In the high-order FIR case, one should also consider "windowed sinc" interpolation (introduced in §4.4 on page 169) as an alternative to Lagrange interpolation. In fact, as discussed in §4.2.16, Lagrange interpolation is a special case of windowed-sinc interpolation in which a scaled binomial window is used. By choosing different windows, optimalities other than "maximally flat at dc" can be achieved.

In the most general Nth-order case, the interpolation-filter impulse response may be designed to achieve any optimality objective, such as Chebyshev optimality (Fig. 4.11 on page 148). That is, design a digital filter (FIR or IIR) that approximates

$$H_\Delta\left(e^{j\omega T}\right) = e^{-j\omega\Delta T}, \quad \Delta = \text{Desired delay in samples}$$

optimally in some sense, with coefficients tabulated over a range of Δ samples (and interpolated on lookup). Tabulated filter-designs of this nature, while generally giving the best interpolation quality, are not included in the above table because the filter-coefficients are not known in closed form.

FIR interpolators have the advantage that they can be used in "random access" mode. IIR interpolators, on the other hand, require a sequential stream of input samples and produce a sequential stream of interpolated signal samples (typically implementing a fractional delay). In IIR fractional-delay filters, the fractional delay must change slowly relative to the IIR duration.

Finally, we note in the last column of the above table that if "order is no object" ($N \rightarrow \infty$), then the ideal bandlimited-interpolator impulse-response is simply a sampled sinc function, as discussed in §4.4.

Chapter 5

Time-Varying Delay Effects

In the category of *delay effects*, variable delay lines are used for

- Doubling
- Phasing
- Flanging
- Chorus
- Leslie
- Reverb

(While reverberators need not be time varying, nowadays they typically are [104, 58].)

In *digital waveguide synthesis*, variable delay lines are used for

- Vibrating strings (guitars, violins, ...)
- Woodwind bores
- Horns
- Tonal percussion (rods, membranes)

After an implementation note on time-varying delay lines, we will look at a number of *time-varying delay effects* requiring variable delay lines. The use of variable delay lines in digital waveguide models will be taken up Chapter 6.

5.1 Variable Delay Lines

Time varying delay lines are fundamental building blocks for delay effects, synthesis algorithms, and computational acoustic models of musical instruments.

Let `A` denote an array of length N. Then we can implement an M-sample variable delay line in the `C` programming language as shown in Fig. 5.1. We require, of course, $M \leq N$.

```
static double A[N];
static double *rptr = A; // read ptr
static double *wptr = A; // write ptr

double setdelay(int M) {
    rptr = wptr - M;
    while (rptr < A) { rptr += N }
}

double delayline(double x)
{
  double y;
  A[wptr++] = x;
  y = A[rptr++];
  if ((wptr-A) >= N) { wptr -= N }
  if ((rptr-A) >= N) { rptr -= N }
  return y;
}
```

Figure 5.1: The M-sample variable delay line using separate read- and write-pointers.

The Synthesis Tool Kit, Version 4 [86] contains the `C++` class "`Delay`" which implements this type of variable (but non-interpolating) delay line. There are additional subclasses which provide *interpolating reads* by various methods. In particular, the class `DelayL` implements continuously variable delay lengths using *linear interpolation*. The code listing in Fig. 5.1 can be modified to use linear interpolation by replacing the line

```
y = A[rptr++];
```

with

```
long rpi = (long)floor(rptr);
double a = rptr - (double)rpi;
y = a * A[rpi] + (1-a) * A[rpi+1];
rptr += 1;
```

To implement a *continuously* varying delay, we add a "delay growth parameter" `g` to the `delayline` function in Fig. 5.1, and change the line

```
rptr += 1; // pointer update
```

above to

```
rptr += 1 - g; // pointer update
```

When `g` is 0, we have a fixed delay line. When `g` > 0, the delay grows `g` samples per sample, which we may also interpret as seconds per second, *i.e.*, $\dot{D}_t =$ `g`. In §5.7.2 on page 192, this will be applied to simulation of the *Doppler effect*.

5.2 Doubling and Slap-Back

The *doubling effect* is a studio recording technique often used to "thicken" vocals in which the same part is sung twice by the same person. In other words, doubling is a "chorus of two", where both parts are sung "in unison" by the same person. As an example, the Beatles used doubling very often, such as on the track "Hard Day's Night". A single variable delay line can simulate doubling very effectively.

The related term *slap back* refers to the use of a single echo on a recorded track. The echo is often placed in a different spatial location in the stereo mix. Normally the echo delay is just large enough to be heard as a discrete echo on careful listening (*e.g.*, on the order of tens of milliseconds). Compared to doubling, slap back uses a larger delay relative to the first voice, and the delay need not vary.

5.3 Flanging

Flanging is a delay effect that has been available in recording studios since at least the 1960s. Surprisingly little literature exists, although there is some [32, 429, 59, 17, 104, 242].[1]

[1]Flanging was apparently invented independently several times. According to `http://www.geofex.com/Article_Folders/phasers/phase.html`, the flanging technique was discovered by Phil Spectre in the 1950s in the context of trying to "thicken" a vocal tract by summing two copies of a track variably delayed in this way. His first use of flanging as an *effect* is said to be in his song "The Big Hurt". According to historian Mark Lewisohn [286], as summarized at `http://en.wikipedia.org/wiki/Flanging` (accessed March 28, 2010), the term "flanging" was coined by John Lennon of the Beatles to describe what EMI engineer Ken Townsend called "Artificial Double Tracking" (ADT). ADT was said to be developed by Townsend in response to John Lennon's request to try to eliminate the work of recording vocal tracks twice in order to "double" them. George Martin has said to have explained the effect as follows: "Now listen, it's very simple. We take the original image and we split it through a double-bifurcated sploshing flange with double negative feedback." [286]. This suggests that negative regenerative feedback was

The term "flanging" is said to have arisen from the way the effect was originally achieved by two tape machines set up to play the same tape in unison, with their outputs added together (mixed equally), as shown in Fig. 5.2. To achieve the flanging effect, the *flange* of one of the supply reels is touched lightly to make it play a littler slower. This causes a *delay* to develop between the two tape machines. The flange is released, and the flange of the other supply reel is touched lightly to slow it down. This causes the delay to gradually disappear and then begin to grow again in the opposite direction. The delay is kept below the threshold of echo perception (*e.g.*, only a few milliseconds in each direction). The process is repeated as desired, pressing the flange of each supply reel in alternation. The flanging effect has been described as a kind of "whoosh" passing subtly through the sound.[2] The effect is also compared to the sound of a jet passing overhead, in which the direct signal and ground reflection arrive at a varying relative delay [59]. If flanging is done rapidly enough, an audible Doppler shift is introduced which approximates the "Leslie" effect commonly used for organs (see §5.9).

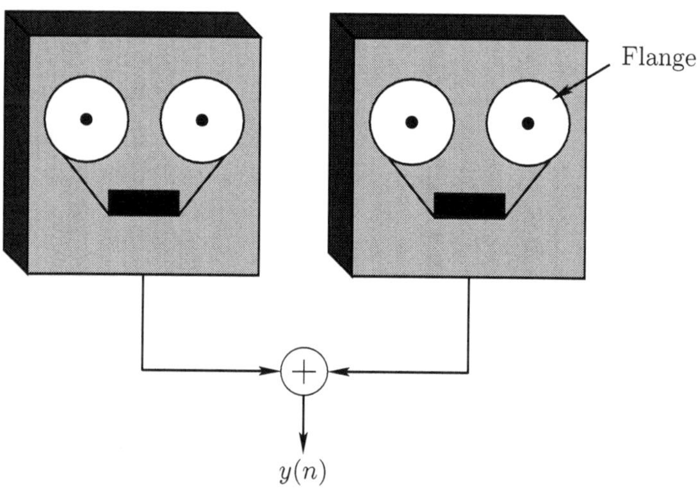

Figure 5.2: Two tape machines configured to produce a *flanging effect*.

used in the original stereo flanging technique (§2.6.2, p. 58). The first Beatles track using flanging is said to have been "Tomorrow Never Knows" on the Revolver album (recorded April 6, 1966). Perhaps the most obvious early example of flanging was in the popular song "Itchykoo Park" by The Small Faces (1967); this instance is said to have originated with engineer George Chkiantz at Olympic Studios in London. The song featured a foreground drum-roll on a snare with strong flanging throughout, and the vocals were flanged as well.

[2]For sound examples, see `http://www.harmony-central.com/Effects/Articles/Flanging/`.

Flanging is modeled quite accurately as a feedforward comb filter, as discussed in §2.6.1, in which the delay M is varied over time. Figure 5.3 depicts such a model. The input-output relation for a basic flanger can be written as

$$y(n) = x(n) + gx[n - M(n)] \tag{5.1}$$

where $x(n)$ is the input signal amplitude at time $n = 0, 1, 2, \ldots$, $y(n)$ is the output at time n, g is the "depth" of the flanging effect, and $M(n)$ is the length of the delay-line at time n. The delay length $M(n)$ is typically varied according to a triangular or sinusoidal waveform. We may say that the delay length is *modulated* by an *"LFO"* (Low-Frequency Oscillator). Since $M(n)$ must vary smoothly over time, it is clearly necessary to use an *interpolated delay line* to provide non-integer values of M in a smooth fashion.

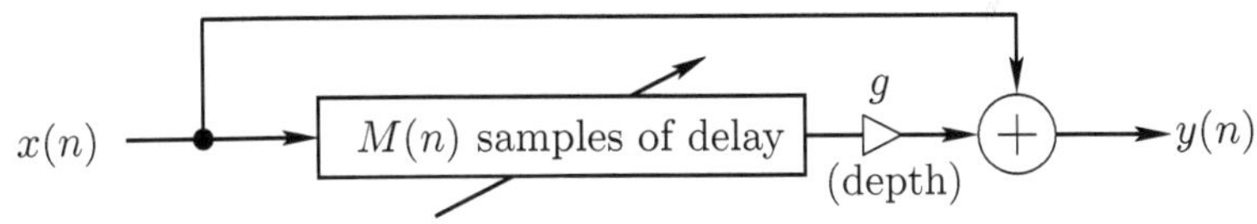

Figure 5.3: The basic flanger effect.

As shown in Fig. 2.25 on page 59, the frequency response of Eq. (5.1) has a "comb" shaped structure. For $g > 0$, there are M *peaks* in the frequency response, centered about frequencies

$$\omega_k^{(p)} = k\frac{2\pi}{M}, \quad k = 0, 1, 2, \ldots, M - 1.$$

For $g = 1$, the peaks are maximally pronounced, with M *notches*[3] occurring between them at frequencies $\omega_k^{(n)} = \omega_k^{(p)} + \pi/M$. As the delay length M is varied over time, these "comb teeth" squeeze in and out like the pleats of an accordion. As a result, the spectrum of any sound passing through the flanger is "massaged" by a variable comb filter.

As is evident from Fig. 2.25 on page 59, at any given time there are $M(n)$ *notches* in the flanger's amplitude response (counting positive- and negative-frequency notches separately). The notches are thus spaced at intervals of f_s/M Hz, where f_s denotes the sampling rate. In particular, the *notch spacing is inversely proportional to delay-line length.*

The time variation of the delay-line length $M(n)$ results in a "sweeping" of uniformly-spaced notches in the spectrum. The flanging effect is

[3]The term *notch* here refers to the elimination of sound energy at a single frequency or over a narrow frequency interval. Another term for this is *"null"*.

thus created by *moving notches* in the spectrum. Notch motion is essential for the flanging effect. Static notches provide some coloration to the sound, but an isolated notch may be inaudible [139]. Since the steady-state sound field inside an undamped *acoustic tube* has a similar set of uniformly spaced notches (except at the ends), a static row of notches tends to sound like being inside an acoustic tube.

5.3.1 Flanger Speed and Excursion

As mentioned above, the delay-line length $M(n)$ in a digital flanger is typically modulated by a *low-frequency oscillator (LFO)*. The oscillator waveform is usually triangular, sinusoidal, or exponential (triangular on a log-frequency scale). In the sinusoidal case, we have the following delay variation:

$$M(n) = M_0 \cdot [1 + A\sin(2\pi f n T)]$$

where f is the "*speed*" (or "*rate*") of the flanger in cycles per second, A is the "excursion" or "sweep" (maximum delay swing) which is often not brought out as a user-controllable parameter, and M_0 is the average delay length controlling the average notch density (also not normally brought out as a user-controllable parameter).

5.3.2 Flanger Depth Control

To obtain a maximum effect, the depth control, g in Fig. 5.3, should be set to 1. A depth of $g = 0$ gives no effect.

5.3.3 Flanger Inverted Mode

A different type of maximum depth is obtained for $g = -1$. In this case, the peaks and notches of the $g = 1$ case trade places. In practice, the depth control g is usually constrained to the interval $[0, 1]$, and a sign inversion for g is controlled separately using a "phase inversion" switch.

In inverted mode, unless the delay M is very large, the bass response will be weak, since the first notch is at dc. This case usually sounds high-pass filtered relative to the "in-phase" case ($g > 0$).

As the notch spacing grows very large (M shrinks), the amplitude response approaches that of a first-order difference $y(n) = x(n) - x(n-1)$, which approximates a differentiator $y(t) = \frac{d}{dt}x(t)$. An ideal differentiator eliminates dc and provides a progressive high-frequency boost rising 6 dB per octave (specifically, the amplitude response is $|H(\omega)| = |\omega|$).

5.3.4 Flanger Feedback Control

Many modern commercial flangers have a control knob labeled "feedback" or "regen." This control sets the level of feedback from the output to the input of the delay line, thereby creating a *feedback comb filter* in addition to the feedforward comb filter, in the same manner as in the creation of a Schroeder allpass filter in §2.8.1 (see Fig. 2.30 on page 70).

5.3.5 Summary of Flanging

In view of the above, we may define a *flanger* in general as any filter which modulates the frequencies of a set of *uniformly* spaced notches and/or peaks in the frequency response. The main parameters are

- Depth $g \in [0, 1]$ — controlling notch depth
- Speed f — speed of notch movement
- Phase — switch to subtract instead of adding the direct signal with the delayed signal

Possible additional parameters include

- Average Delay M_0
- Excursion or Sweep A — amount by which the delay-line grows or shrinks
- Feedback or Regeneration $a_M \in (-1, 1)$ — feedback coefficient from output to input

Note that flanging provides *only* uniformly spaced notches. This can be considered non-ideal for several reasons. First, the ear processes sound over a frequency scale that is more nearly logarithmic than linear [460]. Therefore, exponentially spaced notches (uniformly spaced on a log frequency scale) should sound more uniform perceptually. Secondly, the uniform peaks and notches of the flanger can impose a discernible "resonant pitch" on the program material, giving the impression of being inside a resonant tube. Third, when $g < 0$ (inverted flanging), it is possible for a periodic tone to be completely annihilated by harmonically spaced notches if the harmonics of the tone are unlucky enough to land exactly on a subset of the harmonic notches. In practice, exact alignment is unlikely; however, the signal loudness can be modulated to a possibly undesirable degree as the notches move through alignment with the signal spectrum. For this reason, flangers are best used with noise-like or inharmonic sounds. For harmonic signals, it makes sense to consider methods for creating *non-uniform* moving notches.

A Faust software implementation of flanging may be found in the file `effect.lib` within the Faust distribution [154, 170]. The Faust programming example `phaser_flanger.dsp` may be run to hear the effect on a test signal and experiment with its parameters in real time.

5.4 Phasing

The *phaser*, or *phase shifter*, is closely related to the flanger in that it also works by sweeping notches through the spectrum of the input signal. While the term phasing is sometimes used synonymously with flanging [32],[4] typical commercial phase shifters have been observed to implement *nonuniformly spaced* notches.[5] Furthermore, phasers such as the *Univibe* (1960s) were intended to simulate the *Leslie rotating speaker* effect (§5.9, p. 197) as opposed to being a low-cost analog flanging approximation. In other words, the conceptual unification of phasing and flanging seems to be technical in nature; *i.e.*, based on the fact that both effects operate by sweeping notches through the spectrum. Flangers are constrained to have an "infinite series" of harmonically spaced notches (§2.6.3 on page 59), while phasers have a limited number of nonuniformly spaced notches. In both cases motion of the notches over time is essential to the effect, and this motion is classically periodic. We will therefore define a *phaser* as any linear filter which modulates the frequencies of a set of *non-uniformly spaced notches*, while a *flanger* will remain any device which modulates *uniformly spaced notches.*

Digital implementations of phase shifters are discussed in §8.9 on page 325. A Faust phase-shifter may be found in the file `effect.lib` within the Faust software distribution [154, 170]. The Faust programming example `phaser_flanger.dsp` demonstrates the phasing effect.

5.5 Vibrato Simulation

The term *vibrato* refers to small, quasi-periodic variations in the *pitch* of a tone. On a violin, for example, vibrato is produced by wiggling the finger stopping the string on the fingerboard; a violin vibrato frequency can be very slow, or a bit faster than 6 Hz. A typical vibrato *depth* is on the order of 1 percent (a *semitone* is $2^{1/12} \approx 6$ percent). In the singing

[4] According to `http://www.eventide.com/About/History.aspx`, a dual 200 ms delay-line for simulating flanging was called the *Instant Phaser*. This was the first commercial product circa 1970 made by Eventide Corp. (better known as makers of the *Harmonizer*).

[5] The author discovered this first-hand by looking at the circuit for the MXR phase shifter in 1975.

voice, vibrato is produced by modulating the tension of the vocal folds. Vibrato is typically accompanied by *tremolo*, which is *amplitude modulation* at the same frequency as the vibrato which causes it. For example, in the violin, the frequency-modulations of the string vibrations are translated into amplitude modulations by the complex variations in the frequency response of the violin body.

To apply vibrato to a sound, it is necessary to apply a quasi-periodic *frequency shift*. This can be accomplish using a *modulated delay line*. This works because a time-varying delay line induces a simulated *Doppler shift* on the signal within it.

The flanger in `effect.lib` (Faust distribution) has a vibrato mode in which it becomes a pure time-varying delay line. This mode can be accessed via a checkbox in the example `phaser_flanger.dsp`.

5.6 Doppler Effect

The *Doppler effect* causes the pitch of a sound source to appear to rise or fall due to *motion* of the source and/or listener relative to each other. You have probably heard the pitch of a horn drop lower as it passes by (*e.g.*, from a moving train). As a pitched sound-source moves toward you, the pitch you hear is raised; as it moves away from you, the pitch is lowered. The Doppler effect has been used to enhance the realism of simulated moving sound sources for compositional purposes [80], and it is an important component of the "Leslie effect" (described in §5.9).

As derived in elementary physics texts, the *Doppler shift* is given by

$$\omega_l = \omega_s \frac{1 + \frac{v_{ls}}{c}}{1 - \frac{v_{sl}}{c}} \tag{5.2}$$

where ω_s is the radian frequency emitted by the source at rest, ω_l is the frequency received by the listener, v_{ls} denotes the *speed* of the listener relative to the propagation medium in the direction of the source, v_{sl} denotes the speed of the source relative to the propagation medium in the direction of the listener, and c denotes sound speed. Note that all quantities in this formula are scalars.

5.6.1 Vector Formulation

Denote the sound-source *velocity* by $\underline{v}_s(t)$ where t is time. Similarly, let $\underline{v}_l(t)$ denote the velocity of the listener, if any. The *position* of source and listener are denoted $\underline{x}_s(t)$ and $\underline{x}_l(t)$, respectively, where $\underline{x} \triangleq (x_1, x_2, x_3)^T$ is

3D position. We have velocity related to position by

$$\underline{v}_s = \frac{d}{dt}\underline{x}_s(t) \qquad \underline{v}_l = \frac{d}{dt}\underline{x}_l(t). \tag{5.3}$$

Consider a Fourier component of the source at frequency ω_s. We wish to know how this frequency is shifted to ω_l at the listener due to the Doppler effect.

The Doppler effect depends only on velocity components along the line connecting the source and listener [349, p. 453]. We may therefore *orthogonally project* the source and listener velocities onto the vector $\underline{x}_{sl} = \underline{x}_l - \underline{x}_s$ pointing from the source to the listener. (See Fig. 5.8 on page 199 for a specific example.)

The *orthogonal projection* of a vector $\underline{x}$ onto a vector $\underline{y}$ is given by [451]

$$\mathcal{P}_{\underline{y}}(\underline{x}) = \frac{\langle \underline{x}, \underline{y} \rangle}{\| \underline{y} \|^2}\underline{y} \;\triangleq\; \frac{\underline{x}^T \underline{y}}{\underline{y}^T \underline{y}}\underline{y}.$$

Therefore, we can write the projected source velocity as

$$\underline{v}_{sl} = \mathcal{P}_{\underline{x}_{sl}}(\underline{v}_s) = \frac{\langle \underline{v}_s, \underline{x}_{sl} \rangle}{\| \underline{x}_{sl} \|^2}\underline{x}_{sl} = \frac{\langle \underline{v}_s, \underline{x}_l - \underline{x}_s \rangle}{\| \underline{x}_l - \underline{x}_s \|^2}\left(\underline{x}_l - \underline{x}_s\right). \tag{5.4}$$

In the *far field* (listener far away), Eq. (5.4) reduces to

$$\underline{v}_{sl} \approx \frac{\langle \underline{v}_s, \underline{x}_l \rangle}{\| \underline{x}_l \|^2}\underline{x}_l = \mathcal{P}_{\underline{x}_l}(\underline{v}_s) \qquad (\| \underline{x}_l \| \gg \| \underline{x}_s \|). \tag{5.5}$$

5.7 Doppler Simulation

It is well known that a time-varying delay line results in a frequency shift. Time-varying delay is often used, for example, to provide *vibrato* and *chorus* effects [17]. We therefore expect a time-varying delay-line to be capable of precise Doppler simulation. This section discusses simulating the Doppler effect using a variable delay line [469].[6]

Consider Doppler shift from a physical point of view. The air can be considered as analogous to a *magnetic tape* which moves from source to listener at speed c (see Fig. 5.4). The source is analogous to the *write-head* of a tape recorder, and the listener corresponds to the *read-head*. When the source and listener are fixed, the listener receives what the source records. When either moves, a Doppler shift is observed by the listener, according to Eq. (5.2).[7]

---

[6] `https://ccrma.stanford.edu/ jos/doppler/`

[7] If the tape travels in a loop, then Fig. 5.4 provides a model for the *Echoplex* (Maestro, 1960), which consists of a tape loop with a fixed write-head and movable read-head.

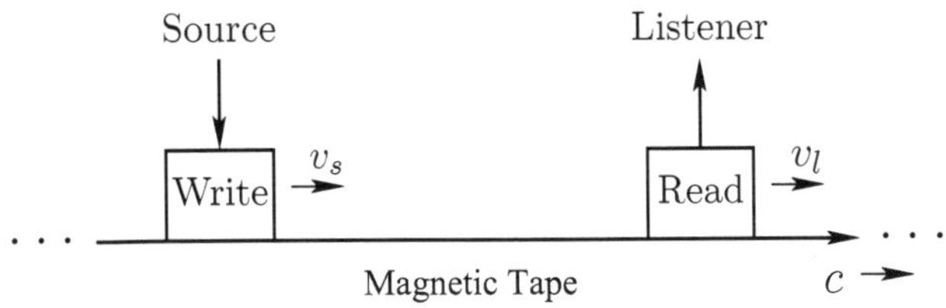

Figure 5.4: Magnetic tape analogy.

5.7.1 Doppler Simulation via Delay Lines

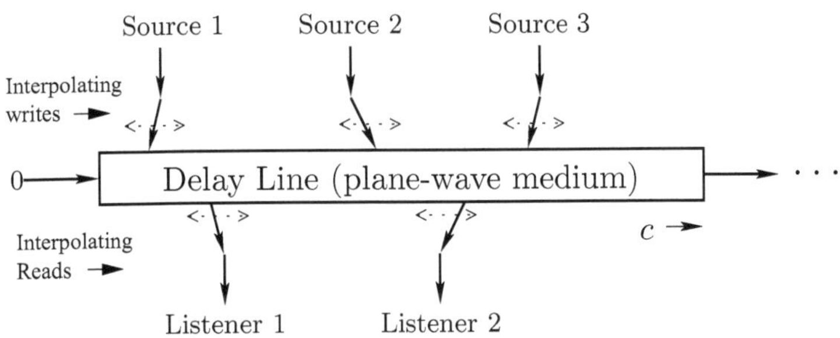

Figure 5.5: Delay line with variable read-pointers and write-pointers.

This analogy also works for a delay-line based computational model, as depicted in Fig. 5.5. The magnetic tape is now the delay line, the tape read-head is the read-pointer of the delay line, and the write-head is the delay-line write-pointer. In this analogy, it is readily verified that modulating delay by changing the read-pointer increment from 1 to $1 + v_{ls}/c$ (thereby requiring interpolated reads) corresponds to listener motion away from the source at speed v_{ls}. It also follows that changing the *write-pointer* increment from 1 to $1 + v_{sl}/c$ corresponds *source motion toward the listener* at speed v_{sl}. When this is done, we must use *interpolating writes* into the delay memory. Interpolating writes may be called *de-interpolation* [503], and they are formally the graph-theoretic *transpose* of interpolating reads (ordinary "interpolation") [333]. A review of time-varying, interpolating, delay-line reads and writes, together with a method using a single shared pointer, are given in [383].

5.7.2 Time-Varying Delay-Line Reads

If $x(t)$ denotes the input to a time-varying delay, the output can be written as

$$y(t) = x(t - D_t).$$

where D_t denotes the time-varying delay in seconds. In discrete-time implementations, when D_t is not an integer multiple of the sampling interval, $x(t - D_t)$ may be approximated to arbitrary accuracy (in a finite band) using *bandlimited interpolation* (see §4.4 on page 169) or other techniques for implementation of *fractional delay* [267, 383].

Let's analyze the frequency shift caused by a time-varying delay by setting $x(t)$ to a complex sinusoid at frequency ω_s:

$$x(t) = e^{j\omega_s t}$$

The output is now

$$y(t) = x(t - D_t) = e^{j\omega_s \cdot (t - D_t)}.$$

The instantaneous phase of this signal is

$$\theta(t) = \angle y(t) = \omega_s \cdot (t - D_t)$$

which can be differentiated to give the instantaneous frequency

$$\omega_l = \omega_s (1 - \dot{D}_t) \tag{5.6}$$

where ω_l denotes the output frequency, and $\dot{D}_t \triangleq \frac{d}{dt} D_t$ denotes the time derivative of the delay D_t. Thus, *the delay growth-rate*, $\dot{D}_t$, equals the *relative frequency downshift*:

$$\dot{D}_t = \frac{\omega_s - \omega_l}{\omega_s}.$$

Comparing Eq. (5.6) to Eq. (5.2), we find that the time-varying delay most naturally simulates Doppler shift caused by a *moving listener*, with

$$\dot{D}_t = -\frac{v_{ls}}{c}. \tag{5.7}$$

That is, the delay growth-rate, $\dot{D}_t$, should be set to the speed of the listener *away* from the source, normalized by sound speed c.

Simulating *source* motion is also possible, but the relation between delay change and desired frequency shift is more complex, *viz.*, from Eq. (5.2) and Eq. (5.6),

$$\dot{D}_t = -\frac{\frac{v_{ls}}{c} + \frac{v_{sl}}{c}}{1 - \frac{v_{sl}}{c}} \approx -\left(\frac{v_{ls}}{c} + \frac{v_{sl}}{c}\right)$$

where the approximation is valid for $v_{sl} \ll c$. In Section 5.7.4, a simplified approach is proposed based on moving the delay *input* instead of its output.

The time-varying delay line was described in §5.1 on page 181. As discussed there, to implement a *continuously* varying delay, we add a "delay growth parameter" `g` to the `delayline` function in Fig. 5.1 on page 182, and change the line

```
rptr += 1; // pointer update
```

to

```
rptr += 1 - g; // pointer update
```

When `g` is 0, we have a fixed delay line, corresponding to $\dot{D}_t = 0$ in Eq. (5.6). When `g` > 0, the delay grows `g` samples per sample, which we may also interpret as seconds per second, *i.e.*, $\dot{D}_t = \mathtt{g}$. By Eq. (5.7), we see that we need

$$\mathtt{g} = -\frac{v_{ls}}{c}$$

to simulate a listener traveling toward the source at speed v_{ls}.

Note that when the read- and write-pointers are driven directly from a model of physical propagation-path geometry, they are always separated by predictable minimum and maximum delay intervals. This implies it is unnecessary to worry about the read-pointer *passing* the write-pointers or vice versa. In generic *frequency shifters* [275], or in a Doppler simulator not driven by a changing geometry, a pointer cross-fade scheme may be necessary when the read- and write-pointers get too close to each other.

5.7.3 Multiple Read Pointers

Using *multiple read pointers*, multiple listeners can be simulated. Furthermore, each read-pointer signal can be *filtered* to simulate propagation losses and radiation characteristics of the source in the direction of the listener. The read-pointers can move independently to simulate the different Doppler shifts associated with different listener motions and relative source directions.

5.7.4 Multiple Write Pointers

It is interesting to consider also what effects can be achieved using multiple interpolating write pointers. From the considerations in §5.7.1, we see that multiple write-pointers correspond to multiple write-heads on a magnetic tape recorder. If they are arranged at a fixed spacing, they are equivalent to multiple read pointers, providing a basic multipath simulation. If, however, the write pointers are moving *independently*, they induce *independent*

Doppler shifts due to source motion. In particular, each write-pointer can lay down a signal from a separate source to a single listener with its own Doppler shift. Furthermore, each write-signal can be passed through its own filter. Such an individualized source filter can implement all filtering incurred along the propagation path from each source to the listener.

When all write pointers have the same input signal, their filters can be implemented using a *series chain* in which the outputs of successive filters in the chain correspond to progressively longer propagation paths (progressively more filtering). Such an implementation can greatly reduce the filter order required for propagation paths longer than the shortest.

The write-pointers may cross each other with no ill effects, since all but the first[8] simply sum into the shared delay line.

We have seen that a single delay line can be used to simulate any number of moving listeners (§5.7.3) or any number of moving sources. However, when simulating *both* multiple listeners and multiple sources, it is not possible to share a single delay line. This is because the different listeners do not see the same Doppler shift for each moving source, and while the listener's read-pointer motion can be adjusted to correct for the Doppler shift seen from any particular source, it cannot correct for more than one in general. Thus, in general, we need as many delay lines as there are sources or listeners, whichever is smaller. More precisely, if there are N moving sources and M moving listeners, simulation requires $\min(N, M)$ delay lines.

5.7.5 Stereo Processing

As a special case, stereo processing of any number of sources can be accomplished using two delay lines, corresponding to left and right stereo channels. The stereo mix may contain a panned mixture of any number sources, each with its own stereo placement, path filtering, and Doppler shift. The two stereo outputs may correspond to "left and right ears," or, more generally, to left- and right-channel microphones in a studio recording set-up.

5.7.6 System Block Diagram

A schematic diagram of a stereo multiple-source simulation is shown in Fig. 5.6. To simplify the layout, the input and output signals are all on the right in the diagram. For further simplicity, only one input source is shown. Additional input sources are handled identically, summing into the same delay lines in the same way.

[8] The "first" write-pointer is defined as the one writing farthest ahead in time; it must *overwrite* memory, instead of summing into it, when a circular buffer is being used, as is typical.

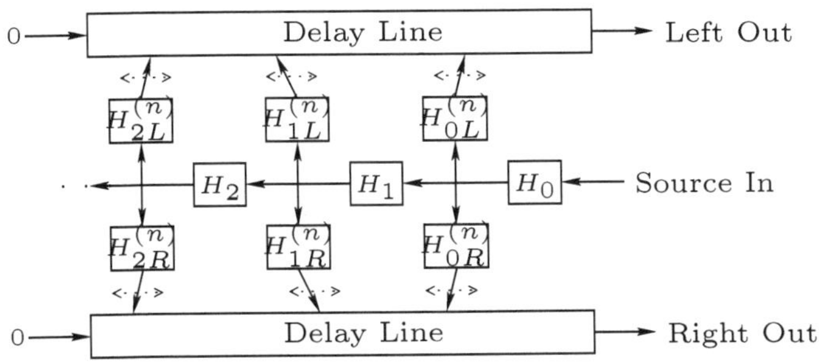

Figure 5.6: Block diagram of a stereo simulator for any number of moving sound sources (from [469]).

The input source signal first passes through filter $H_0(z)$, which provides *time-invariant* filtering common to all propagation paths. The left- and right-channel filters $H^{(n)}_{0L}(z)$ and $H^{(n)}_{0R}(z)$ are typically low-order, linear, *time-varying* filters implementing the time-varying characteristics of the shortest (time-varying) propagation path from the source to each listener. (The (n) superscript here indicates a time-varying filter.) These filter outputs *sum* into the delay lines at arbitrary (time-varying) locations using interpolating writes. The zero signals entering each delay line on the left can be omitted if the left-most filter overwrites delay memory instead of summing into it.

The outputs of $H^{(n)}_{0L}(z)$ and $H^{(n)}_{0R}(z)$ in Fig. 5.6 correspond to the "direct signal" from the moving source, when a direct signal exists. These filters may incorporate modulation of losses due to the changing propagation distance from the moving source to each listener, and they may include dynamic equalization corresponding to the changing radiation strength in different directions from the moving (and possibly turning) source toward each listener.

The next trio of filters in Fig. 5.6, $H_1(z)$, $H^{(n)}_{1L}(z)$, and $H^{(n)}_{1R}(z)$, correspond to the next-to-shortest acoustic propagation path, typically the "first reflection," such as from a wall close to the source. Since a reflection path is longer than the direct path, and since a reflection itself can attenuate (or scatter) an incident sound ray, there is generally more filtering required relative to the direct signal. This additional filtering can be decomposed into its fixed component $H_1(z)$ and time-varying components $H^{(n)}_{1L}(z)$ and $H^{(n)}_{1R}(z)$.

Note that acceptable results may be obtained without implementing all of the filters indicated in Fig. 5.6. Furthermore, it can be convenient to incorporate $H_i(z)$ into $H^{(n)}_{iL}(z)$ and $H^{(n)}_{iR}(z)$ when doing so does not increase their orders significantly.

Note also that the source-filters $H_{iL}^{(n)}(z)$ and $H_{iR}^{(n)}(z)$ may include HRTF filtering [57, 546] in order to impart illusory angles of arrival in 3D space.

5.8 Chorus Effect

The *chorus effect* (or "choralizer") is any signal processor which makes one sound source (such as a voice) sound like *many* such sources singing (or playing) in *unison*. Since performance in unison is never exact, chorus effects simulate this by making independently modified copies of the input signal. Modifications may include

(1) delay,

(2) frequency shift, and

(3) amplitude modulation.

The typical chorus effect today is based on several *time-varying delay lines* which accomplishes (1) and (2) in a qualitative fashion. Reverb generally provides (3) incidentally. Before digital delay lines, analog LC ladder networks were used as an approximation, beginning in the early 1940s in the Hammond organ [59, p. 731].

An efficient chorus-effect implementation may be based on *multiple interpolating taps* working on a single delay line. The taps oscillate back and forth about the positions they would have while implementing a fixed tapped delay line. The tap modulation frequency may be set to achieve a prescribed frequency-shift via the Doppler effect. Each tap should be individually spatialized; in the case of stereo, each tap can be panned to its own stereo position.

5.9 The Leslie

The *Leslie*, named after its inventor, Don Leslie,[9] is a popular audio processor used with electronic organs and other instruments [59, 189]. It employs a rotating horn and rotating speaker port to "choralize" the sound. Since the horn rotates within a cabinet, the listener hears multiple reflections at different Doppler shifts, giving a kind of *chorus effect*. Additionally, the Leslie amplifier distorts at high volumes, producing a pleasing "growl" highly prized by keyboard players. At the time of this writing, there is a nice Leslie Wikipedia page[10], including a stereo sound-example link under the first picture (that is best heard in headphones). Papers on computational audio models of the Leslie include [469, 191].

The Leslie consists primarily of a rotating horn and a rotating speaker port inside a wooden cabinet enclosure [189].

5.9.1 Rotating Horn Simulation

The heart of the Leslie effect is a rotating horn loudspeaker. The rotating horn from a Model 600 Leslie can be seen mounted on a microphone stand in Fig. 5.7. Two horns are apparent, but one is a dummy, serving mainly to cancel the centrifugal force of the other during rotation. The Model 44W horn is identical to that of the Model 600, and evidently standard across all Leslie models [189]. For a circularly rotating horn, the source position can be approximated as

$$\underline{x}_s(t) = \begin{bmatrix} r_s \cos(\omega_m t) \\ r_s \sin(\omega_m t) \end{bmatrix} \tag{5.8}$$

where r_s is the circular radius and ω_m is angular velocity. This expression ignores any *directionality* of the horn radiation, and approximates the horn as an omnidirectional radiator located at the same radius for all frequencies. In the Leslie, a *conical diffuser* is inserted into the end of the horn in order to make the radiation pattern closer to uniform [189], so the omnidirectional assumption is reasonably accurate.[11]

By Eq. (5.3), the source velocity for the circularly rotating horn is

$$\underline{v}_s(t) = \frac{d}{dt}\underline{x}_s(t) = \begin{bmatrix} -r_s\omega_m \sin(\omega_m t) \\ r_s\omega_m \cos(\omega_m t) \end{bmatrix} \tag{5.9}$$

Note that the source velocity vector is always orthogonal to the source position vector, as indicated in Fig. 5.8.

[9] `http://en.wikipedia.org/wiki/Leslie_speaker`

[10] `http://en.wikipedia.org/wiki/Leslie_speaker`

[11] See also the Hammond Leslie FAQ at `http://www.theatreorgans.com/hammond/faq/files/hammond-faq.pdf`

Figure 5.7: Rotating horn recording set up (from [469]).

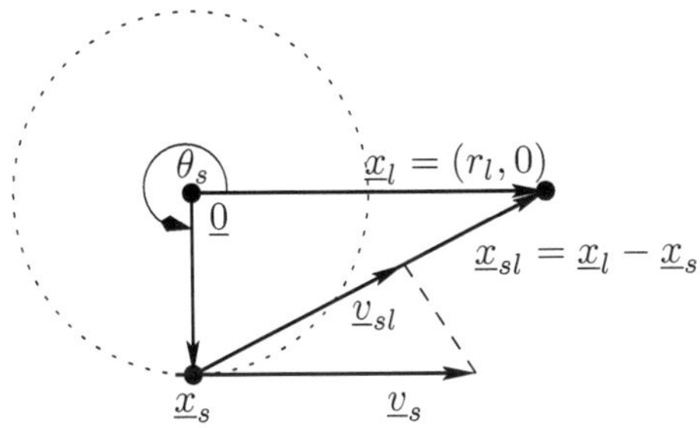

Figure 5.8: Relevant geometry for a rotating horn (from [469]).

Since $\underline{v}_s$ and $\underline{x}_s$ are orthogonal, the projected source velocity Eq. (5.4) simplifies to

$$\underline{v}_{sl} = \mathcal{P}_{\underline{x}_{sl}}(\underline{v}_s) = \frac{\langle \underline{v}_s, \underline{x}_l \rangle}{\| \underline{x}_l - \underline{x}_s \|^2} (\underline{x}_l - \underline{x}_s) \,. \tag{5.10}$$

Arbitrarily choosing $\underline{x}_l = (r_l, 0)$ (see Fig. 5.8), and substituting Eq. (5.8) and Eq. (5.9) into Eq. (5.10) yields

$$\underline{v}_{sl} = \frac{-r_l r_s \omega_m \sin(\omega_m t)}{r_l^2 + 2 r_l r_s \cos(\omega_m t) + r_s^2} \left[\begin{array}{c} r_l - r_s \cos(\omega_m t) \\ -r_s \sin(\omega_m) t \end{array} \right]. \tag{5.11}$$

In the far field, this reduces simply to

$$\underline{v}_{sl} \approx -r_s \omega_m \sin(\omega_m t) \left[\begin{array}{c} 1 \\ 0 \end{array} \right]. \tag{5.12}$$

Substituting into the Doppler expression Eq. (5.2) with the listener velocity v_l set to zero yields

$$\omega_l = \frac{\omega_s}{1 + r_s \omega_m \sin(\omega_m t)/c} \approx \omega_s \left[1 - \frac{r_s \omega_m}{c} \sin(\omega_m t) \right], \tag{5.13}$$

where the approximation is valid for small Doppler shifts. Thus, in the far field, a rotating horn causes an approximately *sinusoidal* multiplicative frequency shift, with the amplitude given by horn length r_s times horn angular velocity ω_m divided by sound speed c. Note that $r_s \omega_m$ is the *tangential speed* of the assumed point of horn radiation.

5.9.2 Rotating Woofer-Port and Cabinet

It is straightforward to extend the above computational model to include the rotating woofer port ("baffle") and wooden cabinet enclosure as follows:

- In [189], it is mentioned that an AM "throb" is the main effect of the rotating woofer port. A modulated lowpass-filter cut-off frequency has been used for this purpose by others. Measured data can be used to construct angle-dependent filtering in a manner analogous to that of the rotating horn, and this "woofer filter" runs in parallel with the rotating horn model.

- The Leslie cabinet multiply-reflects the sound emanating from the rotating horn. The first few early reflections are simply handled as additional sources in Fig. 5.6.

- To qualitatively simulate later, more *reverberant* reflections in the Leslie cabinet, we may feed a portion of the rotating-horn and speaker-port signals to separate states of an *artificial reverberator* (see Chapter 3). This reverberator may be configured as a "very small room" corresponding to the dimensions and scattering characteristics of the Leslie cabinet, and details of the response may be calibrated using measurements of the impulse response of the Leslie cabinet. Finally, in order to emulate the natural spatial diversity of a radiating Leslie cabinet in a room, "virtual cabinet vent outputs" can be extracted from the model and fed into separate states of a *room reverberator*. An alternative time-varying FIR filtering approach based on cabinet impulse-response measurements over a range of horn angles is described in [191].

In summary, we may use multiple interpolating write-pointers to individually simulate the early cabinet reflections, and a "Leslie cabinet" reverberator for handling later reflections more statistically.

5.9.3 Recent Research Modeling the Leslie

As mentioned above, modeling the Leslie via interpolating delay-line writes and cabinet image-sources was described in [469].[12] More recently, Leslie simulation via time-varying FIR filtering has been developed [191]. See these papers and their cited references for further details.

[12] `https://ccrma.stanford.edu/ jos/doppler/`

5.10 Exercises

Delay Line Interpolation Problems

1. As discussed in §4.1.2 on page 133, the frequency response of first-order allpass delay-line interpolation is is

$$H(z) = \frac{\eta + z^{-1}}{1 + \eta z^{-1}}.$$

Find the *phase delay*, and find its dc limit.

2. What range of η corresponds to $\Delta \in [0.5, 1.5]$?

3. What range of η corresponds to a maximum impulse-response duration of 10 ms? (Include the sampling rate f_s as a parameter in your formula.)

4. For the linear-interpolating fraction-delay filter

$$\hat{y}(n-\eta) = \eta \cdot y(n-1) + (1-\eta) \cdot y(n), \quad \eta \in [0, 1],$$

find the

   (a) transfer function,
   (b) poles and zeros,
   (c) amplitude response,
   (d) phase response,
   (e) phase delay, and
   (f) group delay.

5. For the first-order-allpass fraction-delay filter

$$\hat{x}(n-\Delta) \triangleq y(n) = \eta \cdot x(n) + x(n-1) - \eta \cdot y(n-1)$$

find the

   (a) transfer function,
   (b) poles and zeros,
   (c) amplitude response,
   (d) phase response,
   (e) phase delay, and
   (f) group delay.

Delay Line Interpolation Lab

1. Explore using the `sndfile-resample` command provided by the `libsndfile` package, which is included in the Planet CCRMA distribution.[13] Start out by reading the information at

   `http://ccrma.stanford.edu/~jos/resample/`

   and the man page for `resample`. Next, load a standard 44.1 kHz CD audio file and convert it to DAT sampling rate (48 kHz). Try designing your own resampling filter using the `windowfilter` command. Do you notice any difference in the audio quality if you use linear interpolation?

   See `http://ccrma.stanford.edu/realsimple/doppler_flanger_leslie/` for more laboratory exercises related to interpolated delay lines.

[13]`http://ccrma.stanford.edu/planetccrma/software/`

Chapter 6

Digital Waveguide Models

In this chapter, we summarize the basic principles of *digital waveguide models*. Such models are used for efficient synthesis of string and wind musical instruments (and tonal percussion, etc.), as well as for artificial reverberation. They can be further used in modal synthesis by efficiently implementing a quasi harmonic series of modes in a single "filtered delay loop".

We begin with the simplest case of the infinitely long ideal vibrating string, and the model is unified with that of acoustic tubes. The resulting computational model turns out to be a simple *bidirectional delay line*. Next we consider what happens when a finite length of ideal string (or acoustic tube) is rigidly terminated on both ends, obtaining a *delay-line loop*. The delay-line loop provides a basic digital-waveguide synthesis model for (highly idealized) stringed and wind musical instruments. Next we study the simplest possible excitation for a digital waveguide string model, which is to move one of its (otherwise rigid) terminations. Excitation from "initial conditions" is then discussed, including the ideal plucked and struck string. Next we introduce *damping* into the digital waveguide, which is necessary to model realistic losses during vibration. This much modeling yields musically useful results. Another linear phenomenon we need to model, especially for piano strings, is *dispersion*, so that is taken up next. Following that, we consider general excitation of a string or tube model at any point along its length. Methods for calibrating models from recorded data are outlined, followed by modeling of coupled waveguides, and simple memoryless nonlinearities are introduced and analyzed.

6.1 Ideal Vibrating String

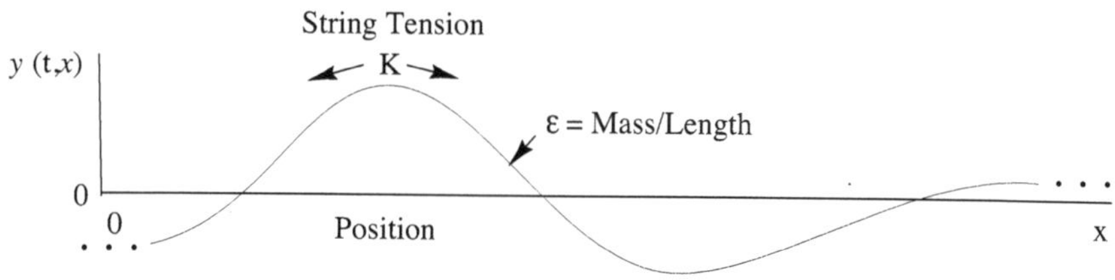

Figure 6.1: The ideal vibrating string.

The ideal vibrating string is depicted in Fig. 6.1. It is assumed to be perfectly flexible and elastic. Once "plucked," it will vibrate forever in one plane as an energy conserving system. The mathematical theory of string vibration is considered in §B.6 and Appendix C. For present purposes, we only need some basic definitions and results.

6.1.1 Wave Equation

The *wave equation* for the ideal vibrating string may be written as

$$Ky'' = \epsilon\ddot{y} \tag{6.1}$$

where we define the following notation:

$$\begin{array}{rcl@{\qquad}rcl} K & \triangleq & \text{string tension} & y & \triangleq & y(t,x) \\ \epsilon & \triangleq & \text{linear mass density} & \dot{y} & \triangleq & \frac{\partial}{\partial t}y(t,x) \\ y & \triangleq & \text{string displacement} & y' & \triangleq & \frac{\partial}{\partial x}y(t,x) \end{array}$$

As discussed in Chapter 1, the wave equation in this form can be interpreted as a statement of Newton's second law,

$$\textit{force} = \textit{mass} \times \textit{acceleration},$$

on a microscopic scale. Since we are concerned with transverse vibrations on the string, the relevant restoring force (per unit length) is given by the string tension (force along the string axis) times the curvature of the string, or $Ky''(t,x)$; the restoring force is balanced at all times by the inertial force per unit length of the string which is equal to mass density (mass per unit length) times transverse acceleration, *i.e.*, $\epsilon\ddot{y}(t,x)$. See Appendix B for a review of basic physical concepts. The wave equation is derived in some detail in §B.6 on page 500.

6.1.2 Wave Equation Applications

The ideal-string wave equation applies to any perfectly elastic medium which is displaced along one dimension. For example, the air column of a clarinet or organ pipe can be modeled using the one-dimensional wave equation by substituting air-pressure deviation for string displacement, and longitudinal volume velocity for transverse string velocity. We refer to the general class of such media as *one-dimensional waveguides.* Extensions to two and three dimensions (and more, for the mathematically curious), are also possible (see §C.14 on page 605) [519, 521, 55].

For a physical string model, at least three coupled waveguide models should be considered. Two correspond to transverse-wave vibrations in the horizontal and vertical planes (two *polarizations* of planar vibration); the third corresponds to *longitudinal waves.* For bowed strings, *torsional waves* should also be considered, since they affect bow-string dynamics [308, 421]. In the piano, for key ranges in which the hammer strikes three strings simultaneously, *nine* coupled waveguides are required per key for a complete simulation (not including torsional waves); however, in a practical, high-quality, virtual piano, one waveguide per coupled string (modeling only the vertical, transverse plane) suffices quite well [42, 43]. It is difficult to get by with fewer than the correct number of strings, however, because their detuning determines the entire amplitude envelope as well as beating and aftersound effects [544].

6.1.3 Traveling-Wave Solution

It can be readily checked (see §C.3 on page 526 for details) that the lossless 1D wave equation

$$Ky'' = \epsilon\ddot{y}$$

(where all terms are defined in Eq. (6.1)) is solved by *any string shape* which travels to the left or right with speed

$$c \triangleq \sqrt{\frac{K}{\epsilon}}.$$

If we denote right-going traveling waves in general by $y_r(t - x/c)$ and left-going traveling waves by $y_l(t + x/c)$, where y_r and y_l are arbitrary twice-differentiable functions, then the general class of solutions to the lossless, one-dimensional, second-order wave equation can be expressed as

$$y(t, x) = y_r(t - x/c) + y_l(t + x/c). \tag{6.2}$$

Note that we have $\ddot{y}_r = c^2 y_r''$ and $\ddot{y}_l = c^2 y_l''$ (derived in §C.3.1) showing that the wave equation is satisfied for all traveling wave shapes y_r and y_l.

However, the derivation of the wave equation itself assumes the string slope $|y'|$ is much less than 1 at all times and positions (see §B.6). An important point to note is that a function of two variables $y(t,x)$ is replaced by two functions of a single (time) variable. This leads to great reductions in computational complexity, as we will see. The traveling-wave solution of the wave equation was first published by d'Alembert in 1747 [100][1]

6.1.4 Sampled Traveling-Wave Solution

As discussed in more detail in Appendix C, the continuous traveling-wave solution to the wave equation given in Eq. (6.2) can be *sampled* to yield

$$\begin{aligned} y(nT,mX) &= y_r(nT-mX/c)+y_l(nT+mX/c) \qquad (\text{set } X=cT) \\ &= y_r(nT-mT)+y_l(nT+mT) \\ &\triangleq y^+(n-m)+y^-(n+m), \end{aligned} \tag{6.3}$$

where T denotes the time sampling interval in seconds, $X = cT$ denotes the spatial sampling interval in meters, and y^+ and y^- are defined for notational convenience.

6.1.5 Wave Impedance

A concept of high practical utility is that of *wave impedance*, defined for vibrating strings as force divided by velocity. As derived in §C.7.2 on page 554, the relevant *force* quantity in this case is minus the string tension times the string slope:

$$f(t,x) \triangleq -Ky'(t,x) \tag{6.4}$$

Physically, this can be regarded as the transverse force *acting to the right* on the string in the vertical direction. (Only transverse vibration is being considered.) In other words, the vertical component of a negative string slope pulls "up" on the segment of string to the right, and "up" is the positive direction for displacement, velocity, and now force. The traveling-wave decomposition of the force into *force waves* is thus given by (see §C.7.2

[1]See §A.1 on page 445 for more about the history of the wave equation and its traveling-wave solution.

on page 554 for a more detailed derivation)[2]

$$\begin{aligned} f(t,x) &= f_r(t-x/c)+f_l(t+x/c) \\ &= -Ky_r'(t-x/c)-Ky_l'(t+x/c) \\ &= \frac{K}{c}\dot{y}_r(t-x/c)-\frac{K}{c}\dot{y}_l(t+x/c) \\ &\triangleq Rv_r(t-x/c)-Rv_l(t+x/c), \end{aligned}$$

where we have defined the new notation $v = \dot{y}$ for transverse velocity, and

$$R \triangleq \frac{K}{c} \triangleq \frac{K}{\sqrt{K/\epsilon}} = \sqrt{K\epsilon},$$

where K is the string tension and ϵ is mass density. The newly defined positive constant $R = \sqrt{K\epsilon}$ is called the *wave impedance* of the string for transverse waves. It is always real and positive for the ideal string. Three expressions for the wave impedance are

$$\boxed{R \triangleq \sqrt{K\epsilon} = \frac{K}{c} = \epsilon c.} \tag{6.5}$$

The wave impedance simply relates force and velocity traveling waves:

$$\boxed{\begin{aligned} f^+(n) &= Rv^+(n) \\ f^-(n) &= -Rv^-(n) \end{aligned}} \tag{6.6}$$

These relations may be called *Ohm's law for traveling waves*. Thus, in a traveling wave, force is always *in phase* with velocity (considering the minus sign in the left-going case to be associated with the direction of travel rather than a 180 degrees phase shift between force and velocity).

The results of this section are derived in more detail in Appendix C. However, all we need in practice for now are the important Ohm's law relations for traveling-wave components given in Eq. (6.6).

---

[2]Note that f_r and f_l, as defined, are traveling-wave components of the force *acting to the right* on the string. That is, their sum $f(t,x) = f_r + f_l$ is physically the force that the string-segment to the *left* of position x applies (in the upward direction) to the string-segment to the *right* of point x. In other words, $f(t,x)$ denotes the vertical force applied by the left segment to the right segment at time t and position x; thus, it "acts to the right", even though its traveling-wave components, f_l and f_r, travel to the left and right, respectively, at speed c. The net physical force acting to the right at each point is exactly canceled by an equal and opposite force acting to the left at each point x. See §C.7.2 on page 554 for a detailed derivation.

6.2 Ideal Acoustic Tube

As discussed in §C.7.3 on page 555, the most commonly used digital waveguide variables ("wave variables") for *acoustic tube* simulation are traveling *pressure* and *volume-velocity* samples. These variables are exactly analogous to the traveling force and transverse-velocity waves used for vibrating string models.

The Ohm's law relations for acoustic-tube wave variables may be written as follows (cf. Eq. (6.6)):

$$\begin{aligned} p^+(n) &= R_{\mathrm{T}}\, u^+(n) \\ p^-(n) &= - R_{\mathrm{T}}\, u^-(n) \end{aligned} \tag{6.7}$$

Here $p^+(n)$ is the right-going traveling *longitudinal pressure wave* component, $p^-(n)$ is the left-going pressure wave, and $u^{\pm}(n)$ are the left- and right-going *volume velocity waves.* For acoustic tubes, the wave impedance R_{T} is given by

$$R_{\mathrm{T}} = \frac{\rho c}{A} \qquad \text{(Acoustic-Tube Wave Impedance)} \tag{6.8}$$

where ρ denotes the density (mass per unit volume) of air, c is sound speed in air, and A is the cross-sectional area of the tube.

In this formulation, the acoustic tube is assumed to contain only *traveling plane waves* to the left and right. This is a reasonable assumption for wavelengths λ much larger than the tube diameter ($\approx \sqrt{A}$). In this case, a change in the tube cross-sectional area A along the tube axis will cause lossless *scattering* of incident plane waves. That is, the plane wave splits into a transmitted and reflected component such that wave energy is conserved (see Appendix C for a detailed derivation).

Figure 6.2 shows a piecewise cylindrical tube model of the vocal tract and a corresponding digital simulation [245, 297]. In the figure, k_1 denotes the *reflection coefficient* associated with the first tube junction (where the cross-sectional area changes), and $1 + k_1$ is the corresponding *transmission coefficient* for traveling *pressure* plane waves. The corresponding reflection and transmission coefficients for *volume velocity* are $-k_1$ and $1 - k_1$, respectively. Again, see Appendix C for a complete derivation.

At higher frequencies, those for which $\lambda \ll \sqrt{A}$, changes in the tube cross-sectional area A give rise to *mode conversion* (which we will neglect in this chapter). Mode conversion means that an incident plane wave (the simplest mode of propagation in the tube) generally scatters into waves traveling in many directions, not just the two directions along the tube axis. Furthermore, even along the tube axis, there are higher orders of mode propagation associated with "node lines" in the transverse plane (such as

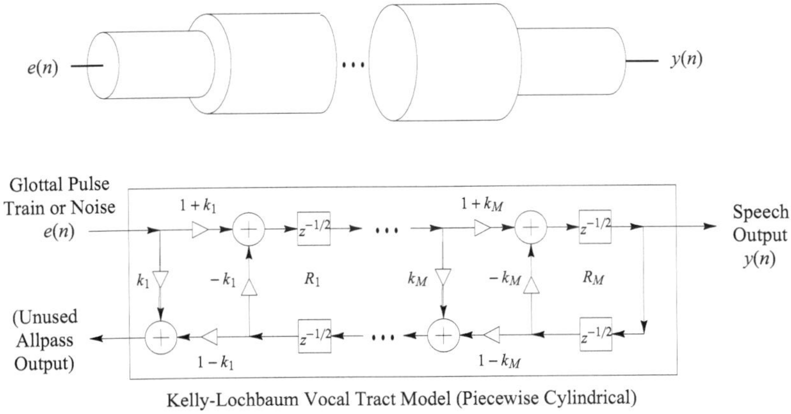

Figure 6.2: Kelly-Lochbaum piecewise cylindrical tube model of the vocal tract (top) and sampled traveling-wave formulation (bottom).

Bessel functions of integer order [542]). When mode conversion occurs, it is necessary to keep track of many components in a more general *modal expansion* of the acoustic field [336, 13, 50]. We may say that when a plane wave encounters a change in the cross-sectional tube area, it is "converted" into a sum of propagation modes. The coefficients (amplitude and phase) of the new modes are typically found by matching boundary conditions. (Pressure and volume-velocity must be continuous throughout the tube.)

As mentioned above, in acoustic tubes we work with *volume velocity*, because it is volume velocity that is conserved when a wave propagates from one tube section to another. For plane waves in open air, on the other hand, we use *particle velocity*, and in this case, the wave impedance of open air is $R_0 = \rho c$ instead. That is, the appropriate wave impedance in open air (not inside an acoustic tube) is pressure divided by particle-velocity for any traveling plane wave. If $u^+(n)$ denotes a sample of the volume-velocity plane-wave traveling to the right in an acoustic tube of cross-sectional area A, and if $v^+(n)$ denotes the corresponding particle velocity, then we have

$$u^+(n) \;=\; v^+(n)\,A.$$

Note that particle velocity is in units of meters per second, while volume velocity is in units of meters-cubed per second (literally a volume of flow per unit time—hence the name). In summary, particle velocity is the appropriate velocity for simulations of waves in open air, while volume velocity is the right choice for acoustic tubes, or "ducts," of varying cross-sectional area. See §B.7.2 on page 504 and Appendix C for further discussion.

6.3 Rigid Terminations

A *rigid termination* is the simplest case of a string (or tube) termination. It imposes the constraint that the string (or air) cannot move at the termination. (We'll look at the more practical case of a *yielding termination* in §9.2.1 on page 343.) If we terminate a length L ideal string at $x = 0$ and $x = L$, we then have the "boundary conditions"

$$y(t, 0) \equiv 0 \qquad y(t, L) \equiv 0 \tag{6.9}$$

where "≡" means "identically equal to," *i.e.*, equal for all t. Let $N \stackrel{\Delta}{=} 2L/X$ denote the time in samples to propagate from one end of the string to the other and back, or the total "string loop" delay. The loop delay N is also equal to twice the number of spatial samples along the string.

Applying the traveling-wave decomposition from Eq. (6.2), we have

$$\begin{aligned} y(nT, 0) &= y^+(n) + y^-(n) \equiv 0 \\ y(nT, NX/2) &= y^+(n - N/2) + y^-(n + N/2) \equiv 0. \end{aligned}$$

Therefore, solving for the reflected waves gives

$$y^+(n) = -y^-(n) \tag{6.10}$$

$$y^-(n + N/2) = -y^+(n - N/2). \tag{6.11}$$

A digital simulation diagram for the rigidly terminated ideal string is shown in Fig. 6.3. A "virtual pickup" is shown at the arbitrary location $x = \xi$.

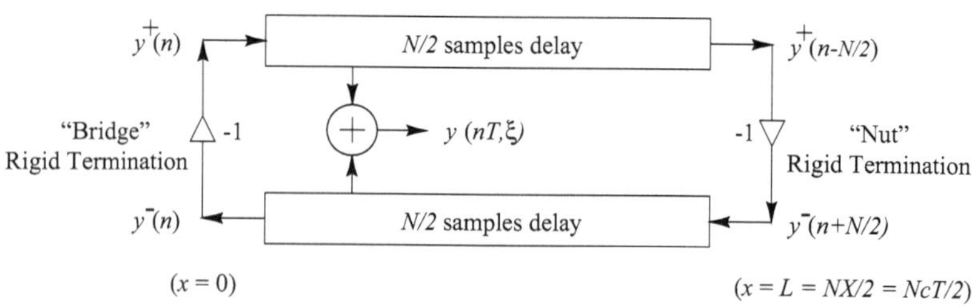

Figure 6.3: The rigidly terminated ideal string, with a displacement output indicated at position $x = \xi$. Rigid terminations reflect traveling displacement, velocity, or acceleration waves with a sign inversion. Slope or force waves reflect with no sign inversion.

6.3.1 Velocity Waves at a Rigid Termination

Since the displacement is always zero at a rigid termination, the velocity is also zero there:

$$v(t,0) \equiv 0 \qquad v(t,L) \equiv 0$$

Therefore, velocity waves reflect from a rigid termination with a sign flip, just like displacement waves:

$$\begin{aligned} v^+(n) &= -v^-(n) \\ v^-(n+N/2) &= -v^+(n-N/2) \end{aligned} \tag{6.12}$$

Such inverting reflections for velocity waves at a rigid termination are identical for models of vibrating strings and acoustic tubes.

6.3.2 Force or Pressure Waves at a Rigid Termination

To find out how force or pressure waves recoil from a rigid termination, we may convert velocity waves to force or velocity waves by means of the Ohm's law relations of Eq. (6.6) for strings (or Eq. (6.7) on page 208 for acoustic tubes), and then use Eq. (6.12), and then Eq. (6.6) again:

$$\begin{aligned} f^+(n) &= Rv^+(n) = -Rv^-(n) = f^-(n) \\ f^-(n+N/2) &= -Rv^-(n+N/2) = Rv^+(n-N/2) = f^+(n-N/2) \end{aligned}$$

Thus, force (and pressure) waves reflect from a rigid termination with *no sign inversion*:[3]

$$\begin{aligned} f^+(n) &= f^-(n) \\ f^-(n+N/2) &= f^+(n-N/2) \end{aligned}$$

The reflections from a rigid termination in a digital-waveguide acoustic-tube simulation are exactly analogous:

$$\begin{aligned} p^+(n) &= p^-(n) \\ p^-(n+N/2) &= p^+(n-N/2) \end{aligned}$$

[3] All wave phenomena involve two physical state variables—one force-like and the other velocity-like. When one of these variables reflects from a termination with a sign inversion, the other reflects with no sign inversion, and vice versa. In acoustic systems, the force-like variable is pressure, and the velocity-like variable is either particle velocity (in open air) or volume velocity (in acoustic tubes), as described in §6.2 above. In electromagnetic systems, the state variables are electric and magnetic field strengths or voltage and current. In a mass-spring oscillator, we may choose the velocity and acceleration of the mass as the coordinates of our state space, or position and velocity. For transverse waves on vibrating strings, it is usually preferable to use force and velocity waves as described above.

Waveguide terminations in acoustic stringed and wind instruments are never perfectly rigid. However, they are typically *passive*, which means that waves at each frequency see a reflection coefficient not exceeding 1 in magnitude. Aspects of passive "yielding" terminations are discussed in §C.11.

6.4 Moving Rigid Termination

It is instructive to study the "waveguide equivalent circuit" of the simple case of a rigidly terminated ideal string with its left endpoint being moved by an external force, as shown in Fig. 6.4. This case is relevant to bowed strings (§9.6) since, during time intervals in which the bow and string are stuck together, the bow provides a termination that divides the string into two largely isolated segments. The bow can therefore be regarded as a moving termination during "sticking".

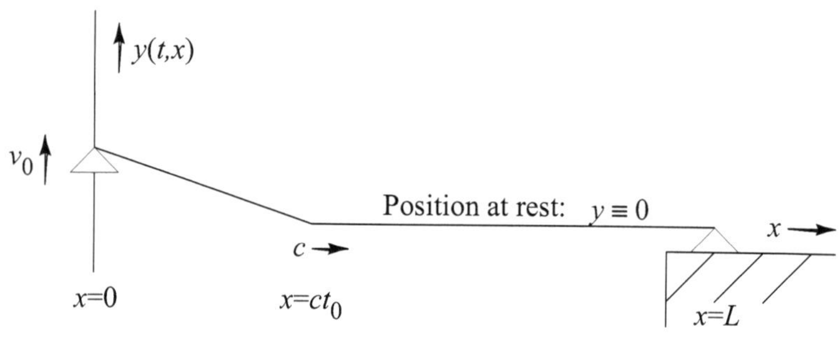

Figure 6.4: Moving rigid termination for an ideal string at time t_0.

Referring to Fig. 6.4, the left termination of the rigidly terminated ideal string is set in motion at time $t = 0$ with a constant velocity v_0. From Eq. (6.5) on page 207, the wave impedance of the ideal string is $R = \sqrt{K\epsilon}$, where K is tension and ϵ is mass density. Therefore, the upward force applied by the moving termination is initially $f_0 = Rv_0$. At time $t_0 < L/c$, the traveling disturbance reaches a distance ct_0 from $x = 0$ along the string. Note that the string slope at the moving termination is given by $-v_0t_0/(ct_0) = -v_0/c = -(f_0/R)/c = -f_0/K$, which derives the fact that force waves are minus tension times slope waves. (See §C.7.2 on page 554 for a fuller discussion.)

6.4.1 Digital Waveguide Equivalent Circuits

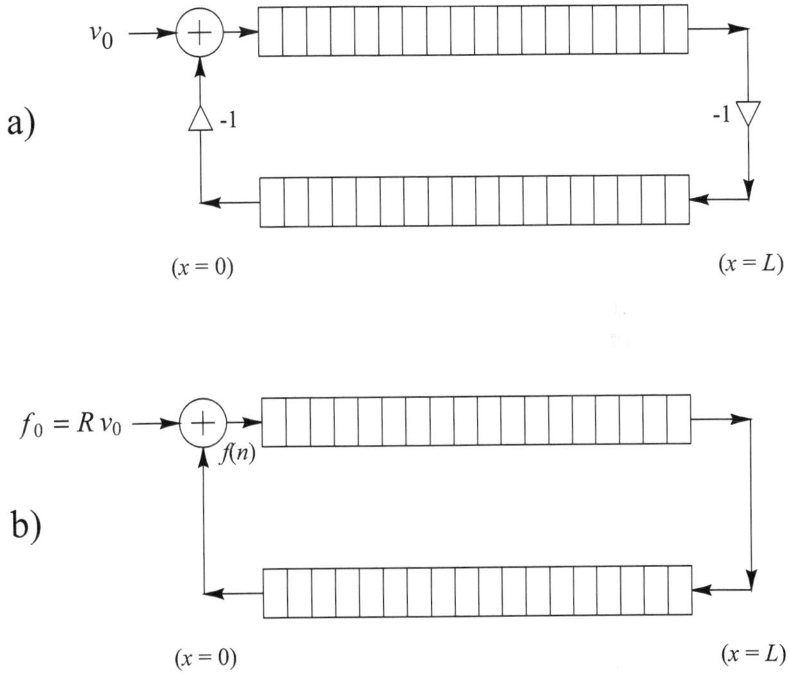

Figure 6.5: Digital waveguide "equivalent circuits" for the rigidly terminated ideal string with a moving termination. a) Velocity-wave simulation. b) Force-wave simulation.

Two digital waveguide "equivalent circuits" are shown in Fig. 6.5. In the velocity-wave case of Fig. 6.5a, the termination motion appears as an additive injection of a constant velocity v_0 at the far left of the digital waveguide. At time 0, this initiates a velocity step from 0 to v_0 traveling to the right. When the traveling step-wave reaches the right termination, it reflects with a sign inversion, thus sending back a "canceling wave" to the left. Behind the canceling wave, the velocity is zero, and the string is not moving. When the canceling step-wave reaches the left termination, it is inverted again and added to the externally injected dc signal, thereby sending an amplitude $2v_0$ positive step-wave to the right, overwriting the amplitude v_0 signal in the upper rail. This can be added to the amplitude $-v_0$ signal in the lower rail to produce a net traveling velocity step of amplitude v_0 traveling to the right. This process repeats forever, resulting in traveling wave components which grow without bound, but whose sum is always either 0 or v_0. Thus, at all times the string can be divided into two segments, where the segment to the left is moving upward with speed v_0, and the segment to the right is motionless.

At this point, it is a good exercise to try to mentally picture the string shape during this process: Initially, since both the left end support and the right-going velocity step are moving with constant velocity c, it is clear that the string shape is piece-wise linear, with a negative-slope segment on the left adjoined to a zero-slope segment on the right. When the velocity step reaches the right termination and reflects to produce a canceling wave, everything to the left of this wave remains a straight line which continues to move upward at speed v_0, while all points to the right of the canceling wave's leading edge are not moving. What is the shape of this part of the string? (The answer is given in the next paragraph, but try to "see" it first.)

6.4.2 Animation of Moving String Termination and Digital Waveguide Models

In the force wave simulation of Fig. 6.5b,[4] the termination motion appears as an additive injection of a constant force $f_0 = Rv_0$ at the far left. At time 0, this initiates a force step from 0 to f_0 traveling to the right. Since force waves are negated slope waves multiplied by tension, *i.e.*, $f^+ = -Ky'^+$, the slope of the string behind the traveling force step is $y' = -f_0/K$. When the traveling step-wave reaches the right termination, it reflects with *no* sign inversion, thus sending back a doubling-wave to the left which elevates the string force from f_0 to $2f_0$. Behind this wave, the slope is then $y' = -2f_0/K$. This answers the question of the previous paragraph: The string is in fact piecewise linear during the first return reflection, consisting of two line segments with slope $-f_0/K$ on the left, and twice that on the right. When the return step-wave reaches the left termination, it is reflected again and added to the externally injected dc force signal, sending an amplitude $2f_0$ positive step-wave to the right (overwriting the amplitude f_0 signal in the upper rail). This can be added to the amplitude f_0 samples in the lower rail to produce a net traveling force step in the string of amplitude $3f_0$ traveling to the right. The slope of the string behind this wave is $y' = -3f_0/K$, and the slope in front of this wave is still $-2f_0/K$. The force applied to the string by the termination has risen to $3f_0$ in order to keep the velocity steady at v_0. (We may interpret the f_0 input as the

[4]This diagram can be seen *animated* along with Figure 6.4 at `http://ccrma.stanford.edu/~jos/swgt/movet.html` .

additional force needed each period to keep the termination moving at speed v_0—see the next paragraph below.) This process repeats forever, resulting in traveling wave components which grow without bound, and whose sum (which is proportional to minus the physical string slope) also grows without bound.[5] The string is always piecewise linear, consisting of at most two linear segments having negative slopes which differ by $-f_0/K$. A sequence of string displacement snapshots is shown in Fig. 6.6.

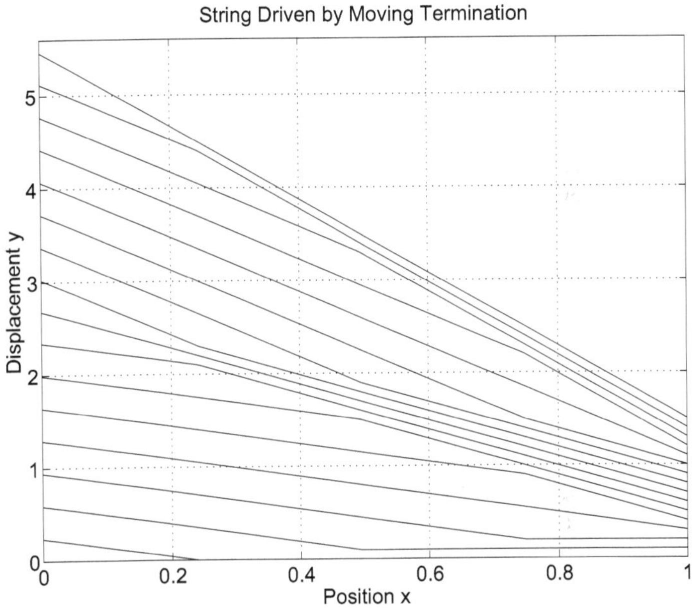

Figure 6.6: Successive snapshots of the rigidly terminated ideal string with a moving termination. For clarity, the string is plotted higher on each successive snapshot. (One can consider both endpoints to be moving at the same speed up to time 0, after which the left termination begins moving faster by a constant velocity offset.)

6.4.3 Terminated String Impedance

Note that the impedance of the *terminated* string, seen from one of its endpoints, is not the same thing as the wave impedance $R = \sqrt{K\epsilon}$ of the

[5] Our model becomes invalid as the slope becomes large. In particular, the string tension K obviously increases as the string length increases. Here we assume K is constant.

string itself. If the string is infinitely long, they are the same. However, when there are *reflections*, they must be included in the impedance calculation, giving it an imaginary part. We may say that the impedance has a "reactive" component. The driving-point impedance of a rigidly terminated string is "purely reactive," and may be called a *reactance* (§7.1). If $f(t)$ denotes the force at the driving-point of the string and $v(t)$ denotes its velocity, then the driving-point impedance is given by (§7.1)

$$R(j\omega) \triangleq \left.\frac{F(s)}{V(s)}\right|_{s=j\omega},$$

where $F(s)$ and $V(s)$ denote the Laplace transforms of $f(t)$ and $v(t)$. In the case of a rigidly terminated string above, as well as in any system in which all energy delivered to the system is ultimately reflected back to the input, the impedance is purely imaginary at every frequency (a "pure reactance"), as is easy to show:

$$R(s) \triangleq \frac{F(s)}{V(s)} = \frac{F^+ + F^-}{V^+ + V^-} = \frac{F^+ + e^{-s2L/c}F^+}{V^+ - e^{-s2L/c}V^-} = R\frac{1+e^{-s2L/c}}{1-e^{-s2L/c}}$$

where L denotes the string length. Let $P = 2L/c$ denote the period of string vibration. Then on the frequency axis $s = j\omega$ we have

$$R(j\omega) = R\frac{1+e^{-j\omega P}}{1-e^{-j\omega P}} = R\frac{2\cos(\omega P/2)}{2j\sin(\omega P/2)} = -jR\cot(\omega P/2).$$

Thus, the driving-point impedance of a rigidly terminated string is purely reactive (imaginary), with alternating poles and zeros along the $j\omega$ axis. Impedance will be discussed further in §7.1 below.

6.5 The Ideal Plucked String

The ideal *plucked string* is defined as an initial string displacement and a zero initial velocity distribution [317]. More generally, the initial displacement along the string $y(0, x)$ and the initial velocity distribution $\dot{y}(0, x)$, for all x, fully determine the resulting motion in the absence of further excitation.

An example of the appearance of the traveling-wave components and the resulting string shape shortly after plucking a doubly terminated string at a point one fourth along its length is shown in Fig. 6.7. The negative traveling-wave portions can be thought of as inverted reflections of the incident waves, or as doubly flipped "images" which are coming from the other side of the terminations.

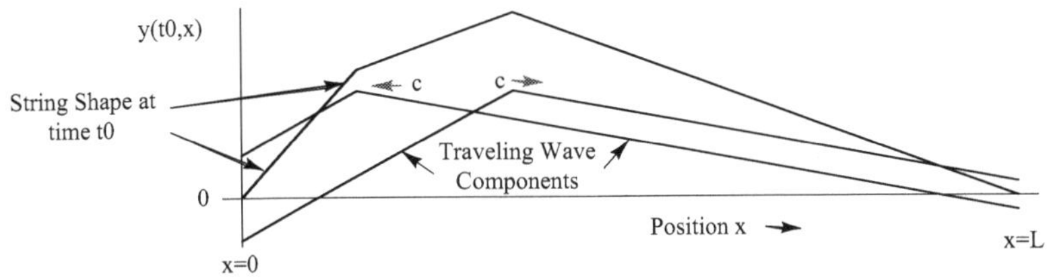

Figure 6.7: A doubly terminated string, "plucked" at 1/4 its length.

An example of an initial "pluck" excitation in a digital waveguide string model is shown in Fig. 6.8. The circulating triangular components in Fig. 6.8 are equivalent to the infinite train of initial images coming in from the left and right in Fig. 6.7.

There is one fine-point to note for the discrete-time case: We cannot admit a sharp corner in the string since that would have infinite bandwidth which would alias when sampled. Therefore, for the discrete-time case, we define the ideal pluck to consist of an arbitrary shape as in Fig. 6.8 *lowpass filtered* to less than half the sampling rate. Alternatively, we can simply require the initial displacement shape to be bandlimited to spatial frequencies less than $f_s/2c$. Since all real strings have some degree of stiffness which prevents the formation of perfectly sharp corners, and since real plectra are never in contact with the string at only one point, and since the frequencies we do allow span the full range of human hearing, the bandlimited restriction is not limiting in any practical sense.

Note that acceleration (or curvature) waves are a simple choice for plucked string simulation, since the ideal pluck corresponds to an initial *impulse* in the delay lines at the pluck point. Of course, since we require a bandlimited excitation, the initial acceleration distribution will be replaced by the impulse response of the anti-aliasing filter chosen. If the anti-aliasing filter chosen is the ideal lowpass filter cutting off at $f_s/2$, the initial acceleration $a(0,x) \triangleq \ddot{y}(0,x)$ for the ideal pluck becomes

$$a(0,x) = \frac{A}{X}\text{sinc}\left(\frac{x-x_p}{X}\right) \tag{6.13}$$

where A is amplitude, x_p is the pick position, and, as we know from §4.4.1 on page 170, $\text{sinc}[(x-x_p)/X]$ is the ideal, bandlimited impulse, centered at x_p and having a rectangular spatial frequency response extending from $-\pi/X$ to π/X. (Recall that $\text{sinc}(\xi) \triangleq \sin(\pi\xi)/(\pi\xi)$). Division by X normalizes the area under the initial shape curve. If x_p is chosen to lie exactly on a

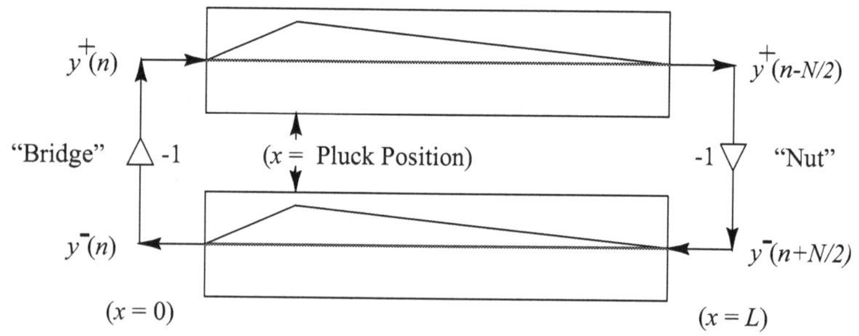

Figure 6.8: Initial conditions for the ideal plucked string. The initial contents of the sampled, traveling-wave delay lines are in effect *plotted* inside the delay-line boxes. The amplitude of each traveling-wave delay line is *half* the amplitude of the initial string displacement. The sum of the upper and lower delay lines gives the physical initial string displacement.

spatial sample $x_m = mX$, the initial conditions for the ideal plucked string are as shown in Fig. 6.9 for the case of acceleration or curvature waves. All initial samples are zero except one in each delay line.

Aside from its obvious simplicity, there are two important benefits of obtaining an impulse-excited model: (1) an extremely efficient "commuted synthesis" algorithm can be readily defined (§8.7), and (2) linear prediction (and its relatives) can be readily used to calibrate the model to recordings of normally played tones on the modeled instrument. Linear Predictive Coding (LPC) has been used extensively in speech modeling [296, 297, 20]. LPC estimates the model filter coefficients under the assumption that the driving signal is *spectrally flat.* This assumption is valid when the input signal is (1) an impulse, or (2) white noise. In the basic LPC model for voiced speech, a periodic impulse train excites the model filter (which functions as the vocal tract), and for unvoiced speech, white noise is used as input.

In addition to plucked and struck strings, simplified *bowed strings* can be calibrated to recorded data as well using LPC [428, 439]. In this simplified model, the bowed string is approximated as a periodically plucked string.

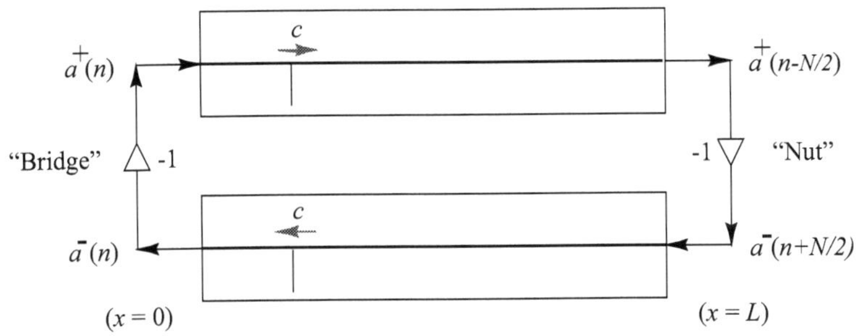

Figure 6.9: Initial conditions for the ideal plucked string when the wave variables are chosen to be proportional to acceleration or curvature. If the bandlimited ideal pluck position is centered on a spatial sample, there is only a single nonzero sample in each of the initial delay lines.

6.6 The Ideal Struck String

The ideal *struck string* [317] can be modeled by a *zero* initial string displacement and a nonzero initial *velocity* distribution. In concept, an inelastic "hammer strike" transfers an "impulse" of momentum to the string at time 0 along the striking face of the hammer. (A more realistic model of a struck string will be discussed in §9.3.1 on page 359.) An example of "struck" initial conditions is shown in Fig. 6.10 for a striking hammer having a rectangular shape. Since $v^{\pm} = \pm f^{\pm}/R = \mp c y'^{\pm}$,[6] the initial velocity distribution can be integrated with respect to x from $x = 0$, divided by c, and negated in the upper rail to obtain equivalent initial displacement waves [317]. Interestingly, the initial displacement waves are not local (see also Appendix E).

The hammer strike itself may be considered to take zero time in the ideal case. A finite spatial width must be admitted for the hammer, however, even in the ideal case, because a zero width and a nonzero momentum transfer sends one (massless) point of the string immediately to infinity under infinite acceleration. In a discrete-time simulation, one sample rep-

[6] In somewhat more detail,

$$\begin{aligned} v^+ &= \frac{f^+}{R} \triangleq \frac{-Ky'^+}{\sqrt{K\epsilon}} = -cy'^+ \\ v^- &= -\frac{f^-}{R} \triangleq \frac{Ky'^-}{\sqrt{K\epsilon}} = cy'^- \end{aligned}$$

so that $v = v^+ + v^- = c(y'^- - y'^+)$, and $y = c\int(y'^- - y'^+)dx$.

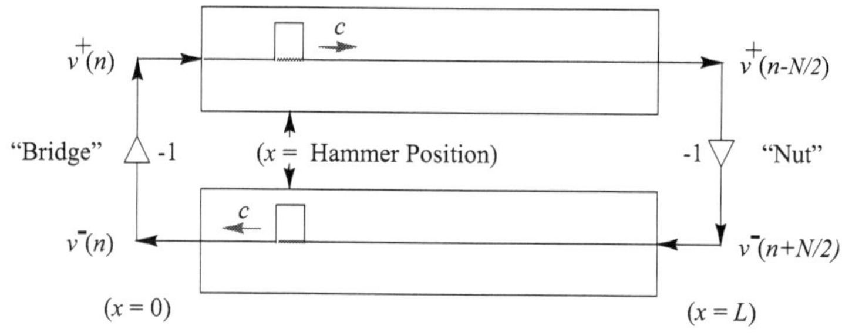

Figure 6.10: Initial conditions for the ideal struck string in a *velocity wave* simulation.

resents an entire sampling interval, so a one-sample hammer width is well defined.

If the hammer velocity is v_h, then the force against the hammer due to pushing against the string wave impedance is $-2Rv_h$. The factor of 2 arises because driving a point in the string's interior is equivalent to driving two string endpoints in "series," *i.e.*, the reaction forces sum. If the hammer is itself a dynamic system which has been "thrown" into the string, as discussed in §9.3.1 below, the reaction force slows the hammer over time, and the interaction is not impulsive, but rather the momentum transfer takes place over a period of time.

The hammer-string collision is ideally *inelastic* since the string provides a reaction force that is equivalent to that of a dashpot. In the case of a pure mass striking a single point on the ideal string, the mass velocity decays exponentially, and an exponential wavefront emanates in both directions. In the musical acoustics literature for the piano, the hammer is often taken to be a nonlinear spring in series with a mass, as discuss further in §9.3.2. A commuted waveguide piano model including a linearized piano hammer is described in §9.4-§9.4.4. "Wave digital hammer" models, which employ a traveling-wave formulation of a lumped model and therefore analogous to a wave digital filter [136], are described in [524, 56, 42]. The "wave digital" modeling approach is introduced in §F.1 on page 701.

6.7 The Damped Plucked String

Without damping, the ideal plucked string sounds more like a cheap electronic organ than a string because the sound is perfectly periodic and never decays. Static spectra are very boring to the ear. The discrete Fourier transform (DFT) of the initial "string loop" contents gives the Fourier series coefficients for the periodic tone produced.

The simplest change to the ideal wave equation of Eq. (6.1) on page 204 that provides damping is to add a term proportional to velocity:

$$Ky'' = \epsilon\ddot{y} + \mu\dot{y} \tag{6.14}$$

Here, $\mu > 0$ can be thought of as a very simple friction coefficient, or resistance. As derived in §C.5 on page 538, solutions to this wave equation can be expressed as sums of left- and right-going *exponentially decaying* traveling waves. When $\mu = 0$, we get non-decaying traveling waves as before. As discussed in §2.2.2 on page 39, propagation losses may be introduced by the substitution

$$z^{-1} \leftarrow gz^{-1}, \quad |g| \leq 1$$

in each delay element (or wherever one sample of delay models one spatial sample of wave propagation). By commutativity of LTI systems, making the above substitution in a delay line of length N is equivalent to simply scaling the output of the delay line by g^N. This *lumping* of propagation loss at one point along the waveguide serves to minimize both computational cost and round-off error. In general finite difference schemes, such a simplification is usually either not possible or nonobvious.

Computational Savings

To illustrate how significant the computational savings can be, consider the simulation of a "damped guitar string" model in Fig. 6.11. For simplicity, the length L string is rigidly terminated on both ends. Let the string be "plucked" by initial conditions so that we need not couple an input mechanism to the string. Also, let the output be simply the signal passing through a particular delay element rather than the more realistic summation of opposite elements in the bidirectional delay line. (A comb filter corresponding to pluck position can be added in series later.)

In this string simulator, there is a loop of delay containing $N = 2L/X = f_s/f_1$ samples where f_1 is the desired pitch of the string. Because there is no input/output coupling, we may lump *all* of the losses at a single point in the delay loop. Furthermore, the two reflecting terminations (gain factors of -1) may be commuted so as to cancel them. Finally, the right-going delay may be combined with the left-going delay to give a single, length N, delay line. The result of these inaudible simplifications is shown in Fig. 6.12.

If the sampling rate is $f_s = 50$ kHz and the desired pitch is $f_1 = 100$ Hz, the loop delay equals $N = 500$ samples. Since delay lines are efficiently

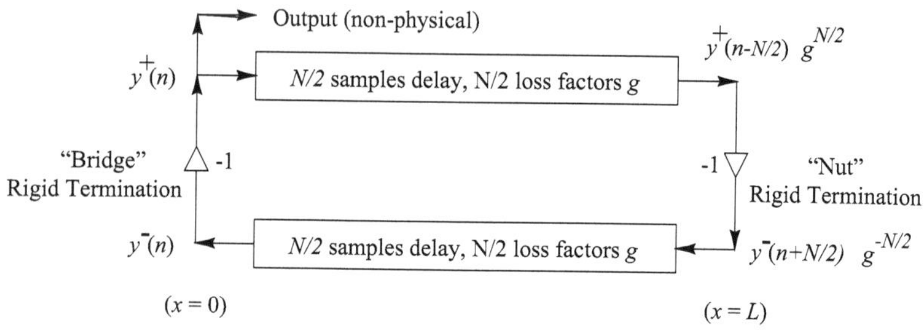

Figure 6.11: Discrete simulation of the rigidly terminated string with distributed resistive losses. The N loss factors g are embedded between the delay-line elements.

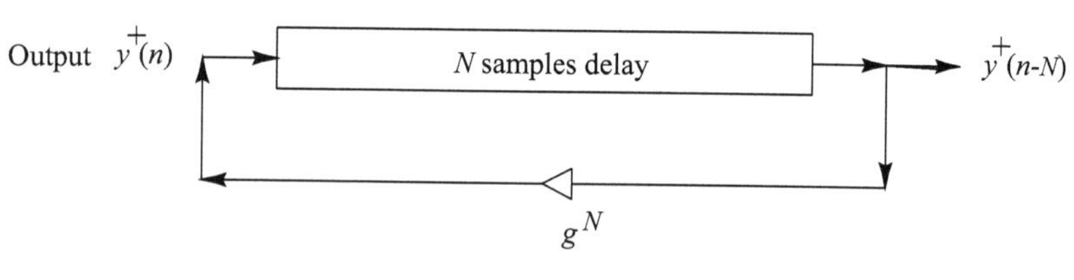

Figure 6.12: Discrete simulation of the rigidly terminated string with *consolidated* losses (frequency-independent). All N loss factors g have been "pushed" through delay elements and combined at a *single* point.

implemented as circular buffers, the cost of implementation is normally dominated by the loss factors, each one requiring a multiply every sample, in general. (Losses of the form $1-2^{-k}$, $1-2^{-k}-2^{-l}$, etc., can be efficiently implemented using shifts and adds.) Thus, the consolidation of loss factors has reduced computational complexity by *three orders of magnitude, i.e.*, by a factor of 500 in this case. However, the physical accuracy of the simulation has not been compromised. In fact, the *accuracy is improved* because the N round-off errors per period arising from repeated multiplication by g have been replaced by a single round-off error per period in the multiplication by g^N.

6.8 Frequency-Dependent Damping

In real vibrating strings, damping typically increases with frequency for a variety of physical reasons [73, 77]. A simple modification [392] to Eq. (6.14) on page 221 yielding frequency-dependent damping is

$$Ky'' = \epsilon\ddot{y} + \mu\dot{y} + \mu_2\dot{y}''.$$

See §C.5.2 on page 541 for some analysis of damped PDEs of this nature. A book-chapter by Vallette containing a thorough analysis of damped vibrating strings is given in [518].

The result of adding such damping terms to the wave equation is that traveling waves on the string decay at *frequency-dependent rates*. This means the loss factors g of the previous section should really be *digital filters* having gains which decrease with frequency (and never exceed 1 for stability of the loop). These filters *commute* with delay elements because they are *linear* and *time invariant* (LTI) [449]. Thus, following the reasoning of the previous section, they can be lumped at a single point in the digital waveguide. Let $\hat{G}(z)$ denote the resulting *string loop filter* (replacing g^N in Fig. 6.12[7]). We have the stability (passivity) constraint $|\hat{G}(e^{j\omega T})| \leq 1$, and making the filter *linear phase* (constant delay at all frequencies) will restrict consideration to *symmetric* FIR filters only.

Restriction to FIR filters yields the important advantage of keeping the approximation problem *convex* in the weighted least-squares norm. Convexity of a norm means that gradient-based search techniques can be used to find a global miminizer of the error norm without exhaustive search [64],[428, Appendix A].

The linear-phase requirement halves the degrees of freedom in the filter coefficients. That is, given $\hat{g}(n)$ for $n \geq 0$, the coefficients $\hat{g}(-n) = \hat{g}(n)$

[7]We should use the notation $\hat{G}_N(z)$ for this loop-filter, since it depends on the string length $N/2$ (in samples). The dependence of $\hat{G}(z)$ on N is suppressed for simplicity of notation.

are also determined. The loss-filter frequency response can be written in terms of its (impulse response) coefficients as

$$\hat{G}(e^{j\omega T}) = \hat{g}(0) + 2\sum_{n=1}^{(N_{\hat{g}}-1)/2} \hat{g}(n)\cos(\omega nT)$$

where $\omega T = 2\pi fT = 2\pi f/f_s$, and the impulse-response length $N_{\hat{g}}$ is assumed odd.

A further degree of freedom is eliminated from the loss-filter approximation by assuming all losses are insignificant at 0 Hz so that $G(0) = 1 \Rightarrow \hat{G}(1) = 1$ (taking the approximation error to be zero at $\omega = 0$). This means the coefficients of the FIR filter must sum to 1. If the length of the filter is $N_{\hat{g}}$, we have

$$1 = \sum_n \hat{g}(n) = \hat{g}(0) + 2\sum_{n=1}^{(N_{\hat{g}}-1)/2} \hat{g}(n)$$

6.9 The Stiff String

Stiffness in a vibrating string introduces a restoring force proportional to the *bending angle* of the string. As discussed further in §C.6 on page 545, the usual *stiffness term* added to the wave equation for the ideal string yields

$$\epsilon\ddot{y} = Ky'' - \kappa y''''.$$

When this wave equation is solved in terms of traveling waves (§C.6), it emerges that high-frequency wave components travel *faster* than low-frequency components. In other words, wave propagation in stiff strings is *dispersive.* (See §C.6 for details.)

Stiff-string models are commonly used in *piano synthesis*. In §9.4, further details of string models used in piano synthesis are described (§9.4.1 on page 383).

Experiments with modified recordings of acoustic classical guitars indicate that overtone inharmonicity due to string-stiffness is generally not audible in nylon-string guitars, although just-noticeable-differences are possible for the 6th (lowest) string [225]. Such experiments may be carried out by retuning the partial overtones in a recorded sound sample so that they become exact harmonics. Such retuning is straightforward using *sinusoidal modeling* techniques [359, 457].

6.9.1 Stiff String Synthesis Models

An ideal stiff-string synthesis model is drawn in Fig. 6.13 [10]. See §C.6 on page 545 for a detailed derivation. The delay-line length N is the number

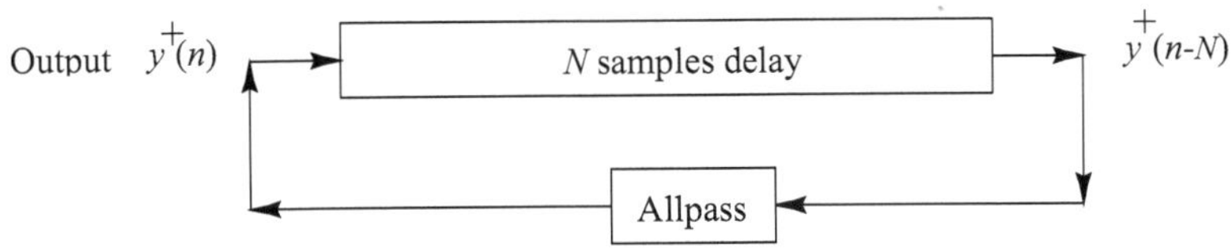

Figure 6.13: Commuted simulation of a rigidly terminated, stiff string (no damping).

of samples in K periods at frequency f_K, where K is the number of the highest partial supported (normally the last one before $f_s/2$). This is the counterpart of Fig. 6.12 which depicted ideal-string damping which was *lumped* at a single point in the delay-line loop. For the ideal stiff string, however, (no damping), it is *dispersion filtering* that is lumped at a single point of the loop. Dispersion can be lumped like damping because it, too, is a linear, time-invariant (LTI) filtering of a propagating wave. Because it is LTI, dispersion-filtering *commutes* with other LTI systems in series, such as delay elements. The allpass filter in Fig. C.9 corresponds to filter $H_s(z)$ in Fig. 9.2 on page 336 for the Extended Karplus-Strong algorithm. In practice, losses are also included for realistic string behavior (filter $H_d(z)$ in Fig. 9.2).

Allpass filters were introduced in §2.8 on page 69, and a fairly comprehensive summary is given in Book II of this series [449, Appendix C].[8] The general transfer function for an allpass filter is given (in the real, single-input, single-output case) by

$$H_s(z) = z^{-K}\frac{\tilde{A}(z)}{A(z)}$$

where $K \geq 0$ is an integer pure-delay in samples (all delay lines are allpass filters),

$$A(z) \triangleq 1 + a_1 z^{-1} + a_2 z^{-2} + \cdots + a_N z^{-N},$$

and

$$\tilde{A}(z) \triangleq z^{-N}\overline{A}(z^{-1}) = \overline{a_N} + \overline{a_{N-1}}z^{-1} + \overline{a_{N-2}}z^{-2} + \cdots + \overline{a_1}z^{-N+1} + z^{-N}.$$

We may think of $\tilde{A}(z)$ as the *flip* of $A(z)$. For example, if $A(z) = 1 + 1.4z^{-1} + 0.49z^{-2}$, we have $\tilde{A}(z) = 0.49 + 1.4z^{-1} + z^{-2}$. Thus, $\tilde{A}(z)$ is obtained from $A(z)$ by simply reversing the order of the coefficients (and

[8] `http://ccrma.stanford.edu/~jos/filters/Allpass_Filters.html`

conjugating them if they are complex, but normally they are real in practice). For an allpass filter $H_s(z)$ simulating stiffness, we would normally have $K = 0$, since the filter is already in series with a delay line.

Section 6.11 below discusses some methods for designing stiffness allpass filters $H_s(z)$ from measurements of stiff vibrating strings, and §9.4.1 gives further details for the case of piano string modeling.

6.10 The Externally Excited String

Sections 6.5 and 6.6 illustrated plucking or striking the string by means of *initial conditions:* an initial displacement for plucking and an initial velocity distribution for striking. Such a description parallels that found in many textbooks on acoustics, such as [317]. However, if the string is already in motion, as it often is in normal usage, it is more natural to excite the string externally by the equivalent of a "pick" or "hammer" as is done in the real world instrument.

Figure 6.14 depicts a rigidly terminated string with an external excitation input. The wave variable w can be set to acceleration, velocity, or displacement, as appropriate. (Choosing force waves would require eliminating the sign inversions at the terminations.) The external input is denoted Δw to indicate that it is an additive incremental input, superimposing with the existing string state.

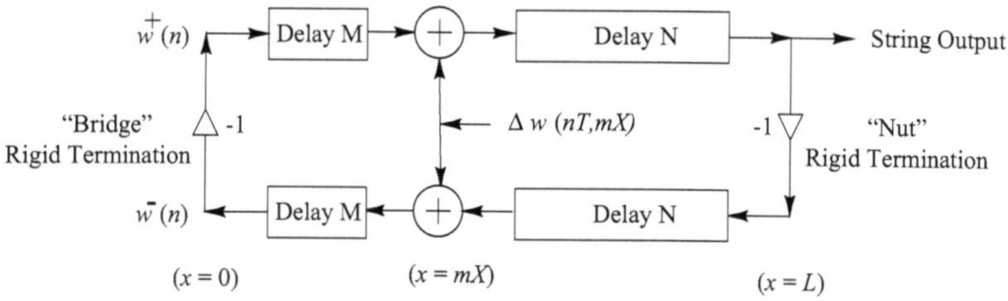

Figure 6.14: Discrete simulation of the rigidly terminated string with an external excitation.

For idealized plucked strings, we may take $w = a$ (acceleration), and Δw can be a single nonzero sample, or impulse, at the plucking instant. As always, bandlimited interpolation can be used to provide a non-integer time or position. In the latter case, there would be two or more summers along both the upper and lower rails, separated by unit delays. More generally, the string may be plucked by a *force distribution* $f_p(t_n, x_m)$. The applied force at a point can be translated to the corresponding velocity increment via the wave impedance R:

$$\Delta v = \frac{f_p}{2R} \tag{6.15}$$

where $R = \sqrt{K\epsilon}$ as before. The factor of two comes from the fact that two string endpoints are being driven physically in parallel. (Physically, they are in parallel, but as impedances, they are formally in series. See §7.2 on page 257 for the theory.)

Note that the force applied by a rigid, stationary pick or hammer varies with the state of a vibrating string. Also, when a pick or hammer makes contact with the string, it partially *terminates* the string, resulting in reflected waves in each direction. A simple model for the termination would be a mass affixed to the string at the excitation point. (This model is pursued in §9.3.1, p. 359.) A more general model would be an arbitrary impedance and force source affixed to the string at the excitation point during the excitation event. This is a special case of the "loaded waveguide junction," discussed in §C.12 on page 598. In the waveguide model for bowed strings (§9.6), the bow-string interface is modeled as a nonlinear scattering junction.

6.10.1 Equivalent Forms

In a physical piano string, as a specific example, the hammer strikes the string between its two inputs, some distance from the agraffe and far from the bridge. This corresponds to the diagram in Fig. 6.15, where the delay lines are again arranged for clarity of physical interpretation. Figure 6.15 is almost identical to Fig. 6.14, except that the delay lines now contain samples of traveling *force waves*, and the bridge is allowed to vibrate, resulting in a *filtered reflection* at the bridge (see §9.2.1 for a derivation of the bridge filter). The hammer-string interaction force-pulse is summed into both the left- and right-going delay lines, corresponding to sending the same pulse toward both ends of the string from the hammer. Force waves are discussed further in §C.7.2 on page 554.

By commutativity of linear, time-invariant elements, Figure 6.15 can be immediately simplified to the form shown in Fig. 6.16, in which each delay line corresponds to the travel time in *both* directions on each string segment. From a structural point of view, we have a conventional filtered delay loop plus a second input which sums into the loop somewhere inside the delay line. The output is shown coming from the middle of the larger delay line, which gives physically correct timing, but in practice, the output can be taken from anywhere in the feedback loop. It is probably preferable in practice to take the output from the loop-delay-line input. That way, other response latencies in the overall system can be compensated.

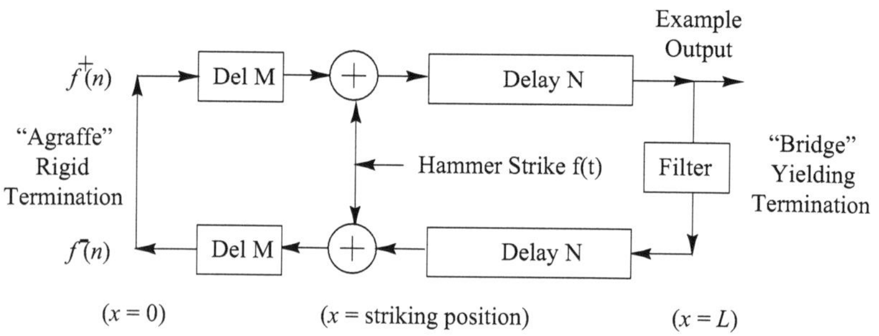

Figure 6.15: Model of a piano string struck in its interior by a hammer.

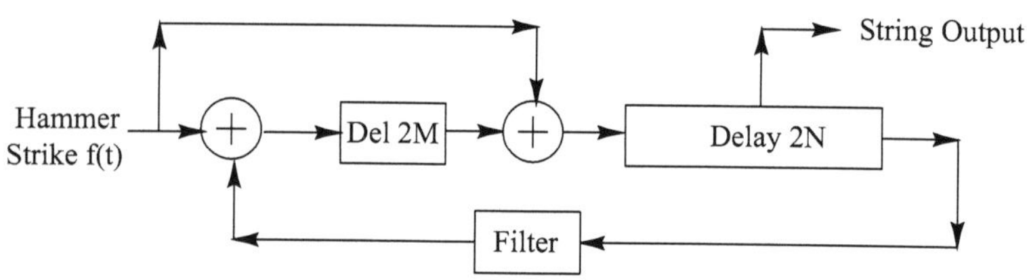

Figure 6.16: Diagram equivalent to Fig. 6.15, obtained by combining upper- and lower-rail delay lines.

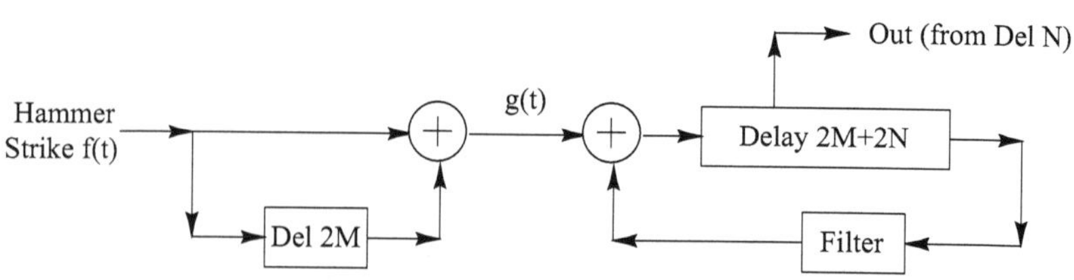

Figure 6.17: Diagram equivalent to Fig. 6.15, obtained by replacing the second string input by a separate comb-filter applied to a single input.

An alternate structure equivalent to Fig. 6.16 is shown in Fig. 6.17, in which the second input injection is factored out into a separate comb-filtering of the input. The comb-filter delay equals the delay between the two inputs in Fig. 6.16, and the delay in the feedback loop equals the sum of both delays in Fig. 6.16. In this case, the string is modeled using a simple filtered delay loop, and the striking-force signal is separately filtered by a comb filter corresponding to the striking-point along the string.

Algebraic derivation

The above equivalent forms are readily verified by deriving the transfer function from the striking-force input $f_i(n)$ to the output force signal $f_o(n)$

Referring to Fig. 6.15 on page 228, denote the input hammer-strike z transform by $F_i(z)$ and the output signal z transform by $F_o(z)$. Also denote the loop-filter transfer function by $H_l(z)$. By inspection of the figure, we can write

$$F_o(z) = z^{-N}\left\{F_i(z) + z^{-2M}\left[F_i(z) + z^{-N}H_l(z)F_o(z)\right]\right\}.$$

Solving for the input-output transfer function yields

$$\begin{aligned} H(z) \triangleq \frac{F_o(z)}{F_i(z)} &= z^{-N}\frac{1+z^{-2M}}{1-H_l(z)\,z^{-(2M+2N)}} \\ &= \left(1+z^{-2M}\right)\frac{z^{-N}}{1-H_l(z)\,z^{-(2M+2N)}} \end{aligned}$$

The final factored form above corresponds to the final equivalent form shown in Fig. 6.17 on page 228.

Related Forms

We see from the preceding example that a filtered-delay loop (a feedback comb-filter using filtered feedback, with delay-line length $2M + 2N$ in the above example) simulates a vibrating string in a manner that is *independent* of where the excitation is applied. To simulate the effect of a particular excitation point, a feedforward comb-filter may be placed in series with the filtered delay loop. Such a "pluck position" illusion may be applied to any basic string synthesis algorithm, such as the EKS [428, 207].

By an exactly analogous derivation, a single feedforward comb filter can be used to simulate the location of a linearized *magnetic pickup* [200] on a simulated electric guitar string. An ideal pickup is formally the *transpose* of an excitation. For a discussion of filter transposition (using Mason's gain theorem [301, 302]), see, *e.g.*, [333, 449].[9]

[9] `http://ccrma.stanford.edu/~jos/filters/Transposed_Direct_Forms.html`

The comb filtering can of course also be implemented *after* the filtered delay loop, again by commutativity. This may be desirable in situations in which comb filtering is one of many options provided for in the "effects section" of a synthesizer. Post-processing comb filters are often used in reverberator design and in virtual pickup simulation.

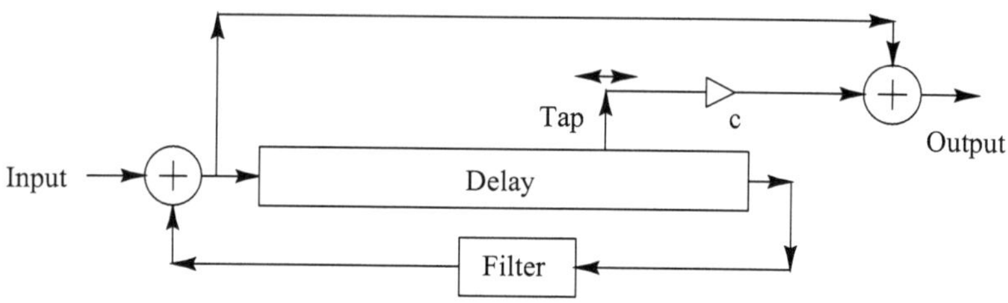

Figure 6.18: Use of a second delay-line tap to implement comb filtering corresponding to hammer-strike echoes returning from the far end of the string.

The comb-filtering can also be conveniently implemented using a second tap from the appropriate delay element in the filtered delay loop simulation of the string, as depicted in Fig. 6.18. The new tap output is simply summed (or differenced, depending on loop implementation) with the filtered delay loop output. Note that making the new tap a moving, interpolating tap (*e.g.*, using linear interpolation), a *flanging* effect is available. The tap-gain c can be brought out as a musically useful timbre control that goes beyond precise physical simulation (*e.g.*, it can be made negative). Adding more moving taps and summing/differencing their outputs, with optional scale factors, provides an economical *chorus* or *Leslie* effect. These extra delay effects cost no extra memory since they utilize the memory that's already needed for the string simulation. While such effects are not traditionally applied to piano sounds, they are applied to electric piano sounds which can also be simulated using the same basic technique.

Summary

In summary, two feedforward comb filters and one feedback comb filter arranged in series (in any order) can be interpreted as a physical model of a vibrating string driven at one point along the string and observed at a different point along the string. The two feedforward comb-filter delays correspond to the excitation and pickup locations along the string, while the amount of feedback-loop delay controls the fundamental frequency of vibration. A filter in the feedback loop determines the decay rate and fine tuning of the partial overtones.

6.11 Loop Filter Identification

In §6.7 we discussed damping filters for vibrating string models, and in §6.9 we discussed dispersion filters. For vibrating strings which are well described by a linear, time-invariant (LTI) partial differential equation, damping and dispersion filtering are the only deviations possible from the ideal string string discussed in §6.1.

The ideal damping filter is "zero phase" (or linear phase) [449],[10] while the ideal dispersion filter is "allpass" (as described in §6.9.1 on page 224). Since every desired frequency response can be decomposed into a zero-phase frequency-response in series with an allpass frequency-response, we may design a single loop filter whose amplitude response gives the desired damping as a function of frequency, and whose phase response gives the desired dispersion vs. frequency. The next subsection summarizes some methods based on this approach. The following two subsections discuss methods for the design of damping and dispersion filters separately.

6.11.1 General Loop-Filter Design

For general loop-filter design in vibrating string models (as well as in woodwind and brass instrument bore models), we wish to minimize [428, pp. 182–184]:

1. Errors in partial overtone *decay times*
2. Errors in partial overtone *tuning*

The partial decay times are controlled by the filter's amplitude response, while the partial tunings are determined by the filter's phase response. There also may be other filters in the loop (such as a delay-line interpolation filter) which need to be considered when designing the main loop filter.

There are numerous methods for designing the string loop filter $\hat{G}(z)$ based on measurements of real string behavior. In [428], a variety of methods for *system identification* [288] were explored for this purpose, including "periodic linear prediction" in which a linear combination of a small group of samples is used to predict a sample one period away from the midpoint of the group. An approach based on a genetic algorithm is described in [378]; in that work, the error measure used with the genetic algorithm is based on properties of human perception of short-time spectra, as is now standard

[10] `http://ccrma.stanford.edu/~jos/filters/Filters_Preserving_Phase.html`

practice in digital audio coding [62]. Overviews of other approaches appear in [29] and [509].

Below is an outline of a simple and effective method used (ca. 1995) to design loop filters for some of the Sondius sound examples:[11]

1. Estimate the fundamental frequency (see §6.11.4 on page 234 below)
2. Set a Hamming FFT-window length to approximately four periods
3. Compute the short-time Fourier transform (STFT)
4. Perform a sinusoidal modeling analysis [467] to
   - detect peaks in each spectral frame, and
   - connect peaks through time to form *amplitude envelopes*
5. Fit an exponential to each amplitude envelope
6. Prepare the desired frequency-response, sampled at the harmonic frequencies of the delay-line loop *without* the loop filter. At each harmonic frequency,
   - the nearest-partial decay-rate gives the desired loop-filter gains,
   - the nearest-partial peak-frequency give the desired loop-filter phase delay.
7. Use a phase-sensitive filter-design method such as `invfreqz` in matlab to design the desired loop filter from its frequency-response samples (further details below).

Physically, amplitude envelopes are expected to decay *exponentially*, although coupling phenomena may obscure the overall exponential trend. On a dB scale, exponential decay is a *straight line*. Therefore, a simple method of estimating the exponential decay time-constant for each overtone frequency is to fit a straight line to its amplitude envelope and use the slope of the fitted line to compute the decay time-constant. For example, the matlab function `polyfit` can be used for this purpose (where the requested polynomial order is set to 1). Since "beating" is typical in the amplitude envelopes, a refinement is to replace the raw amplitude envelope by a piecewise linear envelope that connects the upper local maxima in the raw amplitude envelope. The estimated decay-rate for each overtone determines a sample of the loop-filter amplitude response at the overtone frequency. Similarly, the overtone frequency determines a sample of the loop-filter phase response.

[11] `http://ccrma.stanford.edu/~jos/pasp/Sound_Examples.html`

Taken together, the measured overtone decay rates and tunings determine samples of the complex frequency response of the desired loop filter. The matlab function `invfreqz`[12] can be used to convert these complex samples into recursive filter coefficients (see §8.6.4, p. 297 for a related example). A weighting function inversely proportional to frequency is recommended. Additionally, Steiglitz-McBride iterations can improve the results [287], [428, pp. 101–103]. Matlab's version of `invfreqz` has an iteration-count argument for specifying the number of Steiglitz-McBride iterations. The maximum filter-gain versus frequency should be computed, and the filter should be renormalized, if necessary, to ensure that its gain does not exceed 1 at any frequency; one good setting is that which matches the overall decay rate of the original recording.

6.11.2 Damping Filter Design

When dispersion can be neglected (as it typically can in many cases, such as for guitar strings), we may use *linear-phase filter design* methods, such as provided by the functions `remez`, `firls`, and `fir2` in matlab. Since strings are usually very lightly damped, such linear-phase filter designs give high quality at very low orders. Another approach is to fit a low-order IIR filter, such as by applying `invfreqz` to a *minimum-phase* version of the desired amplitude response, and then subtract the phase-response of the resulting filter from the desired phase-response used in a subsequent allpass design. (Alternatively, the phase response of the loop filter can simply be neglected, as long as tuning is unaffected. If tuning is affected, the tuning allpass can be adjusted to compensate.)

6.11.3 Dispersion Filter Design

A pure *dispersion filter* is an ideal allpass filter. That is, it has a gain of 1 at all frequencies and only delays a signal in a frequency-dependent manner. The need for such filtering in *piano string models* is discussed in §9.4.1 on page 384.

There is a good amount of literature on the topic of *allpass filter design*. Generally, they fall into the categories of *optimized parametric*, *closed-form parametric*, and *nonparametric* methods. Optimized parametric methods can produce allpass filters with optimal group-delay characteristics in some sense [272, 271]. Closed-form parametric methods provide coefficient formulas as a function of a desired parameter such as "inharmonicity" [368]. Nonparametric methods are generally based on measured signals and/or

[12]In Matlab, the Signal Processing Tool Box is required, and in Octave, the `octave-signal` MacPort package is needed. The Linux `octave` package already includes this (at least on Fedora 13).

spectra, and while they are suboptimal, they can be used to design very large-order allpass filters, and the errors can usually be made arbitrarily small by increasing the order [552, 369, 42, 41, 1], [428, pp. 60,172]. In music applications, it is usually the case that the "optimality" criterion is unknown because it depends on aspects of sound perception (see, for example, [211, 384]). As a result, perceptually weighted nonparametric methods can often outperform optimal parametric methods in terms of cost/performance [2].

In historical order, some of the allpass filter-design methods are as follows: A modification of the method in [552] was suggested for designing allpass filters having a phase delay corresponding to the delay profile needed for a stiff string simulation [428, pp. 60,172]. The method of [552] was streamlined in [369]. In [77], piano strings were modeled using finite-difference techniques. An update on this approach appears in [45]. In [340], high quality stiff-string sounds were demonstrated using high-order allpass filters in a digital waveguide model. In [384], this work was extended by applying a least-squares allpass-design method [272] and a spectral Bark-warping technique [460] to the problem of calibrating an allpass filter of arbitrary order to recorded piano strings. They were able to correctly tune the first several tens of partials for any natural piano string with a total allpass order of 20 or less. Additionally, minimization of the L^∞ norm [271] has been used to calibrate a series of allpass-filter sections [42, 41], and a dynamically tunable method, based on Thiran's closed-form, maximally flat group-delay allpass filter design formulas (§4.3 on page 166), was proposed in [368]. An improved closed-form solution appears in [1] based on an elementary method for the robust design of very high-order allpass filters.

6.11.4 Fundamental Frequency Estimation

As mentioned in §6.11.2 above, it is advisable to estimate the fundamental frequency of vibration (often called "*F0*") in order that the partial overtones are well resolved while maintaining maximum time resolution for estimating the decay time-constant.

Below is a summary of the F0 estimation method used in calibrating loop filters with good results [472]:

- Take an FFT of the middle third of a recorded plucked string tone.
- Find the frequencies and amplitudes of the largest K peaks, where K is chosen so that the K retained peaks all have a reasonable signal-to-noise ratio.
- Form a histogram of peak spacing Δf_i

- The pitch estimate $\hat{f}_0$ is defined as the most common spacing Δf_i in the histogram.

The F0 estimate so formed is sufficiently accurate for purposes of calibrating a peak tracker for decay-rate estimation. (The F0 value is only used to set the FFT window length in the STFT to four nominal periods under the window, so it is not critical.)

Approximate Maximum Likelihood F0 Estimation

In applications for which the fundamental frequency F0 must be measured very accurately in a periodic signal, the estimate $\hat{f}_0$ obtained by the above algorithm can be refined using a gradient search which matches a so-called "harmonic comb" to the magnitude spectrum of an interpolated FFT $X(\omega)$:

$$\hat{f}_0 \triangleq \arg\max_{\hat{f}_0} \sum_{k=1}^{K} \log\left[\left|X(k\hat{f}_0)\right| + \epsilon\right] = \arg\max_{\hat{f}_0} \prod_{k=1}^{K} \left[\left|X(k\hat{f}_0)\right| + \epsilon\right]$$

where

$$\begin{aligned} K &= \text{number of peaks} \\ k &= \text{harmonic number of } k\text{th peak (valid method for non-stiff strings)} \\ \epsilon &= \text{small value on the order of spectral magnitude } \textit{noise floor level} \end{aligned}$$

Note that freely vibrating strings are not exactly periodic due to exponenential decay, coupling effects, and stiffness (which stretches harmonics into quasiharmonic overtones, as explained in §6.9 on page 224). However, non-stiff strings can often be analyzed as having approximately harmonic spectra ($\leftrightarrow$ periodic time waveform) over a limited time frame.

Since string spectra typically exhibit harmonically spaced *nulls* associated with the excitation and/or observation points, as well as from other phenomena such as recording multipath and/or reverberation, it is advisable to restrict K to a range that does not include any spectral nulls (or simply omit index k when $k\hat{f}_0$ is too close to a spectral null), since even one spectral null can push the product of peak amplitudes to a very small value. As a practical matter, it is important to inspect the magnitude spectra of the data manually to ensure that a robust row of peaks is being matched by the harmonic comb. For example, a display of the frame magnitude spectrum overlaid with vertical lines at the optimized harmonic-comb frequencies yields an effective picture of the F0 estimate in which typical problems (such as octave errors) are readily seen.

References on F0 Estimation

An often-cited book on classical methods for pitch detection, particularly for voice, is that by Hess [192]. The harmonic comb method can be considered an approximate maximum-likelihood pitch estimator, and more accurate maximum-likelihood methods have been worked out [114, 548, 376, 377]. More recently, Klapuri has been developing some promising methods for multiple pitch estimation [254, 253, 252].[13] A comparison of real-time pitch-tracking algorithms applied to guitar is given in [260], with consideration of latency (time delay).

Extension to Stiff Strings

An advantage of the harmonic-comb method, as well as other frequency-domain maximum-likelihood pitch-estimation methods, is that it is easily extended to accommodate stiff strings. For this, the stretch-factor in the spectral-peak center-frequencies can be estimated—the so-called coefficient of inharmonicity, and then the harmonic-comb (or other maximum-likelihood spectral-matching template) can be stretched by the same amount, so that when set to the correct pitch, the template matches the data spectrum more accurately than if harmonicity is assumed.

6.11.5 EDR-Based Loop-Filter Design

This section discusses use of the Energy Decay Relief (EDR) (§3.2.2 on page 91) to measure the decay times of the partial overtones in a recorded vibrating string.

First we derive what to expect in the case of a simplified string model along the lines discussed in §6.7 above. Assume we have the synthesis model of Fig. 6.12 on page 222, where now the loss factor g^N is replaced by the digital filter $H_l(z)$ that we wish to design. Let $\underline{x}(n)$ denote the contents of the delay line as a vector at time n, with $\underline{x}(0)$ denoting the initial contents of the delay line.

For simplicity, we define the EDR based on a length N DFT of the delay-line vector $\underline{x}$, and use a rectangular window with a "hop size" of N samples, *i.e.*,

$$\underline{X}_m(\omega_k) \isdef \text{DFT}_{N,\omega_k}\{\underline{x}_m\}, \quad m = 0, 1, 2, \ldots,$$

where $\underline{x}_m(n) \isdef \underline{x}(mN)$. That is $\underline{x}_m$ is simply the mth successive snapshot of the delay-line contents, where the snapshots are taken once every N samples. We may interpret $\underline{X}_m$ as mth short-time spectrum of the output

[13] Klapuri's publication home page: `http://www.cs.tut.fi/~klap/iiro/`

signal $y^+(n)$ shown in Fig. 6.12. Due to the special structure of our synthesis model, we have

$$\underline{X}_m(\omega_k) = H_l^m(\omega_k)\underline{X}_0(\omega_k)$$

for each DFT bin number $k \in [0, N-1]$.

Applying the definition of the EDR (§3.2.2) to this short-time spectrum gives

$$\begin{aligned} E_m(\omega_k) &\triangleq \sum_{\nu=m}^{\infty} \left|\underline{X}_\nu(\omega_k)\right|^2 = \sum_{\nu=m}^{\infty} \left|H_l^\nu(\omega_k)\underline{X}_0(\omega_k)\right|^2 \\ &= \left|\underline{X}_0(\omega_k)\right|^2 \sum_{\nu=m}^{\infty} \left|H_l(\omega_k)\right|^{2\nu} = \left|\underline{X}_0(\omega_k)\right|^2 \frac{\left|H_l(\omega_k)\right|^{2m}}{1-\left|H_l(\omega_k)\right|^2} \\ &= \frac{\left|\underline{X}_0(\omega_k)\right|^2}{1-\left|H_l(\omega_k)\right|^2} \left|H_l(\omega_k)\right|^{2m}. \end{aligned}$$

We therefore have the following recursion for successive EDR time-slices:[14]

$$E_{m+1}(\omega_k) = \left|H_l(\omega_k)\right|^2 E_m(\omega_k)$$

Since we normally try to fit straight-line decays to the EDR on a log scale (typically a decibel scale), we will see the relation

$$\log(E_{m+1}) = \log(E_m) + \log(|H_l|^2),$$

where the common argument ω_k is dropped for notational simplicity. Since we require $|H_l(\omega_k)| < 1$ for stability of the filtered-delay loop, the EDR decays monotonically in this example. Thus, the measured slope of the partial overtone decays will be found to be proportional to $\log(|H_l|)$.

This analysis can be generalized to a time-varying model in which the loop filter H_l is allowed to change once per "period" N.[15]

An online laboratory exercise covering the practical details of measuring overtone decay-times and designing a corresponding loop filter is given in [280].

---

[14] Note that this derivation also holds if the power '2' is replaced by an arbitrary p, thereby supporting a generalization from the EDR to what might be called the "LPDR" using a kind of L^p norm on the remaining decay, where the EDR is regarded as the L^2 case.

[15] The delay-line length N is only the "quasi-period" in samples when the phase delay associated with $H_l(z)$ can be neglected. N is never a true "period" because the synthesized signal is decaying from one block to the next.

6.12 String Coupling Effects

It turns out that a single digital waveguide provides a relatively static-sounding string synthesizer. This is because several *coupling* mechanisms exist in natural string instruments. The overtones in coupled strings exhibit more interesting amplitude envelopes over time. Coupling between different strings is not the only important coupling phenomenon. In a real string, there are *two* orthogonal planes of transverse vibration which are intrinsically coupled to each other [181]. There is also intrinsic coupling between transverse vibrational waves and longitudinal waves (see §B.6 on page 500).

6.12.1 Horizontal and Vertical Transverse Waves

The transverse waves considered up to now represent string vibration only in a single two-dimensional plane. One such plane can be chosen as being perpendicular to the top plate of a stringed musical instrument. We will call this the xy plane and refer to it as the *vertical plane of polarization* for transverse waves on a string (or simply the *vertical component* of the transverse vibration). To more fully model a real vibrating string, we also need to include transverse waves in the xz plane, *i.e.*, a *horizontal plane of polarization* (or *horizontal component* of vibration). Any polarization for transverse traveling waves can be represented as a linear combination of horizontal and vertical polarizations, and general transverse string vibration in 3D can be expressed as a linear superposition of vibration in any two distinct polarizations.

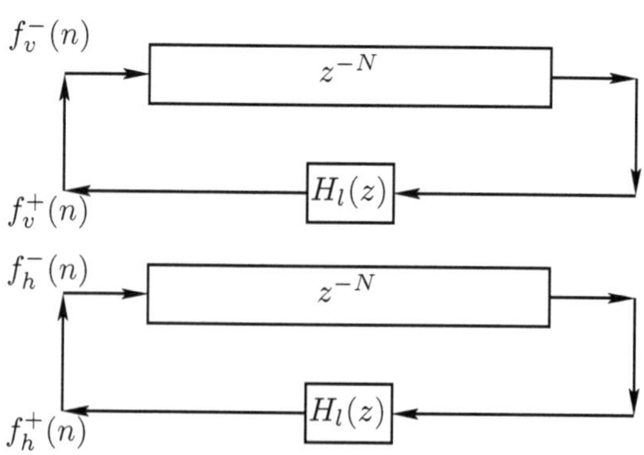

Figure 6.19: Digital waveguide model of a rigidly terminated string vibrating transversally in three-dimensional space (two uncoupled planes of vibration).

If string terminations were perfectly rigid, the horizontal polarization would be largely *independent* of the vertical polarization, and an accurate model would consist of two identical, uncoupled, filtered delay loops (FDL), as depicted in Fig. 6.19. One FDL models vertical force waves while the other models horizontal force waves. This model neglects the small degree of nonlinear coupling between horizontal and vertical traveling waves along the length of the string—valid when the string slope is much less than unity (see §B.6).

Note that the model for two orthogonal planes of vibration on a single string is identical to that for a single plane of vibration on two different strings.

6.12.2 Coupled Horizontal and Vertical Waves

No vibrating string in musical acoustics is truly rigidly terminated, because such a string would produce no sound through the body of the instrument.[16] Yielding terminations result in *coupling* of the horizontal and vertical planes of vibration. In typical acoustic stringed instruments, nearly all of this coupling takes place at the *bridge* of the instrument.

Figure 6.20 illustrates the more realistic case of two planes of vibration which are *linearly coupled* at one end of the string (the "bridge"). Denoting the traveling force waves entering the bridge from the vertical and horizontal vibration components by $F_v^+(z)$ and $F_h^+(z)$, respectively, the outgoing waves in each plane are given by

$$\left[\begin{array}{c} F_v^-(z) \\ F_h^-(z) \end{array}\right] = \left[\begin{array}{cc} H_{vv}(z) & H_{vh}(z) \\ H_{hv}(z) & H_{hh}(z) \end{array}\right] \left[\begin{array}{c} F_v^+(z) \\ F_h^+(z) \end{array}\right] \tag{6.16}$$

as shown in the figure.

In physically symmetric situations, we expect $H_{vh}(z) = H_{hv}(z)$. That is, the transfer function from horizontal to vertical waves is normally the same as the transfer function from vertical to horizontal waves.

If we consider a single frequency ω, then the coupling matrix with $z = e^{j\omega T}$ is a constant (generally complex) matrix (where T denotes the sampling interval as usual). An *eigenanalysis* of this matrix gives information about the *modes* of the coupled system and the *damping* and *tuning* of these modes [544].

As a simple example, suppose the coupling matrix $\mathbf{H}(e^{j\omega T})$ at some frequency has the form

$$\mathbf{H}(e^{j\omega T}) = \left[\begin{array}{cc} A & B \\ B & A \end{array}\right]$$

[16] Electric guitars with magnetic pickups have nearly rigid terminations, but even then, coupling phenomena are clearly observed, especially above the sixth partial or so.

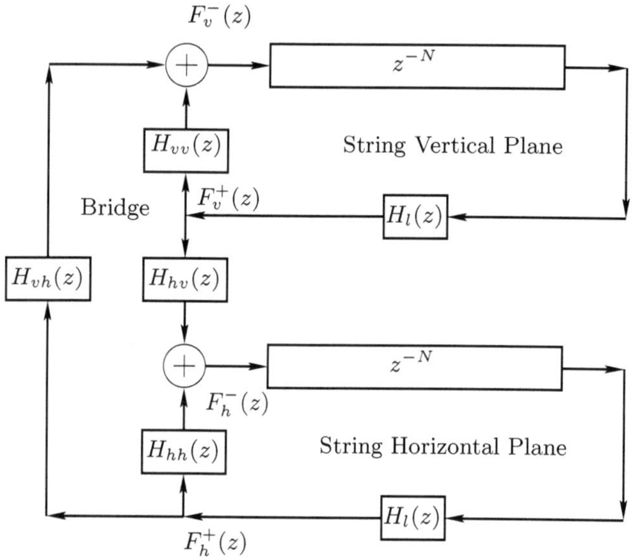

Figure 6.20: Digital waveguide model of a string in which vertical and horizontal planes of vibration are coupled linearly at the bridge.

where A and B are any complex numbers. This means both string terminations are identical, and the coupling is symmetric (the simplest case in practice). The eigenvectors are easily calculated to be[17]

$$\underline{e}_1 = \left[\begin{array}{c} 1 \\ 1 \end{array}\right] \qquad \underline{e}_2 = \left[\begin{array}{c} 1 \\ -1 \end{array}\right],$$

and the eigenvalues are $A+B$ and $A-B$, respectively.

The eigenvector $\underline{e}_T = [1, 1]$ corresponds to "in phase" vibration of the two string endpoints, *i.e.*, $F_v(e^{j\omega T}) = F_h(e^{j\omega T})$, while $\underline{e}_T = [1, -1]$ corresponds to "opposite phase" vibration, for which $F_v(e^{j\omega T}) = -F_h(e^{j\omega T})$. If it happens to be the case that

$$|A+B| < |A-B|$$

then the in-phase vibration component will decay faster than the opposite-phase vibration. This situation applies to coupled piano strings [544], as discussed further below.

More generally, the two *eigenvectors* of the coupling frequency-response

[17] `http://ccrma.stanford.edu/~jos/filters/Finding_Eigenvalues_Practice.html`

matrix

$$\mathbf{H}(e^{j\omega}) \triangleq \begin{bmatrix} H_{vv}(e^{j\omega}) & H_{vh}(e^{j\omega}) \\ H_{hv}(e^{j\omega}) & H_{hh}(e^{j\omega}) \end{bmatrix}$$

correspond to two *decoupled polarization planes*. That is, at each frequency there are two *eigenpolarizations* in which incident vibration reflects in the same plane. In general, the eigenplanes change with frequency. A related analysis is given in [544].

By definition of the eigenvectors of $\mathbf{H}(e^{j\omega T})$, we have

$$\mathbf{H}(e^{j\omega T})\underline{e}_i(e^{j\omega T}) = \lambda_i(e^{j\omega T})\underline{e}_i(e^{j\omega T})$$

where $\lambda_i(e^{j\omega T})$ denotes the ith *eigenvalue* of the coupling-matrix $\mathbf{H}(e^{j\omega T})$ at frequency ω, where $i = 1, 2$. Since the eigenvector $\underline{e}_i(e^{j\omega T})$ holds the Fourier transform of the incoming waves for mode i of the coupled-string system, we see that the eigenvalues have the following interpretation:

> The ith eigenvalue of the coupling matrix equals the *frequency response* seen by the ith eigenpolarization.

In particular, the *modulus* of the eigenvalue gives the reflectance magnitude (affecting mode *damping*), and the *angle* of the eigenvalue is the phase shift of the reflection, for that mode (affecting *tuning* of the mode). Use of coupling matrix eigenanalysis to determine mode damping and tuning is explored further in §C.13 on page 602.

6.12.3 Asymmetry of Horizontal/Vertical Terminations

It is common in real stringed instruments that horizontal and vertical transverse waves are *transduced differently* at the bridge. For example, the bridge on a guitar is typically easier to "push" into the top plate than it is to "shear" sidewise along the top plate. In terms of Eq. (6.16), we have $\left|H_{hh}(e^{j\omega})\right| \gg \left|H_{vv}(e^{j\omega})\right|$ (at most frequencies). This unequal terminating impedance causes the horizontal component of string vibration to *decay slower* than the vertical component of vibration. We can say that this happens because the vertical bridge admittance is much greater than the horizontal admittance, giving rise to a faster rate of energy transfer from the vertical string polarization into the bridge—in other words, the bridge is more "yielding" in the vertical direction. The audible consequence of this unequal rate of decay is a *two-stage amplitude envelope*. The initial fast decay gives a strong onset to the note, while the slower late decay provides a long-lasting sustain—two normally opposing but desirable features.

6.12.4 Coupled Strings

We have just discussed the coupling between vertical and horizontal planes of vibration along a single string. There is also important coupling among different strings on the same instrument. For example, modern pianos are constructed having up to three physical strings associated with each key. These strings are slightly mistuned in order to sculpt the shape of the decay envelope, including its beating characteristics and two-stage decay. A two-stage decay is desired in piano strings in order to provide a strong initial attack followed by a long-sustaining "aftersound" [544], [18, Weinreich chapter].

A simple approximation to the effect of coupled strings is obtained by simply summing two or more slightly detuned strings. While this can provide a realistic beating effect in the amplitude envelope, it does not provide a true two-stage decay. A more realistic simulation of coupling requires signal to flow from each coupled string into all others.

When the bridge moves in response to string vibrations, traveling waves are generated along all other strings attached to the bridge. In the simplest case of a bridge modeled as a rigid body, the generated wave is identical on all strings. In §C.13 on page 602, an efficient scattering formulation of string coupling at a bridge is derived for this case [439]. It can be seen as a simplification of the general coupling matrix shown in Fig. 6.20 for the two-string (or two-polarization) case. Additionally, an eigenanalysis of the coupling matrix is performed, thereby extending the analysis of §6.12.2 above.

6.12.5 Longitudinal Waves

In addition to transverse waves on a string, there are always *longitudinal waves* present as well. In fact, longitudinal waves hold all of the *potential energy* associated with the transverse waves, and they carry the forward *momentum* in the direction of propagation associated with transverse traveling waves [122, 391]. Longitudinal waves in a string typically travel an order of magnitude faster than transverse waves on the same string and are only weakly affected by changes in string tension.

Longitudinal waves are often neglected, *e.g.*, in violin acoustics, because they couple inefficiently to the body through the bridge, and because they are "out of tune" anyway. However, there exist stringed instruments, such as the Finnish Kantele [231], in which longitudinal waves are too important to neglect. In the piano, longitudinal waves are quite audible; to bring this out in a striking way, sound example 5 provided in [18, Conklin chapter] plays *Yankee Doodle* on the longitudinal modes of three piano strings all tuned to the same (transversal) pitch. The nonlinear nature of the coupling from transverse to longitudinal waves has been demonstrated in [163]. Longitudinal waves have been included in some piano synthesis models [30, 28, 24, 23].

6.13 Nonlinear Elements

Many musical instrument models require nonlinear elements, such as

- Amplifier distortion (electric guitar)
- Reed model (woodwinds)
- Bowed string contact friction

Since a nonlinear element generally *expands signal bandwidth*, it can cause *aliasing* in a discrete-time implementation. In the above examples, the nonlinearity also appears inside a *feedback loop*. This means the bandwidth expansion *compounds* over time, causing more and more aliasing.

The topic of nonlinear systems analysis is vast, in part because there is no single analytical approach which is comprehensive. The situation is somewhat analogous to an attempt to characterize "all non-bacterial life". As a result, the only practical approach is to identify useful *classes* of nonlinear systems which are amenable to certain kinds of analysis and characterization. In this chapter, we will look at certain classes of *memoryless* and *passive* nonlinear elements which are often used in digital waveguide modeling.

It is important to keep in mind that a nonlinear element may not be characterized by its impulse response, frequency response, transfer function, or the like. These concepts are only defined, in general, for linear time-invariant systems. However, it is possible to generalize these notions for nonlinear systems using constructs such as Volterra series expansions [52]. However, rather than getting involved with fairly general analysis tools, we will focus instead on approaching each class of nonlinear elements in the manner that best fits that class, with the main goal being to understand its audible effects on discrete-time signals.

6.13.1 Memoryless Nonlinearities

Memoryless or *instantaneous* nonlinearities form the simplest and most commonly implemented form of nonlinear element. Furthermore, many complex nonlinear systems can be broken down into a linear system containing a memoryless nonlinearity.

Given a sampled input signal $x(n)$, the output of any memoryless nonlinearity can be written as

$$y(n) = f(x(n))$$

where $f(\cdot)$ is "some function" which maps numbers to real numbers. We exclude the special case $f(x) = \alpha x$ which defines a simple *linear gain* of α.

The fact that a *function* may be used to describe the nonlinearity implies that each input value is mapped to a unique output value. If it is also true that each output value is mapped to a unique input value, then the function is said to be *one-to-one*, and the mapping is *invertible*. If the function is instead "many-to-one," then the inverse is *ambiguous*, with more than one input value corresponding to the same output value.

Clipping Nonlinearity

A simple example of a *noninvertible* (many-to-one) memoryless nonlinearity is the *clipping* nonlinearity, well known to anyone who records or synthesizes audio signals. In normalized form, the clipping nonlinearity is defined by

$$f(x) = \begin{cases} -1, & x \le -1 \\ x, & -1 \le x \le 1 \\ 1, & x \ge 1 \end{cases} \tag{6.17}$$

Since the clipping nonlinearity abruptly transitions from linear to hard-clipped in a non-invertible, heavily aliasing manner, it is usually desirable to use some form of *soft-clipping* before entering the hard-clipping range.

Arctangent Nonlinearity

A simple example of an *invertible* (one-to-one) memoryless nonlinearity is the arctangent mapping:

$$f(x) = \frac{2}{\pi}\arctan(\alpha x), \quad x \in [-1, 1]$$

where normally $\alpha \gg 1$. This function is graphed for $\alpha = 10$ in Fig. 6.21. (Recall that $\arctan(x)$ is defined as the angle whose tangent is x. Only angles between $-\pi/2$ and $\pi/2$ are needed to cover all real values of x.)

Cubic Soft Clipper

In §9.1.6 on page 337, we used the cubic soft-clipper to simulate amplifier distortion:

$$f(x) = \begin{cases} -\frac{2}{3}, & x \le -1 \\ x - \frac{x^3}{3}, & -1 \le x \le 1 \\ \frac{2}{3}, & x \ge 1 \end{cases}$$

This function is diagrammed in Fig. 6.22. It is non-invertible when driven into "hard clipping".

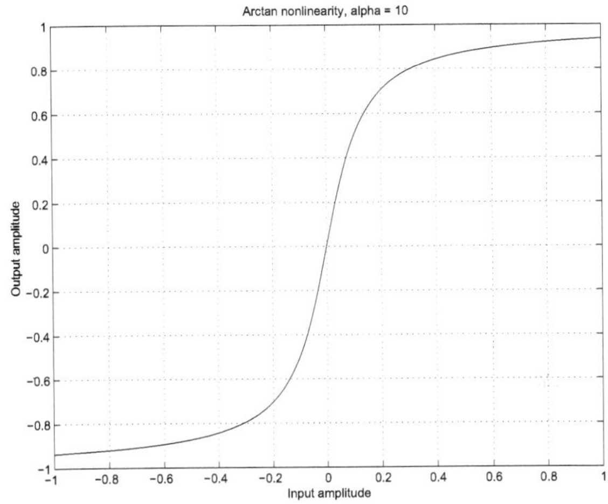

Figure 6.21: Arctangent nonlinearity.

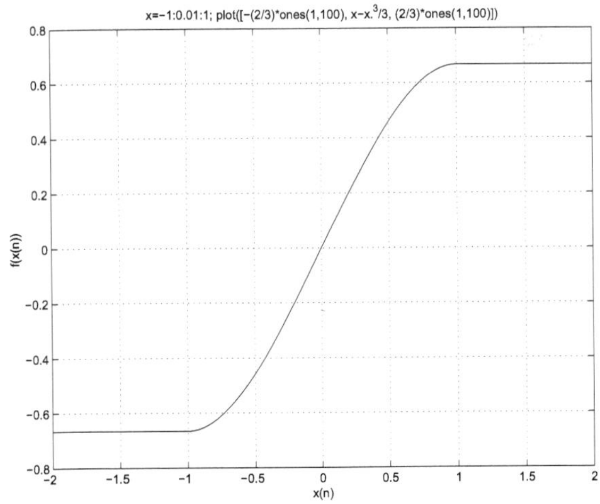

Figure 6.22: Cubic nonlinearity.

Series Expansions

Any "smooth" function $f(x)$ can be expanded as a Taylor series expansion:

$$f(x) = f(0) + \frac{f'(0)}{1}(x) + \frac{f''(0)}{1\cdot 2}(x)^2 + \frac{f'''(0)}{1\cdot 2\cdot 3}(x)^3 + \cdots, \tag{6.18}$$

where "smooth" means that derivatives of all orders must exist over the range of validity. Derivatives of all orders are obviously needed at $x=0$ by the above expansion, and for the expansion to be valid everywhere, the function $f(x)$ must be smooth for all x.

Arctangent Series Expansion

For example, the arctangent function used above can be expanded as

$$\arctan(x) = x - \frac{x^3}{3} + \frac{x^5}{5} - \frac{x^7}{7} + \cdots$$

Note that all even-order terms are zero. This is always the case for *odd functions*, *i.e.*, functions satisfying $f(-x) = -f(x)$. For any smooth function, the odd-order terms of its Taylor expansion comprise the odd part of the function, while the even-order terms comprise the *even part*. The original function is clearly given by the sum of its odd and even parts.[18]

The clipping nonlinearity in Eq. (6.17) is not so amenable to a series expansion. In fact, it is its own series expansion! Since it is not differentiable at $x = \pm 1$, it must be represented as three separate series over the intervals $(-\infty, 1]$, $[-1, 1]$, and $[1, \infty)$, and the result obtained over these intervals is precisely the definition of $f(x)$ in Eq. (6.17).

Spectrum of a Memoryless Nonlinearities

The series expansion of a memoryless nonlinearity is a useful tool for quantifying the *aliasing* caused by that nonlinear mapping when introduced into the signal path of a discrete-time system.

Square Law Series Expansion

When viewed as a Taylor series expansion such as Eq. (6.18), the simplest nonlinearity is clearly the *square law nonlinearity*:

$$f(x) = x + \alpha x^2$$

[18] It is not necessary to perform a Taylor series expansion to separate out the even and odd parts of a function. Instead, the even part of $f(x)$ can be computed as $f_e(x) = [f(x)+f(-x)]/2$ and the odd part as $f_o(x) = [f(x) - f(-x)]/2$. It is easily checked that f_e is even, f_o is odd, and $f = f_e + f_o$.

where α is a parameter of the mapping.[19]

Consider a simple signal processing system consisting only of the square-law nonlinearity:

$$y(n) = x(n) + \alpha x^2(n)$$

The Fourier transform of the output signal is easily found using the dual of the convolution theorem:[20]

$$Y(\omega) = X(\omega) + \alpha(X * X)(\omega)$$

where "$*$" denotes convolution. In general, the bandwidth of $X * X$ is *double* that of X.

Power Law Spectrum

More generally,

$$x^k(n) \longleftrightarrow \underbrace{(X * X * \cdots * X)}_{k \text{ times}}(\omega)$$

so that the spectral bandwidth of $x^k(n)$ is k times that of $x(n)$, in general.

In summary, the spectrum at the output of the square-law nonlinearity can be written as

$$Y(\omega) = X(\omega) + \alpha(X * X)(\omega)$$

This illustrates the great value of series expansions for nonlinearities. In audio signal processing systems, it is essential to keep track of the spectral representation of all output signals, since we hear directly in terms of signal spectra.[21]

Arctangent Spectrum

Since the series expansion of the arctangent nonlinearity is

$$\arctan(x) = x - \frac{x^3}{3} + \frac{x^5}{5} - \frac{x^7}{7} + \cdots$$

[19] Note that memoryless nonlinearities used for distortion simulation are typically *odd functions* of instantaneous amplitude, such as $f(x) = x + \alpha x^3$. (A good practical choice is given in Eq. (6.19).) However, in guitar-amplifier distortion simulation, even-order terms are considered quite important [395].

[20] The convolution theorem for Fourier transforms states that convolution in the time domain corresponds to pointwise multiplication in the frequency domain [449]. The *dual* of the convolution theorem states just the opposite: Pointwise multiplication in the time domain corresponds to convolution in the frequency domain. Thus, if the spectrum of $x(n)$ is $X(\omega)$, then the spectrum of $x^2(n)$ is $(X * X)(\omega)$, where "$*$" denotes the convolution operation.

[21] The basilar membrane of the ear (which is rolled up inside the snail-shaped cochlea of the ear) effectively performs a real-time Fourier analysis which is "felt" by nerve cells leading to the brain.

bandwidth expansion is *infinite* (in continuous time). Thus, we expect problems with *aliasing*.

Cubic Soft-Clipper Spectrum

The cubic soft-clipper, like any polynomial nonlinearity, is defined directly by its series expansion:

$$f(x) = \begin{cases} -\frac{2}{3}, & x \leq -1 \\ x - \frac{x^3}{3}, & -1 \leq x \leq 1 \\ \frac{2}{3}, & x \geq 1 \end{cases} \tag{6.19}$$

In the absence of hard-clipping ($|x| \leq 1$), bandwidth expansion is limited to a factor of *three*. This is the slowest aliasing rate obtainable for an *odd* nonlinearity. Note that smoothing the "corner" in the clipping nonlinearity can reduce the severe bandwidth expansion associated with hard-clipping.

Stability of Nonlinear Feedback Loops

In general, placing a memoryless nonlinearity $f(x)$ in a stable feedback loop preserves stability provided the gain of the nonlinearity is less than one, *i.e.*, $|f(x)| \leq |x|$. A simple proof for the case of a loop consisting of a continuous-time delay-line and memoryless-nonlinearity is as follows.

The delay line can be interpreted as a waveguide model of an ideal string or acoustic pipe having wave impedance $R > 0$ and a noninverting reflection at its midpoint. A memoryless nonlinearity is a special case of an arbitrary time-varying gain [449]. By hypothesis, this gain has magnitude less than one. By routing the output of the delay line back to its input, the gain plays the role of a reflectance g at the "other end" of the ideal string or acoustic pipe. We can imagine, for example, a terminating dashpot with randomly varying positive resistance $\mu(t) > 0$. The set of all $\mu > 0$ corresponds to the set of real reflection coefficients $g = (\mu - R)/(\mu + R)$ in the open interval $(-1, 1)$. Thus, each instantaneous nonlinearity-gain $-1 < g < 1$ corresponds to some instantaneously positive resistance $\mu > 0$. The whole system is therefore passive, even as $\mu(t)$ changes arbitrarily (while remaining positive). (It is perhaps easier to ponder a charged capacitor C terminated on a randomly varying resistor $R(t)$.) This proof method immediately extends to nonlinear feedback around any transfer function that can be interpreted as the reflectance of a passive physical system, *i.e.*, any transfer function $H(s)$ for which the gain is bounded by 1 at each frequency, *viz.*, $|H(j\omega)| \leq 1$.

The finite-sampling-rate case can be embedded in a passive infinite-sampling-rate case by replacing each sample with a constant pulse lasting

T seconds (in the delay line). The continuous-time memoryless nonlinearity $g[x(t)]$ is similarly a held version of the discrete-time case $g[x(nT)]$. Since the discrete-time case is a simple sampling of the (passive) continuous-time case, we are done.

Practical Advice

In summary, the following pointers can be offered regarding nonlinear elements in a digital waveguide model:

- Verify that aliasing can be heard and sounds bad before working to get rid of it.
- Aliasing (bandwidth expansion) is reduced by smoothing "corners" in the nonlinearity.
- Consider an *oversampling factor* for nonlinear subsystems sufficient to accommodate the bandwidth expansion caused by the nonlinearity.
- Make sure there is adequate lowpass filtering in a feedback loop containing a nonlinearity.

As a specific example, consider the cubic nonlinearity used in a feedback loop (as in §9.1.6). This can be done with *no aliasing* at low levels (*i.e.*, at levels below hard clipping) provided we use

- $3\times$ oversampling, and
- a lowpass filter to $\omega \in [-\pi/3, \pi/3]$ after the nonlinearity.

To avoid $3\times$ oversampling in the entire feedback loop, we may downsample by 3 after the lowpass filter and upsample by 3 just before the nonlinearity. If the lowpass filter is good, the downsampling by 3 is trivially accomplished by throwing away every 2 out of 3 samples. For upsampling, however, an additional third-band lowpass-filter is needed for the interpolation (§4.4).

Another variation is to oversample by *two*, in which case there is aliasing, but that aliasing does not reach the "base band." Therefore, a half-band lowpass filter rejects both the second spectral image and the third, which is aliased onto the second.

6.14 Exercises

String Modeling Problems

1. A guitar E string is two feet long (0.61 meters) and, when tuned to 330 Hz, has a tension $K = 20$ pounds (89 Newtons). Find the mass density ϵ of the string in pounds per foot and in kilograms per meter.

2. Describe the details of a digital waveguide model for a vibrating string with pitch 100 Hz at a sampling rate of 10 kHz.

3. Find the decay time at each harmonic in the previous problem, assuming the loop filter has a gain given by $1/k$, where $k = 1, 2, 3, \ldots$ is the harmonic number.

4. Find the precise tunings and decay times for the fundamental frequency and all overtones in a "filtered delay loop" consisting of a 100-sample delay line and a two-point-average lowpass filter $H(z) = (1 + z^{-1})/2$.

5. Consider a basic digital waveguide model for a vibrating electric guitar string with pitch 125 Hz and length 26 inches at a sampling rate of 10 kHz. Assume displacement waves are to be simulated.

   (a) Describe the parameter settings and structure of the model.

   (b) Draw an "efficient" form of the previous model but now with a plucked position specified to be 1 inch from the bridge and a magnetic pick-up position 3 inches from the bridge.

6. A digital waveguide model for a moving rigid string termination was shown in Fig. 6.5 on page 213. In the force-wave model, the "input force" is always a constant f_0, while the string slope grows without bound to $y^+ \to -\infty$. Explain this apparent paradox. [Hint: Read §7.2 on page 257.]

7. In Fig. 6.18 on page 230, the total delay line length N is twice the travel time along the entire length L of the string, *i.e.*, $NT = 2L/c$. Denote the hammer position by $x_h = \beta L$. (a) Determine the tap delay n_h, rounded to the nearest integer, for simulating this hammer position. (b) Assuming traveling force-wave samples in the delay line, describe how to extract the string excitation force as an output signal.

Synthesized Electric Guitar

See `http://ccrma.stanford.edu/realsimple/electric_guitar/` for laboratory exercises related to stringed-instrument synthesis.

Chapter 7

Lumped Models

This chapter introduces modeling of "lumped" physical systems, such as configurations of masses, springs, and "dashpots".

The term "lumped" comes from electrical engineering, and refers to *lumped-parameter analysis*, as opposed to *distributed-parameter analysis.* Examples of "distributed" systems in musical acoustics include ideal strings, acoustic tubes, and anything that propagates waves. In general, a *lumped-parameter approach* is appropriate when the physical object has dimensions that are *small relative to the wavelength of vibration.* Examples from musical acoustics include brass-players' lips (modeled using one or two masses attached to springs—see §9.7), and the piano hammer (modeled using a mass and nonlinear spring, as discussed in §9.4). In contrast to these lumped-modeling examples, the vibrating string is most efficiently modeled as a sampled distributed-parameter system, as discussed in Chapter 6, although lumped models of strings (using, *e.g.*, a *mass-spring-chain* [318]) work perfectly well, albeit at a higher computational expense for a given model quality [69, 145]. In the realm of electromagnetism, distributed-parameter systems include electric transmission lines and optical waveguides, while the typical lumped-parameter systems are ordinary RLC circuits (connecting resistors, capacitors, and inductors). Again, whenever the oscillation wavelength is large relative to the geometry of the physical component, a lumped approximation may be considered. As a result, there is normally a high-frequency limit on the validity of a lumped-parameter model. For the same reason, there is normally an upper limit on physical size as well.

We begin with the fundamental concept of *impedance*, and discuss the elementary lumped impedances associated with *springs*, *mass*, and *dashpots.* These physical objects are analogous to *capacitors*, *inductors*, and *resistors* in lumped-parameter electrical circuits. Next, we discuss general interconnections of such elements, characterized at a single input/output location by means of *one-port network theory*. In particular, we will see that all *passive* networks present a *positive real* impedance at any port (input/output point). A network diagram may be replaced by an *impedance diagram*, which may then be translated into its *equivalent circuit* (replacing springs by capacitors, masses by inductors, and dashpots by resistors).

In the following chapter, we discuss *digitization* of lumped networks by various means, including *finite differences* and the *bilinear transformation*.

7.1 Impedance

Impedance is defined for mechanical systems as force divided by velocity, while the inverse (velocity/force) is called an *admittance*. For dynamic systems, the impedance of a "driving point" is defined for each frequency ω, so that the "force" in the definition of impedance is best thought of as the peak amplitude of a sinusoidal applied force, and similarly for the velocity. Thus, if $F(\omega)$ denotes the Fourier transform of the applied force at a driving point, and $V(\omega)$ is the Fourier transform of the resulting velocity of the driving point, then the *driving-point impedance* is given by

$$R(\omega) \triangleq \frac{F(\omega)}{V(\omega)}.$$

In the lossless case (no dashpots, only masses and springs), all driving-point impedances are purely imaginary, and a purely imaginary impedance is called a *reactance*. A purely imaginary admittance is called a *susceptance*. The term *immittance* refers to either an impedance or an admittance [35]. In mechanics, force is typically in units of newtons (kilograms times meters per second squared) and velocity is in meters per second.

In acoustics [317, 318], force takes the form of *pressure* (*e.g.*, in physical units of newtons per meter squared), and velocity may be either *particle velocity* in open air (meters per second) or *volume velocity* in acoustic tubes (meters cubed per second) (see §B.7.1 on page 503 for definitions). The *wave impedance* (also called the *characteristic impedance*) in open air is the ratio of pressure to particle velocity in a sound wave traveling through air, and it is given by $R = \sqrt{\gamma P_0 \rho} = \rho c$, where ρ is the density (mass per unit volume) of air, c is the speed of sound propagation, P_0 is ambient pressure, and $\gamma_c = 1.4$ is the ratio of the specific heat of air at constant pressure to that at constant volume. In a vibrating string, the wave impedance is given by $R = \sqrt{K\epsilon} = \epsilon c$, where ϵ is string density (mass per unit length) and K is the tension of the string (stretching force), as discussed further in §C.1 on page 523 and §B.5.2 on page 500.

In circuit theory [110], force takes the form of electric potential in volts, and velocity manifests as electric current in amperes (coulombs per second). In an electric transmission line, the characteristic impedance is given by $R = \sqrt{L/C} = Lc$ where L and C are the inductance and capacitance, respectively, per unit length along the transmission line. In free space, the wave impedance for light is $R = \sqrt{\mu_0/\epsilon_0} = \mu_0 c$, where μ_0 and ϵ_0 are the permeability and permittivity, respectively, of free space. One might be led from this to believe that there must exist a medium, or 'ether', which sustains wave propagation in free space; however, this is one instance in which "obvious" predictions from theory turn out to be wrong.

7.1.1 Dashpot

The elementary impedance element in mechanics is the *dashpot* which may be approximated mechanically by a plunger in a cylinder of air or liquid, analogous to a shock absorber for a car. A constant impedance means that the velocity produced is always linearly proportional to the force applied, or $f(t) = \mu v(t)$, where μ is the dashpot impedance, $f(t)$ is the applied force at time t, and $v(t)$ is the velocity. A diagram is shown in Fig. 7.1.

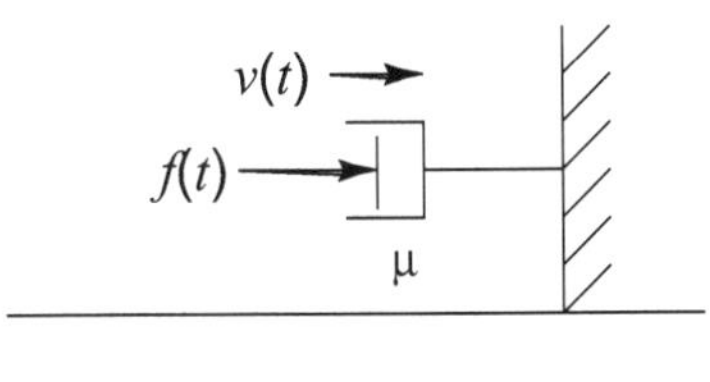

Figure 7.1: The ideal dashpot characterized by a constant impedance μ. For all applied forces $f(t)$, the resulting velocity $v(t)$ obeys $f(t) = \mu v(t)$.

In circuit theory, the element analogous to the dashpot is the *resistor* R, characterized by $v(t) = Ri(t)$, where v is voltage and i is current. In an analog equivalent circuit, a dashpot can be represented using a resistor $R = \mu$.

Over a specific velocity range, *friction force* can also be characterized by the relation $f(t) = \mu v(t)$. However, friction is very complicated in general [419], and as the velocity goes to zero, the coefficient of friction μ may become much larger. The simple model often presented is to use a *static* coefficient of friction when starting at rest ($v(t) = 0$) and a *dynamic* coefficient of friction when in motion ($v(t) \neq 0$). However, these models are too simplified for many practical situations in musical acoustics, *e.g.*, the frictional force between the bow and string of a violin [308, 550], or the internal friction losses in a vibrating string [73].

7.1.2 Ideal Mass

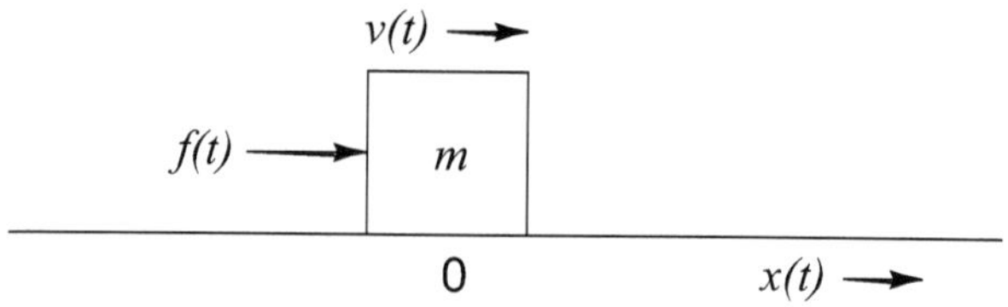

Figure 7.2: The ideal mass characterized by $f(t) = m\dot{v}(t) = m\ddot{x}(t)$.

The concept of impedance extends also to masses and springs. Figure 7.2 illustrates an ideal mass of m kilograms sliding on a frictionless surface. From Newton's second law of motion, we know force equals mass times acceleration, or

$$f(t) = ma(t) \triangleq m\dot{v}(t) \triangleq m\ddot{x}(t).$$

Since impedance is defined in terms of force and velocity, we will prefer the form $f(t) = m\dot{v}(t)$. By the differentiation theorem for Laplace transforms [284],[1] we have

$$F(s) = m[sV(s) - v(0)].$$

If we assume the initial velocity of the mass is zero, we have

$$F(s) = msV(s),$$

and the impedance $F(s)/V(s)$ of the mass in the frequency domain is simply

$$R_m(s) \triangleq ms.$$

The admittance of a mass m is therefore

$$\Gamma_m(s) \triangleq \frac{1}{ms}$$

This is the transfer function of an *integrator*. Thus, an ideal mass integrates the applied force (divided by m) to produce the output velocity. This is just a "linear systems" way of saying force equals mass times acceleration.

[1] For a short online introduction to Laplace transforms, see, *e.g.*, `http://ccrma.stanford.edu/~jos/filters/Laplace_Transform_Analysis.html`.

Since we normally think of an applied force as an *input* and the resulting velocity as an *output*, the corresponding *transfer function* is $H(s) = \Gamma(s) = V(s)/F(s)$. The system diagram for this view is shown in Fig. 7.3.

The *impulse response* of a mass, for a force input and velocity output, is defined as the inverse Laplace transform of the transfer function:

$$\gamma_m(t) \triangleq \mathcal{L}^{-1}\{\Gamma_m(s)\} = \frac{1}{m}u(t)$$

In this instance, setting the input to $\delta(t)$ corresponds to transferring a *unit momentum* to the mass at time 0. (Recall that momentum is the integral of force with respect to time.) Since momentum is also equal to mass m times its velocity $v(t)$, it is clear that the unit-momentum velocity must be $v(t) = 1/m$.

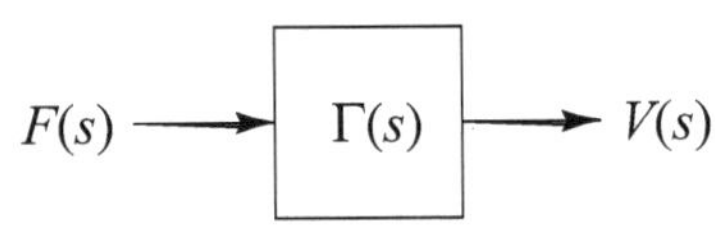

Figure 7.3: Input/output description of a general impedance, with force $F(s)$ as the input, velocity $V(s)$ as the output, and admittance $\Gamma(s)$ as the transfer function.

Once the input and output signal are defined, a transfer function is defined, and therefore a *frequency response* is defined [486]. The frequency response is given by the transfer function evaluated on the $j\omega$ axis in the s plane, *i.e.*, for $s = j\omega$. For the ideal mass, the force-to-velocity frequency response is

$$\Gamma_m(j\omega) = \frac{1}{mj\omega}$$

Again, this is just the frequency response of an integrator, and we can say that the amplitude response rolls off -6 dB per octave, and the phase shift is $-\pi/2$ radians at all frequencies.

In circuit theory, the element analogous to the mass is the *inductor*, characterized by $v(t) = Ldi/dt$, or $V(s) = LsI(s)$. In an analog equivalent circuit, a mass can be represented using an inductor with value $L = m$.

7.1.3 Ideal Spring

Figure 7.4 depicts the ideal spring.

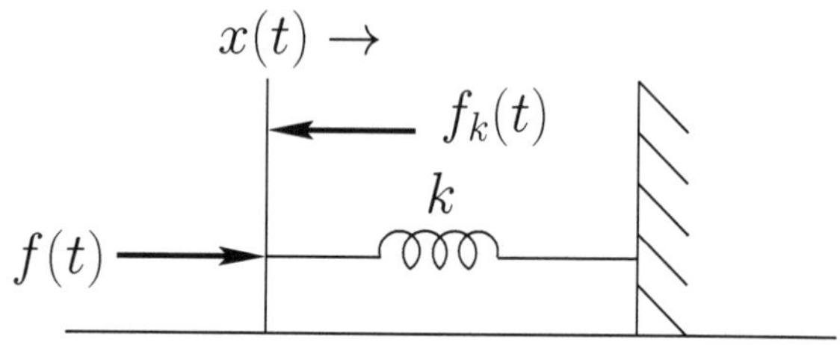

Figure 7.4: The ideal spring characterized by $f(t) = kx(t)$.

From Hooke's law, we have that the applied force is proportional to the *displacement* of the spring:

$$f(t) \;=\; k\,x(t) \;\triangleq\; k\int_0^t v(\tau)d\tau$$

where it is assumed that $x(0) = 0$. The spring constant k is sometimes called the *stiffness* of the spring. Taking the Laplace transform gives

$$F(s) = kV(s)/s$$

so that the impedance of a spring is

$$R_k(s) \triangleq \frac{k}{s}$$

and the admittance is

$$\Gamma_k(s) \triangleq \frac{s}{k}$$

This is the transfer function of a *differentiator*. We can say that the ideal spring differentiates the applied force (divided by k) to produce the output velocity.

The *frequency response* of the ideal spring, given the applied force as input and resulting velocity as output, is

$$\Gamma_k(j\omega) = \frac{j\omega}{k}$$

In this case, the amplitude response grows +6 dB per octave, and the phase shift is $+\pi/2$ radians for all ω. Clearly, there is no such thing as an ideal spring which can produce arbitrarily large gain as frequency goes to infinity; there is always some mass in a real spring.

We call $v(t)$ the *compression velocity* of the spring. In more complicated configurations, the compression velocity is defined as the difference between the velocity of the two spring endpoints, with positive velocity corresponding to spring compression.

In circuit theory, the element analogous to the spring is the *capacitor*, characterized by $i(t) = Cdv/dt$, or $I(s) = CsV(s)$. In an equivalent analog circuit, we can use the value $C = 1/k$. The inverse $1/k$ of the spring stiffness is sometimes called the *compliance* of the spring.

Don't forget that the definition of impedance requires *zero initial conditions* for elements with "memory" (masses and springs). This means we can only use impedance descriptions for *steady state* analysis. For a complete analysis of a particular system, including the transient response, we must go back to full scale Laplace transform analysis.

7.2 One-Port Network Theory

The basic idea of a one-port network [525] is shown in Fig. 7.5. The one-port is a "black box" with a single pair of input/output terminals, referred to as a *port.* A force is applied at the terminals and a velocity "flows" in the direction shown. The admittance "seen" at the port is called the *driving point admittance.* Network theory is normally described in terms of circuit theory elements, in which case a voltage is applied at the terminals and a current flows as shown. However, in our context, mechanical elements are preferable.

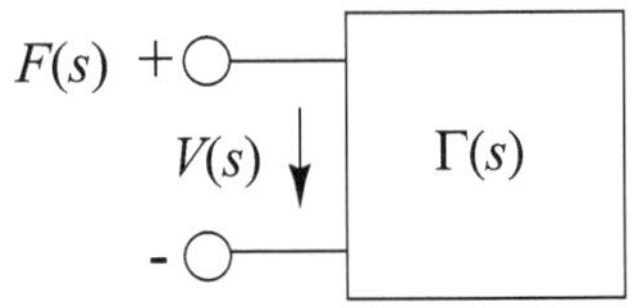

Figure 7.5: A one-port network characterized by its driving point admittance $\Gamma(s)$. For any applied force $F(s)$, the observed velocity is $V(s) = \Gamma(s)F(s)$.

7.2.1 Series Combination of One-Ports

Figure 7.6 shows the *series* combination of two one-ports.

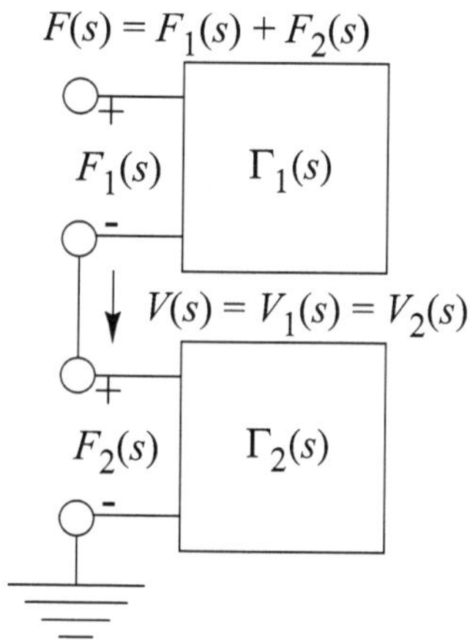

Figure 7.6: Two one-port networks combined in series. The impedance of the series combination is $R(s) = F(s)/V(s) = R_1(s) + R_2(s)$.

Impedances add in series, so the aggregate impedance is $R(s) = R_1(s) + R_2(s)$, and the admittance is

$$\Gamma(s) = \frac{1}{\frac{1}{\Gamma_1(s)} + \frac{1}{\Gamma_2(s)}} = \frac{\Gamma_1(s)\Gamma_2(s)}{\Gamma_1(s) + \Gamma_2(s)}$$

The latter expression is the handy product-over-sum rule for combining admittances in series.

7.2.2 Mass-Spring-Wall System

In a physical situation, if two elements are connected in such a way that they share a common velocity, then they are in series. An example is a mass connected to one end of a spring, where the other end is attached to a rigid support, and the force is applied to the mass, as shown in Fig. 7.7.

Figure 7.8 shows the electrical equivalent circuit corresponding to Fig. 7.7.

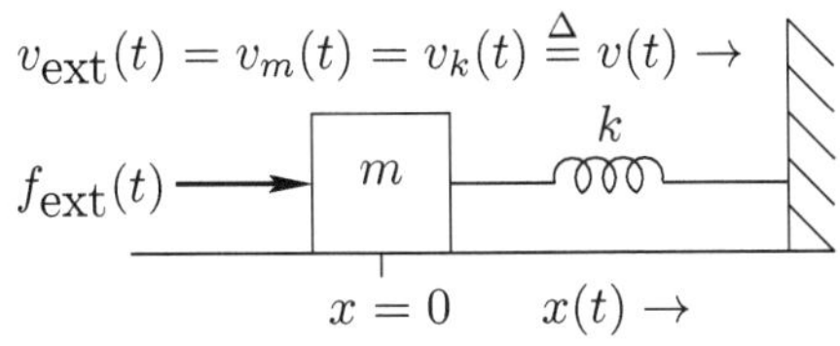

Figure 7.7: A mass and spring combined as one-ports in series.

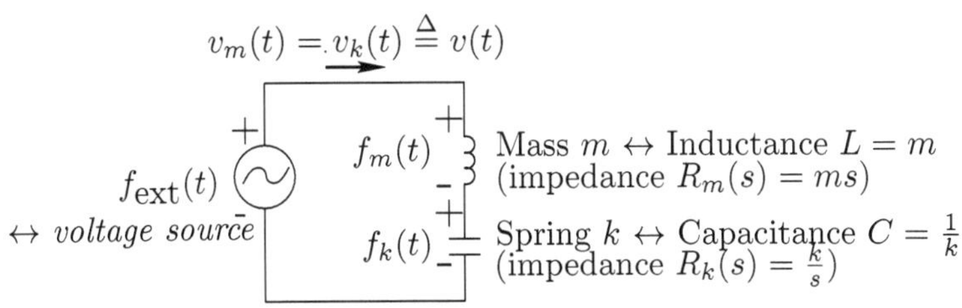

Figure 7.8: Electrical equivalent circuit of the series mass-spring driven by an external force diagrammed in Fig. 7.7.

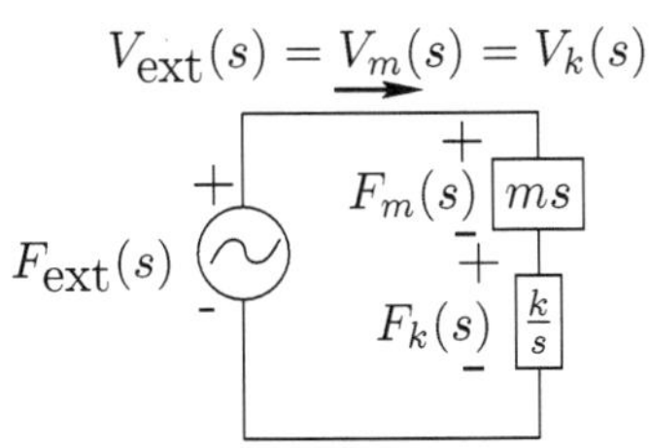

Figure 7.9: Impedance diagram for the force-driven, series arrangement of mass and spring shown in Fig. 7.7.

7.2.3 Parallel Combination of One-Ports

Figure Fig. 7.10 shows the parallel combination of two one-ports.

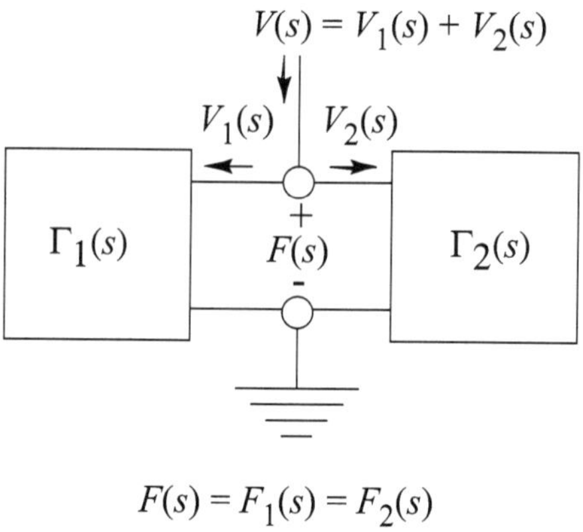

Figure 7.10: Two one-port networks combined in parallel. The admittance of the parallel combination is $\Gamma(s) = \Gamma_1(s) + \Gamma_2(s)$.

Admittances add in parallel, so the combined admittance is $\Gamma(s) = \Gamma_1(s) + \Gamma_2(s)$, and the impedance is

$$R(s) = \frac{1}{\frac{1}{R_1(s)} + \frac{1}{R_2(s)}} = \frac{R_1(s)R_2(s)}{R_1(s) + R_2(s)}$$

which is the familiar product-over-sum rule for combining impedances in parallel. This operation is often denoted by

$$R = R_1 || R_2$$

7.2.4 Spring-Mass System

When two physical elements are driven by a *common force* (yet have independent velocities, as we'll soon see is quite possible), they are formally in *parallel.* An example is a mass connected to a spring in which the driving force is applied to one end of the spring, and the mass is attached to the other end, as shown in Fig. 7.11. The compression force on the spring is equal at all times to the rightward force on the mass. However, the spring compression velocity $v_k(t)$ does not always equal the mass velocity $v_m(t)$. We do have that the sum of the mass velocity and spring compression velocity gives the velocity of the driving point, *i.e.*, $v(t) = v_m(t) + v_k(t)$. Thus, in a parallel connection, forces are equal and velocities sum.

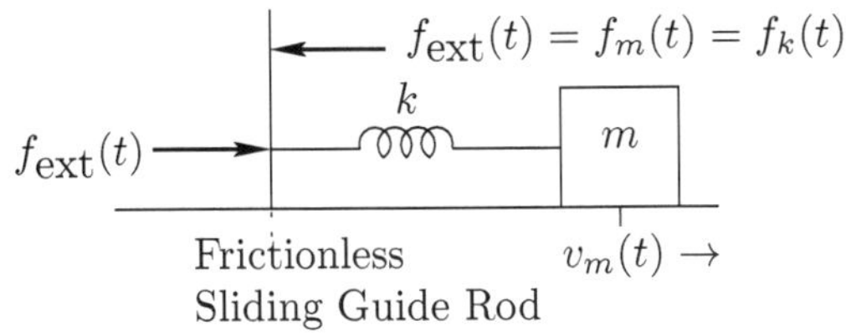

Figure 7.11: A mass and spring combined as one-ports in parallel.

Figure 7.12 shows the electrical equivalent circuit corresponding to Fig. 7.11.

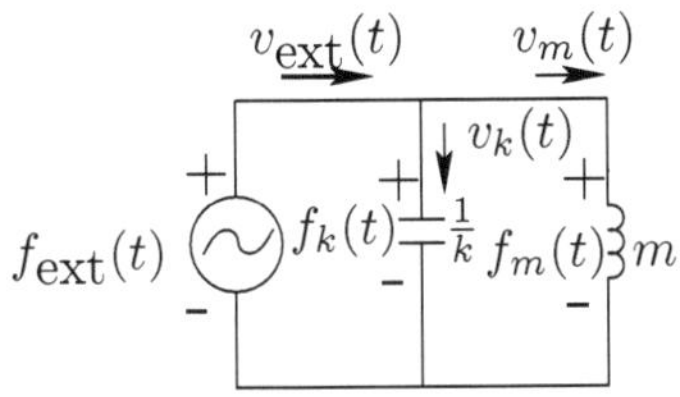

Figure 7.12: Electrical equivalent circuit of the *parallel* mass-spring combination driven by an external force, as diagrammed in Fig. 7.11.

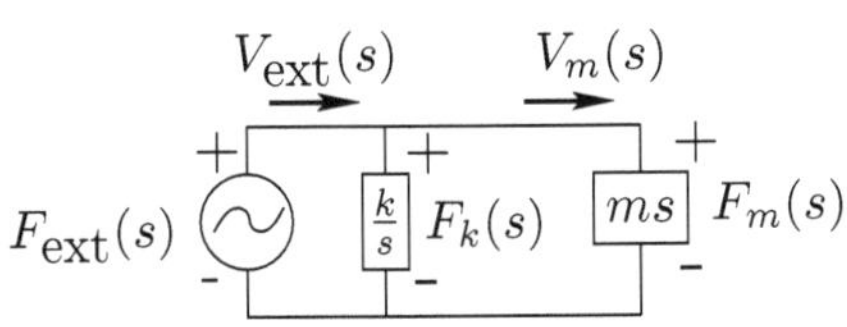

Figure 7.13: Impedance diagram for the force-driven, *parallel* mass-spring arrangement shown in Fig. 7.11.

7.2.5 Mechanical Impedance Analysis

Impedance analysis is commonly used to analyze electrical circuits [110]. By means of equivalent circuits, we can use the same analysis methods for mechanical systems.

For example, referring to Fig. 7.9, the Laplace transform of the force on the spring k is given by the so-called *voltage divider* relation:[2]

$$F_k(s) = F_{\text{ext}}(s)\frac{R_k(s)}{R_m(s)+R_k(s)} = F_{\text{ext}}(s)\frac{\frac{k}{s}}{ms+\frac{k}{s}}$$

Similarly, the Laplace transform of the force on the mass m is given by

$$F_m(s) = F_{\text{ext}}(s)\frac{R_m(s)}{R_m(s)+R_k(s)} = F_{\text{ext}}(s)\frac{ms}{ms+\frac{k}{s}}. \tag{7.1}$$

As a simple application, let's find the motion of the mass m, after time zero, given that the input force is an impulse at time 0:

$$f_{\text{ext}}(t) = \delta(t) \leftrightarrow F_{\text{ext}}(s) = 1$$

Then, by the "voltage divider" relation Eq. (7.1), the Laplace transform of the mass force $f_m(t)$ after time 0 is given by

$$F_m(s) = \frac{ms}{ms+\frac{k}{s}} = \frac{s^2}{s^2+\frac{k}{m}} \stackrel{\Delta}{=} \frac{s^2}{s^2+\omega_0^2},$$

where we have defined $\omega_0^2 \stackrel{\Delta}{=} k/m$. The mass *velocity* Laplace transform is then

$$\begin{aligned} V_m(s) &= \frac{F_m(s)}{ms} = \frac{1}{m}\cdot\frac{s}{s^2+\omega_0^2} \\ &= \frac{1}{m}\left[\frac{1/2}{s+j\omega_0}+\frac{1/2}{s-j\omega_0}\right] \\ &\leftrightarrow \frac{1}{m}\cos(\omega_0 t). \end{aligned}$$

Thus, the impulse response of the mass oscillates sinusoidally with radian frequency $\omega_0 = \sqrt{k/m}$, and amplitude $1/m$. The velocity starts out maximum at time $t = 0$, which makes physical sense. Also, the momentum transferred to the mass at time 0 is $m\,v(0+) = 1$; this is also expected physically because the time-integral of the applied force is 1 (the area under any impulse $\delta(t)$ is 1).

---

[2]Of course, here we should call it the "force divider" relation.

7.2.6 General One-Ports

An arbitrary interconnection of N impedances and admittances, with input and output force and/or velocities defined, results in a one-port with admittance expressible as

$$\Gamma(s) = \frac{b_0 s^N + b_1 s^{N-1} + \cdots + b_N}{s^N + a_1 s^{N-1} + \cdots + a_N} \triangleq \frac{B(s)}{A(s)}$$

In any mechanical situation we have $b_0 = 0$, in principle, since at sufficiently high frequencies, every mechanical system must "look like a mass."[3] However, for purposes of *approximation* to a real physical system, it may well be best to allow $b_0 \neq 0$ and consider the above expression to be a *rational approximation* to the true admittance function.

7.2.7 Passive One-Ports

It is well known that the impedance of every passive one-port is *positive real* (see §C.11.2 on page 588). The reciprocal of a positive real function is positive real, so every passive impedance corresponds also to a passive admittance.

A complex-valued function of a complex variable $\Gamma(s)$ is said to be *positive real* (PR) if

1) $\Gamma(s)$ is real whenever s is real

2) $\Re\{\Gamma(s)\} \geq 0$ whenever $\Re\{s\} \geq 0$.

A particularly important property of positive real functions is that the phase is bounded between plus and minus 90 degrees, *i.e.*,

$$-\frac{\pi}{2} \leq \angle\Gamma(j\omega) \leq \frac{\pi}{2}$$

This is a significant constraint on the rational function $\Gamma(s)$. One implication is that in the lossless case (no dashpots, only masses and springs—a *reactance*) all poles and zeros *interlace* along the $j\omega$ axis, as depicted in Fig. 7.14.

[3]We say that the driving point of every mechanical system must "looks like a mass" at sufficiently high frequency because every mechanical system has at least some mass, and the driving-point impedance of a mass goes to infinity with frequency. However, in a completely detailed model, the contact force between objects should really be the *Coulombe force*, which "looks like a spring". In other words, mechanical interactions are ultimately *electromagnetic* interactions, and it is theoretically possible for a driving force on an atom to be so small and fast that it can vibrate the outermost electron orbital without moving the nucleus appreciably, thus "looking like a spring" in the high-frequency limit. We will not be concerned with atomic-scale models in this book, and will persist in treating masses and springs in idealized form.

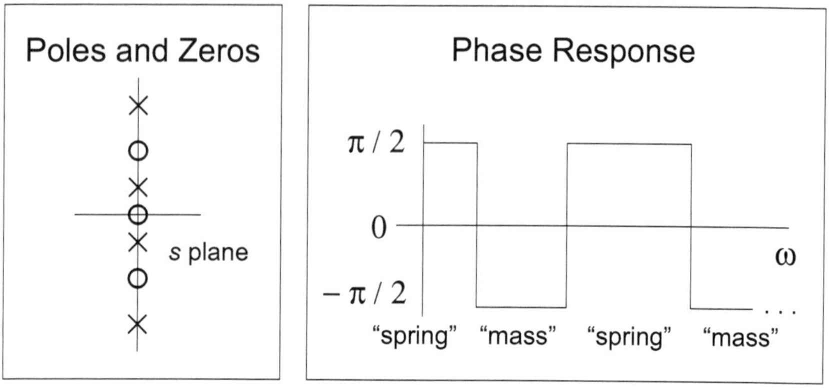

Figure 7.14: Poles and zeros of a lossless immittance (*reactance* or *suseptance*) must *interlace* along the $j\omega$ Axis. Left: Pole-zero plot. Right: Phase response. The "spring/mass" labels along the frequency axis correspond to the case of a lossless *admittance* (susceptance) in which a spring admittance ($\Gamma_k(j\omega) = j\omega/k$) gives a $+\pi/2$ phase shift, while that of a mass ($\Gamma_m(j\omega) = -j/(m\omega)$) gives a $-\pi/2$ phase shift between the input driving-force and output velocity.

Referring to Fig. 7.14, consider the graphical method for computing phase response of a reactance from the pole zero diagram [449].[4] Each zero on the positive $j\omega$ axis contributes a net 90 degrees of phase at frequencies above the zero. As frequency crosses the zero going up, there is a switch from -90 to $+90$ degrees. For each pole, the phase contribution switches from $+90$ to -90 degrees as it is passed going up in frequency. In order to keep phase in $[-\pi/2, \pi/2]$, it is clear that the poles and zeros must strictly alternate. Moreover, all poles and zeros must be simple, since a multiple poles or zero would swing the phase by more than 180 degrees, and the reactance could not be positive real.

The positive real property is fundamental to passive immittances and comes up often in the study of measured resonant systems. A practical modeling example (passive digital modeling of a guitar bridge) is discussed in §9.2.1 on page 343.

[4]Available online at
`http://ccrma.stanford.edu/~jos/filters/Graphical_Amplitude_Response.html`

7.3 Digitization of Lumped Models

Since lumped models are described by *differential equations*, they are digitized (brought into the digital-signal domain) by converting them to corresponding *finite-difference equations* (or simply "*difference equations*"). General aspects of finite difference schemes are discussed in Appendix D. This chapter introduces a couple of elementary methods in common use:

1. Finite difference approximation: $sT \leftarrow 1 - z^{-1}$
2. Bilinear transform: $sT \leftarrow 2\,(1 - z^{-1})/(1 + z^{-1})$

Note that digitization by the bilinear transform is closely related to the Wave Digital Filter (WDF) method introduced in Appendix F. Section 9.3.1 discusses a bilinearly transformed mass colliding with a digital waveguide string (an idealized struck-string example).

7.3.1 Finite Difference Approximation

A *finite difference approximation* (FDA) approximates derivatives with finite differences, *i.e.*,

$$\frac{d}{dt}x(t) \;\triangleq\; \lim_{\delta\to 0}\frac{x(t)-x(t-\delta)}{\delta} \;\approx\; \frac{x(nT)-x[(n-1)T]}{T} \tag{7.2}$$

for sufficiently small T.[5]

Equation (7.2) is also known as the *backward difference* approximation of differentiation.

See §C.2.1 on page 525 for a discussion of using the FDA to model ideal vibrating strings.

FDA in the Frequency Domain

Viewing Eq. (7.2) in the frequency domain, the ideal differentiator transfer-function is $H(s) = s$, which can be viewed as the Laplace transform of the operator d/dt (left-hand side of Eq. (7.2)). Moving to the right-hand side, the z transform of the first-order difference operator is $(1 - z^{-1})/T$. Thus, in the frequency domain, the finite-difference approximation may be performed by making the substitution

$$s \leftarrow \frac{1 - z^{-1}}{T} \tag{7.3}$$

[5]To avoid the introduction of half a sample of delay by the approximation, the first-order finite difference may be defined instead as $x[(n+1)T] - x[(n-1)T]/(2T)$.

in any continuous-time transfer function (Laplace transform of an integro-differential operator) to obtain a discrete-time transfer function (z transform of a finite-difference operator).

The inverse of substitution Eq. (7.3) is

$$z = \frac{1}{1-sT} = 1 + sT + (sT)^2 + \cdots .$$

As discussed in §8.3.1 on page 284, the FDA is a special case of the matched z transformation applied to the point $s = 0$.

Note that the FDA does not alias, since the conformal mapping $s = 1 - z^{-1}$ is one to one. However, it does warp the poles and zeros in a way which may not be desirable, as discussed further on p. 268 below.

Delay Operator Notation

It is convenient to think of the FDA in terms of *time-domain difference operators* using a *delay operator notation.* The *delay operator* d is defined by

$$d^k x(n) = x(n-k).$$

Thus, the first-order difference (derivative approximation) is represented in the time domain by $(1-d)/T$. We can think of d as z^{-1} since, by the *shift theorem* for z transforms, $z^{-k}X(z)$ is the z transform of x delayed (right shifted) by k samples.

The obvious definition for the second derivative is

$$\hat{\ddot{x}}(n) = \frac{(1-d)^2}{T^2} x(n). \tag{7.4}$$

However, a better definition is the *centered finite difference*

$$\hat{\ddot{x}}(n) \triangleq \frac{(d^{-1}-1)(1-d)}{T^2} x(n) = \frac{d^{-1}-2+d}{T^2} x(n) \tag{7.5}$$

where d^{-1} denotes a unit-sample *advance.* This definition is preferable as long as one sample of look-ahead is available, since it avoids an operator delay of one sample. Equation (7.5) is a *zero phase filter,* meaning it has no delay at any frequency, while (7.4) is a *linear phase filter* having a delay of 1 sample at all frequencies.

7.3.2 Bilinear Transformation

The bilinear transform is defined by the substitution

$$s = c\frac{1-z^{-1}}{1+z^{-1}}, \quad c > 0,\ c = 2/T \text{ (typically)} \tag{7.6}$$

$$\Rightarrow z = \frac{1+s/c}{1-s/c} = 1 + 2(s/c) + 2(s/c)^2 + 2(s/c)^3 + \cdots \tag{7.7}$$

where c is some positive constant [83, 326]. That is, given a continuous-time transfer function $H_a(s)$, we apply the bilinear transform by defining

$$H_d(z) = H_a\left(c\frac{1-z^{-1}}{1+z^{-1}}\right)$$

where the "d" subscript denotes "digital," and "a" denotes "analog."

It can be seen that analog dc ($s = 0$) maps to digital dc ($z = 1$) and the highest analog frequency ($s = \infty$) maps to the highest digital frequency ($z = -1$). It is easy to show that the entire $j\omega$ axis in the s plane (where $s \triangleq \sigma + j\omega$) is mapped exactly *once* around the unit circle in the z plane (rather than summing around it infinitely many times, or "aliasing" as it does in ordinary sampling). With c real and positive, the left-half s plane maps to the interior of the unit circle, and the right-half s plane maps outside the unit circle. This means *stability is preserved* when mapping a continuous-time transfer function to discrete time.

Another valuable property of the bilinear transform is that *order is preserved*. That is, an Nth-order s-plane transfer function carries over to an Nth-order z-plane transfer function. (*Order* in both cases equals the maximum of the degrees of the numerator and denominator polynomials [449]).[6]

The constant c provides one remaining degree of freedom which can be used to map any particular finite frequency from the $j\omega_a$ axis in the s plane to a particular desired location on the unit circle $e^{j\omega_d}$ in the z plane. All other frequencies will be *warped*. In particular, approaching half the sampling rate, the frequency axis compresses more and more. Note that at most one resonant frequency can be preserved under the bilinear transformation of a mass-spring-dashpot system. On the other hand, filters having a single transition frequency, such as lowpass or highpass filters, map beautifully under the bilinear transform; one simply uses c to map the cut-off frequency where it belongs, and the response looks great. In particular, "equal ripple" is preserved for optimal filters of the elliptic and Chebyshev types because the values taken on by the frequency response are identical in both cases; only the frequency axis is warped.

The bilinear transform is often used to design digital filters from analog prototype filters [343]. An on-line introduction is given in [449].

---

[6] In continuous time, the order is incremented once for each independently moving mass or spring. In discrete time, the order is increased by one when a sample of delay is added to the system state, and the number of multiplies needed to implement a digital simulation is bounded by twice the order plus one.

Finite Difference Approximation vs. Bilinear Transform

Recall that the Finite Difference Approximation (FDA) defines the elementary differentiator by $y(n) = x(n) - x(n-1)$ (ignoring the scale factor T for now), and this approximates the ideal transfer function $H(s) = s$ by $H_d(z) = 1 - z^{-1}$. The bilinear transform calls instead for the transfer function $H_d'(z) = (1 - z^{-1})/(1 + z^{-1})$ (again dropping the scale factor) which introduces a pole at $z = -1$ and gives us the recursion $y(n) = x(n) - x(n-1) - y(n-1)$. Note that this new pole is right on the unit circle and is therefore undamped. Any signal energy at half the sampling rate will circulate forever in the recursion, and due to round-off error, it will tend to grow. This is therefore a potentially problematic revision of the differentiator. To get something more practical, we need to specify that the filter frequency response approximate $H(j\omega) = j\omega$ over a *finite range* of frequencies $[-\omega_c, \omega_c]$, where $\omega_c \ll \pi f_s$, above which we allow the response to "roll off" to zero. This is how we pose the differentiator problem in terms of general purpose filter design (see §8.6 on page 290) [362].

To understand the properties of the finite difference approximation in the frequency domain, we may look at the properties of its s-plane to z-plane mapping

$$s = \frac{1 - z^{-1}}{T}.$$

We see the FDA is actually a portion of the bilinear transform, since following the FDA mapping by the mapping $s = (c/T)/(1 + z^{-1})$ would convert it to the bilinear transform. Like the bilinear transform, the FDA does not alias, since the mapping $s = 1 - z^{-1}$ is one-to-one.

Setting T to 1 for simplicity and solving the FDA mapping for z gives

$$z = \frac{1}{1 - s}.$$

We see that analog dc ($s = 0$) maps to digital dc ($z = 1$) as desired, but higher frequencies unfortunately map *inside* the unit circle rather than onto the unit circle in the z plane. Solving for the image in the z plane of the $j\omega$ axis in the s plane gives

$$z = \frac{1}{1 - j\omega} = \frac{1 + j\omega}{1 + \omega^2}$$

From this it can be checked that the FDA maps the $j\omega$ axis in the s plane to the circle of radius $1/2$ centered at the point $z = 1/2$ in the z plane, as shown in Fig. 7.15

Under the FDA, analog and digital frequency axes coincide well enough at very low frequencies (high sampling rates), but at high frequencies relative to the sampling rate, *artificial damping* is introduced as the image of the $j\omega$ axis diverges away from the unit circle.

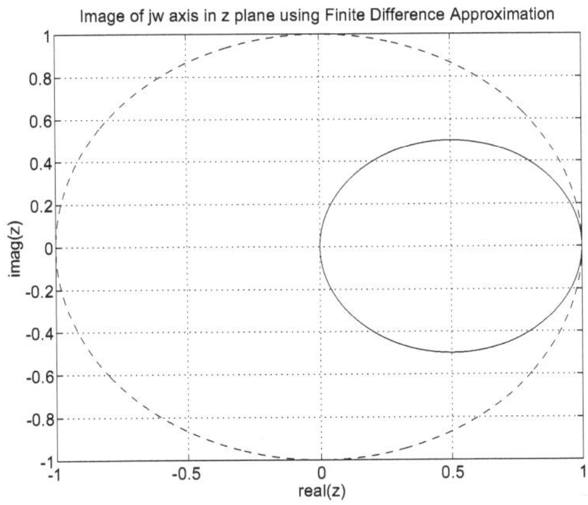

Figure 7.15: Image of the $j\omega$ axis in the z plane: a circle of radius 1/2 centered at the point $z = 1/2$.

While the bilinear transform "warps" the frequency axis, we can say the FDA "doubly warps" the frequency axis: It has a progressive, compressive warping in the direction of increasing frequency, like the bilinear transform, but unlike the bilinear transform, it also warps *normal* to the frequency axis.

Consider a point traversing the upper half of the unit circle in the z plane, starting at $z = 1$ and ending at $z = -1$. At dc, the FDA is perfect, but as we proceed out along the unit circle, we diverge from the $j\omega$ axis image and carve an arc somewhere out in the image of the right-half s plane. This has the effect of introducing an artificial damping.

Consider, for example, an undamped mass-spring system. There will be a complex conjugate pair of poles on the $j\omega$ axis in the s plane. After the FDA, those poles will be inside the unit circle, and therefore damped in the digital counterpart. The higher the resonance frequency, the larger the damping. It is even possible for unstable s-plane poles to be mapped to stable z-plane poles.

In summary, both the bilinear transform and the FDA preserve order, stability, and positive realness. They are both free of aliasing, high frequencies are compressively warped, and both become ideal at dc, or as f_s approaches ∞. However, at frequencies significantly above zero relative to the sampling rate, only the FDA introduces artificial damping. The bilinear transform maps the continuous-time frequency axis in the s (the $j\omega$ axis) plane precisely to the discrete-time frequency axis in the z plane (the unit circle).

7.3.3 Application of the Bilinear Transform

The impedance of a mass in the frequency domain is

$$R_M(s) = Ms.$$

In the s plane, we have

$$F_a(s) = (Ms)V_a(s)$$

where the "a" subscript denotes "analog". For simplicity, let's choose the free constant c in the bilinear transform such that 1 rad/sec maps to one fourth the sampling rate, *i.e.*, $s = j$ maps to $z = j$ which implies $c = 1$. Then the impedance relation maps across as

$$F_d(z) = \left(M\frac{1-z^{-1}}{1+z^{-1}}\right)V_d(z)$$

where the "d" subscript denotes "digital. Multiplying through by the denominator and applying the shift theorem for z transforms gives the corresponding *difference equation*

$$\begin{aligned}(1+z^{-1})F_d(z) &= M(1-z^{-1})V_d(z)\\ \longleftrightarrow f_d(n)+f_d(n-1) &= M[v_d(n)-v_d(n-1)]\\ \Rightarrow f_d(n) &= M[v_d(n)-v_d(n-1)]-f_d(n-1).\end{aligned}$$

This difference equation is diagrammed in Fig. 7.16. We recognize this recursive digital filter as the *direct form I* structure. The direct-form II structure is obtained by commuting the feedforward and feedback portions and noting that the two delay elements contain the same value and can therefore be shared [449]. The two other major filter-section forms are obtained by *transposing* the two direct forms by exchanging the input and output, and reversing all arrows. (This is a special case of Mason's Gain Formula which works for the single-input, single-output case.) When a filter structure is transposed, its summers become branching nodes and vice versa. Further discussion of the four basic filter section forms can be found in [333].

Practical Considerations

While the digital mass simulator has the desirable properties of the bilinear transform, it is also not perfect from a practical point of view: (1) There is a pole at half the sampling rate ($z = -1$). (2) There is a delay-free path from input to output.

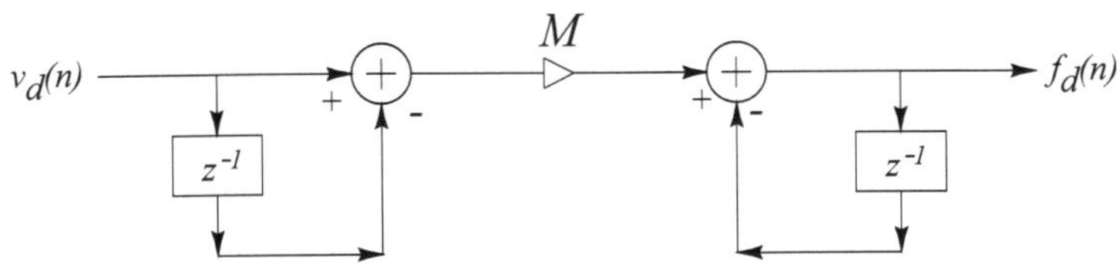

Figure 7.16: A direct-form-I digital filter simulating a mass M created using the bilinear transform $s = (1 - z^{-1})/(1 + z^{-1})$.

The first problem can easily be circumvented by introducing a loss factor g, moving the pole from $z = -1$ to $z = -g$, where $g \in [0, 1)$ and $g \approx 1$. This is sometimes called the "leaky integrator".

The second problem arises when making loops of elements (*e.g.*, a mass-spring chain which forms a loop). Since the individual elements have no delay from input to output, a loop of elements is not computable using standard signal processing methods. The solution proposed by Alfred Fettweis is known as "wave digital filters," a topic taken up in §F.1.

7.3.4 Limitations of Lumped Element Digitization

Model discretization by the FDA (§7.3.1) and bilinear transform (§7.3.2) methods are *order preserving*. As a result, they suffer from significant approximation error, especially at high frequencies relative to half the sampling rate. By allowing a *larger order* in the digital model, we may obtain arbitrarily accurate transfer-function models of LTI subsystems, as discussed in Chapter 8. Of course, in higher-order approximations, the state variables of the simulation no longer have a direct physical intepretation, and this can have implications, particularly when trying to extend to the nonlinear case. The benefits of a physical interpretation should not be given up lightly. For example, one may consider *oversampling* in place of going to higher-order element approximations.

7.4 More General Finite-Difference Methods

The FDA and bilinear transform of the previous sections can be viewed as first-order conformal maps from the analog s plane to the digital z plane. These maps are one-to-one and therefore non-aliasing. The FDA performs well at low frequencies relative to the sampling rate, but it introduces artificial damping at high frequencies. The bilinear transform preserves the frequency axis exactly, but over a warped frequency scale. Being first order, both maps preserve the number of poles and zeros in the model.

We may only think in terms of mapping the s plane to the z plane for *linear, time-invariant* systems. This is because Laplace transform analysis is not defined for nonlinear and/or time-varying differential equations (no s plane). Therefore, such systems are instead digitized by some form of *numerical integration* to produce solutions that are ideally sampled versions of the continuous-time solutions. It is often necessary to work at sampling rates much higher than the desired audio sampling rate, due to the bandwidth-expanding effects of nonlinear elements in the continuous-time system.

A tutorial review of numerical solutions of Ordinary Differential Equations (ODE), including nonlinear systems, with examples in the realm of audio effects (such as a diode clipper), is given in [556]. Finite difference schemes specifically designed for nonlinear discrete-time simulation, such as the energy-preserving "Kirchoff-Carrier nonlinear string model" and "von Karman nonlinear plate model", are discussed in [53].

The remaining sections here summarize a few of the more elementary techniques discussed in [556].

7.4.1 General Nonlinear ODE

In state-space form (§1.3.7 on page 16) [449],[7] a general class of Nth-order Ordinary Differential Equations (ODE), can be written as

$$\underline{\dot{x}}(t) \;=\; f(t, \underline{x}, \underline{u}) \tag{7.8}$$

where t denotes time in seconds, $\underline{x}(t)$ denotes a vector of N *state variables* at time t,

$$\underline{\dot{x}}(t) \;\isdef\; \frac{d}{dt}\underline{x}(t)$$

denotes the time derivative of $\underline{x}(t)$, and $\underline{u}(t)$ is a vector (any length) of the system input signals, if any. Thus, Eq. (7.8) says simply that the time-derivative of the state vector is some function f depending on time t, the current state $\underline{x}(t)$, and the current input signals $\underline{u}(t)$. The basic problem is to solve for the state trajectory $\underline{x}(t)$ given its initial condition $\underline{x}(0)$, the system definition function f, and the input signals $\underline{u}(t)$ for all $t \geq 0$.

In the linear, time-invariant (LTI) case, Eq. (7.8) can be expressed in the usual state-space form for LTI continuous-time systems:

$$\frac{d}{dt}\underline{x}(t) \;=\; A\,\underline{x}(t) + B\,\underline{u}(t) \tag{7.9}$$

[7] http://ccrma.stanford.edu/~jos/filters/State_Space_Filters.html

In this case, standard methods for converting a filter from continuous to discrete time may be used, such as the FDA (§7.3.1) and bilinear transform (§7.3.2).[8]

7.4.2 Forward Euler Method

The finite-difference approximation (Eq. (7.2) on page 265) with the derivative evaluated at time $n-1$ yields the *forward Euler method* of numerical integration:

$$\underline{\hat{x}}(n) \triangleq \underline{\hat{x}}(n-1) + T\,\dot{\underline{\hat{x}}}(n-1) \;=\; \underline{\hat{x}}(n-1) + T\,f[n-1, \underline{\hat{x}}(n-1), \underline{u}(n-1)] \tag{7.10}$$

where $\underline{\hat{x}}(n)$ denotes the *approximation* to $\underline{x}(nT)$ computed by the forward Euler method. Note that the "driving function" f is evaluated at time $n-1$, not n. As a result, given, $\underline{\hat{x}}(0) = \underline{x}(0)$ and the input vector $\underline{u}(n)$ for all $n \geq 0$, Eq. (7.10) can be iterated forward in time to compute $\underline{\hat{x}}(n)$ for all $n > 0$. Since f is an arbitrary function, we have a solver that is applicable to nonlinear, time-varying ODEs Eq. (7.8).

Because each iteration of the forward Euler method depends only on past quantities, it is termed an *explicit method*. In the LTI case, an explicit method corresponds to a *causal* digital filter [449]. Methods that depend on current and/or future solution samples (*i.e.*, $\underline{\hat{x}}(n)$ for $n \geq 0$) are called *implicit methods*. When a nonlinear numerical-integration method is implicit, each step forward in time typically uses some number of iterations of Newton's Method (see §7.4.5 below).

7.4.3 Backward Euler Method

An example of an implicit method is the *backward Euler* method:

$$\underline{\hat{x}}(n) \triangleq \underline{\hat{x}}(n-1) + T\dot{\underline{\hat{x}}}(n) \;=\; \underline{\hat{x}}(n-1) + Tf[n, \underline{\hat{x}}(n), \underline{u}(n)] \tag{7.11}$$

Because the derivative is now evaluated at time n instead of $n-1$, the backward Euler method is implicit. Notice, however, that if time were reversed, it would become explicit; in other words, backward Euler is implicit in forward time and explicit in reverse time.

[8] Normally one or more output signals $\underline{y}$ are defined as linear combinations of the state vector $\underline{x}$, *viz.*, $\underline{y}(t) = C\,\underline{x}(t)$. However, we can define the state itself as the output for now, and form any desired linear combinations separately.

7.4.4 Trapezoidal Rule

The *trapezoidal rule* is defined by

$$\underline{\hat{x}}(n) \triangleq \underline{\hat{x}}(n-1) + T\,\frac{\dot{\underline{\hat{x}}}(n) + \dot{\underline{\hat{x}}}(n-1)}{2}. \tag{7.12}$$

Thus, the trapezoidal rule is driven by the *average* of the derivative estimates at times n and $n-1$. The method is implicit in either forward or reverse time.

The trapezoidal rule gets its name from the fact that it approximates an integral by summing the areas of trapezoids. This can be seen by writing Eq. (7.12) as

$$\underline{\hat{x}}(n) = \underline{\hat{x}}(n-1) + T\,\dot{\underline{\hat{x}}}(n-1) + \frac{1}{2}T\left[\dot{\underline{\hat{x}}}(n) - \dot{\underline{\hat{x}}}(n-1)\right]$$

Imagine a plot of $\dot{\underline{\hat{x}}}(n)$ versus n, and connect the samples with linear segments to form a sequence of trapezoids whose areas must be summed to yield an approximation to $\underline{\hat{x}}(n)$. Then the integral at time n, $\underline{\hat{x}}(n)$, is given by the integral at time $n-1$, $\underline{\hat{x}}(n-1)$, plus the area of the next rectangle, $T\,\dot{\underline{\hat{x}}}(n-1)$, plus the area of the new triangular piece atop the new rectangle,$T\,[\dot{\underline{\hat{x}}}(n) - \dot{\underline{\hat{x}}}(n-1)]/2$. In other words, the integral at time n equals the integral at time $n-1$ plus the area of the next trapezoid in the sum.

An interesting fact about the trapezoidal rule is that it is equivalent to the *bilinear transform* in the linear, time-invariant case. Carrying Eq. (7.12) to the frequency domain gives

$$\begin{aligned} X(z) &= z^{-1}X(z) + T\,\frac{sX(z) + z^{-1}sX(z)]}{2} \\ \Longleftrightarrow \quad s &= \frac{2}{T}\frac{1-z^{-1}}{1+z^{-1}}. \end{aligned}$$

7.4.5 Newton's Method of Nonlinear Minimization

Newton's method [162],[166, p. 143] finds the minimum of a nonlinear (scalar) function of several variables by locally approximating the function by a quadratic surface, and then stepping to the bottom of that "bowl", which generally requires a matrix inversion. Newton's method therefore requires the function to be "close to quadratic", and its effectiveness is directly tied to the accuracy of that assumption. For smooth functions, Newton's method gives very rapid *quadratic convergence* in the last stages of iteration. Quadratic convergence implies, for example, that the number of significant digits in the minimizer approximately doubles each iteration.

Newton's method may be derived as follows: Suppose we wish to minimize the real, positive function $J(\underline{x})$ with respect to $\underline{x}$. The vector Taylor expansion [547] of $J(\underline{x})$ about $\underline{\hat{x}}$ gives

$$J(\underline{\hat{x}}^*) = J(\underline{\hat{x}}) + J'(\underline{\hat{x}})\left(\underline{\hat{x}}^* - \underline{\hat{x}}\right) + \frac{1}{2}\left(\underline{\hat{x}}^* - \underline{\hat{x}}\right)^T J''\left(\lambda\underline{\hat{x}}^* + \overline{\lambda}\underline{\hat{x}}\right)\left(\underline{\hat{x}}^* - \underline{\hat{x}}\right),$$

for some $0 \le \lambda \le 1$, where $\overline{\lambda} \triangleq 1 - \lambda$. It is now necessary to assume that $J''\left(\lambda\underline{\hat{x}}^* + \overline{\lambda}\underline{\hat{x}}\right) \approx J''(\underline{\hat{x}})$. Differentiating with respect to $\underline{\hat{x}}^*$, where $J(\underline{\hat{x}}^*)$ is presumed to be minimized, this becomes

$$0 = 0 + J'(\underline{\hat{x}}) + J''(\underline{\hat{x}})\left(\underline{\hat{x}}^* - \underline{\hat{x}}\right).$$

Solving for $\underline{\hat{x}}^*$ yields

$$\underline{\hat{x}}^* = \underline{\hat{x}} - \left[J''(\underline{\hat{x}})\right]^{-1} J'(\underline{\hat{x}}). \tag{7.13}$$

Applying Eq. (7.13) iteratively, we obtain Newton's method:

$$\underline{\hat{x}}_{n+1} = \underline{\hat{x}}_n - \left[J''(\underline{\hat{x}}_n)\right]^{-1} J'(\underline{\hat{x}}_n), \quad n = 1, 2, \ldots, \tag{7.14}$$

where $\underline{\hat{x}}_0$ is given as an initial condition.

When the $J(\underline{\hat{x}})$ is any *quadratic form* in $\underline{\hat{x}}$, then $J''\left(\lambda\underline{\hat{x}}^* + \overline{\lambda}\underline{\hat{x}}\right) = J''(\underline{\hat{x}})$, and Newton's method produces $\underline{\hat{x}}^*$ in one iteration; that is, $\underline{\hat{x}}_1 = \underline{\hat{x}}^*$ for every $\underline{\hat{x}}_0$.

7.4.6 Semi-Implicit Methods

A *semi-implicit* method for numerical integration is based on an implicit method by using only one iteration of Newton's method [354, 556]. This effectively converts the implicit method into a corresponding explicit method. Best results are obtained for highly oversampled systems (*i.e.*, $f_s = 1/T$ is larger than typical audio sampling rates).

7.4.7 Semi-Implicit Backward Euler

The *semi-implicit backward Euler* method is defined by [556]

$$\underline{\hat{x}}(n) \triangleq \underline{\hat{x}}(n-1) + T\,\frac{f[n, \underline{\hat{x}}(n-1)]}{1 - T\,\ddot{\underline{\hat{x}}}(n-1)} \tag{7.15}$$

where $\ddot{\underline{\hat{x}}}(n-1)$ denotes an estimate of the second time derivative $\ddot{\underline{x}}(t)$.

7.4.8 Semi-Implicit Trapezoidal Rule

The *semi-implicit trapezoidal rule* method is given by [556]

$$\underline{\hat{x}}(n) \triangleq \underline{\hat{x}}(n-1) + \frac{T}{2}\,\frac{f[n,\underline{\hat{x}}(n-1)] + f[n-1,\underline{\hat{x}}(n-1)]}{1 - \frac{T}{2}\,\ddot{\underline{\hat{x}}}(n-1)}. \tag{7.16}$$

7.4.9 Summary

We have looked at a number of methods for solving nonlinear ordinary differential equations, including explicit, implicit, and semi-implicit numerical integration methods. Specific methods included the explicit forward Euler (similar to the finite difference approximation of §7.3.1), backward Euler (implicit), trapezoidal rule (implicit, and equivalent to the bilinear transform of §7.3.2 in the LTI case), and semi-implicit variants of the backward Euler and trapezoidal methods.

As demonstrated and discussed further in [556], implicit methods are generally more accurate than explicit methods for nonlinear systems, with semi-implicit methods (§7.4.6) typically falling somewhere in between. Semi-implicit methods therefore provide a source of improved explicit methods. See [556] and the references therein for a discussion of accuracy and stability of such schemes, as well as applied examples.

7.4.10 Further Reading in Nonlinear Methods

Other well known numerical integration methods for ODEs include second-order backward difference formulas (commonly used in circuit simulation [556]), the fourth-order Runge-Kutta method [99], and their various explicit, implicit, and semi-implicit variations. See [556] for further discussion of these and related finite-difference schemes, and for application examples in the *virtual analog* area (digitization of musically useful analog circuits). Specific digitization problems addressed in [556] include electric-guitar distortion devices [554, 557], the classic "tone stack" [553] (an often-used bass, midrange, and treble control circuit in guitar amplifiers), the Moog VCF, and other electronic components of amplifiers and effects. Also discussed in [556] is the "K Method" for nonlinear system digitization, with comparison to nonlinear wave digital filters (see Appendix F for an introduction to *linear* wave digital filters).

The topic of *real-time finite difference schemes* for virtual analog systems remains a lively research topic [555, 338, 293, 84, 264, 364, 397].

For *Partial Differential Equations* (PDEs), in which spatial derivatives are mixed with time derivatives, the finite-difference approach remains fundamental. An introduction and summary for the LTI case appear in Appendix D. See [53] for a detailed development of finite difference schemes for solving PDEs, both linear and nonlinear, applied to digital sound synthesis. Physical systems considered in [53] include bars, stiff strings, bow coupling, hammers and mallets, coupled strings and bars, nonlinear strings and plates, and acoustic tubes (voice, wind instruments). In addition to numerous finite-difference schemes, there are chapters on finite-element methods and spectral methods.

7.5 Summary of Lumped Modeling

In this chapter, we looked at the fundamentals of lumped modeling elements such as masses, springs, and dashpots. The important concept of driving-point impedance was defined and discussed, and electrical equivalent circuits were developed, along with associated elementary (circuit) network theory. Finally, we looked at basic ways of digitizing lumped elements and more complex ODEs and PDEs, including a first glance at some nonlinear methods.

Practical examples of lumped models begin in §9.3.1 on page 359. In particular, piano-like models require a "hammer" to strike the string, and §9.3.1 explicates the simplest case of an ideal point-mass striking an ideal vibrating string. In that model, when the mass is in contact with the string, it creates a *scattering junction* on the string having reflection and transmission coefficients that are first-order filters. These filters are then digitized via the bilinear transform. The ideal string itself is of course modeled as a digital waveguide. A detailed development of wave scattering at impedance-discontinuities is presented for digital waveguide models in §C.8 on page 560, and for wave digital filters in Appendix F.

Chapter 8

Transfer Function Models

For *linear time-invariant systems*, rather than build an explicit discrete-time model as in §7.3 for each mass, spring, and dashpot, (or inductor, capacitor, and resistor for virtual analog models), we may instead choose to model only the *transfer function* between selected inputs and outputs of the physical system. It should be emphasized that this is an option only when the relevant portion of the system is *linear* and *time invariant* (LTI), or at least sufficiently close to LTI. Transfer-function modeling can be considered a kind of "large-scale" or "macroscopic" modeling in which an entire physical subsystem, such as a guitar body, is modeled by a single transfer function relating specific inputs and outputs. (A transfer function can also of course be a matrix relating a vector of inputs to a vector of outputs [220].) Such models are used extensively in the field of *control system design* [150].[1]

Transfer-function modeling is often the most cost-effective way to incorporate LTI *lumped* elements (Ch. 7) in an otherwise physical computational model. For wave-propagating *distributed* systems, on the other hand, such as vibrating strings and acoustic tubes, digital waveguides models (Ch. 6) are more efficient than transfer-function models, in addition to having a precise physical interpretation that transfer-function coefficients lack. In models containing lumped elements, or distributed components that are not characterized by wave propagation, maximum computational efficiency is typically obtained by deciding which LTI portions of the model can be "frozen" as "black boxes" characterized only by their transfer functions. In return for increased computational efficiency, we sacrifice the ability to access the interior of the black box in a physically meaningful way.

[1] Estimating transfer functions based on input-output measurements is called "system identification" [288, 428]—used in advanced automatic control applications.

An example where such "macroscopic" transfer-function modeling is normally applied is the trumpet bell (§9.7.2, p. 438). A fine-grained model might use a piecewise cylindrical or piecewise conical approximation to the flaring bell [71]. However, there is normally no need for an explicit bell model in a practical virtual instrument when the bell is assumed to be LTI and spatial directivity variations are neglected. In such cases, the transmittance and reflectance of the bell can be accurately summarized by digital filters having frequency responses that are optimal approximations to the measured (or theoretical) bell responses (§9.7). However, it is then not so easy to insert a moveable virtual "mute" into the bell reflectance/transmittance filters. This is an example of the general trade-off between physical extensibilty and computational efficiency/parsimony.

Outline

In this chapter, we will look at a variety of ways to digitize macroscopic point-to-point transfer functions $\Gamma(s)$ corresponding to a desired impulse response $\gamma(t)$:

1. Sampling $\gamma(t)$ to get $\gamma(nT), n = 0, 1, 2, \ldots$
2. Pole mappings (such as $z_i = e^{s_i T}$ followed by Prony's method)
3. Modal expansion
4. Frequency-response matching using digital filter design methods

Next, we'll look at the more specialized technique known as *commuted synthesis*, in which computational efficiency may be greatly increased by interchanging ("commuting") the series order of component transfer functions. Commuted synthesis delivers large gains in efficiency for systems with a short excitation and high-order resonators, such plucked and struck strings. In Chapter 9, commuted synthesis is applied to piano modeling.

Extracting the least-damped modes of a transfer function for separate parametric implementation is often used in commuted synthesis. We look at a number of ways to accomplish this goal toward the end of this chapter.

We close the chapter with a simple example of transfer-function modeling applied to the digital *phase shifter* audio effect. This example classifies as *virtual analog* modeling, in which a valued analog device is converted to digital form in a way that preserves all valued features of the original. Further examples of transfer-function models appear in Chapter 9.

8.1 Sampling the Impulse Response

Sampling the impulse response of a system is of course quite elementary. The main thing to watch out for is aliasing, and the main disadvantage is high computational complexity when the impulse-response is long.

Since we have defined (in §7.2 on page 257) the driving-point admittance $\Gamma(s)$ as the nominal transfer function of a system port, corresponding to defining the input as driving force and the output as resulting velocity (see Fig. 7.3 on page 255), we have that $\gamma(t)$ is defined as the system impulse response. Note, however, that the driving force and observed velocity need not be at the same physical point, and in general we may freely define any physical input and output points. Nevertheless, if the outputs are in velocity units and the inputs are in force units, then the transfer-function matrix will have units of admittance, and we will assume this for simplicity.

Sampling the impulse response can be expressed mathematically as $\gamma(t) \to T\gamma(nT) \to \gamma(n)$.[2] In practice, we can only record a finite number of impulse-response samples. Usually a graceful taper (*e.g.*, using the right half of an FFT window, such as the Hann window) yields better results than simple truncation. The system model is then implemented as a Finite Impulse Response (FIR) digital filter (§2.5.4, p. 56). The next section describes the related *impulse-invariant method* for digital filter design which derives an *infinite* impulse response (IIR) digital filter that matches the analog filter impulse response exactly at the sampling times.

Sampling the impulse response has the advantage of preserving resonant frequencies (see next section), but its big disadvantage is *aliasing* of the frequency response. No "system" is truly bandlimited. For example, even a simple mass and dashpot with a nonzero initial condition produces a continuous decaying exponential response that is not bandlimited.

Before a continuous impulse response is sampled, a *lowpass filter* should be considered for eliminating all frequency components at half the sampling rate and above. In other words, the system itself should be "lowpassed" to avoid aliasing in many applications. (On the other hand, there are also many applications in which the frequency-response aliasing is not objectional to the ear.) If the system is linear and time invariant, and if we excite the system with input signals and initial conditions that are similarly bandlimited to less than half the sampling rate, no signal inside the system or appearing at the outputs will be aliased. In other words, these conditions yield an ideal bandlimited system simulation that remains exact (for the bandlimited signals) at the sampling instants.

Note, however, that time variation or nonlinearity (both common in practical instruments), together with feedback, will "pump" the signal spectrum higher and higher until aliasing is ultimately encountered (see §6.13 on page 243). For this reason, feedback loops in the digital system may need additional lowpass filtering to attenuate newly generated high frequencies.

---

[2]For convenience, we typically define the discrete-time counterpart of a continuous-time function as though the sampling interval were $T = 1$.

A sampled impulse response is an example of a *nonparametric representation* of a linear, time-invariant system. It is not usually regarded as a physical model, even when the impulse-response samples have a physical interpretation (such as when no anti-aliasing filter is used).

8.2 Impulse Invariant Method

The *impulse-invariant method* converts analog filter transfer functions to digital filter transfer functions in such a way that the impulse response is the same (invariant) at the sampling instants [343], [362, pp. 216–219]. Thus, if $\gamma(t)$ denotes the impulse-response of an analog (continuous-time) filter, then the digital (discrete-time) filter given by the impulse-invariant method will have impulse response $\gamma(nT)$, where T denotes the sampling interval in seconds. Moreover, the order of the filter is preserved, and IIR analog filters map to IIR digital filters. However, the digital filter's frequency response is an *aliased* version of the analog filter's frequency response.[3]

To derive the impulse-invariant method, we begin with the analog transfer function

$$\Gamma_a(s) \triangleq \frac{B_a(s)}{A_a(s)} \triangleq \frac{b_a(0)s^{N-1} + b_a(1)s^{N-2} + \cdots + b_a(N-2)s + b_a(N-1)}{s^N + a_a(1)s^{N-1} + \cdots + a_a(N-1)s + a_a(N)} \tag{8.1}$$

and perform a partial fraction expansion (PFE) down to first-order terms [449]:[4]

$$\Gamma_a(s) = \sum_{i=1}^{N} \frac{K_i}{s - s_i},$$

where s_i is the ith pole of the analog system, and K_i is its residue [449]. Assume that the system is at least marginally stable [449] so that there are no poles in the right-half plane ($\text{re}\{s_i\} \leq 0$). Such a PFE is always possible when $\Gamma(s)$ is a *strictly proper transfer function* (more poles than zeros [449]).[5] Performing the inverse Laplace transform on the partial fraction expansion we obtain the impulse response in terms of the system poles and

[3] Frequency-domain aliasing is discussed in [451].

[4] `http://ccrma.stanford.edu/~jos/filters/Partial_Fraction_Expansion.html`

[5] The *strictly proper* constraint is natural in practice because the frequency responses of typical real-world systems generally roll off at least -6 dB per octave. Notice that Eq. (8.1) becomes $b_a(0)/s$ as $s \to \infty$, which is a -6 dB/octave roll off. Setting $b_a(0) = 0$ and $b_a(1) \neq 0$ yields a -12 dB/octave roll-off, and so on. However, this depends on the physical units chosen for the input and output signals of the system. For example, consider an ideal nonzero mass driven by a force; while the resulting velocity and displacement go to zero as frequency goes to infinity (for any finite applied force), the acceleration does not roll off, being proportional to applied force by Newton's 2nd law.

residues:

$$\gamma_a(t) = \sum_{i=1}^{N} K_i e^{s_i t}, \quad t \geq 0.$$

We now sample at intervals of T seconds to obtain the digital impulse response

$$\gamma_d(n) \triangleq \gamma_a(nT) = \sum_{i=1}^{N} K_i e^{s_i nT}, \quad n = 0, 1, 2, \ldots.$$

Taking the z transform gives the digital filter transfer function designed by the impulse-invariant method:

$$\Gamma_d(z) = \sum_{i=1}^{N} \frac{K_i}{1 - e^{s_i T} z^{-1}} \triangleq \frac{B_d(z)}{A_d(z)}$$

We see that the s-plane poles s_i have mapped to the z-plane poles

$$\boxed{z_i \triangleq e^{s_i T}} \tag{8.2}$$

and the residues have remained unchanged. Clearly we must have $-\pi < \text{im}\{s_i\}T < \pi$, *i.e.*, the analog poles must lie within the bandwidth spanned by the digital sampling rate $f_s = 1/T$. Otherwise, the pole angle $\text{im}\{s_i\}T$ will be aliased into the interval $[-\pi, \pi)$. Stability is preserved since $\text{re}\{s_i\} \leq 0 \Leftrightarrow |z_i| \leq 1$.

Note that the series combination of two digital filters designed by the impulse-invariant method is not impulse invariant. In other terms, the convolution of two sampled analog signals is not the same as the sampled convolution of those analog signals. This is easy to see when aliasing is considered. For example, let one signal be the impulse response of an ideal lowpass filter cutting off below half the sampling rate. Then this signal will not alias when sampled, and its convolution with any second signal will similarly not alias when sampled. However, if the second signal does alias upon sampling, then this aliasing is gone when the convolution precedes the sampling, and the results cannot be the same in the two cases.

8.3 Matched Z Transformation

The *matched z transformation* uses the same pole-mapping Eq. (8.2) as in the impulse-invariant method, but the zeros are handled differently. Instead of only mapping the poles of the partial fraction expansion and letting the zeros fall where they may, the matched z transformation maps *both* the poles and zeros in the *factored form* of the transfer function [362, pp. 224–226].

The factored form [449] of a transfer function

$$H(s) \triangleq \frac{B(s)}{A(s)} \triangleq \frac{b_M s^M + \cdots b_1 s + b_0}{a_N s^N + \cdots a_1 s + a_0} \tag{8.3}$$

can be written as

$$H(s) = \left(\frac{b_M}{a_N}\right) \frac{\prod_{i=1}^{M}(s-\xi_i)}{\prod_{i=1}^{N}(s-p_i)} \tag{8.4}$$

The matched z transformation is carried out by replacing each first-order term of the form $(s+a)$ by its digital equivalent $1-e^{-aT}z^{-1}$, *i.e.*,

$$\boxed{s+a \;\rightarrow\; 1-e^{-aT}z^{-1}} \tag{8.5}$$

to get

$$H(s) = \left(\frac{b_M}{a_N}\right) \frac{\prod_{i=1}^{M}(1-e^{\xi_i T}z^{-1})}{\prod_{i=1}^{N}(1-e^{p_i T}z^{-1}} \tag{8.6}$$

Thus, the matched z transformation normally yields different digital zeros than the impulse-invariant method. The impulse-invariant method is generally considered superior to the matched z transformation [343].

8.3.1 Relation to Finite Difference Approximation

The Finite Difference Approximation (FDA) (§7.3.1, p. 265) is a special case of the matched z transformation applied to the point $s=0$. To see this, simply set $a=0$ in Eq. (8.5) to obtain

$$s \;\rightarrow\; 1-z^{-1} \tag{8.7}$$

which is the FDA definition in the frequency domain given in Eq. (7.3) on page 265.

Since the FDA equals the match z transformation for the point $s=0$, it maps analog dc ($s=0$) to digital dc ($z=1$) exactly. However, that is the only point on the frequency axis that is perfectly mapped, as shown in Fig. 7.15 on page 269.

8.4 Pole Mapping with Optimal Zeros

We saw in the preceding sections that both the impulse-invariant and the matched-z transformations map poles from the left-half s plane to the interior of the unit circle in the z plane via

$$z_i = e^{s_i T} \tag{8.8}$$

where s_i is the location of the ith pole in the s plane (assumed to lie in the strip $|\text{im}\{s\}| < \pi/T$ to avoid aliasing). The zeros, on the other hand, were different because the impulse-invariant method started with the partial fraction expansion while the matched-z transformation started with the factored form of the transfer function.

Therefore, an obvious generalization is to map the poles according to Eq. (8.8), but compute the zeros in some optimal way, such as by *Prony's method* [449, p. 393],[273, 297].

It is hard to do better Eq. (8.8) as a pole mapping from s to z, when aliasing is avoided, because it preserves both the resonance frequency and bandwidth for a complex pole [449]. Therefore, good practical modeling results can be obtained by optimizing the zeros (residues) to achieve audio criteria given these fixed poles. Alternatively, only the least-damped poles need be constrained in this way, *e.g.*, to fix and preserve the most important resonances of a stringed-instrument body or acoustic space.

8.5 Modal Expansion

A well known approach to transfer-function modeling is called *modal synthesis*, introduced in §1.3.9 on page 20 [5, 299, 6, 145, 381, 30]. Modal synthesis may be defined as constructing a source-filter synthesis model in which the filter transfer function is implemented as a sum of first- and/or second-order filter sections (*i.e.*, as a parallel filter bank in which each filter is at most second-order—this was reviewed in §1.3.9 on page 20). In other words, the physical system is represented as a superposition of individual modes driven by some external excitation (such as a pluck or strike).

In acoustics, the term *mode of vibration*, or *normal mode*, normally refers to a single-frequency spatial eigensolution of the governing wave equation. For example, the modes of an ideal vibrating string are the harmonically related sinusoidal string-shapes having an integer number of uniformly spaced zero-crossings (or *nodes*) along the string, including its endpoints. As first noted by Daniel Bernoulli (§A.2, p. 447), acoustic vibration can be expressed as a superposition of component sinusoidal vibrations, *i.e.*, as a superposition of modes.[6]

When a single mode is excited by a sinusoidal driving force, all points of the physical object vibrate at the same temporal frequency (cycles per second), and the mode shape becomes proportional to the spatial amplitude envelope of the vibration. The sound emitted from the top plate of a guitar,

[6]Modal synthesis could well be renamed "Bernoulli synthesis," as Daniel Bernoulli was quite alone in advocating the concept of seeing general vibration as a superposition of "simple" sinusoidal vibrations. This view was resisted by Euler and d'Alembert at the time [103].

for example, can be represented as a weighted sum of the radiation patterns of the respective modes of the top plate, where the weighting function is constructed according to how much each mode is excited (typically by the guitar bridge) [143, 390, 109, 205, 209].

The impulse-invariant method (§8.2, p. 282), can be considered a special case of modal synthesis in which a continuous-time s-plane transfer function is given as a starting point. More typically, modal synthesis starts with a measured *frequency response*, and a second-order parallel filter bank is fit to that in some way. In particular, any filter-design technique may be used (§8.6 on page 290), followed by a conversion to second-order parallel form.

Modal expansions find extensive application in industry for determining parametric frequency responses (superpositions of second-order modes) from measured vibration data [299]. Each mode is typically parametrized in terms of its resonant frequency, bandwidth (or damping), and gain (most generally complex gain, to include phase).

Modal synthesis can also be seen as a special case of *source-filter synthesis*, which may be defined as any signal model based on factoring a sound-generating process into a filter driven by some (usually relatively simple) excitation signal. An early example of source-filter synthesis is the *vocoder* (§A.6.1, p. 451). Whenever the filter in a source-filter model is implemented as a parallel second-order filter bank, we can call it modal synthesis (although, strictly speaking, each filter section should correspond to a resonant mode of the modeled resonant system).

Also related to modal synthesis is so-called *formant synthesis*, used extensively in speech synthesis [219, 255, 389, 40, 81, 492, 491, 313, 363, 297]. A *formant* is simply a filter resonance having some center-frequency, bandwidth, and (sometimes specified) gain. Thus, a formant corresponds to a single mode of vibration in the vocal tract. Many text-to-speech systems in use today are based on the Klatt formant synthesizer for speech [255].

Since the importance of spectral formants in sound synthesis has more to do with the way we hear than with the physical parameters of a system, formant synthesis is probably best viewed as a *spectral modeling* synthesis method [457], as opposed to a physical modeling technique [435]. An exception to this rule may occur when the system consists physically of a parallel bank of second-order resonators, such as an array of tuning forks or Helmholtz resonators. In such a case, the mode parameters correspond to physically independent objects; this is of course rare in practice.

In the linear, time-invariant case, only modes in the range of human hearing need to be retained in the model. Also, any "uncontrollable" or "unobservable" modes should obviously be left out as well. With these simplificiation, the modal representation is generally more efficient than an explicit mass-spring-dashpot digitization (§7.3). On the other hand, since the modal representation is usually not directly physical, nonlinear extensions may be difficult and/or behave unnaturally.

As in the impulse-invariant method (§8.2), starting with an order N transfer function $H(s)$ describing the input-output behavior of a physical system, a modal description can be obtained immediately from the *partial fraction expansion* (PFE):

$$H(s) \triangleq \frac{B(s)}{A(s)} \triangleq \frac{b_M s^M + \cdots b_1 s + b_0}{a_N s^N + \cdots a_1 s + a_0} = \sum_{i=1}^{N} \frac{r_i}{s - p_i} \tag{8.9}$$

where p_i denotes the ith pole, ξ_i is the ith zero, and r_i is the residue of the ith pole [449]. (For simplicity of notation, Eq. (8.9) is written for the case $M < N$ and p_i distinct. See [449] for the details when this is not the case.) Since the system is real, complex poles will occur in complex conjugate pairs. This means that for each complex term $r_i/(s - p_i)$, the PFE will also include the complex-conjugate term $\overline{r}_i/(s - \overline{p}_i)$. Such terms can be summed to create half as many real second-order terms:

$$\frac{r_i}{s - p_i} + \frac{\overline{r}_i}{s - \overline{p}_i} = \frac{(r_i + \overline{r}_i)s - (r_i\overline{p}_i + \overline{r}_i p_i)}{s^2 - (p_i + \overline{p}_i)s + |p_i|^2} \triangleq \frac{b_{i1}s + b_{i0}}{s^2 - a_{i1}s + a_{i0}} \tag{8.10}$$

Let N_c denote the number of complex pole pairs, and $N_r = N - 2N_c$ be the number of real poles. Then the modal expansion can be written as

$$H(s) = \sum_{i=1}^{N_r} \frac{r_i}{s - p_i} + \sum_{i=1}^{N_c} \frac{b_{i1}s + b_{i0}}{s^2 - a_{i1}s + a_{i0}}. \tag{8.11}$$

The real poles, if any, can be paired off into a sum of second-order sections as well, with one first-order section left over when N_r is odd. Note that each component second-order mode consists of one zero and two poles. The zero is necessary to adjust the phase of the resonant mode. The mode phase affects the frequency response at frequencies between resonances (from smoothly varying to the appearance of a deep notch somewhere between them).

A typical practical implementation of each mode in the sum is by means of second-order filter, or *biquad section* [449]. A biquad section is easily converted from continuous- to discrete-time form, such as by the impulse-invariant method of §8.2. The final discrete-time model then consist of a parallel bank of second-order digital filters

$$H_d(z) = \sum_{i=1}^{N_d} G_i \frac{1 + b_i z^{-1}}{1 + a_{i1}z^{-1} + a_{i2}z^{-2}} \tag{8.12}$$

where $N_d = N/2$ when the number of real poles is even, or $(N+1)/2$ otherwise. The filter coefficients b_i and a_{ij} are now digital second-order section coefficients which can be computed from the continuous-time filter coefficients, if they are known, or they can be directly estimated from discrete-time data.

Expressing complex poles of $H_d(z)$ in polar form as $p_i \triangleq R_i e^{j\theta_i}$, (where now we assume p_i denotes a pole in the z plane), we obtain the classical relations [449]

$$\begin{aligned} a_{i1} &= p_i + \overline{p}_i = 2R_i\cos(\theta_i) \\ a_{i2} &= |p_i|^2 = R_i^2. \end{aligned} \tag{8.13}$$

These parameters are in turn related to the modal parameters resonance frequency f_i (Hz) and bandwidth B_i (Hz) by

$$\begin{aligned} \theta_i &= 2\pi f_i/f_s \\ R_i &= e^{-\pi B_i/f_s}. \end{aligned}$$

Often f_i and B_i are estimated from a measured log-magnitude frequency-response (see §8.8 on page 309 for a discussion of various methods).

It remains to determine the overall section gain G_i in Eq. (8.12), and the coefficient b_i determining the zero location. If $b_i = 0$ (as is common in practice, since the human ear is not very sensitive to spectral nulls), then G_i can be chosen to match the measured peak height. A particularly simple automatic method is to use least-squares minimization to compute G_i using Eq. (8.12) with a_{i1} and a_{i2} fixed (and $b_i = 0$). Generally speaking, least-squares minimization is always immediate when the error to be minimized is linear in the unknown parameter (in this case G_i).

To estimate b_i in Eq. (8.12), an effective method is to work initially with a *complex*, first-order partial fraction expansion, and use least squares to determine complex gain coefficients r_i for first-order (complex) terms of the form

$$\frac{r_i}{s - p_i}$$

as introduced in Eq. (8.9) above. The angle and radius of p_i can still be determined by measuring peaks in the log-magnitude frequency response. When the first-order terms are paired and summed, the b_i coefficients will "fall out."

Sections implementing two real poles are not conveniently specified in terms of resonant frequency and bandwidth. In such cases, it is more common to work with the filter coefficients b_i and a_i directly, as they are often computed by system-identification software [288, 428] or general purpose filter design utilities, such as in Matlab or Octave [443].

To summarize, numerous techniques exist for estimating modal parameters from measured frequency-response data, and some of these are discussed in §8.8 below. Specifically, resonant modes can be estimated from the magnitude, phase, width, and location of peaks in the Fourier transform of a recorded (or estimated) impulse response Another well known technique is Prony's method [449, p. 393], [297]. There is also a large literature on other methods for estimating the parameters of exponentially decaying sinusoids; examples include the *matrix-pencil method* [274], and *parametric spectrum analysis* in general [237].

State Space Approach to Modal Expansions

The preceding discussion of modal synthesis was based primarily on fitting a sum of biquads to measured frequency-response peaks. A more general way of arriving at a modal representation is to first form a *state space model* of the system [449], and then convert to the modal representation by *diagonalizing* the state-space model. This approach has the advantage of preserving system behavior between the given inputs and outputs. Specifically, the similarity transform used to diagonalize the system provides new input and output gain vectors which properly excite and observe the system modes precisely as in the original system. This procedure is especially more convenient than the transfer-function based approach above when there are multiple inputs and outputs. For some mathematical details, see [449][7] For a related worked example, see §C.17.6 on page 627.

Delay Loop Expansion

When a subset of the resonating modes are nearly harmonically tuned, it can be much more computationally efficient to use a *filtered delay loop* (see §2.6.5 on page 62) to generate an entire *quasi-harmonic series of modes* rather than using a biquad for each modal peak [439]. In this case, the resonator model becomes

$$H(z) \;=\; \sum_{k=1}^{N} \frac{a_k}{1 - H_k(z) z^{-N_k}},$$

where N_k is the length of the delay line in the kth comb filter, and $H_k(z)$ is a low-order filter which can be used to adjust finely the amplitudes and frequencies of the resonances of the kth comb filter [428]. Recall (Chapter 6) that a single filtered delay loop efficiently models a *distributed* 1D propagation medium such as a vibrating string or acoustic tube. More abstractly, a superposition of such quasi-harmonic mode series can provide a

[7] `http://ccrma.stanford.edu/~jos/filters/Modal_Representation.html`

computationally efficient *psychoacoustic equivalent approximation* to arbitrary collections of modes in the range of human hearing.

Note that when $H_k(z)$ is close to -1 instead of $+1$, primarily only odd harmonic resonances are produced, as has been used in modeling the clarinet [431].

8.6 Frequency-Response Matching Using Digital Filter Design Methods

Given force inputs and velocity outputs, the *frequency response* of an ideal mass was given in Eq. (7.1.2) on page 255 as

$$\Gamma_m(j\omega) = \frac{1}{mj\omega},$$

and the frequency response for a spring was given by Eq. (7.1.3) as

$$\Gamma_k(j\omega) = \frac{j\omega}{k}.$$

Thus, an ideal mass is an *integrator* and an ideal spring is a *differentiator*. The modeling problem for masses and springs can thus be posed as a problem in *digital filter design* given the above desired frequency responses. More generally, the admittance frequency response "seen" at the port of a general Nth-order LTI system is, from Eq. (8.3) on page 284,

$$H(s) \triangleq \frac{B(s)}{A(s)} \triangleq \frac{b_M s^M + \cdots b_1 s + b_0}{a_N s^N + \cdots a_1 s + a_0} \tag{8.14}$$

where we assume $M < N$. Similarly, point-to-point "trans-admittances" can be defined as the velocity Laplace transform at one point on the physical object divided by the driving-force Laplace transform at some other point. There is also of course no requirement to always use driving force and observed velocity as the physical variables; velocity-to-force, force-to-force, velocity-to-velocity, force-to-acceleration, etc., can all be used to define transfer functions from one point to another in the system. For simplicity, however, we will prefer admittance transfer functions here.

8.6.1 Ideal Differentiator (Spring Admittance)

Figure 8.1 shows a graph of the frequency response of the ideal differentiator (spring admittance). In principle, a *digital differentiator* is a filter whose frequency response $H(e^{j\omega T})$ optimally approximates $j\omega$ for ωT between

$-\pi$ and π. Similarly, a *digital integrator* must match $1/j\omega$ along the unit circle in the z plane. The reason an exact match is not possible is that the ideal frequency responses $j\omega$ and $1/j\omega$, when wrapped along the unit circle in the z plane, are not "smooth" functions any more (see Fig. 8.1). As a result, there is no filter with a rational transfer function (*i.e.*, finite order) that can match the desired frequency response exactly.

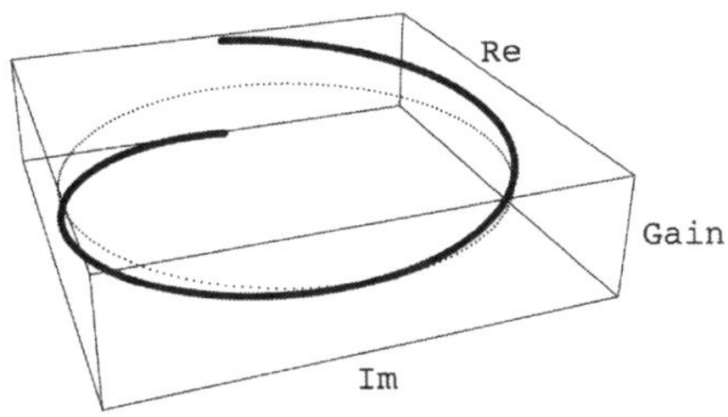

Figure 8.1: Imaginary part of the frequency response $H(e^{j\omega T}) = j\omega$ of the ideal digital differentiator plotted over the unit circle in the z plane (the real part being zero).

The discontinuity at $z = -1$ alone is enough to ensure that no finite-order digital transfer function exists with the desired frequency response. As with bandlimited interpolation (§4.4), it is good practice to reserve a "guard band" between the highest needed frequency $f_{\max}$ (such as the limit of human hearing) and half the sampling rate $f_s/2$. In the guard band $[f_{\max}, f_s/2]$, digital filters are free to smoothly vary in whatever way gives the best performance across frequencies in the audible band $[0, f_{\max}]$ at the lowest cost. Figure 8.2 on page 296 shows an example. Note that, as with filters used for bandlimited interpolation, a small increment in oversampling factor yields a much larger decrease in filter cost (when the sampling rate is near $2f_{\max}$).

In the general case of Eq. (8.14) with $s = j\omega$, digital filters can be designed to implement arbitrarily accurate admittance transfer functions by finding an optimal rational approximation to the complex function of a single real variable ω

$$H(e^{j\omega}) = \frac{B(j\omega)}{A(j\omega)} = \frac{b_M(j\omega)^M + \cdots b_1 j\omega + b_0}{a_N(j\omega)^N + \cdots a_1 j\omega + a_0}$$

over the interval $-\omega_{\max} \leq \omega \leq \omega_{\max}$, where $\omega_{\max}T < \pi$ is the upper limit of human hearing. For small guard bands $\delta \triangleq \pi - \omega_{\max}T$, the filter order required for a given error tolerance is approximately inversely proportional to δ.

8.6.2 Digital Filter Design Overview

This section (adapted from [428]), summarizes some of the more commonly used methods for digital filter design aimed at matching a *nonparametric* frequency response, such as typically obtained from input/output measurements. This problem should be distinguished from more classical problems with their own specialized methods, such as designing lowpass, highpass, and bandpass filters [343, 362], or peak/shelf equalizers [560, 449], and other utility filters designed from *a priori* mathematical specifications.

The problem of fitting a digital filter to a prescribed frequency response may be formulated as follows. To simplify, we set $T = 1$.

Given a continuous complex function $H(e^{j\omega})$, $-\pi < \omega \leq \pi$, corresponding to a causal desired frequency response,[8] find a stable digital filter of the form

$$\hat{H}(z) \triangleq \frac{\hat{B}(z)}{\hat{A}(z)},$$

where

$$\hat{B}(z) \triangleq \hat{b}_0 + \hat{b}_1 z^{-1} + \cdots + \hat{b}_{\hat{N}_b} z^{-\hat{N}_b} \quad (8.15)$$

$$\hat{A}(z) \triangleq 1 + \hat{a}_1 z^{-1} + \cdots + \hat{a}_{\hat{N}_a} z^{-\hat{N}_a}, \quad (8.16)$$

with $\hat{N}_b, \hat{N}_a$ given, such that some *norm* of the error

$$J(\hat{\theta}) \triangleq \left\| H(e^{j\omega}) - \hat{H}(e^{j\omega}) \right\| \quad (8.17)$$

is minimum with respect to the filter coefficients

$$\hat{\theta}^K \triangleq [\hat{b}_0, \hat{b}_1, \ldots, \hat{b}_{\hat{N}_b}, \hat{a}_1, \hat{a}_2, \ldots, \hat{a}_{\hat{N}_a}].$$

The filter coefficients are constrained to lie in some subset $\hat{\Theta} \subset \Re^{\hat{N}}$, where $\hat{N} \triangleq \hat{N}_a + \hat{N}_b + 1$. The filter coefficients may also be complex, in which case $\hat{\Theta} \subset \mathbf{C}^{\hat{N}}$.

The approximate filter $\hat{H}$ is typically *constrained* to be stable, and since $\hat{B}(z)$ is causal (no positive powers of z), stability implies causality. Consequently, the impulse response of the model $\hat{h}(n)$ is zero for $n < 0$.

The filter-design problem is then to find a (strictly) stable $\hat{N}_a$-pole, $\hat{N}_b$-zero digital filter which minimizes some norm of the error in the frequency-response. This is fundamentally *rational approximation of a complex function of a real (frequency) variable, with constraints on the poles.*

[8] A "causal frequency response" $H(e^{j\omega})$ is the Fourier transform of a causal impulse response $h(n)$ (*i.e.*, $h(n) = 0$ for all $n < 0$). Extensions to finitely noncausal spectra are straightforward: Time-shift the desired impulse response to make it causal, perform the filter design, then reverse-time-shift the filter numerator if needed.

While the filter-design problem has been formulated quite naturally, it is difficult to solve in practice. The strict stability assumption yields a compact space of filter coefficients $\hat{\Theta}$, leading to the conclusion that a best approximation $\hat{H}^*$ *exists* over this domain.[9] Unfortunately, the norm of the error $J(\hat{\theta})$ typically is not a *convex*[10] function of the filter coefficients on $\hat{\Theta}$. This means that algorithms based on gradient descent may fail to find an optimum filter due to their premature termination at a suboptimal local minimum of $J(\hat{\theta})$.

Fortunately, there is at least one norm whose global minimization may be accomplished in a straightforward fashion without need for initial guesses or ad hoc modifications of the complex (phase-sensitive) IIR filter-design problem—the *Hankel norm* [155, 428, 177, 36]. Hankel norm methods for digital filter design deliver a spontaneously *stable* filter of any desired order without imposing coefficient constraints in the algorithm.

An alternative to Hankel-norm approximation is to reformulate the problem, replacing Eq. (8.17) with a modified error criterion so that the resulting problem can be solved by *linear least-squares* or *convex optimization* techniques. Examples include

- *Pseudo-norm* minimization: (Pseudo-norms can be zero for nonzero functions.) For example, Padé approximation falls in this category. In Padé approximation, the first $\hat{N}_a + \hat{N}_b + 1$ samples of the impulse-response $h(n)$ of H are matched exactly, and the error in the remaining impulse-response samples is ignored.

- *Ratio Error*: Minimize $\| H(e^{j\omega})/\hat{H}(e^{j\omega}) \|$ subject to $\hat{B}(z) = 1$. Minimizing the L^2 norm of the ratio error yields the class of methods known as *linear prediction* techniques [20, 296, 297]. Since, by the definition of a norm, we have $\| e^{j\theta} E(e^{j\omega}) \| = \| E(e^{j\omega}) \|$, it follows that $\| H/\hat{H} \| = \| |H|/|\hat{H}| \|$; therefore, *ratio error methods ignore the phase of the approximation.* It is also evident that ratio error is minimized by making $|\hat{H}(e^{j\omega})|$ larger than $|H(e^{j\omega})|$.[11] For this reason, ratio-error methods are considered most appropriate for modeling the *spectral envelope* of $|H(e^{j\omega})|$. It is well known that these methods are

[9] A *stability margin* may be specified, for example, by requiring all poles p_i to satisfy $|p_i| \leq 1 - \delta$, where $1 \gg \delta > 0$ determines the stability margin. In particular, with this specification on the poles, the impulse response $h(n)$ must decay asymptotically at least as fast as $(1-\delta)^n$ [449].

[10] Since convex functions bulge upward, like the top of a circle, it makes sense to *maximize* them in general. Since our error-surface formulation is always *minimized* with respect to the filter parameters, we could call it "concave" or "convex from below". In the optimization domain, however, a "convex" region is one that contains all of its chords, thus it applies to either case.

[11] $|\hat{H}|$ cannot go to infinity since the constraint $\hat{B}(z) = 1$ and the stability constraint imply that $\ln|\hat{H}(e^{j\omega})|$ is zero-mean by the argument principle [297, 326].

fast and exceedingly robust in practice, and this explains in part why they are used almost exclusively in some data-intensive applications such as speech modeling and other spectral-envelope applications. In some applications, such as adaptive control or forecasting, the fact that linear prediction error is minimized can justify their choice.

- *Equation error*: Minimize

$$\left\| \hat{A}(e^{j\omega})H(e^{j\omega}) - \hat{B}(e^{j\omega}) \right\| = \left\| \hat{A}(e^{j\omega}) \left[H(e^{j\omega}) - \hat{H}(e^{j\omega}) \right] \right\|.$$

  When the L^2 norm of equation-error is minimized, the problem becomes solving a set of $\hat{N} = \hat{N}_a + \hat{N}_b + 1$ linear equations.

  The above expression makes it clear that equation-error can be seen as a frequency-response error weighted by $|\hat{A}(e^{j\omega})|$. Thus, relatively large errors can be expected where the poles of the optimum approximation (roots of $\hat{A}(z)$) approach the unit circle $|z| = 1$. While this may make the frequency-domain formulation seem ill-posed, in the time-domain, *linear prediction error* is minimized in the L^2 sense, and in certain applications this is ideal. (Equation-error methods thus provide a natural extension of ratio-error methods to include zeros.) Using so-called *Steiglitz-McBride iterations* [287, 449, 288], the equation-error solution iteratively approaches the norm-minimizing solution of Eq. (8.17) for the L^2norm.

  Examples of minimizing equation error using the matlab function `invfreqz` are given in §8.6.3 and §8.6.4 below. See [449, Appendix I] (based on [428, pp. 48–50]) for a discussion of equation-error IIR filter design and a derivation of a fast equation-error method based on the Fast Fourier Transform (FFT) (used in `invfreqz`).

- Conversion to *real-valued approximation*: For example, *power spectrum matching*, i.e., minimization of $\| \, |H(e^{j\omega})|^2 - |\hat{H}(e^{j\omega})|^2 \, \|$, is possible using the Chebyshev or L^∞ norm [428]. Similarly, *linear-phase* filter design can be carried out with some guarantees, since again the problem reduces to real-valued approximation on the unit circle. The essence of these methods is that the *phase-response* is eliminated from the error measure, as in the norm of the ratio error, in order to convert a complex approximation problem into a real one. Real rational approximation of a continuous curve appears to be solved in principle only under the L^∞ norm [373, 374].

- *Decoupling poles and zeros*: An effective example of this approach is Kopec's method [428] which consists of using ratio error to find the poles, computing the error spectrum $E = H/\hat{H}$, inverting it, and fitting poles again (to $1/E(e^{j\omega})$). There is a wide variety of methods which first fit poles and then zeros. None of these methods produce optimum filters, however, in any normal sense.

In addition to these modifications, sometimes it is necessary to reformulate the problem in order to achieve a different goal. For example, in some audio applications, it is desirable to minimize the *log-magnitude* frequency-response error. This is due to the way we hear spectral distortions in many circumstances. A technique which accomplishes this objective to the first order in the L^∞ norm is described in [428].

Sometimes the most important spectral structure is confined to an interval of the frequency domain. A question arises as to how this structure can be accurately modeled while obtaining a cruder fit elsewhere. The usual technique is a weighting function versus frequency. An alternative, however, is to frequency-warp the problem using a first-order *conformal map*. It turns out a first-order conformal map can be made to approximate very well frequency-resolution scales of human hearing such as the *Bark scale* or *ERB scale* [460]. Frequency-warping is especially valuable for providing an effective weighting function connection for filter-design methods, such as the Hankel-norm method, that are intrinsically do not offer choice of a *weighted* norm for the frequency-response error.

There are several methods which produce $\hat{H}(z)\hat{H}(z^{-1})$ instead of $\hat{H}(z)$ directly. A *fast spectral factorization* technique is useful in conjunction with methods of this category [428]. Roughly speaking, a size $2\hat{N}_a$ polynomial factorization is replaced by an FFT and a size $\hat{N}_a$ system of linear equations.

8.6.3 Digital Differentiator Design

We saw the ideal digital differentiator frequency response in Fig. 8.1, where it was noted that the discontinuity in the response at $\omega = \pm\pi$ made an ideal design unrealizable (infinite order). Fortunately, such a design is not even needed in practice, since there is invariably a guard band between the highest supported frequency $f_{\max}$ and half the sampling rate $f_s/2$.

Figure 8.2 illustrates a more practical design specification for the digital differentiator as well as the performance of a tenth-order FIR fit using `invfreqz` (which minimizes equation error) in Octave.[12] The weight function passed to `invfreqz` was 1 from 0 to 20 kHz, and zero from 20 kHz to half the sampling rate (24 kHz). Notice how, as a result, the amplitude

[12]Octave's `remez` function and Matlab's `firpm` function (Signal Processing Toolbox) have a special mode for designing FIR differentiators that are optimal in the Chebeshev sense, while `invfreqz` minimizes equation error, which is arguably not as good. We focus on `invfreqz` in this chapter because it offers the level of generality needed for fitting digital filters to arbitrary measured frequency responses.

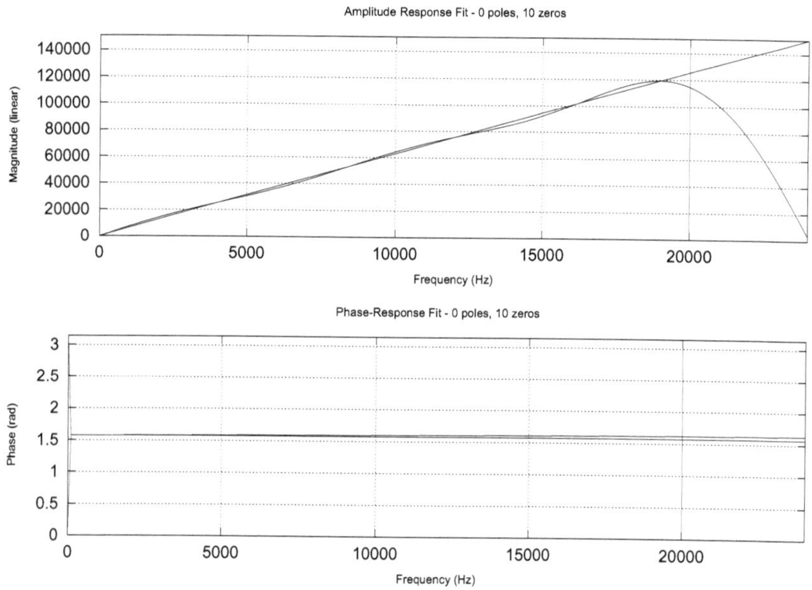

Figure 8.2: FIR differentiator designed by the matlab function **invfreqz** (Octave). Top: Overlay of the ideal amplitude response ($|j2\pi f|$), fitted filter amplitude response, and guard-band limit (at 20 kHz). Bottom: Overlay of ideal phase response ($\pi/2$ radians), fitted filter phase response, and guard-band limit (20 kHz).

response follows that of the ideal differentiator until 20 kHz, after which it rolls down to a gain of 0 at 24 kHz, as it must (see Fig. 8.1). Higher order fits yield better results. Using poles can further improve the results, but the filter should be checked for stability since `invfreqz` designs filters in the frequency domain and does not enforce stability.[13]

8.6.4 Fitting Filters to Measured Amplitude Responses

The preceding filter-design example digitized an ideal differentiator, which is an example of converting an LTI lumped modeling element into a digital filter while maximally preserving its frequency response over the audio band. Another situation that commonly arises is the need for a digital filter that matches a *measured frequency response* over the audio band.

Measured Amplitude Response

Figure 8.3 shows a plot of simulated amplitude-response measurements at 10 frequencies equally spread out between 100 Hz and 3 kHz on a log frequency scale. The "measurements" are indicated by circles. Each circle plots, for example, the output amplitude divided by the input amplitude for a sinusoidal input signal at that frequency [449]. These ten data points are then extended to dc and half the sampling rate, interpolated, and resampled to a uniform frequency grid (solid line in Fig. 8.3), as needed for FFT processing. The details of these computations are listed in Fig. 8.4 on page 299. We will fit a four-pole, one-zero, digital-filter frequency-response to these data.[14]

[13]When working with `invfreqz`, it is often helpful to multiply the desired spectrum by a linear phase term [449] corresponding to an added delay in the impulse response [428]. The filter designed by `invfreqz` will typically be stable if sufficient delay is added to the desired frequency response in this way. Thus, if a filter designed by `invfreqz` is found to be unstable, multiply the desired frequency response by a steeper negative-slope linear-phase term and repeat the design.

[14]The number of poles and zeros needed for a reasonable fit were determined empirically. No zeros gives a poor overall match, and only two poles yields a significantly less sharp resonant peak.

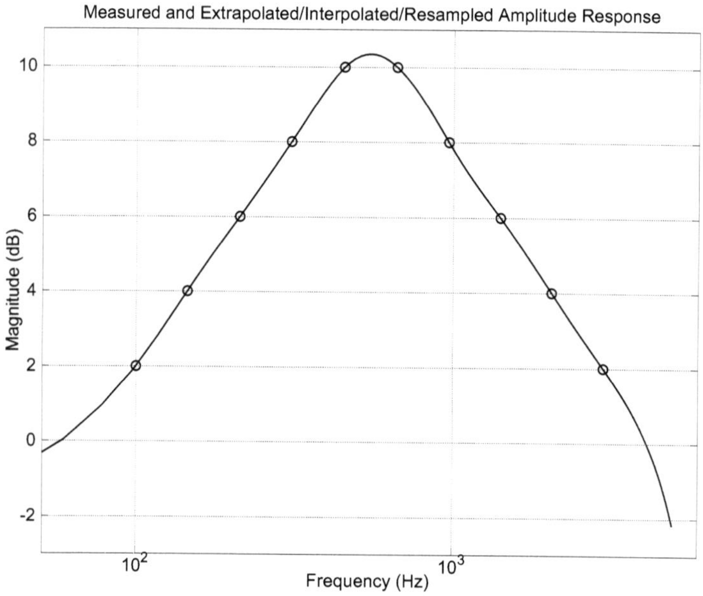

Figure 8.3: Example measured amplitude-response samples at 10 exponentially spaced frequencies. Circles: Measured amplitude-response points. Solid: Extrapolated, spline-interpolated, and uniformly resampled amplitude response, ready for `ifft`.

```
NZ = 1;       % number of ZEROS in the filter to be designed
NP = 4;       % number of POLES in the filter to be designed
NG = 10;      % number of gain measurements
fmin = 100;   % lowest measurement frequency (Hz)
fmax = 3000;  % highest measurement frequency (Hz)
fs = 10000;   % discrete-time sampling rate
Nfft = 512;   % FFT size to use
df = (fmax/fmin)^(1/(NG-1)); % uniform log-freq spacing
f = fmin * df .^ (0:NG-1);   % measurement frequency axis

% Gain measurements (synthetic example = triangular amp response):
Gdb = 10*[1:NG/2,NG/2:-1:1]/(NG/2); % between 0 and 10 dB gain

% Must decide on a dc value.
% Either use what is known to be true or pick something "maximally
% smooth".  Here we do a simple linear extrapolation:
dc_amp = Gdb(1) - f(1)*(Gdb(2)-Gdb(1))/(f(2)-f(1));

% Must also decide on a value at half the sampling rate.
% Use either a realistic estimate or something "maximally smooth".
% Here we do a simple linear extrapolation. While zeroing it
% is appealing, we do not want any zeros on the unit circle here.
Gdb_last_slope = (Gdb(NG) - Gdb(NG-1)) / (f(NG) - f(NG-1));
nyq_amp = Gdb(NG) + Gdb_last_slope * (fs/2 - f(NG));

Gdbe = [dc_amp, Gdb, nyq_amp];
fe = [0,f,fs/2];
NGe = NG+2;

% Resample to a uniform frequency grid, as required by ifft.
% We do this by fitting cubic splines evaluated on the fft grid:
Gdbei = spline(fe,Gdbe); % say 'help spline'
fk = fs*[0:Nfft/2]/Nfft; % fft frequency grid (nonneg freqs)
Gdbfk = ppval(Gdbei,fk); % Uniformly resampled amp-resp

figure(1);
semilogx(fk(2:end-1),Gdbfk(2:end-1),'-k'); grid('on');
axis([fmin/2 fmax*2 -3 11]);
hold('on'); semilogx(f,Gdb,'ok');
xlabel('Frequency (Hz)');   ylabel('Magnitude (dB)');
title(['Measured and Extrapolated/Interpolated/Resampled ',...
       'Amplitude Response']);
```

Figure 8.4: Script (matlab) for simulating a measured amplitude response at 10 exponentially spaced frequencies and extrapolating/interpolating/resampling to obtain a complete, nonparametric amplitude response, uniformly sampled at FFT frequencies. This script generated Fig. 8.3.

Desired Impulse Response

It is good to check that the desired impulse response is not overly aliased in the time domain. The impulse-response for this example is plotted in Fig. 8.5. We see that it appears quite short compared with the inverse FFT used to compute it. The script in Fig. 8.6 gives the details of this computation, and also prints out a measure of "time-limitedness" defined as the L^2 norm of the outermost 20% of the impulse response divided by its total L^2 norm—this measure was reported to be 0.02% for this example.

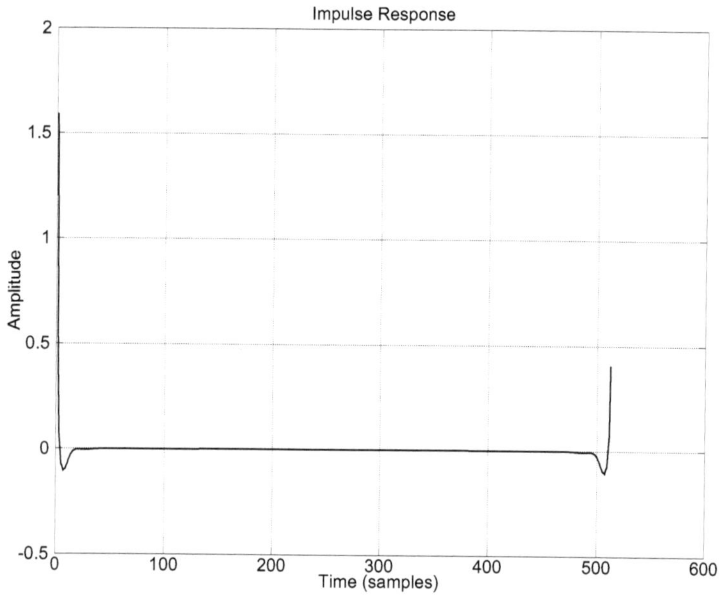

Figure 8.5: Desired impulse response obtained from a length 512 inverse FFT of the interpolated measured amplitude response in Fig. 8.3.

Note also that the desired impulse response is *noncausal.* In fact, it is zero phase [449]. This is of course expected because the desired frequency response was real (and nonnegative).

```
Ns = length(Gdbfk); if Ns~=Nfft/2+1, error("confusion"); end
Sdb = [Gdbfk,Gdbfk(Ns-1:-1:2)]; % install negative-frequencies

S = 10 .^ (Sdb/20); % convert to linear magnitude
s = ifft(S); % desired impulse response
s = real(s); % any imaginary part is quantization noise
tlerr = 100*norm(s(round(0.9*Ns:1.1*Ns)))/norm(s);
disp(sprintf(['Time-limitedness check: Outer 20%% of impulse ' ...
              'response is %0.2f %% of total rms'],tlerr));
% = 0.02 percent
if tlerr>1.0 % arbitrarily set 1% as the upper limit allowed
  error('Increase Nfft and/or smooth Sdb');
end

figure(2);
plot(s,'-k'); grid('on');   title('Impulse Response');
xlabel('Time (samples)');   ylabel('Amplitude');
```

Figure 8.6: Script (matlab) for checking the desired impulse-response for time aliasing. If it wraps around in the time buffer, one can increase the inverse FFT length (`Nfft`) and/or smooth the desired amplitude response (`Sdb`).

Converting the Desired Amplitude Response to Minimum Phase

Phase-sensitive filter-design methods such as the equation-error method implemented in **invfreqz** are normally constrained to produce filters with *causal* impulse responses.[15] In cases such as this (phase-sensitive filter design when we don't care about phase—or don't *have* it), it is best to compute the *minimum phase* corresponding to the desired amplitude response [449].

As detailed in Fig. 8.8, the minimum phase is constructed by the *cepstral method* [449].[16]

The four-pole, one-zero filter fit using **invfreqz** is shown in Fig. 8.7.

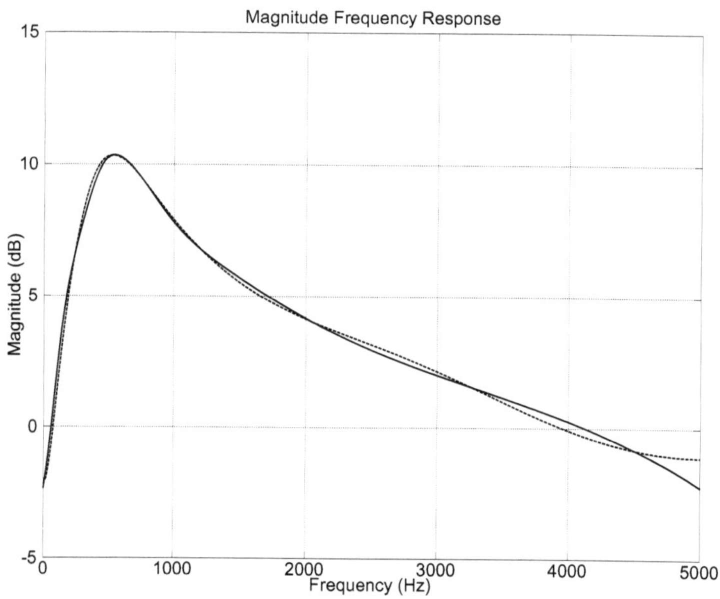

Figure 8.7: Overlay of desired amplitude response (solid) and that of a fourth-order filter fit (dashed) using **invfreqz**.

[15]Ideally, they are also constrained to produce *stable* filters as well, but **invfreqz** does not offer a stability constraint, and so stability should be checked whenever specifying one or more poles.

[16]`https://ccrma.stanford.edu/~jos/filters/Creating_Minimum_Phase_Filters.html`

```
c = ifft(Sdb); % compute real cepstrum from log magnitude spectrum
% Check aliasing of cepstrum (in theory there is always some):
caliaserr = 100*norm(c(round(Ns*0.9:Ns*1.1)))/norm(c);
disp(sprintf(['Cepstral time-aliasing check: Outer 20%% of ' ...
    'cepstrum holds %0.2f %% of total rms'],caliaserr));
% = 0.09 percent
if caliaserr>1.0 % arbitrary limit
  error('Increase Nfft and/or smooth Sdb to shorten cepstrum');
end
% Fold cepstrum to reflect non-min-phase zeros inside unit circle:
% If complex:
% cf=[c(1),c(2:Ns-1)+conj(c(Nfft:-1:Ns+1)),c(Ns),zeros(1,Nfft-Ns)];
cf = [c(1), c(2:Ns-1)+c(Nfft:-1:Ns+1), c(Ns), zeros(1,Nfft-Ns)];
Cf = fft(cf); % = dB_magnitude + j * minimum_phase
Smp = 10 .^ (Cf/20); % minimum-phase spectrum

Smpp = Smp(1:Ns); % nonnegative-frequency portion
wt = 1 ./ (fk+1); % typical weight fn for audio
wk = 2*pi*fk/fs;
[B,A] = invfreqz(Smpp,wk,NZ,NP,wt);
Hh = freqz(B,A,Ns);

figure(3);
plot(fk,db([Smpp(:),Hh(:)])); grid('on');
xlabel('Frequency (Hz)');
ylabel('Magnitude (dB)');
title('Magnitude Frequency Response');
% legend('Desired','Filter');
```

Figure 8.8: Script (matlab) for converting the (real) desired amplitude response to minimum-phase form for `invfreqz`. This script generated Fig. 8.7.

8.6.5 Further Reading on Digital Filter Design

This section provided only a "surface scratch" into the large topic of digital filter design based on an arbitrary frequency response. The main goal here was to provide a high-level orientation and to underscore the high value of such an approach for encapsulating linear, time-invariant subsystems in a computationally efficient yet accurate form. Applied examples will appear in later chapters. We close this section with some pointers for further reading in the area of digital filter design.

Some good books on digital filter design in general include [343, 362, 289]. Also take a look at the various references in the `help/type` info for Matlab/Octave functions pertaining to filter design. Methods for FIR filter design (used in conjunction with FFT convolution) are discussed in Book IV [457], and the equation-error method for IIR filter design was introduced in Book II [449]. See [281, 282] for related techniques applied to guitar modeling. See [454] for examples of using matlab functions `invfreqz` and `invfreqs` to fit filters to measured frequency-response data (specifically the *wah pedal* design example). Other filter-design tools can be found in the same website area.

The overview of methods in §8.6.2 above is elaborated in [428], including further method details, application to violin modeling, and literature pointers regarding the methods addressed. Some of this material was included in [449, Appendix I].

In Octave or Matlab, say `lookfor filter` to obtain a list of filter-related functions. Matlab has a dedicated filter-design toolbox (say `doc filterdesign` in Matlab). In many matlab functions (both Octave and Matlab), there are literature citations in the source code. For example, `type invfreqz` in Octave provides a URL to a Web page (from [449]) describing the FFT method for equation-error filter design.

8.7 Commuted Synthesis

In acoustic stringed musical instruments such as guitars and pianos, the strings couple via a "bridge" to some resonating acoustic structure (typically made of wood) that is required for efficient transduction of string vibration to acoustic propagation in the surrounding air. The resonator also imposes its own characteristic frequency response on the radiated sound. Spectral characteristics of the string excitation, the string resonances, and body/soundboard/enclosure resonator, are thus combined multiplicatively in the radiated sound, as depicted in Fig. 8.9.

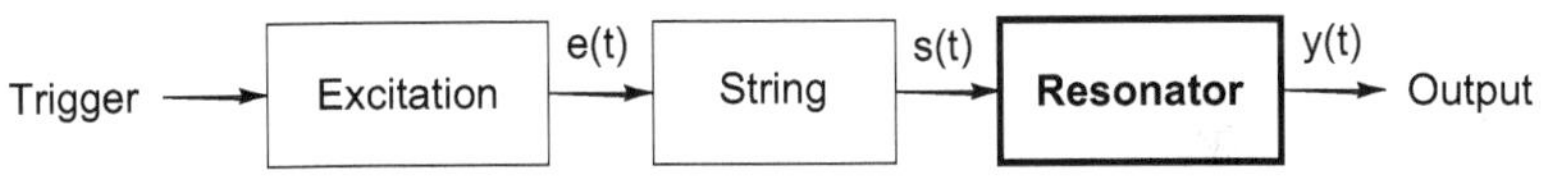

Figure 8.9: Schematic diagram of a stringed musical instrument.

The idea of commuted synthesis is that, because the string and body are close to linear and time-invariant, we may *commute the string and resonator,* as shown in Fig. 8.10.

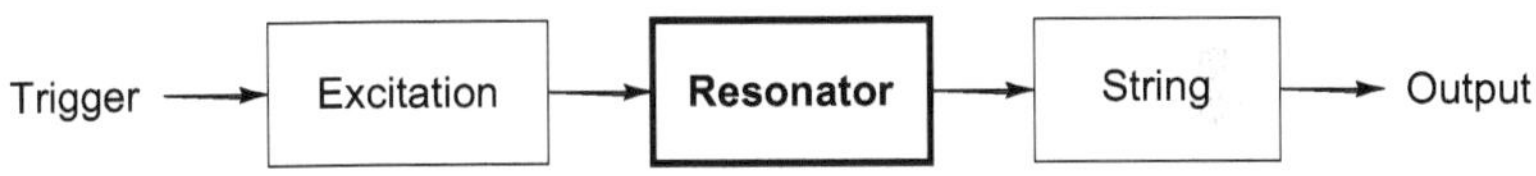

Figure 8.10: Equivalent diagram in the linear, time-invariant case.

The excitation can now be convolved with the resonator impulse response to provide a single, aggregate, excitation table, as depicted in Fig. 8.11. This is the basic idea behind *commuted synthesis,* and it greatly reduces the complexity of stringed instrument implementations, since the body filter is replaced by an inexpensive lookup table [439, 230]. These simplifications are presently important in single-processor polyphonic synthesizers such as multimedia DSP chips.

In the simplest case, the string is "plucked" using the (half-windowed) impulse response of the body.

An example of an excitation is the force applied by a pick or a finger at some point, or set of points, along the string. The input force per sample at each point divided by $4R$ gives the velocity to inject additively at that point in both traveling-wave directions. (The factor of 4 comes from

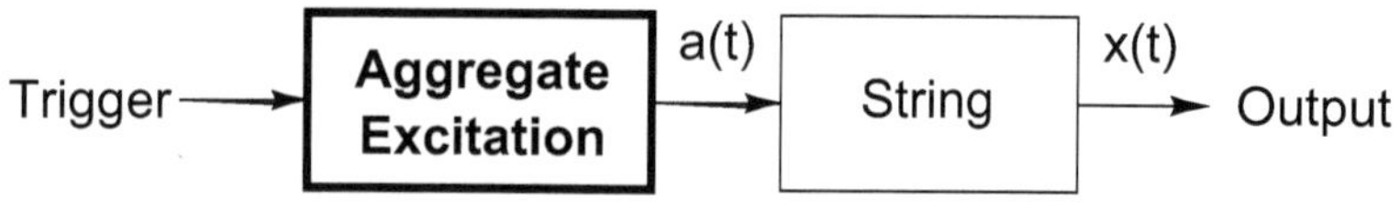

Figure 8.11: Use of an aggregate excitation given by the convolution of original excitation with the resonator impulse response.

splitting the injected velocity into two traveling-wave components, and from the fact that two string end-points are being driven.) Equal injection in the left- and right-going directions corresponds to an excitation force which is stationary with respect to the string.

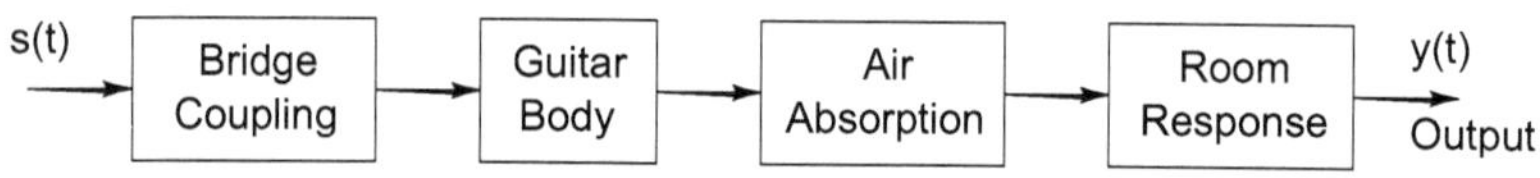

Figure 8.12: Possible components of a guitar resonator.

In a practical instrument, the "resonator" is determined by the choice of output signal in the physical scenario, and it generally includes filtering downstream of the body itself, as shown in Fig. 8.12. A typical example for the guitar or violin would be to choose the output signal at a point a few feet away from the top plate of the body. In practice, such a signal can be measured using a microphone held at the desired output point and recording the response at that point to the striking of the bridge with a force hammer. It is useful to record simultaneously the output of an accelerometer mounted on the bridge in order to also obtain experimentally the driving-point impedance at the bridge. In general, it is desirable to choose the output close to the instrument so as to keep the resonator response as short as possible. The resonator components need to be linear and time invariant, so they will be commutative with the string and combinable with the string excitation signal via convolution.

The string should also be linear and time invariant in order to be able to commute it with the generalized resonator. However, the string is actually the least linear element of most stringed musical instruments, with the main effect of nonlinearity being often a slight increase of the fundamental vibration frequency with amplitude. A secondary effect is to introduce coupling between the two polarizations of vibration along the length of the string. In practice, however, the string can be considered sufficiently close to linear to permit commuting with the body. The string is also time varying in the presence of vibrato, but this too can be neglected in practice. While commuting a live string and resonator may not be identical mathematically, the sound is substantially the same.

There are various options when combining the excitation and resonator into an aggregate excitation, as shown in Fig. 8.11. For example, a wave-table can be prepared which contains the convolution of a particular point excitation with a particular choice of resonator. Perhaps the simplest choice of excitation is the impulse signal. Physically, this would be natural when the wave variables in the string are taken to be acceleration waves for a plucked string; in this case, an ideal pluck gives rise to an impulse of acceleration input to the left and right in the string at the pluck point. If loss of perceived pick position is unimportant, the impulse injection need only be in a single direction. (The comb filtering which gives rise to the pick-position illusion can be restored by injecting a second, negated impulse at a delay equal to the travel time to and from the bridge.) In this simple case of a single impulse to pluck the string, the aggregate excitation is simply the impulse *response* of the resonator. Many excitation and resonator variations can be simulated using a collection of aggregate excitation tables. It is useful to provide for interpolation of excitation tables so as to provide intermediate points along a parameter dimension. In fact, all the issues normally associated with sampling synthesis arise in the context of the string excitation table. A disadvantage of combining excitation and resonator is the loss of multiple output signals from the body simulation, but the timbral effects arising from the mixing together of multiple body outputs can be obtained via a mixing of corresponding excitation tables.

If the aggregate excitation is too long, it may be shortened by a variety of techniques. It is good to first convert the signal $a(n)$ to *minimum phase* so as to provide the maximum shortening consistent with the original magnitude spectrum. Secondly, $a(n)$ can be windowed using the right wing of any window function typically used in spectrum analysis. An interesting choice is the exponential window, since it has the interpretation of increasing the resonator damping in a uniform manner, *i.e.*, all the poles and zeros of the resonator are contracted radially in the z plane by the same factor.

8.7.1 Body-Model Factoring

In commuted synthesis, it is often helpful to *factor out the least-damped resonances* in the body model for implementation in *parametric form* (*e.g.*, as second-order filters, or "biquads"). The advantages of this are manifold, including the following:

- The excitation table is shortened (now containing only the most damped modal components).
- The excitation table signal-to-quantization-noise ratio is improved.
- The most important resonances remain *parametric*, facilitating real-time control. In other words, the parametric resonances can be independently *modulated* to produce interesting effects.
- Multiple body outputs become available from the parametric filters.
- Resonators may be already available in a separate effects unit, making them "free."
- A memory vs. computation trade-off is available for cost optimization.

Techniques for carrying out this factorization in practice are discussed in §8.8 [229].

8.7.2 Further Reading in Commuted Synthesis

Laurson et al. [278, 277] have developed very high quality synthesis of classical guitar from an extended musical score using the commuted synthesis approach. A key factor in the quality of these results is the great attention to detail in the area of musical control when driving the model from a written score. In contrast to this approach, most of the sound examples in [472] were "played" in real time, using a MIDI guitar controller feeding the serial port of a NeXT Computer running SynthBuilder [353].[17]

[17]SynthBuilder is now a proprietary in-house tool at Analog Devices, Inc. See [86] for a description of the STK software prototyping environment that many of us use today. In particular, the `Mandolin.cpp` patch in the STK distribution is based on commuted synthesis. Some of us are also using the STK package in conjunction with `pd` [355] in order to obtain drag-and-drop graphical patch construction and a large library of MIDI processing tools.

8.8 Extracting Parametric Resonators from a Nonparametric Impulse Response

As mentioned in §8.7.1 above, a valuable way of shortening the excitation table in commuted synthesis is to *factor* the body resonator into its *most-damped* and *least-damped* modes. The most-damped modes are then commuted and combined with the excitation in impulse-response form. The least-damped modes can be left in parametric form as recursive digital filter sections.

Commuted synthesis is a technique in which the body resonator is commuted with the string model, as shown in Fig. 8.10 on page 305, in order to avoid having to implement a large body filter at all [439, 232, 229].[18] In commuted synthesis, the excitation (*e.g.*, plucking force versus time) can be convolved with the resonator impulse response to provide a single aggregate excitation signal. This signal is short enough to store in a look-up table, and a note is played by simply summing it into the string.

8.8.1 Mode Extraction Techniques

The goal of resonator factoring is to identify and remove the least-damped resonant modes of the impulse response. In principle, this means ascertaining the precise resonance frequencies and bandwidths associated with each of the narrowest "peaks" in the resonator frequency response, and dividing them out via inverse filtering, so they can be implemented separately as resonators in series. If in addition the amplitude and phase of a resonance peak are accurately measurable in the complex frequency response, the mode can be removed by complex spectral subtraction (equivalent to subtracting the impulse-response of the resonant mode from the total impulse response); in this case, the parametric modes are implemented in a parallel bank as in [66]. However, in the parallel case, the residual impulse response is not readily commuted with the string.

In the inverse-filtering case, the factored resonator components are in cascade (series) so that the damped modes left behind may be commuted with the string and incorporated in the excitation table by convolving the residual impulse response with the desired string excitation signal. In the parallel case, the damped modes do not commute with the string since doing so would require somehow canceling them in the parallel filter sections. In principle, the string would have to be duplicated so that one instance can be driven by the residual signal with no body resonances at all, while the other is connected to the parallel resonator bank and driven only by the natural string excitation without any commuting of string and resonator. Since

[18]Extracts of this material were published previously in [229].

duplicating the string is unlikely to be cost-effective, the impulse response of the high-frequency modes can be commuted and convolved with the string excitation as in the series case to obtain *qualitative* results. The error in doing this is that the high-frequency modes are being multiplied by the parallel resonators rather than being added to them.

Various methods are available for estimating the mode parameters for inverse filtering:

- Amplitude response peak measurement
- Weighted digital filter design
- Linear prediction
- Sinusoidal modeling
- Late impulse-response analysis

Amplitude response peak measurement

The longest ringing modes are associated with the narrowest bandwidths. When they are important resonances in the frequency response, they also tend to be the tallest peaks in the frequency response magnitude. (If they are not the tallest near the beginning of the impulse response, they will be the tallest near the end.) Therefore, one effective technique for measuring the least-damped resonances is simply to find the precise location and width of the narrowest and tallest spectral peaks in the measured amplitude response of the resonator. The center-frequency and bandwidth of a narrow frequency-response peak determine two poles in the resonator to be factored out. Expressing a filter in terms of its poles and zeros is one type of "parametric" filter representation, as opposed to "nonparametric" representations such as the impulse response or frequency response. Prony's method [449, 297, 273] is one well known technique for estimating the frequencies and bandwidths of sums of exponentially decaying sinusoids (two-pole resonator impulse responses).

In the factoring example presented in §8.8.6 on page 318, the frequency and bandwidth of the main Helmholtz air mode are measured manually using an interactive spectrum analysis tool. However, it is a simple matter to automate peak-finding in FFT magnitude data. (See, for example, the peak finders used in sinusoidal modeling, discussed a bit further in §8.8.1 below.)

Weighted digital filter design

Many methods for digital filter design support spectral weighting functions that can be used to focus in on the least-damped modes in the frequency response. One is the *weighted equation-error* method which is available in

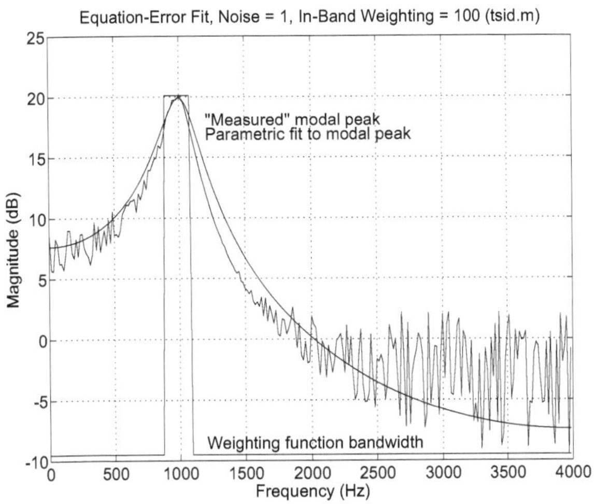

Figure 8.13: Illustration of one way to determine the parameters of a least-damped resonant mode.

the matlab `invfreqz()` function (§8.6.4, p. 297). Figure 8.13 illustrates use of it. For simplicity, only one frequency-response peak plus noise is shown in this synthetic example. First, the peak center-frequency is measured using a quadratically interpolating peak finder operating on the dB spectral magnitude. This is used to set the spectral weighting function. Next, `invfreqz()` is called to design a two-pole filter having a frequency response that approximates the measured data as closely as possible. The weighting function is also shown in Fig. 8.13, renormalized to overlay on the scale of the plot. Finally, the amplitude response of the two-pole filter designed by the equation-error method is shown overlaid in the figure. Note that the agreement is quite good near the peak which is what matters most. The interpolated peak frequency measured initially in the nonparametric spectral magnitude data can be used to fine-tune the pole-angles of the designed filter, thus rendering the equation-error method a technique for measuring only the peak bandwidth in this case. There are of course many, many techniques in the signal processing literature for measuring spectral peaks.

Linear prediction

Another well known method for converting the least-damped modes into parametric form is Linear Predictive Coding (LP) followed by polynomial factorization to obtain resonator poles. LP is particularly good at fitting spectral peaks due to the nature of the error criterion it minimizes [428]. The poles closest to the unit circle in the z plane can be chosen for the "ringy" part of the resonator. It is well known that when using techniques like LP to model spectral peaks for extraction, orders substantially higher than twice the number of spectral peaks should be used. The extra degrees of freedom in the LP fit give more poles for modeling spectral detail other than the peaks, allowing the poles modeling the peaks to fit them with less distraction. On the other hand, if the order chosen is too high, sometimes more than two poles will home in on the same peak. In some cases, this may be appropriate since the body resonances are not necessary resolvable so as to separate the peaks, especially at high frequencies.

Sinusoidal modeling

Another way to find the least-damped mode parameters is by means of an intermediate *sinusoidal model* of the body impulse response, or, more appropriately, the *energy decay relief* (EDR) computed from the body impulse response (see §3.2.2 on page 91). Such sinusoidal models have been used to determine the string loop filter in digital waveguide strings models. In the case of string loop-filter estimation, the sinusoidal model is applied to the impulse response (or "pluck" response) of a vibrating string or acoustic tube. In the present application, it is ideally applied to the EDR of the body impulse response (or "hammer-strike" response).

Since sinusoidal modeling software (*e.g.* [424]) typically quadratically interpolates the peak frequencies, the resonance frequencies are generally quite accurately estimated provided the frame size is chosen large enough to span many cycles of the underlying resonance.

The sinusoidal amplitude envelopes yield a particularly robust measurement of resonance bandwidth. Theoretically, the modal decay should be exponential. Therefore, on a dB scale, the amplitude envelope should decay *linearly*. Linear regression can be used to fit a straight line to the measured log-amplitude envelope of the impulse response of each long-ringing mode. Note that even when amplitude modulation is present due to modal couplings, the ripples tend to average out in the regression and have little effect on the slope measurement. This robustness can be enhanced by starting and ending the linear regression on local maxima in the amplitude envelope. A method for estimating modal decay parameters in the presence of noise is given in [125, 234, 235].

Below is a section of matlab code which performs linear regression:

```
function [slope,offset] = fitline(x,y);
%FITLINE    fit line 'y = slope * x + offset'
```

```
%           to column vectors x and y.

      phi = [x, ones(length(x),1)];
      p = phi' * y;
      r = phi' * phi;
      t = r\p;
      slope = t(1);
      offset = t(2);
```

Late impulse-response analysis

All methods useable with inverse filtering can be modified based on the observation that late in the impulse response, the damped modes have died away, and the least-damped modes dominate. Therefore, by discarding initial impulse-response data, the problem in some sense becomes "easier" at the price of working closer to the noise floor. This technique is most appropriate in conjunction with the inverse filtering method for mode extraction (discussed below), since for subtraction, the modal impulse response must be extrapolated back to the beginning of the data record. However, methods used to compute the filter numerator in variations on Prony's method can be used to scale and phase-align a mode for subtraction [428, 297].

One simple approximate technique based on looking only at the late impulse response is to take a zero-padded FFT of the latest Hanning-windowed data segment. The least-damped modes should give very clearly dominant peaks in the FFT magnitude data. As discussed above, the peak(s) can be interpolated to estimate the mode resonance frequency, and the bandwidth can be measured to determine the time-constant of decay. Alternatively, the time-constant of decay can be measured in the time domain by measuring the decay slope of the log amplitude envelope across the segment (this time using a rectangular window). Since the least-damped mode is assumed to be isolated in the late decay, it is easy to form a pitch-synchronous amplitude envelope.

8.8.2 Inverse Filtering

Whatever poles are chosen for the least-damped part, and however they are computed (provided they are stable), the damped part can be computed from the full impulse response and parametric part using *inverse filtering*, as illustrated in the computed examples above. The inverse filter is formed from zeros equal to the estimated resonant poles. When the inverse filter is applied to the full resonator impulse, a "residual" signal is formed which is defined as the impulse response of the leftover, more damped modes. The residual is in exactly the nonparametric form needed for commuting with the string and convolving with the string excitation signal, such as a "pluck" signal. Feeding the residual signal to the parametric resonator gives the original resonator impulse response to an extremely high degree of accuracy. The error is due only to numerical round-off error during the inverse and forward filtering computations. In particular, the least-damped resonances need not be accurately estimated for this to hold. When there is parametric estimation error, the least-damped components will fail to be completely removed from the residual signal; however, the residual signal through the parametric resonator will always give an exact reconstruction of the original body impulse response, to within roundoff error. This is similar to the well known feature of linear predictive coding that feeding the prediction error signal to the LP model always gives back the original signal [297].

The parametric resonator need not be restricted to all-pole filters, however, although all-pole filters (plus perhaps zeros set manually to the same angles but contracted radii) turn out to be very convenient and simple to work with. Many filter design techniques exist which can produce a parametric part having any prescribed number of poles and zeros, and weighting functions can be used to "steer" the methods toward the least-damped components of the impulse response. The equation-error method illustrated in Fig. 8.13 is an example of a method which can also compute zeros in the parametric part as well as poles. However, for inverse filtering to be an option, the zeros must be constrained to be *minimum phase* so that their inverses will be stable poles.

Empirical Notes on Inverse Filtering

In experiments factoring guitar body impulse responses, it was found that the largest benefit per section comes from pulling out the main Helmholtz air resonance. Doing just this shortens the impulse response (excitation table) by a very large factor, and because the remaining impulse response is noise-like, it can be truncated more aggressively without introducing artifacts.

It also appears that the bandwidth estimate is not very critical in this case. If it is too large, or if "isolation zeros" are not installed behind the poles, as shown in Figs. 8.18b and 8.21b, the inverse filtering serves partially as a *preemphasis* which tends to flatten the guitar body frequency response overall or cause it to rise with frequency. This has a good effect on the signal-to-quantization-noise ratio versus frequency. To maximize the worst-case signal-to-quantization-noise versus frequency, the residual spectrum should be flat since the quantization noise spectrum is normally close to flat. A *preemphasis* filter for flattening the overall spectrum is commonly used in speech analysis [363, 297]. A better preemphasis in this context is an *inverse equal-loudness* preemphasis, taking the inverse of an equal-loudness contour near the threshold of hearing in the Fletcher-Munson curves [476]. This corresponds to psychoacoustic "noise shaping" so that the quantization noise floor is perceptually uniform, and decreasing playback volume until it falls below the threshold of hearing results in all of the noise disappearing across the entire spectrum at the same volume.[19]

Since in some fixed-point implementations, narrow bandwidths may be difficult to achieve, good results are obtained by simply setting the bandwidth of the single resonator to any minimum robust value. As a result, there may still be some main-air-mode response in the residual signal, but it is typically very small, and early termination of it using a half-window for table shortening is much less audible than if the original impulse response were similarly half-windowed. The net effect on the instrument is to introduce *artificial damping* the main air mode in the guitar body. However, since this mode rings so much longer than the rest of the modes in the guitar body, shortening it does not appear to be detrimental to the overall quality of the instrument. In general, it is not desirable for isolated modes to ring longer than all others. Why would a classical guitarist want an audible "ringing" of the guitar body near 100 Hz?

In computing figures 8.16 and Fig. 8.16b, the estimated Q of the main Helmholtz air mode was only 10. As a result, it is still weakly present in the inverse filter output (residual) spectrum Fig. 8.16b.

Matlab Code for Inverse Filtering

Below is the matlab source code used to extract the main Helmholtz air mode from the guitar body impulse response in Figures 8.14 through 8.17:

```
freq = 104.98;   % estimated peak frequency in Hz
bw = 10;         % peak bandwidth estimate in Hz

R = exp( - pi * bw / fs);                  % pole radius
z = R * exp(j * 2 * pi * freq / fs); % pole itself
B = [1, -(z + conj(z)), z * conj(z)] % numerator
r = 0.9;      % zero/pole factor (notch isolation)
A = B .* (r .^ [0 : length(B)-1]);    % denominator

residual = filter(B,A,bodyIR); % apply inverse filter
```

[19]These remarks apply to playback of the excitation only. The string provides further filtering which could be taken into account. However, to a first approximation, the string filtering is like a "sampling" of the excitation spectrum, with a gentle roll-off at high frequencies due to the lowpass loop filtering. Also, the string is a highly variable filter with many settings. It is therefore reasonable to ignore the string filtering when determining an optimal preemphasis for excitation-table quantization.

8.8.3 Sinusoidal Modeling of Mode Decays

When the amplitude envelope, frequency, phase, and onset time are all accurately estimated (for all time), it is possible to *subtract* the synthesized modal impulse response from the measured impulse response. (This contrasts with purely spectral modal parameters which are amplitude, frequency, bandwidth, and phase.) This method of "sinusoidal track removal" is used in sines-plus-noise spectral modeling. (See [424] for further details and supporting C software). In this approach, the resonant mode is subtracted out rather than divided out of the frequency response.

There are some disadvantages of subtraction relative to inverse filtering. First, more parameters must be accurately measured; the precise gain and phase of the resonance are needed in addition to its frequency and bandwidth. Inverse filtering on the other hand requires only estimation of frequency and bandwidth (or frequency and time-constant of decay). In addition, the residual impulse response after subtraction cannot be precisely commuted with the string for commuted synthesis.

The advantages of subtraction over inverse filtering are that amplitude modulation due to mode coupling can be retained in the measured modal decay and subtracted out, whereas a second-order inverse filter cannot subtract out the modulation due to coupling. Also, if the system is time varying (as happens, for example, when the performer's hand is pressing against the resonating body in a time-varying way), the subtraction method can potentially track the changing modal parameters and still successfully remove the modal decay. To compete with this, the inverse filtering method would have to support a time-varying filter model. As another example, if the guitar body is moving, the measured response is a time-varying linear combination of fixed resonant modes (although some Doppler shift is possible). The subtraction method can follow a time-varying gain (and phase) so that the subtraction still will take out the mode.

The inverse filtering method seems most natural when the physical resonator is time-invariant and is well modeled as a *series* of resonant sections. It is also the only one strictly valid for use in commuted synthesis. The subtraction method seems most natural when the physical resonator is best modeled as a *sum* of resonating modes. As a compromise between the two approaches, all parametric modes may be separated from the nonparametric modes by means of inverse filtering, and the parametric part can then be split into parallel form.

8.8.4 Parallel Body Filterbank Design

In [66], a method is described for producing a parallel bank of fourth-order IIR body filters from an arbitrary starting impulse response. Initially, the body impulse is approximated by the impulse response of a fourth-order IIR filter using an algorithm of second author Cheng; the method is said to pick out the most resonant modes of the sampled impulse response, and fourth-order sections were found experimentally to best characterize the resonant modes of the guitar body. The impulse response of the filter so designed is subtracted from the measured impulse response to form a residual impulse response which is given again to the IIR design algorithm. This process is repeated to obtain a parallel bank of fourth-order filters, plus a final residual which is discarded. In the present context, the final residual should be incorporated at least qualitatively into the string excitation signal, or else the method should be applied only to the total parametric impulse response computed by other means and separated from the nonparametric impulse response by inverse filtering.

8.8.5 Approximating Shortened Excitations as Noise

Figure 8.14b suggests that the many damped modes remaining in the shortened body impulse response may not be clearly audible as resonances since their decay time is so short. This is confirmed by listening to shortened and spectrally flattened body responses which sound somewhere between a click and a noise burst. These observations suggest that the shortened, flattened, body response can be replaced by a psychoacoustically equivalent *noise burst*, as suggested for modeling piano soundboards [520]. Thus, the signal of Fig. 8.14b can be synthesized qualitatively by a white noise generator multiplied by an amplitude envelope. In this technique, it is helpful to use LP on the residual signal to flatten it. As a refinement, the noise burst can be time-varying filtered so that high frequencies decay faster [520]. Thus, the stringed instrument model may consist of *noise generator* $\rightarrow$ *excitation amplitude-shaping* $\rightarrow$ *time-varying lowpass* $\rightarrow$ *string model* $\rightarrow$ *parametric resonators* $\rightarrow$ *multiple outputs*. In addition, properties of the physical excitation may be incorporated, such as comb filtering to obtain a virtual pick or hammer position. Multiple outputs provide for individual handling of the most important resonant modes and facilitate simulation of directional radiation characteristics [512].

It is interesting to note that by using what is ultimately a noise-burst excitation, we have almost come full circle back to the original extended Karplus-Strong algorithm [236, 207]. Differences include (1) the amplitude shaping of the excitation noise to follow the envelope of the impulse-response of the highly damped modes of the guitar body (convolved with the physical string excitation), (2) more sophisticated noise filtering which effectively shapes the noise in the frequency domain to follow the frequency response of the highly damped body modes (times the excitation spectrum), and (3) the use of factored-out body resonances which give real resonances such as the main Helmholtz air mode. The present analysis also provides a theoretical explanation as to why use of a noise burst in the Karplus-Strong algorithm is generally considered preferable to a theoretically motivated initial string shape such as the asymmetric triangular wave which leans to the left according to pick position [437, p. 82].

From a psychoacoustical perspective, it may be argued that the excitation noise burst described above is not perceptually uniform. The amplitude envelope for the noise burst provides noise-shaping in the time domain, and the LP filter provides noise-shaping in the frequency domain, but only at *uniform resolution* in time and frequency. Due to properties of hearing, it can be argued that *multi-resolution* noise shaping should be used. Thus, the LP fit for obtaining the noise-shaping filter should be carried out over a Bark frequency axis as in Fig. 8.19b. Since LP operates on the autocorrelation function, a warped autocorrelation can be computed simply as the inverse FFT of the warped squared-magnitude spectrum.

8.8.6 Body Factoring Example

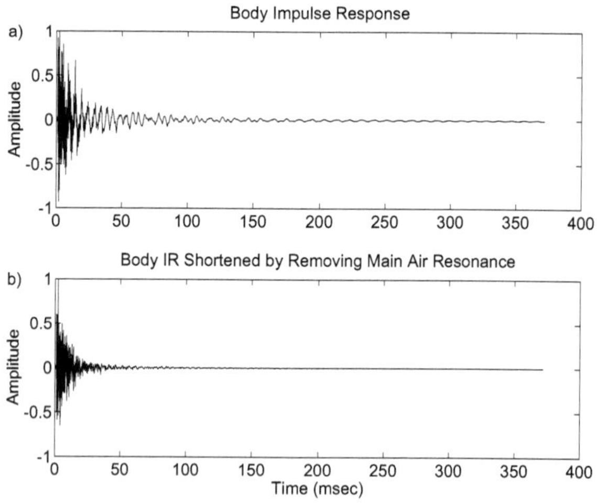

Figure 8.14: Impulse response of a classical guitar body before and after removing the first peak (main air resonance) via the inverse filter defined by Eq. (8.18) on page 323, with $a_1 = -1.9963$ and $a_2 = 0.9972$.

Figure 8.14a shows the impulse response of a classical guitar body sampled at 22050 kHz. It was determined empirically that at least the first

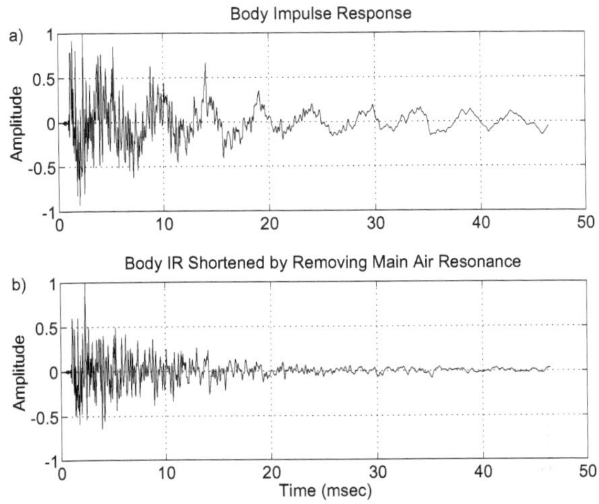

Figure 8.15: First eighth of Figure 8.14.

100 msec of this impulse response needs to be stored in the excitation table to produce a high quality synthetic guitar. Figure 8.14b shows the same impulse response after factoring out a single resonating mode near 100 Hz (the main Helmholtz air mode). A close-up of the initial response is shown in Fig. 8.15. As can be seen, the residual response is considerably shorter than the original.

Figure 8.16a shows the guitar-body amplitude response, and Fig. 8.16b shows the response after the main Helmholtz air mode is removed by inverse filtering with a two-pole, two-zero filter. Figure 8.18 shows the same thing but with only a two-zero inverse filter; in this case the overall spectral shape is more affected.

Figure 8.19a shows the measured guitar-body amplitude response after warping to an approximate Bark frequency axis. Figure 8.19b shows the Bark-warped amplitude response after the main Helmholtz air mode is removed by inverse filtering. Fig. 8.20 shows the low-frequency close-up. The warped amplitude response was computed as the FFT of the impulse response of the FIR filter given by the original impulse response with each unit delay being replaced by a first-order allpass filter, as originally suggested in [334] and described further in [229].

Fig. 8.21 shows the corresponding results if a two-zero inverse filter is used rather than a two-pole, two-zero inverse filter, *i.e.*, without the "isolation poles," ($r = 0$ in Eq. (8.18) below). In this case, there is an overall "EQ" boosting high frequencies and attenuating low frequencies.

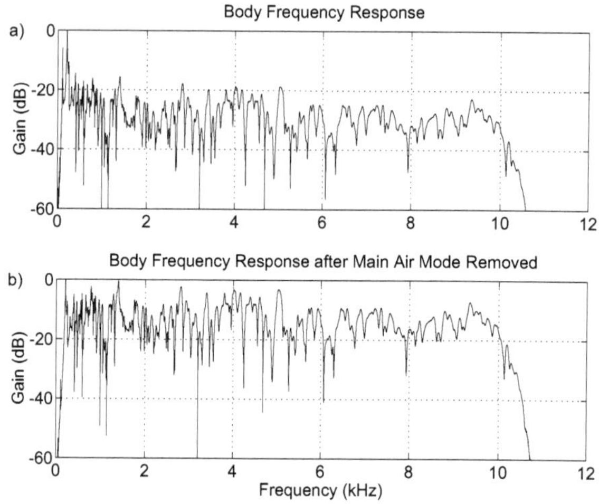

Figure 8.16: Normalized amplitude response of a classical guitar body before and after inverse filtering via Eq. (8.18) on page 323, with $a_1 = -1.9963$ and $a_2 = 0.9972$.

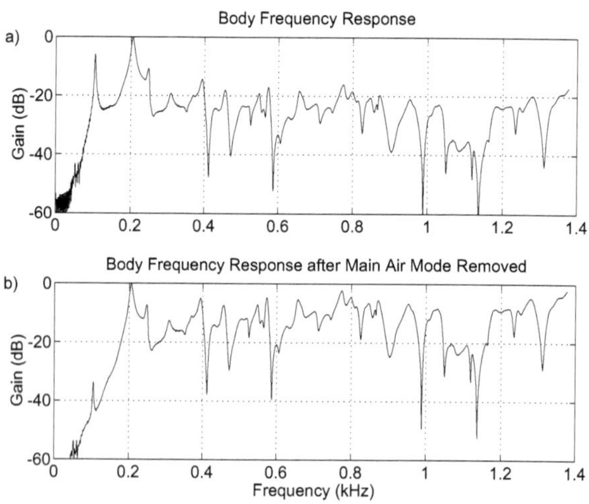

Figure 8.17: First eighth of Figure 8.16.

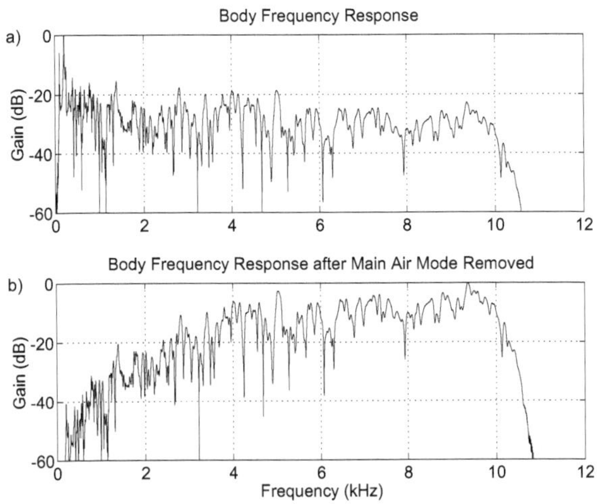

Figure 8.18: Normalized amplitude response of a classical guitar body before and after inverse filtering via Eq. (8.18) on page 323, with $a_1 = -1.9963$, $a_2 = 0.9972$, and $r = 0$.

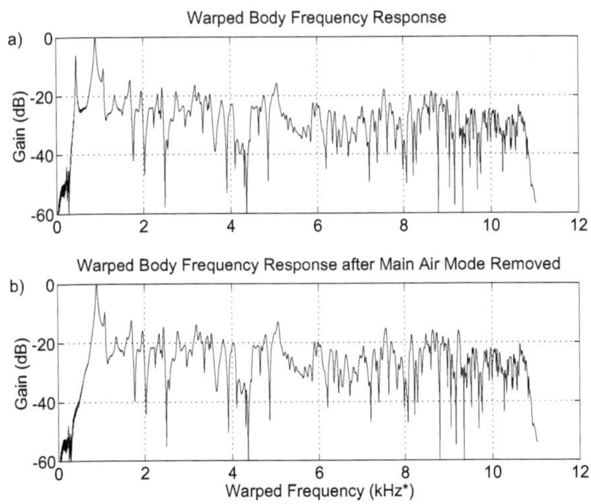

Figure 8.19: Normalized Bark-warped amplitude response of a classical guitar body before and after removing the first peak (main air mode) via Eq. (8.18) on page 323, with $a_1 = -1.9801$ and $a_2 = 0.9972$.

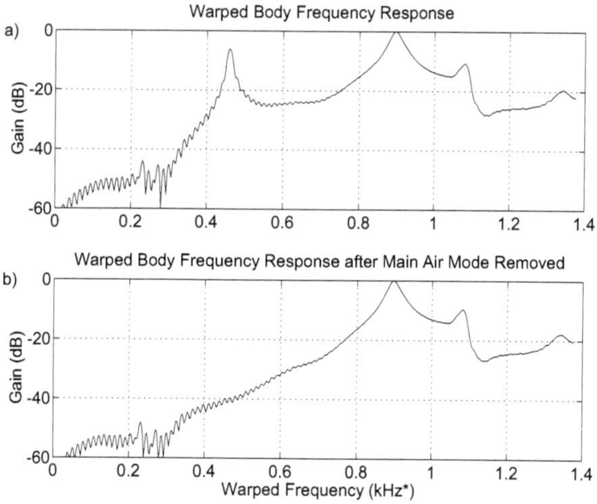

Figure 8.20: First eighth of Figure 8.19.

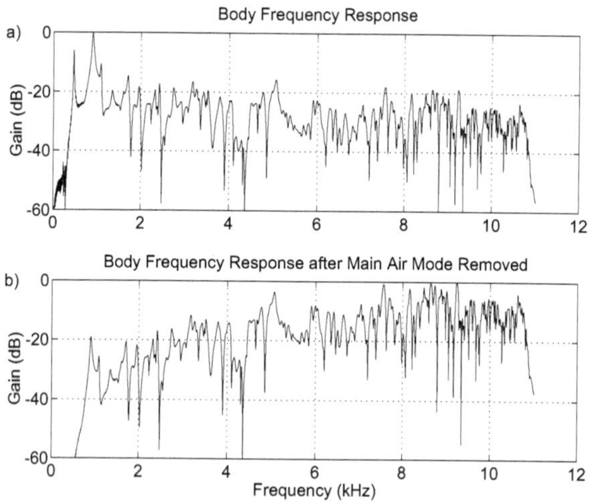

Figure 8.21: Normalized Bark-warped amplitude response of a classical guitar body before and after removing the first peak (main air mode) via Eq. (8.18) on page 323, with $a_1 = -1.9801$, $a_2 = 0.9972$, and $r = 0$.

However, comparing Figs. 8.18b and 8.21b, we see that the global EQ effect is less pronounced in the Bark-warped case. On the Bark frequency scale, it is much easier numerically to eliminate the main air mode.

The modal bandwidth used in the inverse filtering was chosen to be 10 Hz which corresponds to a Q of 46 for the main air mode. If the Bark-warping is done using a first-order conformal map [459], its inverse preserves filter order [428, pp. 61–67]. Applying the inverse warping to the parametric resonator drives its pole radius from 0.99858 in the Bark-warped z plane to 0.99997 in the unwarped z plane.

Note that if the inverse filter consists only of two zeros determined by the spectral peak parameters, other portions of the spectrum will be modified by the inverse filtering, especially at the next higher resonance, and in the linear trend of the overall spectral shape. To obtain a more *localized* mode extraction (useful when the procedure is to be repeated), we define the inverse filter as

$$H_r(z) \triangleq \frac{A(z)}{A(z/r)} \triangleq \frac{1 + a_1 z^{-1} + a_2 z^{-2}}{1 + a_1 r z^{-1} + a_2 r^2 z^{-2}} \qquad (8.18)$$

where $A(z)$ is the inverse filter determined by the peak frequency and bandwidth, and $A(z/r)$ is the same polynomial with its roots contracted by the factor r. If r is close to but less than 1, the poles and zeros substantially cancel far away from the removed modal frequency so that the inverse filter has only a local effect on the frequency response. In the computed examples, r was arbitrarily set to 0.9, but it is not critical.

Because the main air mode is extremely narrow, the probability of overflow can be reduced in fixed-point implementations by artificially dampening it. Reducing the Q of the main Helmholtz air mode from 46 to 10 corresponds to a decay time of about $Q/f_c \approx 10/100 = 0.1$ sec. This is consistent with the original desire to retain the first 100 msec of the body impulse response.

For completeness, the Bark-warped impulse-responses are also shown in Figs. 8.22. Figure 8.22a shows the complete Bark-warped impulse response obtained by taking the inverse FFT of Fig. 8.19a, and Fig. 8.22b shows the shortened Bark-warped impulse response defined as the inverse FFT of Fig. 8.19b. We see that given a Bark-warped frequency axis (which more accurately represents what we hear), the time duration of the high-frequency modes is extended while the low-frequency modes are contracted in time duration. Thus, the modal decay times show less of a spread versus frequency. This also accounts for the reduced apparent shortening by the inverse filtering in the Bark-warped case.

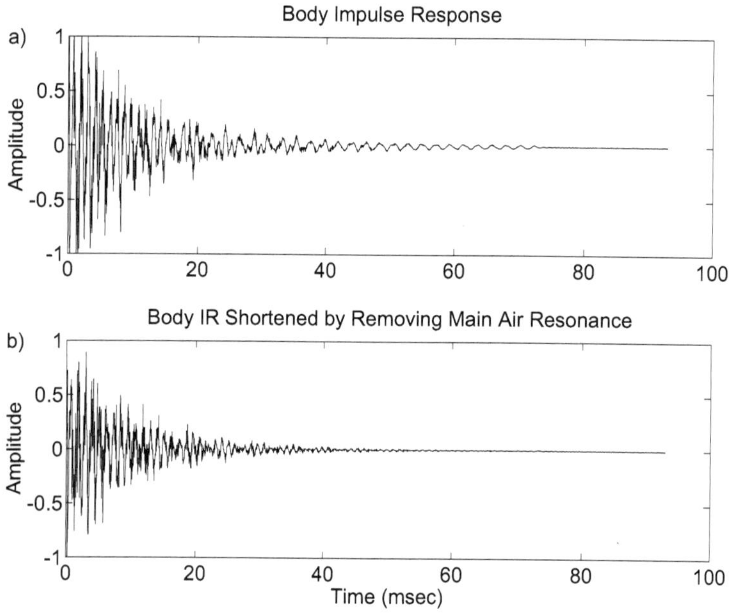

Figure 8.22: Impulse response of a Bark-warped classical guitar body before and after inverse filtering by $1 - 1.9963z^{-1} + 0.9972z^{-2}$.

8.9 Virtual Analog Example: Phasing

As mentioned in §5.4 on page 188, the *phaser*, or *phase shifter* works by sweeping *notches* through a signal's short-time spectrum. The notches are classically spaced nonuniformly, while flangers employ uniformly spaced notches (§5.3). In this section, we look at using a chain of allpass filters to implement the phasing effect.

8.9.1 Phasing with First-Order Allpass Filters

The block diagram of a typical inexpensive *phase shifter* for guitar players is shown in Fig. 8.23.[20] It consists of a series chain of first-order allpass filters,[21] each having a single time-varying parameter $g_i(n)$ controlling the pole and zero location over time, plus a feedforward path through gain g which is a fixed *depth control.* Thus, the delay line of the flanger is replaced by a string of allpass filters. (A delay line is of course an allpass filter itself.)

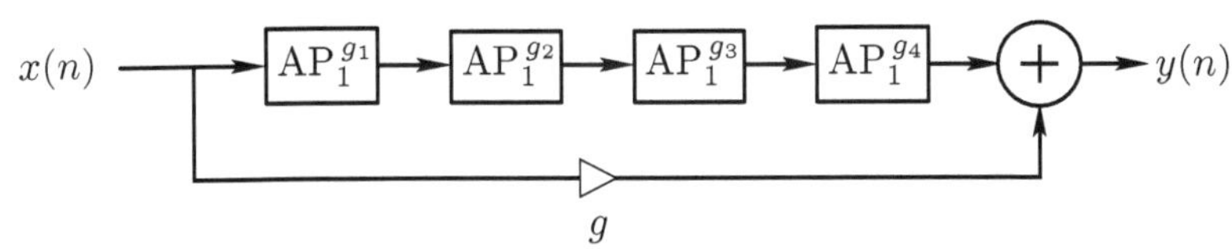

Figure 8.23: Structure of a phaser based on four first-order allpass filters.

In analog hardware, the first-order allpass transfer function [449, Appendix E, Section 8][22] is

$$\text{AP}_1^{\omega_b} \triangleq \frac{s-\omega_b}{s+\omega_b}. \tag{8.19}$$

(In classic phaser circuits such as the Univibe, $-\text{AP}_1^{(}\omega_b)$ is used, but since there is an even number (four) of allpass stages, there is no difference.) In discrete time, the general first-order allpass has the transfer function

$$\text{AP}_1^{g_i} \triangleq \frac{g_i + z^{-1}}{1 + g_i z^{-1}}.$$

We now consider the analog and digital cases, respectively.

[20]This is the basic architecture of the MXR phase shifter as well as the *Univibe* used by Jimi Hendrix [241], and described in detail at (`http://www.geofex.com/Article_Folders/univibe/uvfrindx.htm`).

[21]Moog has built a 12-stage phaser of this type [59] and up to 20 stages (10 notches) have been noted [242].

[22]Available online at
`http://ccrma.stanford.edu/~jos/filters/Analog_Allpass_Filters.html`

Classic Analog Phase Shifters

Setting $s = j\omega$ in Eq. (8.19) gives the frequency response of the analog-phaser transfer function to be

$$H_a(j\omega) \;=\; \frac{j\omega - \omega_b}{j\omega + \omega_b},$$

where the 'a' subscript denotes "analog" as opposed to "digital". The phase response is readily found to be

$$\Theta_a(\omega) \;=\; \pi - 2\tan^{-1}\left(\frac{\omega}{\omega_b}\right).$$

Note that the phase is always π at dc ($\omega = 0$), meaning each allpass section inverts at dc. Also, at $\omega = \infty$ (remember we're talking about analog here), we get a phase of zero. In between, the phase falls from π to 0 as frequency goes from 0 to ∞. In particular, at $\omega = \omega_b$, the phase has fallen exactly half way, to $\pi/2$. We will call $\omega = \omega_b$ the *break frequency* of the allpass section.[23]

Figure 8.24a shows the phase responses of four first-order analog allpass filters with ω_b set to $2\pi[100, 200, 400, 800]$. Figure 8.24b shows the resulting normalized amplitude response for the phaser, for $g = 1$ (unity feedfoward gain). The amplitude response has also been normalized by dividing by 2 so that the maximum gain is 1. Since there is an even number (four) of allpass sections, the gain at dc is $[1 + (-1)(-1)(-1)(-1)]/2 = 1$. Put another way, the initial phase of each allpass section at dc is π, so that the total allpass-chain phase at dc is 4π. As frequency increases, the phase of the allpass chain decreases. When it comes down to 3π, the net effect is a sign inversion by the allpass chain, and the phaser has a notch. There will be another notch when the phase falls down to π. Thus, four first-order allpass sections give two notches. For each notch in the desired response we must add two new first-order allpass sections.

From Fig. 8.24b, we observe that the first notch is near $f = 100$ Hz. This happens to be the frequency at which the first allpass pole "breaks,"

[23]Saying that the frequency ω_b is the *break frequency* for the one-pole term $H(s) = b/(s + \omega_b)$ is terminology from *classical control theory*. Below the break frequency, $H(s) \approx b/\omega_b$, and above, $H(s) \approx b/s$. On a log-log plot, the amplitude response $|H(j\omega)|$ may be approximated by a slope-zero line at height $G_0 = 20\log_{10}(|b/\omega_b|)$ from dc to ω_b, followed by an intersecting line with negative slope of 20 dB per decade for all higher frequencies. At the break frequency, the true gain is down 3 dB from G_0, but far away from the break frequency, the piecewise-linear approximation is very accurate. Such a plot of the log-frequency-response versus log-frequency (the real part being log-magnitude and the imaginary part being phase, plotted separately) is called a *Bode plot*. Bode plots are covered in any introductory course on control systems design (also called the design of "servomechanisms"), and see also the Wikipedia page for "Bode plot."

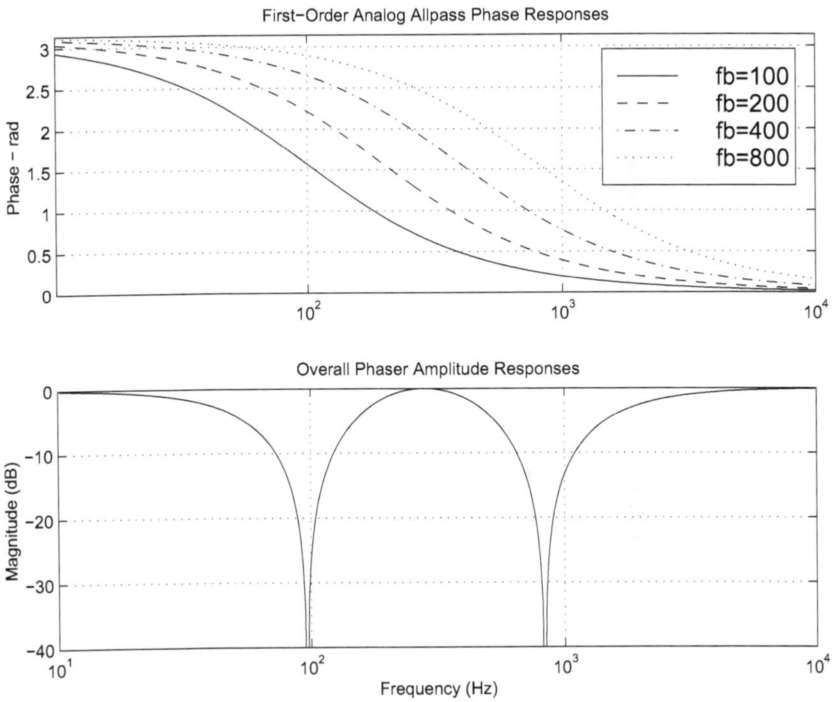

Figure 8.24: (a) Phase responses of first-order analog allpass sections with break frequencies at 100, 200, 400, and 800 Hz. (b) Corresponding phaser amplitude response.

i.e., $\omega = g_1$. Since the phase of a first-order allpass section at its break frequency is $\pi/2$, the sum of the other three sections must be approximately $2\pi + \pi/2$. Equivalently, since the first section has "given up" $\pi/2$ radians of phase at $\omega = g_1 = 2\pi 100$, the other three allpass sections *combined* have given up $\pi/2$ radians as well (with the second section having given up more than the other two).

In practical operation, the break frequencies must *change dynamically*, usually periodically at some rate.

Classic Virtual Analog Phase Shifters

To create a *virtual analog* phaser, following closely the design of typical analog phasers, we must translate each first-order allpass to the digital domain. Working with the transfer function, we must map from s plane to the z plane. There are several ways to accomplish this goal [362]. However, in this case, an excellent choice is the *bilinear transform* (see §7.3.2 on page 266), defined by

$$s \leftarrow c\frac{z-1}{z+1} \tag{8.20}$$

where c is chosen to map one particular frequency to exactly where it belongs. In this case, c can be chosen for each section to map the *break frequency* of the section to exactly where it belongs on the digital frequency axis. The relation between analog frequency ω_a and digital frequency ω_d follows immediately from Eq. (8.20) as

$$\begin{aligned} j\omega_a &= c\frac{e^{j\omega_d T}-1}{e^{j\omega_d T}+1} = c\frac{e^{j\omega_d T/2}\left(e^{j\omega_d T/2}-e^{-j\omega_d T/2}\right)}{e^{j\omega_d T/2}\left(e^{j\omega_d T/2}+e^{-j\omega_d T/2}\right)} \\ &= jc\frac{\sin(\omega_d T/2)}{\cos(\omega_d T/2)} = jc\tan(\omega_d T/2). \end{aligned}$$

Thus, given a particular desired break-frequency $\omega_a = \omega_d = \omega_b$, we can set

$$c = \omega_b \cot\left(\frac{\omega_b T}{2}\right).$$

Recall from Eq. (8.19) on page 325 that the transfer function of the first-order *analog* allpass filter is given by

$$H_a(s) = \frac{s-\omega_b}{s+\omega_b}$$

where ω_b is the break frequency. Applying the general bilinear transformation Eq. (8.20) gives

$$H_d(z) = H_a\left(c\frac{1-z^{-1}}{1+z^{-1}}\right) = \frac{c\left(\frac{1-z^{-1}}{1+z^{-1}}\right) - \omega_b}{c\left(\frac{1-z^{-1}}{1+z^{-1}}\right) + \omega_b} = \frac{p_d - z^{-1}}{1 - p_d z^{-1}}$$

where we have denoted the pole of the digital allpass by

$$p_d \triangleq \frac{c-\omega_b}{c+\omega_b} = \frac{1-\tan(\omega_b T/2)}{1+\tan(\omega_b T/2)} \approx \frac{1-\omega_b T/2}{1+\omega_b T/2} \approx 1-\omega_b T.$$

Figure 8.25 shows the digital phaser response curves corresponding to the analog response curves in Fig. 8.24. While the break frequencies are preserved by construction, the notches have moved slightly, although this is not visible from the plots. An overlay of the total phase of the analog and digital allpass chains is shown in Fig. 8.26. We see that the phase responses of the analog and digital alpass chains diverge visibly only above 9 kHz. The analog phase response approaches zero in the limit as $\omega_a \to \infty$, while the digital phase response reaches zero at half the sampling rate, 10 kHz in this case. This is a good example of when the bilinear transform performs very well.

In general, the bilinear transform works well to digitize feedforward analog structures in which the high-frequency warping is acceptable. When frequency warping is excessive, it can be alleviated by the use of *oversampling*; for example, the slight visible deviation in Fig. 8.26 below 10 kHz can be largely eliminated by increasing the sampling rate by 15% or so. See the case of digitizing the Moog VCF for an example in which the presence of feedback in the analog circuit leads to a delay-free loop in the digitized system [480, 478].

8.9.2 Phasing with 2nd-Order Allpass Filters

The allpass structure proposed in [429] provides a convenient means for generating nonuniformly spaced notches that are independently controllable to a high degree. An advantage of the allpass approach even in the case of uniformly spaced notches (which we call flanging, as introduced in §5.3 on page 183) is that no interpolating delay line is needed.

The architecture of the phaser based on second-order allpasses is shown in Fig. 8.27. It is identical to that in Fig. 8.23 on page 325 with each first-order allpass being replaced by a second-order allpass. *I.e.*, replace $\text{AP}_1^{g_i}$ in Fig. 8.23 by $\text{AP}_2^{g_i}$, for $i = 1, 2, 3, 4$, to get Fig. 8.27. The phaser will have a notch wherever the phase of the allpass chain is at π (180 degrees). It can be shown that these frequencies occur very close to the resonant frequencies of the allpass chain [429]. It is therefore convenient to use a single conjugate

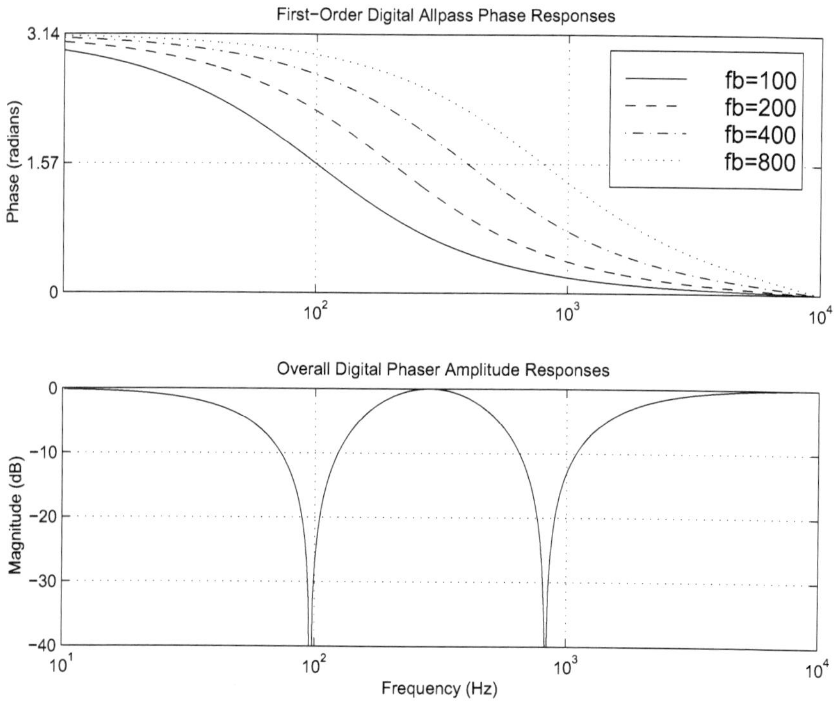

Figure 8.25: (a) Phase responses of first-order digital allpass sections with break frequencies at 100, 200, 400, and 800 Hz, with the sampling rate set to 20,000 Hz. (b) Corresponding phaser amplitude response.

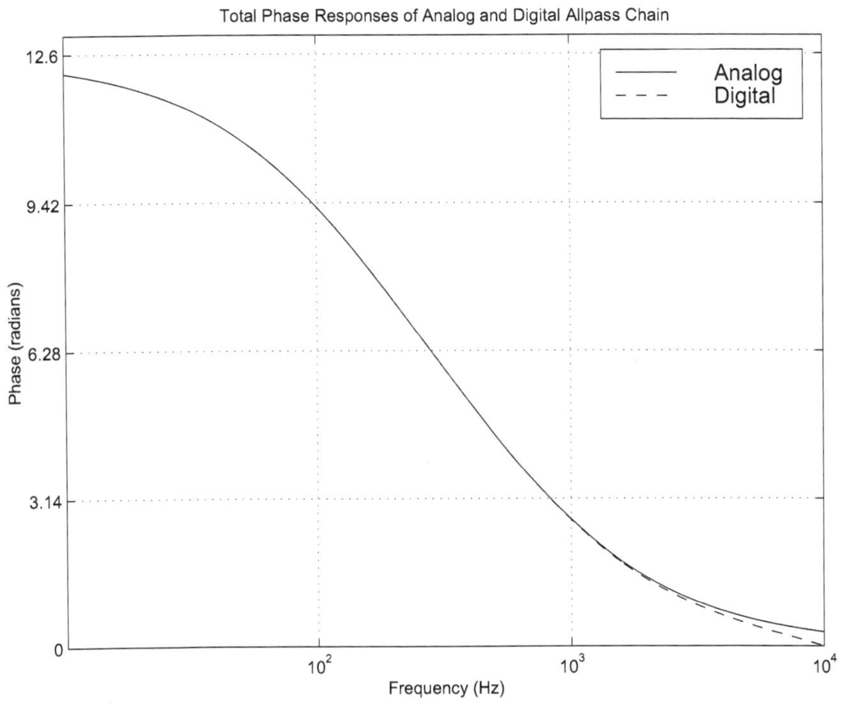

Figure 8.26: Phase response of four first-order allpass sections in series — analog and digital cases overlaid.

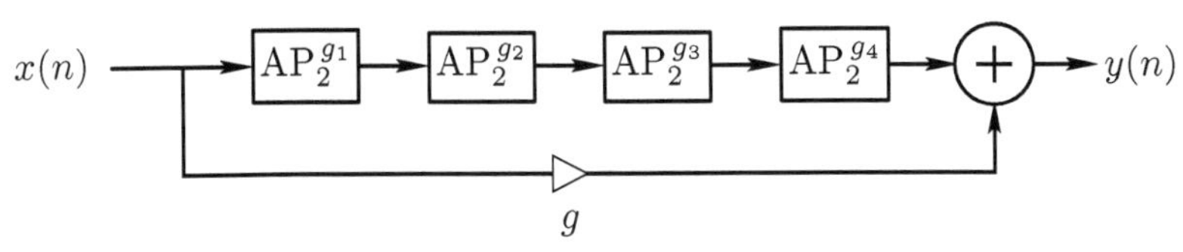

Figure 8.27: Structure of a phaser based on four second-order allpass filters.

pole pair in each allpass section, *i.e.*, use second-order allpass sections of the form

$$H(z) = \frac{a_2 + a_1 z^{-1} + z^{-2}}{1 + a_1 z^{-1} + a_2 z^{-2}}$$

where

$$\begin{aligned} a_1 &= -2R\cos(\theta) \\ a_2 &= R^2 \end{aligned}$$

and R is the radius of each pole in the complex-conjugate pole pair, and pole angles are $\pm\theta$. The pole angle can be interpreted as $\theta = \omega_c T$ where ω_c is the resonant frequency and T is the sampling interval.

Phaser Notch Parameters

To move just one notch, the tuning of the pole-pair in the corresponding section is all that needs to be changed. Note that tuning affects only one coefficient in the second-order allpass structure. (Although the coefficient a_1 appears twice in the transfer function, it only needs to be used once per sample in a slightly modified direct-form implementation [449].)

The depth of the notches can be varied together by changing the gain of the feedforward path.

The bandwidth of individual notches is mostly controlled by the distance of the associated pole-pair from the unit circle. So to widen the notch associated with a particular allpass section, one may increase the "damping" of that section.

Finally, since the gain of the allpass string is unity (by definition of allpass filters), the gain of the entire structure is strictly bounded between 0 and 2. This property allows arbitrary notch controls to be applied without fear of the overall gain becoming ill-behaved.

Phaser Notch Distribution

As mentioned above, it is desirable to avoid exact harmonic spacing of the notches, but what is the ideal non-uniform spacing? One possibility is to space the notches according to the *critical bands of hearing*, since essentially this gives a uniform notch density with respect to "place" along the basilar membrane in the ear. There is no need to follow closely the critical-band structure, so that simple exponential spacing may be considered sufficiently perceptually uniform (corresponding to uniform spacing on a log frequency scale). Due to the immediacy of the relation between notch characteristics and the filter coefficients, the notches can easily be placed under musically meaningful control.

Chapter 9

Virtual Musical Instruments

This chapter is discusses aspects of synthesis models for specific musical instruments such as guitars, pianos, woodwinds, bowed strings, and brasses. Initially we look at model components for electric guitars, followed by some special considerations for acoustic guitars. Next we focus on basic models for string excitation, first by a mass (a simplified piano hammer), then by a damped spring (plectrum model). Piano modeling is then considered, using commuted synthesis for the acoustic resonators and a parametric signal-model for the force-pulse of the hammer. Finally, elementary synthesis of single-reed woodwinds, such as the clarinet, and bowed-string instruments, such as the violin, are summarized, and some pointers are given regarding other wind and percussion instruments.

9.1 Electric Guitars

For electric guitars, which are mainly about the vibrating string and subsequent effects processing, we pick up where §6.7 left off, and work our way up to practical complexity step by step. We begin with practical enhancements for the waveguide string model that are useful for all kinds of guitar models.

9.1.1 Length Three FIR Loop Filter

The simplest nondegenerate example of the loop filters of §6.8 on page 223 is the three-tap FIR case ($N_{\hat{g}} = 3$). The symmetry constraint leaves two degrees of freedom in the frequency response:[1]

$$\hat{G}(e^{j\omega T}) = \hat{g}(0) + 2\hat{g}(1)\cos(\omega T)$$

If the dc gain is normalized to unity, then $\hat{g}(0) + 2\hat{g}(1) = 1$, and there is only one remaining degree of freedom which can be interpreted as a *damping* control. As damping is increased, the duration of free vibration is reduced at all nonzero frequencies, and the decay of higher frequencies is accelerated relative to lower frequencies, provided

$$\hat{g}(0) \geq 2\hat{g}(1) > 0.$$

In this coefficient range, the string-loop amplitude response can be described as a "raised cosine" having a unit-magnitude peak at dc, and minimum gains $\hat{g}(0) - 2\hat{g}(1) \geq 0$ at plus and minus half the sampling rate ($\omega T = \pm\pi$).

9.1.2 Length 3 FIR Loop Filter Controlled by "Brightness" and "Sustain"

Another convenient parametrization of the second-order symmetric FIR case is when the dc normalization is relaxed so that two degrees of freedom are retained. It is then convenient to control them as *brightness* B and *sustain* S according to the formulas

$$\begin{aligned} g_0 &= \exp(-6.91P/S) & (9.1)\\ \hat{g}(0) &= g_0(1+B)/2 & (9.2)\\ \hat{g}(1) &= g_0(1-B)/4 & (9.3) \end{aligned}$$

where P is the period in seconds (total loop delay), S is the desired sustain time in seconds, and B is the brightness parameter in the interval $[0, 1]$. The sustain parameter S is defined here as the time to decay by -60 dB (or ≈ 6.91 time-constants) when brightness B is maximum ($B = 1$) in which case the loop gain is g_0 at all frequencies, or $\hat{G}(e^{j\omega T}) = g_0$. As the brightness is lowered, the dc gain remains fixed at g_0 while higher frequencies decay faster. At the minimum brightness, the gain at half the sampling rate reaches zero, and the loop-filter amplitude-response assumes the form

$$\hat{G}(e^{j\omega T}) = g_0\frac{1+\cos(\omega T)}{2} = g_0\cos^2\left(\frac{\omega T}{2}\right).$$

[1] The corresponding impulse response is $h = [\ldots, 0, \hat{g}(1), \hat{g}(0), \hat{g}(1), 0, \ldots]$, where $\hat{g}(0)$ occurs at time 0, and the transfer function is $\hat{G}(z) = \hat{g}(0) + \hat{g}(1)[z + z^{-1}]$. In practice, this filter is typically implemented in causal form, *i.e.*, $h = [\hat{g}(1), \hat{g}(0), \hat{g}(1), 0, \ldots]$ starting at time 0, and one sample of delay is subtracted from a delay-line preceding or following the filter.

A Faust function implementing this FIR filter as the damping filter in the Extended Karplus Strong (EKS) algorithm is described in [454].

9.1.3 One-Zero Loop Filter

If we relax the constraint that $N_{\hat{g}}$ be odd, then the simplest case becomes the *one-zero* digital filter:

$$\hat{G}(z) = \hat{g}(0) + \hat{g}(1)z^{-1}$$

When $\hat{g}(0) = \hat{g}(1)$, the filter is *linear phase,* and its phase delay and group delay are equal to 1/2 sample [362]. In practice, the half-sample delay must be compensated elsewhere in the filtered delay loop, such as in the delay-line interpolation filter [207]. Normalizing the dc gain to unity removes the last degree of freedom so that $\hat{g}(0) = \hat{g}(1) = 1/2$, and $\hat{G}(e^{j\omega T}) = \cos(\omega T/2)$, $|\omega| \leq \pi f_s$.

See [454] for related discussion from a software implementation perspective.

9.1.4 The Karplus-Strong Algorithm

The simulation diagram for the ideal string with the simplest frequency-dependent loss filter is shown in Fig. 9.1. Readers of the computer music literature will recognize this as the structure of the *Karplus-Strong algorithm* [236, 207, 490].

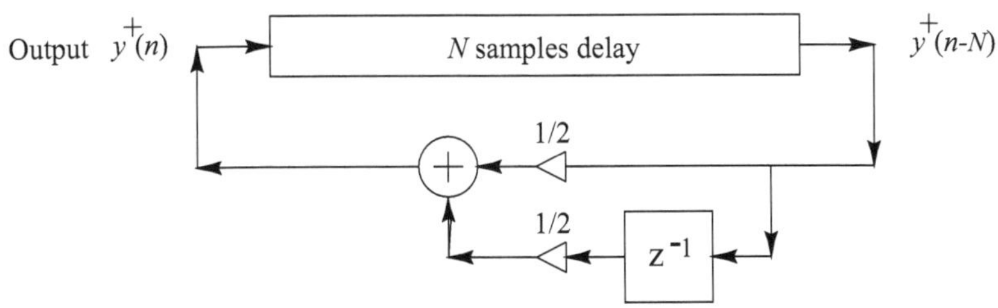

Figure 9.1: Rigidly terminated string with the simplest frequency-dependent loss filter. All N loss factors (possibly including losses due to yielding terminations) have been consolidated at a single point and replaced by a one-zero filter approximation.

The Karplus-Strong algorithm, *per se,* is obtained when the delay-line initial conditions used to "pluck" the string consist of random numbers, or "white noise." We know the initial *shape* of the string is obtained by *adding* the upper and lower delay lines of Fig. 6.11, *i.e.*, $y(t_n, x_m) = y^+(n-m) + y^-(n+m)$. It is shown in §C.7.4 on page 556 that the initial *velocity* distribution along the string is determined by the *difference* between the upper and lower delay lines. Thus, in the Karplus-Strong algorithm, the string is "plucked" by a *random initial displacement and initial velocity distribution.* This is a very energetic excitation, and usually in practice the white noise is lowpass filtered; the lowpass cut-off frequency gives an effective *dynamic level* control since natural stringed instruments are typically brighter at louder dynamic levels [428, 207].

Karplus-Strong sound examples[2] are available on the Web.

An implementation of the Karplus-Strong algorithm in the Faust programming language is described (and provided) in [454].

9.1.5 The Extended Karplus-Strong Algorithm

Figure 9.2 shows a block diagram of the *Extended Karplus-Strong* (EKS) algorithm [207].

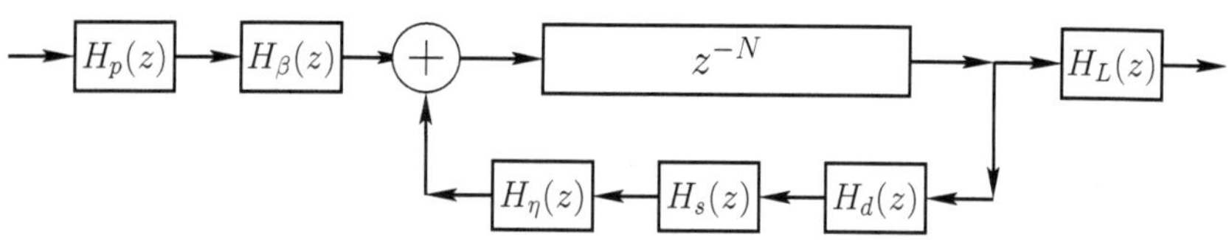

Figure 9.2: Extended Karplus-Strong (EKS) algorithm.

The EKS adds the following features to the KS algorithm:

$$
\begin{aligned}
H_p(z) &= \frac{1-p}{1-p\,z^{-1}} = \text{pick-direction lowpass filter} \\
H_\beta(z) &= 1 - z^{-\lfloor \beta N + 1/2 \rfloor} = \text{pick-position comb filter, } \beta \in (0,1) \\
H_d(z) &= \text{string-damping filter (one/two poles/zeros typical)} \\
H_s(z) &= \text{string-stiffness allpass filter (several poles and zeros)} \\
H_\eta(z) &= -\frac{\eta(N) - z^{-1}}{1-\eta(N)\,z^{-1}} = \text{first-order string-tuning allpass filter} \\
H_L(z) &= \frac{1-R_L}{1-R_L\,z^{-1}} = \text{dynamic-level lowpass filter}
\end{aligned}
$$

[2] http://ccrma.stanford.edu/~jos/Mohonk05/Karplus_Strong_Algorithm.html

where

$$\begin{aligned}
N &= \text{pitch period } (2\times \text{ string length) in samples} \\
\beta &= \text{normalized pick position} \in (0,1) \\
p &= 0 \text{ for one pick direction, } 0 < d < 1 \text{ for opposite direction} \\
R_L &= e^{-\pi LT}, \text{ where } L > \text{is “desired bandwidth” in Hz} \\
T &= \text{Sampling interval (Hz)} \\
\eta &\in [-1/11, 2/3] \text{ for tuning-delays in the range } [0.2, 1.2] \text{ samples} \\
\left|H_d(e^{j\omega T})\right| &\leq 1 \text{ required for stability}
\end{aligned}$$

Note that while $\eta \in [0,1)$ can be used in the tuning allpass, it is better to offset it to $[\epsilon, 1+\epsilon)$ to avoid delays close to zero in the tuning allpass. (A zero delay is obtained by a pole-zero cancellation on the unit circle.) First-order allpass interpolation of delay lines was discussed in §4.1.2 on page 133.

A history of the Karplus-Strong algorithm and its extensions is given in §A.8 on page 457. EKS sound examples[3] are also available on the Web. Techniques for designing the string-damping filter $H_d(z)$ and/or the string-stiffness allpass filter $H_s(z)$ are summarized below in §6.11 on page 231.

An implementation of the Extended Karplus-Strong (EKS) algorithm in the Faust programming language is described (and provided) in [454].

9.1.6 Nonlinear Distortion

In §6.13 on page 243, *nonlinear elements* were introduced in the context of general digital waveguide synthesis. In this section, we discuss specifically virtual electric guitar distortion, and mention other instances of audible nonlinearity in stringed musical instruments.

As discussed in Chapter 6, typical vibrating strings in musical acoustics are well approximated as linear, time-invariant systems, there are special cases in which *nonlinear* behavior is desired.

Tension Modulation

In every freely vibrating string, the fundamental frequency declines over time as the amplitude of vibration decays. This is due to *tension modulation*, which is often audible at the beginning of plucked-string tones, especially for low-tension strings. It happens because higher-amplitude vibrations *stretch* the string to a longer average length, raising the average string tension ⇒ faster wave propagation ⇒ higher fundamental frequency.

[3] http://ccrma.stanford.edu/~jos/Mohonk05/EKS_Sound_Example.html

The are several methods in the literature for simulating tension modulation in a digital waveguide string model [495, 233, 509, 513, 514, 496, 283], as well as in membrane models [298]. The methods can be classified into two categories, *local* and *global.*

Local tension-modulation methods modulate the speed of sound locally as a function of amplitude. For example, opposite delay cells in a force-wave digital waveguide string can be summed to obtain the instantaneous vertical force across that string sample, and the length of the adjacent propagation delay can be modulated using a first-order allpass filter. In principle the string slope reduces as the local tension increases. (Recall from Chapter 6 or Appendix C that force waves are minus the string tension times slope.)

Global tension-modulation methods [496, 495] essentially modulate the string delay-line length as a function of the total energy in the string.

String Length Modulation

A number of stringed musical instruments have a "nonlinear sound" that comes from modulating the physical termination of the string (as opposed to its acoustic length in the case of tension modulation).

The Finnish Kantele [231, 514] has a different effective string-length in the vertical and horizontal vibration planes due to a loose knot attaching the string to a metal bar. There is also nonlinear feeding of the second harmonic due to a nonrigid tuning peg.

Perhaps a better known example is the Indian *sitar*, in which a curved "jawari" (functioning as a nonlinear bridge) serves to shorten the string gradually as it displaces toward the bridge.

The Indian *tambura* also employs a *thread* perpendicular to the strings a short distance from the bridge, which serves to shorten the string whenever string displacement toward the bridge exceeds a certain distance.

Finally, the *slap bass* playing technique for bass guitars involves hitting the string hard enough to cause it to beat against the neck during vibration [263, 366].

In all of these cases, the string length is physically modulated in some manner each period, at least when the amplitude is sufficiently large.

Hard Clipping

A widespread class of distortion used in *electric guitars*, is *clipping* of the guitar waveform. it is easy to add this effect to any string-simulation algorithm by passing the output signal through a nonlinear *clipping function.*

For example, a *hard clipper* has the characteristic (in normalized form)

$$f(x) = \begin{cases} -1, & x \le -1 \\ x, & -1 \le x \le 1 \\ 1, & x \ge 1 \end{cases} \tag{9.4}$$

where x denotes the current input sample $x(n)$, and $f(x)$ denotes the output of the nonlinearity.

Soft Clipping

A *soft clipper* is similar to a hard clipper, but with the corners smoothed. A common choice of soft-clipper is the *cubic nonlinearity*, *e.g.* [490],

$$f(x) = \begin{cases} -\frac{2}{3}, & x \le -1 \\ x - \frac{x^3}{3}, & -1 \le x \le 1 \\ \frac{2}{3}, & x \ge 1. \end{cases} \tag{9.5}$$

This particular soft-clipping characteristic is diagrammed in Fig. 9.3. An analysis of its spectral characteristics, with some discussion of aliasing it may cause, was given in in §6.13 on page 243. An input gain may be used to set the desired degree of distortion.

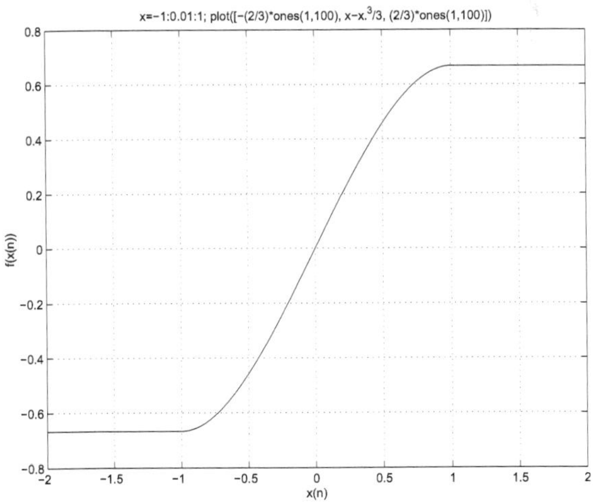

Figure 9.3: Soft-clipper defined by Eq. (9.5).

Enhancing Even Harmonics

A cubic nonlinearity, as well as *any* odd distortion law,[4] generates only odd-numbered harmonics (like in a square wave). For best results, and in particular for *tube distortion* simulation [31, 395], it has been argued that some amount of even-numbered harmonics should also be present. Breaking the odd symmetry in any way will add even-numbered harmonics to the output as well. One simple way to accomplish this is to add an *offset* to the input signal, obtaining

$$y(n) = f[x(n) + c],$$

where c is some small constant. (Signals $x(n)$ in practice are typically constrained to be zero mean by one means or another.)

Another method for breaking the odd symmetry is to add some square-law nonlinearity to obtain

$$f(x) = \alpha x^3 + \beta x^2 + \gamma x + \delta \tag{9.6}$$

where β controls the amount of square-law distortion in the more general third-order polynomial. The square-law is the most gentle nonlinear distortion in existence, adding only some second harmonic to a sinusoidal input signal. The constant δ can be set to zero the mean, on average; if the input signal $x(n)$ is zero-mean with variance is 1, then $\delta = -\beta$ will cancel the nonzero mean induced by the squaring term βx^2. Typically, the output of any audio effect is *mixed* with the original input signal to allow easy control over the amount of effect. The term γ can be used for this, provided the constant gains for $x > 1$ and $x < -1$ are modified accordingly, or x is hard-clipped to the desired range at the input.

Software for Cubic Nonlinear Distortion

The function `cubicnl` in the Faust software distribution (file `effect.lib`) implements cubic nonlinearity distortion [454]. The Faust programming example `cubic_distortion.dsp` provides a real-time demo with adjustable parameters, including an offset for bringing in even harmonics. The free, open-source `guitarix` application (for Linux platforms) uses this type of distortion effect along with many other guitar effects.

In the Synthesis Tool Kit (STK) [91], the class `Cubicnl.cpp` implements a general cubic distortion law.

[4] A function $f(x)$ is said to be *odd* if $f(-x) = -f(x)$.

9.1.7 Amplifier Feedback

A more extreme effect used with distorted electric guitars is *amplifier feedback*. In this case, the amplified guitar waveforms couple back to the strings with some gain and delay, as depicted schematically in Fig. 9.4 [490].

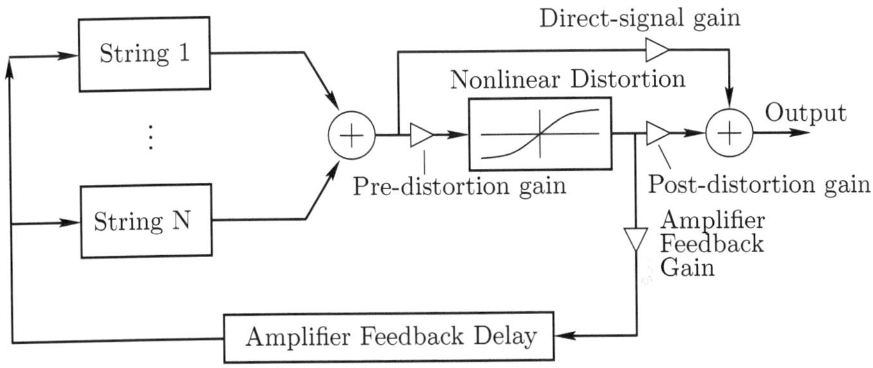

Figure 9.4: Simulation of a basic distorted electric guitar with amplifier feedback.

The Amplifier Feedback Delay in the figure can be adjusted to emphasize certain partial overtones over others. If the loop gain, controllable via the Amplifier Feedback Gain, is greater than 1 at any frequency, a sustained "feedback howl" will be produced. Note that in commercial devices, the Pre-distortion gain and Post-distortion gain are frequency-dependent, *i.e.*, they are implemented as pre- and post-*equalizers* (typically only a few bands, such as three). Another simple choice is an integrator $g/(1 - rz^{-1})$ for the pre-distortion gain, and a differentiator $(1 - rz^{-1})$ for the post-distortion gain.

Faust software implementing electric-guitar amplifier feedback may be found in [454].

9.1.8 Cabinet Filtering

The distortion output signal is often further filtered by an *amplifier cabinet filter*, representing the enclosure, driver responses, etc. It is straightforward to measure the impulse response (or frequency response) of a speaker cabinet and fit it with a recursive digital filter design function such as `invfreqz` in matlab (§8.6.4, p. 297). Individual speakers generally have one major resonance—the "cone resonance"—which is determined by the mass, compliance, and damping of the driver subassembly. The enclosing cabinet introduces many more resonances in the audio range, generally tuned to even out the overall response as much as possible.

Faust software implementing an idealized speaker filter may be found in [454].

9.1.9 Duty-Cycle Modulation

For class A tube amplifier simulation, there should be level-dependent *duty-cycle modulation* as a function:[5]

- 50% at low amplitude levels (no duty-cycle modulation)
- 55-65% duty cycle at high levels

A simple method for implementing duty-cycle modulation is to offset the input to nonlinearity Eq. (9.5) on page 339 by a constant, such as input RMS level times some scale factor.

9.1.10 Vacuum Tube Modeling

The elementary nonlinearities discussed above are aimed at approximating the sound of *amplifier distortion*, such as often used by guitar players. Generally speaking, the distortion characteristics of vacuum-tube amplifiers are considered superior to those of solid-state amplifiers. A real-time numerically solved nonlinear differential algebraic model for class A single-ended guitar power amplification is presented in [84]. Real-time simulation of a tube guitar-amp (triode preamp, inverter, and push-pull amplifier) based on a precomputed approximate solution of the nonlinear differential equations is presented in [293]. Other references in this area include [338, 228, 339].

9.1.11 Lab Exercises and Software

1. See `http://ccrma.stanford.edu/realsimple/electric_guitar/` for laboratory exercises related to virtual electric guitars.

2. See `http://ccrma.stanford.edu/realsimple/faust_strings/` for Faust software for virtual electric guitars.

3. See `http://guitarix.sourceforge.net/` for `guitarix` — a full-featured, free, open-source, guitar-effects application for Linux platforms (many effects written in the Faust language).

[5]See `http://www.trueaudio.com/at_eetjlm.htm` for further discussion.

9.2 Acoustic Guitars

This section addresses some modeling considerations specific to *acoustic* guitars. Section 9.2.1 discusses special properties of acoustic guitar *bridges*, and §9.2.1 gives ways to simulate passive bridges, including a look at some laboratory measurements. For further reading in this area, see, *e.g.*, [281, 128].

9.2.1 Bridge Modeling

In §6.3 on page 210 we analyzed the effect of *rigid string terminations* on traveling waves. We found that waves derived by time-derivatives of displacement (displacement, velocity, acceleration, and so on) reflect with a sign inversion, while waves defined in terms of the first spatial derivative of displacement (force, slope) reflect with no sign inversion. We now look at the more realistic case of *yielding terminations* for strings. This analysis can be considered a special case of the loaded string junction analyzed in §C.12 on page 598.

Yielding string terminations (at the bridge) have a large effect on the sound produced by acoustic stringed instruments. Rigid terminations can be considered a reasonable model for the *solid-body electric guitar* in which *maximum sustain* is desired for played notes. Acoustic guitars, on the other hand, must transduce sound energy from the strings into the body of the instrument, and from there to the surrounding air. All audible sound energy comes from the string vibrational energy, thereby reducing the sustain (decay time) of each played note. Furthermore, because the bridge vibrates more easily in one direction than another, a kind of "chorus effect" is created from the detuning of the horizontal and vertical planes of string vibration (as discussed further in §6.12.1 on page 238). A perfectly rigid bridge, in contrast, cannot transmit any sound into the body of the instrument, thereby requiring some other transducer, such as the magnetic pickups used in electric guitars, to extract sound for output.[6]

Passive String Terminations

When a traveling wave reflects from the bridge of a real stringed instrument, the bridge moves, transmitting sound energy into the instrument body. How far the bridge moves is determined by the *driving-point impedance* of the bridge, denoted $R_b(s)$. The driving point impedance is the ratio of Laplace

[6]So-called "electric-acoustic" guitars, such as the Godin electric-acoustic, use piezoelectric crystals in the bridge to measure string force at the string endpoint. Electric violins, such as the Zeta Violin, typically use this approach as well.

transform of the force on the bridge $F_b(s)$ to the velocity of motion that results $V_b(s)$. That is, $R_b(s) \triangleq F_b(s)/V_b(s)$.

For passive systems (*i.e.*, for all unamplified acoustic musical instruments), the driving-point impedance $R_b(s)$ is *positive real* (a property defined and discussed in §C.11.2 on page 588). Being positive real has strong implications on the nature of $R_b(s)$. In particular, the phase of $R_b(j\omega)$ cannot exceed plus or minus 90 degrees at any frequency, and in the lossless case, all poles and zeros must *interlace* along the $j\omega$ axis. Another implication is that the *reflectance* of a passive bridge, as seen by traveling waves on the string, is a so-called *Schur function* (defined and discussed in §C.11 on page 584); a Schur reflectance is a stable filter having gain not exceeding 1 at any frequency. In summary, a guitar bridge is passive if and only if its driving-point impedance is positive real and (equivalently) its reflectance is Schur. See §C.11 for a fuller discussion of this point.

At $x = 0$, the force on the bridge is given by (§C.7.2, p. 554)

$$f_b(t) \;=\; Ky'(t,0) \;=\; -f(t,0)$$

where K is the string tension as in Chapter 6, and $y'(t,0)$ is the *slope* of the string at $x = 0$. In the Laplace frequency domain, we have

$$F_b(s) \;=\; KY'(s,0) \;=\; -F(s,0),$$

due to linearity, and the velocity of the string endpoint is therefore

$$V(s,0) \equiv V_b(s) \;\triangleq\; \frac{F_b(s)}{R_b(s)} \;=\; -\frac{F(s,0)}{R_b(s)}.$$

A Terminating Resonator

Suppose a guitar bridge couples an ideal vibrating string to a single resonance, as depicted schematically in Fig. 9.5. This is often an accurate model of an acoustic bridge impedance in a narrow frequency range, especially at low frequencies where the resonances are well separated. Then, as developed in Chapter 7, the driving-point impedance seen by the string at the bridge is

$$R_b(s) \;=\; ms + \mu + k/s.$$

That is, the driving-point impedance is the series combination of a mass m, spring k, and dashpot μ (§7.2). More general bridge impedances can be modeled as a sum of such terms. Since the bridge is passive, $R_b(s)$ is *positive real* (§C.11.2).

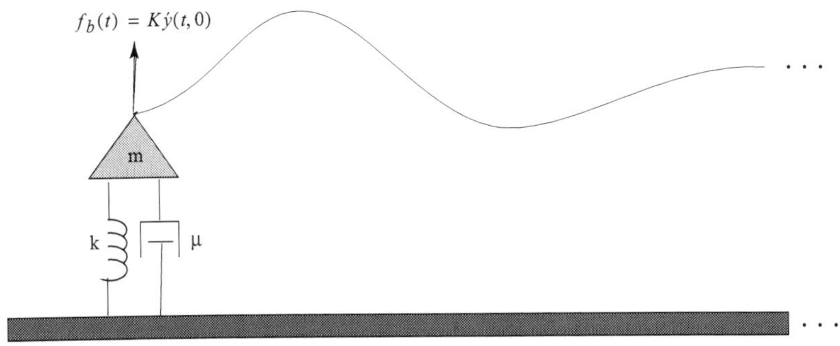

Figure 9.5: Ideal vibrating string terminated by a second-order driving-point impedance consisting of a mass m, spring k, and dashpot μ.

Bridge Reflectance

The bridge *reflectance* is needed as part of the loop filter in a digital waveguide model (Chapter 6).

As derived in §C.11.1 on page 584, the force-wave reflectance of $R_b(s)$ seen on the string is

$$\hat{\rho}_b(s) = \frac{R_b(s) - R_0}{R_b(s) + R_0} = \frac{s^2 + \frac{1}{m}(\mu - R_0)s + \omega_0^2}{s^2 + \frac{1}{m}(\mu + R_0)s + \omega_0^2} \qquad (9.7)$$

where R_0 denotes the wave impedance of the ideal string, and $\omega_0 \triangleq \sqrt{k/m}$ denotes the resonance frequency in radians per second. The velocity reflectance is simply minus the force reflectance (§C.11.1).

Bridge Transmittance

The bridge *transmittance* is the filter needed for the signal path from the vibrating string to the resonant acoustic body.

Since the bridge velocity equals the string endpoint velocity (a "series" connection), the velocity transmittance is simply

$$\hat{\tau}_b(s) = \frac{V_b(s)}{V^+(s)} = \frac{V^+(s) + V^-(s)}{V^+(s)} = 1 - \hat{\rho}_b(s).$$

If the bridge is rigid, then its motion becomes a velocity input to the acoustic resonator. In principle, there are three such velocity inputs for each point along the bridge. However, it is typical in stringed instrument models to consider only the vertical transverse velocity on the string as significant, which results in one (vertical) driving velocity at the base of the bridge. In violin models (§9.6), the use of a "sound post" on one side of the bridge and "bass bar" on the other strongly suggests supporting a rocking motion along the bridge.

Digitizing Bridge Reflectance

Converting continuous-time transfer functions such as $\hat{\rho}_b(s)$ and $\hat{\tau}_b(s)$ to the digital domain is analogous to converting an analog electrical filter to a corresponding digital filter—a problem which has been well studied [343]. For this task, the *bilinear transform* (§7.3.2, p. 266) is a good choice. In addition to preserving order and being free of aliasing, the bilinear transform preserves the positive-real property of passive impedances (§C.11.2).

Digitizing $\hat{\rho}_b(s)$ via the bilinear transform (§7.3.2) transform gives

$$\hat{\rho}_b^d(z) \triangleq \hat{\rho}_b\left(c\frac{1-z^{-1}}{1+z^{-1}}\right)$$

which is a second-order digital filter having gain less than one at all frequencies—*i.e.*, it is a *Schur* filter that becomes an allpass as the damping μ approaches zero. The choice of bilinear-transform constant $c = 1/\tan(\omega_0 T/2)$ maps the peak-frequency ω_0 without error (see Problem 4).

A Two-Resonance Guitar Bridge

Now let's consider a two-resonance guitar bridge, as shown in Fig. 9.6.

Like all mechanical systems that don't "slide away" in response to a constant applied input force, the bridge must "look like a spring" at zero frequency. Similarly, it is typical for systems to "look like a mass" at very high frequencies, because the driving-point typically has mass (unless the driver is spring-coupled by what seems to be massless spring). This implies the driving point admittance should have a zero at dc and a pole at infinity. If we neglect losses, as frequency increases up from zero, the first thing we encounter in the admittance is a pole (a "resonance" frequency at which energy is readily accepted by the bridge from the strings). As we pass the admittance peak going up in frequency, the phase switches around from being near $\pi/2$ ("spring like") to being closer to $-\pi/2$ ("mass like"). (Recall the *graphical method* for calculating the *phase response* of a linear system [449].) Below the first resonance, we may say that the system is *stiffness controlled* (admittance phase $\approx \pi/2$), while above the first resonance, we say it is *mass controlled* (admittance phase $\approx -\pi/2$). This qualitative description is typical of any lightly damped, linear, dynamic system. As we proceed up the $j\omega$ axis, we'll next encounter a near-zero, or "anti-resonance," above which the system again appears "stiffness controlled,"

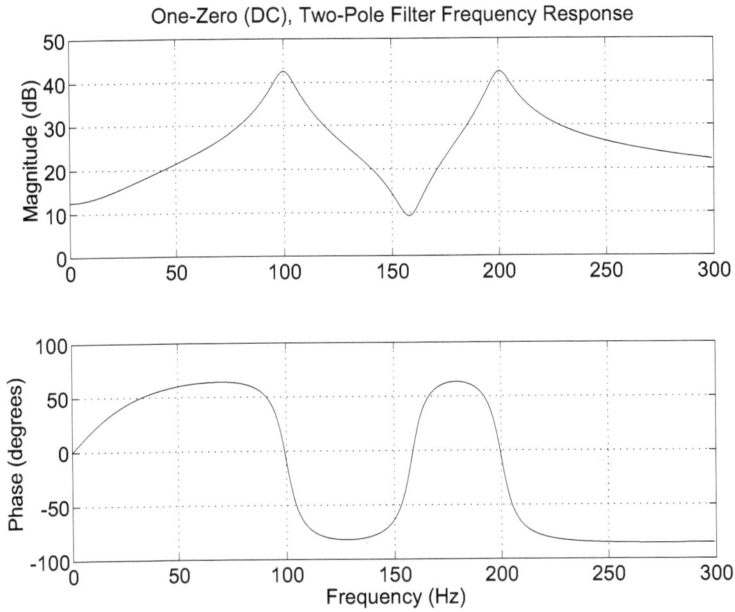

Figure 9.6: Synthetic input admittance of a passive, linear, dynamic system using a pair of resonating two-pole filters, a pair of zeros between the resonances, and a zero near dc.

or spring-like, and so on in alternation to infinity. The strict alternation of poles and zeros near the $j\omega$ axis is required by the *positive real* property of all passive admittances (§C.11.2).

Measured Guitar-Bridge Admittance

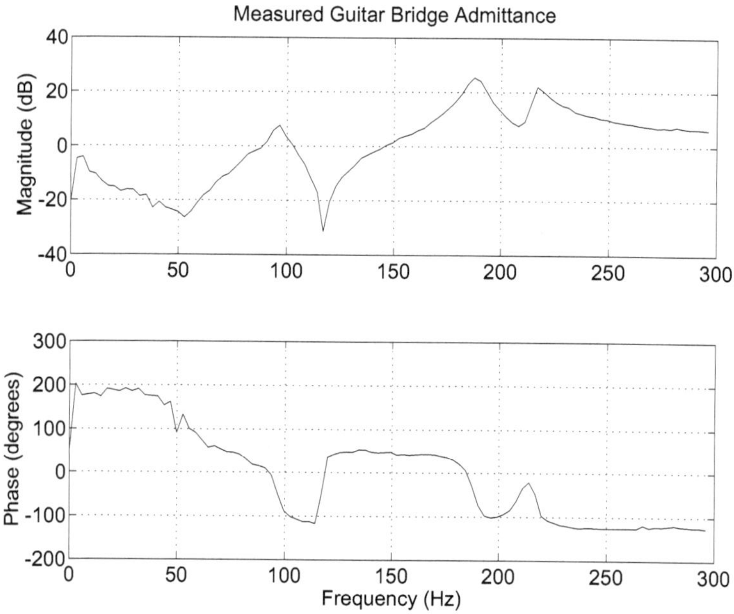

Figure 9.7: Input admittance of a real classical guitar bridge measured by striking the bridge with a force hammer in the vertical direction.

A *measured* driving-point admittance [269] for a real guitar bridge is shown in Fig. 9.7. Note that at very low frequencies, the phase information does not appear to be bounded by $\pm\pi/2$ as it must be for the admittance to be positive real (passive). This indicates a poor signal-to-noise ratio in the measurements at very low frequencies. This can be verified by computing and plotting the *coherence* function between the bridge input and output signals using multiple physical measurements.[7]

Figures 9.8 and 9.9 show overlays of the admittance magnitudes and phases, and also the input-output coherence, for three separate measure-

[7]The matlab function `cohere()` can be used to compute the coherence function between two signals across multiple physical measurements (one per non-overlapping frame).

ments. A coherence of 1 means that all the measurements are identical, while a coherence less than 1 indicates variation from measurement to measurement, implying a low signal-to-noise ratio. As can be seen in the figures, at frequencies for which the coherence is very close to 1, successive measurements are strongly in agreement, and the data are reliable. Where the coherence drops below 1, successive measurements disagree, and the measured admittance is not even positive real at very low frequencies.

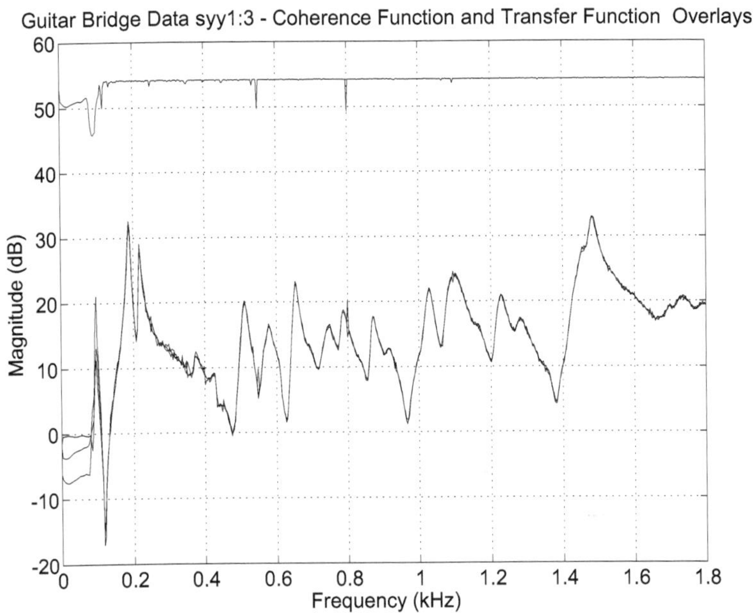

Figure 9.8: Overlay of three successively measured input admittances of a classical guitar bridge, displayed over a wider frequency range, and showing the coherence function, normalized to lie on the same dB scale as the admittance magnitude data.

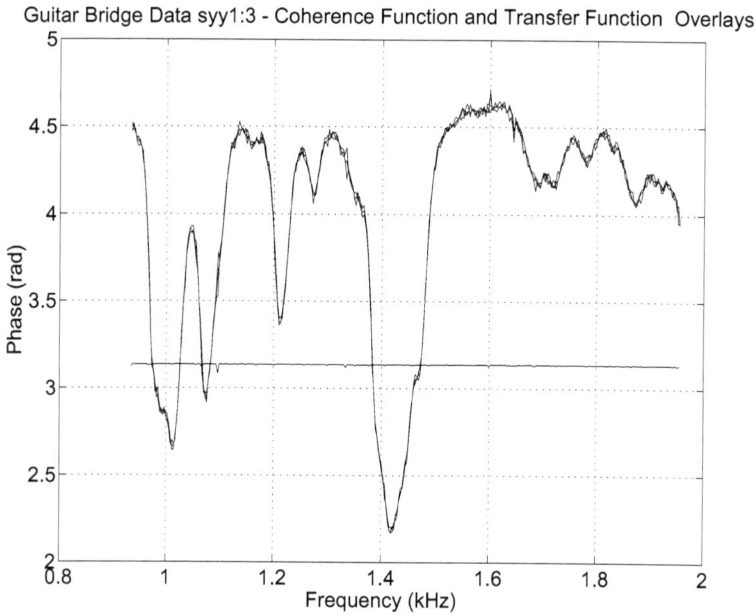

Figure 9.9: Overlay of the phase data for three successively measured input admittances of a classical guitar bridge, displayed over the frequency interval for which the coherence is very close to 1 (also shown).

Building a Synthetic Guitar Bridge Admittance

To construct a synthetic guitar bridge model, we can first measure empirically the admittance of a real guitar bridge, or we can work from measurements published in the literature, as shown in Fig. 9.7. Each peak in the admittance curve corresponds to a *resonance* in the guitar body that is well coupled to the strings via the bridge. Whether or not the corresponding vibrational mode *radiates efficiently* depends on the geometry of the vibrational mode, and how well it couples to the surrounding air. Thus, a complete bridge model requires not only a synthetic bridge admittance which determines the reflectance "seen" on the string, but also a separate filter which models the *transmittance* from the bridge to the outside air; the transmittance filter normally contains the same poles as the reflectance filter, but the zeros are different. Moreover, keep in mind that each string sees a slightly different reflectance and transmittance because it is located at a slightly different point relative to the guitar top plate; this changes the coupling coefficients to the various resonating modes to some extent. (We normally ignore this for simplicity and use the same bridge filters for all the strings.)

Finally, also keep in mind that each string excites the bridge in *three dimensions*. The two most important are the *horizontal and vertical planes of vibration*, corresponding to the two planes of transverse vibration on the string. The vertical plane is normal to the guitar top plate, while the horizontal plane is parallel to the top plate. Longitudinal waves also excite the bridge, and they can be important as well, especially in the piano. Since longitudinal waves are much faster in strings than transverse waves, the corresponding overtones in the sound are normally inharmonically related to the main (nearly harmonic) overtones set up by the transverse string vibrations.

The frequency, complex amplitude, and width of each peak in the measured admittance of a guitar bridge can be used to determine the parameters of a second-order digital resonator in a parallel bank of such resonators being used to model the bridge impedance. This is a variation on *modal synthesis* [5, 299]. However, for the bridge model to be *passive* when attached to a string, its transfer function must be *positive real*, as discused previously. Since strings are very nearly lossless, passivity of the bridge model is actually quite critical in practice. If the bridge model is even slightly active at some frequency, it can either make the whole guitar model unstable, or it might objectionably lengthen the decay time of certain string overtones.

We will describe two methods of constructing passive "bridge filters" from measured modal parameters. The first is guaranteed to be positive real but has some drawbacks. The second method gives better results, but it has to be checked for passivity and possibly modified to give a positive real admittance. Both methods illustrate more generally applicable signal processing methods.

Passive Reflectance Synthesis—Method 1

The first method is based on constructing a *passive reflectance* $\hat{\rho}(z)$ having the desired poles, and then converting to an admittance via the fundamental relation

$$\Gamma(z) = \Gamma_0 \frac{1-\hat{\rho}(z)}{1+\hat{\rho}(z)}$$

where Γ_0 is an arbitrary real, positive number which can be interpreted as the wave admittance of the string on which waves enter and return from the bridge.

As we saw in §C.11.1, every passive impedance corresponds to a passive reflectance which is a Schur function (stable and having gain not exceeding 1 around the unit circle). Since damping is light in a guitar bridge impedance (otherwise the strings would not vibrate very long, and sustain is a highly prized feature of real guitars), we can expect the bridge reflectance to be close to an allpass transfer function $H_A(z)$.

It is well known that every allpass transfer function can be expressed as

$$H_A(z) \triangleq \frac{\tilde{A}(z)}{A(z)}$$

where

$$\begin{aligned}
A(z) &= 1 + a_1 z^{-1} + a_2 z^{-2} + \cdots + a_{M-1} z^{-(M-1)} + a_M z^{-M} \\
\tilde{A}(z) &= a_M + a_{M-1} z^{-1} + \cdots a_1 z^{-(M-1)} + z^{-M} \\
&= z^{-M} A\left(z^{-1}\right) = \text{Flip}(A)(z)
\end{aligned}$$

We will then construct a Schur function as

$$\hat{\rho}(z) = g H_A(z) F(z), \quad 0 < g < 1$$

where $F(z) = f_0 + f_1(z + z^{-1}) + f_2(z^2 + z^{-2} + \cdots + f_N(z^N + z^{-N})$ is a zero-phase FIR filter which can be used to adjust peak heights without altering phase.[8] We require $\angle F(e^{j\omega T}) = 0$ which means its frequency response is real and positive. Note that being a symmetric FIR filter is not sufficient; there also can be no zero-crossings in the frequency response, which tends to limit how far the impedance magnitudes can be shaped. (For simplicity, feel

[8] Zero-phase filters can normally be implemented in practice because there are pure delay lines preceding and following the reflectance filter, and taps can be introduced in the delay-line preceding the reflectance to implement noncausal terms in the impulse response.

free to forget about $F(z)$ and accept that all the admittance resonances will be constrained to the same height. The second method is easier anyway.)

Recall that in every allpass filter with real coefficients, to every pole at radius R_i there corresponds a zero at radius $1/R_i$.[9]

Because the impedance is lightly damped, the poles and zeros of the corresponding reflectance are close to the unit circle. This means that at points along the unit circle between the poles, the poles and zeros tend to cancel. It can be easily seen using the graphical method for computing the phase of the frequency response that the pole-zero angles in the allpass filter are very close to the resonance frequencies in the corresponding passive impedance [429]. Furthermore, the distance of the allpass poles to the unit circle controls the bandwidth of the impedance peaks. Therefore, to a first approximation, we can treat the allpass pole-angles as the same as those of the impedance pole angles, and the pole radii in the allpass can be set to give the desired impedance peak bandwidth. The zero-phase shaping filter $F(z)$ gives the desired mode height.

From the measured peak frequencies F_i and bandwidths B_i in the guitar bridge admittance, we may approximate the pole locations $z_i = R_i e^{j\theta_i}$ as

$$\begin{aligned} R_i &\approx e^{-\pi B_i T} \\ \theta_i &= 2\pi F_i T \end{aligned}$$

where T is the sampling interval as usual. Next we construct the allpass denominator as the product of elementary second-order sections:

$$A(z) = \prod_{i=1}^{M} \frac{1}{1 + a_1(i)z^{-1} + a_2(i)z^{-2}}$$

where

$$\begin{aligned} a_1(i) &\triangleq -2R_i \cos(\theta_i) \\ a_2(i) &\triangleq R_i^2 \end{aligned}$$

Now that we've constructed a Schur function, a passive admittance can be computed using (9.2.1). While it is guaranteed to be positive real, the modal frequencies, bandwidths, and amplitudes are only indirectly controlled and therefore approximated. (Of course, this would provide a good initial guess for an iterative procedure which computes an optimal approximation directly.)

A simple example of a synthetic bridge constructed using this method with $F(z) = 1$ and $g = 0.9$ is shown in Fig. 9.10.

[9] To prove this, note that the roots of $A(z^{-1})$ are the reciprocals of the roots of $A(z)$,

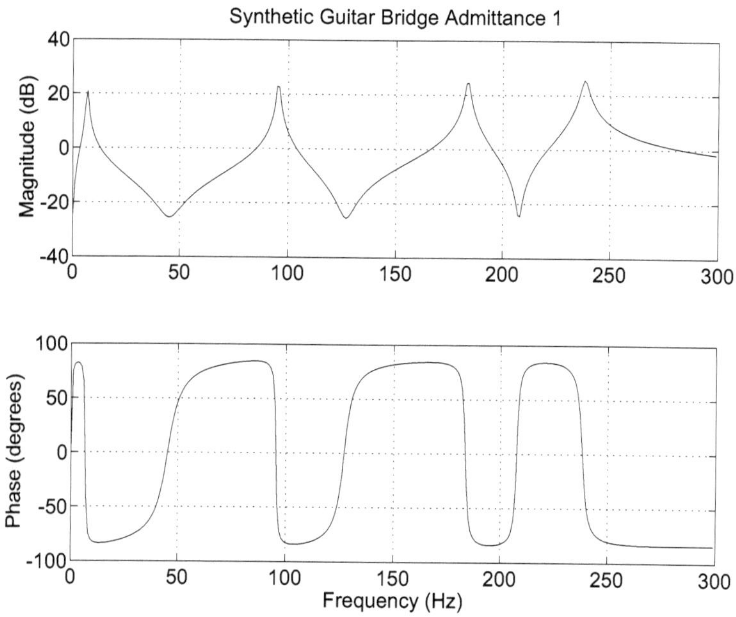

Figure 9.10: Synthetic guitar-bridge admittance using method 1.

Passive Reflectance Synthesis—Method 2

The second method is based on constructing a partial fraction expansion of the admittance directly:

$$\Gamma(z) = \Gamma_0(1 - z^{-1}) \sum_{i=1}^{M/2} \frac{1}{1 + a_1(i)z^{-1} + a_2(i)z^{-2}}$$

with $a_m(i)$ as above. While such a construction is not guaranteed to be positive real, (please feel free to find general conditions for which the positive-real condition can be guaranteed), we now have direct control over the bandwidths and modal gains (pole residues in the admittance). The reason the construction *tends* to be positive real is that by using the same phase for each section (the 1 in all the section numerators), we are sure to get a zero forming at some frequency near the middle between the resonance frequencies, and at a similar distance from the unit circle. This means we are constructing interlacing poles and zeros by simply adding the resonators in parallel. The extra zero near dc is to ensure that the admittance looks like a lightly damped spring at zero frequency. Since half the sampling rate merely "cuts off" the frequency response, there is no corresponding requirement to add a pole near $z = -1$ as if it were the point at infinity. However, the phase should be checked not to exceed plus or minus 90 degrees there (or at any other frequency), and a pole or zero added if necessary to pull it into the positive-real range.

A simple example of a synthetic bridge constructed using this method with is shown in Fig. 9.11.

Matlab for Passive Reflectance Synthesis Method 1

```
fs = 8192;  % Sampling rate in Hz (small for display convenience)
fc = 300;   % Upper frequency to look at
nfft = 8192;% FFT size (spectral grid density)

nspec = nfft/2+1;
nc = round(nfft*fc/fs);
f = ([0:nc-1]/nfft)*fs;

% Measured guitar body resonances
F = [4.64 96.52 189.33 219.95]; % frequencies
B = [ 10   10     10     10  ]; % bandwidths
```

since the conformal map $z \leftarrow z^{-1}$ exchanges interior of the unit circle with the exterior of the unit circle.

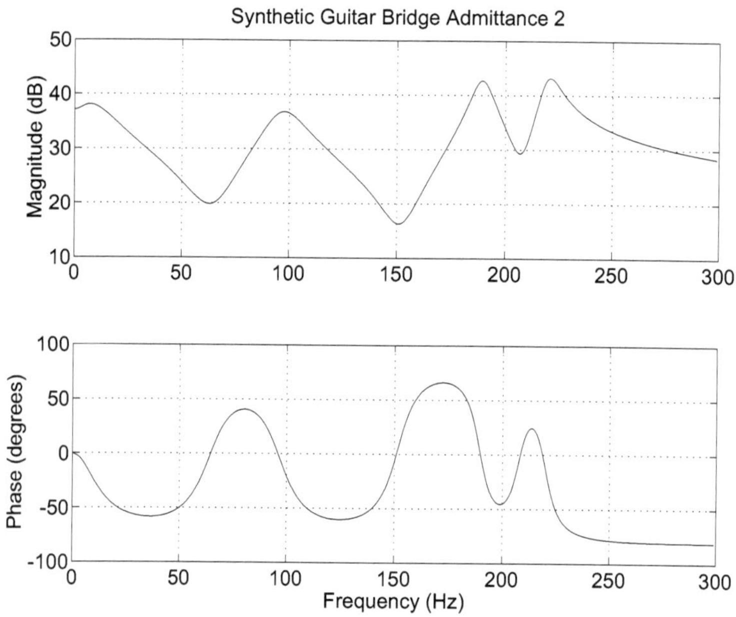

Figure 9.11: Synthetic guitar-bridge admittance using method 2.

```
nsec = length(F);

R = exp(-pi*B/fs);      % Pole radii
theta = 2*pi*F/fs;      % Pole angles
poles = R .* exp(j*theta);
A1 = -2*R.*cos(theta);  % 2nd-order section coeff
A2 = R.*R;              % 2nd-order section coeff
denoms = [ones(size(A1)); A1; A2]'
A = [1,zeros(1,2*nsec)];
for i=1:nsec,
    % polynomial multiplication = FIR filtering:
    A = filter(denoms(i,:),1,A);
end;
```

Now A contains the (stable) denominator of the desired bridge admittance. We want now to construct a numerator which gives a positive-real result. We'll do this by first creating a passive reflectance and then computing the corresponding PR admittance.

```
g = 0.9;            % Uniform loss factor

B = g*fliplr(A); % Flip(A)/A = desired allpass

Badm = A - B;
Aadm = A + B;
Badm = Badm/Aadm(1); % Renormalize
Aadm = Aadm/Aadm(1);

% Plots

fr = freqz(Badm,Aadm,nfft,'whole');

nc = round(nfft*fc/fs);
spec = fr(1:nc);
f = ([0:nc-1]/nfft)*fs;
dbmag = db(spec);
phase = angle(spec)*180/pi;

subplot(2,1,1);
plot(f,dbmag); grid;
title('Synthetic Guitar Bridge Admittance');
ylabel('Magnitude (dB)');
subplot(2,1,2);
```

```
plot(f,phase); grid;
ylabel('Phase (degrees)');
xlabel('Frequency (Hz)');
```

Matlab for Passive Reflectance Synthesis Method 2

```
... as in Method 1 for poles ...

% Construct a resonator as a sum of arbitrary modes
% with unit residues, adding a near-zero at dc.

B = zeros(1,2*nsec+1);
impulse = [1,zeros(1,2*nsec)];
for i=1:nsec,
    % filter() does polynomial division here:
    B = B + filter(A,denoms(i,:),impulse);
end;

% add a near-zero at dc
B = filter([1 -0.995],1,B);

... as in Method 1 for display ...
```

Matrix Bridge Impedance

A passive *matrix* bridge admittance (*e.g.*, taking all six strings of the guitar together) can be conveniently synthesized as a vector of positive-real scalar admittances multiplied by a positive definite matrix [25]. In [25], a dual-polarization guitar-string model is connected to a matrix reflectance computed from the matrix driving-point impedance at the bridge.

9.2.2 Body Modeling

As introduced in §8.7 on page 305, the *commuted synthesis* technique involves *commuting* the LTI filter representing acoustic resonator (guitar body) with the string, so that the string is excited by the *pluck response* of the body in place of the plectrum directly. The techniques described in §8.8 on page 309 can be used to build practical computational synthesis models of acoustic guitar bodies and the like [229].

9.3 String Excitation

In §2.4 and §6.10 we looked at the basic architecture of a digital waveguide string excited by some external disturbance. We now consider the specific cases of string excitation by a *hammer* (mass) and plectrum (spring).

9.3.1 Ideal String Struck by a Mass

In §6.6 on page 219, the ideal struck string was modeled as a simple initial velocity distribution along the string, corresponding to an instantaneous transfer of linear momentum from the striking hammer into the transverse motion of a string segment at time zero. (See Fig. 6.10 on page 220 for a diagram of the initial traveling velocity waves.) In that model, we neglected any effect of the striking hammer after time zero, as if it had bounced away at time 0 due to a so-called *elastic collision*. In this section, we consider the more realistic case of an *inelastic collision*, *i.e.*, where the mass hits the string and *remains in contact* until something, such as a wave, or gravity, causes the mass and string to separate.

For simplicity, let the string length be infinity, and denote its wave impedance by R. Denote the colliding mass by m and its speed prior to collision by v_0. It will turn out in this analysis that we may approximate the size of the mass by zero (a so-called *point mass*). Finally, we neglect the effects of gravity and drag by the surrounding air. When the mass collides with the string, our model must switch from two separate models (mass-in-flight and ideal string), to that of two ideal strings joined by a mass m at $x = 0$, as depicted in Fig. 9.12. The "force-velocity port" connections of the mass and two semi-infinite string endpoints are formally in *series* because they all move together; that is, the mass velocity equals the velocity of each of the two string endpoints connected to the mass (see §7.2 on page 257 for a fuller discussion of impedances and their parallel/series connection).

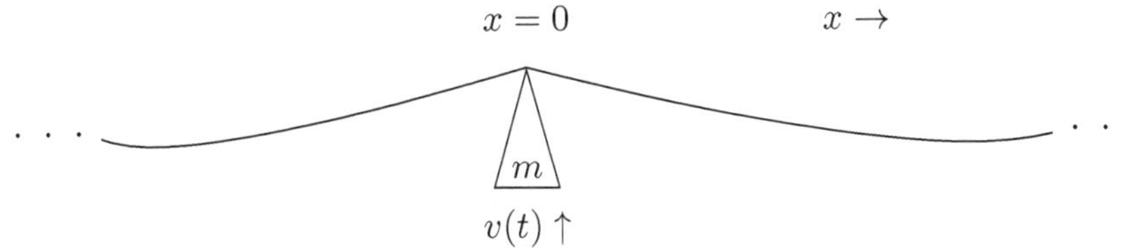

Figure 9.12: Physical model of mass-string collision after time 0.

The equivalent circuit for the mass-string assembly after time zero is shown in Fig. 9.13. Note that the string wave impedance R appears twice, once for each string segment on the left and right. Also note that there is a single common velocity $v(t)$ for the two string endpoints and mass. LTI circuit elements in series can be arranged in any order.

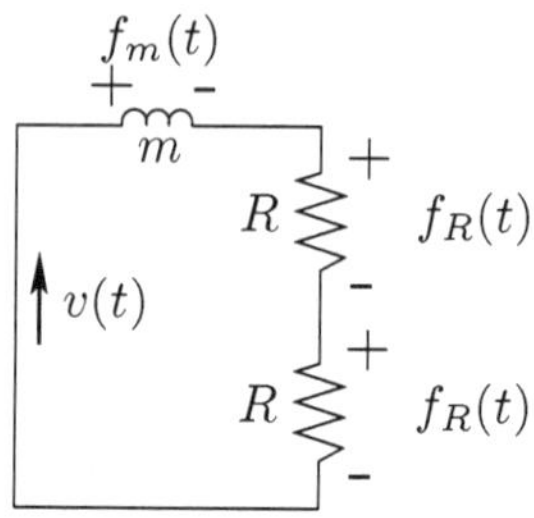

Figure 9.13: Electrical equivalent circuit for the mass and two string endpoints after time zero. The mass is represented by an inductor of m Henrys, while each string endpoint is represented by a resistor of R Ohms (impedance R).

From the equivalent circuit, it is easy to solve for the velocity $v(t)$. Formally, this is accomplished by applying Kirchoff's Loop Rule, which states that the sum of voltages ("forces") around any series loop is zero:

$$0 = f_m(t) + f_R(t) + f_R(t) \qquad (9.8)$$

All of the signs are '+' in this equation because the "current" (velocity $v(t)$) flows into the '+' sign of each element. These *reference directions* indicated by '+/−' on each element in Fig. 9.13 may be chosen arbitrarily, but it is convenient to adopt the convention that $v(t)$ flows into the '+' sign for each *reaction force*, and into the '−' sign for each *action force* (or *driving force* or *voltage/current source*, etc.). With this convention, Kirchoff's Loop Rule effectively states "the sum of all action forces equals the sum of all reaction forces," where the forces are understood to act on a common point. This is essentially Newton's third law of motion "for every action there is an equal and opposite reaction" (§B.1). In our mass-string example, all three forces are defined to be reaction forces because there is no external driving force on the mass, and no incoming waves from the string segments. External driving forces on the mass-string junction will be formulated later below.[10] We could equally well have defined the mass inertial force as a driving force on the two string segments, or the string segments could comprise a pair of driving forces for the mass. Ultimately, any force polarities may be assigned and consistently handled.

[10]In particular, as illustrated in Fig. 9.14 below, we can formulate the initial mass momentum as being supplied by an external impulsive force at time zero.

Taking the Laplace transform[11] of Eq. (9.8) yields, by linearity,

$$0 = F_m(s) + 2F_R(s), \tag{9.9}$$

where $F_m(s)$ and $F_R(s)$ denote the Laplace transforms of $f_m(t)$ and $f_R(t)$, respectively. As discussed above, the impedance relation for each string endpoint is given by

$$f_R(t) = R\,v(t) \quad \longleftrightarrow \quad F_R(s) = R\,V(s),$$

where $V(s)$ denotes the Laplace transform of $v(t)$.

For the mass, we have

$$f_m(t) = m\,a(t) = m\,\frac{d}{dt}v(t) \quad \longleftrightarrow \quad F_m(s) = m\left[s\,V(s) - v_0\right],$$

where we used the differentiation theorem for Laplace transforms [449, Appendix D].[12] Note that the mass is characterized by its *impedance* $F_m(s)/V(s) = ms$ when the initial velocity v_0 is zero (§7.1).

Substituting these relations into Eq. (9.9) yields

$$m\,v_0 = m\,s\,V(s) + 2R\,V(s). \tag{9.10}$$

We see that the initial momentum $m\,v_0$ of the mass effectively provides an *impulsive external driving force*

$$f_{\text{ext}}(t) = m\,v_0\,\delta(t).$$

That is, an equivalent problem formulation is to start with the mass at rest and in contact with the string, followed by striking the mass with an ideal hammer (impulse) that imparts momentum $m\,v_0$ to the mass at time zero. This formulation is diagrammed in Fig. 9.14.

An advantage of the external-impulse formulation is that the system has a zero initial state, so that an *impedance description* (§7.1) is complete. In other words, the system can be fully described as a series combination of the three impedances ms, R (on the left), and R (on the right), driven by an external force-source $f_{\text{ext}}(t)$.

Solving Eq. (9.10) for $V(s)$ yields

$$V(s) = \frac{m\,v_0}{ms + 2R}.$$

Since the Laplace transform of $e^{-at}u(t)$ is $1/(s+a)$, where $u(t)$ denotes the Heaviside unit step function,[13] the velocity of the contact point is

$$v(t) = v_0\,e^{-\frac{2R}{m}t}, \quad t \geq 0.$$

[11] http://ccrma.stanford.edu/~jos/filters/Introduction_Laplace_Transform_Analysis.html

[12] http://ccrma.stanford.edu/~jos/filters/Differentiation.html

[13] $u(t) = 0$ for $t < 0$ and 1 for $t \geq 0$.

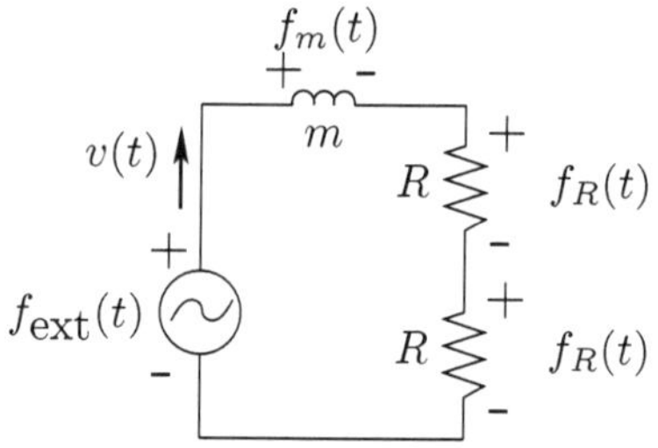

Figure 9.14: Electrical equivalent circuit for the mass and two string endpoints after time zero, using an impulsive driving force in place of a nonzero initial velocity. (*Cf.* Fig. 9.13.)

We see that at time zero the mass velocity is v_0, as it must be, and after that it decays exponentially to zero with time-constant $m/2R$. The decay rate depends on the ratio of the mass to the string wave impedance. In particular, the heavier the mass, the slower the mass velocity decays to zero. Since $R = \sqrt{K\epsilon}$, the greater the string tension K or mass-density ϵ, the faster the mass velocity decays to zero.

The displacement of the string at $x = 0$ is given by the integral of the velocity:

$$y(t,0) = \int_0^t v(\tau)\,d\tau = v_0\,\frac{m}{2R}\,\left[1 - e^{-\frac{2R}{m}t}\right]$$

where we defined the initial transverse displacement as $y(0,0) = 0$. The final displacement of the string is

$$y(\infty,0) = v_0\,\frac{m}{2R}.$$

Thus, the final string displacement is proportional to both the "hammer mass" and the initial striking velocity; it is inversely proportional to the string wave impedance R.

The momentum of the mass before time zero is mv_0, and after time zero it is

$$m\,v(t) = m\,v_0\,e^{-\frac{2R}{m}t}.$$

The force applied to the two string endpoints by the mass is given by $f_m(t) = 2Rv(t)$. From Newton's Law, $f = ma = m\dot{v}$, we have that momentum mv, delivered by the mass to the string, can be calculated as the *time integral* of applied force:

$$\int_0^t f_m(\tau)\,d\tau = 2R\int_0^t v(\tau)\,d\tau = 2Rv_0\frac{m}{2R}\left(1 - e^{-\frac{2R}{m}t}\right) = m\,v_0\left(1 - e^{-\frac{2R}{m}t}\right).$$

Thus, the momentum delivered to the string by the mass starts out at zero, and grows as a relaxing exponential to mv_0 at time infinity. We see that an ideal string struck inelastically by a mass does not at all correspond to an instantaneous momentum transfer, as considered in §6.6. Instead, the mass's momentum is transferred over a period of time (in this case infinite time). This is why it is possible to approximate the mass width by zero in this analysis, unlike in the excitation-by-initial-velocity in §6.6.

In a real piano, the hammer, which strikes in an upward (grand) or sideways (upright) direction, falls away from the string a short time after collision, but it may remain in contact with the string for a substantial fraction of a period (see §9.4 on piano modeling).

Mass Termination Model

The previous discussion solved for the motion of an ideal mass striking an ideal string of infinite length. We now investigate the same model from the string's point of view. As before, we will be interested in a digital waveguide (sampled traveling-wave) model of the string, for efficiency's sake (Chapter 6), and we therefore will need to know what the mass "looks like" at the end of each string segment. For this we will find that the impedance description (§7.1) is especially convenient.

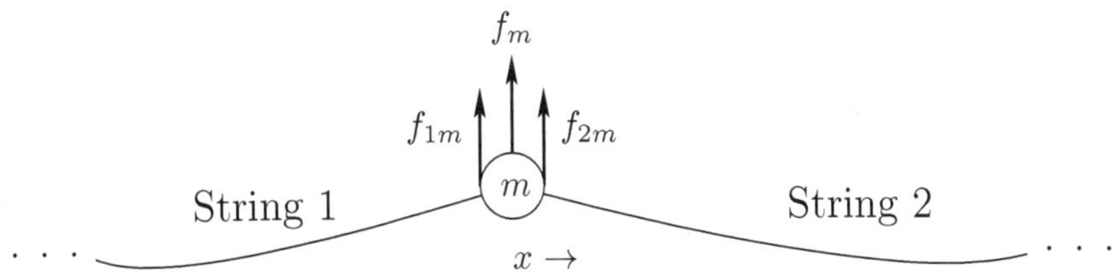

Figure 9.15: Physical model of mass-string collision after time 0. The mass is drawn as having a finite diameter for conceptual clarity. However, the model is formulated for the limit as the diameter approaches zero in the figure (bringing all three forces together to act on a single mass-string junction point). In other words, we assume a point mass.

Let's number the string segments to the left and right of the mass by 1 and 2, respectively, as shown in Fig. 9.15. Then Eq. (9.8) on page 360 above may be written

$$0 = f_m(t) + f_{1m}(t) + f_{2m}(t), \tag{9.11}$$

where $f_{1m}(t)$ denotes the force applied by string-segment 1 to the mass (defined as positive in the "up", or positive-y direction), $f_{2m}(t)$ is the force applied by string-segment 2 to the mass (again positive upwards), and $f_m(t)$ denotes the inertial force applied by the mass to both string endpoints (where again, a positive force points up).

To derive the traveling-wave relations in a digital waveguide model, we want to use the force-wave variables $f_1 = f_1^+ + f_1^-$ and $f_2 = f_2^+ + f_2^-$ that we defined for vibrating strings in §6.1.5 on page 206; *i.e.*, we defined $f_i \triangleq -Ky_i'$, where K is the string tension and y_i' is the string slope, $i = 1, 2$.

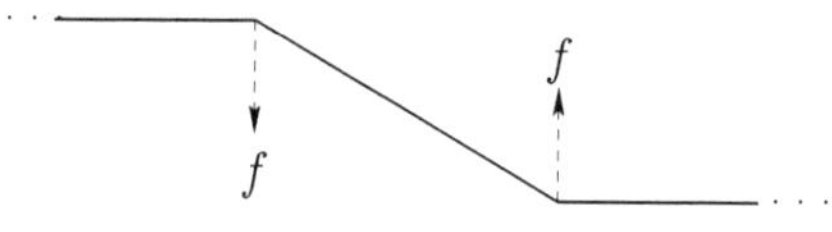

Figure 9.16: Depiction of a string segment with negative slope (center), pulling up to the right and down to the left. (Horizontal force components are neglected.)

As shown in Fig. 9.16, a negative string slope pulls "up" to the right. Therefore, at the mass point we have $f_{1m}(t) = f_1(t, x_m) = -Ky_1'(t, x_m)$, where x_m denotes the position of the mass along the string. On the other hand, the figure also shows that a negative string slope pulls "down" to the left, so that $f_{2m}(t) = -f_2(t, x_m) = Ky_2'(t, x_m)$. In summary, relating the forces we have defined for the mass-string junction to the force-wave variables in the string, we have

$$\begin{aligned} f_{1m}(t) &= f_1(t, x_m) \\ f_{2m}(t) &= -f_2(t, x_m) \end{aligned}$$

where x_m denotes the position of the mass along the string.

Thus, we can rewrite Eq. (9.11) in terms of string wave variables as

$$0 = f_m(t) + f_1(t, x_m) - f_2(t, x_m), \tag{9.12}$$

or, substituting the definitions of these forces,

$$0 = -m\dot{v}(t) - K\, y_1'(t, 0) + K\, y_2'(t, 0). \tag{9.13}$$

The inertial force of the mass is $f_m(t) = -m\dot{v}(t)$ because the mass must be accelerated downward in order to produce an upward reaction force. The signs of the two string forces follow from the definition of force-wave variables on the string, as discussed above.

The force relations can be checked individually. For string 1,

$$m\dot{v}(t) + K\, y_1'(t, 0) = 0$$

states that a positive slope in the string-segment to the left of the mass corresponds to a negative acceleration of the mass by the endpoint of that string segment. Similarly, for string 2,

$$m\dot{v}(t) - K\, y_2'(t,0) = 0$$

says that a positive slope on the right accelerates the mass upwards. Similarly, a negative slope pulls "up" to the right and "down" to the left, as shown in Fig. 9.16 above.

Now that we have expressed the string forces in terms of string force-wave variables, we can derive digital waveguide models by performing the traveling-wave decompositions $f_1 = f_1^+ + f_1^-$ and $f_2 = f_2^+ + f_2^-$ and using the Ohm's law relations $f_i^+ = Rv_i^+$ and $f_i^- = -Rv_i^-$ for $i = 1, 2$ (introduced above near Eq. (6.6) on page 207).

Mass Reflectance from Either String

Let's first consider how the mass looks from the viewpoint of string 1, assuming string 2 is at rest. In this situation (no incoming wave from string 2), string 2 will appear to string 1 as a simple resistor (or dashpot) of R Ohms in series with the mass impedance ms. (This observation will be used as the basis of a rapid solution method in §9.3.1 below.)

When a wave from string 1 hits the mass, it will cause the mass to move. This motion carries both string endpoints along with it. Therefore, both the reflected and transmitted waves include this mass motion. We can say that we see a "dispersive transmitted wave" on string 2, and a dispersive reflection back onto string 1. Our object in this section is to calculate the transmission and reflection *filters* corresponding to these transmitted and reflected waves.

By physical symmetry the *velocity* reflection and transmission will be the same from string 1 as it is from string 2. We can say the same about force waves, but we will be more careful because the sign of the transverse force flips when the direction of travel is reversed.[14] Thus, we expect a scattering junction of the form shown in Fig. 9.17 (recall the discussion of physically interacting waveguide inputs in §2.4.3 on page 52). This much invokes the superposition principle (for simultaneous reflection and transmission), and imposes the expected symmetry: equal reflection filters $\hat{\rho}(s)$ and equal transmission filters $\hat{\tau}(s)$ (for either force or velocity waves).

Let's begin with Eq. (9.12) above, restated as follows:

$$0 \;=\; f_m(t) + f_1(t, x_m) - f_2(t, x_m). \tag{9.14}$$

[14] In acoustic tubes, it is the longitudinal-velocity wave variable that changes sign when the direction of propagation is changed, because the air particles then move in the opposite direction along the tube axis x. The pressure, on the other hand, being a scalar field, does not fundamentally change sign when the propagation direction toggles.

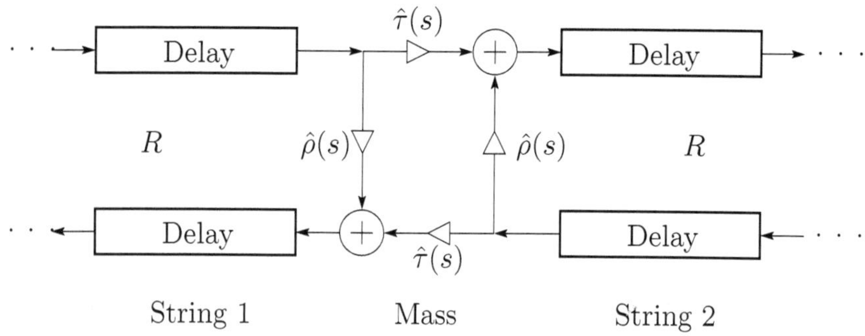

Figure 9.17: Expected form of the scattering junction representing a mass m attached to the string. This form is expected for both force and velocity waves. The LTI filter transfer function $\hat{\rho}(s)$ models the *reflectance* of the mass on the spring, while $\hat{\tau}(s)$ models its *transmittance*.

The traveling-wave decompositions can be written out as

$$\begin{aligned} f_1(t, x_m) &= f_1^+(t - x_m/c) + f_1^-(t + x_m/c) \\ f_2(t, x_m) &= f_2^+(t + x_m/c) + f_2^-(t - x_m/c) \end{aligned} \tag{9.15}$$

where a "+" superscript means "right-going" and a "-" superscript means "left-going" on either string.[15]

Let's define the mass position x_m to be zero, so that Eq. (9.14) with the substitutions Eq. (9.15) becomes

$$0 \;=\; f_m(t) + [f_1^+(t) + f_1^-(t)] - [f_2^+(t) + f_2^-(t)] = 0.$$

Let's now omit the common "(t)" arguments and write more simply

$$f_m + f_1^+ + f_1^- - f_2^+ - f_2^- \;=\; 0.$$

Let $v = v_m = v_1 = v_2$ denote the common velocity of the mass and string endpoints. Then we have $f_m = -m\dot{v}$ for the mass (as discussed above at Eq. (9.13)), and the Ohm's-law relations for the string are $f_i^+ = Rv_i^+$ and $f_i^- = -Rv_i^-$. Making these substitutions gives

$$-m\dot{v} + Rv_1^+ - Rv_1^- - Rv_2^+ + Rv_2^- \;=\; 0.$$

In the Laplace domain, dropping the common "(s)" arguments,

$$\begin{aligned} F_m + F_1^+ + F_1^- - F_2^+ - F_2^- &= 0 \\ \Longleftrightarrow \quad -msV + RV_1^+ - RV_1^- - RV_2^+ + RV_2^- &= 0. \end{aligned}$$

[15]Since we are in continuous time, a notation more consistent with Chapter 6 would be $f_{i,l}(t - x/c)$ and $f_{i,r}(t - x/c)$, for $i = 1, 2$, as we reserved the $\pm$ superscripts for the discrete-time case in that chapter. Notation has thus been changed for this example.

To compute the reflection coefficient of the mass seen on string 1, we may set $V_2^- = 0$, which means $V = V_2^+ + V_2^- = V_2^+$, so that we have

$$-msV + RV_1^+ - RV_1^- - RV \;=\; 0.$$

Also, since $V = V_1^+ + V_1^-$, we can substitute $V_1^- = V - V_1^+$ and solve for the common velocity to get

$$\boxed{V(s) \;=\; \frac{2R}{ms+2R} V_1^+(s).} \tag{9.16}$$

From this, the reflected velocity is immediate:

$$\boxed{V_1^-(s) \;=\; V(s) - V_1^+(s) \;=\; -\frac{ms}{ms+2R} V_1^+(s).}$$

The transmitted velocity is of course $Vp_2(s) = V(s)$. We have thus found the *velocity reflection transfer function* (or velocity *reflectance*) of the mass as seen from string 1:

$$\boxed{\hat{\rho}_v(s) \;\triangleq\; \frac{V_1^-(s)}{V_1^+(s)} \;=\; -\frac{ms}{ms+2R}}$$

By physical symmetry, we also have $V_2^+(s)/V_2^-(s) = \hat{\rho}_v(s)$, *i.e.*, the transverse-velocity reflectance is the same on either side of the mass. Thus we have found both velocity reflection filters at the mass-string junction. In summary, the velocity reflectance of the mass is $-ms/(ms+2R)$ from either string segment. This is a simple first-order filter model of what the mass (and string beyond) looks like dynamically from either string.

It is always good to check that our answers make physical sense in limiting cases. For this problem, easy cases to check are $m = 0$ and $m = \infty$. When the mass is $m = 0$, the reflectance goes to zero (no reflected wave at all). When the mass goes to infinity, the reflectance approaches $\hat{\rho}_v = -1$, corresponding to a rigid termination, which also makes sense.

The results of this section can be more quickly obtained as a special case of the main result of §C.12 on page 598, by choosing $N = 2$ waveguides meeting at a load impedance $R_J(s) = ms$. The next section gives another fast-calculation method based on a standard formula.

Simplified Impedance Analysis

The above results are quickly derived from the general reflection-coefficient for force waves (or voltage waves, pressure waves, etc.):

$$\boxed{\rho = \frac{R_2 - R_1}{R_2 + R_1} = \frac{\text{Impedance Step}}{\text{Impedance Sum}}} \tag{9.17}$$

where ρ is the reflection coefficient of impedance R_2 as "seen" from impedance R_1. If a force wave f^+ traveling along in impedance R_1 suddenly hits a new impedance R_2, the wave will split into a reflected wave $f^- = \rho f^+$, and a transmitted wave $(1+\rho)f^+$. It therefore follows that a velocity wave v^+ will split into a reflected wave $v^- = -\rho v^+$ and transmitted wave $(1-\rho)v^+$. This rule is derived in §C.8.4 on page 569 (and implicitly above as well).

In the mass-string-collision problem, we can immediately write down the *force reflectance* of the mass as seen from either string:

$$\boxed{\hat{\rho}_f(s) = \frac{(ms+R)-R}{(ms+R)+R} = \frac{ms}{ms+2R}}$$

That is, waves in the string are traveling through wave impedance R, and when they hit the mass, they are hitting the series combination of the mass impedance ms and the wave impedance R of the string on the other side of the mass. Thus, in terms of Eq. (9.17) above, $R_1 = R$ and $R_2 = ms + R$.

Since, by the Ohm's-law relations,

$$\hat{\rho}_f(s) \stackrel{\Delta}{=} \frac{F^-}{F^+} = \frac{-RV^-}{RV^+} = -\frac{V^-}{V^+} \stackrel{\Delta}{=} -\hat{\rho}_v(s).$$

we have that the *velocity reflectance* is simply

$$\boxed{\hat{\rho}_v(s) \stackrel{\Delta}{=} \frac{V^-}{V^+} = -\hat{\rho}_f(s) \stackrel{\Delta}{=} -\frac{F^-}{F^+}.}$$

Mass Transmittance from String to String

Referring to Fig. 9.15 on page 363, the *velocity transmittance* from string 1 to string 2 may be defined as

$$\hat{\tau}_v(s) = \frac{V_2^+(s)}{V_1^+(s)}.$$

By physical symmetry, we expect the transmittance to be the same in the opposite direction: $\hat{\tau}_v(s) = \frac{V_1^-(s)}{V_2^-(s)}$. Assuming the incoming wave V_2^- on string 2 is zero, we have $V_2^+ = V$, which we found in Eq. (9.16) on page 367:

$$V = \frac{2R}{ms+2R}V_1^+$$

Thus, the mass transmittance for velocity waves is

$$\boxed{\hat{\tau}_v(s) = \frac{2R}{ms+2R} = 1 - \hat{\rho}_v(s)}$$

We see that $m \to \infty$ corresponds to $\hat{\tau}_v(s) \to 0$, as befits a rigid termination. As $m \to 0$, the transmittance becomes 1 and the mass has no effect, as desired.

We can now refine the picture of our scattering junction Fig. 9.17 to obtain the form shown in Fig. 9.18.

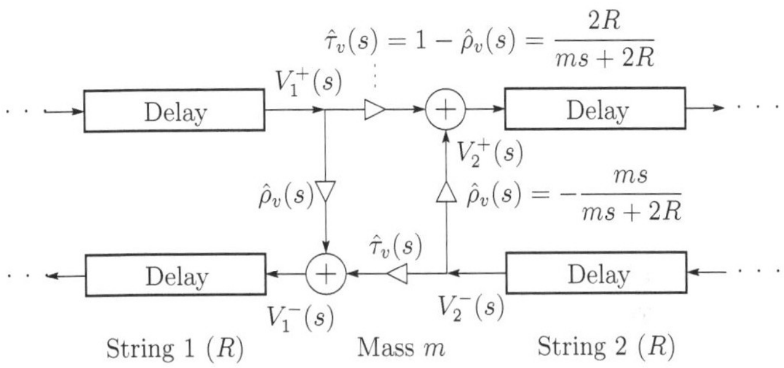

Figure 9.18: Velocity-wave scattering junction for a mass m (impedance ms) attached to an ideal string having wave impedance R.

Force Wave Mass-String Model

The velocity transmittance is readily converted to a *force transmittance* using the Ohm's-law relations:

$$\hat{\tau}_f(s) \triangleq \frac{F_2^+}{F_1^+} = \frac{RV_2^+}{RV_1^+} = \hat{\tau}_v(s)$$

Calculating it for the other direction as a check gives

$$\frac{F_1^-}{F_2^-} = \frac{-RV_1^+}{-RV_2^+} = \hat{\tau}_v(s).$$

Thus, while the reflectance of the mass toggles sign when going from force to velocity waves, the transmittance of the mass is the same in all cases. We therefore have the force-wave scattering junction shown in Fig. 9.19.

Checking as before, we see that $m \to \infty$ corresponds to $\hat{\tau}_v(s) \to 0$, which means no force is transmitted through an infinite mass, which is reasonable. As $m \to 0$, the force transmittance becomes 1 and the mass has no effect, as desired.

Summary of Mass-String Scattering Junction

In summary, we have characterized the mass on the string in terms of its reflectance and transmittance from either string. For force waves, we have

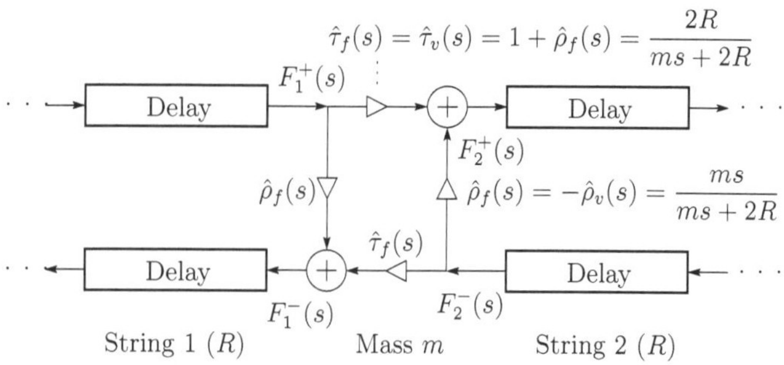

Figure 9.19: Force-wave scattering junction representing a mass m (impedance ms) attached to an ideal string having wave impedance R.

outgoing waves given by

$$\begin{aligned} F_1^-(s) &= \hat{\rho}_f(s)F_1^+(s) + \hat{\tau}_f(s)F_2^-(s) \\ F_2^+(s) &= \hat{\tau}_f(s)F_1^+(s) + \hat{\rho}_f(s)F_2^-(s) \end{aligned}$$

or

$$\begin{bmatrix} F_2^+ \\ F_1^- \end{bmatrix} = \begin{bmatrix} \hat{\tau}_f(s) & \hat{\rho}_f(s) \\ \hat{\rho}_f(s) & \hat{\tau}_f(s) \end{bmatrix} \begin{bmatrix} F_1^+ \\ F_2^- \end{bmatrix}$$

in terms of the incoming waves F_1^+ and F_2^-, the force reflectance $\hat{\rho}_f(s) = ms/(ms+2R)$, and the force transmittance $\hat{\tau}_f(s) = 1 + \hat{\rho}_f(s) = 2R/(ms+2R)$. We may say that the mass creates a *dynamic scattering junction* on the string. (If there were no dependency on s, such as when a dashpot is affixed to the string, we would simply call it a *scattering junction*.) The above form of the dynamic scattering junction is analogous to the *Kelly-Lochbaum scattering junction* (§C.8.4, p. 569). The general relation $\hat{\tau}_f = 1+\hat{\rho}_f$ can be used to simplify the Kelly-Lochbaum form to a *one-filter scattering junction* analogous to the one-multiply scattering junction (§C.8.5 on page 570):

$$\begin{aligned} F_1^- &= \hat{\rho}_f F_1^+ + (1+\hat{\rho}_f)F_2^- \;=\; F_2^- + \hat{\rho}_f \cdot (F_1^+ + F_2^-) \\ F_2^+ &= (1+\hat{\rho}_f)F_1^+ + \hat{\rho}_f F_2^- \;=\; F_1^+ + \hat{\rho}_f \cdot (F_1^+ + F_2^-) \end{aligned}$$

The one-filter form follows from the observation that $\hat{\rho}_f \cdot (F_1^+ + F_2^-)$ appears in both computations, and therefore need only be implemented once:

$$\begin{aligned} F^+ &\triangleq \hat{\rho}_f \cdot (F_1^+ + F_2^-) \\ F_1^- &= F_2^- + F^+ \\ F_2^+ &= F_1^+ + F^+ \end{aligned}$$

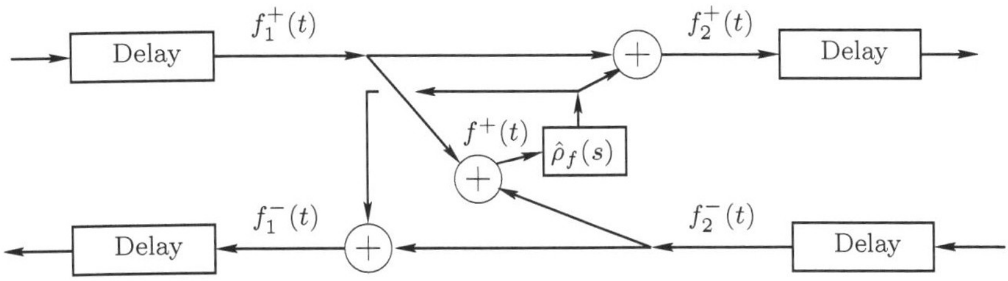

Figure 9.20: Continuous-time force-wave simulation diagram, in one-filter form, for an ideal string with a point mass attached.

This structure is diagrammed in Fig. 9.20.

Again, the above results follow immediately from the more general formulation of §C.12 on page 598.

Digital Waveguide Mass-String Model

To obtain a force-wave digital waveguide model of the string-mass assembly after the mass has struck the string, it only remains to digitize the model of Fig. 9.20. The delays are obviously to be implemented using digital delay lines. For the mass, we must digitize the force reflectance appearing in the one-filter model of Fig. 9.20 on page 371:

$$\hat{\rho}_f(s) = \frac{ms}{ms + 2R} \tag{9.18}$$

A common choice of digitization method is the *bilinear transform* (§7.3.2 on page 266) because it preserves losslessness and does not alias. This will effectively yield a *wave digital filter* model for the mass in this context (see Appendix F for a tutorial on wave digital filters).

The bilinear transform is typically scaled as

$$s = \frac{2}{T}\frac{1 - z^{-1}}{1 + z^{-1}}$$

where T denotes the sampling interval. This choice optimizes the low-frequency match between the digital and analog mass frequency responses. Rewriting Eq. (9.18) as

$$\hat{\rho}_f(s) = \frac{1}{1 + \frac{2R}{ms}}$$

the bilinear transform gives

$$\begin{aligned}\hat{\rho}_d(z) &= \frac{1}{1+\frac{2R}{m}\frac{T}{2}\frac{1+z^{-1}}{1-z^{-1}}} \\ &= g\frac{1-z^{-1}}{1-pz^{-1}}\end{aligned}$$

where the gain coefficient g and pole p are given by

$$\begin{aligned}g &\triangleq \frac{1}{1+\frac{RT}{m}} < 1 \\ p &\triangleq \frac{1-\frac{RT}{m}}{1+\frac{RT}{m}} < 1.\end{aligned}$$

Thus, the reflectance of the mass is a one-pole, one-zero filter. The zero is exactly at dc, the real pole is close to dc, and the gain at half the sampling rate is 1. We may recognize this as the classic *dc-blocking filter* [449]. Comparing with Eq. (9.18), we see that the behavior at dc is correct, and that the behavior at infinite frequency ($\hat{\rho}(\infty) = 1$) is now the behavior at half the sampling rate ($\hat{\rho}_d(-1) = 1$).

Physically, the mass reflectance is zero at dc because sufficiently slow waves can freely move a mass of any finite size. The reflectance is 1 at infinite frequency because there is no time for the mass to start moving before it is pushed in the opposite direction. In short, a mass behaves like a rigid termination at infinite frequency, and a free end (no termination) at zero frequency. The reflectance of a mass is therefore a "dc blocker".

The final digital waveguide model of the mass-string combination is shown in Fig. 9.21.

Additional examples of lumped-element modeling, including masses, springs, dashpots, and their various interconnections, are discussed the Wave Digital Filters (WDF) appendix (Appendix F). A nice feature WDFs is that they employ traveling-wave input/output signals which are ideal for interfacing to digital waveguides. The main drawback is that the WDFs operate over a *warped frequency axis* (due to the bilinear transform), while digital delay lines have a normal (unwarped) frequency axis. On the plus side, WDFs cannot *alias*, while digital waveguides do alias in the frequency domain for signals that are not bandlimited to less than half the sampling rate. At low frequencies (or given sufficient oversampling), the WDF frequency warping is minimal, and in such cases, WDF "lumped element models" may be connected directly to digital waveguides, which are "sampled-wave distributed parameter" models.

Even when the bilinear-transform frequency-warping is severe, it is often well tolerated when the frequency response has only one "important

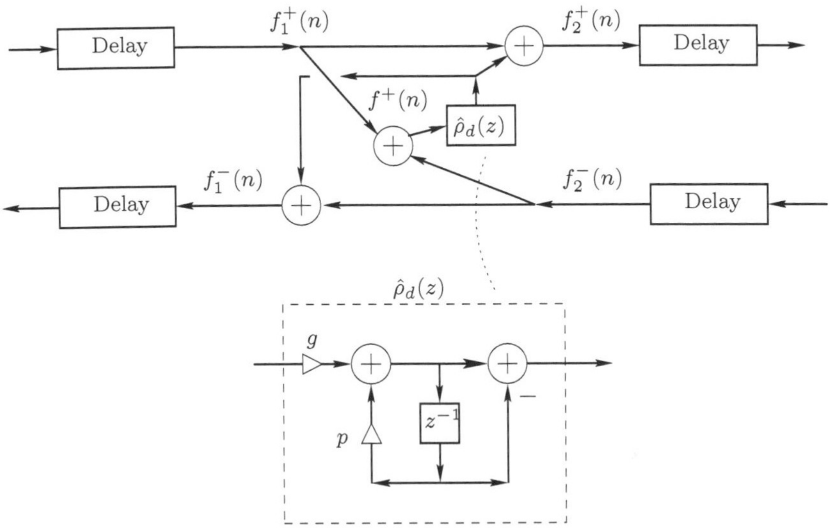

Figure 9.21: Digital waveguide model of an ideal string with a point mass attached.

frequency", such as a second-order resonator, lowpass, or highpass response. In other words, the bilinear transform can be scaled to map any single analog frequency to any desired corresponding digital frequency (see §7.3.2 on page 266 for details), and the frequency-warped responses above and below the exactly mapped frequency may "sound as good as" the unwarped responses for musical purposes. If not, higher order filters can be used to model lumped elements (Chapter 7).

Displacement-Wave Simulation

As discussed in [121], displacement waves are often preferred over force or velocity waves for guitar-string simulations, because such strings often hit obstacles such as frets or the neck. To obtain displacement from velocity at a given x, we may time-integrate velocity as above to produce displacement at any spatial sample along the string where a collision might be possible. However, all these integrators can be eliminated by simply going to a displacement-wave simulation, as has been done in nearly all papers to date on plucking models for digital waveguide strings.

To convert our force-wave simulation to a displacement-wave simulation, we may first convert force to velocity using the Ohm's law relations $f_i^+ = Rv_i^+$ and $f_i^- = -Rv_i^-$ and then conceptually integrate all signals with respect to time (in advance of the simulation). R is the same on both sides of the finger-junction, which means we can convert from force to velocity by simply negating all left-going signals. (Conceptually, all signals are converted from force to velocity by the Ohm's law relations and then divided by R, but the common scaling by $1/R$ can be omitted (or postponed) unless signal values are desired in particular physical units.) An all-velocity-wave simulation can be converted to displacement waves even more easily by simply changing v to y everywhere, because velocity and displacement waves scatter identically. In more general situations, we can go to the Laplace domain and replace each occurrence of $V^+(s)$ by $sY^+(s)$, each $V^-(s)$ by $sY^-(s)$, divide all signals by s, push any leftover s around for maximum simplification, perhaps absorbing it into a nearby filter. In an all-velocity-wave simulation, each signal gets multiplied by s in this procedure, which means it cancels out of all definable transfer functions. All filters in the diagram (just $\hat{\rho}_f(s)$ in this example) can be left alone because their inputs and outputs are still force-valued in principle. (We expressed each force wave in terms of velocity and wave impedance without changing the signal flow diagram, which remains a force-wave simulation until minus signs, scalings, and s operators are moved around and combined.) Of course, one can absorb scalings and sign reversals into the filter(s) to change the physical input/output units as desired. Since we routinely assume zero initial conditions in an impedance description, the integration constants obtained by time-integrating velocities to get displacements are all defined to be zero. Additional considerations regarding the choice of displacement waves over velocity (or force) waves are given in §E.3.3 on page 680. In particular, their initial conditions can be very different, and traveling-wave components tend not to be as well behaved for displacement waves.

9.3.2 Piano Hammer Modeling

The previous section treated an ideal point-mass striking an ideal string. This can be considered a simplified piano-hammer model. The model can be improved by adding a damped spring to the point-mass, as shown in Fig. 9.22 (*cf.* Fig. 9.12, p. 359).

The impedance of this plucking system, as seen by the string, is the parallel combination of the mass impedance ms and the damped spring impedance $\mu+k/s$. (The damper μ and spring k/s are formally in series—see §7.2, p. 257, for a refresher on series versus parallel connection.) Denoting the driving-point impedance of the hammer at the string contact-point by $R_h(s)$, we have

$$R_h(s) \;=\; ms \left\| \left(\mu + \frac{k}{s}\right) \right. \;=\; \frac{\mu s^2 + ks}{s^2 + \frac{\mu}{m}s + \frac{k}{m}}. \qquad (9.19)$$

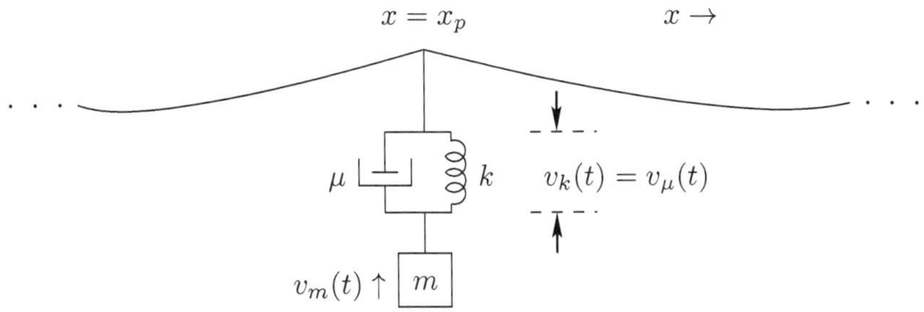

Figure 9.22: Ideal string excited by a mass and damped spring (a more realistic piano-hammer model).

Thus, the scattering filters in the digital waveguide model are second order (biquads), while for the string struck by a mass (§9.3.1) we had first-order scattering filters. This is expected because we added another energy-storage element (a spring).

The impedance formulation of Eq. (9.19) assumes all elements are linear and time-invariant (LTI), but in practice one can normally modulate element values as a function of time and/or state-variables and obtain realistic results for low-order elements. For this we must maintain filter-coefficient formulas that are explicit functions of physical state and/or time. For best results, state variables should be chosen so that any nonlinearities remain *memoryless* in the digitization [361, 348, 555, 556].

Nonlinear Spring Model

In the musical acoustics literature, the piano hammer is classically modeled as a *nonlinear spring* [494, 63, 178, 76, 60, 487, 164].[16] Specifically, the piano-hammer damping in Fig. 9.22 is typically approximated by $\mu = 0$, and the spring k is *nonlinear* and *memoryless* according to a simple power law:

$$k(x_k) \approx Q_0 \, x_k^{p-1}$$

where $p = 1$ for a linear spring, and generally $p > 2$ for pianos. A fairly complete model across the piano keyboard (based on acoustic piano mea-

[16] http://paws.kettering.edu/~drussell/Piano/NonlinearHammer.html

surements) is as follows [488]:

$$\begin{aligned} Q_0 &= 183\, e^{0.045\, n} \\ p &= 3.7 + 0.015\, n \\ n &= \text{piano key number } n \in [1, 88] \\ x_k &= \text{hammer-felt (nonlinear spring) compression} \\ v_k &= \dot{x}_k \end{aligned}$$

The upward force applied to the string by the hammer is therefore

$$f_h(t) \;=\; Q_0\, x_k^p. \tag{9.20}$$

This force is balanced at all times by the downward string force (string tension times slope difference), exactly as analyzed in §9.3.1 above.

Including Hysteresis

Since the compressed hammer-felt (wool) on real piano hammers shows significant *hysteresis memory*, an improved piano-hammer felt model is

$$f_h(t) \;=\; Q_0 \left[x_k^p + \alpha \frac{d(x_k^p)}{dt} \right], \tag{9.21}$$

where $\alpha = 248 + 1.83\, n - 5.5 \cdot 10^{-2} n^2$ (μs), and again n denotes piano key number [488].

Equation (9.21) is said to be a good approximation under normal playing conditions. A more complete hysteresis model is [488]

$$f_h(t) \;=\; f_0 \left[x_k^p(t) - \frac{\epsilon}{\tau_0} \int_0^t x_k^p(\xi) \exp\left(\frac{\xi - t}{\tau_0} \right) d\xi \right]$$

where

$$\begin{aligned} f_0 &= \text{instantaneous hammer stiffness} \\ \epsilon &= \text{hysteresis parameter} \\ \tau_0 &= \text{hysteresis time constant.} \end{aligned}$$

Relating to Eq. (9.21) above, we have $Q_0 = f_0 \cdot (1 - \epsilon)$ (N/mm^p).

Piano Hammer Mass

The piano-hammer *mass* may be approximated across the keyboard by [488]

$$m = 11.074 - 0.074\, n + 0.0001\, n^2.$$

where n is piano key number as before.

9.3.3 Pluck Modeling

The piano-hammer model of the previous section can also be configured as a *plectrum* by making the mass and damping small or zero, and by releasing the string when the contact force exceeds some threshold $f_{\max}$. That is, to a first approximation, a plectrum can be modeled as a *spring* (linear or nonlinear) that disengages when either it is far from the string or a maximum spring-force is exceeded. To avoid discontinuities when the plectrum and string engage/disengage, it is good to taper both the damping and spring-constant to zero at the point of contact (as shown below).

Starting with the piano-hammer impedance of Eq. (9.19) on page 374 and setting the mass m to infinity (the plectrum holder is immovable), we define the *plectrum impedance* as

$$R_p(s) \triangleq \mu + \frac{k}{s} = \frac{\mu s + k}{s}. \tag{9.22}$$

The force-wave reflectance of impedance $R_p(s)$ in Eq. (9.22), as seen from the string, may be computed exactly as in §9.3.1 on page 367:

$$\begin{aligned} \hat{\rho}_f(s) &= \frac{\text{Impedance Step}}{\text{Impedance Sum}} = \frac{[R_p(s)+R]-R}{[R_p(s)+R]+R} = \frac{R_p(s)}{R_p(s)+2R} \\ &= \frac{\mu}{\mu+2R}\cdot\frac{s+\frac{k}{\mu}}{s+\frac{k}{\mu+2R}} \end{aligned} \tag{9.23}$$

If the spring damping is much greater than twice the string wave impedance ($\mu \gg 2R$), then the plectrum looks like a rigid termination to the string (force reflectance $\hat{\rho}_f(s) = 1$), which makes physical sense.

Again following §9.3.1, the transmittance for force waves is given by

$$\hat{\tau}_f(s) = 1 + \hat{\rho}_f(s),$$

and for velocity and displacement waves, the reflectance and transmittance are respectively $-\hat{\rho}_f(s)$ and $1 - \hat{\rho}_f(s)$.

If the damping μ is set to zero, *i.e.*, if the plectrum is to be modeled as a simple linear spring, then the impedance becomes $R_k(s) = k/s$, and the force-wave reflectance becomes [128]

$$\hat{\rho}_f(s) = \frac{\frac{k}{2R}}{s+\frac{k}{2R}}. \tag{9.24}$$

Digital Waveguide Plucked-String Model

When plucking a string, it is necessary to detect "collisions" between the plectrum and string. Also, more complete plucked-string models will allow the string to "buzz" on the frets and "slap" against obstacles such as the fingerboard. For these reasons, it is convenient to choose *displacement waves* for the waveguide string model. The reflection and transmission filters for displacement waves are the same as for velocity, namely, $\hat{\rho}_y(s) = -\hat{\rho}_f(s)$ and $\hat{\tau}_y(s) = 1 - \hat{\rho}_f(s)$.

As in the mass-string collision case, we obtain the one-filter scattering-junction implementation shown in Fig. 9.23. The filter $\hat{\rho}_f(s)$ may now be digitized using the bilinear transform as previously (§9.3.1, p. 371).

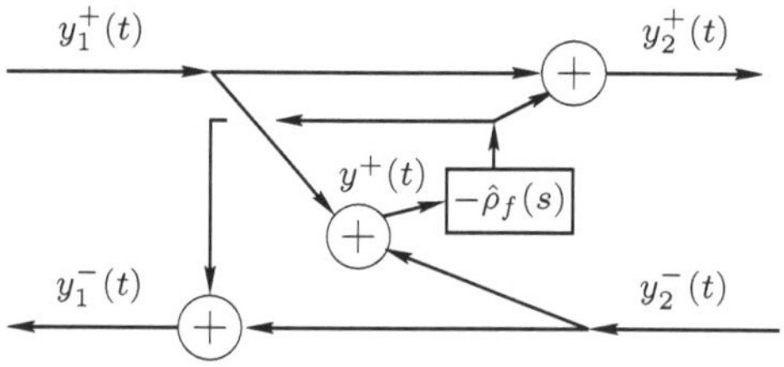

Figure 9.23: Displacement-wave scattering model for a spring.

Incorporating Control Motion

Let $y_m(t)$ denote the vertical position of the *mass* in Fig. 9.22 on page 375. (We still assume $m = \infty$.) We can think of y_m as the position of the *control point* on the plectrum, *e.g.*, the position of the "pinch-point" holding the plectrum while plucking the string. In a harpsichord, y_m can be considered the *jack* position [347].

Also denote by L the *rest length* of the spring k in Fig. 9.22, and let $y_e \triangleq y_m + L$ denote the position of the "end" of the spring while not in contact with the string. Then the plectrum makes *contact* with the string when

$$y_e(t) \geq y(t)$$

where $y(t)$ denotes string vertical position at the plucking point x_p. This may be called the *collision detection* equation.

Let the subscripts 1 and 2 each denote one side of the scattering system, as indicated in Fig. 9.23. Then, for example, $y_1 = y_1^- + y_1^+$ is the displacement of the string on the left (side 1) of plucking point, and y_2 is on the right side of x_p (but still located at point x_p). By continuity of the string, we have

$$y(t) = y_1(t) = y_2(t).$$

When the spring engages the string ($y_e > y$) and begins to compress, the upward force on the string at the contact point is given by

$$f_k = k \cdot (y_e - y)$$

where again $y_e = y_m + L$. The force f_k is applied given $y_e \geq y$ (spring is in contact with string) and given $f_k < f_{\max}$ (the force at which the pluck releases in a simple max-force model).[17] For $y_e < y$ or $f_k \geq f_{\max}$ the applied force is zero and the entire plucking system disappears to leave $y_1^- = y_2^-$ and $y_2^+ = y_1^+$, or equivalently, the force reflectance becomes $\hat{\rho}_f = 0$ and the transmittance becomes $\hat{\tau}_f = 1$.

During contact, force equilibrium at the plucking point requires (*cf.* §9.3.1 on page 363)

$$0 = f_1 + f_k - f_2 \tag{9.25}$$

where $f_i \triangleq -Ky_i' \triangleq -K\partial y_i/\partial x$ as usual (§6.1, p. 204), with K denoting the string tension. Using Ohm's laws for traveling-wave components (p. 204), we have

$$f_i = f_i^+ + f_i^- = Rv_i^+ - Rv_i^-,$$

where $R = \sqrt{K\epsilon}$ denotes the string wave impedance (p. 204). Solving Eq. (9.25) for the velocity at the plucking point yields

$$v = v_1^+ + v_2^- + \frac{1}{2R} f_k,$$

or, for displacement waves,

$$y = y_1^+ + y_2^- + \frac{1}{2R}\int_t f_k. \tag{9.26}$$

Substituting $f_k = k \cdot (y_e - y)$ and taking the Laplace transform yields

$$Y(s) = Y_1^+(s) + Y_2^-(s) + \frac{1}{2R}\frac{F_k(s)}{s} = Y_1^+(s) + Y_2^-(s) + \frac{k}{2Rs}\left[Y_e(s) - Y(s)\right].$$

Solving for $Y(s)$ and recognizing the force reflectance $\hat{\rho}_f(s)$ gives

$$\begin{aligned} Y(s) &= \left[1 - \hat{\rho}_f(s)\right] \cdot \left[Y_1^+(s) + Y_2^-(s)\right] + \hat{\rho}_f(s) Y_e(s) \\ &= \left[Y_1^+(s) + Y_2^-(s)\right] + \hat{\rho}_f(s) \cdot \left\{Y_e(s) - \left[Y_1^+(s) + Y_2^-(s)\right]\right\}, \end{aligned}$$

where, as first noted at Eq. (9.24) above,

$$\hat{\rho}_f(s) \triangleq \frac{\left(\frac{k}{s} + R\right) - R}{\left(\frac{k}{s} + R\right) + R} = \frac{\frac{k}{s}}{2R + \frac{k}{s}} = \frac{\frac{k}{2R}}{s + \frac{k}{2R}}.$$

---

[17] See, *e.g.*, [347] for a more elaborate model in which release is calculated from the computed geometry of the plectrum.

We can thus formulate a one-filter scattering junction as follows:

$$\begin{aligned} Y_d^+ &= Y_e - (Y_1^+ + Y_2^-) \\ Y_1^- &= Y_2^- + \hat{\rho}_f Y_d^+ \\ Y_2^+ &= Y_1^+ + \hat{\rho}_f Y_d^+. \end{aligned}$$

This system is diagrammed in Fig. 9.24. The manipulation of the minus signs relative to Fig. 9.23 makes it convenient for restricting $y_d^+(t)$ to positive values only (as shown in the figure), corresponding to the plectrum engaging the string going up. This uses the approximation $y_1(t) = y_2(t) \approx y_1^+(t) + y_2^-(t)$, which is exact when $\hat{\rho}_f = 0$, *i.e.*, when the plectrum does not affect the string displacement at the current time. It is therefore exact at the time of collision and also applicable just after release. Similarly, $y_d^+(t) > f_{\max}/k$ can be used to trigger a release of the string from the plectrum.

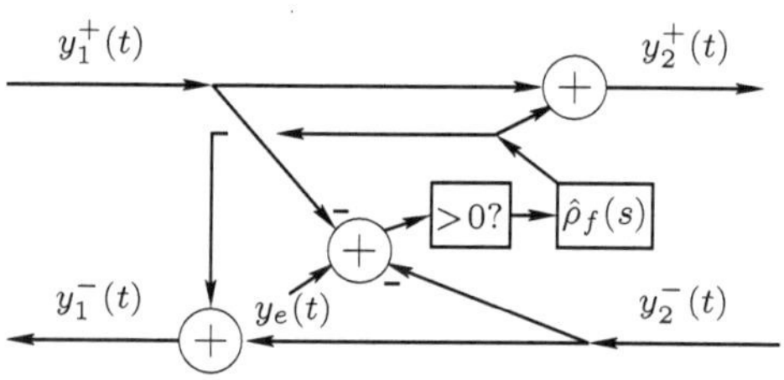

Figure 9.24: Instantaneous spring displacement-wave scattering model driven by the spring edge $y_e(t) = y_m(t) + L$.

Successive Pluck Collision Detection

As discussed above, in a simple 1D plucking model, the plectrum comes up and engages the string when $y_e(t) \geq y(t)$, and above some maximum force the plectrum releases the string. At this point, it is "above" the string. To pluck again in the same direction, the collision-detection must be disabled until we again have $y_e < y$, requiring one bit of state to keep track of that.[18] The harpsichord jack plucks the string only in the "up" direction due to its asymmetric behavior in the two directions [143]. If only "up picks" are supported, then engagement can be suppressed after a release until $y_e(t)$ comes back down below the envelope of string vibration (*e.g.*, $y_e(t) < -y_{\max}$). Note that intermittent disengagements as a plucking cycle

[18]Of course, a truly complete plucking model would allow the plectrum to move in 3D space, at least within a plucking plane, and the collision detection would determine when the string (a point whirling about in the plucking plane) intersects the plectrum.

begins are normal; there is often an audible "buzzing" or "chattering" when plucking an already vibrating string.

When plucking up and down in alternation, as in the *tremolo* technique (common on mandolins), the collision detection alternates between $y_e < y$ and $y_e > y$, and again a bit of state is needed to keep track of which comparison to use.

Plectrum Damping

To include damping μ in the plectrum model, the load impedance $R_p(s) = k/s$ goes back to Eq. (9.22):

$$R_p(s) = \frac{k}{s} + \mu$$

The corresponding force reflectance is then given by Eq. (9.23) on page 377.

Digitization of the Damped-Spring Plectrum

Applying the bilinear transformation (§7.3.2) to the reflectance $\hat{\rho}_f(s)$ in Eq. (9.23) (including damping) yields the following first-order digital force-reflectance filter:

$$\hat{\rho}_f(z) = \frac{\mu}{\mu + 2R} \cdot \frac{\frac{2}{T}\frac{1-z^{-1}}{1+z^{-1}} + \frac{K}{\mu}}{\frac{2}{T}\frac{1-z^{-1}}{1+z^{-1}} + \frac{K}{\mu+2R}} = g\frac{1-\zeta z^{-1}}{1-pz^{-1}}$$

where

$$p = \frac{1 - \frac{KT}{2(\mu+2R)}}{1 + \frac{KT}{2(\mu+2R)}} \quad \text{(digital pole)} \tag{9.27}$$

$$\zeta = \frac{1 - \frac{KT}{2\mu}}{1 + \frac{KT}{2\mu}} \quad \text{(digital zero)} \tag{9.28}$$

$$g = \frac{1-p}{1-\zeta} \quad \text{(gain term)} \tag{9.29}$$

The transmittance filter is again $1 + \hat{\rho}_f(z)$, and there is a one-filter form for the scattering junction as usual.

Feathering

Since the pluck model is linear, the parameters are not signal-dependent. As a result, when the string and spring separate, there is a discontinuous change in the reflection and transmission coefficients. In practice, it is useful to "feather" the switch-over from one model to the next [471]. In this instance, one appealing choice is to introduce a *nonlinear spring*, as is commonly used for piano-hammer models (see §9.3.2 on page 375 for details).

Let the nonlinear spring model take the form

$$f_k(y_d) = k y_d^p,$$

where $p = 1$ corresponds to a linear spring. The spring constant linearized about zero displacement y_d is

$$k(y_d) = f_k'(y_d) = pk y_d^{p-1}$$

which, for $p > 1$, approaches zero as $y_d \to 0$. In other words, the spring-constant itself goes to zero with its displacement, instead of remaining a constant. This behavior serves to "feather" contact and release with the string. We see from Eq. (9.23) above that, as displacement goes to zero, the reflectance approaches a frequency-independent reflection coefficient $\hat{\rho}_f = \mu/(\mu+2r)$, resulting from the damping μ that remains in the spring model. As a result, there is still a discontinuity when the spring disengages from the string.

The foregoing suggests a nonlinear tapering of the damping μ in addition to the tapering the stiffness k as the spring compression approaches zero. One natural choice would be

$$\mu(y_d) = p\mu y_d^{p-1}$$

so that $\mu(y_d)$ approaches zero at the same rate as $k(y_d)$. It would be interesting to estimate p for the spring and damper from measured data. In the absence of such data, $p = 2$ is easy to compute (requiring a single multiplication). More generally, an interpolated lookup of y_d^p values can be used.

In summary, the engagement and disengagement of the plucking system can be "feathered" by a nonlinear spring and damper in the plectrum model.

9.4 Piano Synthesis

The main elements of the piano we need to model are the

- hammer (§9.3.2),
- string (Chapter 6, §6.6),
- bridge (§9.2.1), and
- soundboard, enclosure, acoustic space, mic placement, etc.

Since piano strings are audibly *stiff*, we need to include a *dispersion allpass filter* in the string model, as introduced in (§6.11.3). The next section provides additional details specific to piano strings.

From the bridge (which runs along the soundboard) to each "virtual microphone" (or "virtual ear"), the soundboard, enclosure, and listening space can be well modeled as a high-order LTI filter characterized by its impulse response (Chapter 8). Such long impulse responses can be converted to more efficient recursive digital filters by various means (Chapter 3,§8.6.2).

A straightforward piano model along the lines indicated above turns out to be relatively expensive computationally. Therefore, we also discuss, in §9.4.4 below, the more specialized *commuted piano synthesis* technique [468], which is capable of high quality piano synthesis at low cost relative to other model-based approaches. It can be seen as a *hybrid* method in which the string is *physically* modeled, the hammer is represented by a *signal* model, and the acoustic resonators (soundboard, enclosure, etc.) are handled by ordinary *sampling* (of their impulse response).

9.4.1 Stiff Piano Strings

Piano strings are audibly *inharmonic* due to stiffness [211, 210]. General stiff-string modeling was introduced in §6.9 on page 224. In this section, we discuss further details specific to piano strings.

The main effect of string stiffness is to *stretch* the partial overtone series, so that piano tones are not precisely harmonic [144]. As a result, piano tuners typically stretch the tuning of the piano slightly. For example, the total amount of tuning stretch from the lowest to highest note has been measured to be approximately 35 cents on a Kurzweil PC88, 45 cents on a Steinway Model M, and 60 cents on a Steinway Model D.[19]

Piano String Wave Equation

A wave equation suitable for modeling linearized piano strings is given by [77, 45, 317, 518]

$$f(t,x) = \epsilon\ddot{y} - Ky'' + EIy'''' + R_0\dot{y} + R_2\dot{y}' \tag{9.30}$$

[19]The PC88 stretching begins in the fourth octave, while the measured Steinway stretching is greatest in the first octave (reaching nearly -20 cents relative to "nominal tuning" for the first note A0) and picking up again, going sharp, in the sixth octave. See `http://www.precisionstrobe.com/apps/stretchdata/stretchdata.html` for the measured curves.

where the partial derivative notation y' and $\dot{y}$ are defined on page 523, and

$$\begin{aligned} f(t,x) &= \text{driving force density (N/m) at position } x \text{ and time } t \\ \epsilon &= \text{mass density (kg/m)} \\ K &= \text{tension force along the string axis (N)} \\ E &= \text{Young's modulus (N/m}^2\text{)} \\ I &= \text{radius of gyration of the string cross-section (m).} \end{aligned}$$

Young's modulus and the radius of gyration are defined in Appendix B.

The first two terms on the right-hand side of Eq. (9.30) come from the ideal string wave equation (see Eq. (C.1) on page 523), and they model transverse acceleration and transverse restoring force due to tension, respectively. The term EIy'''' approximates the transverse restoring force exerted by a stiff string when it is bent. In an ideal string with zero diameter, this force is zero; in an *ideal rod* (or *bar*), this term is dominant [317, 261, 169]. The final two terms provide *damping*. The damping associated with R_0 is frequency-independent, while the damping due R_2 increases with frequency.

Damping-Filter Design

In [46], the damping in real piano strings was modeled using a length 17 FIR filter for the lowest strings, and a length 9 FIR filter for the remaining strings. Such FIR filters (or nonparametric measured frequency-response data) can be converted to lower-order IIR filters by the usual methods (§8.6.2). It is convenient to have separate damping and dispersion filters in the string model. The damping filter in piano strings is significantly less demanding than the dispersion filter.

Dispersion Filter-Design

In the context of a digital waveguide string model, dispersion associated with stiff strings can be supplied by an *allpass filter* in the basic feedback loop. Methods for designing dispersion allpass filters were summarized in §6.11.3 on page 233. In this section, we are mainly concerned with how to *specify* the desired dispersion allpass filter for piano strings.

Perceptual studies regarding the audibility of inharmonicity in stringed instrument sounds [211] indicate that the just noticeable coefficient of inharmonicity is given approximately by

$$B_{\text{thresh}} = \exp[2.54\log(f_0) - 24.6] \tag{9.31}$$

where f_0 is the fundamental frequency of the string vibration in hertz, and B—the so-called *coefficient of inharmonicity*—affects the nth partial overtone tuning via

$$f_n = nf_0\sqrt{1+Bn^2}. \tag{9.32}$$

For a stiff string with Young's modulus E, radius a, length L, and tension K, the coefficient of inharmonicity B is predicted from theory [144, p. 65],[211] to be

$$B = \frac{\pi^2 ESI^2}{KL^2} = \frac{\pi^3 Ea^4}{16L^2K}, \tag{9.33}$$

where $S = \pi a^2$ is the string cross-sectional area, and $I = a/2$ is the radius of gyration of the string cross-section (see §B.4.9 on page 476).

In general, when designing dispersion filters for vibrating string models, it is highly cost-effective to obtain an allpass filter which correctly tunes only the lowest-frequency partial overtones, where the number of partials correctly tuned is significantly less than the total number of partials present, as in [384].

Application of the method of [2] to piano-string dispersion-filter design is reported in [1].

A Faust implementation of a closed-form expression [367] for dispersion allpass coefficients as a function of inharmonicity coefficient B may be found in the function `piano_dispersion_filter` within `effect.lib`.

9.4.2 Nonlinear Piano Strings

It turns out that piano strings exhibit audible nonlinear effects, especially in the first three octaves of its pitch range at fortissimo playing levels and beyond [26]. As a result, for highest quality piano synthesis, we need more than what is obtainable from a linearized wave equation such as Eq. (9.30).

As can be seen from a derivation of the wave equation for an ideal string vibrating in 3D space (§B.6 on page 500), there is fundamentally *nonlinear coupling* between transverse and longitudinal string vibrations. It turns out that the coupling from transverse-to-longitudinal is much stronger than vice versa, so that piano synthesis models can get by with one-way coupling at normal dynamic playing levels [30, 163]. As elaborated in §B.6 and the references cited there, the longitudinal displacement ξ is driven by longitudinal changes in the *squared slope* of the string:

$$\epsilon\ddot{\xi} = ES\xi'' + \frac{1}{2}ES[(y')^2]'$$

where

$$[(y')^2]' \triangleq \frac{\partial}{\partial x}\left[\frac{\partial}{\partial x}\,y(t,x)\right]^2.$$

Since longitudinal waves travel an order of magnitude faster than transverse waves, this coupling gives rise to *high-frequency inharmonic overtones* (corresponding to longitudinal modes of vibration) in the sound. Since the nonlinear coupling is *distributed* along the length of the string, the longitudinal modes are continuously being excited by the transverse vibration across time t and position x along the string.

In addition to the excitation of longitudinal modes, the nonlinear transverse-to-longitudinal coupling results in a powerful *longitudinal attack pulse*, which is the leading component of the initial "shock noise" audible in a piano tone. This longitudinal attack pulse hits the bridge well before the first transverse wave and is therefore quite significant perceptually. A detailed simulation of both longitudinal and transverse waves in an ideal string excited by a Gaussian pulse is given in [391].

Another important (*i.e.*, audible) effect due to nonlinear transverse-to-longitudinal coupling is so-called *phantom partials*. Phantom partials are ongoing *intermodulation products* from the transverse partials as they transduce (nonlinearly) into longitudinal waves. The term "phantom partial" was coined by Conklin [85]. The Web version of [18] includes some illuminating sound examples by Conklin.

Nonlinear Piano-String Synthesis

When one-way transverse-to-longitudinal coupling is sufficiently accurate, we can model its effects *separately* based on *observations* of transverse waves. For example [30, 28],

- longitudinal modes can be implemented as *second-order resonators* ("modal synthesis"), with driving coefficients given by
- *orthogonally projecting* [451] the spatial derivative of the squared string slope onto the longitudinal mode shapes (both functions of position x).
- If tension variations along the string are neglected (reasonable since longitudinal waves are so much faster than transverse waves), then the longitudinal force on the bridge can be derived from the estimated *instantaneous tension* in the string, and efficient methods for this have been developed for guitar-string synthesis, particularly by Tolonen (§9.1.6, p. 337).

An excellent review of nonlinear piano-string synthesis (emphasizing the modal synthesis approach) is given in [30].

Regimes of Piano-String Vibration

In general, more complex synthesis models are needed at higher dynamic playing levels. The main three regimes of vibration are as follows [27]:

1. The simplest piano-string vibration regime is characterized by *linear superposition* in which transverse and longitudinal waves *decouple* into separate modes, as implied by Eq. (9.30), p. 383. In this case, transverse and longitudinal waves can be simulated in separate digital waveguides (Ch. 6). The longitudinal waveguide is of course an order of magnitude shorter than the transverse waveguide(s).

2. As dynamic playing level is increased, transverse-to-longitudinal coupling becomes audible [26].

3. At very high dynamic levels, the model should also include longitudinal-to-transverse coupling. However, this is usually neglected.

Other regimes arise depending on whether tension can be assumed to be uniform along the string [27].

Efficient Waveguide Synthesis of Nonlinear Piano Strings

The following nonlinear waveguide piano-synthesis model is under current consideration [427]:

- Initial striking force determines the starting regime (1, 2, or 3 above).
- The string model is simplified as it decays.

At pianissimo playing levels, for example, a linear string model (§9.4.1) can suffice. At higher playing levels (*e.g.*, fortissimo for the first three octaves of the piano keyboard [26]), the string model starts in regime 2.

Because the energy stored in a piano string decays monotonically after a hammer strike (neglecting coupling from other strings), we may *switch to progressively simpler models* as the energy of vibration falls below corresponding thresholds. Since the purpose of model-switching is merely to save computation, it need not happen immediately, so it may be triggered by a string-energy estimate based on observing the string at a *single point* x over the past period or so of vibration. Perhaps most simply, the model-regime classifier can be based on the maximum magnitude of the *bridge force* over at least one period. If the regime 2 model includes an instantaneous string-length (tension) estimate, one may simply compare that to a threshold to determine when the simpler model can be used. If the longitudinal components and/or phantom partials are not completely inaudible when the model switches, then standard cross-fading techniques should be applied so that inharmonic partials are faded out rather than abruptly and audibly cut off.

To obtain a realistic initial shock noise in the tone for regime 1 (and for any model that does not compute it automatically), the appropriate shock pulse, computed as a function of hammer striking force (or velocity, displacement, etc.), can be summed into the longitudinal waveguide during the hammer contact period.

The longitudinal bridge force may be generated from the estimated string length (§9.1.6). This force should be exerted on the bridge in the direction coaxial with the string at the bridge (a direction available from the two transverse displacements one sample away from the bridge).

Phantom partials may be generated in the longitudinal waveguide as explicit intermodulation products based on the transverse-wave overtones known to be most contributing; for example, the Goetzel algorithm [451] could be used to track relevant partial amplitudes for this purpose. Such explicit synthesis of phantom partials, however, makes modal synthesis more compelling for the longitudinal component [30]; in a modal synthesis model, the longitudinal attack pulse can be replaced by a one-shot (per hammer strike) table playback, scaled and perhaps filtered as a function of the hammer striking velocity.

On the high end for regime 2 modeling, a full nonlinear coupling may be implemented along the three waveguides (two transverse and one longitudinal). At this level of complexity, a wide variety of finite-difference schemes should also be considered (§7.3.1, p. 265) [53, 556].

Checking the Approximations

The preceding discussion considered *several* possible approximations for nonlinear piano-string synthesis. Other neglected terms in the stiff-string wave equation were not even discussed, such as terms due to *shear deformation* and *rotary inertia* that are included in the (highly accurate) *Timoshenko beam theory* formulation [261, 169]. The following questions naturally arise:

- How do we know for sure our approximations are inaudible?
- We can listen, but could we miss an audible effect?
- Could a difference become audible after more listening?

To answer these questions, it is helpful to have a *truth reference*—a "perceptually exact" model.

Note that there are software tools (*e.g.*, from the world of *perceptual audio coding* [62, 473]) that can be used to measure the *audible equivalence* of two sounds [457]. For an audio coder, these tools predict the audibility of the difference between original and compressed sounds. For sound synthesis applications, we want to compare our "exact" and "computationally efficient" synthesis models.

9.4.3 High-Accuracy Piano-String Modeling

In [265, 266], an extension of the mass-spring model of [391] was presented for the purpose of high-accuracy modeling of nonlinear piano strings struck by a hammer model such as described in §9.3.2. This section provides a brief overview.

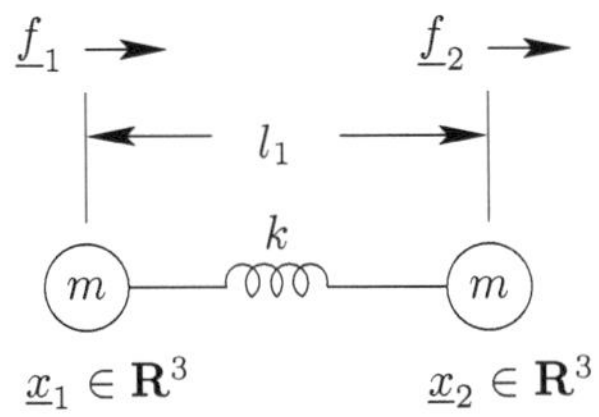

Figure 9.25: Mass-spring model in 3D space.

Figure 9.25 shows a mass-spring model in 3D space. From *Hooke's Law* (§B.1.3, p. 462), we have

$$||\,\underline{f}_1\,|| \;=\; k\cdot|l_1 - l_0| \;=\; ||\,\underline{f}_1\,||$$

where l_0 denotes the rest-length of the spring k, and $||\,\underline{f}_i\,||$ denotes the vector norm (length) of the 3D vector $\underline{f}_i \in \mathbf{R}^3$ [451]. The *vector equation of motion* for mass 1 is given by Newton's second law $f = ma$:

$$\begin{aligned} m_1\,\ddot{\underline{x}}_1 \;=\; \underline{f}_1 &= k\cdot(||\,\underline{x}_2 - \underline{x}_1\,|| - l_0)\cdot\frac{\underline{x}_2 - \underline{x}_1}{||\,\underline{x}_2 - \underline{x}_1\,||} \\ &= k\left[1 - \frac{l_0}{||\,\underline{x}_2 - \underline{x}_1\,||}\right](\underline{x}_2 - \underline{x}_1) \end{aligned}$$

and similarly for mass 2, where $\underline{x}_i \in \mathbf{R}^3$ is the vector position of mass i in 3D space.

Generalizing to a *chain* of masses and spring is shown in Fig. 9.26. Mass-spring chains—also called *beaded strings*—have been analyzed in numerous textbooks (*e.g.*, [295, 318]), and numerical software simulation is described in [391].

The force on the ith mass can be expressed as

$$\begin{aligned} \underline{f}_i &= \alpha_i\cdot(\underline{x}_{i+1} - \underline{x}_i) + \alpha_{i-1}\cdot(\underline{x}_{i-1} - \underline{x}_i) \\ &= \alpha_{i-1}\,\underline{x}_{i-1} - (\alpha_{i-1} + \alpha_i)\,\underline{x}_i + \alpha_i\,\underline{x}_{i+1} \end{aligned} \tag{9.34}$$

where

$$\alpha_i \;\stackrel{\Delta}{=}\; k\cdot\left[1 - \frac{l_0}{||\,\underline{x}_{i+1} - \underline{x}_i\,||}\right].$$

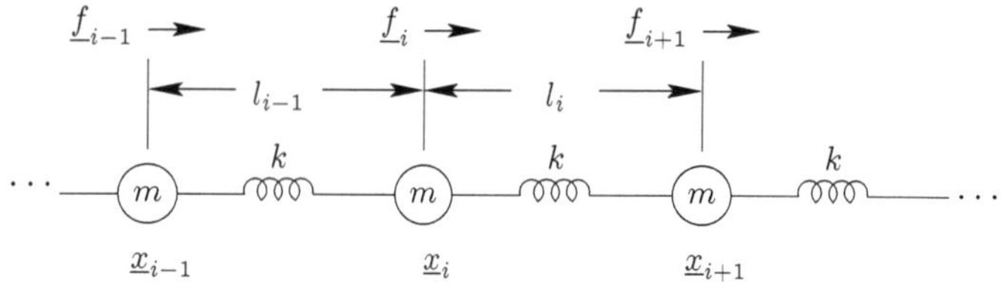

Figure 9.26: Mass-spring string model

A Stiff Mass-Spring String Model

Following the classical derivation of the stiff-string wave equation [317, 144], an obvious way to introduce *stiffness* in the mass-spring chain is to use a *bundle* of mass-spring chains to form a kind of "lumped stranded cable". One section of such a model is shown in Fig. 9.27. Each mass is now modeled as a 2D *mass disk*. Complicated rotational dynamics can be avoided by assuming *no torsional waves* (no "twisting" motion) (§B.4.20, p. 497).

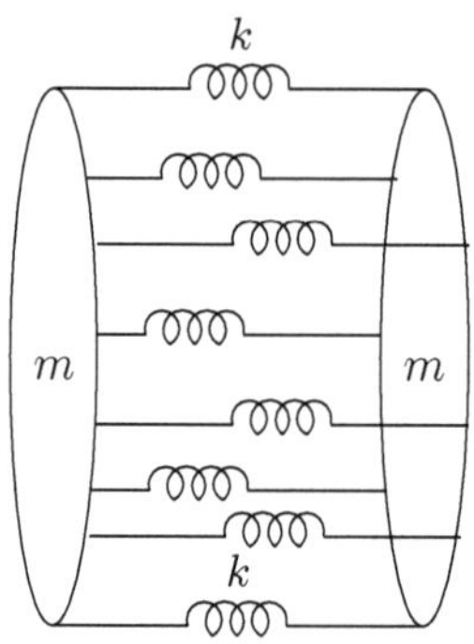

Figure 9.27: Adding bending stiffness to the mass-spring string model.

A three-spring-per-mass model is shown in Fig. 9.28 [266]. The spring positions alternate between angles $(0, 2\pi/3, 4\pi/3)$, say, on one side of a mass disk and $(\pi/3, \pi, 5\pi/3)$ on the other side in order to provide effectively *six* spring-connection points around the mass disk for only three connecting springs per section. This improves *isotropy* of the string model with respect to bending direction.

A problem with the simple mass-spring-chain-bundle is that there is no resistance whatsoever to *shear deformation*, as is clear from Fig. 9.29. To rectify this problem (which does not arise due implicit assumptions when classically deriving the stiff-string wave equation), diagonal springs can be added to the model, as shown in Fig. 9.30.

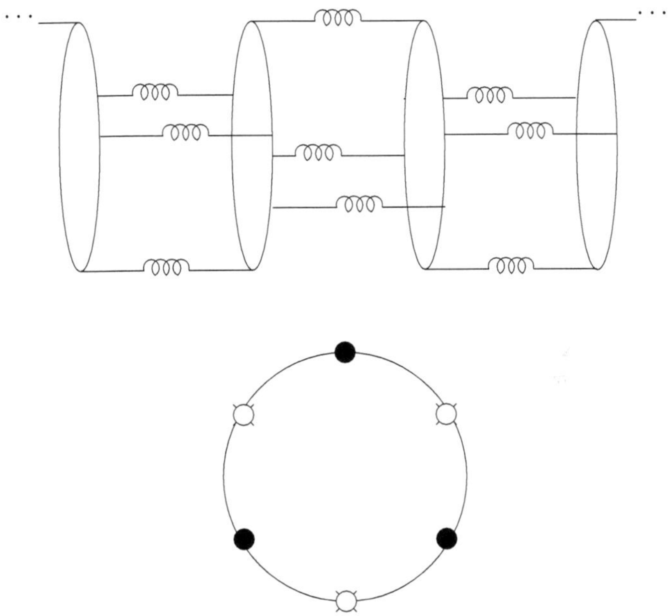

Figure 9.28: Stiff mass-spring chain with alternating three-spring placement.

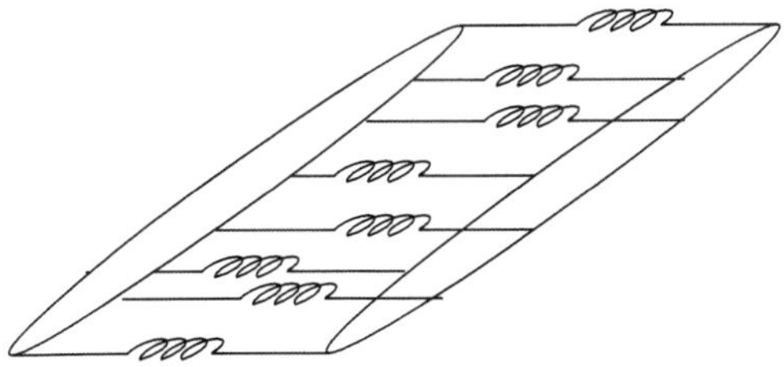

Figure 9.29: Illustration of the need for shear stiffness in the model.

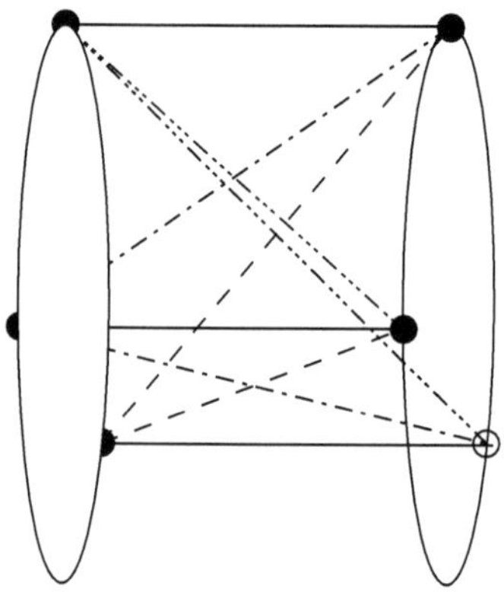

Figure 9.30: Geometry of added shear springs (non-solid lines).

In the simulation results reported in [266], the spring-constants of the shear springs were chosen so that their stiffness in the longitudinal direction would equal that of the longitudinal springs.

Nonlinear Piano-String Equations of Motion in State-Space Form

For the flexible (non-stiff) mass-spring string, referring to Fig. 9.26 and Eq. (9.34), p. 389, we have the following equations of motion:

$$\begin{aligned}
\underline{f}_1 &= m_1\,\underline{\ddot{x}}_1 = \alpha_1\cdot(\underline{x}_2-\underline{x}_1)\\
\vdots \quad & \quad \vdots\\
\underline{f}_i &= m_i\,\underline{\ddot{x}}_i = \alpha_{i-1}\,\underline{x}_{i-1}-(\alpha_{i-1}+\alpha_i)\,\underline{x}_i+\alpha_i\,\underline{x}_{i+1}\\
\vdots \quad & \quad \vdots\\
\underline{f}_M &= m_M\,\underline{\ddot{x}}_M = \alpha_{M-1}\cdot(\underline{x}_{M-1}-\underline{x}_M)
\end{aligned}$$

or, in $3M\times 1$ vector form,

$$\underline{F} = \mathbf{M}\underline{\ddot{X}} = \mathbf{A}\,\underline{X}.$$

Here the string terminations (bridge and agraffe) are modeled simply as very large masses m_1 and m_M.

Finite Difference Implementation

Digitizing $\underline{\ddot{x}}_i(t)$ via the centered second-order difference [Eq. (7.5), p. 266]

$$\underline{\ddot{x}}_n \isdef \underline{x}_{n+1} - 2\underline{x}_n + \underline{x}_{n-1}$$

we obtain the following *explicit finite-difference scheme* (§D.1):

$$\underline{X}_{n+1} = \left[2\mathbf{I} + \mathbf{M}^{-1}\mathbf{A}\right] \underline{X}_n - \underline{X}_{n-1} + B\underline{u}_n,$$

where $\underline{u}_n$ is the external input vector for exciting the string (driving three adjacent masses in simulations to date).

Note that requiring three adjacent spatial string samples to be in contact with the piano hammer during the attack (which helps to suppress aliasing of spatial frequencies on the string during the attack) implies a sampling rate in the vicinity of 6 megahertz [265]. Thus, the model is expensive to compute! However, results to date show a high degree of accuracy, as desired. In particular, the stretching of the partial overtones in the stiff-string model of Fig. 9.30 has been measured to be highly accurate despite using only three spring attachment points on one side of each mass disk [265].

See [53] for alternative finite-difference formulations that better preserve physical energy and have other nice properties worth considering.

9.4.4 Commuted Piano Synthesis

The general technique of *commuted synthesis* was introduced in §8.7 on page 305. This method is enabled by the *linearity* and *time invariance* of both the vibrating string and its acoustic resonator, allowing them to be interchanged, in principle, without altering the overall transfer function [449].

While the piano-hammer is a strongly nonlinear element (§9.3.2), it is nevertheless possible to synthesize a piano using commuted synthesis with high fidelity and low computational cost given the approximation that the string itself is linear and time-invariant (§9.4.2) [468, 520, 30]. The key observation is that the interaction between the hammer and string is essentially *discrete* (after deconvolution) at only one or a few time instants per hammer strike. The deconvolution needed is a function of the hammer-string collision velocity v_c. As a result, *the hammer-string interaction can be modeled as one or a few discrete impulses that are filtered in a v_c-dependent way.*

Figure 9.31 illustrates a typical series of interaction forces-pulses at the contact point between a piano-hammer and string. The vertical lines indicate the locations and amplitudes of three single-sample impulses passed through three single-pulse filters to produce the overlapping pulses shown.

Force-Pulse Synthesis

The creation of a single force-pulse for a given hammer-string collision velocity v_c (a specific "dynamic level") is shown in Fig. 9.32. The filter input

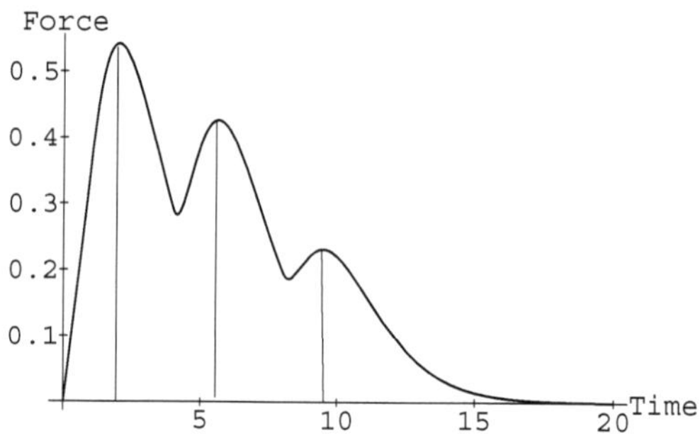

Figure 9.31: Overlapping hammer-string interaction force pulses.

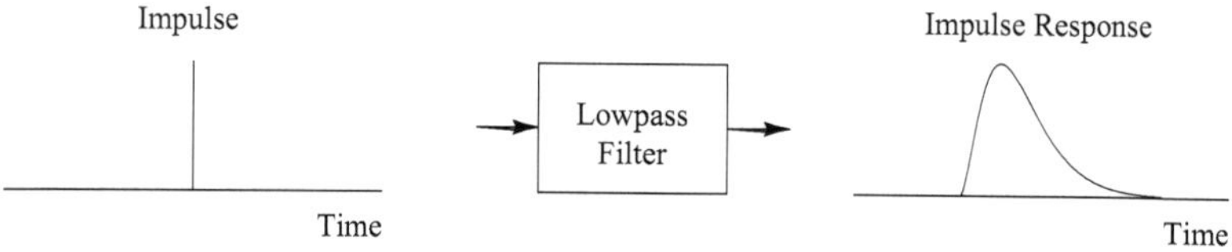

Figure 9.32: Creating a single hammer-string interaction force-pulse as the impulse response of a filter. The filter depends on the hammer-string collision velocity v_c, but it is LTI while v_c is fixed.

is an impulse, and the output is the desired hammer-string force pulse. As v_c increases, the output pulse increases in amplitude *and decreases in width*, which means the filter is nonlinear. In other words, the force pulse gets "brighter" as its amplitude (dynamic level) increases. In a real piano, this brightness increase is caused by the nonlinear felt-compression in the piano hammer. Recall from §9.3.2 that piano-hammer felt is typically modeled as a nonlinear spring described by $f(x) = k\,x^p$, where x is felt compression. Here, the brightness is increased by shrinking the duration of the filter impulse response as v_c increases. The key property enabling commuted synthesis is that, when v_c is constant, the filter operates as a normal LTI filter. In this way, the entire piano has been "linearized" with respect to a given collision velocity v_c.

Multiple Force-Pulse Synthesis

One method of generating multiple force pulses as shown in Fig. 9.31 is to sum several single-pulse filters, as shown in Fig. 9.33. Furthermore, the three input impulses can be generated using a single impulse into a tapped delay line (§2.5, p. 54). The sum of the three filter outputs gives the desired superposition of three hammer-string force pulses. As the collision velocity v_c increases, the output pulses become taller and thinner, showing less overlap. The filters LPF1–LPF3 can be given v_c as side information, or the input impulse amplitudes can be set to v_c, or the like.

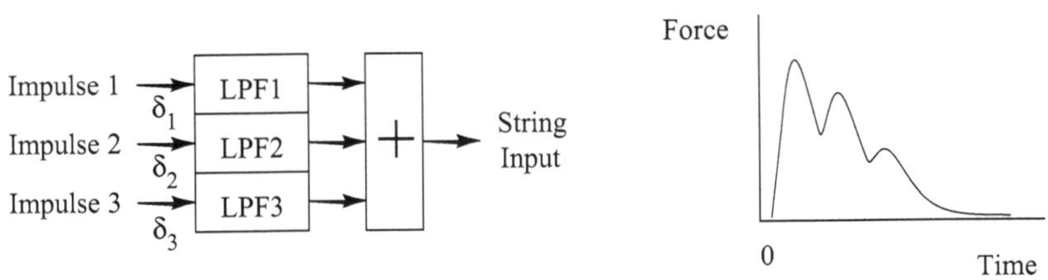

Figure 9.33: Multiple hammer-string interaction force-pulse as the superposition of impulse-responses of a parallel lowpass-filter (LPF) bank.

Commuted Piano Synthesis Architecture

Figure 9.34 shows a complete piano synthesis system along the lines discussed. At a fixed dynamic level, we have the critical feature that the model is *linear* and *time invariant.* Therefore we may *commute* the "Soundboard & Enclosure" filter with not only the string, but with the hammer filter-bank as well. The result is shown in Fig. 9.35.

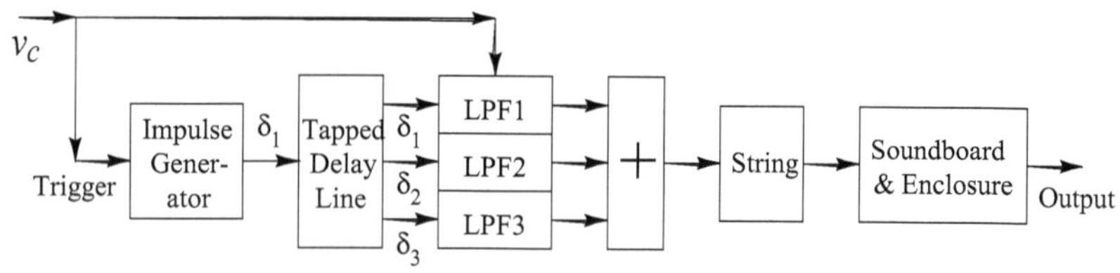

Figure 9.34: Piano synthesis using natural ordering of all elements.

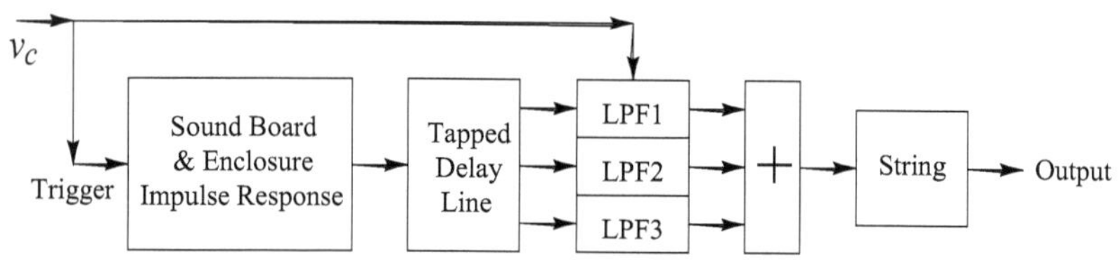

Figure 9.35: Piano synthesis using commuted ordering.

As in the case of commuted guitar synthesis (§8.7, p. 305), we replace a high-order digital filter (at least thousands of poles and zeros [229]) with a simple *excitation table* containing that filter's impulse response. Thus, the large digital filter required to implement the soundboard, piano enclosure, and surrounding acoustic space, has been eliminated. At the same time, we still have explicit models for the hammer and string, so physical variations can be implemented, such as harder or softer hammers, more or less string stiffness, and so on.

Even some resonator modifications remain possible, however, such as changing the virtual mic positions (§2.2.7). However, if we now want to "open the top" of our virtual piano, we have to measure the impulse response that case separately and have a separate table for it, which is not a problem, but it means we're doing "sampling" instead of "modeling" of the various resonators.

An approximation made in the commuted technique is that detailed reflections generated on the string during hammer-string contact are neglected; that is, the hammer force-pulse depends only on the hammer-string velocity at the time of their initial contact, and the string velocity is treated as remaining constant throughout the contact duration.

Progressive Filtering

To save computation in the filtering, we can make use of the observation that, under the assumption of a string initially at rest, each interaction pulse is smoother than the one before it. That suggests applying the force-pulse filtering *progressively*, as was done with Leslie cabinet reflections in §5.7.6 on page 194. In other words, the second force-pulse is generated as a filtering of the first force-pulse. This arrangement is shown in Fig. 9.36.

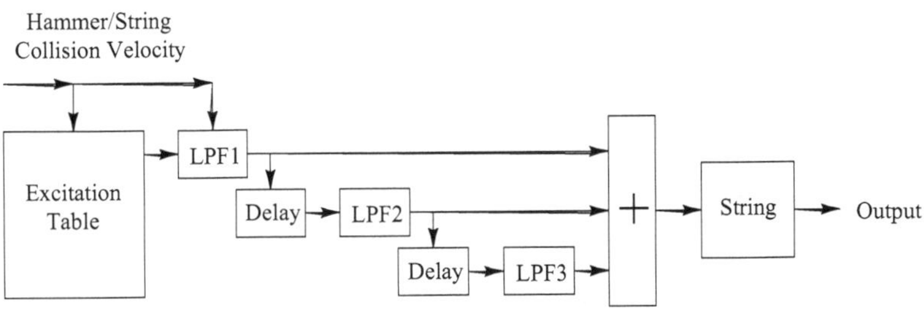

Figure 9.36: Commuted piano synthesis supporting three hammer-string interaction pulses using separate filters for each pulse and implementing the filters successively. Each new delay is equal to the travel from the hammer, to the agraffe, and back to the hammer.

With progressive filtering, each filter need only supply the mild smoothing (and perhaps dispersion) associated with traveling from the hammer to the agraffe and back, plus the mild attenuation associated with reflection from the felt-covered hammer (a nonlinear mass-spring system as described in §9.3.2).

Referring to Fig. 9.36, The first filter LPF1 can shape a velocity-independent excitation signal to obtain the appropriate "shock spectrum" for that hammer velocity. Alternatively, the Excitation Table itself can be varied with velocity to produce the needed signal. In this case, filter LPF1 can be eliminated entirely by applying it in advance to the excitation signal. It is possible to interpolate between tables for two different striking velocities; in this case, the tables should be pre-processed to eliminate phase cancellations during cross-fade.

Assuming the first filter in Fig. 9.36 is "weakest" at the highest hammer velocity (MIDI velocity 127), that filtering can be applied to the excitation table in advance, and the first filter then becomes no computation for MIDI velocity 127, and as velocity is lowered, the filter only needs to make up the difference between what was done in advance to the table and what is desired at that velocity.

Since, for most keys, only a few interaction impulses are observed, per hammer strike, in real pianos, this computational model of the piano achieves a high degree of realism for a price comparable to the cost of the strings only. The soundboard and enclosure filtering have been eliminated and replaced by a look-up table using a few read-pointers per note, and the hammer costs only one or a few low-order filters which in principle convert the interaction impulse into an accurate force pulse.

Excitation Factoring

As another refinement, it is typically more efficient to implement the highest Q resonances of the soundboard and piano enclosure using actual digital filters (see §8.8 on page 309). By factoring these out, the impulse response is shortened and thus the required excitable length is reduced. This provides a classical computation vs. memory trade-off which can be optimized as needed in a given implementation. For lack of a better name, let us refer to such a resonator bank as an "equalizer" since it can be conveniently implemented using parametric equalizer sections, one per high-Q resonance.

A possible placement of the comb filter and equalizer is shown in Fig. 9.37. The resonator/eq/reverb/comb-filter block may include filtering for partially implementing the response of the soundboard and enclosure, equalization sections for piano color variations, reverberation, comb-filter(s) for flanging, chorus, and simulated hammer-strike echoes on the string. Multiple outputs having different spectral characteristics can be extracted at various points in the processing.

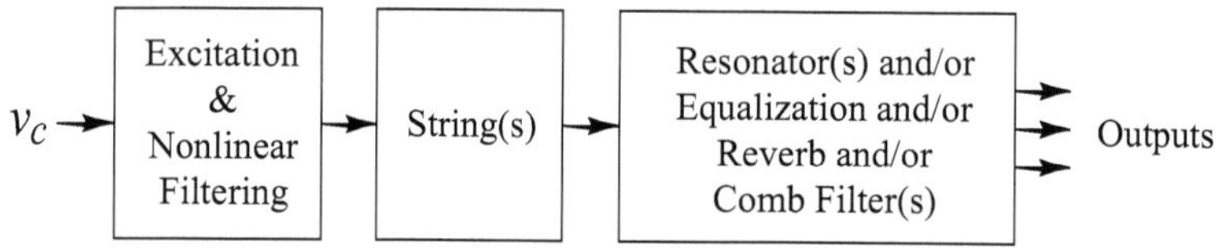

Figure 9.37: Example block diagram of a more complete commuted-piano synthesis system.

Excitation Synthesis

For a further reduction in cost of implementation, particularly with respect to memory usage, it is possible to *synthesize* the excitation using a noise signal through a time-varying lowpass filter [440]. The synthesis replaces the recorded soundbard/enclosure sound. It turns out that the sound produced by tapping on a piano soundboard sounds very similar to noise which is lowpass-filtered such that the filter bandwidth contracts over time. This approach is especially effective when applied only to the high-frequency residual excitation after factoring out all long-ringing modes as biquads.

Coupled Piano Strings

Perhaps it should be emphasized that for good string-tone synthesis, *multiple* filtered delay loops should be employed for each note rather than the single-loop case so often used for simplicity (§6.12, p. 238; §C.13 on page 602). For piano, two or three delay loops can correspond to the two or three strings hit by the same hammer in the plane orthogonal to both the string and bridge. This number of delay loops is *doubled* from there if the transverse vibration planes *parallel* to the bridge are added; since these are slightly detuned, they are worth considering. Finally, another 2–3 (shorter) delay loops are needed to incorporate longitudinal waves for each string, and all loops should be appropriately coupled. In fact, for full accurancy, longitudinal waves and transverse waves should be *nonlinearly* coupled along the length of the string [391] (see §B.6.2 on page 503).

Commuted Synthesis of String Reverberation

The sound of all strings ringing can be summed with the excitation to simulate the effect of many strings resonating with the played string when the sustain pedal is down [520]. The string loop filters out the unwanted frequencies in this signal and selects only the overtones which would be excited by the played string.

High Piano Key Numbers

At very high pitches, the delay-line length used in the string may become too short for the implementation used, especially when using vectorized module signals (sometimes called "chunks" in place of samples [355]). In this case, good results can be obtained by replacing the filtered-delay-loop string model by a simpler model consisting only of a few parallel, second-order filter sections or enveloped sinusoidal oscillators, etc. In other words, *modal synthesis* (§8.5, p. 285) is a good choice for very high key numbers.

Force-Pulse Filter Design

In the commuted piano synthesis technique, it is necessary to fit a filter impulse response to the desired hammer-string force-pulse shape *for each key* and *for each key velocity* v_c, as indicated in Figures 9.32 and 9.33. Such a desired curve can be found in the musical acoustics literature [489] or it can be easily generated from a good piano-hammer model (§9.3.2, p. 374) striking the chosen piano string model.

For psychoacoustic reasons, it is preferable to optimize the force-pulse filter-coefficients in the *frequency domain* §8.6.2 on page 292. Some additional suggestions are given in [468, 520].

9.4.5 Literature on Piano Acoustics and Synthesis

There has been much research into the musical acoustics of the piano, including models of piano strings [544, 18, 75, 76, 144, 42] and the piano hammer [494, 63, 178, 60, 487, 164, 488]. Additionally, there has been considerable work to date on the problem of piano synthesis by digital waveguide methods [468, 520, 524, 56, 42].

A careful computational model of the piano, partially based on physical modeling and suitable for high-quality real-time sound synthesis, was developed by Bensa in the context of his thesis work [42]. Related publications include [43, 45, 46].

An excellent simulation of the *clavichord*, also using the commuted waveguide synthesis technique, is described in [502]. A detailed simulation of the *harpsichord*, along similar lines, but with some interesting new variations, may be found in [515].

The 2003 paper of Bank et al. [29] includes good summaries of prior research on elements of piano modeling such as efficient computational string models, loss-filter design, dispersion filter design, and soundboard models. A more recent paper in this area is [30].

9.4.6 Piano Modeling Exercises

1. Given a measured decay time (t_{60}) of 5 seconds at a frequency of A4 $=$ 440 Hz, find the desired loop-filter amplitude response at that frequency, for a string having pitch A0 and a sampling rate of $f_s = 44$ kHz.

2. Compute the displacement of an infinite string with wave impedance R when it is struck with an ideal mass-spring system at velocity v_0. The linear spring is described by $F = kx$, where k is the spring constant and x is the displacement. For what range of values does the mass m oscillate and for what range doesn't it oscillate? Make a sketch of the displacement of the string at the contact point.

3. The formula for partial stretching in a piano string is given, for small stiffness, by

$$f_k \approx k f_0 \sqrt{1 + Bk^2}, \quad k = 1, 2, 3, \ldots$$

where B is known as the *inharmonicity coefficient.* On a real piano string, B has been measured to be near 10^{-4} for the lowest 30 or so semitones of a piano, then climbing smoothly to around 10^{-3} over the next 25 or so semitones.

(a) For the lowest note A0 = 27.5 Hz of the piano, assuming $B = 10^{-4}$ exactly, find samples $P(\omega_k)$ of the desired phase delay of an allpass filter which can be used in a waveguide model of a piano string at that pitch. Assume that the loss filter in the model is just a scalar, so that it has no contribution to the phase of the system. Plot your result using $f_s = 44.1$ kHz.

(b) Repeat for a higher note, C7 = 2093.0 Hz of the piano, assuming $B = 10^{-3}$ exactly.

4. Design a dispersion filter to be used in a waveguide model of a piano string, using frequency data from the same recorded piano tone (`D4.wav`[20]). The dispersion filter is to be modeled as a cascade of first-order allpass filters. Assuming that the waveguide loop can be represented by a length N delay line and a block of M cascaded one-pole allpass filters $H(z) = \frac{a+z^{-1}}{1+az^{-1}}$, the transfer function once around the loop is $L(z) \triangleq z^{-N}H(z)^M$. Its phase response is $\theta(\omega) \triangleq \angle L(e^{j\omega T}) = -N\omega T + M\phi(\omega)$, where $\phi(\omega) = -\omega T + 2\ \arctan\{\frac{a\sin(\omega T)}{1+a\cos(\omega T)}\}$.

The coefficient a of the allpass filters can be solved by constructing a set of linear equations and solving for a:

$$a\ \sin(\beta_i + \omega_i T) = -\sin(\beta_i)$$

where

$$\beta_i = -\frac{1}{2}\left\{\frac{\theta_{desired}(\omega_i) + N\omega_i T}{M} + \omega_i T\right\},$$

ω_i are the radian frequencies of the partials analyzed from the piano tone, $i = 1, \ldots, K$, and K is the number of partials analyzed. This system ($\mathbf{Ax} = \mathbf{b}$) is overdetermined, so the solution is obtained as

$$\mathbf{x} = [\mathbf{A}^T\mathbf{W}\mathbf{A}]^{-1}\mathbf{A}^T\mathbf{W}\mathbf{b},$$

where $\mathbf{x}$ is the length 1 vector holding the allpass coefficient a, $\mathbf{b}$ is the length K vector $-[\sin(\beta_1)\ \sin(\beta_2)\ \ldots\ \sin(\beta_K)]^T$, $\mathbf{A}$ is the $K \times 1$ matrix $[\sin(\beta_1 + \omega_1)\ \sin(\beta_2 + \omega_2)\ \ldots\ \sin(\beta_K + \omega_K)]^T$, and $\mathbf{W}$ is a $K \times K$ diagonal matrix of weights which we can choose according to various criteria (e.g., following a curve of ear frequency sensitivity). For now, we can set $\mathbf{W} = \mathbf{I}$, the identity matrix.

[20] `http://ccrma.stanford.edu/realsimple/piano/D4.wav`

Find the best delay line and allpass cascade combination using different values of M and N, according to the sum of squares error between the actual partial frequencies and the ones resulting from the waveguide loop ($e = \sum_{i=1}^{K}(f_{i,desired} - f_{i,loop})^2$). Document a, M, N, and the resulting error e. Plot the frequency difference (in Hz) between the measured frequencies and the waveguide loop partial frequencies.

5. While commuted synthesis for the piano has advantages in being computationally efficient, modeling the excitation directly allows for a more precise and parametric hammer force realization. Draw the structure of a basic digital waveguide piano model which uses *displacement waves*. Include appropriate reflection and/or filtering modules to account for the agraffe, bridge, and soundboard. (For simplification, don't worry about coupled strings or dispersion from string stiffness. Assume transverse wave propagation in the vertical polarization only.) Label the wave variables in the system and work out the relationship between the hammer and the string, e.g., how to calculate the input into the waveguide string. The hammer is modeled as a nonlinear spring, such that $F_h = k(\Delta y)^p = k(y_h - y_s)^p$, where y_h is the displacement of the hammer and y_s is the displacement of the string. However, when $\Delta y \leq 0$, $F_h = 0$. For background regarding converting to and from different wave types, see p. 551.

6. Suppose a mass m attached to a spring k is flying along at constant speed v_0, and the assembly collides "spring-first" into an ideal wall at time 0.

   (a) Find and sketch the mass motion $x_m(t)$, defining $x_m = 0$ when the spring first contacts the wall. (9 points)

   (b) How long is the spring in contact with the wall? (3 points)

   (c) How much momentum is delivered to the wall, and what is its direction? (3 points)

Commuted Piano Synthesis Lab

See `http://ccrma.stanford.edu/realsimple/piano/` for laboratory exercises related to piano synthesis.

9.5 Woodwinds

We now consider models of woodwind instruments, focusing on the single-reed case. In §9.5.4, tonehole modeling is described in some detail.

9.5.1 Single-Reed Instruments

A simplified model for a single-reed woodwind instrument is shown in Fig. 9.38 [431].

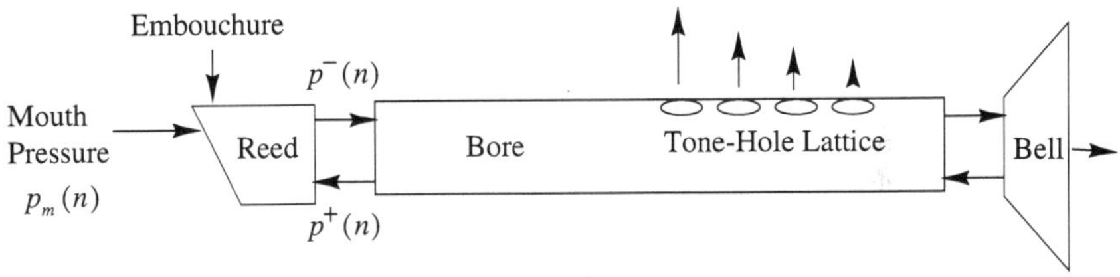

Figure 9.38: A schematic model for woodwind instruments.

If the bore is cylindrical, as in the clarinet, it can be modeled quite simply using a bidirectional delay line. If the bore is conical, such as in a saxophone, it can still be modeled as a bidirectional delay line, but interfacing to it is slightly more complex, especially at the mouthpiece [37, 7, 160, 436, 507, 508, 503, 527, 406, 529], Because the main control variable for the instrument is air pressure in the mouth at the reed, it is convenient to choose *pressure wave variables.*

To first order, the bell passes high frequencies and reflects low frequencies, where "high" and "low" frequencies are divided by the wavelength which equals the bell's diameter. Thus, the bell can be regarded as a simple "cross-over" network, as is used to split signal energy between a woofer and tweeter in a loudspeaker cabinet. For a clarinet bore, the nominal "cross-over frequency" is around 1500 Hz [38]. The flare of the bell lowers the cross-over frequency by decreasing the bore characteristic impedance toward the end in an approximately non-reflecting manner [51]. Bell flare can therefore be considered analogous to a transmission-line *transformer.*

Since the length of the clarinet bore is only a quarter wavelength at the fundamental frequency, (in the lowest, or "chalumeau" register), and since the bell diameter is much smaller than the bore length, most of the sound energy traveling into the bell reflects back into the bore. The low-frequency energy that makes it out of the bore radiates in a fairly omnidirectional pattern. Very high-frequency traveling waves do not "see" the enclosing bell and pass right through it, radiating in a more directional beam. The directionality of the beam is proportional to how many wavelengths fit along the bell diameter; in fact, many wavelengths away from the bell, the radiation pattern is proportional to the two-dimensional spatial Fourier transform of the exit aperture (a disk at the end of the bell) [308].

The theory of the single reed is described, *e.g.*, in [102, 249, 308]. In the digital waveguide clarinet model described below [431], the reed is modeled as a signal- and embouchure-dependent *nonlinear reflection coefficient* terminating the bore. Such a model is possible because the reed mass is neglected. The player's embouchure controls damping of the reed, reed aperture width, and other parameters, and these can be implemented as parameters on the contents of the lookup table or nonlinear function.

Digital Waveguide Single-Reed Implementation

A diagram of the basic clarinet model is shown in Fig. 9.39. The delay-lines carry left-going and right-going *pressure* samples p_b^+ and p_b^- (respectively) which sample the traveling pressure-wave components within the bore.

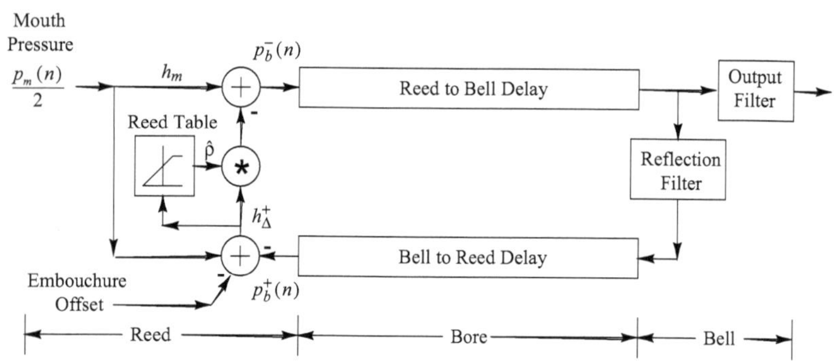

Figure 9.39: Waveguide model of a single-reed, cylindrical-bore woodwind, such as a clarinet.

The reflection filter at the right implements the bell or tone-hole losses as well as the round-trip attenuation losses from traveling back and forth in the bore. The bell output filter is highpass, and *power complementary* with respect to the bell reflection filter [501]. Power complementarity follows from the assumption that the bell itself does not vibrate or otherwise absorb sound. The bell is also *amplitude complementary*. As a result, given a reflection filter $H_r(z)$ designed to match measured mode decay-rates in the bore, the transmission filter can be written down simply as $H_t(z) = 1 - H_r(z)$ for velocity waves, or $H_t(z) = 1 + H_r(z)$ for pressure waves. It is easy to show that such amplitude-complementary filters are also power complementary by summing the transmitted and reflected power waves:

$$\begin{aligned} P_t U_t + P_r U_r &= (1+H_r)P \cdot (1-H_r)U + H_r P \cdot (-H_r)(-U) \\ &= [1 - H_r^2 + H_r^2]PU \;=\; PU, \end{aligned}$$

where P denotes the z transform transform of the incident pressure wave, and U denotes the z transform of the incident volume-velocity. (All z transform have omitted arguments $(\exp(j\omega T)$, where T denotes the sampling interval in seconds.)

At the far left is the reed mouthpiece controlled by *mouth pressure* p_m. Another control is *embouchure,* changed in general by modifying the *reflection-coefficient* function $\rho(h_\Delta^+)$, where $h_\Delta^+ \triangleq p_m/2 - p_b^+$. A simple choice of embouchure control is an offset in the reed-table address. Since the main feature of the reed table is the pressure-drop where the reed begins to open, a simple embouchure offset can implement the effect of biting harder or softer on the reed, or changing the reed stiffness.

9.5.2 A View of Single-Reed Oscillation

To start the oscillation, the player applies a pressure at the mouthpiece which "biases" the reed in a "negative-resistance" region. (The pressure drop across the reed tends to close the air gap at the tip of the reed so that an increase in pressure will result in a net decrease in volume velocity—this is negative resistance.) The high-pressure front travels down the bore at the speed of sound until it encounters an open air hole or the bell. To a first approximation, the high-pressure wave reflects with a sign inversion and travels back up the bore. (In reality a lowpass filtering accompanies the reflection, and the complementary highpass filter shapes the spectrum that emanates away from the bore.)

As the negated pressure wave travels back up the bore, it cancels the elevated pressure that was established by the passage of the first wave. When the negated pressure front gets back to the mouthpiece, it is reflected again, this time with no sign inversion (because the mouthpiece looks like a closed end to a first approximation). Therefore, as the wave travels back down to the bore, a *negative* pressure zone is left behind. Reflecting from the open end again with a sign inversion brings a return-to-zero wave traveling back to the mouthpiece. Finally the positive traveling wave reaches the mouthpiece and starts the second "period" of oscillation, after four trips along the bore.

So far, we have produced oscillation without making any use of the negative-resistance of the reed aperture. This is merely the start-up transient. Since in reality there are places of pressure loss in the bore, some mechanism is needed to feed energy back into the bore and prevent the oscillation just described from decaying exponentially to zero. This is the function of the reed: When a traveling pressure-drop reflects from the mouthpiece, making pressure at the mouthpiece switch from high to low, the reed changes from open to closed (to first order). The closing of the reed increases the reflection coefficient "seen" by the impinging traveling wave, and so as the pressure falls, it is amplified by an increasing gain (whose maximum is unity when the reed shuts completely). This process sharpens the falling edge of the pressure drop. But this is not all. The closing of the reed also cuts back on the steady incoming flow from the mouth. This causes the pressure to drop even more, potentially providing effective *amplification* by more than unity.

An analogous story can be followed through for a rising pressure appearing at the mouthpiece. However, in the rising pressure case, the reflection coefficient falls as the pressure rises, resulting in a progressive attenuation of the reflected wave; however, the increased pressure let in from the mouth amplifies the reflecting wave. It turns out that the reflection of a positive wave is boosted when the incoming wave is below a certain level and it is attenuated above that level. When the oscillation reaches a very high amplitude, it is limited on the negative side by the shutting of the reed, which sets a maximum reflective amplification for the negative excursions, and it is limited on the positive side by the attenuation described above. *Unlike* classical negative-resistance oscillators, in which the negative-resistance device is terminated by a simple resistance instead of a lossy transmission line, a *dynamic equilibrium* is established between the amplification of the negative excursion and the dissipation of the positive excursion.

In the first-order case, where the reflection-coefficient varies linearly with pressure drop, it is easy to obtain an exact quantitative description of the entire process. In this case it can be shown, for example, that amplification occurs only on the positive half of the cycle, and the amplitude of oscillation is typically close to half the incoming mouth pressure (when losses in the bore are small). The threshold blowing pressure (which is relatively high in this simplified case) can also be computed in closed form.

9.5.3 Single-Reed Theory

A simplified diagram of the clarinet mouthpiece is shown in Fig. 9.40. The pressure in the mouth is assumed to be a constant value p_m, and the bore pressure p_b is defined located at the mouthpiece. Any pressure drop $p_\Delta = p_m - p_b$ across the mouthpiece causes a flow u_m into the mouthpiece through the reed-aperture impedance $R_m(p_\Delta)$ which changes as a function of p_Δ since the reed position is affected by p_Δ. To a first approximation,

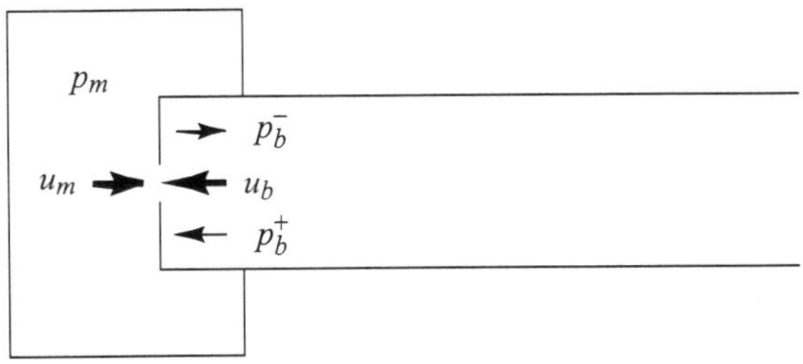

Figure 9.40: Schematic diagram of mouth cavity, reed aperture, and bore.

the clarinet reed can be regarded as a spring flap regulated Bernoulli flow (§B.7.5 on page 506), [249]). This model has been verified well experimentally until the reed is about to close, at which point viscosity effects begin to appear [102]. It has also been verified that the mass of the reed can be neglected to first order,[21] so that $R_m(p_\Delta)$ is a positive real number for all values of p_Δ. Possibly the most important neglected phenomenon in this model is sound generation due to turbulence of the flow, especially near reed closure. Practical synthesis models have always included a noise component of some sort which is modulated by the reed [431], despite a lack of firm basis in acoustic measurements to date.

The fundamental equation governing the action of the reed is *continuity of volume velocity, i.e.,*

$$u_b + u_m = 0 \tag{9.35}$$

where

$$u_m(p_\Delta) \stackrel{\Delta}{=} \frac{p_\Delta}{R_m(p_\Delta)} \tag{9.36}$$

and

$$u_b \stackrel{\Delta}{=} u_b^+ + u_b^- = \frac{p_b^+ - p_b^-}{R_b} \tag{9.37}$$

is the volume velocity corresponding to the incoming pressure wave p_b^+ and outgoing pressure wave p_b^-. (The physical pressure in the bore at the mouthpiece is of course $p_b = p_b^+ + p_b^-$.) The wave impedance of the bore air-column is denoted R_b (computable as the air density times sound speed c divided by cross-sectional area).

In operation, the mouth pressure p_m and incoming traveling bore pressure p_b^+ are given, and the reed computation must produce an outgoing bore

[21] It is valid to neglect the reed mass when the first reed resonance is well above the fundamental frequency of the played note, as is normally the case.

pressure p_b^- which satisfies (9.35), *i.e.*, such that

$$\begin{aligned} 0 &= u_m + u_b = \frac{p_\Delta}{R_m(p_\Delta)} + \frac{p_b^+ - p_b^-}{R_b}, \\ p_\Delta &\stackrel{\Delta}{=} p_m - p_b = p_m - (p_b^+ + p_b^-) \end{aligned} \tag{9.38}$$

Solving for p_b^- is not immediate because of the dependence of R_m on p_Δ which, in turn, depends on p_b^-. A graphical solution technique was proposed [151, 244, 308] which, in effect, consists of finding the intersection of the two terms of the equation as they are plotted individually on the same graph, varying p_b^-. This is analogous to finding the operating point of a transistor by intersecting its operating curve with the "load line" determined by the load resistance.

It is helpful to normalize (9.38) as follows: Define $G(p_\Delta) = R_b u_m(p_\Delta) = R_b p_\Delta / R_m(p_\Delta)$, and note that $p_b^+ - p_b^- = 2p_b^+ - p_m - (p_b^+ + p_b^- - p_m) \stackrel{\Delta}{=} p_\Delta - p_\Delta^+$, where $p_\Delta^+ \stackrel{\Delta}{=} p_m - 2p_b^+$. Then (9.38) can be multiplied through by R_b and written as $0 = G(p_\Delta) + p_\Delta - p_\Delta^+$, or

$$G(p_\Delta) = p_\Delta^+ - p_\Delta, \qquad p_\Delta^+ \stackrel{\Delta}{=} p_m - 2p_b^+ \tag{9.39}$$

The solution is obtained by plotting $G(x)$ and $p_\Delta^+ - x$ on the same graph, finding the point of intersection at (x, y) coordinates $(p_\Delta, G(p_\Delta))$, and computing finally the outgoing pressure wave sample as

$$p_b^- = p_m - p_b^+ - p_\Delta(p_\Delta^+) \tag{9.40}$$

An example of the qualitative appearance of $G(x)$ overlaying $p_\Delta^+ - x$ is shown in Fig. 9.41.

Scattering-Theoretic Formulation

Equation (9.38) can be solved for p_b^- to obtain

$$\begin{aligned} p_b^- &= \frac{1-r}{1+r} p_b^+ + \frac{r}{1+r} p_m && (9.41) \\ &= \rho p_b^+ + \frac{1-\rho}{2} p_m && (9.42) \\ &= \frac{p_m}{2} - \rho \frac{p_\Delta^+}{2} && (9.43) \end{aligned}$$

where

$$\rho(p_\Delta) \stackrel{\Delta}{=} \frac{1 - r(p_\Delta)}{1 + r(p_\Delta)}, \qquad r(p_\Delta) \stackrel{\Delta}{=} \frac{R_b}{R_m(p_\Delta)} \tag{9.44}$$

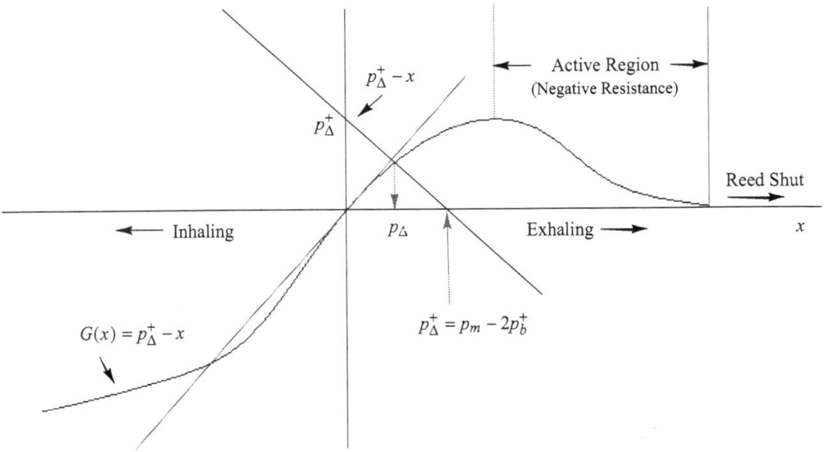

Figure 9.41: Normalized reed impedance $G(p_\Delta) = R_b u_m(p_\Delta)$ overlaid with the "bore load line" $p_\Delta^+ - p_\Delta = R_b u_b$.

We interpret $\rho(p_\Delta)$ as a *signal-dependent reflection coefficient.*

Since the mouthpiece of a clarinet is nearly closed, $R_m \gg R_b$ which implies $r \approx 0$ and $\rho \approx 1$. In the limit as R_m goes to infinity relative to R_b, (9.42) reduces to the simple form of a rigidly capped acoustic tube, *i.e.*, $p_b^- = p_b^+$. If it were possible to open the reed wide enough to achieve matched impedance, $R_m = R_b$, then we would have $r = 1$ and $\rho = 0$, in which case $p_b^- = p_m/2$, with no reflection of p_b^+, as expected. If the mouthpiece is removed altogether to give $R_m = 0$ (regarding it now as a tube section of infinite radius), then $r = \infty$, $\rho = -1$, and $p_b^- = -p_b^+ + p_m$.

Computational Methods

Since finding the intersection of $G(x)$ and $p_\Delta^+ - x$ requires an expensive iterative algorithm with variable convergence times, it is not well suited for real-time operation. In this section, fast algorithms based on precomputed nonlinearities are described.

Let h denote *half-pressure* $p/2$, *i.e.*, $h_m \triangleq p_m/2$ and $h_\Delta^+ \triangleq p_\Delta^+/2$. Then (9.43) becomes

$$p_b^- = h_m - \rho(p_\Delta) \cdot h_\Delta^+ \qquad (9.45)$$

Subtracting this equation from p_b^+ gives

$$\begin{aligned} p_b^+ + \rho h_\Delta^+ - h_m &= p_b^+ - p_b^- = p_\Delta - p_\Delta^+ \\ \Rightarrow \rho h_\Delta^+ &= \underbrace{h_m - p_b^+}_{h_\Delta^+} + p_\Delta - 2h_\Delta^+ = p_\Delta - h_\Delta^+ \\ \Rightarrow \rho &= \frac{p_\Delta}{h_\Delta^+} - 1 \end{aligned} \tag{9.46}$$

The last expression above can be used to precompute ρ as a function of $h_\Delta^+ \triangleq h_m - p_b^+ = p_m/2 - p_b^+$. Denoting this newly defined function as

$$\hat{\rho}(h_\Delta^+) = \rho(p_\Delta(h_\Delta^+)) \tag{9.47}$$

(9.45) becomes

$$p_b^- = h_m - \hat{\rho}(h_\Delta^+) \cdot h_\Delta^+ \tag{9.48}$$

This is the form chosen for implementation in Fig. 9.39 [431]. The control variable is mouth half-pressure h_m, and $h_\Delta^+ = h_m - p_b^+$ is computed from the incoming bore pressure using only a single subtraction. The table is indexed by h_Δ^+, and the result of the lookup is then multiplied by h_Δ^+. Finally, the result of the multiplication is subtracted from h_m. The cost of the reed simulation is only two subtractions, one multiplication, and one table lookup per sample.

Because the table contains a coefficient rather than a signal value, it can be more heavily quantized both in address space and word length than a direct lookup of a signal value such as $p_\Delta(p_\Delta^+)$ or the like. A direct signal lookup, though requiring much higher resolution, would eliminate the multiplication associated with the scattering coefficient. For example, if p_b^+ and p_b^- are 16-bit signal samples, the table would contain on the order of 64K 16-bit p_Δ^* samples. Clearly, some compression of this table would be desirable. Since $p_\Delta^*(p_\Delta^+/2)$ is smoothly varying, significant compression is in fact possible. However, because the table is directly in the signal path, comparatively little compression can be done while maintaining full audio quality (such as 16-bit accuracy).

In the field of computer music, it is customary to use simple piecewise linear functions for functions other than signals at the audio sampling rate, *e.g.*, for amplitude envelopes, FM-index functions, and so on [382, 380]. Along these lines, good initial results were obtained [431] using the simplified *qualitatively* chosen reed table

$$\hat{\rho}(h_\Delta^+) = \begin{cases} 1 - m(h_\Delta^c - h_\Delta^+), & -1 \le h_\Delta^+ < h_\Delta^c \\ 1, & h_\Delta^c \le h_\Delta^+ \le 1 \end{cases} \tag{9.49}$$

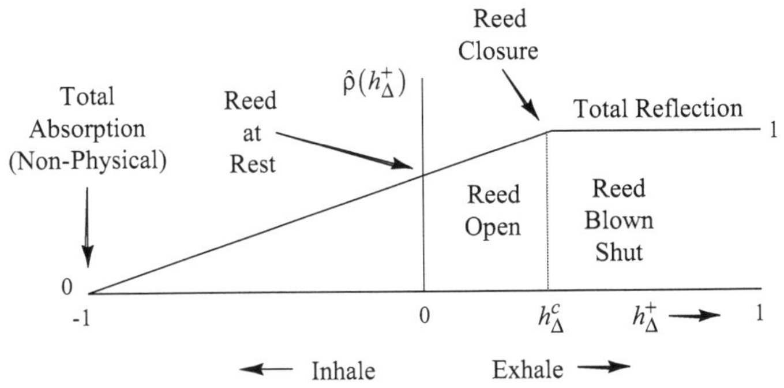

Figure 9.42: Simple, qualitatively chosen reed table for the digital waveguide clarinet.

depicted in Fig. 9.42 for $m = 1/(h^c_\Delta + 1)$. The corner point h^c_Δ is the smallest pressure difference giving reed closure.[22] Embouchure and reed stiffness correspond to the choice of offset h^c_Δ and slope m. Brighter tones are obtained by increasing the curvature of the function as the reed begins to open; for example, one can use $\hat{\rho}^k(h^+_\Delta)$ for increasing $k \geq 1$.

Another variation is to replace the table-lookup contents by a piecewise polynomial approximation. While less general, good results have been obtained in practice [89, 91, 92]. For example, one of the SynthBuilder [353] clarinet patches employs this technique using a cubic polynomial.

An intermediate approach between table lookups and polynomial approximations is to use interpolated table lookups. Typically, linear interpolation is used, but higher order polynomial interpolation can also be considered (see Chapter 4).

Clarinet Synthesis Implementation Details

To finish off the clarinet example, the remaining details of the SynthBuilder clarinet patch `Clarinet2.sb` are described.

The input mouth pressure is summed with a small amount of white noise, corresponding to turbulence. For example, 0.1% is generally used as a minimum, and larger amounts are appropriate during the attack of

[22]For operation in fixed-point DSP chips, the independent variable $h^+_\Delta \triangleq p_m/2 - p^+_b$ is generally confined to the interval $[-1, 1)$. Note that having the table go all the way to zero at the maximum negative pressure $h^+_\Delta = -1$ is not physically reasonable (0.8 would be more reasonable), but it has the practical benefit that when the lookup-table input signal is about to clip, the reflection coefficient goes to zero, thereby opening the feedback loop.

a note. Ideally, the turbulence level should be computed automatically as a function of pressure drop p_Δ and reed opening geometry [141, 531]. It should also be lowpass filtered as predicted by theory.

Referring to Fig. 9.39, the reflection filter is a simple one-pole with transfer function

$$H(z) = \frac{1 + a_1(t)}{1 + a_1(t)z^{-1}} \tag{9.50}$$

where $a_1(t) = v(t) - 0.642$, $v(t) = A_v \sin(2\pi f_v t)$, A_v is vibrato amplitude (*e.g.*, 0.03), and f_v is vibrato frequency (*e.g.*, 5 Hz). Further loop filtering occurs as a result of using simple linear interpolation of the delay line. (There is only one delay line in the actual implementation since the lower delay line of Fig. 9.39 can be commuted with the reflection filter and combined with the upper delay line, ignoring the path to the output filter since a pure delay of less than a period in the final output sound is inconsequential.) There is no transmission filter or tone-hole modeling.

Legato note transitions are managed using two delay line taps and cross-fading from one to the other during a transition [208, 441, 405]. In general, legato problems arise when the bore length is changed suddenly while sounding, corresponding to a new fingering. The reason is that really the model itself should be changed during a fingering change from that of a statically terminated bore to that of a bore with a new scattering junction appearing where each "finger" is lifting, and with disappearing scattering junctions where tone holes are being covered. In addition, if a hole is covered abruptly (especially when there are large mechanical caps, as in the saxophone), there will also be new signal energy injected in both directions on the bore in superposition with the signal scattering. As a result of this ideal picture, is difficult to get high quality legato performance using only a single delay line.

A reduced-cost, approximate solution for obtaining good sounding note transitions in a single delay-line model was proposed in [208]. In this technique, the bore delay line is "branched" during the transition, *i.e.*, a second feedback loop is formed at the new loop delay, thus forming two delay lines sharing the same memory, one corresponding to the old pitch and the other corresponding to the new pitch. A cross-fade from the old-pitch delay to the new-pitch delay sounds good if the cross-fade time and duration are carefully chosen. Another way to look at this algorithm is in terms of "read pointers" and "write pointers." A normal delay line consists of a single write pointer followed by a single read pointer, delayed by one period. During a legato transition, we simply cross-fade from a read-pointer at the old-pitch delay to a read-pointer at the new-pitch delay. In this type of implementation, the write-pointer always traverses the full delay memory corresponding to the minimum supported pitch in order that read-pointers may be instantiated at any pitch-period delay at any time. Conceptually, this simplified model of note transitions can be derived from the more rigorous model by replacing the tone-hole scattering junction by a single reflection coefficient.

STK software implementing a model as in Fig. 9.39 can be found in the file `Clarinet.cpp`.

9.5.4 Tonehole Modeling

Toneholes in woodwind instruments are essentially cylindrical holes in the bore. One modeling approach would be to treat the tonehole as a small waveguide which connects to the main bore via one port on a three-port junction. However, since the tonehole length is small compared with the distance sound travels in one sampling instant ($cT = 1125/44100 = 0.3$in, *e.g.*), it is more straightforward to treat the tonehole as a lumped load along the bore, and most modeling efforts have taken this approach.

The musical acoustics literature contains experimentally verified models of tone-hole acoustics, such as by Keefe [238]. Keefe's tonehole model is formulated as a "transmission matrix" description, which we may convert to a traveling-wave formulation by a simple linear transformation (described in §9.5.4 below) [466]. For typical fingerings, the first few open tone holes jointly provide a bore termination [38]. Either the individual tone holes can be modeled as (interpolated) scattering junctions, or the whole ensemble of terminating tone holes can be modeled in aggregate using a single reflection and transmission filter, like the bell model. Since the tone hole diameters are small compared with most audio frequency wavelengths, the reflection and transmission coefficients can be implemented to a reasonable approximation as constants, as opposed to cross-over filters as in the bell. Taking into account the inertance of the air mass in the tone hole, the tone hole can be modeled as a two-port loaded junction having load impedance equal to the air-mass inertance [143, 510]. At a higher level of accuracy, adapting transmission-matrix parameters from the existing musical acoustics literature leads to first-order reflection and transmission filters [238, 406, 403, 404, 466]. The individual tone-hole models can be simple lossy two-port junctions, modeling only the internal bore loss characteristics, or three-port junctions, modeling also the transmission characteristics to the outside air. Another approach to modeling toneholes is the "wave digital" model [528] (see §F.1 for a tutorial introduction to this approach). The subject of tone-hole modeling is elaborated further in [406, 503]. For simplest practical implementation, the bell model can be used unchanged for all tunings, as if the bore were being cut to a new length for each note and the same bell were attached. However, for best results in dynamic performance, the tonehole model should additionally include an explicit valve model for physically accurate behavior when slowly opening or closing the tonehole [405].

The Clarinet Tonehole as a Two-Port Junction

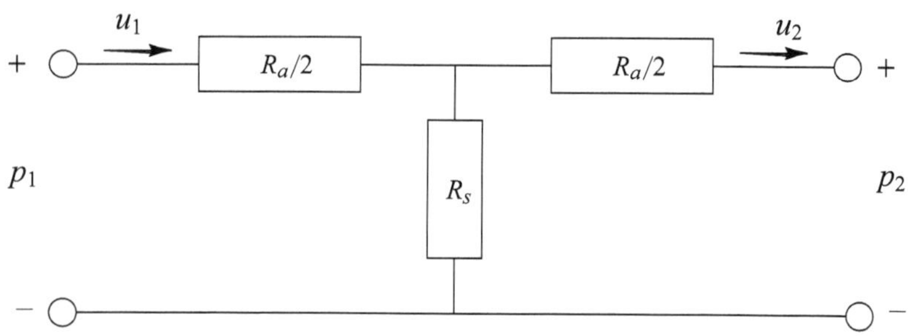

Figure 9.43: Lumped-parameter description of the clarinet tonehole.

The clarinet tonehole model developed by Keefe [240] is parametrized in terms of series and shunt resistance and reactance, as shown in Fig. 9.43. The *transmission matrix* description of this two-port is given by the product of the transmission matrices for the series impedance $R_a/2$, shunt impedance R_s, and series impedance $R_a/2$, respectively:

$$\begin{aligned}\begin{bmatrix} P_1 \\ U_1 \end{bmatrix} &= \begin{bmatrix} 1 & R_a/2 \\ 0 & 1 \end{bmatrix}\begin{bmatrix} 1 & 0 \\ R_s^{-1} & 1 \end{bmatrix}\begin{bmatrix} 1 & R_a/2 \\ 0 & 1 \end{bmatrix}\begin{bmatrix} P_2 \\ U_2 \end{bmatrix} \\ &= \begin{bmatrix} 1+\frac{R_a}{2R_s} & R_a[1+\frac{R_a}{4R_s}] \\ \frac{1}{R_s} & 1+\frac{R_a}{2R_s} \end{bmatrix}\begin{bmatrix} P_2 \\ U_2 \end{bmatrix}\end{aligned} \tag{9.51}$$

where all quantities are written in the frequency domain, and the impedance parameters are given by

$$\begin{array}{rrcl} \text{(open-hole shunt impedance)} & R_s^o & = & R_b(jkt_e+\xi_e) \\ \text{(closed-hole shunt impedance)} & R_s^c & = & -jR_b\cot(kt_h) \\ \text{(open-hole series impedance)} & R_a^o & = & -jR_bkt_a^o \\ \text{(closed-hole series impedance)} & R_a^c & = & -jR_bkt_a^c \end{array} \tag{9.52}$$

where $R_b = \rho c/(\pi b^2)$ is the wave impedance of the tonehole entrance, *i.e.*, that of an acoustic tube of cross-sectional area πb^2 (ρ is air density and

c is sound speed as usual), b is the tonehole radius, $k = \omega/c = 2\pi/\lambda$ is the wavenumber (radian spatial frequency), t_e is the open-tonehole effective length (which is slightly greater than its physical length due to the formation of a small air-piston inside the open tonehole), ξ_e is the "specific resistance" of the open tonehole due to air viscosity in and radiation from the hole, t_h is the closed-tonehole height, defined such that its product times the cross-sectional area of the tonehole exactly equals the geometric volume V_h of the closed tonehole. Finally, t_a^o and t_a^c are the equivalent *series* lengths of the open and closed tonehole, respectively, and are given by

$$\begin{aligned} t_a^o &= \frac{0.47b(b/a)^4}{\tanh(1.84t_h/b) + 0.62(b/a)^2 + 0.64(b/a)} \\ t_a^c &= \frac{0.47b(b/a)^4}{\coth(1.84t_h/b) + 0.62(b/a)^2 + 0.64(b/a)} \end{aligned}$$

where a is the radius of the main bore. The closed-tonehole height $V_h/(\pi b^2)$ can be estimated as [240]

$$t_h = t_w + \frac{1}{8}\frac{b^2}{a}\left[1 + 0.172\left(\frac{b}{a}\right)^2\right]$$

where t_w is the physical tonehole chimney height at its center.

Note that the specific resistance of the open tonehole, ξ_e, is the only real impedance and therefore the only source of wave energy loss at the tonehole. It is given by [240]

$$\xi_e = 0.25(kb)^2 + \alpha t_h + (1/4)kd_v \ln(2b/r_c),$$

where r_c is the radius of curvature of the tonehole, d_v is the viscous boundary layer thickness which expressible in terms of the shear viscosity η of air as

$$d_v = \sqrt{\frac{2\eta}{\rho\omega}}$$

and α is the real part of the propagation wavenumber (or minus the imaginary part of complex spatial frequency k). In [239], for the large-tube limit (*i.e.*, when the tube radius is large compared with the viscous boundary layer), α is given by

$$\alpha = \frac{1}{2bc}\left[\sqrt{\frac{2\eta\omega}{\rho}} + (\gamma - 1)\sqrt{\frac{2\kappa\omega}{\rho C_p}}\right]$$

where $\gamma = 1.4$ is the adiabatic gas constant for air [318], κ is the thermal conductivity of air, and C_p is the specific heat of air at constant pressure.

In [239], the following values are given for air at 300° Kelvin (26.85° C), and valid within ±10 degrees of that temperature:

$$
\begin{aligned}
\rho &= 1.1769 \times 10^{-3}(1 - 0.00335\Delta T)\,\mathrm{g/cm}^3 \\
\eta &= 1.846 \times 10^{-4}(1 + 0.0025\Delta T)\,\mathrm{g/sec/cm} \\
\gamma &= 1.4017(1 - 0.00002\Delta T) \\
\nu &= \sqrt{\eta C_p/\kappa} = 0.8418(1 - 0.0002\Delta T) \\
c &= 3.4723 \times 10^4(1 + 0.00166\Delta T)\,\mathrm{cm/sec} \\
\alpha &= \frac{\omega}{c}\left(\frac{1.045}{r_v} + \frac{1.080}{r_v^2} + \frac{0.750}{r_v^3}\right) \quad \text{(valid for } r_v > 2)
\end{aligned}
$$

where

$$r_v = b\sqrt{\frac{\rho\omega}{\eta}} = \sqrt{2}\frac{b}{d_v}$$

can be interpreted as $\sqrt{2}$ times the ratio of the tonehole radius b to the viscous boundary layer thickness d_v [239]. The constant ν^2 is referred to as the Prandtl number, and η is the shear viscosity coefficient [239]. In [71], it is noted that r_v is greater than 8 under practical conditions in musical acoustics, and so it is therefore sufficient to keep only the first and second-order terms in the expression above for α.

The open-hole effective length t_e, assuming no pad above the hole, is given in [240] as

$$t_e = \frac{(1/k)\tan(kt) + b[1.40 - 0.58(b/a)^2]}{1 - 0.61kb\tan(kt)}$$

See [240] for the case in which a pad lies above the open hole. In [405], a unified tonehole model is given which supports continuous opening and closing of the tonehole.

For implementation in a digital waveguide model, the lumped parameters above must be converted to scattering parameters. Such formulations of toneholes have appeared in the literature: Vesa Välimäki [510, 503] developed tonehole models based on a "three-port" digital waveguide junction loaded by an inertance, as described in Fletcher and Rossing [143], and also extended his results to the case of interpolated digital waveguides. It should be noted in this context, however, that in the terminology of Appendix C, Välimäki's tonehole representation is a loaded 2-port junction rather than a three-port junction. (A load can be considered formally equivalent to a "waveguide" having wave impedance given by the load impedance.) Scavone and Smith [402] developed digital waveguide tonehole models based on the more rigorous "symmetric T" acoustic model of Keefe

[240], using general purpose digital filter design techniques to obtain rational approximations to the ideal tonehole frequency response. A detailed treatment appears in Scavone's CCRMA Ph.D. thesis [406]. This section, adapted from [466], considers an exact translation of the Keefe tonehole model, obtaining two one-filter implementations: the "shared reflectance" and "shared transmittance" forms. These forms are shown to be stable without introducing an approximation which neglects the series inertance terms in the tonehole model.

By substituting $k = \omega/c$ in (9.53) to convert spatial frequency to temporal frequency, and by substituting

$$P_i = P_i^+ + P_i^- \tag{9.53}$$

$$U_i = \frac{P_i^+ - P_i^-}{R_0} \tag{9.54}$$

for $i = 1, 2$, into (9.51) to convert physical variables to wave variables, ($R_0 = \rho c/(\pi a^2)$ is the bore wave impedance), we may solve for the outgoing waves P_1^-, P_2^- in terms of the incoming waves P_1^+, P_2^+. Mathematica code for obtaining the general conversion formula from lumped parameters to scattering parameters is as follows:

```
Clear["t*", "p*", "u*", "r*"]
transmissionMatrix = {{t11, t12}, {t21, t22}};
leftPort = {{p2p+p2m}, {(p2p-p2m)/r2}};
rightPort = {{p1p+p1m}, {(p1p-p1m)/r1}};
Format[t11, TeXForm] := "{T_{11}}"
Format[p1p, TeXForm] := "{P_1^+}"
... (etc. for all variables) ...
TeXForm[Simplify[Solve[leftPort ==
      transmissionMatrix . rightPort, {p1m, p2p}]]]
```

The above code produces the following formulas:

$$\begin{aligned} P_1^- &= \frac{2P_2^- R_1 - P_1^+ R_1 T_{11} - P_1^+ T_{12} + P_1^+ R_1 R_2 T_{21} + P_1^+ R_2 T_{22}}{R_1 T_{11} - T_{12} - R_1 R_2 T_{21} + R_2 T_{22}}, \\ P_2^+ &= \frac{P_2^- R_1 T_{11} - P_2^- T_{12} + P_2^- R_1 R_2 T_{21} - 2P_1^+ R_2 T_{12} T_{21} - P_2^- R_2 T_{22} + 2P_1^+ R_2 T_{11} T_{22}}{R_1 T_{11} - T_{12} - R_1 R_2 T_{21} + R_2 T_{22}} \end{aligned} \tag{9.55}$$

Substituting relevant values for Keefe's tonehole model, we obtain, in matrix notation,

$$\begin{aligned} \begin{bmatrix} P_1^- \\ P_2^+ \end{bmatrix} &= \begin{bmatrix} S & T \\ T & S \end{bmatrix} \begin{bmatrix} P_1^+ \\ P_2^- \end{bmatrix} \\ &= \frac{1}{(2R_0 + R_a)(2R_0 + R_a + 4R_s)} \times \\ &\quad \begin{bmatrix} 4R_a R_s + R_a^2 - 4R_0^2 & 8R_0 R_s \\ 8R_0 R_s & 4R_a R_s + R_a^2 - 4R_0^2 \end{bmatrix} \begin{bmatrix} P_1^+ \\ P_2^- \end{bmatrix} \end{aligned} \tag{9.56}$$

We thus obtain the scattering formulation depicted in Fig. 9.44, where

$$S(\omega) = \frac{4R_aR_s + R_a^2 - 4R_0^2}{(2R_0 + R_a)(2R_0 + R_a + 4R_s)} \approx -\frac{R_0}{R_0 + 2R_s} \qquad (9.57)$$

is the *reflectance* of the tonehole (the same from either direction), and

$$T(\omega) = \frac{8R_0R_s}{(2R_0 + R_a)(2R_0 + R_a + 4R_s)} \approx \frac{2R_s}{R_0 + 2R_s} \qquad (9.58)$$

is the *transmittance* of the tonehole (also the same from either direction). The notation "S" for reflectance is chosen because every reflectance is a *Schur function* (stable and not exceeding unit magnitude on the unit circle in the z plane) [428, p. 221].

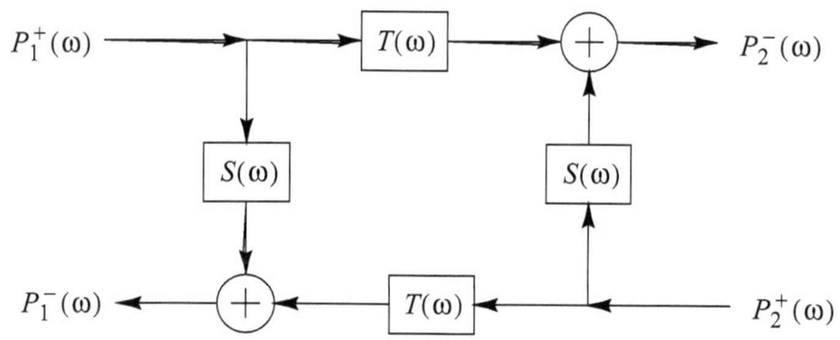

Figure 9.44: Frequency-domain, traveling-wave description of the clarinet tonehole.

The approximate forms in (9.57) and (9.58) are obtained by neglecting the negative series inertance R_a which serves to adjust the effective length of the bore, and which therefore can be implemented elsewhere in the interpolated delay-line calculation as discussed further below. The open and closed tonehole cases are obtained by substituting $\{R_a = R_a^o, R_s = R_s^o\}$ and $\{R_a = R_a^c, R_s = R_s^c\}$, respectively, from (9.53).

In a manner analogous to converting the four-multiply Kelly-Lochbaum (KL) scattering junction [245] into a one-multiply form (cf. (C.60) and (C.62) on page 570), we may pursue a "one-filter" form of the waveguide tonehole model. However, the series inertance gives some initial trouble, since

$$[1 + S(\omega)] - T(\omega) = \frac{2R_a}{2R_0 + R_a} \triangleq L(\omega)$$

instead of zero as in the KL junction. In the scattering formulas (C.100) and (C.101) on page 600 for the general loaded waveguide junction, the reflectance seen on any branch is always the transmittance from that branch to any other branch minus 1. *I.e.*, if α_i denotes the transmittance from branch i to all other branches meeting at the junction, then $\alpha_i - 1$ is the reflectance seen on branch i. Substituting

$$T = 1 + S - L$$

into the basic scattering relations (9.56), and factoring out S, we obtain, in the frequency domain,

$$\begin{aligned} P_1^-(\omega) &= SP_1^+ + TP_2^+ \\ &= SP_1^+ + [1 + S - L]P_2^+ \\ &= S[P_1^+ + P_2^+] + [1 - L]P_2^+ \\ &\triangleq S[P_1^+ + P_2^+] + AP_2^+ \end{aligned} \tag{9.59}$$

and, similarly,

$$P_2^-(\omega) = S[P_1^+ + P_2^+] + AP_1^+ \tag{9.60}$$

The resulting tonehole implementation is shown in Fig. 9.45. We call this the "shared reflectance" form of the tonehole junction.

In the same way, an alternate form is obtained from the substitution

$$S = T - 1 + L$$

which yields the "shared transmittance" form:

$$P_1^- = T[P_1^+ + P_2^+] - AP_1^+ \tag{9.61}$$

$$P_2^- = T[P_1^+ + P_2^+] - AP_2^+ \tag{9.62}$$

shown in Fig. 9.46.

Since $L(\omega) \approx 0$, it can be neglected to first order, and $A(\omega) \approx 1$, reducing both of the above forms to an approximate "one-filter" tonehole implementation.

Since $R_a = -jR_b\omega t_a/c$ is a pure negative reactance, we have

$$A(\omega) = 1 - L(\omega) = \frac{R_0 - R_a/2}{R_0 + R_a/2} = \frac{p + j\omega}{p - j\omega}, \quad p = \frac{R_0 c}{R_b t_a} \tag{9.63}$$

In this form, it is clear that $A(\omega)$ is a first-order *allpass* filter with a single pole-zero pair near infinity. Unfortunately, the pole is in the right-half-plane and hence *unstable*. We cannot therefore implement it as shown in Fig. 9.45 or Fig. 9.46. Using elementary manipulations, the unstable allpasses in Figs. 9.45 and Fig. 9.46 can be moved to the configuration

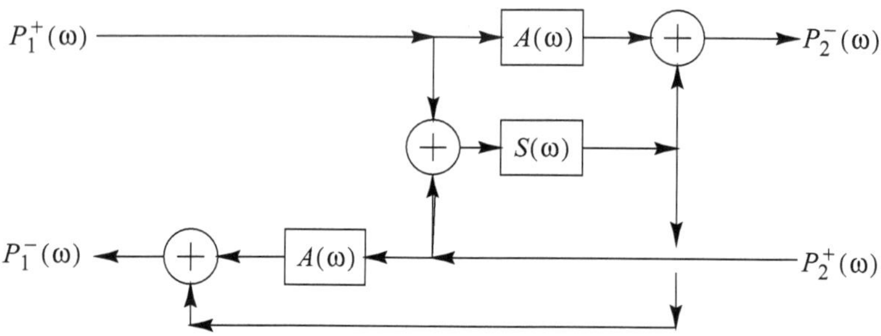

Figure 9.45: "Shared-reflectance" implementation of the clarinet tonehole model.

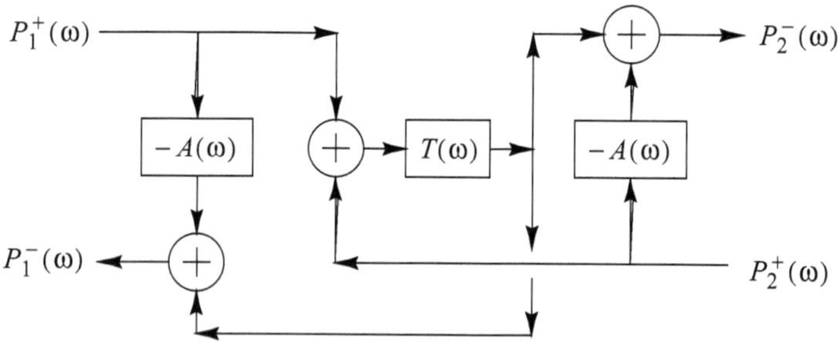

Figure 9.46: "Shared-transmittance" implementation of the clarinet tonehole model.

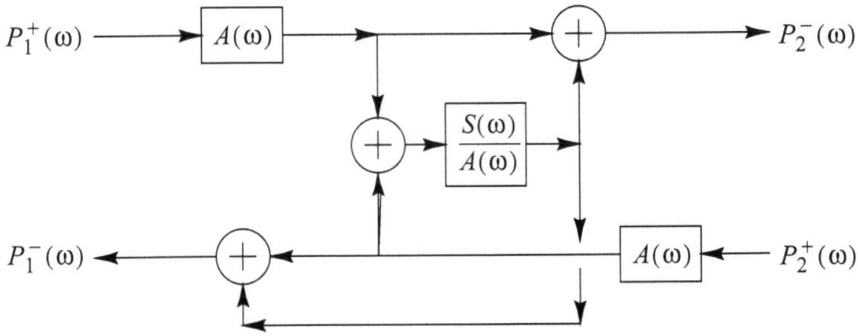

Figure 9.47: "Shared-reflectance" tonehole model with unstable all-passes pulled out to the inputs.

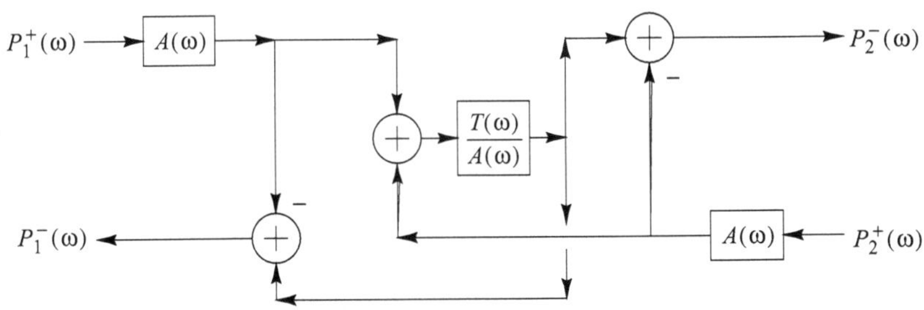

Figure 9.48: "Shared-transmittance" tonehole model with unstable all-passes pulled out to inputs.

shown in Figs. 9.47 and 9.48, respectively. Note that $T(\omega)/A(\omega)$ is stable whenever T is stable. The unstable allpasses now operate only on the two incoming wave variables, and they can be implemented implicitly by slightly reducing the (interpolated) delay-lines leading to the junction from either side. The tonehole then requires only one filter S/A or T/A.

We now see precisely how the negative series inertance R_a provides a *negative, frequency-dependent, length correction* for the bore. From (9.63), the phase delay of $A(\omega)$ can be computed as

$$D_A(\omega) \triangleq -\frac{\angle A(\omega)}{\omega} = -2\tan^{-1}(\omega/p) = -2\tan^{-1}(kt_a R_b/R_0)$$

Thus, the negative delay correction goes to zero with frequency $k = \omega/c$, series tonehole length t_a, tonehole impedance R_b, or main bore admittance $\Gamma_0 = 1/R_0$.

In practice, it is common to combine all delay corrections into a single "tuning allpass filter" for the whole bore [428, 207]. Whenever the desired allpass delay goes negative, we simply add a sample of delay to the desired allpass phase-delay and subtract it from the nearest delay. In other words, negative delays have to be "pulled out" of the allpass and used to shorten an adjacent interpolated delay line. Such delay lines are normally available in practical modeling situations.

Tonehole Filter Design

The tone-hole reflectance and transmittance must be converted to discrete-time form for implementation in a digital waveguide model. Figure 9.49 plots the responses of second-order discrete-time filters designed to approximate the continuous-time magnitude and phase characteristics of the reflectances for closed and open toneholes, as carried out in [403, 406]. These filter designs assumed a tonehole of radius $b = 4.765$ mm, minimum tonehole height $t_w = 3.4$ mm, tonehole radius of curvature $r_c = 0.5$ mm, and air column radius $a = 9.45$ mm. Since the measurements of Keefe do not extend to 5 kHz, the continuous-time responses in the figures are extrapolated above this limit. Correspondingly, the filter designs were weighted to produce best results below 5 kHz.

The closed-hole filter design was carried out using weighted L^2 equation-error minimization [428, p. 47], *i.e.*, by minimizing $||\, W(e^{j\omega})[\hat{A}(e^{j\omega})H(e^{j\omega}) - \hat{B}(e^{j\omega})]\,||_2$, where W is the weighting function, $H(e^{j\omega})$ is the desired frequency response, Ω denotes discrete-time radian frequency, and the designed filter response is $\hat{H}(e^{j\omega}) = \hat{B}(e^{j\omega})/\hat{A}(e^{j\omega})$. Note that both phase and magnitude are matched by equation-error minimization, and this error criterion is used extensively in the field of system identification [288] due to its ability to design optimal IIR filters via quadratic minimization. In the spirit of the well-known Steiglitz-McBride algorithm [287], equation-error minimization can be iterated, setting the weighting function at iteration $i+1$ to the inverse of the inherent weighting $\hat{A}_i$ of the previous iteration, *i.e.*, $W_{i+1}(e^{j\omega}) = 1/\hat{A}_i(e^{j\omega})$. However, for this study, the weighting was used only to increase accuracy at low frequencies relative to high frequencies. Weighted equation-error minimization is implemented in the matlab function `invfreqz()` (§8.6.4, p. 297).

The open-hole discrete-time filter was designed using Kopec's method [297], [428, p. 46] in conjunction with weighted equation-error minimization. Kopec's method is based on linear prediction:

- Given a desired complex frequency response $H(e^{j\omega})$, compute an allpole model $1/\hat{A}(z)$ using linear prediction
- Compute the error spectrum $\hat{E}(e^{j\omega}) \triangleq \hat{A}(e^{j\omega})H(e^{j\omega})$.
- Compute an allpole model $1/\hat{B}(z)$ for $\hat{E}^{-1}(e^{j\omega})$ by minimizing

$$\left\| \hat{B}(e^{j\omega})\hat{E}^{-1}(e^{j\omega}) \right\|_2 = \left\| \frac{\hat{B}(e^{j\omega})}{\hat{A}(e^{j\omega})} H^{-1}(e^{j\omega}) \right\|_2 .$$

Use of linear prediction is equivalent to minimizing the L^2 *ratio error*

$$\left\| \hat{E}(e^{j\omega}) \right\|_2 = \left\| \hat{A}(e^{j\omega})H(e^{j\omega}) \right\|_2 .$$

This optimization criterion causes the filter to fit the *upper spectral envelope* of the desired frequency-response. Since the first step of Kopec's method captures the upper spectral envelope, the "nulls" and "valleys" are largely "saved" for the next step which computes zeros. When computing the zeros, the spectral "dips" become "peaks," thereby receiving more weight under the L^2 ratio-error norm. Thus, in Kopec's method, the poles model the upper spectral envelope, while the zeros model the lower spectral envelope. To apply Kopec's method to the design of an open-tonehole filter, a one-pole model $\hat{H}_1(z)$ was first fit to the continuous-time response, $H(e^{j\Omega})$. Subsequently, the inverse error spectrum, $\hat{H}_1(e^{j\Omega})/H(e^{j\Omega})$ was modeled with a two-pole digital filter, $\hat{H}_2(z)$. The discrete-time approximation to $H(e^{j\Omega})$ was then given by $\hat{H}_1(z)/\hat{H}_2(z)$.

The reasonably close match in both phase and magnitude by second-order filters indicates that there is in fact only one important tonehole resonance and/or anti-resonance within the audio band, and that the measured frequency responses can be modeled with very high audio accuracy using only second-order filters.

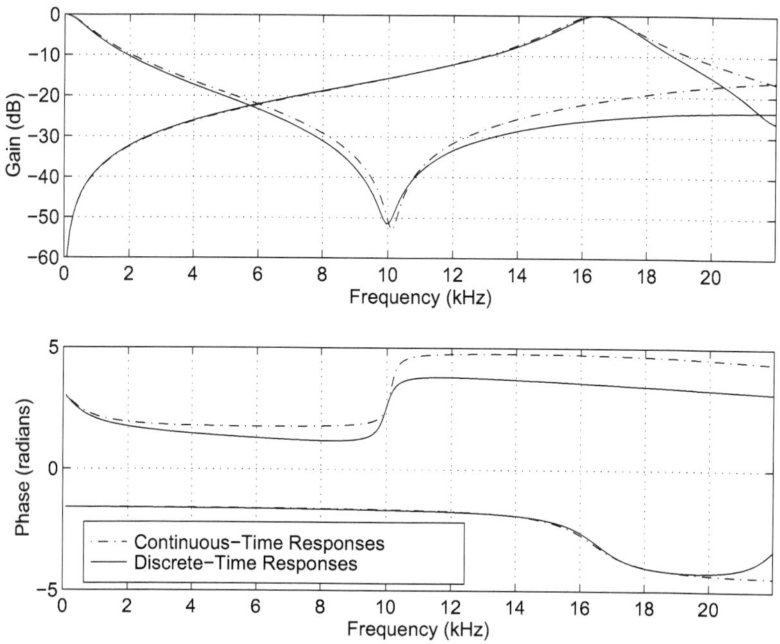

Figure 9.49: Two-port tonehole junction closed-hole and open-hole reflectances based on Keefe's acoustic measurements (dashed) versus second-order digital filter approximations (solid). Top: Reflectance magnitude; Bottom: Reflectance phase. The closed tonehole has one resonance in the audio band just above 16 kHz. The open tonehole has one anti-resonance in the audio band near 10 kHz. At dc, the open tonehole fully reflects, while the closed tonehole reflects close to nothing (from [403]).

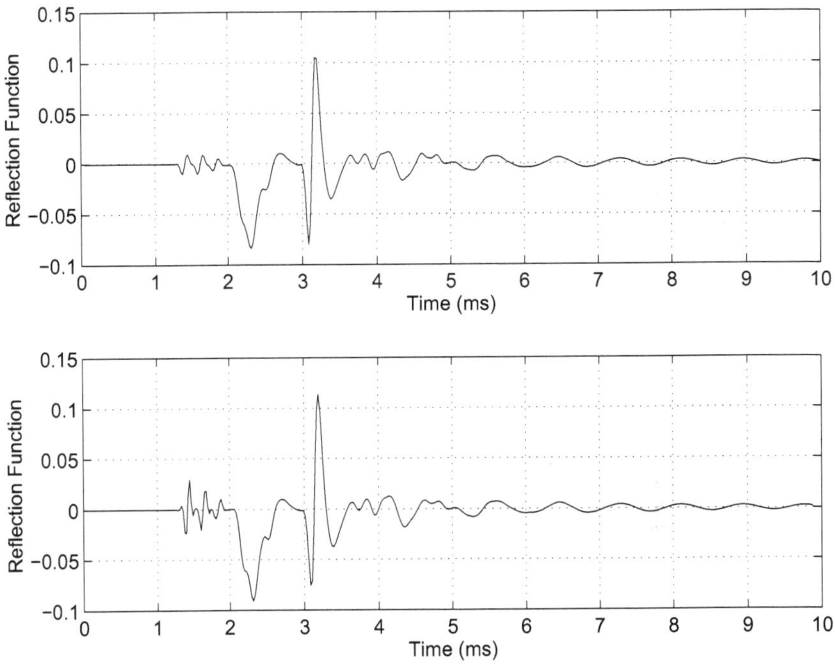

Figure 9.50: Reflection functions for note G (three finger holes closed, three finger holes open) on a simple flute (from [403]). (top) Transmission-line calculation; (bottom) Digital waveguide two-port tonehole implementation.

Figure 9.50 plots the reflection function calculated for a six-hole flute bore, as described in [240].

The upper plot was calculated using Keefe's frequency-domain transmission matrices, such that the reflection function was determined as the inverse Fourier transform of the corresponding reflection coefficient. This response is equivalent to that provided by [240], though scale factor discrepancies exist due to differences in open-end reflection models and lowpass filter responses. The lower plot was calculated from a digital waveguide model using two-port tonehole scattering junctions. Differences between the continuous- and discrete-time results are most apparent in early, high-frequency, closed-hole reflections. The continuous-time reflection function was low-pass filtered to remove time-domain aliasing effects incurred by the inverse Fourier transform operation and to better correspond with the plots of [240]. By trial and error, a lowpass filter with a cutoff frequency around 4 kHz was found to produce the best match to Keefe's results. The digital waveguide result was obtained at a sampling rate of 44.1 kHz and then lowpass filtered to a 10 kHz bandwidth, corresponding to that of [240]. Further lowpass filtering is inherent from the first-order Lagrangian, delay-line length interpolation technique used in this model [503]. Because such filtering is applied at different locations along the "bore," a cumulative effect is difficult to accurately determine. The first tonehole reflection is affected by only two interpolation filters, while the second tonehole reflection is affected by four of these filtering operations. This effect is most responsible for the minor discrepancies apparent in the plots.

The Tonehole as a Two-Port Loaded Junction

It seems reasonable to expect that the tonehole should be representable as a *load* along a waveguide bore model, thus creating a loaded two-port junction with two identical bore ports on either side of the tonehole. From the relations for the loaded parallel junction (C.101), in the two-port case with $R_1 = R_2 = R_0$, and considering pressure waves rather than force waves, we have

$$\begin{aligned} P_J(s) &= \alpha P_1^+ + \alpha P_2^+, \quad \alpha = 2\Gamma_0/[G_J(s)+2\Gamma_0] && (9.64)\\ P_1^-(s) &= P_J(s) - P_1^+ = (\alpha-1)P_1^+ + \alpha P_2^+ = \alpha(P_1^+ + P_2^+) - P_1^+ && (9.65)\\ P_2^-(s) &= P_J(s) - P_2^+ = \alpha P_1^+ + (\alpha-1)P_2^+ = \alpha(P_1^+ + P_2^+) - P_2^+ && (9.66)\end{aligned}$$

Thus, the loaded two-port junction can be implemented in "one-filter form" as shown in Fig. 9.48 with $A(\omega) = 1$ ($L(\omega) = 0$) and

$$T(\omega) = \alpha = \frac{2\Gamma_0}{2\Gamma_0 + G_J(s)} = \frac{2R_J(s)}{2R_J(s) + R_0}$$

Comparing with (9.58), we see that the simplified Keefe tonehole model with the negative series inertance removed ($R_a = 0$), is equivalent to a loaded two-port waveguide junction with $R_J = R_s$, *i.e.*, the parallel load impedance is simply the shunt impedance in the tonehole model.

Each series impedance $R_a/2$ in the split-T model of Fig. 9.43 can be modeled as a *series* waveguide junction with a load of $R_a/2$. To see this, set the transmission matrix parameters in (9.55) to the values $T_{11} = T_{22} = 1$, $T_{12} = R_a/2$, and $T_{21} = 0$ from (9.51) to get

$$\begin{aligned} P_1^- &= (1-\alpha)P_1^+ + \alpha P_2^- \\ P_2^+ &= \alpha P_1^+ + (1-\alpha)P_2^- \end{aligned} \tag{9.67}$$

where $\alpha = 2R_0/(2R_0 + R_a/2)$ is the alpha parameter for a series loaded waveguide junction involving two impedance R_0 waveguides joined in series with each other and with a load impedance of $R_a/2$, as can be seen from (C.99). To obtain exactly the loaded series scattering relations (C.100), we first switch to the more general convention in which the "+" superscript denotes waves traveling *into* a junction of any number of waveguides. This exchanges "+" with "−" at port 2 to yield

$$\begin{aligned} P_1^- &= (1-\alpha)P_1^+ + \alpha P_2^+ \\ P_2^- &= \alpha P_1^+ + (1-\alpha)P_2^+ \end{aligned} \tag{9.68}$$

Next we convert pressure to velocity using $P_i^+ = R_0 U_i^+$ and $P_i^- = -R_0 U_i^-$ to obtain

$$\begin{aligned} U_1^- &= (\alpha-1)U_1^+ - \alpha U_2^+ \\ U_2^- &= -\alpha U_1^+ + (\alpha-1)U_2^+ \end{aligned} \tag{9.69}$$

Finally, we toggle the reference direction of port 2 (the "current" arrow for u_2 on port 2 in Fig. 9.43) so that velocity is positive flowing *into* the junction on both ports (which is the convention used to derive (C.100) and which is typically followed in circuit theory). This amounts to negating $U_2^\pm$, giving

$$\begin{aligned} U_1^- &= U_J - U_1^+ \\ U_2^- &= U_J - U_2^+ \end{aligned} \tag{9.70}$$

where $U_J \triangleq (\alpha U_1^+ + \alpha U_2^+)$. This is then the canonical form (C.100).

9.5.5 Exercises

See `http://ccrma.stanford.edu/realsimple/reeds-problems/` for laboratory exercises related to reed-instrument synthesis.

9.6 Bowed Strings

A schematic block diagram for bowed strings is shown in Fig. 9.51 [431]. The bow divides the string into two sections, so the bow model is a nonlinear two-port, in contrast with the reed which was a nonlinear one-port terminating the bore at the mouthpiece. In the case of bowed strings, the primary control variable is bow velocity, so *velocity waves* are the natural choice for the delay lines.

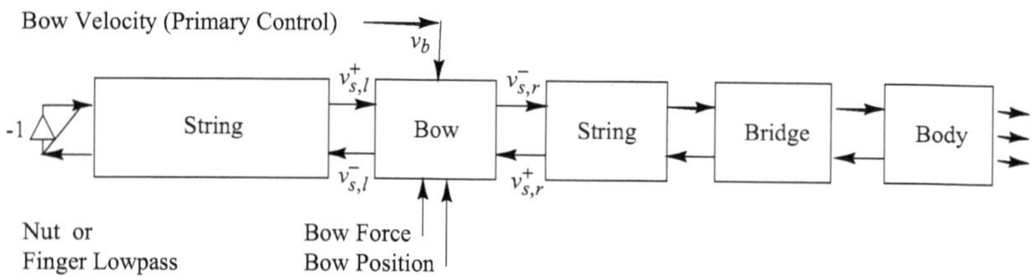

Figure 9.51: A schematic model for bowed-string instruments.

9.6.1 Digital Waveguide Bowed-String

A more detailed diagram of the digital waveguide implementation of the bowed-string instrument model is shown in Fig. 9.52. The right delay-line pair carries left-going and right-going velocity waves samples v^+_{sr} and v^-_{sr}, respectively, which sample the traveling-wave components within the string to the right of the bow, and similarly for the section of string to the left of the bow. The '+' superscript refers to waves traveling *into* the bow.

String velocity at any point is obtained by adding a left-going velocity sample to the right-going velocity sample immediately opposite in the other delay line, as indicated in Fig. 9.52 at the bowing point. The reflection filter at the right implements the losses at the bridge, bow, nut or finger-terminations (when stopped), and the round-trip attenuation/dispersion from traveling back and forth on the string. To a very good degree of approximation, the nut reflects incoming velocity waves (with a sign inversion) at all audio wavelengths. The bridge behaves similarly to a first order, but there are additional (complex) losses due to the finite bridge driving-point impedance (necessary for transducing sound from the string into the resonating body). According to [95, page 27], the bridge of a violin can be modeled up to about 5 kHz, for purposes of computing string loss, as a single spring in parallel with a frequency-independent resistance

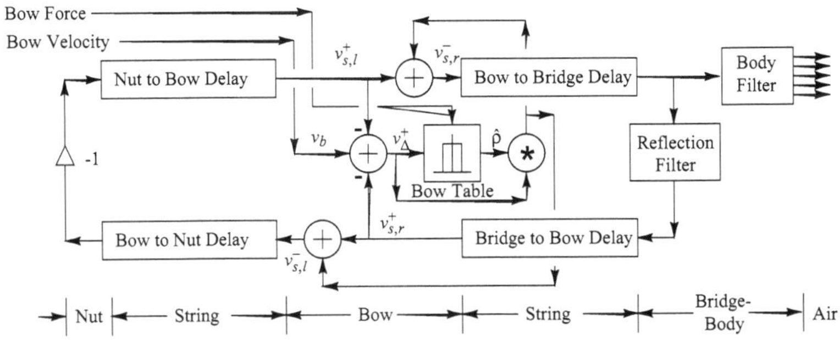

Figure 9.52: Waveguide model for a bowed string instrument, such as a violin.

("dashpot"). Bridge-filter modeling is discussed further in §9.2.1.

Figure 9.52 is drawn for the case of the lowest note. For higher notes the delay lines between the bow and nut are shortened according to the distance between the bow and the finger termination. The bow-string interface is controlled by *differential velocity* v_Δ^+ which is defined as the bow velocity minus the total incoming string velocity. Other controls include *bow force* and *angle* which are changed by modifying the reflection-coefficient $\rho(v_\Delta^+)$. Bow position is changed by taking samples from one delay-line pair and appending them to the other delay-line pair. Delay-line interpolation can be used to provide continuous change of bow position [267].

9.6.2 The Bow-String Scattering Junction

The theory of bow-string interaction is described in [95, 151, 244, 307, 308]. The basic operation of the bow is to reconcile the nonlinear bow-string friction curve $R_b(v_d)$ with the string wave impedance R_s:

$$\begin{aligned}\text{Applied Force} &= \text{Bow-String Friction Curve} \times \text{Differential Velocity}\\ \text{Reaction Force} &= \text{String Wave Impedance} \times \text{Velocity Change}\end{aligned}$$

or, equating these equal and opposite forces, we obtain

$$R_b(v_\Delta) \times v_\Delta = R_s\left[v_\Delta^+ - v_\Delta\right]$$

where $v_\Delta = v_b - v_s$ is the velocity of the bow minus that of the string, $v_s = v_{sl}^+ + v_{sl}^- = v_{sr}^+ + v_{sr}^-$ is the string velocity in terms of traveling waves, R_s is the wave impedance of the string (equal to the geometric mean of tension and density), and $R_b(v_\Delta)$ is the friction coefficient for the bow against the string, *i.e.*, bow force $F_b(v_\Delta) = R_b(v_\Delta)\cdot v_\Delta$. (Force and velocity point in the same direction when they have the same sign.) Here, v_{sr} denotes transverse velocity on the segment of the bowed string to the *right* of the bow, and v_{sl} denotes velocity waves to the *left* of the bow. The corresponding normalized functions to be used in the Friedlander-Keller graphical solution technique are depicted in Fig. 9.53.

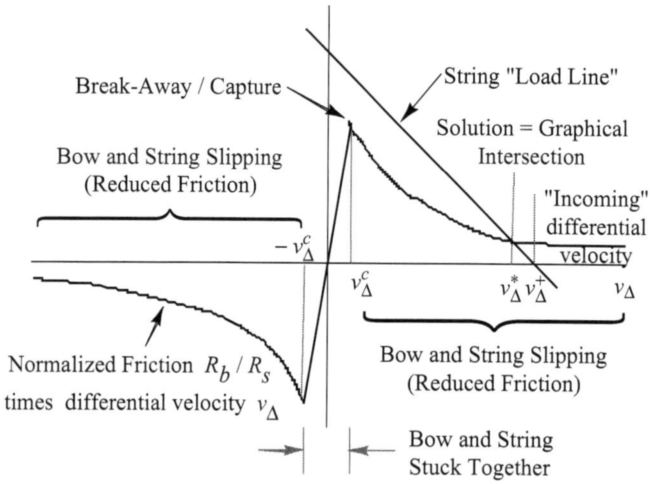

Figure 9.53: Overlay of normalized bow-string friction curve $R_b(v_\Delta)/R_s$ with the string "load line" $v_\Delta^+ - v_\Delta$. The "capture" and "break-away" differential velocity is denoted v_Δ^c. Note that increasing the bow force increases v_Δ^c as well as enlarging the maximum force applied (at the peaks of the curve).

In a bowed string simulation as in Fig. 9.51, a velocity input (which is injected equally in the left- and right-going directions) must be found such that the transverse force of the bow against the string is balanced by the reaction force of the moving string. If bow-hair dynamics are neglected [176], the bow-string interaction can be simulated using a memoryless table lookup or segmented polynomial in a manner similar to single-reed woodwinds [431].

A derivation analogous to that for the single reed is possible for the simulation of the bow-string interaction. The final result is as follows.

$$\begin{aligned} v_{sr}^- &= v_{sl}^+ + \hat{\rho}(v_\Delta^+)\cdot v_\Delta^+ \\ v_{sl}^- &= v_{sr}^+ + \hat{\rho}(v_\Delta^+)\cdot v_\Delta^+ \end{aligned}$$

where $v_\Delta^+ \triangleq v_b - (v_{sr}^+ + v_{sl}^+)$, v_b is bow velocity, and

$$\hat{\rho}(v_\Delta^+) = \frac{r(v_\Delta(v_\Delta^+))}{1 + r(v_\Delta(v_\Delta^+))}$$

The impedance ratio is defined as $r(v_\Delta) = 0.25 R_b(v_\Delta)/R_s$,

Nominally, $R_b(v_\Delta)$ is constant (the so-called static coefficient of friction) for $|v_\Delta| \le v_\Delta^c$, where v_Δ^c is both the capture and break-away differential velocity. For $|v_\Delta| > v_\Delta^c$, $R_b(v_\Delta)$ falls quickly to a low dynamic coefficient of friction. It is customary in the bowed-string physics literature to assume that the dynamic coefficient of friction continues to approach zero with increasing $|v_\Delta| > v_\Delta^c$ [308, 95].

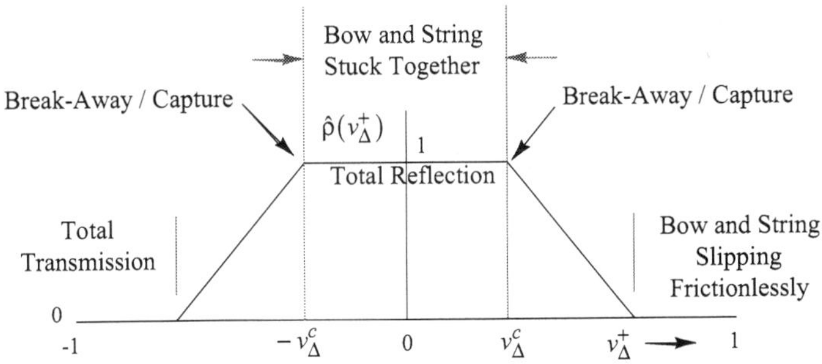

Figure 9.54: Simple, qualitatively chosen bow table for the digital waveguide violin.

Figure 9.54 illustrates a simplified, piecewise linear bow table $\hat{\rho}(v_\Delta^+)$. The flat center portion corresponds to a fixed reflection coefficient "seen" by a traveling wave encountering the bow stuck against the string, and the outer sections of the curve give a smaller reflection coefficient corresponding to the reduced bow-string interaction force while the string is slipping under the bow. The notation v_Δ^c at the corner point denotes the capture or break-away differential velocity. Note that hysteresis is neglected.

9.6.3 Bowed String Synthesis Extensions

For maximally natural interaction between the bow and string, bow-hair dynamics should be included in a proper bow bodel [176]. In addition, the width of the bow should be variable, depending on bow "angle". A finite difference model for the finite width bow was reported in [351, 352]. The bow-string friction characteristic should employ a *thermodynamic model* of bow friction in which the bow rosin has a *time-varying viscosity* due to temperature variations within a period of sound [550]. It is well known by bowed-string players that rosin is sensitive to temperature. Thermal models of dynamic friction in bowed strings are described in [426], and they have been used in virtual bowed strings for computer music [419, 423, 21].

Given a good model of a bowed-string instrument, it is most natural to interface it to a physical bow-type controller, with sensors for force, velocity, position, angle, and so on [420, 294].

A real-time software implementation of a bowed-string model, similar to that shown in Fig. 9.52, is available in the Synthesis Tool Kit (STK) distribution as `Bowed.cpp`. It provides a convenient starting point for more refined bowed-string models [276].

9.6.4 Linear Commuted Bowed Strings

The commuted synthesis technique (§8.7, p. 305) can be extended to bowed strings in special case of "ideal" bowed attacks [440]. Here, an ideal attack is defined as one in which *Helmholtz motion*[23] is instantly achieved. This technique will be called "linear commuted synthesis" of bowed strings.

Additionally, the linear commuted-synthesis model for bowed strings can be driven by a separate nonlinear model of bowed-string dynamics. This gives the desirable combination of a full range of complex bow-string interaction behavior together with an efficiently implemented body resonator.

Bowing as Periodic Plucking

The "leaning sawtooth" waveforms observed by Helmholtz for steady state bowed strings can be obtained by periodically "plucking" the string in only one direction along the string [428]. In principle, a traveling impulsive excitation is introduced into the string in the right-going direction for a "down bow" and in the left-going direction for an "up bow." This simplified bowing simulation works best for smooth bowing styles in which the notes have slow attacks. More varied types of attack can be achieved using the more physically accurate McIntyre-Woodhouse theory [307, 431].

Commuting the string and resonator means that the string is now plucked by a *periodically repeated resonator impulse response*. A nice simplified vibrato implementation is available by varying the impulse-response retriggering period, *i.e.*, the vibrato is implemented in the excitation oscillator and not in the delay loop. The string loop delay need not be modulated at all. While this departs from being a physical model, the vibrato quality is satisfying and qualitatively similar to that obtained by a rigorous physical model. Figure 9.55 illustrates the overall block diagram of the simplified

[23] `https://ccrma.stanford.edu/realsimple/travelingwaves/Helmholtz_Motion.html`

bowed string and its commuted and response-excited versions.

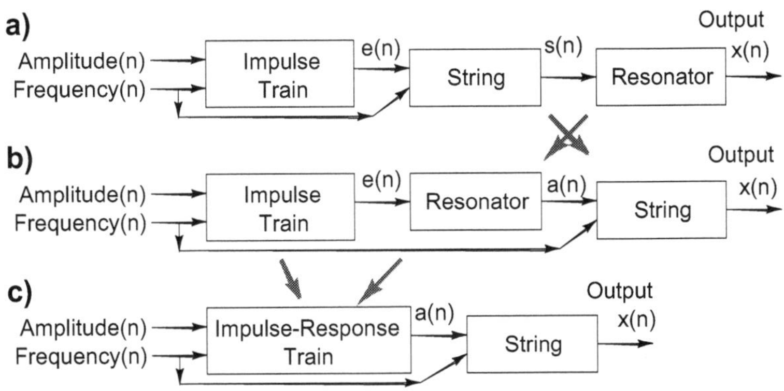

Figure 9.55: a) The simplified bowed string, including amplitude, pitch, and vibrato controls. The frequency control is also used by the string. b) Equivalent diagram with resonator and string commuted. c) Equivalent diagram in which the resonator impulse response is played into the string each pitch period.

In current technology, it is reasonable to store one recording of the resonator impulse response in digital memory as one of many possible *string excitation tables.* The excitation can contribute to many aspects of the tone to be synthesized, such as whether it is a violin or a cello, the force of the bow, and where the bow is playing on the string. Also, graphical equalization and other time-invariant filtering can be provided in the form of alternate excitation-table choices.

During the synthesis of a single bowed-string tone, the excitation signal is played into the string quasi-periodically. Since the excitation signal is typically longer than one period of the tone, it is necessary to either (1) interrupt the excitation playback to replay it from the beginning, or (2) start a new playback which overlaps with the playback in progress. Variant (2) requires a separate incrementing pointer and addition for each instance of the excitation playback; thus it is more expensive, but it is preferred from a quality standpoint. This same issue also arises in the Chant synthesis technique for the singing voice [389]. In the Chant technique, a sum of three or more enveloped sinusoids (called FOFs) is periodically played out to synthesize a sung vowel tone. In the unlikely event that the excitation table is less than one period long, it is of course extended by zeros, as is done in the VOSIM voice synthesis technique [219] which can be considered a simplified forerunner of Chant.

Of course, ordinary wavetable synthesis [303, 199, 327] or any other type of synthesis can also be used as an excitation signal in which case the string loop behaves as a pitch-synchronous comb filter following the wavetable oscillator. Interesting effects can be obtained by slightly detuning the wavetable oscillator and delay loop; tuning the wavetable oscillator to a harmonic of the delay loop can also produce an ethereal effect.

More General Quasi-Periodic Synthesis

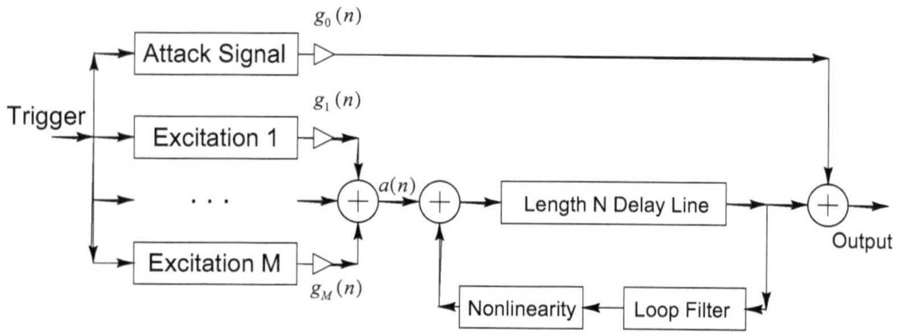

Figure 9.56: More general table-excited filtered-delay loop [440].

Figure 9.56 illustrates a more general version of the table-excited, filtered delay loop synthesis system [440]. The generalizations help to obtain a wider class of timbres. The multiple excitations summed together through time-varying gains provide for timbral evolution of the tone. For example, a violin can transform smoothly into a cello, or the bow can move smoothly toward the bridge by interpolating among two or more tables. Alternatively, the tables may contain "principal components" which can be scaled and added together to approximate a wider variety of excitation timbres. An excellent review of multiple wavetable synthesis appears in [199]. The nonlinearity is useful for obtaining distortion guitar sounds and other interesting evolving timbres [387].

Finally, the "attack signal" path around the string has been found to be useful for reducing the cost of implementation: the highest frequency components of a struck string, say, tend to emanate immediately from the string to the resonator with very little reflection back into the string (or pipe, in the case of wind instrument simulation). Injecting them into the delay loop increases the burden on the loop filter to quickly filter them out. Bypassing the delay loop altogether alleviates requirements on the loop filter and even allows the filtered delay loop to operate at a lower sampling rate; in this case, a signal interpolator would appear between the string output and the summer which adds in the scaled attack signal in Fig. 9.56. For example, it was found that the low E of an electric guitar (Gibson Les Paul) can be synthesized quite well using a filtered delay loop running at a sampling rate of 3 kHz. (The pickups do not pick up much energy above 1.5 kHz.) Similar savings can be obtained for any instrument having a high-frequency content which decays much more quickly than its low-frequency content.

Stochastic Excitation for Quasi-Periodic Synthesis

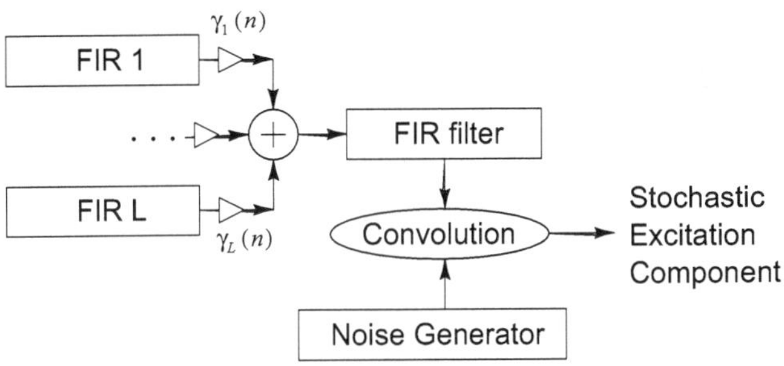

Figure 9.57: Example of a filtered noise excitation implementation.

For good generality, at least one of the excitation signals should be a filtered noise signal [440]. An example implementation is shown in Fig. 9.57, in which there is a free-running bandlimited noise generator filtered by a finite impulse response (FIR) digital filter. As noted in §9.4.4, p. 398, such a filtered-noise signal can synthesize the perceptual equivalent of the impulse response of many high-frequency modes that have been separated from the lower frequency modes in commuted synthesis (§8.7.1, p. 308). It can also handle pluck models in which successive plucking variations are imposed by the FIR filter coefficients.

In a simple implementation, only two gains might be used, allowing simple interpolation from one filter to the next, and providing an overall amplitude control for the noise component of the excitation signal.

9.6.5 Exercises

See `http://ccrma.stanford.edu/realsimple/bs-problems/` for laboratory exercises related to virtual bowed strings.

9.7 Brasses

Brass instruments include the trumpet, trombone, French horn, tuba, and other instruments consisting of a horn driven by lip vibrations at one end. The mouthpiece of a brass instrument is generally cup-shaped, and the lips vibrate at a frequency near that of the sounding note.

There are three main components needed in synthesis models for brass instruments: the mouthpiece, bore, and bell. Both cylindrical and conical bores are accurately modeled using digital waveguide techniques, as discussed in previous chapters (see §C.18 on page 635 for further discussion). However, bore models may need to simulate nonlinear shock wave formation at high dynamic levels [198, 341], which is relatively expensive to do [537]. The bell, which flares more rapidly than conical (typically *exponentially*[24]) does not support true traveling waves [357], and so it must be modeled using a more general simulation technique, such as by means of reflection and transmission filters (discussed in §9.7.2 below).

9.7.1 Modeling the Lips and Mouthpiece

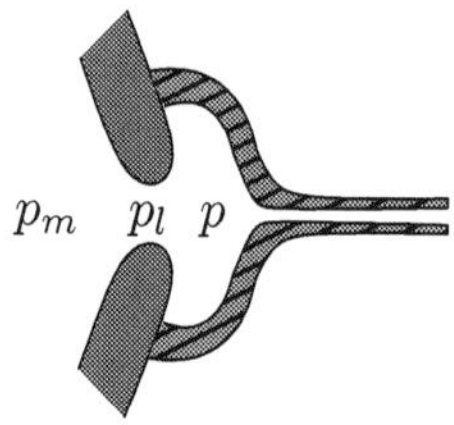

Figure 9.58: Schematic of a brass player's lips at a mouthpiece.

Figure 9.58 depicts a schematic of a brass-player's lips positioned within a mouthpiece. The air pressure[25] is denoted p_m inside the mouth, p_l between the lips, and p within the mouthpiece cup.

A simple first approximation is to assume that the air flow through the lips is *inviscid* ("frictionless") and *incompressible*. In this case, the pressure between the lips is governed by *Bernoulli's equation* (see §B.7.4 on page 505):

$$p_m = p_l + \frac{1}{2}\rho u_l^2$$

[24] If the flare of the bell is expressed as $r(x) = ae^{\alpha x} + b$, where $r(x)$ denotes the horn radius at position x along the bore axis, then α is called the *flare constant* of the bell.

[25] As always in this book, by "air pressure" we mean the *excess air pressure* above ambient air pressure. In the case of brass instruments, excess air pressure is created by the muscles of the lungs.

where ρ is the air density (mass per unit volume) and u_l denotes the *particle velocity* (see §B.7.1 on page 503) (distance per unit time) between the lips. The particle velocity u_m in the mouth is so small that it is usually neglected (approximated by zero).

Following Cullen et al. [98] and previous researchers [159, 161], one may assume that the flow between the lips creates a jet in the mouthpiece having velocity u_l. The pressure within the jet is therefore p_l, the same as between the lips. The jet is assumed to mix turbulently with the air in the mouthpiece so that it dissipates its energy without raising pressure in the mouthpiece.[26]

One- and two-mass models were also developed by Rodet and Vergez [388, 534]. An early one-mass, "swinging door" model that works surprisingly well is in Cook's `HosePlayer` [89] and and `TBone` [88]. It is also used in the `Brass.cpp` patch in the Synthesis Tool Kit [91]; in this simplified model, the lip-valve is modeled as a second-order resonator whose output is squared and hard-clipped to a maximum magnitude.

The acoustic tube in a horn can be modeled as a simple digital waveguide [22, 37, 436, 507, 529], whether it is cylindrical or conical. More complicated, flaring horn sections may be modeled as two-port digital filters which provide the appropriate reflection and transmission from the bell [39, 50, 71, 118, 406, 529]. Section 9.7.2 below summarizes a particularly economical approach to brass bell modeling.

It is known that nonlinear *shock waves* occur in the bore for large amplitude vibrations [198, 341]. A simple nonlinear waveguide simulation of this effect can be implemented by shortening the delay elements slightly when the instantaneous amplitude is large. (The first-order effect of air nonlinearity in large-amplitude wave propagation is an increase in sound speed at very high pressures [52].)

It is also known that the bores of "bent horns," such as the trumpet, do not behave as ideal waveguides [425]. This is because sharp bends in the metal tube cause some scattering, and mode conversion at high frequencies. Such points along the horn can be modeled using high-frequency loss and reflection.

---

[26]When the kinetic energy of a jet is converted back into air pressure, this is called *pressure recovery*. We assume, following [98] and others, that pressure recovery does not occur in this model. Instead, the kinetic energy of the ket is assumed to be dissipated in the form of turbulence (vortices and ultimately heat). Note that air flow is therefore assumed inviscid within the mouth and between the lips, but viscous around the jet in the mouthpiece.

9.7.2 Bell Models

The flaring bell of a horn cannot be accurately modeled as a sparse digital waveguide, because traveling pressure waves only propagate without reflection in conical bores (which include cylindrical bores as a special case) [357].[27] Digital waveguides are "sparse" (free of internal scattering) only when there are long sections at a constant wave impedance.

The most cost-effective bell filters (and, more generally, "flare filters") to date appears to be the use of *truncated IIR* (TIIR) digital filters [541]. These filters use an unstable pole to produce exponentially rising components in the impulse response, but the response is cut off after a finite time, as is needed in the case of a bell impulse response. By fitting a piecewise polynomial/exponential approximation to the reflection impulse response of the trumpet bell, very good approximations can be had for the computational equivalent of approximately a 10th order IIR filter (but using more memory in the form of a delay line, which costs very little computation).

In more detail, the most efficient computational model for flaring bells in brass instruments seems to be one that consists of one or more sections having an impulse response given by the sum of a growing exponential and a constant, *i.e.*,

$$y(n) = \begin{cases} ae^{cn} + b, & n = 0, 1, 2, \ldots, N-1 \\ 0, & \text{otherwise.} \end{cases}$$

The truncated constant b can also be generated using a one-pole TIIR filter, with its pole set to $z = 1$. The remaining reflection impulse response has a decaying trend, and can therefore be modeled accurately using one of many conventional filter design techniques. In [529, 530], the Steiglitz-McBride IIR filter design algorithm [287] yielded good results from pulse-reflectometry data [425], as shown in Fig. 9.60.

The C++ class in the Synthesis Tool Kit (STK) implementing a basic brass synthesis model is called `Brass.cpp`.

Literature Relevant to Brasses

A nice summary of the musical acoustics of brass instruments is given by Campbell [70].

Research by Cullen et al. [98] and Gilbert et al. [159, 161] has been concerned with artificial mouth experiments to verify a theoretical model of vibrating lips, and to calibrate the model to realistic playing conditions. These data can be used in the construction of improved lip-valve models

[27] For velocity waves, the flare may be *hyperbolic* [50].

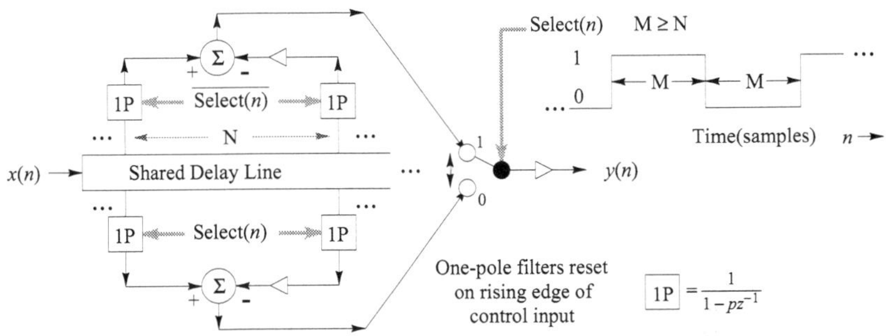

Figure 9.59: Example of a TIIR filter for generating a growing exponential or constant segment (from [529]).

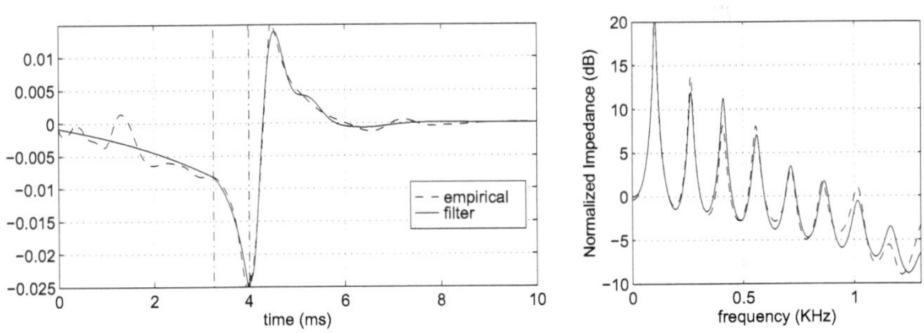

Figure 9.60: Impulse-response and driving-point-impedance fit for the trumpet bell using two offset exponentials and two biquads designed as a 4th-order IIR filter using the Steiglitz-McBride algorithm (from [529]). The dot-dashed lines show the model segment boundaries.

for virtual brass instruments. Additional literature relevant to brass instruments includes [187, 186, 188, 247, 530, 319, 331, 336, 535, 536, 537, 445, 93].

9.8 Other Instruments

One of the most interesting aspects of building virtual musical instruments is that every instrument has its own unique characteristics and modeling considerations. One disadvantage of this, however, is that it is not feasible to cover all cases in a single book. This section provides some literature pointers that can hopefully serve as a starting point for further research in some of the instrument areas not discussed here. While these pointers are neither complete nor up to date, the most recent papers and dissertations by the cited researchers can help fill in the gaps. (This is true of the preceding sections as well.)

9.8.1 Singing Voice

An historical summary of voice modeling and synthesis appears in §A.6.3 on page 452.

In [290], a model is proposed for the singing voice in which the driving glottal pulse train is estimated jointly with filter parameters describing the shape of the vocal tract (the complete airway from the base of the throat to the lip opening). The model can be seen as an improvement over linear-predictive coding (LPC) of voice in the direction of a more accurate physical model of voice production, while maintaining a low computational cost relative to more complex articulatory models of voice production. In particular, the parameter estimation involves only convex optimization plus a one-dimensional (possibly non-convex) line search over a compact interval. The line search determines the so-called "open quotient" which is fraction of the time there is glottal flow within each period. The glottal pulse parameters are based on the derivative-glottal-wave models of Liljencrants, Fant, and Klatt [133, 257]. Portions of this research have been published in the ICMC-00 [291] and WASPAA-01 [292] proceedings. Related subsequent work includes [250, 213, 251, 214, 212]

Earlier work in voice synthesis, some summarized in Appendix A, includes [40, 81, 87, 90, 206, 257, 389, 493]; see also the KTH "Research Topics" home page.[28]

---

[28]`http://www.speech.kth.se/music/music_research_topics.html`

9.8.2 Flutes, Recorders, and Pipe Organs

A chapter on the fundamental aero-acoustics of wind instruments appears in [196], and a comprehensive treatment of the acoustics of air jets in recorder-like instruments is given in [531].

A comprehensive review article on lumped models for flue instruments appears [130] in a special issue (July/August 2000) of the *Acustica* journal on "musical wind instrument acoustics." An overview of research on woodwinds and organs appears in [129]. Follow-up publications by this research group include papers concerning the influence of mouth geometry on sound production in flue instruments [108, 418].

9.8.3 Percussion Instruments

While sample-playback synthesis is especially effective for percussion instruments that are supposed to play "in the background," some form of parametrization is needed for the more expressive performances, or for highly variable percussion instruments such as the Indian tabla. Highly efficient computational models of 2D membranes and 3D volumes may be built using the *digital waveguide mesh* [146, 521, 519, 522]. More recently, Lauri and Välimäki have developed a frequency-warping approach to compensating for dispersive wave propagation in a variety of mesh types [398, 399, 401]. The 2001 thesis of Bilbao [54] provides a unified view of the digital waveguide mesh and *wave digital filters* [136] as particular classes of energy invariant finite difference schemes [482]. The problem of modeling diffusion at a mesh boundary was addressed by [268].

Chapter 10

Conclusion

We have visited some basic methods for computational acoustic modeling with applications in digital audio effects and musical instrument modeling. Topics visited included acoustic echo simulation, artificial reverberation, chorus and flanging effects, digital waveguide modeling of strings and winds, lumped modeling, transfer-function modeling, plucked and struck strings, electric and acoustic guitars, piano, woodwinds, bowed strings, and brass instruments.

The remainder of this book consists of a series of appendices which extend and supplement the material introduced thus far. First, some historical background is filled in, followed by a tutorial appendix on underlying physical principles. The next appendix discusses further extensions of the digital waveguide modeling paradigm, followed by an introduction to finite difference schemes, and a discussion of equivalences between the two modeling approaches. Finally, the closely related topic of *wave digital filters* is introduced, followed by some pointers on the Internet.

Appendix A

A History of Enabling Ideas Leading to Virtual Musical Instruments

This appendix summarizes some milestones toward the development of virtual musical instruments. Beginning with early musical acoustics, key developments in the literature deemed most enabling for virtual acoustic instruments are traced up to the mid 1980s, with an emphasis on string models (as in the text overall). Selected follow-on efforts are mentioned. A condensed version of this appendix was published in [446], with associated presentation overheads [448].

A.1 Early Musical Acoustics

> *All things which can be known have number; for it is not possible that without number anything can be either conceived or known.*
> — Philolaus (ca. 400 BC)

Vibrating strings were studied by the Pythagoreans (6th–5th century BC). Pythagorus noticed that harmonics were produced by dividing the string length by whole numbers, and he was interested in understanding consonant pitch intervals in terms of simple ratios of string lengths. "Harmony theory" from the Pythagoreans was taught throughout the Middle Ages as one of the seven liberal arts: the *quadrivium*, consisting of arithmetic, geometry, astronomy, and music (harmony theory); and the *trivium*, consisting of grammar, logic, and rhetoric [411]. The correspondence between musical pitch and frequency of physical vibration was not discovered until the seventeenth century [113].

It took until Galileo (1564–1642) to be free of the formulation of Aristotle (384–322 BC) that all motion required an ongoing applied force, thereby opening the way for modern differential equations of motion. The ideas of Galileo were formalized and extended by Newton (1642–1727), whose famous second law of motion "$f = ma$" lies at the foundation of essentially all classical mechanics and acoustics. Newton's *Principia* (1686) describes sound as traveling pressure pulses, and single-frequency sound waves were analyzed.

The first to publish a one-dimensional wave equation for the vibrating string was the applied mathematician Jean Le Rond d'Alembert (1717–1783) [100, 103].[1] The 1D wave equation can be written as

$$K\frac{\partial^2 y}{\partial x^2} = \epsilon\frac{\partial^2 y}{\partial t^2} \tag{A.1}$$

where

$$\begin{array}{rclcrcl} K & \stackrel{\Delta}{=} & \text{string tension} & \quad & y & \stackrel{\Delta}{=} & y(t,x) \\ \epsilon & \stackrel{\Delta}{=} & \text{linear mass density} & & & = & \text{string displacement,} \end{array}$$

t is time in seconds, and x denotes position along the string. (See discussion in §C.1 on page 523.) It can be derived directly from Newton's second law applied to a differential string element by equating the string restoring force about a small element (first described by Brook Taylor in 1713) to Newton's mass times acceleration of that element, which is how d'Alembert in fact proceeded in his derivation [103]. In addition to introducing the 1D wave equation as an equation among differentials, d'Alembert derived its solution in terms of traveling waves:

$$y(t,x) = y_r\left(t - \frac{x}{c}\right) + y_l\left(t + \frac{x}{c}\right) \tag{A.2}$$

where $c = \sqrt{K/\epsilon}$ denotes the wave propagation speed. Earlier, in 1717, Brook Taylor (1685-1731)[2] published what was apparently the first mathematical paper on vibrating strings, but without introducing differential equations. D'Alembert's ideas were developed into essentially modern form primarily by Euler (1707–1783)[3] [65]. Incidentally, d'Alembert also invented the theory of construction of eyeglasses.[4]

[1] For a short biography of d'Alembert, see `http://www-groups.dcs.st-and.ac.uk/~history/Mathematicians/D'Alembert.html`

[2] `http://`<*ibidem*>`/Taylor.html`

[3] `http://`<*ibidem*>`/Euler.html`

[4] `http://www.mathphysics.com/pde/history.html`

A.2 History of Modal Expansion

While the traveling-wave solution for the ideal vibrating string is fully general, and remains the basis for the most efficient known string synthesis algorithms, an important alternate formulation is in terms of *modal expansions*, that is, a superposition sum of orthogonal basis functions which solve the differential equation and obey the boundary conditions. Daniel Bernoulli (1700-1782) developed the notion that string vibrations can be expressed as the superposition of an infinite number of harmonic vibrations [103].[5] This approach ultimately evolved to Hilbert spaces of orthogonal basis functions that are solutions of Hermitian linear operators—a formulation at the heart of quantum mechanics describing what can be observed in nature [112, 540]. In computational acoustic modeling, Sturm-Liouville theory has been used to give improved models of nonuniform acoustic tubes such as horns [50], and to provide spatial transforms analogous to the Laplace transform [360].

In the field of computer music, the introduction of *modal synthesis* is credited to Adrien [5]. More generally, a modal synthesis model is perhaps best formulated via a diagonalized state-space formulation [559, 220, 107].[6] A more recent technique, which has also been used to derive modal synthesis models, is the so-called *functional transformation method* [360, 500, 499]. In this approach, physically meaningful transfer functions are mapped from continuous to discrete time in a way which preserves desired physical parametric controls.

A.3 Mass-Spring Resonators

Since a harmonic oscillator is produced by a simple mass-spring system, a mechanical generator for the harmonic basis functions of Bernoulli is readily obtained by equating Newton's second law $f = m\ddot{x}$ for the reaction force of an ideal mass m, with Robert Hooke's spring force law $f = -kx$ (published in 1676), where k is an empirical spring constant [65]. Hooke (1635–1703) was a contemporary of Newton's who carried out extensive experiments with springs in search of a spring-regulated clock [259, pp. 274–288]. Hooke's law was generalized to 3D by Cauchy (1789–1857) as the familiar linear relationship between six components of *stress* and *strain*.[7]

Elementary mass-spring models have found much use in computational physical models for purposes of sound synthesis [69, 92]. For example, a mass-spring oscillator is typically used to model a brass-player's lips [4],

[5] `http://www.stetson.edu/~efriedma/periodictable/html/B.html`

[6] `http://ccrma.stanford.edu/~jos/PianoString/`

[7] `http://www.efunda.com/formulae/solid_mechanics/mat_mechanics/hooke.cfm`

piano hammers [44], and is sometimes included in woodwind-reed models [406].

A.4 Sampling Theory

Nowadays, audio processing is typically carried out in discrete time. As a result, *sampling theory* is fundamental to digital audio signal processing. The sampling theorem is credited to Harold Nyquist (1928), extending an earlier result by Cauchy (1831) based on series expansions. Claude Shannon is credited with reviving interest in the sampling theorem after World War II when computers became public.[8] As a result, the sampling theorem is often called "Nyquist's sampling theorem," "Shannon's sampling theorem," or the like. The *sampling rate* has been called the "Nyquist rate" in honor of Nyquist's contributions [333]. Often in common usage, however, the term "Nyquist rate" is used to refer instead to *half* the sampling rate. To preserve the historically correct meaning, we might encourage use of the term *Nyquist limit* to mean half the sampling rate, and simply say "sampling rate" instead of "Nyquist rate", so as to minimize confusion.

A.5 Physical Digital Filters

A.5.1 Analog Computers

Analog filters are inherently "physical" since voltages and currents are direct "analogs" for physical variables such as force and velocity. So-called "analog computers" were often used for physical system simulation before digital computers took over.

A.5.2 Finite Difference Methods

One of the first applications of digital computers to numerical simulation of physical systems was the so-called *finite difference* approach [482]. The general procedure is to replace derivatives by finite differences, and there are many variations on how this can be done. For example, the first partial derivative with respect to time in Eq. (A.1) may be approximated by

$$\dot{y}(t,x) \triangleq \frac{\partial}{\partial t}y(t,x) \approx \frac{y(t,x)-y(t-T,x)}{T} \approx \frac{y(t+T,x)-y(t-T,x)}{2T}. \tag{A.3}$$

The second form, called a *centered finite difference*, has the advantage of not introducing a time delay, but at the expense of requiring an extra factor

[8] http: //cnx.rice.edu/content/m0050/latest/

of two oversampling for a given accuracy in its magnitude response (when viewed as a digital filter).

Finite differences are at least as old as Taylor series, since,

$$y(t-T) = y(t) - T\dot{y}(t) + \cdots \;\Rightarrow\; \dot{y}(t) \approx \frac{y(t) - y(t-T)}{T}. \tag{A.4}$$

Interestingly, the general Taylor series, published in 1715 by Taylor in his book *Methodus Incrementorum Directa et Inversa*, was known more than forty years earlier to James Gregory (1668), somewhat earlier to Jean Bernoulli, and to some extent even before 1550 in India [65, pp. 422,469].

Finite differences were used to construct the earliest known digital models of vibrating strings by Pierre Ruiz and Lejaren Hiller ca. 1971 [194].

Finite-difference methods have not historically been aimed at real-time simulation, and they are generally used with very large sampling rates compared with the "band of interest". On the order of ten-times oversampling is needed to obtain reasonably accurate simulation across the entire audio band when using classical finite difference methods. In the finite-difference method literature, accuracy is usually only considered at dc, which is inaudible. Since finite difference models are usually linear, time-invariant digital filters, it is straightforward to improve them by filter-design methods, optimizing perceived audio error over the entire audio band. Such improved (and optionally extended) coefficients can then be used to construct a refined, indirectly estimated, partial differential equation.

Other offline (slower than real time) computational physical modeling methods include finite element [481] and boundary element methods. Such offline simulations can be valuable as a source of "virtual experiments" which enable the testing and calibration of faster algorithms. As an example, a detailed simulation of guitar acoustics, employing both finite difference and finite element modeling approaches, is reported in [109].

A.5.3 Transfer Function Models

As indicated in the previous section, instead of digitizing a differential equation by finite differences, one can often formulate a *filter design problem*. This is ideal when all that matters is the input-output response of the physical system, and the physical system is linear and time-invariant (LTI). When the desired transfer function spans more than one system element, non-physical models are usually obtained, so we will not consider such models further. However, digital filter design methods optimizing perceptually motivated error criteria are extremely effective in spectral modeling and audio compression applications [337]. They are also good choices for subsystems which are to remain fixed over time, such as cello bodies, piano soundboards, and the like.

A.5.4 Wave Digital Filter Models

Perhaps the best known physics-based approach to digital filter design is *wave digital filters*, developed principally by Alfred Fettweis [136].[9] Wave digital filters may be constructed by applying the bilinear transform [343] to the scattering-theoretic formulation of lumped RLC networks introduced in circuit theory by Vitold Belevitch [34]. Fettweis in fact worked with Belevitch.[10] Scattering theory had been in use for many years prior in quantum mechanics.

A key, driving property of wave digital filters is low sensitivity to coefficient round-off error. This follows from the correspondence to passive circuit networks. Wave digital filters also have the nice property of preserving order of the original (analog) system. For example, a "wave digital spring" is simply a unit delay, and a "wave digital mass" is a unit delay with a sign flip. The only approximation aspect is the frequency-warping caused by the bilinear transform. It is interesting to note that when it is possible to frequency-warp input/output signals exactly, a wave digital filter can implement a continuous-time LTI system exactly! See [55] for a discussion of wave digital filters and their relation to finite differences et al.

In computer music, various "wave digital elements" have been proposed, including wave digital toneholes [528], piano hammers [56], and woodwind bores [526].

A.6 Voice Synthesis

Unquestionably, the most extensive prior work in the 20th century relevant to virtual acoustic musical instruments occurred within the field of *speech synthesis* [139, 142, 363, 408, 335, 106, 243].[11] This research was driven by both academic interest and the potential practical benefits of speech compression to conserve telephone bandwidth. It was clear at an early point that the bandwidth of a telephone channel (nominally 200–3200 Hz) was far greater than the "information rate" of speech. It was reasoned, therefore, that instead of encoding the speech waveform, it should be possible to encode instead more slowly varying parameters of a good synthesis model for speech.

Before the 20th century, there were several efforts to simulate the voice *mechanically*, going back at least until 1779 [140].

---

[9] Derivation: `http://ccrma.stanford.edu/~jos/pasp/Wave_Digital_Filters.html`

[10] `http://www.ieee.org/organizations/history_center/oral_histories/-transcripts/fettweis.html`

[11] The overview [306] of early speech production models is freely available online, thanks to the Smithsonian Speech Synthesis History Project.

A.6.1 Dudley's Vocoder

The first major effort to encode speech electronically was Homer Dudley's *vocoder* ("voice coder") [119] developed starting in October of 1928 at AT&T Bell Laboratories [414]. A manually controlled version of the vocoder synthesis engine, called the *Voder* (Voice Operation Demonstrator [140]), was constructed and demonstrated at the 1939 World's Fairs in New York and San Francisco [119]. Pitch was controlled by a foot pedal, and ten fingers controlled the bandpass gains. Buzz/hiss selection was by means of a wrist bar. Three additional keys controlled transient excitation of selected filters to achieve stop-consonant sounds [140]. "Performing speech" on the Voder required on the order of a year's training before intelligible speech could reliably be produced. The Voder was a very interesting performable instrument!

The vocoder and Voder can be considered based on a *source-filter* model for speech which includes a *non-parametric spectral model* of the vocal tract given by the output of a fixed bandpass-filter-bank over time. Later efforts included the *formant vocoder* (Munson and Montgomery 1950)—a type of parametric spectral model—which encoded F_0 and the amplitude and center-frequency of the first three spectral formants. See [306, p. 2452-3] for an overview and references.

While we have now digressed to into the realm of *spectral models*, as opposed to physical models, it seems worthwhile to point out that the early efforts toward speech synthesis were involved with essentially all of the mainstream sound modeling methods in use today (both spectral and physical domains).

A.6.2 Vocal Tract Analog Models

There is one speech-synthesis thread that clearly classifies under computational physical modeling, and that is the topic of *vocal tract analog models*. In these models, the vocal tract is regarded as a *piecewise cylindrical acoustic tube*. The first mechanical analogue of an acoustic-tube model appears to be a hand-manipulated leather tube built by Wolfgang von Kempelen in 1791, reproduced with improvements by Sir Charles Wheatstone [140]. In *electrical* vocal-tract analog models, the piecewise cylindrical acoustic tube is modeled as a cascade of electrical *transmission line* segments, with each cylindrical segment being modeled as a transmission line at some fixed characteristic impedance. An early model employing four cylindrical sections was developed by Hugh K. Dunn in the late 1940s [120]. An even earlier model based on two cylinders joined by a conical section was published by T. Chiba and M. Kajiyama in 1941 [120]. Cylinder cross-sectional areas A_i were determined based on X-ray images of the vocal tract, and the corresponding characteristic impedances were proportional to $1/A_i$. An impedance-based, lumped-parameter approximation to the transmission-line sections was used in order that analog LC ladders could be used to implement the model electronically. By the 1950s, LC vocal-tract analog models included a side-branch for nasal simulation [131].

The theory of transmission lines is credited to applied mathematician Oliver Heaviside (1850-1925), who worked out the *telegrapher's equations* (sometime after 1874) as an application of Maxwell's equations, which he simplified (sometime after 1880) from the original 20 equations of Maxwell to the modern vector formulation.[12] Additionally, Heaviside is credited with introducing complex numbers into circuit analysis, inventing essentially Laplace-transform methods for solving circuits (sometime between 1880 and 1887), and coining the terms 'impedance' (1886), 'admittance' (1887), 'electret', 'conductance' (1885), and 'permeability' (1885). A little later, Lord Rayleigh worked out the theory of waveguides (1897), including multiple propagating modes and the cut-off phenomenon.[13]

A.6.3 Singing Kelly-Lochbaum Vocal Tract

In 1962, John L. Kelly and Carol C. Lochbaum published a software version of a digitized vocal-tract analog model [245, 246]. This may be the first instance of a *sampled traveling-wave model* of the vocal tract, as opposed to a lumped-parameter transmission-line model. In other words, Kelly and Lochbaum apparently returned to the original acoustic tube model (a sequence of cylinders), obtained d'Alembert's traveling-wave solution in each section, and applied Nyquist's sampling theorem to digitize the system. This sampled, bandlimited approach to digitization contrasts with the use of bilinear transforms as in wave digital filters; an advantage is that the frequency axis is not warped, but it is prone to aliasing when the parameters vary over time (or if nonlinearities are present).

At the junction of two cylindrical tube sections, *i.e.*, at area discontinuities, *lossless scattering* occurs.[14] As mentioned in §A.5.4, reflection/transmission at impedance discontinuities was well formulated in classical network theory [34, 35], and in transmission-line theory.

The Kelly-Lochbaum model can be regarded as a kind of *ladder filter* [297] or, more precisely, using later terminology, a *digital waveguide filter* [433]. Ladder and lattice digital filters can be used to realize arbitrary transfer functions [297], and they enjoy low sensitivity to round-off error, guaranteed stability under coefficient interpolation, and freedom from over-

[12] `http: en.wikipedia.org/wiki/Oliver_Heaviside`

[13] `http:www.microwaves101.com/encyclopedia/history.cfm`

[14] Scattering theory:
`http://ccrma.stanford.edu/~jos/pasp/Scattering_Impedance_Changes.html`

flow oscillations and limit cycles under general conditions. Ladder/lattice filters remain important options when designing fixed-point implementations of digital filters, *e.g.*, in VLSI. In the context of wave digital filters, the Kelly-Lochbaum model may be viewed as a digitized *unit element filter* [136], reminiscent of waveguide filters used in microwave engineering. In more recent terminology, it may be called a *digital waveguide model* of the vocal tract in which the digital waveguides are degenerated to single-sample width [433, 442, 452].

In 1961, Kelly and Lochbaum collaborated with Max Mathews to create what was most likely the first digital physical-modeling synthesis example by any method.[15] The voice was computed on an IBM 704 computer using speech-vowel data from Gunnar Fant's recent book [132]. Interestingly, Fant's vocal-tract shape data were obtained (via x-rays) for Russian vowels, not English, but they were close enough to be understandable. Arthur C. Clarke, visiting John Pierce at Bell Labs, heard this demo, and he later used it in "2001: A Space Odyssey,"—the HAL9000 computer slowly sang its "first song" ("Bicycle Built for Two") as its disassembly by astronaut Dave Bowman neared completion.[16]

Perhaps due in part to J. L. Kelly's untimely death afterward, research on vocal-tract analog models tapered off thereafter, although there was some additional work [306]. Perhaps the main reason for the demise of this research thread was that spectral models (both nonparametric models such as the vocoder, and parametric source-filter models such as linear predictive coding (discussed below)) proved to be more effective when the application was simply speech coding at acceptably low bit rates and high fidelity levels. In telephone speech-coding applications, there was no requirement that a physical voice model be retained for purposes of expressive musical performance. In fact, it was desired to automate and minimize the "performance expertise" required to operate the voice production model. One could go so far as to say that the musical expressivity of voice synthesis models reached their peak in the 1939 Voder and related (manual) systems (§A.6.1).

In computer music, the Kelly-Lochbaum vocal tract model was revived for singing-voice synthesis in the thesis research of Perry Cook [87].[17] In addition to the basic vocal tract model with side branch for the nasal tract, Cook included neck radiation (*e.g.*, for 'b'), and damping extensions. Additional work on incorporating damping within the tube sections was carried out by Amir et al. [15]. Other extensions include sparse acoustic tube modeling [149] and extension to piecewise *conical* acoustic tubes [508]. The

[15] "Bicycle Built for Two" by Kelly, Lochbaum, and Matthews, 1961: `http://ccrma.stanford.edu/~jos/wav/daisy-klm.wav`

[16] `http://www.bell-labs.com/news/1997/march/5/2.html`

[17] `http: www.cs.princeton.edu/~prc/SingingSynth.html` (includes sound examples)

digital waveguide modeling framework [430, 431, 433] can be viewed as an adaptation of extremely sparse acoustic-tube models for artificial reverberation, vibrating strings, and wind-instrument bores.

A.6.4 Linear Predictive Coding of Speech

Approximately a decade after the Kelly-Lochbaum voice model was developed, *Linear Predictive Coding (LPC) of speech* began [20, 296, 297]. The linear-prediction voice model is best classified as a parametric, spectral, source-filter model, in which the short-time spectrum is decomposed into a flat *excitation* spectrum multiplied by a smooth spectral *envelope* capturing primarily vocal *formants* (resonances).

LPC has been used quite often as a spectral transformation technique in computer music, as well as for general-purpose audio spectral envelopes [381], and it remains much used for low-bit-rate speech coding in the variant known as *Codebook Excited Linear Prediction (CELP)* [337].[18] When applying LPC to audio at high sampling rates, it is important to carry out some kind of auditory *frequency warping*, such as according to mel, Bark, or ERB frequency scales [182, 460, 483].

Interestingly, it was recognized from the beginning that the all-pole LPC vocal-tract model could be interpreted as a modified piecewise-cylindrical acoustic-tube model [20, 297], and this interpretation was most explicit when the vocal-tract filters (computed by LPC in direct form) were realized as ladder filters [297]. The physical interpretation is not really valid, however, unless the vocal-tract filter parameters are estimated jointly with a realistic *glottal pulse shape.* LPC demands that the vocal tract be driven by a *flat* spectrum—either an impulse (or low-pitched impulse train) or white noise—which is not physically accurate. When the glottal pulse shape (and lip radiation characteristic) are "factored out", it becomes possible to convert LPC coefficients into vocal-tract shape parameters (area ratios). Approximate results can be obtained by assuming a simple roll-off characteristic for the glottal pulse spectrum (*e.g.*, -12 dB/octave) and lip-radiation frequency response (nominally +6dB /octave), and compensating with a simple preemphasis characteristic (*e.g.*, $12-6 = 6$ dB/octave) [297]. More accurate glottal pulse estimation in terms of parameters of the *derivative-glottal-wave* models by Liljencrants, Fant, and Klatt [133, 257] (still assuming +6dB/octave for lip radiation) was carried out in the thesis research of Vicky Lu [290], and further extension of that work appears in [250, 213, 251].

[18] In particular, CELP is used in the free, open-source speech codec called *Speex* (`http://www.speex.org`).

A.6.5 Formant Synthesis Models

A *formant synthesizer* is a source-filter model in which the source models the glottal pulse train and the filter models the formant resonances of the vocal tract. Constrained linear prediction can be used to estimate the parameters of formant synthesis models, but more generally, formant peak parameters may be estimated directly from the short-time spectrum (*e.g.*, [255]). The filter in a formant synthesizer is typically implemented using cascade or parallel second-order filter sections, one per formant. Most modern rule-based text-to-speech systems descended from software based on this type of synthesis model [255, 256, 257].

Another type of formant-synthesis method, developed specifically for singing-voice synthesis is called the *FOF* method [386]. It can be considered an extension of the *VOSIM* voice synthesis algorithm [219]. In the FOF method, the formant filters are implemented *in the time domain* as parallel second-order sections; thus, the vocal-tract impulse response is modeled a sum of three or so exponentially decaying sinusoids. Instead of driving this filter with a glottal pulse wave, a simple impulse is used, thereby greatly reducing computational cost. A convolution of an impulse response with an impulse train is simply a periodic superposition of the impulse response. In the VOSIM algorithm, the impulse response was trimmed to one period in length, thereby avoiding overlap and further reducing computations.

The FOF method also tapers the beginning of the impulse-response using a rising half-cycle of a sinusoid. This qualitatively reduces the "buzziness" of the sound, and compensates for having replaced the glottal pulse with an impulse. In practice, however, the synthetic signal is matched to the desired signal in the frequency domain, and the details of the onset taper are adjusted to optimize audio quality more generally, including to broaden the formant resonances.

One of the difficulties of formant synthesis methods is that formant parameter estimation is not always easy [408]. The problem is particularly difficult when the fundamental frequency F_0 is so high that the formants are not adequately "sampled" by the harmonic frequencies, such as in high-pitched female voice samples. Formant ambiguities due to insufficient spectral sampling can often be resolved by incorporating additional physical constraints to the extent they are known.

Formant synthesis is an effective combination of physical and spectral modeling approaches. It is a physical model in that there is an explicit division between glottal-flow wave generation and the formant-resonance filter, despite the fact that a physical model is rarely used for either the glottal waveform or the formant resonator. On the other hand, it is a spectral modeling method in that its parameters are estimated by explicitly matching short-time audio spectra of desired sounds. It is usually most effective for any synthesis model, physical or otherwise, to be optimized in the "audio perception" domain to the extent it is known how to do this [312, 165]. For an illustrative example, see, *e.g.*, [201].

A.6.6 Further Reading in Speech Synthesis

More on the history and practice of speech synthesis can be found in, e.g., [139, 140, 142, 408, 335, 106]. A surprisingly complete collection of sound examples[19] spanning the history of speech synthesis can be heard on the CD-ROM accompanying a JASA-87 review article by Dennis Klatt [256]. Sound examples of these and related models in music applications may be found online as well [448].

A.7 String Models

As mentioned in §A.5.2, the first known computational models of vibrating strings were based on finite difference modeling by Ruiz and Hiller [194]. In the field of musical acoustics, software simulations of bowed strings were being carried out by Michael E. McIntyre and James Woodhouse as early as the mid 1970s [307]. In these simulations, the string is represented by the "Green's function" (impulse response) seen by a traveling wave traversing the string once in both directions (*i.e.*, one round trip). This model was evidently based on investigations in the 1960s by John Schelleng, Lothar Cremer, and Cremer's co-worker Hans Lazarus into the behavior of the so-called "corner rounding function" (round-trip impulse response) in the context of bowed-string dynamics [409, 410, 95].[20] The bow-string junction was based on theory worked out by Friedlander [151] and Keller [244], both published in 1953. These mathematical models of bowed-string dynamics were in turn preceded by influential investigations by Raman, published in 1918 [365], and Hermann von Helmholtz, published in 1863 [539]. An important enabling scientific measurement was that of the *friction curve* describing the bow-string contact, and many models, such as in [307, 308] used a hyperbolic friction curve, approximating measurements by Lazarus [95]. Interestingly, while researchers in musical acoustics would often use their models to produce example waveforms illustrating the successful modeling of various visible physical effects, they apparently never listened to them as sound.

Robert T. Schumacher, who had been interested primarily in woodwind musical acoustics, collaborated with McIntyre and Woodhouse, and the

[19] `http://www.cslu.ogi.edu/tts/research/history/`

[20] `http://www.zainea.com/Oscilationsofbowedstring.htm`

result was a generalization of the Friedlander-Keller bowed-string model to wind instruments [308].

Incidentally, electrical equivalent circuits for bowed-string instruments appeared as early as 1952 [332, p. 122], with perhaps the most cited model being Schelleng's published in 1963 [409].

A.8 Karplus-Strong Algorithms

In 1983 the Karplus-Strong [236] and Extended Karplus-Strong (EKS) [207, 428] algorithms were published (in the same issue of the Computer Music Journal). Kevin Karplus and Alex Strong were computer science students at Stanford trying out wavetable synthesis algorithms on an 8-bit microcomputer. The sounds were rather boring, as any precisely periodic sound tends to be, so Strong had the idea of filtering the wavetable by a two-point average, $H(z) = 0.5 + 0.5\,z^{-1}$, on each reading pass (which requires no multiplies), in order to change the timbre dynamically. The result sounded curiously like a string, almost no matter what initial table was used. After some experimentation, they settled on random numbers as the standard excitation. Strong played in a string quartet with David Jaffe, a CCRMA Composition PhD student, and Jaffe decided to use the algorithm in his piece "Silicon Valley Breakdown" (1982) which became a tour de force demonstration of the EKS algorithm. In preparation for the piece, Jaffe approached the author regarding how to obtain better musical control over tuning, brightness, dynamic level, and the like, and the resulting extensions were described in [207, 428]. Having studied the McIntyre-Woodhouse papers, the author recognized the filtered-delay-loop in the Karplus-Strong algorithm as having the same transfer function as that of an idealized vibrating string [428]. This equivalence later led to deriving the Karplus-Strong algorithm as a special case of a digital waveguide string model (see Chapter 6 up to §9.1.4 on page 335 for the derivation). When viewed as a digital waveguide model, the musical quality and expressivity of the EKS algorithm suggested (ca. 1985) that extremely simplified traveling-wave models with sparsely lumped losses and dispersion could capture the necessary dimensions of musical expressivity at very low computational cost.

A.9 Digital Waveguide Models

The *digital waveguide* modeling paradigm was developed in 1985-86 as a guaranteed-stable "construction kit" for lossless reverberator prototypes supporting general feedback topologies [430]. The following year, digital waveguide building blocks were extended to vibrating strings and single-cylinder acoustic tubes (principally for the clarinet family) [431]. A modular synthesis architecture was developed in which various "nonlinear junctions" could be used to excite digital waveguide networks of general design. The iterative Friedlander-Keller solver of [308] was replaced by a look-up table plus a couple of additions and a multiply, facilitating sound synthesis in real time. This computational model was presented, with sound examples, at the 1986 ICMC, and it so happened that Yamaha's chief engineer was in the audience. Perhaps significantly, the FM synthesis patent was nearing the end of its life. Yamaha soon hired some consultants to evaluate waveguide synthesis, and in 1989 they began a strenuous development effort culminating in the VL1 ("Virtual Lead") synthesizer family, introduced in the U.S. at the 1994 NAMM show in LA. The jazz trio demonstrating the VL1 at NAMM created quite a "buzz", and the cover of Keyboard Magazine soon proclaimed it as "The Next Big Thing" in synthesis technology [375]. As it happened, physical modeling synthesis did not immediately meet with large commercial success, perhaps because wavetable synthesis required less computation and was so much easier to "voice", and because memory was fairly inexpensive. The market at large simply has not yet demanded more expressive sound synthesis algorithms than what wavetable synthesis can provide. However, the quality of the VL1 in the hands of skilled players constituted proof-of-concept for many of us, and helped stimulate further academic interest in the approach. Trained performing musicians generally prefer the expressive richness of a computational physical model, and find wavetable synthesis to be comparatively limiting, as typically implemented.

A.10 Summary

This appendix has attempted to trace some of the early technical developments which enabled and shaped virtual acoustic musical instrument technology as addressed in this book, particularly for the case of virtual stringed instruments. For the early period covered (roughly up to the mid 1980s), many lines of developments specifically in the fields of acoustics and musical acoustics *per se* have been omitted. Some recent review articles include [445, 516]. For sound examples and few other recent developments, see the companion overheads for this appendix [448].

Appendix B

Elementary Physics, Mechanics, and Acoustics

This appendix derives some basic results from the field of *physics*, particularly *mechanics* and *acoustics*, which are referenced elsewhere in this book, or which are known to be needed in practical synthesis models.

B.1 Newton's Laws of Motion

Perhaps the most heavily used equation in physics is *Newton's second law of motion*:

$$\boxed{\textit{Force} = \textit{Mass} \times \textit{Acceleration}}$$

That is, when a force is applied to a mass, the mass experiences an acceleration proportional to the applied force. Denoting the mass by m, force at time t by $f(t)$, and acceleration by

$$a(t) \triangleq \ddot{x}(t) \triangleq \frac{d^2x(t)}{dt^2},$$

we have

$$\boxed{f(t) = m\,a(t) = m\,\ddot{x}(t).} \tag{B.1}$$

In this formulation, the applied force $f(t)$ is considered positive in the direction of positive mass-position $x(t)$. The force $f(t)$ and acceleration $a(t)$ are, in general, *vectors* in three-dimensional space $x \in \mathbf{R}^3$. In other words, force and acceleration are generally vector-valued functions of time t. The mass m is a scalar quantity, and can be considered a measure of the *inertia* of the physical system (see §B.1.1 below).

Newton's Three Laws of Motion

Newton's three laws of motion may be stated as follows:

1. Every object in a state of uniform motion will remain in that state of motion unless an external force acts on it.

2. Force equals mass times acceleration [$f(t) = m\,a(t)$].

3. For every action there is an equal and opposite reaction.

The first law, also called the *law of inertia*, was pioneered by Galileo. This was quite a conceptual leap because it was not possible in Galileo's time to observe a moving object without at least some frictional forces dragging against the motion. In fact, for over a thousand years before Galileo, educated individuals believed Aristotle's formulation that, wherever there is motion, there is an external force producing that motion.

The second law, $f(t) = m\,a(t)$, actually implies the first law, since when $f(t) = 0$ (no applied force), the acceleration $a(t)$ is zero, implying a constant velocity $v(t)$. (The velocity is simply the integral with respect to time of $a(t) = \dot{v}(t)$.)

Newton's third law implies *conservation of momentum* [137]. It can also be seen as following from the second law: When one object "pushes" a second object at some (massless) point of contact using an applied force, there must be an equal and opposite force from the second object that cancels the applied force. Otherwise, there would be a nonzero net force on a massless point which, by the second law, would accelerate the point of contact by an infinite amount.

In summary, Newton's laws boil down to $f = ma$. An enormous quantity of physical science has been developed by applying this simple[1] mathematical law to different physical situations.

B.1.1 Mass

Mass is an intrinsic property of matter. From Newton's second law, $f(t) = m\,a(t)$, we have that the amount of force required to accelerate an object, by a given amount, is proportional to its mass. Thus, the mass of an object quantifies its *inertia*—its resistance to a change in velocity.

We can measure the mass of an object by measuring the *gravitational force* between it and another known mass, as described in the next section. This is a special case of measuring its acceleration in response to a known

[1]While this formula seems fairly simple now, in Newton's day, it was necessary to invent *calculus* before it could be stated in this way.

force. Whatever the force f, the mass m is given by f divided by the resulting acceleration a, again by Newton's second law $f = ma$.

The usual mathematical model for an ideal mass is a dimensionless *point* at some location in space. While no real objects are dimensionless, they can often be treated mathematically as dimensionless points located at their *center of mass*, or *centroid* (§B.4.1, p. 470).

The *physical state* of a mass m at time t consists of its *position* $x(t)$ and *velocity* $\dot{x}(t)$ in 3D space. The amount of mass itself, m, is regarded as a fixed parameter that does not change. In other words, the *state* $(x, \dot{x})$ of a physical system typically changes over time, while any *parameters* of the system, such as mass m, remain fixed over time (unless otherwise specified).

B.1.2 Gravitational Force

We are all familiar with the *force of gravity*. It is a fundamental observed property of our universe that any two masses m_1 and m_2 experience an *attracting force* f given by the formula

$$f(t) = G\frac{m_1 m_2}{r^2(t)} \tag{B.2}$$

where $r(t)$ is the distance between the centroids of the masses m_1 and m_2 at time t, and G is the *gravitation constant.*[2]

The law of gravitation Eq. (B.2) can be accepted as an *experimental fact* which defines the concept of a *force*.[3] The giant conceptual leap taken by Newton was that the law of gravitation is *universal*—applying to celestial

[2]The gravitation constant is given in *International Standard Units ("SI units")* by

$$G = 6.672 \times 10^{-11}\ \mathrm{N\ m^2\ kg^{-2}} \quad (= \mathrm{m^3\ kg^{-1}\ s^{-2}})$$

where the following *physical units* abbreviations are used:

$$\begin{aligned}
\mathrm{m} &\triangleq \text{meters (length)}\\
\mathrm{kg} &\triangleq \text{kilograms} = 1000\ \mathrm{g}\ \text{(mass)}\\
\mathrm{g} &\triangleq \text{grams (mass)}\\
\mathrm{s} &\triangleq \text{seconds (time)}\\
\mathrm{N} &\triangleq \text{newtons} = \mathrm{kg\ m\ s^{-2}}\ \text{(force)}
\end{aligned}$$

The physical units for newtons follow from Newton's second law of motion $f = ma$, and the physical units for G follow immediately from Eq. (B.2).

[3]From a modern point of view, all forces are "mediated" by some particle, and there are only four basic forces: The *electromagnetic force* is mediated by the *photon*, the *strong nuclear force* is mediated by the *gluon*, the *weak nuclear force* is mediated the W and Z bosons, and *gravity* is thought to be mediated by the *graviton*, although gravity is not yet incorporated into the *Standard Model* of theoretical physics. Only the electromagnetic and gravitational forces are encountered in everyday life, since the strong and weak nuclear forces decay exponentially on the scale of an atomic nucleus.

bodies as well as objects on earth. When a mass is "dropped" and allowed to "fall" in a gravitational field, it is observed to experience a *uniform acceleration* proportional to its mass. Newton's second law of motion (§B.1 on page 459) quantifies this result.

B.1.3 Hooke's Law

Consider an ideal spring suspending a mass from a rigid ceiling, as depicted in Fig. B.1. Assume the mass is at rest, and that its distance $x(t)$ from the ceiling is fixed.

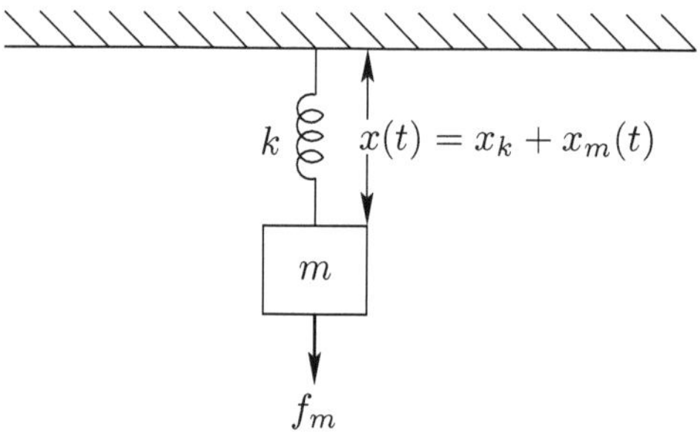

Figure B.1: Mass hung by a spring from a rigid ceiling.

If M denotes the mass of the earth, and r is the distance of mass m's center from the earth's center of mass, then the downward force on the mass m is given by Eq. (B.2) as

$$f_m = gm,$$

where

$$g = G\frac{M}{r^2}$$

is called the *acceleration due to gravity*. Changes in r due to the motion of the mass are assumed negligible relative to the radius of the earth (about 4000 miles), and so g is treated as a constant for most practical purposes near the earth's surface. We see that if we double the mass m, we double the force f_m pulling on the spring. It is an *experimental fact* that typical springs exhibit a displacement x_m that is approximately proportional to the applied force f_m for a wide range of applied forces. This is *Hooke's law* for ideal springs:

$$\boxed{f(t) = k\,x_m(t),} \quad \text{(Hooke's Law)} \tag{B.3}$$

where $x_m(t)$ is the *displacement* of the spring from its natural length. We call k the *spring constant*, or *stiffness* of the spring. In terms of our previous notation, we have

$$x(t) = x_k + x_m(t),$$

where x_k is the length of the spring with no mass attached.

Note that the force on the spring in Fig. B.1 is gravitational force. Equal and opposite to the force of gravity is the *spring force* exerted upward by the spring on the mass (which is not moving). We know that the spring force is equal and opposite to the gravitational force because the mass would otherwise be accelerated by the net force.[4] Therefore, like gravity, a displaced spring can be regarded as a definition of an applied force. That is, whenever you have to think of an applied force, you can always consider it as being delivered by the end of some ideal spring attached to some external physical system.

Note, by the way, that normal interaction forces when objects touch arise from the *Coulomb force* (electrostatic force, or repulsion of like charges) between electron orbitals. This electrostatic force obeys an "inverse square law" like gravity, and therefore also behaves like an ideal spring for small displacements.[5]

The specific value of k depends on the physical units adopted as well as the "stiffness" of the spring. What is most important in this definition of force is that a doubling of spring displacement doubles the force. That is, the spring force is a *linear* function of spring displacement (compression or stretching).

B.1.4 Applying Newton's Laws of Motion

As a simple example, consider a mass m driven along a frictionless surface by an ideal spring k, as shown in Fig. B.2. Assume that the mass position

---

[4]In this example, the mass-spring system is in *equilibrium* (not moving), so all forces in the system must sum to zero. Equilibrium also must hold if the whole system is traveling with a constant velocity; in other words, it is the lack of *relative* motion that matters.

[5]To show this quantitatively, consider two electrons, each having electric charge e, separated by r meters. Then the Coulomb force between the electrons is proportional to e^2/r^2. Adding a small perturbation Δr to r yields a new force proportional to

$$\frac{e^2}{(r+\Delta r)^2} = \frac{e^2}{r^2}\left(1+\frac{\Delta r}{r}\right)^{-2} = \frac{e^2}{r^2}\left[1-2\frac{\Delta r}{r}+\mathcal{O}\left(\frac{\Delta r}{r}\right)^2\right] \approx \frac{e^2}{r^2} + \underbrace{\left(-\frac{2}{r^3}\right)}_{k}\Delta r,$$

where we used the *binomial expansion* to obtain the approximation. (The notation $\mathcal{O}(\epsilon^2)$ means "terms order of ϵ^2", and such terms approach zero as fast as ϵ^2 approaches zero.) Thus, the effective spring constant connecting the two electrons is $k = -2/r^3$, when $|\Delta r| \ll r$.

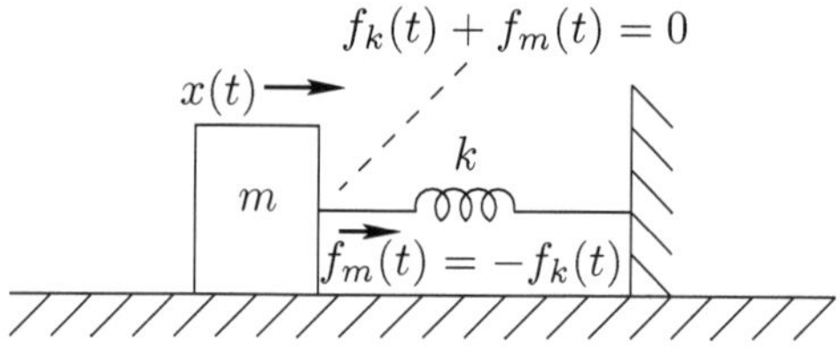

Figure B.2: Mass-spring system.

$x = 0$ corresponds to the spring at rest, *i.e.*, not stretched or compressed. The force necessary to compress the spring by a distance x is given by *Hooke's law* (§B.1.3):

$$f_k(t) = -k\,x(t)$$

This force is balanced at all times by the inertial force $f_m(x) = -m\ddot{x}$ of the mass m, *i.e.* $f_k + f_m = 0$, yielding[6]

$$m\ddot{x}(t) + k\,x(t) = 0 \quad \forall t \geq 0, \quad x(0) = A, \quad \dot{x}(0) = 0, \tag{B.4}$$

where we have defined A as the initial displacement of the mass along x. This is a *differential equation* whose solution gives the equation of motion of the mass-spring junction for all time:[7]

$$x(t) = A\cos(\omega_0 t), \quad \forall t \geq 0, \tag{B.5}$$

where $\omega_0 \triangleq \sqrt{k/m}$ denotes the *frequency of oscillation* in radians per second. More generally, the complete space of solutions to Eq. (B.4), corresponding to all possible initial displacements $x(0)$ and initial velocities $\dot{x}(0)$, is the set of all sinusoidal oscillations at frequency ω_0:

$$x(t) = A\cos(\omega_0 t + \phi), \quad \forall A, \phi \in \mathbf{R}.$$

The amplitude of oscillation A and phase offset ϕ are determined by the *initial conditions*, *i.e.*, the initial position $x(0)$ and initial velocity $\dot{x}(0)$ of the mass (its *initial state*) when we "let it go" or "push it off" at time $t = 0$.

[6]A more formal methodology for arriving at differential equations by applying Newton's law and Hooke's law is described in Appendix F. In the present example, consideration of the underlying physics should convince you that the signs in Eq. (B.4) are correct. For example, when x is positive, the spring must push the mass back to the left. Therefore, $m\ddot{x} = -kx$ means the mass will be accelerated to the left.

[7]See, *e.g.*, [449, p. 313–316] for a derivation, and Appendix F, specifically §F.3.6 on page 737, for related analysis.

B.2 Work = Force times Distance = Energy

Work is defined as *force times distance*. Work is a measure of the *energy* expended in applying a force to move an object.[8]

The work required to compress a spring k through a displacement of x meters, starting from rest, is then

$$W_k(x) = \int_0^x k\,\xi\,d\xi = \frac{1}{2}k\,x^2. \tag{B.6}$$

Work can also be negative. For example, when uncompressing an ideal spring, the (positive) work done by the spring on its moving end support can be interpreted also as saying that the end support performs negative work on the spring as it allows the spring to uncompress. When negative work is performed, the driving system is always accepting energy from the driven system. This is all simply accounting. Physically, one normally considers the *driver* as the agent performing the positive work, *i.e.*, the one expending energy to move the driven object. Thus, when allowing a spring to uncompress, we consider the spring as performing (positive) work on whatever is attached to its moving end.

During a sinusoidal mass-spring oscillation, as derived in §B.1.4, each period of the oscillation can be divided into equal sections during which either the mass performs work on the spring, or vice versa.

Gravity, spring forces, and electrostatic forces are examples of *conservative forces*. Conservative forces have the property that the work required to move an object from point a to point b, either with or against the force, depends only on the locations of points a and b in space, not on the path taken from a to b.

B.2.1 Potential Energy in a Spring

When compressing an ideal spring, *work* is performed, and this work is *stored* in the spring in the form of what we call *potential energy*. Equation (B.6) above gives the quantitative formula for the potential energy $W_k(x)$ stored in an ideal spring after it has been compressed x meters from rest.

B.2.2 Kinetic Energy of a Mass

Kinetic energy is energy associated with *motion*. For example, when a spring uncompresses and accelerates a mass, as in the configuration of

[8] In SI units, *work* is in units called *joules* (abbreviated J). Thus, joules are newtons times meters. Power in watts is defined as joules per second. One joule is the energy dissipated in one second by an electric current of one ampere (coulomb per second) through a resistance of one ohm. For a list of physical units and their abbreviations, see, for example, `http://physics.nist.gov/cuu/Units/units.html`.

Fig. B.2, work is performed on the mass by the spring, and we say that the potential energy of the spring is converted to *kinetic energy* of the mass.

Suppose in Fig. B.2 we have an initial spring compression by A meters at time $t = 0$, and the mass velocity is zero at $t = 0$. Then from the equation of motion Eq. (B.5), we can calculate when the spring returns to rest ($x(t) = 0$). This first happens at the first zero of $\cos(\omega_0 t)$, which is time $t = (\pi/2)/\omega_0 = (\pi/2)\sqrt{m/k}$. At this time, the velocity, given by the time-derivative of Eq. (B.5),

$$v(t) = -A\omega_0 \sin(\omega_0 t),$$

can be evaluated at $t = (\pi/2)/\omega_0$ to yield the mass velocity $v[(\pi/2)/\omega_0] = -A\omega_0 = -A\sqrt{k/m}$, which is when all potential energy from the spring has been converted to kinetic energy in the mass. The square of this value is

$$v_{\mathrm{max}}^2 = A^2\omega_0^2 = A^2\frac{k}{m},$$

and we see that if we multiply v_{max}^2 by $m/2$, we get

$$\frac{1}{2}m\,v_{\mathrm{max}}^2 = \frac{1}{2}k\,A^2,$$

which is the initial potential energy stored in the spring. We require this result. Therefore, the kinetic energy of a mass must be given by

$$E_m(v) = \frac{1}{2}m\,v^2$$

in order that the kinetic energy of the mass when spring compression is zero equals the original potential energy in the spring when the kinetic energy of the mass was zero. In the next section we derive this result in a more general way.

B.2.3 Mass Kinetic Energy from Virtual Work

From Newton's second law, $f = ma = m\ddot{x}$ (introduced in Eq. (B.1) on page 459), we can use d'Alembert's idea of *virtual work* to derive the formula for the kinetic energy of a mass given its speed $v = \dot{x}$. Let dx denote a small (infinitesimal) displacement of the mass in the x direction. Then we have, using the calculus of differentials,

$$\begin{aligned} f(t) &= m\,\ddot{x}(t) \\ \Rightarrow \quad dW \triangleq f\,dx &= m\,\ddot{x}\,dx = m\,d\left(\frac{1}{2}\dot{x}^2\right) \\ &= d\left(\frac{1}{2}m\,v^2\right). \end{aligned}$$

Thus, by Newton's second law, a differential of work dW applied to a mass m by force f through distance dx boosts the kinetic energy of the mass by $d(m\,v^2/2)$. The kinetic energy of a mass moving at speed v is then given by the integral of all such differential boosts from 0 to v:

$$E_m(v) = \int_0^v dW = \int_0^v d\left(\frac{1}{2}m\nu^2\right) = \frac{1}{2}mv^2 = \frac{1}{2}m\,\dot{x}^2,$$

where $E_m(v)$ denotes the *kinetic energy* of mass m traveling at speed v.

The quantity $dW = f\,dx$ is classically called the *virtual work* associated with force f, and dx a *virtual displacement* [545].

B.2.4 Energy in the Mass-Spring Oscillator

Summarizing the previous sections, we say that a compressed spring holds a *potential energy* equal to the *work* required to compress the spring from rest to its current displacement. If a compressed spring is allowed to expand by pushing a mass, as in the system of Fig. B.2 on page 464, the potential energy in the spring is converted to *kinetic energy* in the moving mass.

We can draw some inferences from the oscillatory motion of the mass-spring system written in Eq. (B.5):

$$x(t) = A\cos(\omega_0 t), \quad t \geq 0$$

- From a global point of view, we see that *energy is conserved*, since the oscillation never decays.

- At the *peaks* of the displacement $x(t)$ (when $\cos(\omega_0 t)$ is either 1 or -1), all energy is in the form of *potential* energy, *i.e.*, the spring is either maximally compressed or stretched, and the mass is momentarily stopped as it is changing direction.

- At the *zero-crossings* of $x(t)$, the spring is momentarily relaxed, thereby holding no potential energy; at these instants, all energy is in the form of *kinetic* energy, stored in the motion of the mass.

- Since total energy is conserved (§B.2.5), the kinetic energy of the mass at the displacement zero-crossings is exactly the amount needed to stretch the spring to displacement $-A$ (or compress it to $+A$) before the mass stops and changes direction. At all times, the total energy E is equal to the sum of the potential energy $E_k(t)$ stored in the spring, and the kinetic energy $E_m(t)$ stored in the mass:

$$E = E_k(t) + E_m(t) = \text{constant } [E_k(0)]$$

Regarding the last point, the potential energy, $E_k(t) = k\,x^2(t)/2$ was defined as the work required to displace the spring by x meters, where work was defined in Eq. (B.6). The kinetic energy of a mass m moving at speed v was found to be $E_m(t) = m\,v^2(t)/2 = m\,\dot{x}^2/2$. The constance of the potential plus kinetic energy at all times in the mass-spring oscillator is easily obtained from its equation of motion using the trigonometric identity $\cos^2(\theta) + \sin^2(\theta) = 1$ (see Problem 3).

B.2.5 Energy Conservation

It is a remarkable property of our universe that *energy is conserved* under all circumstances. There are no known exceptions to the conservation of energy, even when relativistic and quantum effects are considered.[9]

Energy may be defined as the ability to do *work*, where work may be defined as *force times distance* (§B.2).

B.2.6 Energy Conservation in the Mass-Spring System

Recall that Newton's second law applied to a mass-spring system, as in §B.1.4 on page 463, yields

$$f_m(t) + f_k(t) = 0, \quad \forall t,$$

which led to the differential equation obeyed by the mass-spring system:

$$m\ddot{x}(t) + k\,x(t) = 0 \quad \forall t$$

Multiplying through by $\dot{x}(t) = v(t)$ gives

$$\begin{aligned}
0 &= m\ddot{x}(t)\dot{x}(t) + k\,x(t)\dot{x}(t) \\
&= m\dot{v}(t)v(t) + k\,x(t)\dot{x}(t) \\
&= \frac{d}{dt}\left[\frac{1}{2}mv^2(t)\right] + \frac{d}{dt}\left[\frac{1}{2}kx^2(t)\right] \\
&= \frac{d}{dt}\left[E_m(t) + E_k(t)\right] \\
&= \frac{d}{dt}E(t).
\end{aligned}$$

Thus, Newton's second law and Hooke's law imply conservation of energy in the mass-spring system of Fig. B.2.

[9] Ok, there *does* exist energy fluctuation on the scale of "Heisenberg uncertainty". That is, it is possible to have a violation in energy conservation by an amount ΔE over a time duration Δt, provided $\Delta E \cdot \Delta t$ is sufficiently small (on the order of Planck's constant h). This can be considered a form of the Heisenberg uncertainty principle.

B.3 Momentum

The *momentum* of a mass m is usually defined by

$$p = mv,$$

where v is the velocity of the mass. Note that momentum is a vector quantity in general, so we should more clearly write

$$\underline{p} = m\underline{v},$$

where $\underline{p}$ and $\underline{v}$ are *vectors* in 3D space. We will return to vector momentum in §B.4.1 below.

B.3.1 Conservation of Momentum

Like energy, momentum is *conserved* in physical systems. For example, when two masses collide and recoil from each other, the total momentum after the collision equals that before the collision. Since the momentum is a three-dimensional vector in Euclidean space, momentum conservation provides three simultaneous equations, in general.[10]

B.4 Rigid-Body Dynamics

Below are selected topics from *rigid-body dynamics*, a subtopic of *classical mechanics* involving the use of Newton's laws of motion to solve for the motion of *rigid bodies* moving in 1D, 2D, or 3D space.[11] We may think of a rigid body as a *distributed mass*, that is, a mass that has length, area, and/or volume rather than occupying only a single point in space. Rigid body models have application in stiff strings (modeling them as disks of

---

[10] Conservation of energy and momentum are unified in what is called the "four-momentum" of spacetime, $P = [E/c, p_x, p_y, p_z]$, often used in special relativity calculations (see, *e.g.*, `http://en.wikipedia.org/wiki/Four-momentum`). Thus, energy (divided by the speed of light c) is just the 0th component of the four-momentum.

[11] Additionally, mechanics problems are formulated using the *principle of least action*, in which the *action* is defined as the time integral of the *Lagrangian* (usually the kinetic energy minus potential energy). Equations of motion, including Newton's second law, are obtained by finding the path which minimizes this action integral. The two main resulting formulations for the equations of motion (in addition to Newton's formulation) are called the *Lagrangian* and *Hamiltonian* formulations, and they generalize more cleanly to constrained motion, more general *phase-space* coordinate systems, relativistic invariance, and problems in quantum mechanics. Another nice implication of the principle of least action is that the conservation of energy follows from homogeneity of time, and conservation of momentum from isotropy of space [270]. We will be fine with the classical Newtonian formulation here.

mass interconnect by ideal springs), rigid bridges, resonator braces, and so on.

We have already used Newton's $f = ma$ to formulate mathematical dynamic models for the ideal *point-mass* (§B.1.1), *spring* (§B.1.3), and a simple mass-spring system (§B.1.4). Since many physical systems can be modeled as assemblies of masses and (normally damped) springs, we are pretty far along already. However, when the springs interconnecting our point-masses are very stiff, we may approximate them as rigid to simplify our simulations. Thus, rigid bodies can be considered mass-spring systems in which the springs are so stiff that they can be treated as rigid massless rods (infinite spring-constants k, in the notation of §B.1.3).

So, what is new about *distributed* masses, as opposed to the point-masses considered previously? As we will see, the main new ingredient is *rotational dynamics*. The *total momentum* of a rigid body (distributed mass) moving through space will be described as the *sum* of the *linear momentum* of its *center of mass* (§B.4.1 below) plus the *angular momentum* about its center of mass (§B.4.13 below).

B.4.1 Center of Mass

The *center of mass* (or *centroid*) of a rigid body is found by averaging the spatial points of the body $\underline{x}_i \in \mathbf{R}^3$ weighted by the mass m_i of those points:[12]

$$\underline{x}_c \triangleq \sum_{i=1}^{N} m_i \underline{x}_i \bigg/ \sum_{i=1}^{N} m_i$$

Thus, the center of mass is the *mass-weighted average location* of the object. For a continuous mass distribution totaling up to M, we can write

$$\underline{x}_c \triangleq \frac{1}{M}\int_V \underline{x}\, dm(\underline{x}) = \int_V \underline{x}\, \rho(\underline{x})\, dV = \iiint_{\underline{x}\in V} \underline{x}\, \rho(\underline{x})\, dx\, dy\, dz$$

where the volume integral is taken over a volume V of 3D space that includes the rigid body, and $dm(\underline{x}) = m(\underline{x})dV = m(\underline{x})\, dx\, dy\, dz$ denotes the mass contained within the differential volume element dV located at the point $\underline{x} \in \mathbf{R}^3$, with $\rho(\underline{x})$ denoting the *mass density* at the point $\underline{x}$. The total mass is

$$M = \int_V dm(\underline{x}) = \int_V \rho(\underline{x})\, dV.$$

[12] We denote by $\mathbf{R}^3$ the set of all points in 3D Euclidean space. Thus $\underline{x}_i$ has three coordinates, usually denoted x, y, and z, or (as we will use below), x_1, x_2, and x_3, all real scalars ($x_i \in \mathbf{R}$).

A nice property of the center of mass is that gravity acts on a far-away object as if all its mass were concentrated at its center of mass. For this reason, the center of mass is often called the *center of gravity*.

Linear Momentum of the Center of Mass

Consider a system of N point-masses m_i, each traveling with vector velocity $\underline{v}_i$, and not necessarily rigidly attached to each other. Then the total momentum of the system is

$$\underline{p} = \sum_{i=1}^{N} m_i \underline{v}_i = \sum_{i=1}^{N} m_i \dot{\underline{x}}_i = M\frac{d}{dt}\left(\frac{1}{M}\sum_{i=1}^{N} m_i \underline{x}_i\right) = M\frac{d}{dt}\underline{x}_c \stackrel{\Delta}{=} M\underline{v}_c$$

where $M = \sum m_i$ denotes the total mass, and $\underline{v}_c$ is the velocity of the center of mass.

Thus, the *momentum* $\underline{p}$ of any collection of masses m_i (including rigid bodies) equals the total mass M times the *velocity of the center-of-mass.*

Whoops, No Angular Momentum!

The previous result might be surprising since we said at the outset that we were going to *decompose* the total momentum into a sum of linear plus angular momentum. Instead, we found that the total momentum is simply that of the center of mass, which means any angular momentum that might have been present just went away. (The center of mass is just a point that cannot rotate in a measurable way.) Angular momentum does not contribute to linear momentum, so it provides three new "degrees of freedom" (three new energy storage dimensions, in 3D space) that are "missed" when considering only linear momentum.

To obtain the desired decomposition of momentum into linear plus angular momentum, we will choose a fixed reference point in space (usually the center of mass) and then, with respect to that reference point, decompose an arbitrary mass-particle travel direction into the sum of two mutually orthogonal vector components: one will be the vector component pointing *radially* with respect to the fixed point (for the "linear momentum" component), and the other will be the vector component pointing *tangentially* with respect to the fixed point (for the "angular momentum"), as shown in Fig. B.3. When the reference point is the center of mass, the resultant radial force component gives us the force on the center of mass, which creates *linear momentum,* while the net tangential component (times distance from the center-of-mass) give us a resultant *torque* about the reference point, which creates *angular momentum.* As we saw above, because the tangential force component does not contribute to linear momentum, we can simply sum the external force vectors and get the same result as summing their radial components. These topics will be discussed further below, after some elementary preliminaries.

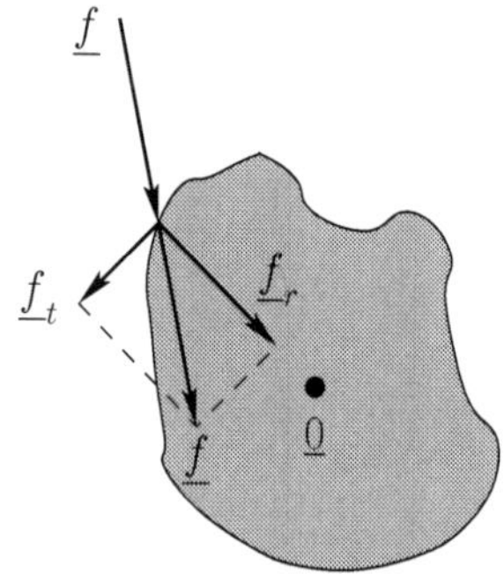

Figure B.3: Rigid body having center of mass at $\underline{0}$, experiencing an external force $\underline{f}$ that can be expressed as the sum of radial and tangential components $\underline{f} = \underline{f}_r + \underline{f}_t$. The radial component $\underline{f}_r$ accelerates the center-of-mass, while the tangential component $\underline{f}_t$ causes rotation about the center of mass.

B.4.2 Translational Kinetic Energy

The *translational kinetic energy* of a collection of masses m_i is given by

$$E_K = \frac{1}{2} M v_c^2$$

where $M = \sum_i m_i$ is the total mass, and v_c denotes the speed of the center-of-mass. We have $v_c \triangleq \| \underline{v}_c \|$, where $\underline{v}_c$ is the velocity of the center of mass.

More generally, the *total energy* of a collection of masses (including distributed and/or rigidly interconnected point-masses) can be expressed as the *sum* of the *translational* and *rotational* kinetic energies [270, p. 98].

B.4.3 Rotational Kinetic Energy

The *rotational kinetic energy* of a rigid assembly of masses (or mass distribution) is the sum of the rotational kinetic energies of the component masses. Therefore, consider a point-mass m rotating[13] in a circular orbit of

[13] We are generally concerned with the rotation of a rigid body about some axis passing through its center of mass. Therefore, we will not distinguish between "rotating," as the earth does to produce night and day, and "revolving," as the earth does around the sun to produce the seasons. Thus, the individual mass particles of a rotating rigid body

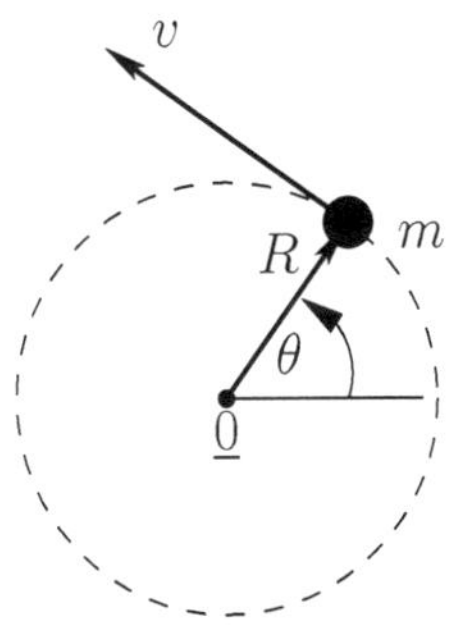

Figure B.4: Point-mass m rotating in a circle of radius R with tangential speed $v = R\omega$, where $\omega = \dot{\theta}$ denotes the angular velocity in rad/s.

radius R and angular velocity ω (radians per second), as shown in Fig. B.4. To make it a closed system, we can imagine an effectively infinite mass at the origin $\underline{0}$. Then the speed of the mass along the circle is $v = R\omega$, and its kinetic energy is $(1/2)mv^2 = (1/2)mR^2\omega^2$. Since this is what we want for the *rotational* kinetic energy of the system, it is convenient to define it in terms of angular velocity ω in radians per second. Thus, we write

$$E_R = \frac{1}{2}I\omega^2, \tag{B.7}$$

where

$$I = mR^2 \tag{B.8}$$

is called the *mass moment of inertia.*

B.4.4 Mass Moment of Inertia

The *mass moment of inertia I* (or simply *moment of inertia*), plays the role of *mass* in rotational dynamics, as we saw in Eq. (B.7) above.

The mass moment of inertia of a rigid body, relative to a given axis of rotation, is given by a weighted sum over its mass, with each mass-point weighted by the square of its distance from the rotation axis. Compare this with the center of mass (§B.4.1) in which each mass-point is weighted by its vector location in space (and divided by the total mass).

Equation (B.8) above gives the moment of inertia for a single point-mass m rotating a distance R from the axis to be $I = mR^2$. Therefore, for a rigid collection of point-masses m_i, $i = 1,\dots,N$,[14] the moment of

will be said to *rotate* about the axis of rotation, even though "revolve" might win more points on a middle-school exam.

[14]To create a rigid collection of point masses, we can imagine them to be interconnected by "massless rigid rods" or, equivalently, ideal springs having infinite stiffness.

inertia about a given axis of rotation is obtained by adding the component moments of inertia:

$$I = m_1 R_1^2 + m_2 R_2^2 + \cdots + m_N R_N^2, \tag{B.9}$$

where R_i is the distance from the axis of rotation to the ith mass.

For a continuous mass distribution, the moment of inertia is given by integrating the contribution of each differential mass element:

$$I = \int_M R^2 dm, \tag{B.10}$$

where R is the distance from the axis of rotation to the mass element dm. In terms of the *density* $\rho(\underline{x})$ of a continuous mass distribution, we can write

$$I = \int_V R^2(\underline{x})\, \rho(\underline{x})\, dV,$$

where $\rho(\underline{x})$ denotes the mass density (kg/m^3) at the point $\underline{x}$, and $dV = dx\,dy\,dz$ denotes a differential volume element located at $\underline{x} \in \mathbf{R}^3$.

Circular Disk Rotating in Its Own Plane

For example, the moment of inertia for a uniform circular disk of total mass M and radius R, rotating in its own plane about a rotation axis piercing its center, is given by

$$I = \frac{M}{\pi R^2}\int_{-\pi}^{\pi}\int_0^R r^2\, r\, dr\, d\theta = \frac{2M}{R^2}\int_0^R r^3 dr = \frac{2M}{R^2}\frac{1}{4}R^4 = \frac{1}{2}MR^2.$$

Circular Disk Rotating About Its Diameter

The moment of inertia for the same circular disk rotating about an axis in the plane of the disk, passing through its center, is given by

$$I = \frac{M}{\pi R^2}\cdot 4\int_0^{\pi/2}\int_0^R [r\cos(\theta)]^2\, r\, dr\, d\theta = \frac{1}{4}MR^2$$

Thus, the uniform disk's moment of inertia in its own plane is twice that about its diameter.

B.4.5 Perpendicular Axis Theorem

In general, for any 2D distribution of mass, the moment of inertia about an axis orthogonal to the plane of the mass equals the sum of the moments of inertia about any two mutually orthogonal axes in the plane of the mass intersecting the first axis. To see this, consider an arbitrary mass element dm having rectilinear coordinates $(x_1, x_2, 0)$ in the plane of the mass. (All three coordinate axes intersect at a point in the mass-distribution plane.) Then its moment of inertia about the axis orthogonal to the mass plane is $dm\,(x_1^2 + x_2^2)$ while its moment of inertia about coordinate axes within the mass-plane are respectively $dm\,x_1^2$ and $dm\,x_2^2$. This, the perpendicular axis theorem is an immediate consequence of the Pythagorean theorem for right triangles.

B.4.6 Parallel Axis Theorem

Let I_0 denote the moment of inertia for a rotation axis passing through the center of mass, and let I_d denote the moment of inertia for a rotation axis parallel to the first but a distance d away from it. Then the *parallel axis theorem* says that

$$I_d = I_0 + Md^2$$

where M denotes the total mass. Thus, the added inertia due to displacement by d meters away from the centroidal axis is equal to that of a *point mass* M rotating a distance d from the center of rotation.

B.4.7 Stretch Rule

Note that the moment of inertia does not change when masses are moved along a vector parallel to the axis of rotation (see, *e.g.*, Eq. (B.9) on page 474). Thus, any rigid body may be "stretched" or "squeezed" parallel to the rotation axis without changing its moment of inertia. This is known as the *stretch rule*, and it can be used to simplify geometry when finding the moment of inertia.

For example, we saw in §B.4.4 that the moment of inertia of a point-mass m a distance R from the axis of rotation is given by $I = mR^2$. By the stretch rule, the same applies to an ideal *rod* of mass m parallel to and distance R from the axis of rotation.

Note that mass can be also be "stretched" along the circle of rotation without changing the moment of inertia for the mass about that axis. Thus, the point mass m can be stretched out to form a mass *ring* at radius R about the axis of rotation without changing its moment of inertia about that axis. Similarly, the ideal rod of the previous paragraph can be stretched tangentially to form a *cylinder* of radius R and mass m, with its axis of symmetry coincident with the axis of rotation. In all of these examples, the moment of inertia is $I = mR^2$ about the axis of rotation.

B.4.8 Area Moment of Inertia

The *area moment of inertia* is the second moment of an area A around a given axis:

$$I_A = \int_A R^2(\underline{x})dA$$

where dA denotes a differential element of the area (summing to A), and $R(\underline{x})$ denotes its distance from the axis of rotation.

Comparing with the definition of *mass moment of inertia* in §B.4.4 above, we see that mass is replaced by area in the area moment of inertia.

In a planar mass distribution with total mass M uniformly distributed over an area A (*i.e.*, a constant mass density of $\rho = M/A$), the mass moment of inertia I_ρ is given by the area moment of inertia I_A times mass-density ρ:

$$I_\rho \isdef \int_M R^2 dM = \int_A R^2 \rho\, dA = \rho I_A$$

B.4.9 Radius of Gyration

For a planar distribution of mass rotating about some axis in the plane of the mass, the *radius of gyration* is the distance from the axis that all mass can be concentrated to obtain the same mass moment of inertia. Thus, the radius of gyration is the "equivalent distance" of the mass from the axis of rotation. In this context, *gyration* can be defined as *rotation* of a planar region about some axis lying in the plane.

For a bar cross-section with area S, the radius of gyration is given by

$$R_g = \sqrt{\frac{I_S}{S}} \tag{B.11}$$

where I_S is the area moment of inertia (§B.4.8) of the cross-section about a given axis of rotation lying in the plane of the cross-section (usually passing through its centroid):

$$I_S = \int_S R^2 dS,$$

where R denotes the distance of the differential area element dS from the axis of gyration.

Rectangular Cross-Section

For a rectangular cross-section of height h and width w, area $S = hw$, the area moment of inertia about the horizontal midline is given by

$$I_w = w\int_{-h/2}^{h/2} y^2 dy = w\,\frac{1}{3}y^3\bigg|_{-h/2}^{h/2} = \frac{Sh^2}{12}.$$

The radius of gyration about this axis is then

$$R_g = \sqrt{\frac{I_w}{S}} = \sqrt{\frac{h^2}{12}} = \frac{h}{2\sqrt{3}}.$$

Similarly, the radius of gyration about a vertical axis passing through the center of the cross-section is $R_g = w/(2\sqrt{3})$.

The radius of gyration can be thought of as the "effective radius" of the mass distribution with respect to its inertial response to rotation ("gyration") about the chosen axis.

Most cross-sectional shapes (*e.g.*, rectangular), have at least two radii of gyration. A circular cross-section has only one, and its radius of gyration is equal to half its radius, as shown in the next section.

Circular Cross-Section

For a circular cross-section of radius a, Eq. (B.11) tells us that the squared radius of gyration about any line passing through the center of the cross-section is given by

$$\begin{aligned}
R_g^2 &= \frac{1}{\pi a^2}\int_{-a}^{a} dx \int_{-\sqrt{a^2-x^2}}^{\sqrt{a^2-x^2}} y^2 dy = \frac{4}{\pi a^2}\int_0^a dx \int_0^{\sqrt{a^2-x^2}} y^2 dy \\
&= \frac{4}{\pi a^2}\int_0^a \left[\frac{1}{3}(a^2-x^2)^{\frac{3}{2}}\right] dx = \frac{4a}{3\pi}\int_0^a \left[1-\left(\frac{x}{a}\right)^2\right]^{\frac{3}{2}} dx \\
&\qquad [\text{let } \sin(\theta) = x/a \Rightarrow dx = a\cos(\theta)d\theta] \\
&= \frac{4a^2}{3\pi}\int_0^{\frac{\pi}{2}} \cos^4(\theta)d\theta.
\end{aligned}$$

Using the elementrary trig identity $\cos(2\theta) = 2\cos^2(\theta)-1$, we readily derive

$$\cos^4(\theta) = \frac{1}{8}\cos(4\theta) + \frac{1}{2}\cos(2\theta) + \frac{3}{8}.$$

The first two terms of this expression contribute zero to the integral from 0 to $\pi/2$, while the last term contributes $3\pi/16$, yielding

$$R_g^2 = \frac{4a^2}{3\pi}\frac{3\pi}{16} = \frac{a^2}{4}.$$

Thus, the radius of gyration about any midline of a circular cross-section of radius a is

$$R_g = \frac{a}{2}.$$

For a circular *tube* in which the mass of the cross-section lies within a circular *annulus* having inner radius b and outer radius a, the radius of gyration is given by

$$R_g = \frac{\sqrt{a^2 - b^2}}{2}. \tag{B.12}$$

B.4.10 Two Masses Connected by a Rod

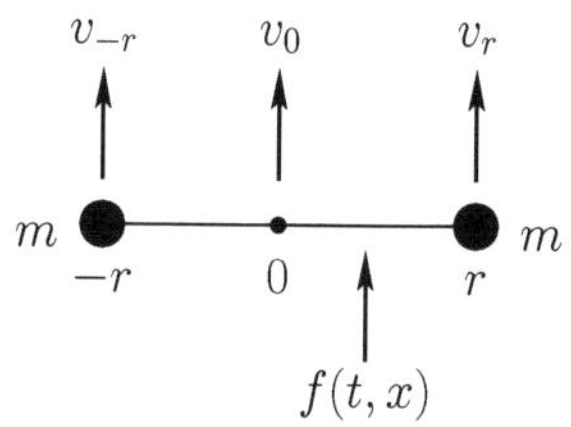

Figure B.5: Two ideal point-masses m connected by an ideal, rigid, massless rod of length $2r$.

As an introduction to the decomposition of rigid-body motion into *translational* and *rotational* components, consider the simple system shown in Fig. B.5. The excitation force density[15] $f(t,x)$ can be applied anywhere between $x = -r$ and $x = r$ along the connecting rod. We will deliver a vertical impulse of momentum to the mass on the right, and show, among other observations, that the total kinetic energy is split equally into (1) the *rotational* kinetic energy about the center of mass, and (2) the *translational* kinetic energy of the total mass, treated as being located at the center of mass. This is accomplished by defining a *new frame of reference* (*i.e.*, a moving coordinate system) that has its origin at the center of mass.

First, note that the driving-point impedance (§7.1, p. 252) "seen" by the driving force $f(t,x)dx$ varies as a function of x. At $x = 0$, The excitation $f(t,0)dx$ sees a "point mass" $2m$, and no rotation is excited by the force (by symmetry). At $x = \pm R$, on the other hand, the excitation $f(t,\pm R)dx$ only sees mass m at time 0, because the vertical motion of either point-mass initially only rotates the other point-mass via the massless connecting rod. Thus, an observation we can make right away is that the *driving point impedance seen by $f(t,x)$ depends on the striking point x and, away from $x = 0$, it depends on time t as well.*

To avoid dealing with a time-varying driving-point impedance, we will use an impulsive force input at time $t = 0$. Since momentum is the time-

[15]We let f be a force *density* (a linear, spatial, force density, in newtons per meter) in order that it can be applied impulsive along x as well as t. We could have instead chosen a force $f(t)$ applied at a particular point x.

integral of force ($f = ma = m\dot{v} \;\Rightarrow\; mv = \int f\,dt$), our excitation will be a *unit momentum* transferred to the two-mass system at time 0.

Striking the Rod in the Middle

First, consider $f(t,x) = \delta(t)\delta(x)$. That is, we apply an upward unit-force impulse at time 0 in the middle of the rod. The total momentum delivered in the neighborhood of $x = 0$ and $t = 0$ is obtained by integrating the applied force density with respect to time and position:

$$p \;=\; \iint f(t,x)\,dt\,dx \;=\; \iint \delta(t)\delta(x)\,dt\,dx \;=\; 1$$

This unit momentum is transferred to the two masses m. By symmetry, we have $v_{-r} = v_r = v_0$. We can also refer to v_0 as the *velocity of the center of mass*, again obvious by symmetry. Continuing to refer to Fig. B.5, we have

$$p \;=\; 1 \;=\; mv_{-r} + mv_r \;=\; (2m)v_0 \;\Rightarrow\; v_0 \;=\; \frac{1}{2m}.$$

Thus, after time zero, each mass is traveling upward at speed $v_0 = 1/(2m)$, and there is no rotation about the center of mass at $x = 0$.

The kinetic energy of the system after time zero is

$$E_K \;=\; \frac{1}{2}mv_{-r}^2 + \frac{1}{2}mv_r^2 \;=\; m\left(\frac{1}{2m}\right)^2 \;=\; \frac{1}{4m}.$$

Note that we can also compute E_K in terms of the total mass $M = 2m$ and the velocity of the center of mass $v_0 = 1/(2m)$:

$$E_K \;=\; \frac{1}{2}Mv_0^2 \;=\; \frac{1}{2}(2m)\left(\frac{1}{2m}\right)^2 \;=\; \frac{1}{4m}$$

Striking One of the Masses

Now let $f(t,x) = \delta(t)\delta(x - r)$. That is, we apply an impulse of vertical momentum to the mass on the right at time 0.

In this case, the unit of vertical momentum is transferred entirely to the mass on the right, so that

$$p \;=\; 1 \;=\; mv_r \;\Rightarrow\; v_r \;=\; \frac{1}{m},$$

which is twice as fast as before. Just after time zero, we have $v_r = 1/m$, $v_{-r} = 0$, and, because the massless rod remains rigid, $v_0 = 1/(2m)$.

Note that the velocity of the center-of-mass $1/(2m)$ is the *same* as it was when we hit the midpoint of the rod. This is an important general equivalence: The sum of all external force vectors acting on a rigid body can be applied as a single resultant force vector to the total mass concentrated at the center of mass to find the linear (translational) motion produced. (Recall from §B.4.1 that such a sum is the same as the sum of all radially acting external force components, since the tangential components contribute only to rotation and not to translation.)

All of the kinetic energy is in the mass on the right just after time zero:

$$E_K = \frac{1}{2}mv_r^2 = \frac{1}{2}m\left(\frac{1}{m}\right)^2 = \frac{1}{2m} \tag{B.13}$$

However, after time zero, things get more complicated, because the mass on the left gets dragged into a rotation about the center of mass.

To simplify ongoing analysis, we can define a *body-fixed frame of reference*[16] having its origin at the center of mass. Let v' denote a velocity in this frame. Since the velocity of the center of mass is $v_0 = 1/(2m)$, we can convert any velocity v' in the body-fixed frame to a velocity v in the original frame by adding v_0 to it, *viz.*,

$$v = v' + v_0 = v' + \frac{1}{2m}.$$

The mass velocities in the body-fixed frame are now

$$v'_{-r} = -\frac{1}{2m}, \qquad v'_r = \frac{1}{2m},$$

and of course $v'_0 = 0$.

In the body-fixed frame, all kinetic energy is *rotational* about the origin. Recall (Eq. (B.9), p. 474) that the moment of inertia for this system, with respect to the center of mass at $x = 0$, is

$$I = m(-r)^2 + mr^2 = 2mr^2.$$

Thus, the *rotational kinetic energy* (§B.4.3, p. 472) is found to be

$$E'_R = \frac{1}{2}I\omega^2 = \frac{1}{2}(2mr^2)\left(\frac{v'_r}{r}\right)^2 = mr^2\left(\frac{1}{2mr}\right)^2 = \frac{1}{4m}.$$

This is *half* of the kinetic energy we computed in the original "space-fixed" frame (Eq. (B.13) on page 480 above). The other half is in the *translational*

[16] At this stage, a better name for "body-fixed frame" would be *center-of-mass frame*. However, later on, we will see that not only do we need to follow the center-of-mass with the origin of our body-fixed frame, but its coordinate axes will also need to remain pointing in the so-called *principal directions* (defined in §B.4.16).

kinetic energy not seen in the body-fixed frame. As we saw in §B.4.2 on page 472 above, we can easily calculate the translational kinetic energy as that of the total mass $M = 2m$ traveling at the center-of-mass velocity $v_0 = 1/(2m)$:

$$E'_K = \frac{1}{2}Mv_0^2 = \frac{1}{2}(2m)\left(\frac{1}{2m}\right)^2 = \frac{1}{4m}$$

Adding this translational kinetic energy to the rotational kinetic energy in the body-fixed frame yields the total kinetic energy, as it must.

In summary, we defined a moving body-fixed frame having its origin at the center-of-mass, and the total kinetic energy was computed to be

$$E_K = E'_K + E'_R = \frac{1}{4m} + \frac{1}{4m} = \frac{1}{2m}$$

in agreement with the more complicated (after time zero) space-fixed analysis in Eq. (B.13).

It is important to note that, after time zero, both the linear momentum of the center-of-mass ($p_0 = Mv_0 = (2m)(v_r/2) = mv_r = m \cdot (1/(2m)) = 1/2$), and the angular momentum in the body-fixed frame ($L' = I\omega = (2mr^2)(v'_r/r) = (2mr^2)(1/(2mr)) = r/2$) remain *constant* over time.[17] In the original space-fixed frame, on the other hand, there is a complex transfer of momentum back and forth between the masses after time zero.

Similarly, the translational kinetic energy of the total mass, treated as being concentrated at its center-of-mass, and the rotational kinetic energy in the body-fixed frame, are both constant after time zero, while in the space-fixed frame, kinetic energy transfers back and forth between the two masses. At all times, however, the total kinetic energy is the same in both formulations.

B.4.11 Angular Velocity Vector

When working with rotations, it is convenient to define the *angular-velocity vector* as a vector $\underline{\omega} \in \mathbf{R}^3$ pointing along the *axis of rotation*. There are two directions we could choose from, so we pick the one corresponding to the *right-hand rule*, *i.e.*, when the fingers of the right hand curl in the direction of the rotation, the thumb points in the direction of the angular velocity

[17] There is no problem with conservation of momentum here. We will see in §B.4.13 that, for a rotating mass m at radius r, the angular momentum $L = I\omega = I(v/r) = (mr^2)(v/r) = mvr$ is always r times the instantaneous linear momentum $p = mv$. Thus, in our problem, converting angular momentum $L = r/2$ to linear momentum $p = 1/2$ (in the body-fixed frame) verifies conservation of momentum when we add it to the linear momentum of the center of mass, which was also found to be 1/2. In other words, the total momentum is still 1, as initially delivered to the system.

vector.[18] The *length* $\|\underline{\omega}\|$ should obviously equal the angular velocity ω. It is convenient also to work with a unit-length variant $\underline{\tilde{\omega}} \triangleq \underline{\omega}/\|\underline{\omega}\|$.

As introduced in Eq. (B.8) above, the mass moment of inertia is given by $I = mR^2$ where R is the distance from the (instantaneous) axis of rotation to the mass m located at $\underline{x} \in \mathbf{R}^3$. In terms of the angular-velocity vector $\underline{\omega}$, we can write this as (see Fig. B.6)

$$\begin{aligned} I &= mR^2 = m \cdot \left\| \underline{x} - \mathcal{P}_{\underline{\omega}}(\underline{x}) \right\|^2 \\ &= m \cdot \left\| \underline{x} - (\underline{\tilde{\omega}}^T \underline{x})\underline{\tilde{\omega}} \right\|^2 \end{aligned} \tag{B.14}$$

where

$$\mathcal{P}_{\underline{\omega}}(\underline{x}) \triangleq \frac{\underline{\omega}^T \underline{x}}{\underline{\omega}^T \underline{\omega}}\underline{\omega} = (\underline{\tilde{\omega}}^T \underline{x})\underline{\tilde{\omega}}$$

denotes the orthogonal projection of $\underline{x}$ onto $\underline{\omega}$ (or $\underline{\tilde{\omega}}$) [451]. Thus, we can project the mass position $\underline{x}$ onto the angular-velocity vector $\underline{\omega}$ and subtract to get the component of $\underline{x}$ that is orthogonal to $\underline{\omega}$, and the length of that difference vector is the distance to the rotation axis R, as shown in Fig. B.6.

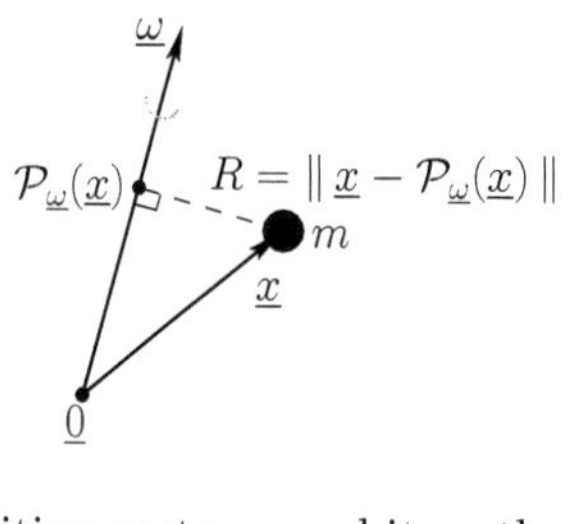

Figure B.6: Mass position vector $\underline{x}$ and its orthogonal projection $\mathcal{P}_{\underline{\omega}}(\underline{x})$ onto the angular velocity vector $\underline{\omega}$ for purposes of finding the distance R of the mass m from the axis of rotation $\underline{\tilde{\omega}}$.

Using the *vector cross product* (defined in the next section), we will show (in §B.4.17) that R can be written more succinctly as

$$R = \left\| \underline{x} - \mathcal{P}_{\underline{\omega}}(\underline{x}) \right\| = \| \underline{\tilde{\omega}} \times \underline{x} \| .$$

B.4.12 Vector Cross Product

The *vector cross product* (or simply *vector product*, as opposed to the *scalar product* (which is also called the *dot product*, or *inner product*)) is commonly used in *vector calculus*—a basic mathematical toolset used in mechanics

[18] Another analogy is the *right-hand screw*, in which the screw drives in the direction of the vector when it is turned clockwise from behind, as is standard.

[270, 258], acoustics [349], electromagnetism [356], quantum mechanics, and more. It can be defined symbolically in the form of a *matrix determinant*:[19]

$$\begin{aligned}
\underline{x} \times \underline{y} &= \begin{vmatrix} \underline{e}_1 & \underline{e}_2 & \underline{e}_3 \\ x_1 & x_2 & x_3 \\ y_1 & y_2 & y_3 \end{vmatrix} \\
&= (x_2y_3 - y_2x_3)\underline{e}_1 + (x_3y_1 - y_3x_1)\underline{e}_2 + (x_1y_2 - y_1x_2)\underline{e}_3 \\
&= \begin{bmatrix} 0 & -x_3 & x_2 \\ x_3 & 0 & -x_1 \\ -x_2 & x_1 & 0 \end{bmatrix} \begin{bmatrix} y_1 \\ y_2 \\ y_3 \end{bmatrix} = \begin{bmatrix} 0 & y_3 & -y_2 \\ -y_3 & 0 & y_1 \\ y_2 & -y_1 & 0 \end{bmatrix} \begin{bmatrix} x_1 \\ x_2 \\ x_3 \end{bmatrix}
\end{aligned} \tag{B.15}$$

where $\underline{e}_i$ denote the *unit vectors* in $\mathbf{R}^3$. The cross-product is a vector in 3D that is orthogonal to the plane spanned by $\underline{x}$ and $\underline{y}$, and is oriented positively according to the *right-hand rule*.[20]

The second and third lines of Eq. (B.15) make it clear that $\underline{y} \times \underline{x} = -\underline{x} \times \underline{y}$. This is one example of a host of identities that one learns in vector calculus and its applications.

Cross-Product Magnitude

It is a straightforward exercise to show that the cross-product magnitude is equal to the product of the vector lengths times the sine of the angle between them:[21]

$$\left\|\underline{x} \times \underline{y}\right\| = \left\|\underline{x}\right\| \cdot \left\|\underline{y}\right\| \cdot \left|\sin(\theta)\right| \tag{B.16}$$

where

$$\sin(\theta) = \sqrt{1 - \cos^2(\theta)}$$

with

$$\cos(\theta) \triangleq \frac{\langle \underline{x}, \underline{y} \rangle}{\left\|\underline{x}\right\| \cdot \left\|\underline{y}\right\|}. = \frac{x_1y_1 + x_2y_2 + x_3y_3}{\sqrt{x_1^2 + x_2^2 + x_3^2} \cdot \sqrt{y_1^2 + y_2^2 + y_3^2}}$$

[19] http://mathworld.wolfram.com/CrossProduct.html

[20] Rotate $\underline{x}$ to $\underline{y}$ by "pushing" with the fingers of the right hand, taking the smallest angle possible, and your thumb points in the positive direction for $\underline{x} \times \underline{y}$.

[21] We are using "·" (optionally, when it looks better) to denote *scalar multiplication* here, while in vector calculus, $\underline{x} \cdot \underline{y}$ would normally denote a *dot product* of the two vectors $\underline{x}$ and $\underline{y}$, but we are writing the dot product as $\underline{x} \cdot \underline{y} = \underline{x}^T \underline{y} = \underline{y}^T \underline{x}$. (We could also use the "inner-product" notation $\langle \underline{x}, \underline{y} \rangle = \underline{x}^T \underline{y}$ for real vectors [451].) We could say that "·" means ordinary multiplication when appearing between two scalars, and it means the dot product when used between two vectors. Between two matrices it would of course mean matrix multiplication.

(Recall that the *vector cosine* of the angle between two vectors is given by their inner product divided by the product of their norms [451].)

To derive Eq. (B.16), let's begin with the cross-product in matrix form as $\underline{x} \times \underline{y} = \mathbf{X}\underline{y}$ using the first matrix form in the third line of the cross-product definition in Eq. (B.15) above. Then

$$\begin{aligned}(\underline{x} \times \underline{y})^2 &= \underline{y}^T\mathbf{X}^T\mathbf{X}\underline{y} = \underline{y}^T\left(\|\underline{x}\|^2 \cdot \mathbf{E} - \underline{x}\underline{x}^T\right)\underline{y} = \underline{y}^T\|\underline{x}\|^2\left(\mathbf{E} - \mathcal{P}_{\underline{x}}\right)\underline{y} \\ &\triangleq \|\underline{x}\|^2\underline{y}^T\mathcal{P}_{\underline{x}^\perp}\underline{y}, = \|\underline{x}\|^2\underline{y}^T\mathcal{P}_{\underline{x}^\perp}^T\mathcal{P}_{\underline{x}^\perp}\underline{y}, \triangleq \|\underline{x}\|^2\underline{y}_{\underline{x}^\perp}^T\underline{y}_{\underline{x}^\perp}\end{aligned}$$

where $\mathbf{E} = [\underline{e}_1, \underline{e}_2, \underline{e}_3]$ denotes the identity matrix in $\mathbf{R}^3$, $\mathcal{P}_{\underline{x}}$ denotes the orthogonal-projection matrix onto $\underline{x}$ [451], $\mathcal{P}_{\underline{x}}$ denotes the projection matrix onto the orthogonal *complement* of $\underline{x}$, $\underline{y}_{\underline{x}^\perp}$ denotes the component of $\underline{y}$ orthogonal to $\underline{x}$, and we used the fact that orthogonal projection matrices $\mathcal{P}$ are *idempotent* (*i.e.*, $\mathcal{P}^2 = \mathcal{P}$) and *symmetric* (when real, as we have here) when we replaced $\mathcal{P}_{\underline{x}^\perp}$ by $\mathcal{P}_{\underline{x}^\perp}^T\mathcal{P}_{\underline{x}^\perp}$ above. Finally, note that the length of $\underline{y}_{\underline{x}^\perp}$ is $\|\underline{y}\|\cdot|\sin(\theta)|$, where θ is the angle between the 1D subspaces spanned by $\underline{x}$ and $\underline{y}$ in the plane including both vectors. Thus,

$$(\underline{x} \times \underline{y})^2 = \|\underline{x}\|^2\underline{y}_{\underline{x}^\perp}^T\underline{y}_{\underline{x}^\perp} = \|\underline{x}\|^2\|\underline{y}_{\underline{x}^\perp}\|^2 = \|\underline{x}\|^2\|\underline{y}\|^2\sin^2(\theta),$$

which establishes the desired result:

$$\left\|\underline{x} \times \underline{y}\right\| = \|\underline{x}\| \cdot \|\underline{y}\| \cdot |\sin(\theta)|$$

Moreover, this proof gives an appealing geometric interpretation of the vector cross-product $\underline{x}\times\underline{y}$ as having magnitude given by the product of $\|\underline{x}\|$ times the norm of the difference between $\underline{x}$ and the *orthogonal projection* of $\underline{x}$ onto $\underline{y}$ ($\|\underline{x} \times \underline{y}\| = \|\underline{x}\| \cdot \|\underline{x} - \mathcal{P}_{\underline{y}}(\underline{x})\|$) or vice versa ($\|\underline{x} \times \underline{y}\| = \|\underline{y}\| \cdot \|\underline{y} - \mathcal{P}_{\underline{x}}(\underline{y})\|$). In this geometric picture it is clear that the cross-product magnitude is maximized when the vectors are orthogonal, and it is zero when the vectors are collinear. It is "length times orthogonal length."

The direction of the cross-product vector is then taken to be orthogonal to both $\underline{x}$ and $\underline{y}$ according to the right-hand rule. This orthogonality can be checked by verifying that $\mathbf{X}\underline{x} = \mathbf{Y}\underline{y} = \underline{0}$. The right-hand-rule parity can be checked by rotating the space so that $\underline{x}' = [\|\underline{x}\|, 0, 0]^T$ and $\underline{y}' = [\|\underline{y}\|\cos(\theta), \|\underline{y}\|\sin(\theta), 0]^T$ in which case $\underline{x}'\times\underline{y}' = \|\underline{x}\|[0, 0, \|\underline{y}\|\sin(\theta)]^T = \|\underline{x}\|\|\underline{y}\|\sin(\theta)\underline{e}_3$. Thus, the cross product points "up" relative to the $(\underline{x}, \underline{y})$ plane for $\theta \in (0, \pi)$ and "down" for $\theta \in (0, -\pi)$.

Mass Moment of Inertia as a Cross Product

In Eq. (B.14) above, the mass moment of inertia was expressed in terms of orthogonal projection as $I = mR^2 = m\cdot\|\underline{x} - \mathcal{P}_{\underline{\omega}}(\underline{x})\|^2 = m\cdot\|\underline{x} - (\underline{\tilde{\omega}}^T\underline{x})\underline{\tilde{\omega}}\|^2$, where $\underline{\tilde{\omega}} \triangleq \underline{\omega}/\|\,\underline{\omega}\,\|$. In terms of the vector cross product, we can now express it as

$$I = m\cdot(\underline{\tilde{\omega}}\times\underline{x})^2 = m\cdot\left[\|\,\underline{\tilde{\omega}}\,\|\cdot\|\,\underline{x}\,\|\cdot\sin(\theta_{\underline{\tilde{\omega}}\underline{x}})\right]^2 = mR^2$$

where $R = \|\,\underline{x}\,\|\sin(\theta_{\underline{\tilde{\omega}}\underline{x}})$ is the distance from the rotation axis out to the point $\underline{x}$ (which equals the length of the vector $\underline{x} - \mathcal{P}_{\underline{\omega}}(\underline{x})$).

Tangential Velocity as a Cross Product

Referring again to Fig. B.4 on page 473, we can write the *tangential velocity vector* $\underline{v}$ as a vector cross product of the angular-velocity vector $\underline{\omega}$ (§B.4.11) and the position vector $\underline{x}$:

$$\underline{v} = \underline{\omega}\times\underline{x} \tag{B.17}$$

To see this, let's first check its direction and then its magnitude. By the right-hand rule, $\underline{\omega}$ points up out of the page in Fig. B.4. Crossing that with $\underline{x}$, again by the right-hand rule, produces a tangential velocity vector $\underline{v}$ pointing as shown in the figure. So, the direction is correct. Now, the magnitude: Since $\underline{\omega}$ and $\underline{x}$ are mutually orthogonal, the angle between them is $\pi/2$, so that, by Eq. (B.16),

$$\|\,\underline{\omega}\times\underline{x}\,\| = \|\,\underline{\omega}\,\|\cdot\|\,\underline{x}\,\| = \omega R = v$$

as desired.

B.4.13 Angular Momentum

The *angular momentum* of a mass m rotating in a circle of radius R with angular velocity ω (rad/s), is defined by

$$L = I\omega$$

where $I = mR^2$ denotes the *mass moment of inertia* of the rigid body (§B.4.4).

Relation of Angular to Linear Momentum

Recall (§B.3) that the momentum of a mass m traveling with velocity v in a straight line is given by

$$p = mv,$$

while the angular momentum of a point-mass m rotating along a circle of radius R at ω rad/s is given by

$$L = I\omega,$$

where $I = mR^2$. The *tangential speed* of the mass along the circle of radius R is given by

$$v = R\omega.$$

Expressing the angular momentum I in terms of v gives

$$L \triangleq I\omega = I\frac{v}{R} \triangleq mR^2\frac{v}{R} = Rmv = Rp. \tag{B.18}$$

Thus, the angular momentum L is R times the linear momentum $p = mv$.

Linear momentum can be viewed as a renormalized special case of angular momentum in which the radius of rotation goes to infinity.

B.4.14 Angular Momentum Vector

Like linear momentum, angular momentum is fundamentally a vector in $\mathbf{R}^3$. The definition of the previous section suffices when the direction does not change, in which case we can focus only on its magnitude $L = I\omega$.

More generally, let $\underline{x}_m \in \mathbf{R}^3$ denote the 3-space coordinates of a point-mass m, and let $\underline{v}_m = \dot{\underline{x}}_m$ denote its velocity in $\mathbf{R}^3$. Then the *instantaneous angular momentum vector* of the mass relative to the origin (not necessarily rotating about a fixed axis) is given by

$$\underline{L} \triangleq m\,\underline{x}_m \times \underline{v}_m = m\,\underline{x}_m \times (\underline{\omega} \times \underline{x}_m) \tag{B.19}$$

where $\times$ denotes the *vector cross product*, discussed in §B.4.12 above. The identity $\underline{v}_m = \underline{\omega} \times \underline{x}_m$ was discussed at Eq. (B.17) on page 485.

For the special case in which $\underline{v}_m$ is *orthogonal* to $\underline{x}_m$, as in Fig. B.4 on page 473, we have that $x_m \times \underline{v}_m$ points, by the right-hand rule, in the direction of the angular velocity vector $\underline{\omega}$ (up out of the page), which is satisfying. Furthermore, its magnitude is given by

$$\|\,\underline{L}\,\| = m\,\|\,\underline{x}_m \times \underline{v}_m\,\| = m\,\|\,\underline{x}_m\,\|\,\|\,\underline{v}_m\,\| = mRv = mR^2\omega = I\omega$$

which agrees with the scalar case.

In the more general case of an arbitrary mass velocity vector $\underline{v}_m$, we know from §B.4.12 that the magnitude of $\underline{x}_m \times \underline{v}_m$ equals the product of the distance from the axis of rotation to the mass, *i.e.*, $\|\,\underline{x}_m\,\|$, times the length of the component of $\underline{v}_m$ that is orthogonal to $\underline{x}$, *i.e.*, $\|\,\underline{v}_m\,\|\sin(\theta)$, as needed.

It can be shown that vector angular momentum, as defined, is *conserved.*[22] For example, in an orbit, such as that of the moon around the earth, or that of Halley's comet around the sun, the orbiting object speeds

[22] Angular momentum is also conserved on the smallest scales, such as *orbital angular momentum* and the *spin* of fundamental particles such as an electron. Thus, conservation

up as it comes closer to the object it is orbiting. (See Kepler's laws of planetary motion.) Similarly, a spinning ice-skater spins faster when pulling in arms to reduce the moment of inertia about the spin axis. The conservation of angular momentum can be shown to result from the principle of least action and the isotrophy of space [270, p. 18].

Angular Momentum Vector in Matrix Form

The two cross-products in Eq. (B.19) can be written out with the help of the vector analysis identity[23]

$$\underline{x} \times (\underline{y} \times \underline{z}) \;=\; \underline{y} \cdot (\underline{z}^T \underline{x}) - \underline{z} \cdot (\underline{x}^T \underline{y}).$$

This (or a direct calculation) yields, starting with Eq. (B.19),

$$\begin{aligned}
\underline{L} \;&=\; m\,\underline{x}_m \times (\underline{\omega} \times \underline{x}_m) \;=\; m\,\underline{\omega} \cdot (\underline{x}^T \underline{x}) - \underline{x} \cdot (\underline{x}^T \underline{\omega}) \\
&=\; m \left(\| \underline{x} \|^2 \mathbf{E} - \underline{x}\underline{x}^T \right) \underline{\omega} \\
&\triangleq\; \mathbf{I}\underline{\omega}
\end{aligned} \tag{B.20}$$

where

$$\mathbf{I}\underline{\omega} \;=\; \left[\begin{array}{ccc} I_{11} & I_{12} & I_{13} \\ I_{21} & I_{22} & I_{23} \\ I_{31} & I_{32} & I_{33} \end{array}\right] \left[\begin{array}{c} \omega_1 \\ \omega_2 \\ \omega_3 \end{array}\right]$$

with $I_{ii} = m\left(\sum_{j=1}^3 x_j^2 - x_i^2\right)$, and $I_{ij} = -m x_i x_j$, for $i \neq j$. That is,

$$\mathbf{I} \;=\; m \left[\begin{array}{ccc} x_2^2 + x_3^2 & -x_1 x_2 & -x_1 x_3 \\ -x_2 x_1 & x_1^2 + x_3^2 & -x_2 x_3 \\ -x_3 x_1 & -x_3 x_2 & x_1^2 + x_2^2 \end{array}\right] \;=\; m \left(\| \underline{x} \|^2 \mathbf{E} - \underline{x}\underline{x}^T \right). \tag{B.21}$$

The matrix $\mathbf{I}$ is the Cartesian representation of the *mass moment of inertia tensor*, which will be explored further in §B.4.15 below.

---

of angular momentum is a fundamental invariant in physics, along with conservation of energy and conservation of linear momentum. Since we live in 3D, that makes *seven* conserved quantities pertaining to motion of a mass through empty space: its kinetic energy, the three components of the linear momentum of its center of mass, and the three components of its angular momentum if it is rotating about its center of mass.

[23] In `maxima`, one can verify this identity with the following input (the output of which is [0, 0, 0]):

```
load(vect);
A:[A1,A2,A3]; B:[B1,B2,B3]; C:[C1,C2,C3];
expand(express((A ~ (B ~ C)) - (B * (C . A) - C * (A . B))));
```

Note that the cross-product is called the *wedge product* in `maxima`, and is denoted by a tilde (~).

The vector angular momentum of a rigid body is obtained by summing the angular momentum of its constituent mass particles. Thus,

$$\underline{L} = \sum_i m_i \left(\| \underline{x}_i \|^2 \mathbf{E} - \underline{x}_i \underline{x}_i^T \right) \underline{\omega} \;\stackrel{\Delta}{=}\; \mathbf{I}\,\underline{\omega}.$$

Since $\underline{\omega}$ factors out of the sum, we see that the mass moment of inertia tensor for a rigid body is given by the sum of the mass moment of inertia tensors for each of its component mass particles.

In summary, the angular momentum vector $\underline{L}$ is given by the mass moment of inertia tensor $\mathbf{I}$ times the angular-velocity vector $\underline{\omega}$ representing the axis of rotation.

Note that the angular momentum vector $\underline{L}$ does *not* in general point in the same direction as the angular-velocity vector $\underline{\omega}$. We saw above that it does in the special case of a point mass traveling orthogonal to its position vector. In general, $\underline{L}$ and $\underline{\omega}$ point in the same direction whenever $\underline{\omega}$ is an *eigenvector* of $\mathbf{I}$, as will be discussed further below (§B.4.16). In this case, the rigid body is said to be *dynamically balanced*.[24]

B.4.15 Mass Moment of Inertia Tensor

As derived in the previous section, the *moment of inertia tensor*, in 3D Cartesian coordinates, is a three-by-three matrix $\mathbf{I}$ that can be multiplied by any angular-velocity vector to produce the corresponding angular momentum vector for either a point mass or a rigid mass distribution. Note that the origin of the angular-velocity vector $\underline{\omega}$ is always fixed at $\underline{0}$ in the space (typically located at the center of mass). Therefore, the moment of inertia tensor $\mathbf{I}$ is defined relative to that origin.

The moment of inertia tensor can similarly be used to compute the *mass moment of inertia* for any normalized angular velocity vector $\underline{\tilde{\omega}} = \underline{\omega}/\|\,\underline{\omega}\,\|$ as

$$I = \underline{\tilde{\omega}}^T \mathbf{I}\, \underline{\tilde{\omega}}. \tag{B.22}$$

Since rotational energy is defined as $(1/2)I\omega^2$ (see Eq. (B.7) on page 473), multiplying Eq. (B.22) by ω^2 gives the following expression for the rotational kinetic energy in terms of the moment of inertia tensor:

$$E_R = \frac{1}{2}\underline{\omega}^T \mathbf{I}\, \underline{\omega} \tag{B.23}$$

We can show Eq. (B.22) starting from Eq. (B.14) on page 482. For a point-

[24] `http://www.ph.man.ac.uk/~mikeb/lecture/pc167/rigidbody/principal.html`

mass m located at $\underline{x}$, we have

$$\begin{aligned}
I &= m\left\|\underline{x} - (\underline{\tilde{\omega}}^T\underline{x})\underline{\tilde{\omega}}\right\|^2 \\
&= m\left[\underline{x}^T - (\underline{\tilde{\omega}}^T\underline{x})\underline{\tilde{\omega}}^T\right]\left[\underline{x} - (\underline{\tilde{\omega}}^T\underline{x})\underline{\tilde{\omega}}\right] \\
&= m\left[\underline{x}^T\underline{x} - \underline{x}^T(\underline{\tilde{\omega}}^T\underline{x})\underline{\tilde{\omega}} - (\underline{\tilde{\omega}}^T\underline{x})\underline{\tilde{\omega}}^T\underline{x} + (\underline{\tilde{\omega}}^T\underline{x})^2\underline{\tilde{\omega}}^T\underline{\tilde{\omega}}\right] \\
&= m\left[\underline{x}^T\underline{x} - (\underline{\tilde{\omega}}^T\underline{x})^2\right] \;=\; m\left[\|x\|^2 - \underline{\tilde{\omega}}^T\underline{xx}^T\underline{\tilde{\omega}}\right] \\
&= \underline{\tilde{\omega}}^T\left[m\left(\|\underline{x}\|^2\mathbf{E} - \underline{xx}^T\right)\right]\underline{\tilde{\omega}} \\
&\triangleq \underline{\tilde{\omega}}^T\mathbf{I}\,\underline{\tilde{\omega}}
\end{aligned}$$

where again $\mathbf{E}$ denotes the three-by-three identity matrix, and

$$\mathbf{I} = m\left(\|\underline{x}\|^2\mathbf{E} - \underline{xx}^T\right), \tag{B.24}$$

which agrees with Eq. (B.20). Thus we have derived the moment of inertia I in terms of the moment of inertia tensor $\mathbf{I}$ and the normalized angular velocity $\underline{\tilde{\omega}}$ for a point-mass m at $\underline{x}$.

For a collection of N masses m_i located at $\underline{x}_i \in \mathbf{R}^3$, we simply sum over their masses to add up the moments of inertia:

$$\mathbf{I} = \sum_{i=1}^{N} m_i\left(\|\underline{x}_i\|^2\mathbf{E} - \underline{x}_i\underline{x}_i^T\right)$$

Finally, for a continuous mass distribution, we integrate as usual:

$$\mathbf{I} = \frac{1}{M}\int_V \rho(\underline{x})\left(\|\underline{x}\|^2\mathbf{E} - \underline{xx}^T\right)dV$$

where $M = \int_V \rho(\underline{x})dV$ is the total mass.

Simple Example

Consider a mass m at $\underline{x} = [x, 0, 0]^T$. Then the mass moment of inertia tensor is

$$\mathbf{I} = m\left(\|\underline{x}\|^2\mathbf{E} - \underline{xx}^T\right) = m\begin{bmatrix} 0 & 0 & 0 \\ 0 & 1 & 0 \\ 0 & 0 & 1 \end{bmatrix}.$$

For the angular-velocity vector $\underline{\omega} = [\omega, 0, 0]^T$, we obtain the moment of inertia

$$I = \underline{\tilde{\omega}}^T \mathbf{I} \underline{\tilde{\omega}} = \begin{bmatrix} 1 & 0 & 0 \end{bmatrix} \begin{bmatrix} 0 & 0 & 0 \\ 0 & x^2 & 0 \\ 0 & 0 & x^2 \end{bmatrix} \begin{bmatrix} 1 \\ 0 \\ 0 \end{bmatrix} m = 0.$$

This makes sense because the axis of rotation passes through the point mass, so the moment of inertia *should* be zero about that axis. On the other hand, if we look at $\underline{\omega} = [0, 1, 0]^T$, we get

$$I = \underline{\tilde{\omega}}^T \mathbf{I} \underline{\tilde{\omega}} = \begin{bmatrix} 0 & 1 & 0 \end{bmatrix} \begin{bmatrix} 0 & 0 & 0 \\ 0 & x^2 & 0 \\ 0 & 0 & x^2 \end{bmatrix} \begin{bmatrix} 0 \\ 1 \\ 0 \end{bmatrix} = mx^2$$

which is what we expected.

Example with Coupled Rotations

Now let the mass m be located at $\underline{x} = [1, 1, 0]^T$ so that

$$\begin{aligned} \mathbf{I} &= m\left(\| \underline{x} \|^2 \mathbf{E} - \underline{x}\underline{x}^T \right) = m\left(2\mathbf{E} - \begin{bmatrix} 1 & 1 & 0 \\ 1 & 1 & 0 \\ 0 & 0 & 0 \end{bmatrix} \right) \\ &= m \begin{bmatrix} 1 & -1 & 0 \\ -1 & 1 & 0 \\ 0 & 0 & 2 \end{bmatrix}. \end{aligned}$$

We expect $\underline{\omega} = [1, 1, 0]$ to yield zero for the moment of inertia, and sure enough $\underline{\tilde{\omega}}^T \mathbf{I} \underline{\tilde{\omega}} = 0$. Similarly, the vector angular momentum is zero, since $\mathbf{I}\underline{\omega} = \underline{0}$.

For $\underline{\omega} = [1, 0, 0]^T$, the result is

$$\mathbf{I} = \begin{bmatrix} 1 & 0 & 0 \end{bmatrix} \begin{bmatrix} 1 & -1 & 0 \\ -1 & 1 & 0 \\ 0 & 0 & 2 \end{bmatrix} \begin{bmatrix} 1 \\ 0 \\ 0 \end{bmatrix} m = m,$$

which makes sense because the distance from the axis $\underline{e}_1$ to $\underline{x}$ is one. The same result is obtained for rotation about $\underline{\omega} = \underline{e}_2$. For $\underline{\omega} = \underline{e}_3$, however, the result is $I = 2m = m\| \underline{x} \|^2$, as expect.

Off-Diagonal Terms in Moment of Inertia Tensor

This all makes sense, but what about those −1 off-diagonal terms in **I**? Consider the vector angular momentum (§B.4.14, p. 486):

$$\underline{L} = \mathbf{I}\,\underline{\omega} = m \begin{bmatrix} I_{11} & I_{12} & I_{13} \\ I_{21} & I_{22} & I_{23} \\ I_{31} & I_{32} & I_{33} \end{bmatrix} \begin{bmatrix} \omega_1 \\ \omega_2 \\ \omega_3 \end{bmatrix}$$

We see that the off-diagonal terms I_{ij} correspond to a *coupling* of rotation about $\underline{e}_i$ with rotation about $\underline{e}_j$. That is, there is a component of moment-of-inertia I_{ij} that is contributed (or subtracted, as we saw above for $\underline{\omega} = [1,1,0]^T$) when *both* ω_i and ω_j are nonzero. These cross-terms can be eliminated by *diagonalizing* the matrix [449],[25] as discussed further in the next section.

B.4.16 Principal Axes of Rotation

A *principal axis of rotation* (or *principal direction*) is an eigenvector of the *mass moment of inertia tensor* (introduced in the previous section) defined relative to some point (typically the center of mass). The corresponding eigenvalues are called the *principal moments of inertia*. Because the moment of inertia tensor is defined relative to the point $\underline{x} = \underline{0}$ in the space, the principal axes all pass through that point (usually the center of mass).

As derived above (§B.4.14), the angular momentum vector is given by the moment of inertia tensor times the angular-velocity vector:

$$\underline{L} = \mathbf{I}\,\underline{\omega}$$

If $\underline{\omega}$ is an eigenvector of $\mathbf{I}$, then we have

$$\underline{L} = \mathbf{I}\,\underline{\omega} = \lambda\underline{\omega}$$

where the (scalar) eigenvalue λ is called a *principal moment of inertia*. If we set the rigid body assocated with $\mathbf{I}$ rotating about the axis $\underline{\omega}$, then λ is the mass moment of inertia of the body for that rotation. As will become clear below, there are always three mutually orthogonal principal axes of rotation, and three corresponding principal moments of inertia (in 3D space, of course).

Positive Definiteness of the Moment of Inertia Tensor

From the form of the moment of inertia tensor introduced in Eq. (B.24) on page 489

$$\mathbf{I} = m\left(\|\,\underline{x}\,\|^2\,\mathbf{E} - \underline{x}\underline{x}^T\right),$$

[25] https://ccrma.stanford.edu/~jos/filters/Diagonalizing_State_Space_Model.html

it is clear that $\mathbf{I}$ is *symmetric.* Moreover, for any normalized angular-velocity vector $\underline{\tilde{\omega}}$ we have

$$\begin{aligned} I &= \underline{\tilde{\omega}}^T \mathbf{I} \underline{\tilde{\omega}} = m\left(\|\underline{x}\|^2 - \underline{\tilde{\omega}}^T \underline{x}\underline{x}^T \underline{\tilde{\omega}}\right) = m\,\|\underline{x}\|^2 \left[1 - \underline{\tilde{\omega}}^T \left(\frac{\underline{x}\,\underline{x}^T}{\underline{x}^T \underline{x}}\,\underline{\tilde{\omega}}\right)\right] \\ &= m\,\|\underline{x}\|^2 \left[1 - \underline{\tilde{\omega}}^T \mathcal{P}_{\underline{x}}(\underline{\tilde{\omega}})\right] \geq 0 \end{aligned}$$

since $\underline{\tilde{\omega}}$ is unit length, and projecting it onto any other vector can only shorten it or leave it unchanged. That is, $|\underline{\tilde{\omega}}^T \mathcal{P}_{\underline{x}}(\underline{\tilde{\omega}})| \leq 1$, with equality occurring for $\underline{x} = \alpha\underline{\tilde{\omega}}$ for any nonzero $\alpha \in \mathbf{R}$. Zooming out, *of course* we expect any moment of inertia I for a positive mass m to be nonnegative. Thus, $\mathbf{I}$ is *symmetric nonnegative definite.* If furthermore $\underline{\tilde{\omega}}$ and $\underline{x}$ are not collinear, *i.e.*, if there is any nonzero angle between them, then $\mathbf{I}$ is *positive definite* (and $I > 0$). As is well known in linear algebra [329], real, symmetric, positive-definite matrices have *orthogonal eigenvectors* and *real, positive eigenvalues.* In this context, the orthogonal eigenvectors are called the *principal axes of rotation.* Each corresponding eigenvalue is the moment of inertia about that principal axis—the corresponding principal moment of inertia. When angular velocity vectors $\underline{\omega}$ are expressed as a linear combination of the principal axes, there are no cross-terms in the moment of inertia tensor—no so-called *products of inertia.*

The three principal axes are *unique* when the eigenvalues of $\mathbf{I}$ (principal moments of inertia) are *distinct.* They are not unique when there are repeated eigenvalues, as in the example above of a disk rotating about any of its diameters (§B.4.4, p. 474). In that example, one principal axis, the one corresponding to eigenvalue $MR^2/2$, was $\underline{e}_3$ (*i.e.*, orthogonal to the disk and passing through its center), while any two orthogonal diameters in the plane of the disk may be chosen as the other two principal axes (corresponding to the repeated eigenvalue $MR^2/4$).

Symmetry of the rigid body about any axis $\underline{\omega}_s$ (passing through the origin) means that $\underline{\omega}_s$ is a principal direction. Such a symmetric body may be constructed, for example, as a *solid of revolution.*[26] In rotational dynamics, this case is known as the *symmetric top* [270]. Note that the center of mass will lie somewhere along an axis of symmetry. The other two principal axes can be arbitrarily chosen as a mutually orthogonal pair in the (circular) plane orthogonal to the $\underline{\omega}_s$ axis, intersecting at the $\underline{\omega}_s$ axis. Because of the circular symmetry about $\underline{\omega}_s$, the two principal moments of inertia in that plane are equal. Thus the moment of inertia tensor can be

[26] `http://mathworld.wolfram.com/SolidofRevolution.html`

diagonalized to look like

$$\mathbf{I} = \begin{bmatrix} I_1 & 0 & 0 \\ 0 & I_2 & 0 \\ 0 & 0 & I_2 \end{bmatrix},$$

where I_1 is the principal moment of inertia about $\underline{\omega}_s$, and I_2 is the (twice repeated) principal moment of inertia about the two axes in the circular-symmetry plane. We saw in §B.4.5 (Perpendicular Axis theorem) that if the mass distribution is planar, then $I_1 = 2I_2$.

B.4.17 Rotational Kinetic Energy Revisited

If a point-mass is located at $\underline{x}$ and is rotating about an axis-of-rotation $\tilde{\underline{\omega}}$ with angular velocity ω, then the distance from the rotation axis to the mass is $r = \|\,\underline{x} - \mathcal{P}_{\tilde{\underline{\omega}}}(\underline{x})\,\| = \|\,(\mathbf{E} - \tilde{\underline{\omega}}\tilde{\underline{\omega}}^T)\underline{x}\,\|$, or, in terms of the vector cross product, $r = \|\,\tilde{\underline{\omega}} \times \underline{x}\,\|$. The tangential velocity of the mass is then $r\omega$, so that the kinetic energy can be expressed as (*cf.* Eq. (B.23) on page 488)

$$E_R \;=\; \frac{1}{2} m\,\|\,\tilde{\underline{\omega}} \times \underline{x}\,\|^2\,\omega^2 \;=\; \frac{1}{2} I\omega^2 \tag{B.25}$$

where

$$I \;=\; m\,\|\,\tilde{\underline{\omega}} \times \underline{x}\,\|^2 .$$

In a collection of N masses m_i having velocities $\underline{v}_i$, we of course sum the individual kinetic energies to get the total kinetic energy.

Finally, we may also write the rotational kinetic energy as half the *inner product* of the angular-velocity vector and the angular-momentum vector:[27]

$$E_R \;=\; \frac{1}{2}\underline{\omega}\cdot\underline{L} \;=\; \frac{1}{2} I\omega^2,$$

where the second form (introduced above in Eq. (B.7) on page 473) derives from the vector-dot-product form by using Eq. (B.20) on page 487 and Eq. (B.22) to establish that $\underline{\omega}\cdot\underline{L} = \underline{\omega}\cdot\mathbf{I}\underline{\omega} = \underline{\omega}^T\mathbf{I}\underline{\omega} = \omega^2\tilde{\underline{\omega}}^T\mathbf{I}\tilde{\underline{\omega}} = I\omega^2$.

B.4.18 Torque

When twisting things, the rotational force we apply about the center is called a *torque* (or *moment*, or *moment of force*). Informally, we think of

[27] Here we are using '$\cdot$' to denote the "dot product" (or scalar product, or inner product), since that is the traditional notation in physics. Since it appears between two vectors, this usage is unambiguous. Thus, we have the following equivalent notations: $\underline{x}\cdot\underline{y} = \underline{x}^T\underline{y} = \langle\underline{x},\underline{y}\rangle$.

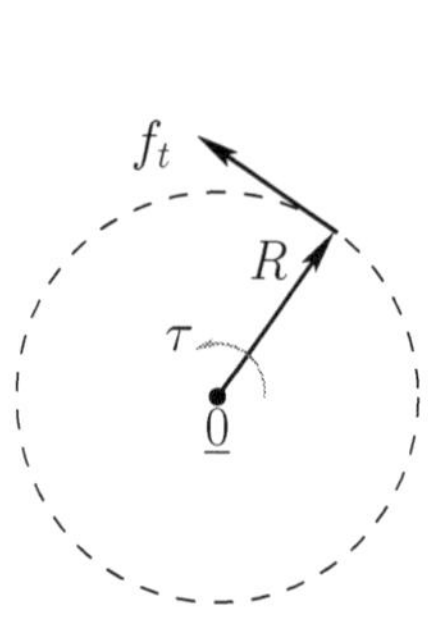

Figure B.7: Application of torque τ about the origin given by a tangential force f_t on a lever arm of length R.

the torque as the *tangential applied force* f_t times the *moment arm* (length of the *lever arm*) R

$$\tau \isdef f_t R \tag{B.26}$$

as depicted in Fig. B.7. The moment arm is the distance from the applied force to the point being twisted. For example, in the case of a wrench turning a bolt, f_t is the force applied at the end of the wrench by one's hand, orthogonal to the wrench, while the moment arm R is the length of the wrench. Doubling the length of the wrench doubles the torque. This is an example of *leverage*. When R is increased, a given twisting angle θ is spread out over a larger arc length $R\theta$, thereby reducing the tangential force f_t required to assert a given torque τ.

For more general applied forces $\underline{f} \in \mathbf{R}^3$, we may compute the tangential component $\underline{f}_t$ by *projecting* $\underline{f}$ onto the tangent direction. More precisely, the *torque* τ about the origin $\underline{0}$ applied at a point $\underline{x} \in \mathbf{R}^3$ may be defined by

$$\underline{\tau} \isdef \underline{x} \times \underline{f} \tag{B.27}$$

where $\underline{f}$ is the applied force (at $\underline{x}$) and $\times$ denotes the cross product, introduced above in §B.4.12 on page 482.

Note that the torque vector $\underline{\tau}$ is orthogonal to both the lever arm and the tangential-force direction. It thus points in the direction of the angular velocity vector (along the axis of rotation).

The torque magnitude is

$$\tau \isdef \|\underline{\tau}\| = \|\underline{x}\| \cdot \|\underline{f}\| \cdot \sin(\theta),$$

where θ denotes the angle from $\underline{x}$ to $\underline{f}$. We can interpret $\|\underline{f}\|\sin(\theta)$ as the length of the projection of $\underline{f}$ onto the tangent direction (the line orthogonal to $\underline{x}$ in the direction of the force), so that we can write

$$\tau = \|\underline{x}\| \cdot \|\underline{f}_t\|$$

where $\|\underline{f}_t\| = \|\underline{f}\|\sin(\theta)$, thus getting back to Eq. (B.26).

B.4.19 Newton's Second Law for Rotations

The rotational version of Newton's law $f = ma$ is

$$\tau = I\alpha, \tag{B.28}$$

where $\alpha \triangleq \dot{\omega}$ denotes the *angular acceleration*. As in the previous section, τ is *torque* (tangential force f_t times a moment arm R), and I is the *mass moment of inertia.* Thus, the net applied torque τ equals the time derivative of angular momentum $L = I\omega$, just as force f equals the time-derivative of linear momentum p:

$$\begin{aligned} \tau &= \dot{L} = I\dot{\omega} \triangleq I\alpha \\ f &= \dot{p} = m\dot{v} \triangleq ma \end{aligned}$$

To show that Eq. (B.28) results from Newton's second law $f = ma$, consider again a mass m rotating at a distance R from an axis of rotation, as in §B.4.3 above, and let f_t denote a *tangential* force on the mass, and a_t the corresponding tangential acceleration. Then we have, by Newton's second law,

$$f_t = ma_t$$

Multiplying both sides by R gives

$$f_t R = ma_t R \triangleq m\dot{v}_t R \triangleq m\dot{\omega}R^2 = I\dot{\omega} = I\alpha.$$

where we used the definitions $\omega = v_t R$ and $I = mR^2$. Furthermore, the left-hand side is the definition of *torque* $\tau = f_t R$. Thus, we have derived

$$\tau = I\alpha$$

from Newton's second law $f_t = ma_t$ applied to the tangential force f_t and acceleration a_t of the mass m.

In summary, force equals the time-derivative of linear momentum, and torque equals the time-derivative of angular momentum. By Newton's laws, the time-derivative of linear momentum is mass times acceleration, and the time-derivative of angular momentum is the mass moment of inertia times angular acceleration:

$$\dot{p}_t = ma_t \quad \Leftrightarrow \quad \dot{L} = I\alpha$$

B.4.20 Equations of Motion for Rigid Bodies

We are now ready to write down the general equations of motion for rigid bodies in terms of $f = ma$ for the center of mass and $\tau = I\alpha$ for the rotation of the body about its center of mass.

As discussed above, it is useful to decompose the motion of a rigid body into

(1) the *linear* velocity $\underline{v}$ of its center of mass, and

(2) its *angular* velocity $\underline{\omega}$ about its center of mass.

The linear motion is governed by Newton's second law $\underline{f} = M\dot{\underline{v}}$, where M is the total mass, $\underline{v}$ is the velocity of the center-of-mass, and $\underline{f}$ is the sum of all external forces on the rigid body. (Equivalently, $\underline{f}$ is the sum of the radial force components pointing toward or away from the center of mass.) Since this is so straightforward, essentially no harder than dealing with a point mass, we will not consider it further.

The angular motion is governed the *rotational* version of Newton's second law introduced in §B.4.19:

$$\underline{\tau} = \dot{\underline{L}} = \mathbf{I}\,\dot{\underline{\omega}} = \mathbf{I}\,\dot{\underline{\omega}} \tag{B.29}$$

where τ is the vector torque defined in Eq. (B.27) on page 494, $\underline{L}$ is the angular momentum, $\mathbf{I}$ is the mass moment of inertia tensor, and $\underline{\omega}$ is the angular velocity of the rigid body about its center of mass. Note that if the center of mass is moving, we are in a moving coordinate system moving with the center of mass (see next section). We may call $\underline{L}$ the *intrinsic momentum* of the rigid body, *i.e.*, that in a coordinate system moving with the center of the mass. We will translate this to the non-moving coordinate system in §B.4.20 below.

The driving torque $\underline{\tau}$ is given by the *resultant moment* of the external forces, using Eq. (B.27) for each external force to obtain its contribution to the total moment. In other words, the external moments (tangential forces times moment arms) sum up for the net torque just like the radial force components summed to produce the net driving force on the center of mass.

Body-Fixed and Space-Fixed Frames of Reference

Rotation is always about some (instantaneous) axis of rotation that is free to change over time. It is convenient to express rotations in a coordinate system having its origin ($\underline{0}$) located at the center-of-mass of the rigid body (§B.4.1), and its coordinate axes aligned along the principal directions for the body (§B.4.16). This *body-fixed frame* then moves within a stationary *space-fixed frame* (or "star frame").

In Eq. (B.29) above, we wrote down Newton's second law for angular motion in the *body-fixed frame*, *i.e.*, the coordinate system having its origin at the center of mass. Furthermore, it is simplest (**I** is diagonal) when its axes lie along principal directions (§B.4.16).

As an example of a local body-fixed coordinate system, consider a spinning top. In the body-fixed frame, the "vertical" axis coincides with the top's axis of rotation (spin). As the top loses rotational kinetic energy due to friction, the top's rotation-axis *precesses* around a circle, as observed in the space-fixed frame. The other two body-fixed axes can be chosen as any two mutually orthogonal axes intersecting each other (and the spin axis) at the center of mass, and lying in the plane orthogonal to the spin axis. The space-fixed frame is of course that of the outside observer's inertial frame[28] in which the top is spinning.

Angular Motion in the Space-Fixed Frame

Let's now consider angular motion in the presence of linear motion of the center of mass. In general, we have [270]

$$\underline{L} = \sum \underline{x} \times \underline{p}$$

where the sum is over all mass particles in the rigid body, and $\underline{p}$ denotes the vector linear momentum for each particle. That is, the angular momentum is given by the tangential component of the linear momentum times the associated moment arm. Using the chain rule for differentiation, we find

$$\underline{\tau} = \dot{\underline{L}} = \frac{d}{dt}\sum \underline{x} \times \underline{p} = \sum (\dot{\underline{x}} \times \underline{p} + \underline{x} \times \dot{\underline{p}}) \triangleq \sum (\underline{v} \times \underline{p} + \underline{x} \times \dot{\underline{p}}).$$

However, $\underline{v} \times \underline{p} = \underline{v} \times m\underline{v} = \underline{0}$, so that

$$\underline{\tau} = \dot{\underline{L}} = \sum \underline{x} \times \dot{\underline{p}} = \sum \underline{x} \times \underline{f}$$

which is the sum of moments of all external forces.

Euler's Equations for Rotations in the Body-Fixed Frame

Suppose now that the body-fixed frame is rotating in the space-fixed frame with angular velocity $\underline{\omega}$. Then the total torque on the rigid body becomes [270]

$$\underline{\tau} = \dot{\underline{L}} + \underline{\omega} \times \underline{L}. \tag{B.30}$$

[28] An *inertial frame* means an *unaccelerated* frame. That is, the coordinate system is not spinning or accelerated in any way. (Even gravity is neglected.)

Similarly, the total external forces on the center of mass become

$$\underline{f} = \dot{\underline{p}} + \underline{\omega} \times \underline{p}.$$

If the body-fixed frame is aligned with the principal axes of rotation (§B.4.16), then the mass moment of inertia tensor is diagonal, say $\mathbf{I} = \text{diag}(I_1, I_2, I_3)$. In this frame, the angular momentum is simply

$$\underline{L} = \begin{bmatrix} I_1\omega_1 \\ I_2\omega_2 \\ I_3\omega_3 \end{bmatrix}$$

so that the term $\underline{\omega} \times \underline{L}$ becomes (*cf.* Eq. (B.15) on page 483)

$$\begin{aligned}
\underline{\omega} \times \underline{L} &= \begin{vmatrix} \underline{e}_1 & \underline{e}_2 & \underline{e}_3 \\ \underline{\omega}_1 & \underline{\omega}_2 & \underline{\omega}_3 \\ I_1\omega_1 & I_2\omega_2 & I_3\omega_3 \end{vmatrix} \\
&= (\underline{\omega}_2 I_3\omega_3 - I_2\omega_2\underline{\omega}_3)\underline{e}_1 + (\underline{\omega}_3 I_1\omega_1 - I_3\omega_3\underline{\omega}_1)\underline{e}_2 + (\underline{\omega}_1 I_2\omega_2 - I_1\omega_1\underline{\omega}_2)\underline{e}_3 \\
&= (I_3 - I_2)\omega_2\omega_3\,\underline{e}_1 + (I_1 - I_3)\omega_3\omega_1\,\underline{e}_2 + (I_2 - I_1)\omega_1\omega_2\,\underline{e}_3.
\end{aligned}$$

Substituting this result into Eq. (B.30), we obtain the following equations of angular motion for an object rotating in the body-fixed frame defined by its three principal axes of rotation:

$$\begin{aligned}
\tau_1 &= I_1\dot{\omega}_1 + (I_3 - I_2)\omega_2\omega_3 \\
\tau_2 &= I_2\dot{\omega}_2 + (I_1 - I_3)\omega_3\omega_1 \\
\tau_3 &= I_3\dot{\omega}_3 + (I_2 - I_1)\omega_1\omega_2
\end{aligned}$$

These are call *Euler's equations:*[29] Since these equations are in the body-fixed frame, I_i is the mass moment of inertia about principal axis i, and ω_i is the angular velocity about principal axis i.

Examples

For a uniform sphere, the cross-terms disappear and the moments of inertia are all the same, leaving $\tau_i = I\omega_i$, for $i = 1, 2, 3$. Since any three orthogonal vectors can serve as eigenvectors of the moment of inertia tensor, we have that, for a uniform sphere, any three orthogonal axes can be chosen as principal axes.

For a cylinder that is not spinning about its axis, we similarly obtain two uncoupled equations $\tau_i = I\omega_i$, for $i = 1, 2$, given $\omega_3 = \tau_3 = 0$ (no spin).

[29] `http://www.physicsforums.com/library.php?do=view_item&itemid=182`

Note, however, that if we replace the circular cross-section of the cylinder by an *ellipse*, then $I_1 \neq I_2$ and there is a coupling term that drives $\dot{\omega}_3$ (unless τ_3 happens to cancel it).

B.5 Properties of Elastic Solids

B.5.1 Young's Modulus

Young's modulus can be thought of as the *spring constant for solids.* Consider an ideal *rod* (or bar) of length L and cross-sectional area S. Suppose we apply a force F to the face of area S, causing a displacement ΔL along the axis of the rod. Then Young's modulus Y is given by

$$Y \triangleq \frac{\text{Stress}}{\text{Strain}} \triangleq \frac{F/S}{\Delta L/L}$$

where

$$\begin{aligned} F &= \text{total applied force} \\ S &= \text{area over which force is applied} \\ F/S &= \textit{stress} = \text{force per unit area} \\ L &= \text{rest length} \\ \Delta L &= \text{displacement} \\ \Delta L/L &= \textit{strain} = \text{displacement per unit length} \end{aligned}$$

For wood, Young's modulus Y is on the order of 10 N/m^2. For aluminum, it is around 70 (a bit higher than glass which is near 65), and structural steel has $Y \approx 200$ [180].

Young's Modulus as a Spring Constant

Recall (§B.1.3) that *Hooke's Law* defines a spring constant k as the applied force F divided by the spring displacement x, or $F = kx$. An elastic solid can be viewed as a *bundle* of ideal springs. Consider, for example, an *ideal bar* (a rectangular solid in which one dimension, usually its longest, is designated its length L), and consider compression by ΔL along the length dimension. The length of each spring in the bundle is the length of the bar, so that each spring constant k must be inversely proportional to L; in particular, each doubling of length L doubles the length of each "spring" in the bundle, and therefore halves its stiffness. As a result, it is useful to normalize displacement ΔL by length L and use *relative displacement*

$\Delta L/L$. We need displacement per unit length because we have a constant spring compliance per unit length.

The *number* of springs in parallel is proportional to the cross-sectional area S of the bar. Therefore, the force applied to each spring is proportional to the total applied force F divided by the cross-sectional area S. Thus, Hooke's law for each spring in the bundle can be written

$$\frac{F}{S} = Y\frac{\Delta L}{L}$$

where Y is Young's modulus.

We may say that Young's modulus is the Hooke's-law spring constant for the spring made from a specifically cut section of the solid material, cut to length 1 and cross-sectional area 1. The shape of the cross-sectional area does not matter since all displacement is assumed to be longitudinal in this model.

B.5.2 String Tension

The *tension* of a vibrating string is the *force* F used to stretch it. It is therefore directed along the axis of the string. A force F must be applied at the endpoint on the right, and a force $-F$ is applied at the endpoint on the left. Each point interior to the string is pulled equally to the left and right, *i.e.*, the net force on an interior point is $F + (-F) = 0$. (A nonzero force on a massless point would produce an infinite acceleration.)

If the cross-sectional area of the string is S, then the tension is given by the stress on the string times S.

B.6 Wave Equation for the Vibrating String

Consider an elastic string under tension which is at rest along the x dimension. Let $\mathbf{i}$, $\mathbf{j}$, and $\mathbf{k}$ denote the unit vectors in the x, y, and z directions, respectively. When a wave is present, a point $\mathbf{p} = (x, 0, 0)$ originally at x along the string is displaced to some point $\mathbf{a} = \mathbf{p} + d\mathbf{p}$ specified by the displacement vector

$$d\mathbf{p} = \mathbf{i}\xi + \mathbf{j}\eta + \mathbf{k}\zeta.$$

Note that typical derivations of the wave equation consider only the displacement η in the y direction. This more general treatment is adapted from [122]. An alternative clear development is given in [391].

The displacement of a neighboring point originally at $\mathbf{q} = (x + dx, 0, 0)$ along the string can be specified as

$$d\mathbf{q} = \mathbf{i}(\xi + d\xi) + \mathbf{j}(\eta + d\eta) + \mathbf{k}(\zeta + d\zeta).$$

Let K denote string tension along x when the string is at rest, and $\mathbf{K}$ denote the vector tension at the point $\mathbf{p}$ in the present displaced scenario under analysis. The net vector force acting on the infinitesimal string element between points $\mathbf{p}$ and $\mathbf{q}$ is given by the vector sum of the force $-\mathbf{K}$ at $\mathbf{p}$ and the force $\mathbf{K}+(\partial\mathbf{K}/\partial x)dx$ at $\mathbf{q}$, that is, $(\partial\mathbf{K}/\partial x)dx$. If the string has stiffness, the two forces will in general not be tangent to the string at these points. The mass of the infinitesimal string element is $\epsilon\, dx$, where ϵ denotes the mass per unit length of the string at rest. Applying Newton's second law gives

$$\frac{\partial \mathbf{K}}{\partial x} = \epsilon \frac{\partial^2 \mathbf{p}}{\partial t^2} \tag{B.31}$$

where dx has been canceled on both sides of the equation. Note that no approximations have been made so far.

The next step is to express the force $\mathbf{K}$ in terms of the tension K of the string at rest, the elastic constant of the string, and geometrical factors. The displaced string element $\mathbf{pq}$ is the vector

$$\begin{aligned} d\mathbf{s} &= \mathbf{i}(dx+d\xi)+\mathbf{j}d\eta+\mathbf{k}d\zeta & \text{(B.32)}\\ &= \left[\mathbf{i}\left(1+\frac{\partial\xi}{\partial x}\right)+\mathbf{j}\frac{\partial\eta}{\partial x}+\mathbf{k}\frac{\partial\zeta}{\partial x}\right]dx & \text{(B.33)}\end{aligned}$$

having magnitude

$$ds = \sqrt{\left(1+\frac{\partial\xi}{\partial x}\right)^2+\left(\frac{\partial\eta}{\partial x}\right)^2+\left(\frac{\partial\zeta}{\partial x}\right)^2}. \tag{B.34}$$

B.6.1 Non-Stiff String

Let's now assume the string is perfectly flexible (zero stiffness) so that the direction of the force vector $\mathbf{K}$ is given by the unit vector $d\mathbf{s}/ds$ tangent to the string. (To accommodate stiffness, it would be necessary to include a force component at right angles to the string which depends on the curvature and stiffness of the string.) The magnitude of $\mathbf{K}$ at any position is the rest tension K plus the incremental tension needed to stretch it the fractional amount

$$\frac{ds-dx}{dx} = \frac{ds}{dx} - 1.$$

If S is the constant cross-sectional area of the string, and Y is the *Young's modulus* (stress/strain—the "spring constant" for solids—see §B.5.1 on page 499), then

$$|\mathbf{K}| = K + SY\left(\frac{ds}{dx}-1\right),$$

so that

$$\mathbf{K} = \left[K + SY\left(\frac{ds}{dx} - 1\right)\right]\frac{d\mathbf{s}}{ds} \tag{B.35}$$

where no *geometrical* limitations have yet been placed on the magnitude of $\mathbf{p}$ and $\partial\mathbf{p}/\partial x$, other than to prevent the string from being stretched beyond its elastic limit.

The four equations (B.31) through (B.35) can be combined into a single *vector* wave equation that expresses the propagation of waves on the string having three displacement components. This differential equation is *nonlinear*, so that *superposition* no longer holds. Furthermore, the three displacement components of the wave are *coupled* together at all points along the string, so that the wave equation is no longer *separable* into three independent 1D wave equations.

To obtain a linear, separable wave equation, it is necessary to assume that the *strains* $\partial\xi/\partial x$, $\partial\eta/\partial x$, and $\partial\zeta/\partial x$ be *small compared with unity.* This is the same assumption ($|\partial\eta/\partial x| \ll 1$) necessary to derive the usual wave equation for transverse vibrations only in the y-x plane.

When (B.35) is expanded into a Taylor series in the strains, and when only the first-order terms are retained, we obtain

$$\mathbf{K} = \mathbf{i}\left(K + SY\frac{\partial\xi}{\partial x}\right) + \mathbf{j}K\frac{\partial\eta}{\partial x} + \mathbf{k}K\frac{\partial\zeta}{\partial x}. \tag{B.36}$$

This is the *linearized* wave equation for the string, based only on the assumptions of elasticity of the string, and strain magnitudes much less than unity. Using this linearized equation for the force $\mathbf{K}$, it is found that (B.31) separates into the three wave equations

$$\frac{\partial^2\xi}{\partial x^2} = \frac{1}{c_b^2}\frac{\partial^2\xi}{\partial t^2} \tag{B.37}$$

$$\frac{\partial^2\eta}{\partial x^2} = \frac{1}{c_t^2}\frac{\partial^2\eta}{\partial t^2} \tag{B.38}$$

$$\frac{\partial^2\zeta}{\partial x^2} = \frac{1}{c_t^2}\frac{\partial^2\zeta}{\partial t^2} \tag{B.39}$$

where $c_b = \sqrt{SY/\epsilon}$ is the longitudinal wave velocity, and $c_t = \sqrt{K/\epsilon}$ is the transverse wave velocity.

In summary, the two transverse wave components and the longitudinal component may be considered *independent* (*i.e.*, "superposition" holds with respect to vibrations in these three dimensions of vibration) provided powers higher than 1 of the strains (relative displacement) can be neglected, *i.e.*,

$$\left|\frac{\partial \xi}{\partial x}\right| \ll 1, \quad \left|\frac{\partial \eta}{\partial x}\right| \ll 1, \quad \text{and} \left|\frac{\partial \zeta}{\partial x}\right| \ll 1.$$

B.6.2 Wave Momentum

The physical forward *momentum* carried by a transverse wave along a string is conveyed by a secondary *longitudinal wave* [391].

A less simplified wave equation which supports longitudinal wave momentum is given by [391, Eqns. 38ab]

$$\epsilon\ddot{\xi} = (SY+K)\,\xi'' + SY\eta'\eta'' \tag{B.40}$$

$$\begin{aligned}\epsilon\ddot{\eta} &= K\eta'' + SY\left(\frac{3}{2}(\eta')^2\eta'' + \xi'\eta'' + \eta'\xi''\right) && \text{(B.41)}\\ &\approx K\eta'', && \text{(B.42)}\end{aligned}$$

where ξ and η denote longitudinal and transverse displacement, respectively, and the commonly used "dot" and "prime" notation for partial derivatives has been introduced, *e.g.*,

$$\dot{\xi} \triangleq \frac{\partial \xi}{\partial t} \tag{B.43}$$

$$\xi' \triangleq \frac{\partial \xi}{\partial x}. \tag{B.44}$$

(See also Eq. (C.1) on page 523.) We see that the term $SY\eta'\eta''$ in the first equation above provides a mechanism for transverse waves to "drive" the generation of longitudinal waves. This coupling cannot be neglected if momentum effects are desired.

Physically, the rising edge of a transverse wave generates a longitudinal displacement in the direction of wave travel that propagates ahead at a much higher speed (typically an order of magnitude faster). The falling edge of the transverse wave then cancels this forward displacement as it passes by. See [391] for further details (including computer simulations).

B.7 Properties of Gases

B.7.1 Particle Velocity of a Gas

The *particle velocity* of a gas flow at any point can be defined as the average velocity (in meters per second, m/s) of the air molecules passing through a plane cutting orthogonal to the flow. The term "velocity" in this book, when referring to air, means "particle velocity."

It is common in acoustics to denote particle velocity by lower-case u.

B.7.2 Volume Velocity of a Gas

The *volume velocity* U of a gas flow is defined as particle velocity u times the cross-sectional area A of the flow, or

$$U(t,x) = u(t,x)A(x)$$

where x denotes position along the flow, and t denotes time in seconds. Volume velocity is thus in physical units of volume per second (m^3/s).

When a flow is confined within an enclosed channel, as it is in an acoustic tube, *volume velocity is conserved* when the tube changes cross-sectional area, assuming the density ρ remains constant. This follows directly from conservation of mass in a flow: The total mass passing a given point x along the flow is given by the mass density ρ times the integral of the volume volume velocity at that point, or

$$M(t_0 : t_f, x) = \rho \int_{t_0}^{t_f} U(t,x)dt.$$

As a simple example, consider a constant flow through two cylindrical acoustic tube sections having cross-sectional areas A_1 and A_2, respectively. If the particle velocity in cylinder 1 is u_1, then the particle velocity in cylinder 2 may be found by solving

$$u_1 A_1 = u_2 A_2$$

for u_2.

It is common in the field of acoustics to denote volume velocity by an upper-case U. Thus, for the two-cylinder acoustic tube example above, we would define $U_1 \triangleq u_1 A_1$ and $U_2 \triangleq u_2 A_2$, so that

$$U_1 = U_2$$

would express the conservation of volume velocity from one tube segment to the next.

B.7.3 Pressure is Confined Kinetic Energy

According the *kinetic theory of ideal gases* [180], *air pressure* can be defined as the *average momentum transfer per unit area per unit time* due to molecular collisions between a confined gas and its boundary. Using Newton's second law, this pressure can be shown to be given by one third of the average kinetic energy of molecules in the gas.

$$p = \frac{1}{3}\rho \left\langle u^2 \right\rangle$$

Here, $\langle u^2 \rangle$ denotes the average squared particle velocity in the gas. (The constant 1/3 comes from the fact that we are interested only in the kinetic energy directed along one dimension in 3D space.)

Proof: This is a classical result from the kinetic theory of gases [180]. Let M be the total mass of a gas confined to a rectangular volume $V = Aw$, where A is the area of one side and w the distance to the opposite side. Let $\overline{u}_x$ denote the average molecule velocity in the x direction. Then the total net molecular momentum in the x direction is given by $M|\overline{u}_x|$. Suppose the momentum $\overline{u}_x$ is directed against a face of area A. A rigid-wall elastic collision by a mass M traveling into the wall at velocity $\overline{u}_x$ imparts a momentum of magnitude $2M\overline{u}_x$ to the wall (because the momentum of the mass is changed from $+M\overline{u}_x$ to $-M\overline{u}_x$, and momentum is conserved). The average momentum-transfer per unit area is therefore $2M\overline{u}_x/A$ at any instant in time. To obtain the definition of pressure, we need only multiply by the average collision rate, which is given by $\overline{u}_x/(2w)$. That is, the average x-velocity divided by the round-trip distance along the x dimension gives the collision rate at either wall bounding the x dimension. Thus, we obtain

$$p \stackrel{\Delta}{=} \frac{2M\overline{u}_x}{A} \cdot \frac{\overline{u}_x}{2w} = \rho\overline{u}_x^2$$

where $\rho = M/V$ is the density of the gas in mass per unit volume. The quantity $\rho\overline{u}_x^2/2$ is the average kinetic energy density of molecules in the gas along the x dimension. The total kinetic energy density is $\rho\overline{u}^2/2$, where $\overline{u} = \sqrt{\overline{u}_x^2 + \overline{u}_y^2 + \overline{u}_z^2}$ is the average molecular velocity magnitude of the gas. Since the gas pressure must be the same in all directions, by symmetry, we must have $\overline{u}_x^2 = \overline{u}_y^2 = \overline{u}_z^2 = \overline{u}^2/3$, so that

$$p = \frac{1}{3}\rho\overline{u}^2.$$

B.7.4 Bernoulli Equation

In an ideal *inviscid, incompressible flow*, we have, by *conservation of energy*,

$$p + \frac{1}{2}\rho u^2 + \rho g h = \text{constant}$$

where

$$\begin{aligned} p &= \text{pressure (newtons/m}^2 = \text{kg /(m s}^2\text{))} \\ u &= \text{particle velocity (m/s)} \\ \rho &= \text{volume density of air (kg/m}^3\text{)} \\ g &= \text{Newton's gravitational constant (m/s}^2\text{)} \\ h &= \text{Height of flow's center-of-mass axis (m)} \\ \text{“Inviscid”} &= \text{“Frictionless”, “Lossless”} \end{aligned}$$

This basic energy conservation law was published in 1738 by Daniel Bernoulli in his classic work *Hydrodynamica.*

From §B.7.3, we have that the pressure of a gas is proportional to the average kinetic energy of the molecules making up the gas. Therefore, when a gas flows at a constant height h, some of its "pressure kinetic energy" must be given to the kinetic energy of the flow as a whole. If the mean height of the flow changes, then kinetic energy trades with potential energy as well.

B.7.5 Bernoulli Effect

The *Bernoulli effect* provides that, when a gas such as air flows, its pressure drops. This is the basis for how aircraft wings work: The cross-sectional shape of the wing, called an *aerofoil* (or *airfoil*), forces air to follow a longer path over the top of the wing, thereby speeding it up and creating a net upward force called *lift.*

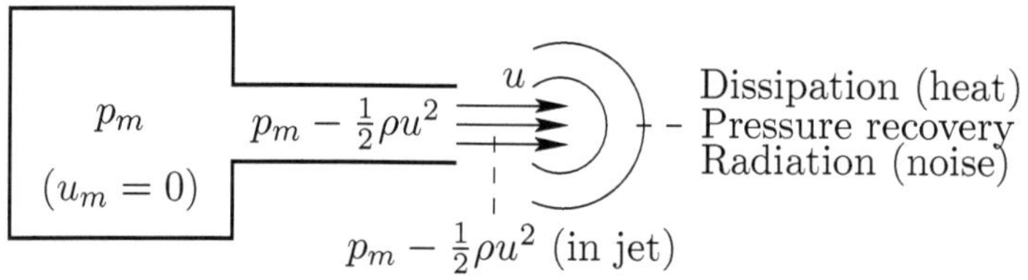

Figure B.8: Illustration of the Bernoulli effect in an acoustic tube.

Figure B.8 illustrates the Bernoulli effect for the case of a reservoir at constant pressure p_m ("mouth pressure") driving an acoustic tube. Any flow inside the "mouth" is neglected. Within the acoustic channel, there is a flow with constant particle velocity u. To conserve energy, the pressure within the acoustic channel must drop down to $p_m - \rho u^2/2$. That is, the flow kinetic energy subtracts from the pressure kinetic energy within the channel.

For a more detailed derivation of the Bernoulli effect, see, *e.g.*, [179]. Further discussion of its relevance in musical acoustics is given in [144, 197].

B.7.6 Air Jets

Referring again to Fig. B.8, the gas flow exiting the acoustic tube is shown as forming a *jet*. The jet "carries its own pressure" until it dissipates in some form, such as any combination of the following:

- heat (now allowing for "friction" in the flow),
- vortices (angular momentum),
- radiation (sound waves), or
- pressure recovery.

Pressure recovery refers to the conversion of flow kinetic energy back to pressure kinetic energy. In situations such as the one shown in Fig. B.8, the flow itself is *driven* by the pressure drop between the confined reservoir (pressure p_m) and the outside air (pressure $p_m - \rho u^2/2$). Therefore, any pressure recovery would erode the pressure drop and hence the flow velocity u.

For a summary of more advanced aeroacoustics, including consideration of vortices, see [196]. In addition, basic textbooks on *fluid mechanics* are relevant [171].

B.7.7 Acoustic Intensity

Acoustic intensity may be defined by

$$\boxed{\underline{I} \triangleq p\underline{v}} \quad \left(\frac{\text{Energy Flux}}{\text{Area} \cdot \text{Time}} = \frac{\text{Power Flux}}{\text{Area}}\right)$$

where

$$\begin{aligned} p &= \text{acoustic pressure} \quad \left(\frac{\text{Force}}{\text{Area}}\right) \\ \underline{v} &= \text{acoustic particle velocity} \quad \left(\frac{\text{Length}}{\text{Time}}\right). \end{aligned}$$

For a *plane traveling wave*, we have

$$\boxed{p = Rv}$$

where

$$\boxed{R \triangleq \rho c}$$

is called the *wave impedance* of air, and

$$\begin{aligned} c &= \text{sound speed}, \\ \rho &= \text{mass density of air} \quad \left(\frac{\text{Mass}}{\text{Volume}}\right), \\ v &\triangleq |\underline{v}|. \end{aligned}$$

Therefore, in a plane wave,

$$\boxed{I = pv = Rv^2 = \frac{p^2}{R}.}$$

B.7.8 Acoustic Energy Density

The two forms of energy in a wave are *kinetic* and *potential.* Denoting them at a particular time t and position $\underline{x}$ by $w_v(t, \underline{x})$ and $w_p(t, \underline{x})$, respectively, we can write them in terms of velocity v and wave impedance $R = \rho c$ as follows:

$$\begin{aligned} w_v &= \frac{1}{2}\rho v^2 = \frac{1}{2c} R v^2 \quad \left(\frac{\text{Energy}}{\text{Volume}}\right) \\ w_p &= \frac{1}{2}\frac{p^2}{\rho c^2} = \frac{1}{2c}\frac{p^2}{R} \quad \left(\frac{\text{Energy}}{\text{Volume}}\right) \end{aligned}$$

More specifically, w_v and w_p may be called the *acoustic kinetic energy density* and the *acoustic potential energy density*, respectively.

At each point in a plane wave, we have $p(t, \underline{x}) = Rv(t, \underline{x})$ (pressure equals wave-impedance times velocity), and so

$$\begin{aligned} w_v &= \frac{1}{2c} R v^2 = \frac{1}{2} \cdot \frac{I}{c} \\ w_p &= \frac{1}{2c}\frac{p^2}{R} = \frac{1}{2} \cdot \frac{I}{c}, \end{aligned}$$

where $I(t, \underline{x}) \triangleq p(t, \underline{x})\, v(t, \underline{x})$ denotes the acoustic intensity (pressure times velocity) at time t and position $\underline{x}$. Thus, half of the acoustic intensity I in a plane wave is kinetic, and the other half is potential:[30]

$$\frac{I}{c} = w = w_v + w_p = 2w_v = 2w_p$$

Note that acoustic intensity I has units of *energy per unit area per unit time* while the acoustic energy density $w = I/c$ has units of *energy per unit volume.*

B.7.9 Energy Decay through Lossy Boundaries

Since the acoustic energy density $w = w_v + w_p$ is the energy per unit volume in a 3D sound field, it follows that the total energy of the field is given by

[30] This was first pointed out by Rayleigh [349, p. 39].

integrating over the volume:

$$E(t) = \iiint\limits_V w(t, \underline{x})\, dV$$

In reverberant rooms and other acoustic systems, the field energy decays over time due to losses. Assuming the losses occur only at the boundary of the volume, we can equate the rate of total-energy change to the rate at which energy exits through the boundaries. In other words, the energy lost by the volume V in time interval Δt must equal the acoustic intensity $\underline{I}(t, \underline{x})$ exiting the volume, times Δt (approximating I as constant between times t and $t + \Delta t$):

$$E(t + \Delta t) - E(t) = -\Delta t \iint\limits_A \underline{I} \cdot \underline{n}\, dA$$

The term $\underline{I}(t, \underline{x}) \cdot \underline{n}(\underline{x})$ is the dot-product of the (vector) intensity $\underline{I}$ with a unit-vector $\underline{n}$ chosen to be normal to the surface at position $\underline{x}$ along the surface. Thus, $\underline{I} \cdot \underline{n}$ is the component of the acoustic intensity $\underline{I}$ exiting the volume normal to its surface. (The tangential component does not exit.) Dividing through by Δt and taking a limit as $\Delta t \to 0$ yields the following conservation law, originally published by Kirchoff in 1867:

$$\frac{d}{dt} E = \frac{d}{dt} \iiint\limits_V w(t, \underline{x})\, dV = -\iint\limits_A \underline{I}(t, \underline{x}) \cdot \underline{n}(\underline{x})\, dA.$$

Thus, the rate of change of energy in an ideal acoustic volume V is equal to the surface integral of the power crossing its boundary. A more detailed derivation appears in [349, p. 37].

Sabine's theory of acoustic energy decay in reverberant room impulse responses can be derived using this conservation relation as a starting point.

B.7.10 Ideal Gas Law

The *ideal gas law* can be written as

$$PV = nRT = NkT \tag{B.45}$$

where

$$\begin{aligned} P &= \text{total pressure (Pascals)} \\ V &= \text{volume (cubic meters)} \\ n &= \text{number of } \textit{moles} \text{ of gas in the volume } V \\ R &= 8.3143 = \textit{ideal gas constant} \text{ (SI units)}, \textit{and} \\ T &= \textit{absolute temperature} \text{ (degrees Kelvin)}. \end{aligned}$$

The alternate form $PV = NkT$ comes from the *statistical mechanics* derivation in which N is the number of gas molecules in the volume, and k is *Boltzmann's constant.* In this formulation (the *kinetic theory of ideal gases*), the *average kinetic energy* of the gas molecules is given by $(3/2)kT$. Thus, *temperature is proportional to average kinetic energy* of the gas molecules, where the kinetic energy of a molecule m with translational speed v is given by $(1/2)mv^2$.

In an ideal gas, the molecules are like little rubber balls (or rubbery assemblies of rubber balls) in a weightless vacuum, colliding with each other and the walls elastically and losslessly (an "ideal rubber"). Electromagnetic forces among the molecules are neglected, other than the electron-orbital repulsion producing the elastic collisions; in other words, the molecules are treated as electrically neutral far away. (Gases of ionized molecules are called *plasmas*.)

The mass m of the gas in volume V is given by $m = nM$, where M is the *molar mass* of the gass (about 29 g per mole for air). The air density is thus $\rho = m/V$ so that we can write

$$P = \frac{R}{M}\rho T.$$

That is, pressure P is proportional to density ρ at constant temperature T (with R/M being a constant).

We normally do not need to consider the (nonlinear) ideal gas law in audio acoustics because it is usually *linearized* about some ambient pressure P_0. The physical pressure is then $P = P_0 + p$, where p is the usual acoustic pressure-wave variable. That is, we are only concerned with small pressure *perturbations* p in typical audio acoustics situations, so that, for example, variations in volume V and density ρ can be neglected. Notable exceptions include brass instruments which can achieve nonlinear sound-pressure regions, especially near the mouthpiece [198, 52]. Additionally, the aeroacoustics of air jets is nonlinear [196, 531, 532, 533, 102, 101].

B.7.11 Isothermal versus Isentropic

If air compression/expansion were *isothermal* (constant temperature T), then, according to the ideal gas law $PV = nRT$, the pressure P would simply be proportional to density ρ. It turns out, however, that heat diffusion is much slower than audio acoustic vibrations. As a result, air compression/expansion is much closer to *isentropic* (constant entropy S) in normal acoustic situations. (An isentropic process is also called a *reversible adiabatic process*.) This means that when air is compressed by shrinking its volume V, for example, not only does the pressure P increase (§B.7.3, p.

504), but the temperature T increases as well (as quantified in the next section). In a constant-entropy compression/expansion, temperature changes are not given time to diffuse away to thermal equilibrium. Instead, they remain largely frozen in place. Compressing air heats it up, and relaxing the compression cools it back down.

B.7.12 Adiabatic Gas Constant

The relative amount of compression/expansion energy that goes into temperature T versus pressure P can be characterized by the *heat capacity ratio*

$$\gamma \triangleq \frac{C_P}{C_V}$$

where C_P is the *specific heat* (also called *heat capacity*) at constant pressure, while C_V is the specific heat at constant volume. The *specific heat*, in turn, is the amount of heat required to raise the temperature of the gas by one degree. It is derived in *statistical thermodynamics* [138] that, for an ideal gas, we have $C_p = C_v + R$, where R is the ideal gas constant (introduced in Eq. (B.45) on page 509). Thus, $\gamma > 1$ for any ideal gas. The extra heat absorption that occurs when heating a gas at constant pressure is associated with the *work* (§B.2) performed on the volume boundary (fore times distance = pressure times area times distance) as it expands to keep pressure constant. Heating a gas at constant volume involves increasing the kinetic energy of the molecules, while heating a gas at constant pressure involves both that *and* pushing the boundary of the volume out. The reason not all gases have the same γ is that they have different *internal degrees of freedom*, such as those associated with spinning and vibrating internally. Each degree of freedom can store energy.

In terms of γ, we have

$$P_1 V_1^\gamma = P_2 V_2^\gamma, \tag{B.46}$$

where $\gamma \approx 1.4$ for dry air at normal temperatures. Thus, if a volume of ideal gas is changed from V_1 to V_2, the pressure change is given by

$$P_2 = P_1 \left(\frac{V_1}{V_2}\right)^{1/\gamma}$$

and the temperature change is

$$T_2 = T_1 \left(\frac{V_1}{V_2}\right)^{\gamma-1}.$$

These equations both follow from Eq. (B.46) and the ideal gas law Eq. (B.45).

The value $\gamma = 1.4$ is typical for any *diatomic gas*.[31] Monatomic inert gases, on the other hand, such as Helium, Neon, and Argon, have $\gamma \approx 1.6$. Carbon dioxide, which is *triatomic*, has a heat capacity ratio $\gamma = 1.28$. We see that more complex molecules have lower γ values because they can store heat in more degrees of freedom.

B.7.13 Heat Capacity of Ideal Gases

In statistical thermodynamics [175, 138], it is derived that each molecular degree of freedom contributes $R/2$ to the molar heat capacity of an ideal gas, where again R is the ideal gas constant.

An ideal *monatomic* gas molecule (negligible spin) has only three degrees of freedom: its kinetic energy in the three spatial dimensions. Therefore, $C_v = (3/2)R$. This means we expect

$$\gamma \stackrel{\Delta}{=} \frac{C_p}{C_v} = \frac{C_v + R}{C_v} = \frac{3/2+1}{3/2} = 5/3,$$

a result that agrees well with experimental measurements [138].

For an ideal *diatomic* gas molecule such as air, which can be pictured as a "bar bell" configuration of two rubber balls, two additional degrees of freedom are added, both associated with spinning the molecule about an axis orthogonal to the line connecting the atoms, and piercing its center of mass. There are two such axes. Spinning about the connecting axis is neglected because the moment of inertia is so much smaller in that case. Thus, for diatomic gases such as dry air, we expect

$$\gamma \stackrel{\Delta}{=} \frac{C_p}{C_v} = \frac{C_v + R}{C_v} = \frac{5/2+1}{5/2} = 7/5 = 1.4,$$

as observed to a good degree of approximation at normal temperatures. At high temperatures, new degrees of freedom appear associated with vibrations in the molecular bonds. (For example, the "bar bell" can vibrate longitudinally.) However, such vibrations are "frozen out" at normal room temperatures, meaning that their (quantized) energy levels are too high and spaced too far apart to be excited by room temperature collisions [138, p. 147].[32]

---

[31] Air has been measured to contain 75.54% nitrogen and 23.10% oxygen, both of which are diatomic gases. Thus, dry air is about 99% diatomic.

[32] This is a rare example in which quantum mechanics must be considered in an acoustic calculation!

B.7.14 Speed of Sound in Air

The *speed of sound* in a gas depends primarily on the temperature, and can be estimated using the following formula from the kinetic theory of gases:[33]

$$c = \sqrt{\gamma R_m T},$$

where, as discussed in the previous section, the adiabatic gas constant is $\gamma = 1.4$ for dry air, $R_m = 286$ is the ideal gas constant for air in meters-squared per second-squared per degrees-Kelvin-squared, and T is absolute temperature in degrees Kelvin (which equals degrees Celsius + 273.15). For example, at zero degrees Celsius (32 degrees Fahrenheit), the speed of sound is calculated to be 1085.1 feet per second. At 20 degrees Celsius, we get 1124.1 feet per second.

B.7.15 Air Absorption

This section provides some further details regarding acoustic air absorption [318]. For a plane wave, the decline of acoustic intensity as a function of propagation distance x is given by

$$I(x) = I_1 e^{-x/\xi},$$

where

$$\begin{aligned} I(x) &= \text{intensity } x \text{ meters from the source} \\ &\quad (\S\text{B.7.7 defines acoustic intensity}), \\ I_1 &= \text{intensity one meter from the source, and} \\ \xi &= \text{intensity decay constant (meters)} \\ &\quad \text{(depends on frequency, temperature, humidity, and pressure).} \end{aligned}$$

Tables B.1 and B.2 (adapted from [314]) give some typical values for air.

There is also a (weaker) dependence of air absorption on *temperature* [183].

Theoretical models of energy loss in a gas are developed in Morse and Ingard [318, pp. 270–285]. Energy loss is caused by *viscosity*, *thermal diffusion*, *rotational relaxation*, *vibration relaxation*, and *boundary losses* (losses due to heat conduction and viscosity at a wall or other acoustic boundary). Boundary losses normally dominate by several orders of magnitude, but in resonant modes, which have nodes along the boundaries, interior losses

[33] `http://www.grc.nasa.gov/WWW/K-12/airplane/sound.html`

Relative Humidity	Frequency in Hz 1000	2000	3000	4000
40	0.0013	0.0037	0.0069	0.0242
50	0.0013	0.0027	0.0060	0.0207
60	0.0013	0.0027	0.0055	0.0169
70	0.0013	0.0027	0.0050	0.0145

Table B.1: Attenuation constant $m = 1/\xi$ (in inverse meters) at 20 degrees Celsius and standard atmospheric pressure

Relative Humidity	Frequency in Hz 1000	2000	3000	4000
40	5.6	16	30	105
50	5.6	12	26	90
60	5.6	12	24	73
70	5.6	12	22	63

Table B.2: Attenuation in dB per kilometer at 20 degrees Celsius and standard atmospheric pressure.

dominate, especially for polyatomic gases such as air.[34] For air having moderate amounts of water vapor (H_2O) and/or carbon dioxide (CO_2), the loss and dispersion due to N_2 and O_2 vibration relaxation hysteresis becomes the largest factor [318, p. 300]. The vibration here is that of the molecule itself, accumulated over the course of many collisions with other molecules. In this context, a diatomic molecule may be modeled as two masses connected by an ideal spring. Energy stored in molecular vibration typically dominates over that stored in molecular rotation, for polyatomic gas molecules [318, p. 300]. Thus, vibration relaxation hysteresis is a loss mechanism that converts wave energy into heat.

In a resonant mode, the attenuation per wavelength due to vibration relaxation is greatest when the sinusoidal period (of the resonance) is equal to 2π times the time-constant for vibration-relaxation. The relaxation time-constant for oxygen is on the order of one millisecond. The presence of water vapor (or other impurities) decreases the vibration relaxation time, yielding loss maxima at frequencies above 1000 rad/sec. The energy loss approaches zero as the frequency goes to infinity (wavelength to zero).

Under these conditions, the speed of sound is approximately that of dry air below the maximum-loss frequency, and somewhat higher above. Thus, the humidity level changes the dispersion cross-over frequency of the

[34] Air is predominately *diatomic* due to nitrogen (N_2, 78%) and oxygen (O_2, 21%) comprising 99% of the Earth's atmosphere.

air in a resonant mode.

B.8 Wave Equation in Higher Dimensions

The *wave equation* in 1D, 2D, or 3D may be written as

$$\left(\nabla^2 - \frac{1}{c^2}\frac{\partial^2}{\partial t^2}\right) z(\underline{x},t) \;=\; 0, \tag{B.47}$$

where, in 3D, $z(\underline{x},t)$ denotes the amplitude of the wave at time t and position $\underline{x} \in \mathbf{R}^3$, and

$$\nabla^2 \;\stackrel{\Delta}{=}\; \nabla\cdot\nabla \;\stackrel{\Delta}{=}\; \nabla^T\nabla \;=\; \frac{\partial^2}{\partial x^2} + \frac{\partial^2}{\partial y^2} + \frac{\partial^2}{\partial z^2}$$

denotes the *Laplacian operator* in Euclidean coordinates. (In general coordinates, it is often denoted by Δ.) To investigate solutions of the wave equation, as pursued in §B.8.3 below, it is useful to first develop some simple expressions and notations for elementary waves in 2D and 3D.

B.8.1 Plane Waves in Air

Figure B.9 shows a 2D xy cross-section of a snapshot (in time) of the sinusoidal plane wave

$$p(x,y,z) = p_0 + \cos(k_x x + k_y y)$$

for $k_x = 2\pi 5$ and $k_y = 2\pi 5/2$, with x and y in the range $[0,1)$.

Figure B.10 depicts a more mathematical schematic of a sinusoidal plane wave traveling toward the upper-right of the figure. The dotted lines indicate the *crests* (peak amplitude location) along the wave.

The direction of travel and spatial frequency are indicated by the *vector wavenumber* $\underline{k}$, as discussed in in the following section.

B.8.2 Vector Wavenumber

Mathematically, a sinusoidal plane wave, as in Fig. B.9 or Fig. B.10, can be written as

$$p(t,\underline{x}) \;=\; p_0 + A\cos\left(\omega t - \underline{k}^T\underline{x} + \phi\right), \quad \underline{x} \in \mathbf{R}^3 \tag{B.48}$$

where p(t,x) is the pressure at time t (seconds) and position $\underline{x} \in \mathbf{R}^3$ (3D Euclidean space). The amplitude A, phase ϕ, and radian frequency ω are

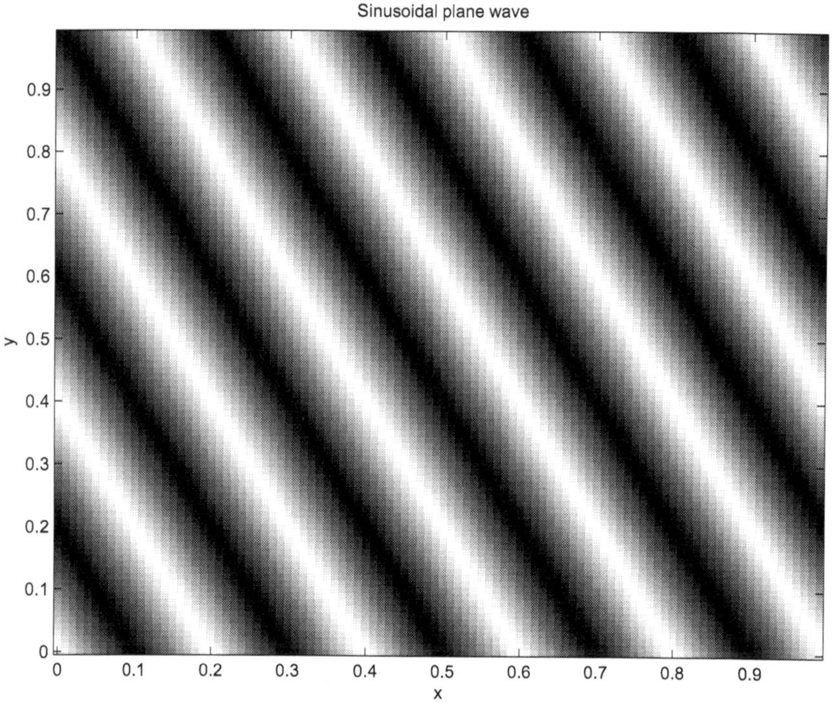

Figure B.9: Gray-scale density plot of the xy cross-section of a sinusoidal plane wave $p(t, \underline{x}) = \cos\left(\omega t - \underline{k}^T \underline{x}\right)$, at $t = 0$ with vector wavenumber $\underline{k}^T = [10\pi, 5\pi, 0]$.

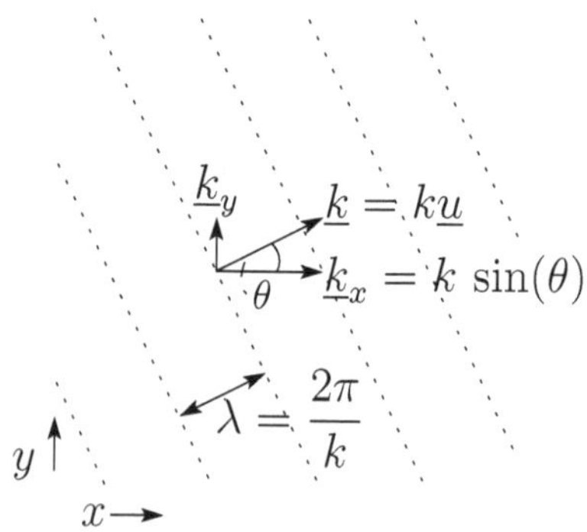

Figure B.10: Wave crests of the sinusoidal traveling plane wave $p(t, \underline{x}) = \cos\left(\omega t - \underline{k}^T \underline{x}\right)$, for some fixed time t and $\underline{x}$ in the $(x, y, 0)$ plane.

ordinary sinusoid parameters [451], and $\underline{k}$ is the *vector wavenumber*:

$$\underline{k} = \begin{bmatrix} k_x \\ k_y \\ k_z \end{bmatrix} = k \begin{bmatrix} k_x/k \\ k_y/k \\ k_z/k \end{bmatrix} \triangleq k \begin{bmatrix} \cos\alpha \\ \cos\beta \\ \cos\gamma \end{bmatrix} \triangleq k\,\underline{u},$$

where

- $\underline{u} =$ (unit) vector of *direction cosines*
- $k = 2\pi/\lambda =$ (scalar) *wavenumber* along travel direction

Thus, the vector wavenumber $\underline{k} = k\,\underline{u}$ contains

- wavenumber along the travel direction in its magnitude $k = \|\,\underline{k}\,\|$
- travel direction in its orientation $\underline{u} = \underline{k}/k$

Note that wavenumber units are *radians per meter* (spatial radian frequency).

To see that the vector wavenumber $\underline{k} = k\,\underline{u}$ has the claimed properties, consider that the *orthogonal projection* of any vector $\underline{x}$ onto a vector collinear with $\underline{u}$ is given by $(\underline{u}^T\underline{x})\underline{u}$ [451].[35] Thus, $(\underline{u}^T\underline{x})\underline{u}$ is the component of $\underline{x}$ lying along the direction of wave propagation indicated by $\underline{u}$. The norm of this component is $\|\,(\underline{u}^T\underline{x})\underline{u}\,\| = |\underline{u}^T\underline{x}|$, since $\underline{u}$ is unit-norm by construction. More generally, $\underline{u}^T\underline{x}$ is the *signed length* (in meters) of the component of $\underline{x}$ along $\underline{u}$. This length times wavenumber k gives the spatial phase advance along the wave, or, $\theta(\underline{x}) = k\cdot(\underline{u}^T\underline{x}) \triangleq \underline{k}^T\underline{x}$.

For another point of view, consider the plane wave $\cos(\underline{k}^T\underline{x})$, which is the varying portion of the general plane-wave of Eq. (B.48) at time $t = 0$, with unit amplitude $A = 1$ and zero phase $\phi = 0$. The spatial phase of this plane wave is given by

$$\theta(\underline{x}) \triangleq \underline{k}^T\underline{x} = k_x x + k_y y + k_z z.$$

Recall that the general equation for a plane in 3D space is

$$\alpha x + \beta y + \gamma z = \text{constant}$$

where α, β, and γ are real constants, and x, y, and z are 3D spatial coordinates. Thus, the set of all points $\underline{x}^T = (x, y, z)$ yielding the same value $\theta(\underline{x}) = \theta_0$ define a *plane of constant phase* θ_0 in $\mathbf{R}^3$.

As we know from elementary vector calculus, the direction of maximum phase advance is given by the *gradient* of the phase $\theta(\underline{x})$:

$$\underline{\nabla}\theta(\underline{x}) \triangleq \begin{bmatrix} \frac{\partial}{\partial x} \\ \frac{\partial}{\partial y} \\ \frac{\partial}{\partial z} \end{bmatrix} \theta(\underline{x}) = \begin{bmatrix} k_x \\ k_y \\ k_z \end{bmatrix} \triangleq \underline{k}$$

[35] `http://ccrma.stanford.edu/~jos/mdft/Projection.html`

This shows that the vector wavenumber $\underline{k}$ is equal to the gradient of the phase $\theta(\underline{x})$, so that $\underline{k}$ points in the direction of *maximum spatial-phase advance.*

Since the wavenumber k is the spatial frequency (in radians per meter) along the direction of travel, we should be able to compute it as the *directional derivative* of $\theta(\underline{x})$ along $\underline{k}$, *i.e.*,

$$k \triangleq d_{\underline{\nabla\theta}}\theta(\underline{x}) \triangleq \lim_{\delta\to 0}\frac{\theta\left(\underline{x}+\delta\underline{\nabla\theta}\right)-\theta(\underline{x})}{\delta\,\|\,\underline{\nabla\theta}\,\|}.$$

An explicit calculation yields

$$k = \|\,\underline{\nabla\theta}\,\| = \sqrt{k_x^2+k_y^2+k_z^2} \triangleq \|\,\underline{k}\,\|$$

as needed.

Scattering of plane waves is discussed in §C.8.1 on page 560.

B.8.3 Solving the 2D Wave Equation

Since solving the wave equation in 2D has all the essential features of the 3D case, we will look at the 2D case in this section.

Specializing Eq. (B.47) on page 515 to 2D, the *2D wave equation* may be written as

$$\left(\nabla^2 - \frac{1}{c^2}\frac{\partial^2}{\partial t^2}\right) z(\underline{x},t) = 0.$$

where

$$\nabla^2 \triangleq \nabla\cdot\nabla \triangleq \nabla^T\nabla = \frac{\partial^2}{\partial x^2}+\frac{\partial^2}{\partial y^2}.$$

The 2D wave equation is obeyed by *traveling sinusoidal plane waves* having any amplitude A, radian frequency ω, phase ϕ, and direction $\underline{u}$:

$$z(\underline{x},t) = A\,e^{j\phi}\,e^{j(\omega t-\underline{k}^T\underline{x})}$$

where $\underline{k} = k\underline{u}$ denotes the vector-wavenumber, $k=\omega/c$ denotes the wavenumber (spatial radian frequency) of the wave along its direction of travel, and $\underline{u}$ is a unit vector of direction cosines. This is the analytic-signal form of a sinusoidal traveling plane wave, and we may define the real (physical) signal as the real part of the analytic signal, as usual [451]. We see that the only constraint imposed by the wave equation on this general traveling-wave is the so-called *dispersion relation*:

$$k = |\underline{k}| = \frac{\omega}{c}$$

In particular, the wave can travel in any direction, with any amplitude, frequency, and phase. The only constraint is that its spatial frequency k is tied to its temporal frequency ω by the dispersion relation.[36]

The sum of two such waves traveling in opposite directions with the same amplitude and frequency produces a *standing wave*. For example, if the waves are traveling parallel to the x axis, we have

$$z(\underline{x},t) \;=\; A\,e^{j\phi}\,e^{j\omega t-kx} + A\,e^{j\phi}\,e^{j\omega t+kx} \;=\; 2A\,e^{j(\omega t+\phi)}\cos(kx) \tag{B.49}$$

which is a standing wave along x.

B.8.4 2D Boundary Conditions

We often wish to find solutions of the 2D wave equation that obey certain known *boundary conditions*. An example is transverse waves on an ideal elastic membrane, rigidly clamped on its boundary to form a rectangle with dimensions $X \times Y$ meters.

Similar to the derivation of Eq. (B.49), we can *subtract* the second sinusoidal traveling wave from the first to yield

$$z(\underline{x},t) \;=\; A\,e^{j\phi}\,e^{j\omega t-kx} - A\,e^{j\phi}\,e^{j\omega t+kx} \;=\; 2A\,e^{j(\omega t+\phi+\pi/2)}\sin(kx)$$

which satisfies the zero-displacement boundary condition along the y axis. If we restrict the wavenumber k to the set $m\pi/X$, where m is any positive integer, then we also satisfy the boundary condition along the line parallel to the y axis at $x = X$. Similar standing waves along y will satisfy both boundary conditions along $(x, 0)$ and (x, Y).

Note that we can also use *products* of horizontal and vertical standing waves

$$z(\underline{x},t) \;=\; A\,e^{j\omega t+\phi}\sin(kx)\sin(ky)$$

because, when taking the partial derivative with respect to x, the term $\sin(ky)$ is simply part of the constant coefficient, and vice versa.

To build solutions to the wave equation that obey all of the boundary conditions, we can form linear combinations of the above standing-wave products having zero displacement ("nodes") along all four boundary lines:

$$z(\underline{x},t) \;=\; e^{j\omega t}\sum_{m=1}^{\infty}\sum_{n=1}^{\infty} A_{mn}e^{j\phi_{mn}}\,W_{mn}(x,y) \tag{B.50}$$

[36]Do not be misled by the name "dispersion relation" into thinking that wave propagation is dispersive in this ase—it is not. Dispersion occurs when $c = \omega/k$ changes as a function of frequency; here it is a constant. Thus, the dispersion relation implies wave propagation dispersion when k is a *nonlinear* function of ω.

where

$$W_{mn}(x,y) \;\stackrel{\Delta}{=}\; \sin\left(m\frac{\pi}{X}x\right)\sin\left(n\frac{\pi}{Y}y\right).$$

By construction, all linear combinations of the form Eq. (B.50) are solutions of the wave equation that satisfy the zero boundary conditions along the rectangle (0–X, 0–Y). Since sinusoids at different frequencies are *orthogonal*, the solution building-blocks $W_{mn}(x,y)$ are orthogonal under the inner product

$$\langle f, g\rangle \;\stackrel{\Delta}{=}\; \int_0^X\int_0^Y f(x,y)\overline{g}(x,y)\,dx\,dy.$$

It remains to be shown that the set of functions $W_{mn}(x,y)$ is *complete*, that is, that they form a *basis* for the set of *all* solutions to the wave equation satisfying the boundary conditions. Given that, we can solve the problem of *arbitrary initial conditions*. That is, given any initial $z(x,y)$ over the membrane (subject to the boundary conditions, of course), we can find the amplitude of each excited mode by simple projection:

$$Z_{mn} \stackrel{\Delta}{=} \frac{\langle z, W\rangle}{\langle W, W\rangle}$$

Showing completeness of the basis $W_{mn}(x,y)$ in the desired solution space is a special case (zero boundary conditions) of the problem of showing that the 2D Fourier series expansion is complete in the space of all continuous rectangular surfaces.

The Wikipedia page (as of 1/31/10) on the Helmholtz equation provides a nice "entry point" on the above topics and further information.

B.8.5 3D Sound

The mathematics of 3D sound is quite elementary, as we will see below. The hard part of the theory of practical systems typically lies in the mathematical approximation to the ideal case. Examples include *Ambisonics* [158] and *wave field synthesis* [49].

Consider a point source at position $\underline{x}_s \in \mathbf{R}^3$. Then the acoustic complex amplitude at position $\underline{x}_l \in \mathbf{R}^3$ is given by

$$p(\underline{x}_l;\underline{x}_s) = p_1(\underline{x}_s)\frac{e^{jk|\underline{x}_l-\underline{x}_s|}}{|\underline{x}_l-\underline{x}_s|}$$

where $p_1(\underline{x}_s)$ denotes the complex amplitude one meter from the point source in any direction, and $k = 2\pi/\lambda$ denotes the wavenumber (spatial radian frequency). Distributed acoustic sources are handled as a superposition of point sources, so the point source is a completely general building block for all types of sources in linear acoustics.

The fundamental approximation problem in 3D sound is to approximate the complex acoustic field at one or more listening points using a finite set of M loudspeakers, which are often modeled as a point source for each speaker.

Physics Theory Problems

1. Prove the parallel axis theorem (§B.4.6 on page 475).

2. Derive the radius of gyration for a cylindrical tube, given by Eq. (B.12) on page 478.

3. For the system of Fig. B.2, given an initial displacement $x(0) = A$ and initial velocity $\dot{x}(0) = 0$, the potential energy stored in the spring at time 0 is

$$E_k(0) = \frac{1}{2}kA^2.$$

   Using the equation of motion

$$x(t) = A\cos(\omega_0 t), \quad \forall t \geq 0,$$

   show that the potential energy in the spring, $E_k(t)$, plus the kinetic energy of the mass, $E_m(t)$, is constant for all times $t > 0$, *i.e.*, show that

$$E_k(t) + E_m(t) = E_k(0), \quad \forall t \geq 0.$$

4. In analogy with the mass-spring-dashpot impedance analyzed in §9.2.1, suppose we have a transfer function

$$H(s) = \frac{1}{ms^2 + \mu s + k}.$$

   (a) Find formulas for ζ and ω_0 in the *normalized form*

$$H(s) = \frac{1}{m}\frac{1}{s^2 + \zeta\omega_0 s + \omega_0^2}.$$

   (b) Find the values of ζ for which there are two complex poles (the so-called *underdamped* case).

   (c) Find the values of ζ for which there are two distinct real poles (the so-called *overdamped* case).

   (d) Find the value(s) of ζ for which there is a single repeated poles (the so-called *critically damped* case).

   (e) Find the frequency ω_p where $|H(j\omega)|$ achieves its maximum, *i.e.*, $|H(j\omega_p)| \geq |H(j\omega)|, \forall \omega \in \mathbf{R}$.

(f) Why can $H(s)$ not be a driving-point impedance or admittance for a passive linear system?

5. Repeat the previous problem for

$$\Gamma(s) = \frac{s}{ms^2 + \mu s + k}.$$

except for the last part show that $\Gamma(s)$ *can* be a driving-point admittance, and give a diagram of a physical system having that admittance.

Appendix C

Digital Waveguide Theory

In this appendix, the basic principles of digital waveguide acoustic modeling are derived from a mathematical point of view. For this, the reader is expected to have some background in linear systems and elementary physics. In particular, facility with Laplace transforms [284], Newtonian mechanics [180], and basic differential equations is assumed.

We begin with the partial differential equation (PDE) describing the ideal vibrating string, which we first digitize by converting partial derivatives to finite differences. This yields a discrete-time recursion which approximately simulates the ideal string. Next, we go back and solve the original PDE, obtaining continuous traveling waves as the solution. These traveling waves are then digitized by ordinary sampling, resulting in the digital waveguide model for the ideal string. The digital waveguide simulation is then shown to be equivalent to a particular finite-difference recursion. (This only happens for the lossless ideal vibrating string with a particular choice of sampling intervals, so it is an interesting case.) Next digital waveguides simulating lossy and dispersive vibrating strings are derived, and alternative choices of wave variables (displacement, velocity, slope, force, power, etc.) are derived. Finally, an introduction to scattering theory for digital waveguides is presented.

C.1 The Ideal Vibrating String

The *wave equation* for the ideal (lossless, linear, flexible) vibrating string, depicted in Fig. C.1, is given by

$$Ky'' = \epsilon\ddot{y} \tag{C.1}$$

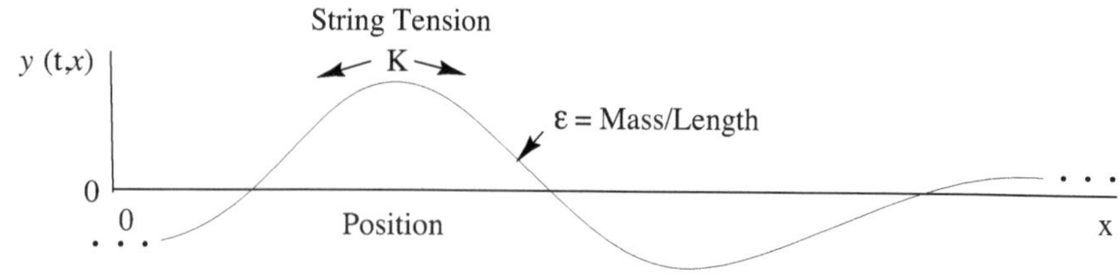

Figure C.1: The ideal vibrating string.

where

$$
\begin{array}{rclcrcl}
K & \triangleq & \text{string tension} & \qquad & y & \triangleq & y(t,x) \\
\epsilon & \triangleq & \text{linear mass density} & & \dot{y} & \triangleq & \frac{\partial}{\partial t}y(t,x) \\
y & \triangleq & \text{string displacement} & & y' & \triangleq & \frac{\partial}{\partial x}y(t,x)
\end{array}
$$

where "$\triangleq$" means "is defined as." The wave equation is derived in §B.6 on page 500. It can be interpreted as a statement of Newton's second law, *"force = mass × acceleration,"* on a microscopic scale. Since we are concerned with transverse vibrations on the string, the relevant restoring force (per unit length) is given by the string tension times the curvature of the string (Ky''); the restoring force is balanced at all times by the inertial force per unit length of the string which is equal to mass density times transverse acceleration ($\epsilon\ddot{y}$).

C.2 The Finite Difference Approximation

In the musical acoustics literature, the normal method for creating a computational model from a differential equation is to apply the so-called *finite difference approximation* (FDA) in which differentiation is replaced by a finite difference (see Appendix D) [482, 311]. For example

$$\dot{y}(t,x) \approx \frac{y(t,x) - y(t-T,x)}{T} \tag{C.2}$$

and

$$y'(t,x) \approx \frac{y(t,x) - y(t,x-X)}{X} \tag{C.3}$$

where T is the time sampling interval to be used in the simulation, and X is a spatial sampling interval. These approximations can be seen as arising directly from the definitions of the partial derivatives with respect to t and x. The approximations become exact in the limit as T and X approach zero. To avoid a delay error, the second-order finite-differences are defined with a compensating time shift:

$$\ddot{y}(t,x) \approx \frac{y(t+T,x) - 2y(t,x) + y(t-T,x)}{T^2} \tag{C.4}$$

$$y''(t,x) \approx \frac{y(t,x+X) - 2y(t,x) + y(t,x-X)}{X^2} \tag{C.5}$$

The odd-order derivative approximations suffer a half-sample delay error while all even order cases can be compensated as above.

C.2.1 FDA of the Ideal String

Substituting the FDA into the wave equation gives

$$K\frac{y(t,x+X) - 2y(t,x) + y(t,x-X)}{X^2} = \epsilon\frac{y(t+T,x) - 2y(t,x) + y(t-T,x)}{T^2}$$

which can be solved to yield the following recursion for the string displacement:

$$\begin{aligned} y(t+T,x) &= \frac{KT^2}{\epsilon X^2}\left[y(t,x+X) - 2y(t,x) + y(t,x-X)\right] \\ &\qquad +2y(t,x) - y(t-T,x) \end{aligned}$$

In a practical implementation, it is common to set $T = 1$, $X = (\sqrt{K/\epsilon})T$, and evaluate on the integers $t = nT = n$ and $x = mX = m$ to obtain the difference equation

$$y(n+1,m) = y(n,m+1) + y(n,m-1) - y(n-1,m). \tag{C.6}$$

Thus, to update the sampled string displacement, past values are needed for each point along the string at time instants n and $n-1$. Then the above recursion can be carried out for time $n+1$ by iterating over all m along the string.

Perhaps surprisingly, it is shown in Appendix E that the above recursion is *exact* at the sample points in spite of the apparent crudeness of the finite difference approximation [442]. The FDA approach to numerical simulation was used by Pierre Ruiz in his work on vibrating strings [392], and it is still in use today [74, 75].

When more terms are added to the wave equation, corresponding to complex losses and dispersion characteristics, more terms of the form $y(n-l,m-k)$ appear in (C.6). These higher-order terms correspond to frequency-dependent *losses* and/or *dispersion* characteristics in the FDA. All linear differential equations with constant coefficients give rise to *some* linear, time-invariant discrete-time system via the FDA. A general subclass of the linear, time-invariant case giving rise to "filtered waveguides" is

$$\sum_{k=0}^{\infty} \alpha_k \frac{\partial^k y(t,x)}{\partial t^k} = \sum_{l=0}^{\infty} \beta_l \frac{\partial^l y(t,x)}{\partial x^l}, \tag{C.7}$$

while the fully general linear, time-invariant 2D case is

$$\sum_{k=0}^{\infty}\sum_{l=0}^{\infty} \alpha_{k,l} \frac{\partial^k \partial^l y(t,x)}{\partial t^k \partial x^l} = \sum_{m=0}^{\infty}\sum_{n=0}^{\infty} \beta_{m,n} \frac{\partial^m \partial^n y(t,x)}{\partial x^m \partial x^n}. \tag{C.8}$$

A nonlinear example is

$$\frac{\partial y(t,x)}{\partial t} = \left(\frac{\partial y(t,x)}{\partial x}\right)^2, \tag{C.9}$$

and a time-varying example can be given by

$$\frac{\partial y(t,x)}{\partial t} = t^2 \frac{\partial y(t,x)}{\partial x}. \tag{C.10}$$

C.3 Traveling-Wave Solution

It is easily shown that the lossless 1D wave equation $Ky'' = \epsilon \ddot{y}$ is solved by any string shape which travels to the left or right with speed $c \triangleq \sqrt{K/\epsilon}$. Denote *right-going* traveling waves in general by $y_r(t-x/c)$ and *left-going* traveling waves by $y_l(t+x/c)$, where y_r and y_l are assumed twice-differentiable.[1] Then a general class of solutions to the lossless, one-dimensional, second-order wave equation can be expressed as

$$y(t,x) = y_r\left(t - \frac{x}{c}\right) + y_l\left(t + \frac{x}{c}\right). \tag{C.11}$$

The next section derives the result that $\ddot{y}_r = c^2 y_r''$ and $\ddot{y}_l = c^2 y_l''$, establishing that the wave equation is satisfied for all traveling wave shapes y_r and y_l. However, remember that the derivation of the wave equation in §B.6 assumes the string slope is much less than 1 at all times and positions. Finally, we show in §C.3.6 that the traveling-wave picture is *general*; that

[1] Any function of $(t - x/c)$, where t denotes time and x denotes position along a "waveshape", may be interpreted as a fixed waveshape traveling to the *right* (positive x direction), with speed c. Similarly, any function of $(t+x/c)$ may be seen as a waveshape traveling to the *left* (negative x direction) at speed c meters per second. In both cases, x denotes a position along the waveshape, and t denotes time. For any fixed t, a "snapshot" of the waveshape may be seen by evaluating along x.

is, any physical state of the string can be converted to a pair of equivalent traveling force- or velocity-wave components.

An important point to note about the traveling-wave solution of the 1D wave equation is that a function of two variables $y(t,x)$ has been replaced by two functions of a single variable in time units. This leads to great reductions in computational complexity.

The traveling-wave solution of the wave equation was first published by d'Alembert in 1747 [100]. See Appendix A for more on the history of the wave equation and related topics.

C.3.1 Traveling-Wave Partial Derivatives

Because we have defined our traveling-wave components $y_r(t-x/c)$ and $y_l(t+x/c)$ as having arguments in units of *time*, the partial derivatives with respect to time t are identical to simple derivatives of these functions. Let $\dot{y}_r$ and $\dot{y}_l$ denote the (partial) derivatives with respect to time of y_r and y_l, respectively. In contrast, the partial derivatives with respect to x are

$$\begin{aligned}\frac{\partial}{\partial x} y_r\left(t-\frac{x}{c}\right) &= -\frac{1}{c}\dot{y}_r\left(t-\frac{x}{c}\right) \\ \frac{\partial}{\partial x} y_l\left(t+\frac{x}{c}\right) &= \frac{1}{c}\dot{y}_l\left(t+\frac{x}{c}\right).\end{aligned}$$

Denoting the *spatial* partial derivatives by y'_r and y'_l, respectively, we can write more succinctly

$$\begin{aligned}y'_r &= -\frac{1}{c}\dot{y}_r \\ y'_l &= \frac{1}{c}\dot{y}_l,\end{aligned}$$

where this argument-free notation assumes the same t and x for all terms in each equation, and the subscript l or r determines whether the omitted argument is $t+x/c$ or $t-x/c$.

Now we can see that the *second* partial derivatives in x are

$$\begin{aligned}y''_r &= \left(-\frac{1}{c}\right)^2 \ddot{y}_r = \frac{1}{c^2}\ddot{y}_r \\ y''_l &= \left(\frac{1}{c}\right)^2 \ddot{y}_l = \frac{1}{c^2}\ddot{y}_l.\end{aligned}$$

These relations, together with the fact that partial differention is a *linear operator*, establish that

$$y(t,x) = y_r\left(t-\frac{x}{c}\right) + y_l\left(t+\frac{x}{c}\right).$$

obeys the ideal wave equation $\ddot{y} = c^2 y''$ for all twice-differentiable functions y_r and y_l.

C.3.2 Use of the Chain Rule

These traveling-wave partial-derivative relations may be derived a bit more formally by means of the *chain rule* from calculus, which states that, for the composition of functions f and g, *i.e.*,

$$y(x) = f(g(x)),$$

the derivative of the composition with respect to x can be expressed according to the chain rule as

$$y'(x) = f'(g(x))g'(x),$$

where $f'(x)$ denotes the derivative of $f(x)$ with respect to x.

To apply the chain rule to the spatial differentiation of traveling waves, define

$$\begin{aligned} g_r(t,x) &= t - \frac{x}{c} \\ g_l(t,x) &= t + \frac{x}{c}. \end{aligned}$$

Then the traveling-wave components can be written as $y_r[g_r(t,x)]$ and $y_l[g_l(t,x)]$, and their partial derivatives with respect to x become

$$\begin{aligned} y_r' \isdef \frac{\partial}{\partial x} y_r\left[g_r(t,x)\right] &= \left(\left.\frac{\partial y_r(\xi)}{\partial \xi}\right|_{\xi=g_r(t,x)}\right)\left(\left.\frac{\partial g_r(t,\xi)}{\partial \xi}\right|_{\xi=x}\right) \\ &= \dot{y}_r[g_r(t,x)] \cdot \frac{\partial g_r(t,x)}{\partial x} \\ &= \dot{y}_r[g_r(t,x)] \cdot \left(-\frac{1}{c}\right) \isdef -\frac{1}{c}\dot{y}_r, \end{aligned}$$

and similarly for y_l'.

C.3.3 String Slope from Velocity Waves

Let's use the above result to derive the *slope* of the ideal vibrating string From Eq. (C.11) on page 526, we have the string displacement given by

$$y(t,x) = y_r(t - x/c) + y_l(t + x/c).$$

By linearity of differentiation, the *string slope* is given by

$$s(t,x) \triangleq \frac{\partial}{\partial x} y(t,x) = \frac{\partial}{\partial x} y_r(t-x/c) + \frac{\partial}{\partial x} y_l(t+x/c).$$

Consider only the right-going component, and define

$$\dot{y}_r(\tau) \triangleq \frac{d}{d\tau} y_r(\tau)$$

with $\tau \triangleq t - x/c$. By the *chain rule*,

$$\frac{\partial}{\partial x} y_r(t-x/c) = \frac{dy_r}{d\tau} \cdot \frac{d\tau}{dx} = \dot{y}_r(t-x/c) \cdot \left(-\frac{1}{c}\right).$$

The left-going component is similar, but with $+1/c$. Thus, the string slope in terms of traveling velocity-wave components can be written as

$$s(t,x) = \frac{1}{c}\dot{y}_l(t+x/c) - \frac{1}{c}\dot{y}_r(t-x/c).$$

C.3.4 Wave Velocity

Because e^{st} is an *eigenfunction* under differentiation (*i.e.*, the exponential function is its own derivative), it is often profitable to replace it with a generalized exponential function, with maximum degrees of freedom in its parametrization, to see if parameters can be found to fulfill the constraints imposed by differential equations.

In the case of the one-dimensional ideal wave equation (Eq. (C.1) on page 523), with no boundary conditions, an appropriate choice of eigensolution is

$$y(t,x) = e^{st+vx} \tag{C.12}$$

Substituting into the wave equation yields

$$\begin{array}{rclcrcl} \dot{y} & = & sy & \quad & y' & = & vy \\ \ddot{y} & = & s^2 y & & y'' & = & v^2 y \end{array}$$

Defining the *wave velocity* (or *phase velocity*[2]) as $c \triangleq s/v$, the wave equation becomes

$$\begin{aligned} Kv^2 y &= \epsilon s^2 y \\ \Rightarrow \frac{K}{\epsilon} &= \frac{s^2}{v^2} \triangleq c^2 \\ \Rightarrow v &= \pm\frac{s}{c}. \end{aligned} \tag{C.13}$$

[2]The term *phase velocity* is normally used when it differs from the *group velocity*, as in stiff, dispersive strings (§C.6). In the present context, the phase velocity and group velocity are the same, so the term "wave velocity" is unambigous here. See the analogous terms *phase delay* and *group delay* in [449] for more details about the difference.

Thus

$$y(t,x) = e^{s(t \pm x/c)}$$

is a solution for all s. By *superposition,*

$$y(t,x) = \sum_i A^+(s_i)e^{s_i(t-x/c)} + A^-(s_i)e^{s_i(t+x/c)}$$

is also a solution, where $A^+(s_i)$ and $A^-(s_i)$ are arbitrary complex-valued functions of arbitrary points s_i in the complex plane.

C.3.5 D'Alembert Derived

Setting $s \triangleq j\omega$, and extending the summation to an integral, we have, by Fourier's theorem,

$$y(t,x) = y_r\left(t - \frac{x}{c}\right) + y_l\left(t + \frac{x}{c}\right) \tag{C.14}$$

for *arbitrary* continuous functions $y_r(\cdot)$ and $y_l(\cdot)$. This is again the traveling-wave solution of the wave equation attributed to d'Alembert, but now derived from the eigen-property of sinusoids and Fourier theory rather than "guessed".

An example of the appearance of the traveling wave components shortly after plucking an infinitely long string at three points is shown in Fig. C.2.

C.3.6 Converting Any String State to Traveling Slope-Wave Components

We verified in §C.3.1 above that traveling-wave components y_r and y_l in Eq. (C.14) satisfy the ideal string wave equation $\ddot{y} = c^2 y''$. By definition, the physical string displacement is given by the sum of the traveling-wave components, or

$$y(t,x) \;=\; y_r\left(t - \frac{x}{c}\right) + y_l\left(t + \frac{x}{c}\right). \tag{C.15}$$

Thus, given any pair of traveling waves y_r and y_l, we can compute a corresponding string displacement y. This leads to the question whether any initial string state can be converted to a pair of equivalent traveling-wave components. If so, then d'Alembert's traveling-wave solution is *complete*, and all solutions to the ideal string wave equation can be expressed in terms of traveling waves.

The *state* of an ideal string at time t is classically specified by its displacement $y(t,x)$ and velocity

$$v(t,x) \;\triangleq\; \dot{y}(t,x) \;\triangleq\; \frac{\partial}{\partial t} y(t,x)$$

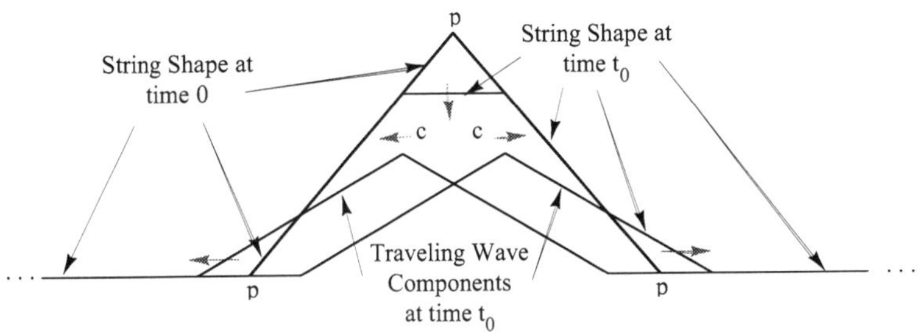

Figure C.2: An infinitely long string, "plucked" simultaneously at *three* points, labeled "p" in the figure, so as to produce an initial triangular displacement. The initial displacement is modeled as the sum of two identical triangular pulses which are exactly on top of each other at time 0. At time t_0 shortly after time 0, the traveling waves centers are separated by $2ct_0$ meters, and their sum gives the trapezoidal physical string displacement at time t_0 which is also shown. Note that only three short string segments are in motion at that time: the flat top segment which is heading to zero where it will halt forever, and two short pieces on the left and right which are the leading edges of the left- and right-going traveling waves. The string is not moving where the traveling waves overlap at the same slope. When the traveling waves fully separate, the string will be at rest everywhere but for two half-amplitude triangular pulses heading off to plus and minus infinity at speed c.

for all x [317]. Equation (C.15) gives us y as a simple sum of the traveling-wave components, and now we need a formula for v in terms of them as well. It will be derived in §C.7.3 (see Equations (C.44–C.46)) that we can write

$$v(t,x) = -cy_r'\left(t-\frac{x}{c}\right)+cy_l'\left(t+\frac{x}{c}\right).$$

where y' denotes the partial derivative with respect to x as usual. We have

$$\left[\begin{array}{c} y(t,x) \\ v(t,x) \end{array}\right] = \left[\begin{array}{cc} 1 & 1 \\ -c\frac{\partial}{\partial x} & c\frac{\partial}{\partial x} \end{array}\right] \left[\begin{array}{c} y_r(t-x/c) \\ y_l(t+x/c) \end{array}\right].$$

Inverting the two-by-two differential operator matrix yields left- and right-going *slope waves* as a function of an arbitrary initial slope and velocity:

$$\left[\begin{array}{c} y'^{+} \\ y'^{-} \end{array}\right] = \frac{1}{2}\left[\begin{array}{c} y' - \frac{v}{c} \\ y' + \frac{v}{c} \end{array}\right]$$

Integrating both sides with respect to x, and choosing the constant of integration to give the correct constant component of y, we obtain the displacement-wave components

$$\left[\begin{array}{c} y^{+} \\ y^{-} \end{array}\right] = \frac{1}{2}\left[\begin{array}{c} y - w \\ y + w \end{array}\right]$$

where

$$w(t,x) \triangleq \frac{1}{c}\int_{-\infty}^{x} v(t,\xi)\,d\xi.$$

Notice that if the initial velocity is zero, each of the initial traveling displacement waves is simply half the initial displacement, as expected. On the other hand, if the initial displacement is zero and there is a uniform initial velocity (the whole string is moving), the initial displacement-wave components are *unbounded* as the string length goes to infinity. Related discussion appears in Appendix E.

It will be seen in §C.7.4 on page 556 that state conversion between physical variables and traveling-wave components is simpler when *force* and *velocity* are chosen as the physical state variables (as opposed to displacement and velocity used here).

C.4 Sampled Traveling Waves

To carry the traveling-wave solution into the "digital domain," it is necessary to *sample* the traveling-wave amplitudes at intervals of T seconds, corresponding to a sampling rate $f_s \triangleq 1/T$ samples per second. For CD-quality audio, we have $f_s = 44.1$ kHz. The natural choice of *spatial sampling interval* X is the distance sound propagates in one temporal sampling interval T, or $X \triangleq cT$ meters. In a lossless traveling-wave simulation, the whole wave moves left or right one spatial sample each time sample; hence, lossless simulation requires only digital delay lines. By lumping losses parsimoniously in a real acoustic model, most of the traveling-wave simulation can in fact be lossless even in a practical application.

Formally, sampling is carried out by the change of variables

$$\begin{aligned} x &\to x_m &= mX, \\ t &\to t_n &= nT. \end{aligned}$$

Substituting into the traveling-wave solution of the wave equation gives

$$\begin{aligned} y(t_n, x_m) &= y_r(t_n - x_m/c) + y_l(t_n + x_m/c) \\ &= y_r(nT - mX/c) + y_l(nT + mX/c) \\ &= y_r\left[(n-m)T\right] + y_l\left[(n+m)T\right]. \end{aligned}$$

Since T multiplies all arguments, let's suppress it by defining

$$y^+(n) \triangleq y_r(nT), \qquad y^-(n) \triangleq y_l(nT). \tag{C.16}$$

This new notation also introduces a "+" superscript to denote a traveling-wave component propagating to the right, and a "−" superscript to denote propagation to the left. This notation is similar to that used for acoustic tubes [297].

C.4.1 Digital Waveguide Model

In this section, we interpret the sampled d'Alembert traveling-wave solution of the ideal wave equation as a digital filtering framework. This is an example of what are generally known as *digital waveguide models* [430, 431, 433, 437, 442].

The term $y_r\left[(n-m)T\right] \triangleq y^+(n-m)$ in Eq. (C.16) can be thought of as the output of an m-sample delay line whose input is $y^+(n)$. In general, subtracting a positive number m from a time argument n corresponds to *delaying* the waveform by m samples. Since y^+ is the right-going component, we draw its delay line with input $y^+(n)$ on the left and its output $y^+(n-m)$ on the right. This can be seen as the upper "rail" in Fig. C.3

Similarly, the term $y_l\left[(n+m)T\right] \triangleq y^-(n+m)$ can be thought of as the *input* to an m-sample delay line whose *output* is $y^-(n)$. Adding m to the time argument n produces an m-sample waveform *advance*. Since y^- is the

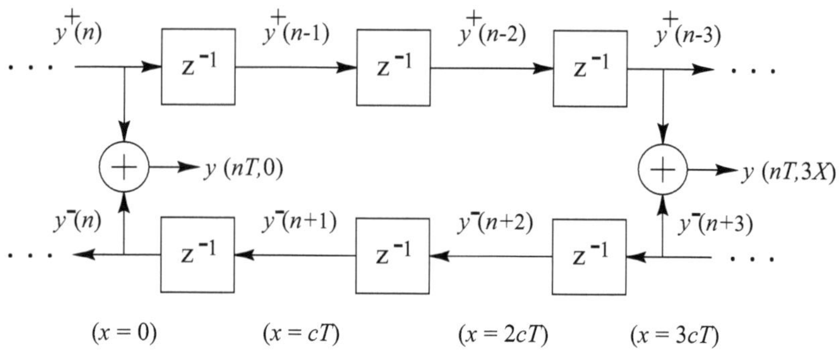

Figure C.3: Digital simulation of the ideal, lossless waveguide with observation points at $x = 0$ and $x = 3X = 3cT$. (The symbol "z^{-1}" denotes a one-sample delay.)

left-going component, it makes sense to draw the delay line with its input $y^-(n+m)$ on the right and its output $y^-(n)$ on the left. This can be seen as the lower "rail" in Fig. C.3.

Note that the position along the string, $x_m = mX = mcT$ meters, is laid out from left to right in the diagram, giving a physical interpretation to the horizontal direction in the diagram. Finally, the left- and right-going traveling waves must be summed to produce a physical output according to the formula

$$y(t_n, x_m) = y^+(n-m) + y^-(n+m). \tag{C.17}$$

We may compute the physical string displacement at any spatial sampling point x_m by simply adding the upper and lower rails together at position m along the delay-line pair. In Fig. C.3, "transverse displacement outputs" have been arbitrarily placed at $x = 0$ and $x = 3X$. The diagram is similar to that of well known ladder and lattice digital filter structures (§C.9.4, p. 579,[297]), except for the delays along the upper rail, the absence of scattering junctions, and the direct physical interpretation. (A scattering junction implements partial reflection and partial transmission in the waveguide.) We could proceed to ladder and lattice filters by (1) introducing a perfectly reflecting (rigid or free) termination at the far right, and (2) commuting the delays rightward from the upper rail down to the lower rail [432, 434]. The absence of scattering junctions is due to the fact that the string has a uniform wave impedance. In acoustic tube simulations, such as for voice [87, 297] or wind instruments [195], lossless scattering junctions are used at changes in cross-sectional tube area and lossy scattering junctions are used to implement tone holes. In waveguide bowed-string synthesis (discussed in a later section), the bow itself creates an active, time-varying, and nonlinear scattering junction on the string at the bowing point.

Any ideal, one-dimensional waveguide can be simulated in this way. It is important to note that the simulation is *exact* at the sampling instants, to within the numerical precision of the samples themselves. To avoid *aliasing* associated with sampling, we require all waveshapes traveling along the string to be initially *bandlimited* to less than half the sampling frequency. In other words, the highest frequencies present in the signals $y_r(t)$ and $y_l(t)$ may not exceed half the temporal sampling frequency $f_s \triangleq 1/T$; equivalently, the highest *spatial* frequencies in the shapes $y_r(x/c)$ and $y_l(x/c)$ may not exceed half the spatial sampling frequency $\nu_s \triangleq 1/X$.

A C program implementing a plucked/struck string model in the form of Fig. C.3 is available at `http://ccrma.stanford.edu/~jos/pmudw/`.

C.4.2 Digital Waveguide Interpolation

A more compact simulation diagram which stands for either sampled or continuous waveguide simulation is shown in Fig. C.4. The figure emphasizes that the ideal, lossless waveguide is simulated by a *bidirectional delay line*, and that bandlimited *spatial* interpolation may be used to construct a displacement output for an arbitrary x not a multiple of cT, as suggested by the output drawn in Fig. C.4. Similarly, bandlimited interpolation across time serves to evaluate the waveform at an arbitrary time not an integer multiple of T (§4.4 on page 169).

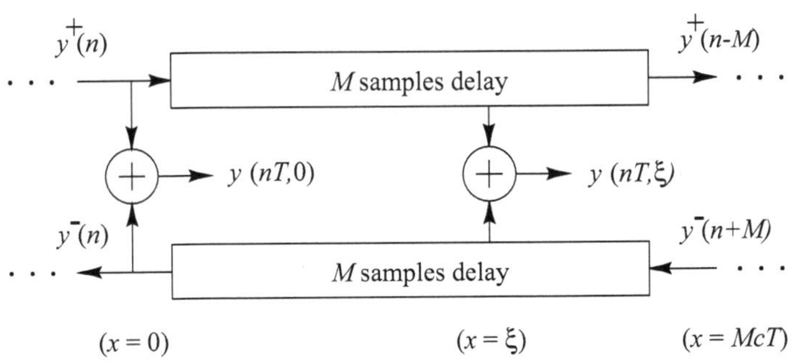

Figure C.4: Simplified picture of ideal waveguide simulation.

Ideally, bandlimited interpolation is carried out by convolving a continuous "sinc function" $\text{sinc}(x) \triangleq \sin(\pi x)/\pi x$ with the signal samples. Specifically, convolving a sampled signal $x(t_n)$ with $\text{sinc}[(t_n - t_0)/T)]$ "evaluates" the signal at an arbitrary continuous time t_0. The sinc function is the impulse response of the ideal lowpass filter which cuts off at half the sampling rate.

In practice, the interpolating sinc function must be *windowed* to a finite duration. This means the associated lowpass filter must be granted a "transition band" in which its frequency response is allowed to "roll off" to zero at half the sampling rate. The interpolation quality in the "pass band" can always be made perfect to within the resolution of human hearing by choosing a sufficiently large product of window-length times transition-bandwidth. Given "audibly perfect" quality in the pass band, increasing the transition bandwidth reduces the computational expense of the interpolation. In fact, they are approximately inversely proportional. This is one reason why *oversampling* at rates higher than twice the highest audio frequency is helpful. For example, at a 44.1 kHz sampling rate, the transition bandwidth above the nominal audio upper limit of 20 kHz is only 2.1 kHz, while at a 48 kHz sampling rate (used in DAT machines) the guard band is 4 kHz wide—nearly double. Since the required window length (impulse response duration) varies inversely with the provided transition bandwidth, we see that increasing the sampling rate by less than ten percent reduces the filter expense by almost fifty percent. Windowed-sinc interpolation is described further in §4.4 on page 169. Many more techniques for digital resampling and delay-line interpolation are reviewed in [267].

C.4.3 Relation to the Finite Difference Recursion

In this section we will show that the digital waveguide simulation technique is *equivalent* to the recursion produced by the finite difference approximation (FDA) applied to the wave equation [442, pp. 430–431]. A more detailed derivation, with examples and exploration of implications, appears in Appendix E. Recall from (C.6) that the time update recursion for the ideal string digitized via the FDA is given by

$$y(n+1,m) = y(n,m+1) + y(n,m-1) - y(n-1,m). \qquad \text{(C.18)}$$

To compare this with the waveguide description, we substitute the traveling-wave decomposition $y(n,m) = y^+(n-m) + y^-(n+m)$ (which is exact in the ideal case at the sampling instants) into the right-hand side of the FDA recursion above and see how good is the approximation to the left-hand side $y(n+1,m) = y^+(n+1-m) + y^-(n+1+m)$. Doing this gives

$$\begin{aligned} y(n+1,m) &= y(n,m+1) + y(n,m-1) - y(n-1,m) \\ &= y^+(n-m-1) + y^-(n+m+1) \\ &\quad +y^+(n-m+1) + y^-(n+m-1) \\ &\quad -y^+(n-m-1) - y^-(n+m-1) \\ &= y^-(n+m+1) + y^+(n-m+1) \\ &= y^+[(n+1)-m] + y^-[(n+1)+m] \\ &\triangleq y(n+1,m). \end{aligned} \tag{C.19}$$

Thus, we obtain the result that the FDA recursion is also *exact* in the lossless case, because it is equivalent to the digital waveguide method which we know is exact at the sampling points. This is surprising since the FDA introduces artificial damping when applied to lumped, mass-spring systems, as discussed earlier.

The last identity above can be rewritten as

$$\begin{aligned} y(n+1,m) &\triangleq y^+[(n+1)-m] + y^-[(n+1)+m] \\ &= y^+[n-(m-1)] + y^-[n+(m+1)] \end{aligned} \tag{C.20}$$

which says the displacement at time $n+1$, position m, is the superposition of the right-going and left-going traveling wave components at positions $m-1$ and $m+1$, respectively, from time n. In other words, the physical wave variable can be computed for the next time step as the sum of incoming traveling wave components from the left and right. This picture also underscores the lossless nature of the computation.

This results extends readily to the digital waveguide mesh (§C.14), which is essentially a lattice-work of digital waveguides for simulating membranes and volumes. The equivalence is important in higher dimensions because the finite-difference model requires less computations per node than the digital waveguide approach.

Even in one dimension, the digital waveguide and finite-difference methods have unique advantages in particular situations, and as a result they are often combined together to form a hybrid traveling-wave/physical-variable simulation [351, 352, 222, 124, 123, 224, 263, 223]. In this hybrid simulations, the traveling-wave variables are called "W variables" (where 'W' stands for "Wave"), while the physical variables are caled "K variables" (where 'K' stands for "Kirchoff"). Each K variable, such as displacement $y(nT, mX)$ on a vibrating string, can be regarded as the sum of two traveling-wave components, or W variables:

$$y(nT, mX) = y_r(nT - mX/c) + y_l(nT + mX/c)$$

Conversion between K variables and W variables can be non-trivial due to the non-local dependence of one set of state variables on the other, in general. A detailed examination of this issue is given in Appendix E.

C.5 A Lossy 1D Wave Equation

In any real vibrating string, there are energy losses due to yielding terminations, drag by the surrounding air, and internal friction within the string. While losses in solids generally vary in a complicated way with frequency, they can usually be well approximated by a small number of odd-order terms added to the wave equation. In the simplest case, force is directly proportional to transverse string velocity, independent of frequency. If this proportionality constant is μ, we obtain the modified wave equation

$$Ky'' = \epsilon\ddot{y} + \mu\dot{y}. \tag{C.21}$$

Thus, the wave equation has been extended by a "first-order" term, *i.e.*, a term proportional to the first derivative of y with respect to time. More realistic loss approximations would append terms proportional to $\dddot{y}$, $\partial^5 y/\partial t^5$, and so on, giving frequency-dependent losses.

Setting $y(t,x) = e^{st+vx}$ in the wave equation to find the relationship between temporal and spatial frequencies in the eigensolution, the wave equation becomes

$$\begin{aligned} K\left(v^2 y\right) &= \epsilon\left(s^2 y\right) + \mu\left(sy\right) \;\Rightarrow\; Kv^2 = \epsilon s^2 + \mu s \\ \Rightarrow\; v^2 &= \frac{\epsilon}{K}s^2 + \frac{\mu}{K}s = \frac{\epsilon}{K}s^2\left(1+\frac{\mu}{\epsilon s}\right) \stackrel{\Delta}{=} \frac{s^2}{c^2}\left(1+\frac{\mu}{\epsilon s}\right) \\ \Rightarrow\; v &= \pm\frac{s}{c}\sqrt{1+\frac{\mu}{\epsilon s}} \end{aligned}$$

where $c \stackrel{\Delta}{=} \sqrt{K/\epsilon}$ is the wave velocity in the lossless case. At high frequencies (large $|s|$), or when the friction coefficient μ is small relative to the mass density ϵ at the lowest frequency of interest, we have the approximation

$$\left(1+\frac{\mu}{\epsilon s}\right)^{\frac{1}{2}} \approx 1 + \frac{1}{2}\frac{\mu}{\epsilon s} \tag{C.22}$$

by the binomial theorem. For this small-loss approximation, we obtain the following relationship between temporal and spatial frequency:

$$v \approx \pm\frac{1}{c}\left(s+\frac{\mu}{2\epsilon}\right) \tag{C.23}$$

The eigensolution is then

$$e^{st+vx} = \exp\left[st \pm \left(s+\frac{\mu}{2\epsilon}\right)\frac{x}{c}\right] = \exp\left[s\left(t\pm\frac{x}{c}\right)\right]\exp\left(\pm\frac{\mu}{2\epsilon}\frac{x}{c}\right). \tag{C.24}$$

The right-going part of the eigensolution is

$$e^{-(\mu/2\epsilon)x/c}e^{s(t-x/c)} \tag{C.25}$$

which *decays exponentially* in the direction of propagation, and the left-going solution is

$$e^{(\mu/2\epsilon)x/c}e^{s(t+x/c)} \tag{C.26}$$

which also decays exponentially in the direction of travel.

Setting $s = j\omega$ and using superposition to build up arbitrary traveling wave shapes, we obtain the general class of solutions

$$y(t,x) = e^{-(\mu/2\epsilon)x/c}y_r\left(t - x/c\right) + e^{(\mu/2\epsilon)x/c}y_l\left(t + x/c\right). \tag{C.27}$$

Sampling these exponentially decaying traveling waves at intervals of T seconds (or $X = cT$ meters) gives

$$\begin{aligned}
y(t_n, x_m) &= e^{-(\mu/2\epsilon)x_m/c}y_r\left(t_n - x_m/c\right) + e^{(\mu/2\epsilon)x_m/c}y_l\left(t_n + x_m/c\right) \\
&= e^{-(\mu/2\epsilon)mX/c}y_r\left(nT - mX/c\right) + e^{(\mu/2\epsilon)mX/c}y_l\left(nT + mX/c\right) \\
&= e^{-\mu mT/2\epsilon}y_r\left[(n-m)T\right] + e^{\mu mT/2\epsilon}y_l\left[(n+m)T\right] \\
&= \left(e^{-\mu T/2\epsilon}\right)^m y^+(n-m) + \left(e^{\mu T/2\epsilon}\right)^m y^-(n+m) \\
&\triangleq g^m y^+(n-m) + g^{-m}y^-(n+m).
\end{aligned}$$

The simulation diagram for the lossy digital waveguide is shown in Fig. C.5.

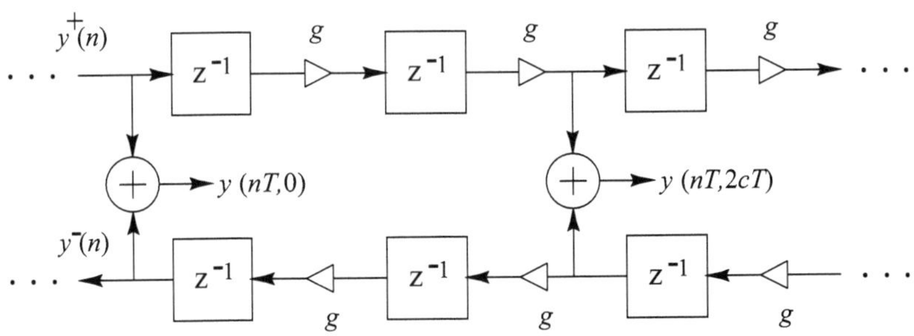

Figure C.5: Discrete simulation of the ideal, lossy waveguide.

Again the discrete-time simulation of the decaying traveling-wave solution is an *exact* implementation of the continuous-time solution at the sampling positions and instants, even though losses are admitted in the wave equation. Note also that the losses which are *distributed* in the continuous solution have been consolidated, or *lumped,* at discrete intervals of cT meters in the simulation. The loss factor $g = e^{-\mu T/2\epsilon}$ *summarizes* the distributed loss incurred in one sampling interval. The lumping of distributed losses does not introduce an approximation error at the sampling points. Furthermore, bandlimited interpolation can yield arbitrarily accurate reconstruction between samples. The only restriction is again that all initial conditions and excitations be bandlimited to below half the sampling rate.

C.5.1 Loss Consolidation

In many applications, it is possible to realize vast computational savings in digital waveguide models by *commuting* losses out of unobserved and undriven sections of the medium and consolidating them at a minimum number of points. Because the digital simulation is linear and time invariant (given constant medium parameters K, ϵ, μ), and because linear, time-invariant elements commute, the diagram in Fig. C.6 is exactly equivalent (to within numerical precision) to the previous diagram in Fig. C.5.

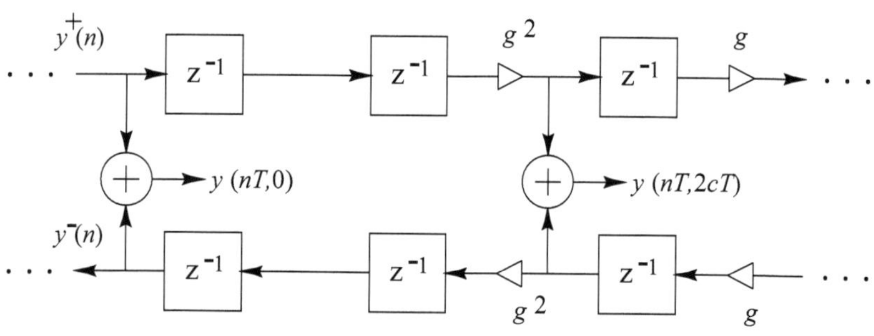

Figure C.6: Discrete simulation of the ideal, lossy waveguide. Each per-sample loss factor g may be "pushed through" delay elements and combined with other loss factors until an input or output is encountered which inhibits further migration. If further consolidation is possible on the other side of a branching node, a loss factor can be pushed *through* the node by pushing a copy into each departing branch. If there are other *inputs* to the node, the *inverse* of the loss factor must appear on each of them. Similar remarks apply to pushing backwards through a node.

C.5.2 Frequency-Dependent Losses

In nearly all natural wave phenomena, losses increase with frequency. Distributed losses due to air drag and internal bulk losses in the string tend to increase monotonically with frequency. Similarly, air absorption increases with frequency, adding loss for sound waves in acoustic tubes or open air [318].

Perhaps the apparently simplest modification to Eq. (C.21) yielding frequency-dependent damping is to add a *third-order* time-derivative term [392]:

$$Ky'' = \epsilon\ddot{y} + \mu\dot{y} + \mu_3\dddot{y} \tag{C.28}$$

While this model has been successful in practice [77], it turns out to go *unstable* at very high sampling rates. The technical term for this problem is that the PDE is *ill posed* [45].

A well posed replacement for Eq. (C.28) is given by

$$Ky'' = \epsilon\ddot{y} + \mu\dot{y} + \mu_2\dot{y}'' \tag{C.29}$$

in which the third-order partial derivative with respect to time, $\dddot{y}$, has been replaced by a third-order *mixed* partial derivative—twice with respect to x and once with respect to t.

The solution of a lossy wave equation containing higher odd-order derivatives with respect to time yields traveling waves which propagate with frequency-dependent attenuation. Instead of scalar factors g distributed throughout the diagram as in Fig. C.5, each g factor becomes a *lowpass filter* having some frequency-response per sample denoted by $G(\omega)$. Because propagation is passive, we will always have $|G(\omega)| \leq 1$.

More specically, As shown in [392], odd-order partial derivatives with respect to time in the wave equation of the form

$$\frac{\partial^n}{\partial t^n}y(t,x), \quad n = 1, 3, 5, \ldots,$$

correspond to attenuation of traveling waves on the string. (The even-order time derivatives can be associated with variations in *dispersion* as a function of frequency, which is considered in §C.6 below.) For $n > 1$, the losses are *frequency dependent*, and the per-sample amplitude-response "rolls off" proportional to

$$G(\omega) \propto \frac{1}{\omega^{n-1}}, \qquad (n \text{ odd}).$$

In particular, if the wave equation (C.21) is modified by adding terms proportional to $\mu_3\dddot{y}$ and $\mu_5\partial^5 y/\partial t^5$, for instance, then the per-sample propagation gain $G(\omega)$ has the form

$$G(\omega) = g_0 + g_2\omega^2 + g_4\omega^4$$

where the g_i are constants depending on the constants μ_3 and μ_5 in the wave equation. Since these per-sample loss filters are linear and time-invariant [449], they may also be consolidated at a minimum number of points in the waveguide without introducing any approximation error, just like the constant gains g in Fig. C.5. This result does not extend precisely to the waveguide mesh (§C.14).

In view of the above, we see that we can add odd-order time derivatives to the wave equation to approximate experimentally observed frequency-dependent damping characteristics in vibrating strings [73]. However, we then have the problem that such wave equations are ill posed, leading to possible stability failures at high sampling rates. As a result, it is generally preferable to use mixed derivatives, as in Eq. (C.29), and try to achieve realistic damping using higher order *spatial derivatives* instead.

C.5.3 Well Posed PDEs for Modeling Damped Strings

A large class of well posed PDEs is given by [45]

$$\ddot{y} + 2\sum_{k=0}^{M} q_k \frac{\partial^{2k+1} y}{\partial x^{2k} \partial t} + \sum_{k=0}^{N} r_k \frac{\partial^{2k} y}{\partial x^{2k}}. \tag{C.30}$$

Thus, to the ideal string wave equation Eq. (C.1) on page 523 we add any number of even-order partial derivatives in x, plus any number of mixed odd-order partial derivatives in x and t, where differentiation with respect to t occurs only once. Because every member of this class of PDEs is only second-order in time, it is guaranteed to be *well posed*, as shown in §D.2.2 on page 664.

C.5.4 Digital Filter Models of Damped Strings

In an efficient digital simulation, lumped loss factors of the form $G^k(\omega)$ are approximated by a rational frequency response $\hat{G}_k(e^{j\omega T})$. In general, the coefficients of the optimal rational loss filter are obtained by minimizing $||\, G^k(\omega) - \hat{G}_k(e^{j\omega T})\, ||$ with respect to the filter coefficients or the poles and zeros of the filter. To avoid introducing frequency-dependent delay, the loss filter should be a *zero-phase, finite-impulse-response* (FIR) filter [362]. Restriction to zero phase requires the impulse response $\hat{g}_k(n)$ to be finite in length (*i.e.*, an FIR filter) and it must be symmetric about time zero, *i.e.*, $\hat{g}_k(-n) = \hat{g}_k(n)$. In most implementations, the zero-phase FIR filter can be converted into a causal, *linear phase* filter by reducing an adjacent delay line by half of the impulse-response duration.

C.5.5 Lossy Finite Difference Recursion

We will now derive a finite-difference model in terms of string displacement samples which correspond to the lossy digital waveguide model of Fig. C.5. This derivation generalizes the lossless case considered in §C.4.3 on page 536.

Figure C.7 depicts a digital waveguide section once again in "physical canonical form," as shown earlier in Fig. C.5 on page 539, and introduces a doubly indexed notation for greater clarity in the derivation below [442, 222, 124, 123].

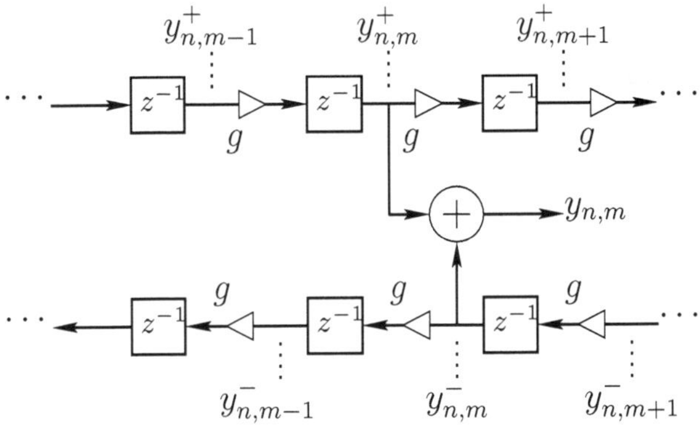

Figure C.7: Lossy digital waveguide—frequency-independent loss-factors g.

Referring to Fig. C.7, we have the following time-update relations:

$$\begin{aligned} y^+_{n+1,m} &= gy^+_{n,m-1} = g\cdot\left(y_{n,m-1} - y^-_{n,m-1}\right) \\ y^-_{n+1,m} &= gy^-_{n,m+1} = g\cdot\left(y_{n,m+1} - y^+_{n,m+1}\right) \end{aligned}$$

Adding these equations gives

$$\begin{aligned} y_{n+1,m} &= g\cdot\left(y_{n,m-1} + y_{n,m+1}\right) - g\cdot\left(\underbrace{y^-_{n,m-1}}_{gy^-_{n-1,m}} + \underbrace{y^+_{n,m+1}}_{gy^+_{n-1,m}}\right) \\ &= g\cdot\left(y_{n,m-1} + y_{n,m+1}\right) - g^2 y_{n-1,m} \end{aligned} \tag{C.31}$$

This is now in the form of the *finite-difference time-domain* (FDTD) scheme analyzed in [222]:

$$y_{n+1,m} = g^+_m y_{n,m-1} + g^-_m y_{n,m+1} + a_m y_{n-1,m},$$

with $g^+_m = g^-_m = g$, and $a_m = -g^2$. In [124], it was shown by von Neumann analysis (§D.4 on page 669) that these parameter choices give rise to a stable finite-difference scheme (§D.2.3), provided $|g| \leq 1$. In the present context, we expect stability to follow naturally from starting with a passive digital waveguide model.

Frequency-Dependent Losses

The preceding derivation generalizes immediately to frequency-dependent losses. First imagine each g in Fig. C.7 to be replaced by $G(z)$, where for passivity we require

$$\left|G(e^{j\omega T})\right| \leq 1.$$

In the time domain, we interpret $g(n)$ as the *impulse response* corresponding to $G(z)$. We may now derive the frequency-dependent counterpart of Eq. (C.31) as follows:

$$\begin{aligned}
y^+_{n+1,m} &= g * y^+_{n,m-1} = g * (y_{n,m-1} - y^-_{n,m-1}) \\
y^-_{n+1,m} &= g * y^-_{n,m+1} = g * (y_{n,m+1} - y^+_{n,m+1}) \\
\Rightarrow \quad y_{n+1,m} &= g * (y_{n,m-1} + y_{n,m+1}) - g * (\underbrace{y^-_{n,m-1}}_{g*y^-_{n-1,m}} + \underbrace{y^+_{n,m+1}}_{g*y^+_{n-1,m}}) \\
&= g * (y_{n,m-1} + y_{n,m+1}) - g * g * y_{n-1,m} \\
&= g * \left[(y_{n,m-1} + y_{n,m+1}) - g * y_{n-1,m}\right]
\end{aligned}$$

where $*$ denotes convolution (in the time dimension only). Define *filtered node variables* by

$$\begin{aligned}
y^f_{n,m} &= g * y_{n,m} \\
y^{ff}_{n,m} &= g * y^f_{n,m}.
\end{aligned}$$

Then the frequency-dependent FDTD scheme is simply

$$y_{n+1,m} = y^f_{n,m-1} + y^f_{n,m+1} - y^{ff}_{n-1,m}.$$

We see that generalizing the FDTD scheme to frequency-dependent losses requires a simple filtering of each node variable $y_{n,m}$ by the per-sample propagation filter $G(z)$. For computational efficiency, two spatial lines should be stored in memory at time n: $y^f_{n,m}$ and $y^{ff}_{n-1,m}$, for all m. These state variables enable computation of $y_{n+1,m}$, after which each sample of $y^f_{n,m}$ $(\forall m)$ is filtered by $G(z)$ to produce $y^{ff}_{n-1,m}$ for the next iteration, and $y_{n+1,m}$ is filtered by $G(z)$ to produce $y^f_{n,m}$ for the next iteration.

The frequency-dependent generalization of the FDTD scheme described in this section extends readily to the digital waveguide mesh. See §C.14.5 on page 611 for the outline of the derivation.

C.6 The Dispersive 1D Wave Equation

In the ideal vibrating string, the only restoring force for transverse displacement comes from the string tension (§C.1 above); specifically, the transverse restoring force is equal the net transverse component of the axial string tension. Consider in place of the ideal string a *bundle* of ideal strings, such as a stranded cable. When the cable is bent, there is now a new restoring force arising from some of the fibers being compressed and others being stretched by the bending. This force sums with that due to string tension. Thus, stiffness in a vibrating string introduces a new restoring force proportional to bending angle. It is important to note that *string stiffness is a linear phenomenon* resulting from the finite diameter of the string.

In typical treatments,[3] bending stiffness adds a new term to the wave equation that is proportional to the *fourth spatial derivative* of string displacement:

$$\epsilon\ddot{y} = Ky'' - \kappa y'''' \tag{C.32}$$

where the moment constant $\kappa = YI$ is the product of Young's modulus Y (the "relative-displacement spring constant per unit cross-sectional area," discussed in §B.5.1 on page 499) and the *area moment of inertia* I (§B.4.8 on page 476); as derived in §B.4.9 on page 477, a cylindrical string of radius a has area moment of inertia equal to $\pi a^2 \cdot (a/2)^2 = \pi a^4/4$. This wave equation works well enough for small amounts of bending stiffness, but it is clearly missing some terms because it predicts that deforming the string into a parabolic shape will incur no restoring force due to stiffness. See §6.9, p. 224 for further discussion of wave equations for stiff strings.

To solve the stiff wave equation Eq. (C.32), we may set $y(t,x) = e^{st+vx}$ to get

$$\epsilon s^2 = Kv^2 - \kappa v^4.$$

At very low frequencies, or when stiffness is negligible in comparison with K/v^2, we obtain again the non-stiff string: $\epsilon s^2 \approx Kv^2 \;\Rightarrow\; v = \pm s/c$.

At very high frequencies, or when the tension K is negligible relative to κv^2, we obtain the *ideal bar* (or rod) approximation:

$$\epsilon s^2 \approx -\kappa v^4 \;\Rightarrow\; v \approx \pm e^{\pm j\frac{\pi}{4}} \left(\frac{\epsilon}{\kappa}\right)^{1/4} \sqrt{s}$$

In an ideal bar, the *only* restoring force is due to bending stiffness. Setting $s = j\omega$ gives solutions $v = \pm j\,(\epsilon/\kappa)^{1/4}\sqrt{\omega}$ and $v = \pm\,(\epsilon/\kappa)^{1/4}\sqrt{\omega}$. In the

[3] A detailed derivation of the usual stiff-string wave equation Eq. (C.32) is given by Morse in [317] or [318]. Derivations of more elaborate wave equations including rotary inertia and shear effects are given in Graff [169, pp. 180–195] ("flexural waves in thin rods"). See also Kolsky [261]. See Fletcher and Rossing [144] regarding stiff piano strings, and Cremer [95] regarding stiffness effects in violin strings (principally the prevention of sharp corners in the string).

first case, the wave velocity becomes proportional to $\sqrt{\omega}$. That is, waves travel faster along the ideal bar as oscillation frequency increases, going up as the square root of frequency. The second solution corresponds to a change in the wave shape which prevents sharp corners from forming due to stiffness [95, 118].

At intermediate frequencies, between the ideal string and the ideal bar, the stiffness contribution can be treated as a correction term [95]. This is the region of most practical interest because it is the principal operating region for strings, such as piano strings, whose stiffness has audible consequences (an inharmonic, stretched overtone series). Assuming $\kappa_0 \triangleq \kappa/K \ll 1$,

$$\begin{aligned} s^2 &= \frac{K}{\epsilon}v^2 - \frac{\kappa}{\epsilon}v^4 = \frac{K}{\epsilon}v^2\left(1 - \frac{\kappa}{K}v^2\right) \triangleq c_0^2 v^2\left(1 - \kappa_0 v^2\right) \\ \Rightarrow v^2 &\approx \frac{s^2}{c_0^2}\left(1 + \kappa_0 v^2\right) \approx \frac{s^2}{c_0^2}\left(1 + \kappa_0 \frac{s^2}{c_0^2}\right) \\ \Rightarrow v &\approx \pm\frac{s}{c_0}\sqrt{1 + \kappa_0\frac{s^2}{c_0^2}} \approx \pm\frac{s}{c_0}\left(1 + \frac{1}{2}\kappa_0\frac{s^2}{c_0^2}\right). \end{aligned}$$

Substituting for v in terms of s in e^{st+vx} gives the general eigensolution

$$e^{st+vx} = \exp\left\{s\left[t \pm \frac{x}{c_0}\left(1 + \frac{1}{2}\kappa_0\frac{s^2}{c_0^2}\right)\right]\right\}.$$

Setting $s = j\omega$ as before, corresponding to driving the medium sinusoidally over time at frequency ω, the medium response is

$$e^{st+vx} = e^{j\omega[t \pm x/c(\omega)]}$$

where

$$c(\omega) \triangleq c_0\left(1 + \frac{\kappa\omega^2}{2Kc_0^2}\right).$$

Because the effective wave velocity depends on ω, we cannot use Fourier's theorem to construct arbitrary traveling shapes by superposition. At $x = 0$, we can construct any function of time, but the waveshape *disperses* as it propagates away from $x = 0$. The higher-frequency Fourier components travel faster than the lower-frequency components.

Since the temporal and spatial sampling intervals are related by $X = cT$, this must generalize to $X = c(\omega)T \;\Rightarrow\; T(\omega) = X/c(\omega) = c_0 T_0/c(\omega)$, where $T_0 = T(0)$ is the size of a unit delay in the absence of stiffness. Thus, a unit delay z^{-1} may be replaced by

$$z^{-1} \rightarrow z^{-c_0/c(\omega)} \qquad \text{(for frequency-dependent wave velocity).}$$

That is, each delay element becomes an *allpass filter* which approximates the required delay versus frequency. A diagram appears in Fig. C.8, where $H_a(z)$ denotes the allpass filter which provides a rational approximation to $z^{-c_0/c(\omega)}$.

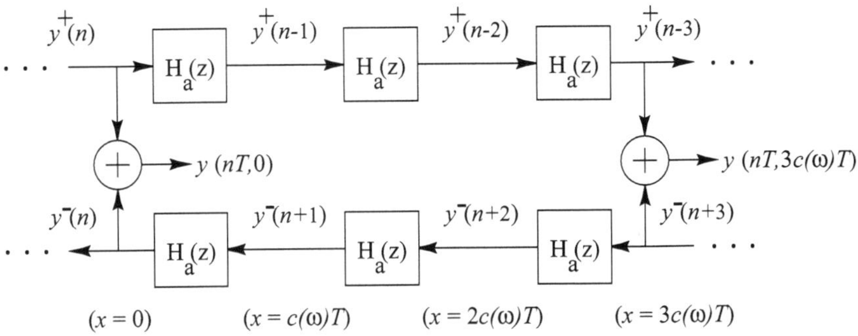

Figure C.8: Section of a stiff string where allpass filters play the role of unit delay elements.

The general, order L, allpass filter[4] is given by [449]

$$H_a(z) \stackrel{\Delta}{=} z^{-L}A(z^{-1})/A(z)$$

where

$$A(z) \stackrel{\Delta}{=} 1 + a_1 z^{-1} + a_2 z^{-2} + \cdots + a_L z^{-L}$$

and the roots of $A(z)$ must all have modulus less than 1. That is, the numerator polynomial is just the reverse of the denominator polynomial. This implies each pole p_i is gain-compensated by a zero at $z_i = 1/p_i$.

For computability of the string simulation in the presence of scattering junctions, there must be at least one sample of pure delay along each uniform section of string. This means for at least one allpass filter in Fig. C.8, we must have $H_a(\infty) = 0$ which implies $H_a(z)$ can be factored as $z^{-1}H'_a(z)$. In a systolic VLSI implementation, it is desirable to have at least one real delay from the input to the output of *every* allpass filter, in order to be able to pipeline the computation of all of the allpass filters in parallel. Computability can be arranged in practice by deciding on a minimum delay, (*e.g.*, corresponding to the wave velocity at a maximum frequency), and using an allpass filter to provide excess delay beyond the minimum.

Because allpass filters are linear and time invariant, they commute like gain factors with other linear, time-invariant components. Fig. C.9

[4] `http://ccrma.stanford.edu/~jos/filters/Allpass_Filters.html`

shows a diagram equivalent to Fig. C.8 in which the allpass filters have been commuted and consolidated at two points. For computability in all possible contexts (*e.g.*, when looped on itself), a single sample of delay is pulled out along each rail. The remaining transfer function, $H_c(z) = zH_a^3(z)$ in the example of Fig. C.9, can be approximated using any allpass filter design technique [1, 2, 267, 272, 552]. Alternatively, both gain and dispersion for a stretch of waveguide can be provided by a single filter which can be designed using any general-purpose filter design method which is sensitive to frequency-response phase as well as magnitude; examples include equation error methods (such as used in the matlab `invfreqz` function (§8.6.4, p. 297), and Hankel norm methods [177, 428, 36].

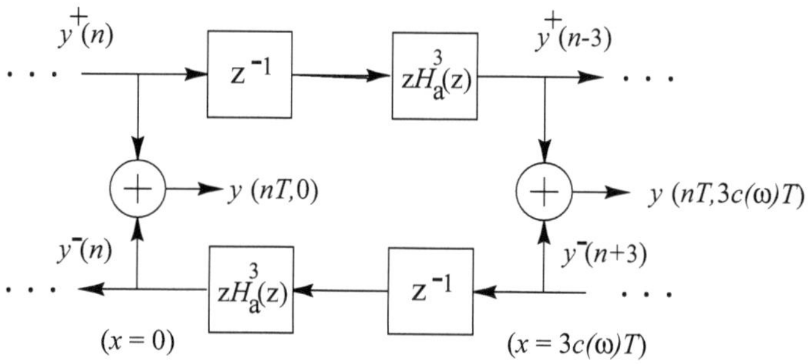

Figure C.9: Section of a stiff string where the allpass delay elements are consolidated at two points, and a sample of pure delay is extracted from each allpass chain.

In the case of a lossless, stiff string, if $H_c(z)$ denotes the consolidated allpass transfer function, it can be argued that the filter design technique used should minimize the *phase-delay error,* where phase delay is defined by [362]

$$P_c(\omega) \triangleq -\frac{\angle H_c\left(e^{j\omega T}\right)}{\omega}. \qquad \text{(Phase Delay)}$$

Minimizing the Chebyshev norm of the phase-delay error,

$$||\, P_c(\omega) - c_0/c(\omega)\, ||_\infty$$

approximates minimization of the error in *mode tuning* for the freely vibrating string [428, pp. 182–184]. Since the stretching of the overtone series is typically what we hear most in a stiff, vibrating string, the worst-case phase-delay error is a good choice in such a case.

Alternatively, a lumped allpass filter can be designed by minimizing *group delay,*

$$D_c(\omega) \triangleq -\frac{d\angle H_c\left(e^{j\omega T}\right)}{d\omega}. \qquad \text{(Group Delay)}$$

The group delay of a filter gives the delay experienced by the amplitude *envelope* of a narrow frequency band centered at ω, while the phase delay applies to the "carrier" at ω, or a sinusoidal component at frequency ω [342]. As a result, for proper *tuning* of overtones, phase delay is what matters, while for precisely estimating (or controlling) the *decay* time in a lossy waveguide, group delay gives the effective filter delay "seen" by the exponential decay envelope.

See §9.4.1 on page 384 for designing allpass filters with a prescribed delay versus frequency. To model stiff strings, the allpass filter must supply a phase delay which *decreases* as frequency increases. A good approximation may require a fairly high-order filter, adding significantly to the cost of simulation. (For low-pitched piano strings, order 8 allpass filters work well perceptually [1].) To a large extent, the allpass order required for a given error tolerance increases as the number of lumped frequency-dependent delays is increased. Therefore, increased dispersion consolidation is accompanied by larger required allpass filters, unlike the case of resistive losses.

The function `piano_dispersion_filter` in the Faust distribution (in `effect.lib`) designs and implements an allpass filter modeling the dispersion due to stiffness in a piano string [154, 170, 368].

C.6.1 Higher Order Terms

The complete, linear, time-invariant generalization of the lossy, stiff string is described by the differential equation

$$\sum_{k=0}^{\infty} \alpha_k \frac{\partial^k y(t,x)}{\partial t^k} = \sum_{l=0}^{\infty} \beta_l \frac{\partial^l y(t,x)}{\partial x^l}. \qquad \text{(C.33)}$$

which, on setting $y(t,x) = e^{st+vx}$, (or taking the 2D Laplace transform with zero initial conditions), yields the algebraic equation,

$$\sum_{k=0}^{\infty} \alpha_k s^k = \sum_{l=0}^{\infty} \beta_l v^l \qquad \text{(C.34)}$$

Solving for v in terms of s is, of course, nontrivial in general. However, in specific cases, we can determine the appropriate attenuation per sample $G(\omega)$ and wave propagation speed $c(\omega)$ by numerical means. For example, starting at $s = 0$, we normally also have $v = 0$ (corresponding to the absence of static deformation in the medium). Stepping s forward by a small differential $j\Delta\omega$, the left-hand side can be approximated by $\alpha_0 + \alpha_1 j\Delta\omega$. Requiring the generalized wave velocity $s/v(s)$ to be continuous, a physically reasonable assumption, the right-hand side can be approximated by $\beta_0 + \beta_1\Delta v$, and the solution is easy. As s steps forward, higher order terms become important one by one on both sides of the equation. Each new term in v spawns a new solution for v in terms of s, since the order of the polynomial in v is incremented. It appears possible that *homotopy continuation methods* [316] can be used to keep track of the branching solutions of v as a function of s. For each solution $v(s)$, let $v_r(\omega)$ denote the real part of $v(j\omega)$ and let $v_i(\omega)$ denote the imaginary part. Then the eigensolution family can be seen in the form $\exp\{j\omega t \pm v(j\omega)x\} = \exp\{\pm v_r(\omega)x\} \cdot \exp\{j\omega\,(t \pm v_i(\omega)x/\omega)\}$. Defining $c(\omega) \triangleq \omega/v_i(\omega)$, and sampling according to $x \to x_m \triangleq mX$ and $t \to t_n \triangleq nT(\omega)$, with $X \triangleq c(\omega)T(\omega)$ as before, (the spatial sampling period is taken to be frequency invariant, while the temporal sampling interval is modulated versus frequency using allpass filters), the left- and right-going sampled eigensolutions become

$$\begin{aligned} e^{j\omega t_n \pm v(j\omega)x_m} &= e^{\pm v_r(\omega)x_m} \cdot e^{j\omega(t_n \pm x_m/c(\omega))} \\ &= G^m(\omega) \cdot e^{j\omega(n\pm m)T(\omega)} \end{aligned} \tag{C.35}$$

where $G(\omega) \triangleq e^{\pm v_r(\omega)X}$. Thus, a general map of v versus s, corresponding to a partial differential equation of any order in the form (C.33), can be translated, in principle, into an accurate, local, linear, time-invariant, discrete-time simulation. The boundary conditions and initial state determine the initial mixture of the various solution branches as usual.

We see that a large class of wave equations with constant coefficients, of any order, admits a decaying, dispersive, traveling-wave type solution. Even-order time derivatives give rise to frequency-dependent dispersion and odd-order time derivatives correspond to frequency-dependent losses. The corresponding digital simulation of an arbitrarily long (undriven and unobserved) section of medium can be simplified via commutativity to at most two pure delays and at most two linear, time-invariant filters.

Every linear, time-invariant filter can be expressed as a zero-phase filter in series with an allpass filter. The zero-phase part can be interpreted as implementing a frequency-dependent gain (damping in a digital waveguide), and the allpass part can be seen as frequency-dependent delay (dispersion in a digital waveguide).

C.7 Alternative Wave Variables

We have thus far considered discrete-time simulation of transverse *displacement* y in the ideal string. It is equally valid to choose *velocity* $v \triangleq \dot{y}$, *acceleration* $a \triangleq \ddot{y}$, *slope* y', or perhaps some other derivative or integral of displacement with respect to time or position. Conversion between various time derivatives can be carried out by means *integrators* and *differentiators*, as depicted in Fig. C.10. Since integration and differentiation are *linear* operators, and since the traveling wave arguments are in units of time, the conversion formulas relating y, v, and a hold also for the traveling wave *components* $y^{\pm}, v^{\pm}, a^{\pm}$.

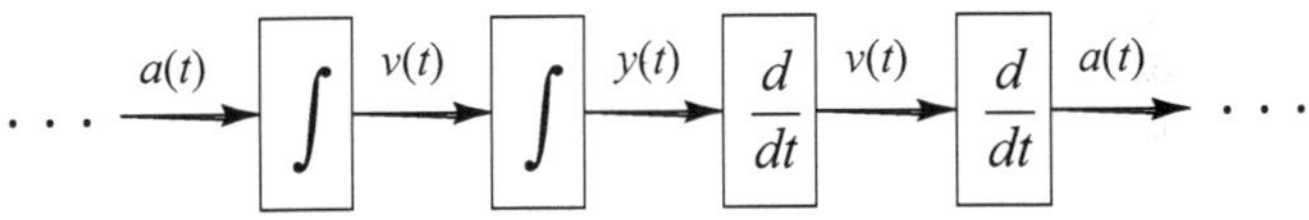

Figure C.10: Conversions between various time derivatives of displacement: $y =$ displacement, $v = \dot{y} =$ velocity, $a = \ddot{y} =$ acceleration, where $\dot{y}$ denotes dy/dt and $\ddot{y}$ denotes d^2y/dt^2.

Differentiation and integration have a simple form in the frequency domain. Denoting the *Laplace Transform* of $y(t,x)$ by

$$Y(s,x) \triangleq \mathcal{L}_s\{y(\cdot,x)\} \triangleq \int_0^{\infty} y(t,x)e^{-st}dt, \tag{C.36}$$

where "$\cdot$" in the time argument means "for all time," we have, according to the *differentiation theorem* for Laplace transforms [284],

$$\mathcal{L}_s\{\dot{y}(\cdot,x)\} = sY(s,x) - y(0,x). \tag{C.37}$$

Similarly, $\mathcal{L}_s\{\dot{y}^+\} = sY^+(s) - y^+(0)$, and so on. Thus, in the frequency domain, the conversions between displacement, velocity, and acceleration appear as shown in Fig. C.11.

In discrete time, integration and differentiation can be accomplished using digital filters [362]. Commonly used first-order approximations are shown in Fig. C.12.

If discrete-time acceleration $a_d(n)$ is defined as the sampled version of continuous-time acceleration, *i.e.*, $a_d(n) \triangleq a(nT,x) = \ddot{y}(nT,x)$, (for some fixed continuous position x which we suppress for simplicity of notation), then the frequency-domain form is given by the z *transform* [486]:

$$A_d(z) \triangleq \sum_{n=0}^{\infty} a_d(n)z^{-n} \tag{C.38}$$

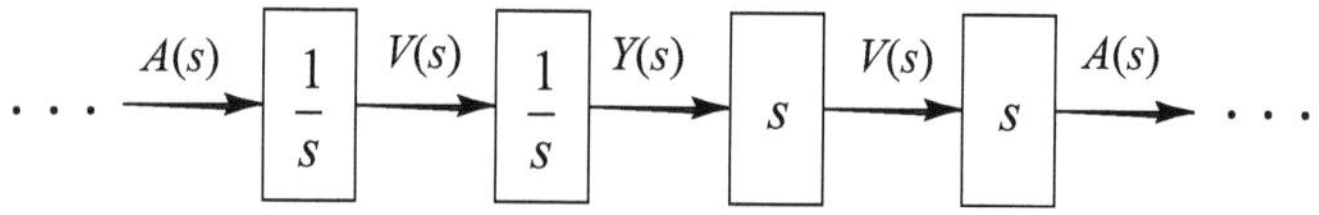

Figure C.11: Conversions between various time derivatives of displacement in the frequency domain.

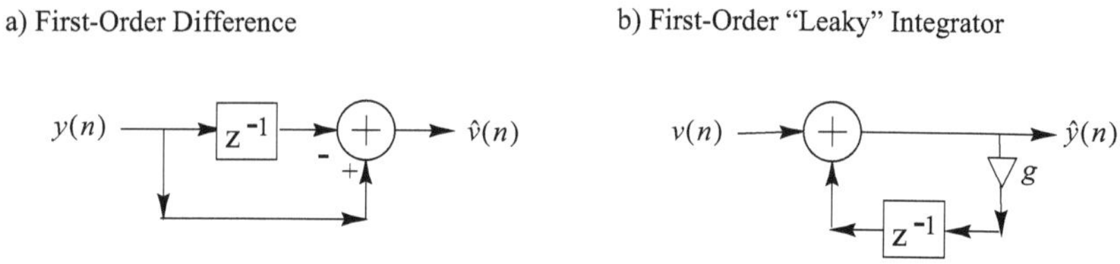

Figure C.12: Simple approximate conversions between time derivatives in the discrete-time case: a) The first-order difference $\hat{v}(n) = y(n) - y(n-1)$. b) The first-order "leaky" integrator $\hat{y}(n) = v(n) + g\hat{y}(n-1)$ with loss factor g (slightly less than 1) used to avoid infinite dc build-up.

In the frequency domain for discrete-time systems, the first-order approximate conversions appear as shown in Fig. C.13.

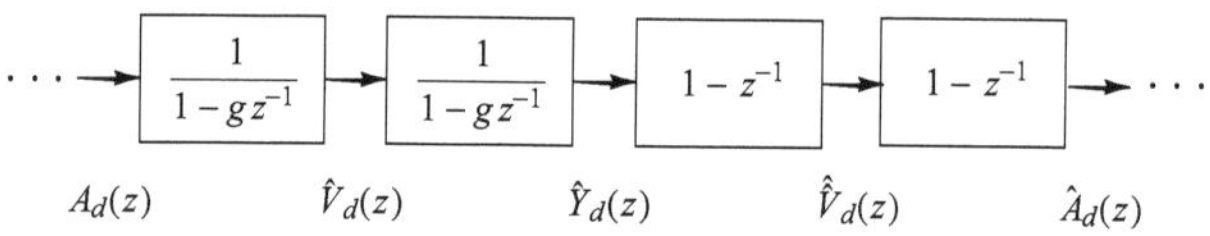

Figure C.13: Frequency-domain description of the approximate conversions between time derivatives in the discrete-time case. The subscript "d" denotes the "digital" case. A "hat" over a variable indicates it is an approximation to the variable without the hat.

The z transform plays the role of the Laplace transform for discrete-time systems. Setting $z = e^{sT}$, it can be seen as a sampled Laplace transform (divided by T), where the sampling is carried out by halting the limit of the rectangle width at T in the definition of a Reimann integral for the Laplace transform. An important difference between the two is that the frequency axis in the Laplace transform is the imaginary axis (the "$j\omega$ axis"), while the frequency axis in the z plane is on the unit circle $z = e^{j\omega T}$. As one would expect, the frequency axis for discrete-time systems has unique information only between frequencies $-\pi/T$ and π/T while the continuous-time frequency axis extends to plus and minus infinity.

These first-order approximations are accurate (though scaled by T) at low frequencies relative to half the sampling rate, but they are not "best" approximations in any sense other than being most like the definitions of integration and differentiation in continuous time. Much better approximations can be obtained by approaching the problem from a *digital filter design* viewpoint, as discussed in §8.6 on page 290.

C.7.1 Spatial Derivatives

In addition to time derivatives, we may apply any number of *spatial derivatives* to obtain yet more wave variables to choose from. The first spatial derivative of string displacement yields *slope waves*

$$\begin{aligned} y'(t,x) &\triangleq \frac{\partial}{\partial x} y(t,x) \\ &= y'_r(t-x/c) + y'_l(t+x/c) \\ &= -\frac{1}{c}\dot{y}_r(t-x/c) + \frac{1}{c}\dot{y}_l(t+x/c), \end{aligned} \tag{C.39}$$

or, in discrete time,

$$\begin{aligned} y'(t_n,x_m) &\triangleq y'(nT,mX) \\ &= y'_r\left[(n-m)T\right] + y'_l\left[(n+m)T\right] \\ &\triangleq y'^{+}(n-m) + y'^{-}(n+m) \\ &= -\frac{1}{c}\dot{y}^{+}(n-m) + \frac{1}{c}\dot{y}^{-}(n+m) \\ &\triangleq -\frac{1}{c}v^{+}(n-m) + \frac{1}{c}v^{-}(n+m) \\ &= \frac{1}{c}\left[v^{-}(n+m) - v^{+}(n-m)\right]. \end{aligned} \tag{C.40}$$

From this we may conclude that $v^- = cy'^-$ and $v^+ = -cy'^+$. That is, traveling slope waves can be computed from traveling velocity waves by dividing by c and negating in the right-going case. Physical string slope can thus be computed from a velocity-wave simulation in a digital waveguide by *subtracting* the upper rail from the lower rail and dividing by c.

By the wave equation, *curvature waves*, $y'' = \ddot{y}/c^2$, are simply a scaling of acceleration waves, in the case of ideal strings.

In the field of acoustics, the state of a vibrating string at any instant of time t_0 is often specified by the displacement $y(t_0,x)$ and velocity $\dot{y}(t_0,x)$ for all x [317]. Since velocity is the *sum* of the traveling velocity waves and displacement is determined by the *difference* of the traveling velocity waves, *viz.*, from Eq. (C.39),

$$y(t,x) = \int_0^x y'(t,\xi)d\xi = -\frac{1}{c}\int_0^x \left[v_r(t-\xi/c) - v_l(t+\xi/c)\right]d\xi,$$

one state description can be converted to the other.

In summary, all traveling-wave variables can be computed from any one, as long as both the left- and right-going component waves are available. Alternatively, any *two* linearly independent *physical* variables, such as displacement and velocity, can be used to compute all other wave variables. Wave variable conversions requiring differentiation or integration are relatively expensive since a large-order digital filter is necessary to do it right (§8.6.1, p. 290). Slope and velocity waves can be computed from each other by simple scaling, and curvature waves are identical to acceleration waves to within a scale factor.

In the absence of factors dictating a specific choice, *velocity waves* are a good overall choice because (1) it is numerically easier to perform digital integration to get displacement than it is to differentiate displacement to get velocity, (2) slope waves are immediately computable from velocity waves. Slope waves are important because they are a simple scaling of force waves.

C.7.2 Force Waves

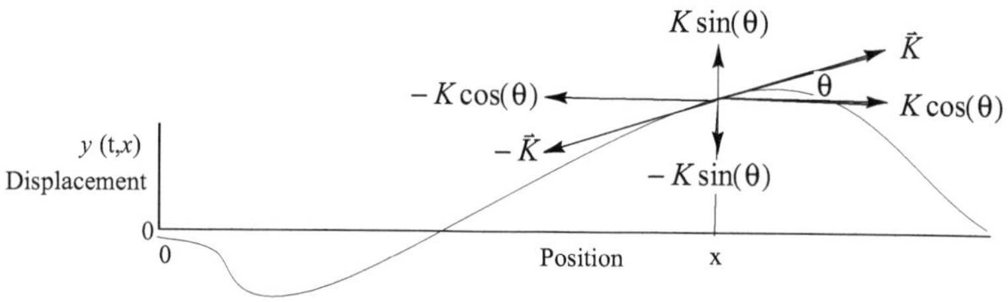

Figure C.14: Transverse force propagation in the ideal string.

Referring to Fig. C.14, at an arbitrary point x along the string, the vertical force applied at time t *to* the portion of string to the left of position x *by* the portion of string to the right of position x is given by

$$f_{rl}(t,x) = K\sin(\theta) \approx K\tan(\theta) = Ky'(t,x) \qquad \text{(C.41)}$$

assuming $|y'(t,x)| \ll 1$, as is assumed in the derivation of the wave equation. Similarly, the force applied *by* the portion to the left of position x *to* the portion to the right is given by

$$f_{lr}(t,x) = -K\sin(\theta) \approx -Ky'(t,x). \tag{C.42}$$

These forces must cancel since a nonzero net force on a massless point would produce infinite acceleration. *I.e.*, we must have $f_{rl} + f_{lr} \equiv 0$ at all times t and positions x.

Vertical force waves propagate along the string like any other transverse wave variable (since they are just slope waves multiplied by tension K). We may choose either f_{rl} or f_{lr} as the string force wave variable, one being the negative of the other. It turns out that to make the description for vibrating strings look the same as that for air columns, we have to pick f_{lr}, the one that *acts to the right.* This makes sense intuitively when one considers longitudinal pressure waves in an acoustic tube: a compression wave traveling to the right in the tube pushes the air in front of it and thus acts to the right. We therefore define the *force wave variable* to be

$$f(t,x) \stackrel{\Delta}{=} f_{lr}(t,x) = -Ky'(t,x). \tag{C.43}$$

Note that a negative slope pulls up on the segment to the right. At this point, we have not yet considered a traveling-wave decomposition.

C.7.3 Wave Impedance

Using the above identities, we have that the force distribution along the string is given in terms of velocity waves by

$$f(t,x) = \frac{K}{c}\left[\dot{y}_r(t-x/c) - \dot{y}_l(t+x/c)\right], \tag{C.44}$$

where $K/c \stackrel{\Delta}{=} K/\sqrt{K/\epsilon} = \sqrt{K\epsilon}$. This is a fundamental quantity known as the *wave impedance* of the string (also called the *characteristic impedance*), denoted as

$$R \stackrel{\Delta}{=} \sqrt{K\epsilon} = \frac{K}{c} = \epsilon c. \tag{C.45}$$

The wave impedance can be seen as the geometric mean of the two resistances to displacement: tension (spring force) and mass (inertial force).

The digitized traveling force-wave components become

$$\begin{aligned} f^+(n) &= \quad Rv^+(n) \\ f^-(n) &= -\ Rv^-(n) \end{aligned} \tag{C.46}$$

which gives us that the right-going force wave equals the wave impedance times the right-going velocity wave, and the left-going force wave equals *minus* the wave impedance times the left-going velocity wave.[5] Thus, in a traveling wave, force is always *in phase* with velocity (considering the minus sign in the left-going case to be associated with the direction of travel rather than a 180 degrees phase shift between force and velocity). Note also that if the left-going force wave were defined as the string force acting to the left, the minus sign would disappear. The fundamental relation $f^+ = Rv^+$ is sometimes referred to as the mechanical counterpart of *Ohm's law for traveling waves*, and R in c.g.s. units can be called *acoustical ohms* [261].

In the case of the *acoustic tube* [317, 297], we have the analogous relations

$$\begin{aligned} p^+(n) &= \quad R_T\, u^+(n) \\ p^-(n) &= -\; R_T\, u^-(n) \end{aligned} \tag{C.47}$$

where $p^+(n)$ is the right-going traveling *longitudinal pressure wave* component, $p^-(n)$ is the left-going pressure wave, and $u^\pm(n)$ are the left and right-going *volume velocity waves*. In the acoustic tube context, the wave impedance is given by

$$R_T = \frac{\rho c}{A} \qquad \text{(Acoustic Tubes)} \tag{C.48}$$

where ρ is the mass per unit volume of air, c is sound speed in air, and A is the cross-sectional area of the tube. Note that if we had chosen *particle velocity* rather than volume velocity, the wave impedance would be $R_0 = \rho c$ instead, the wave impedance in open air. Particle velocity is appropriate in open air, while volume velocity is the conserved quantity in acoustic tubes or "ducts" of varying cross-sectional area.

C.7.4 State Conversions

In §C.3.6 on page 530, an arbitrary string state was converted to traveling displacement-wave components to show that the traveling-wave representation is complete, *i.e.*, that any physical string state can be expressed as a pair of traveling-wave components. In this section, we revisit this topic using *force* and *velocity* waves.

By definition of the traveling-wave decomposition, we have

$$\begin{aligned} f &= f^+ + f^- \\ v &= v^+ + v^-. \end{aligned}$$

Using Eq. (C.46), we can eliminate $v^+ = f^+/R$ and $v^- = -f^+/R$, giving, in matrix form,

$$\begin{bmatrix} f \\ v \end{bmatrix} = \begin{bmatrix} 1 & 1 \\ \frac{1}{R} & -\frac{1}{R} \end{bmatrix} \begin{bmatrix} f^+ \\ f^- \end{bmatrix}.$$

[5] The sum of the left- and right-going components, $f = f^+ + f^-$, equals the *net* force acting to the right.

Thus, the string state (in terms of force and velocity) is expressed as a linear transformation of the traveling force-wave components. Using the Ohm's law relations to eliminate instead $f^+ = Rv^+$ and $f^- = -Rv^-$, we obtain

$$\left[\begin{array}{c} f \\ v \end{array}\right] = \left[\begin{array}{cc} R & -R \\ 1 & 1 \end{array}\right] \left[\begin{array}{c} v^+ \\ v^- \end{array}\right].$$

To convert an arbitrary initial string state (f, v) to either a traveling force-wave or velocity-wave simulation, we simply must be able to *invert* the appropriate two-by-two matrix above. That is, the matrix must be *nonsingular*. Requiring both determinants to be nonzero yields the condition

$$0 < R < \infty.$$

That is, the wave impedance must be a positive, finite number. This restriction makes good physical sense because one cannot propagate a finite-energy wave in either a zero or infinite wave impedance.

Carrying out the inversion to obtain force waves (f^+, f^-) from (f, v) yields

$$\left[\begin{array}{c} f^+ \\ f^- \end{array}\right] = \frac{1}{2}\left[\begin{array}{cc} 1 & R \\ 1 & -R \end{array}\right] \left[\begin{array}{c} f \\ v \end{array}\right] = \left[\begin{array}{c} \frac{f+Rv}{2} \\ \frac{f-Rv}{2} \end{array}\right].$$

Similarly, velocity waves (v^+, v^-) are prepared from (f, v) according to

$$\left[\begin{array}{c} v^+ \\ v^- \end{array}\right] = \frac{1}{2}\left[\begin{array}{cc} \frac{1}{R} & 1 \\ -\frac{1}{R} & 1 \end{array}\right] \left[\begin{array}{c} f \\ v \end{array}\right] = \left[\begin{array}{c} \frac{v+f/R}{2} \\ \frac{v-f/R}{2} \end{array}\right].$$

C.7.5 Power Waves

Basic courses in physics teach us that *power* is *work per unit time*, and *work* is a measure of *energy* which is typically defined as *force times distance*. Therefore, power is in physical units of force times distance per unit time, or force times velocity. It therefore should come as no surprise that traveling *power waves* are defined for strings as follows:

$$\begin{aligned} \mathcal{P}^+(n) &\stackrel{\Delta}{=} \quad f^+(n)v^+(n) \\ \mathcal{P}^-(n) &\stackrel{\Delta}{=} -f^-(n)v^-(n) \end{aligned}$$

From the Ohm's-law relations $f^+ = Rv^+$ and $f^- = -Rv^-$, we also have

$$\begin{aligned} \mathcal{P}^+(n) &= R[v^+(n)]^2 = [f^+(n)]^2/R, \\ \mathcal{P}^-(n) &= R[v^-(n)]^2 = [f^-(n)]^2/R. \end{aligned}$$

Thus, both the left- and right-going components are *nonnegative.* The sum of the traveling powers at a point gives the total power at that point in the waveguide:

$$\mathcal{P}(t_n, x_m) \triangleq \mathcal{P}^+(n-m) + \mathcal{P}^-(n+m) \tag{C.49}$$

If we had left out the minus sign in the definition of left-going power waves, the sum would instead be a *net* power flow.

Power waves are important because they correspond to the actual ability of the wave to do *work* on the outside world, such as on a violin bridge at the end of a string. Because *energy* is conserved in closed systems, power waves sometimes give a simpler, more fundamental view of wave phenomena, such as in conical acoustic tubes. Also, implementing nonlinear operations such as *rounding* and *saturation* in such a way that signal power is not increased, gives suppression of *limit cycles* and *overflow oscillations* [432], as discussed in the section on signal scattering.

For example, consider a waveguide having a wave impedance which increases smoothly to the right. A converging cone provides a practical example in the acoustic tube realm. Then since the energy in a traveling wave must be in the wave unless it has been transduced elsewhere, we expect $\mathcal{P}^+$ to propagate unchanged along the waveguide. However, since the wave impedance is increasing, it must be true that force is increasing and velocity is decreasing according to $\mathcal{P}^+ = R(v^+)^2 = (f^+)^2/R$. Looking only at force or velocity might give us the mistaken impression that the wave is getting stronger (looking at force) or weaker (looking at velocity), when really it was simply sailing along as a fixed amount of energy. This is an example of a *transformer* action in which force is converted into velocity or vice versa. It is well known that a conical tube acts as if it's open on both ends even though we can plainly see that it is closed on one end. A tempting explanation is that the cone acts as a transformer which exchanges pressure and velocity between the endpoints of the tube, so that a closed end on one side is equivalent to an open end on the other. However, this view is oversimplified because, while spherical pressure waves travel nondispersively in cones, velocity propagation is dispersive [22, 50].

C.7.6 Energy Density Waves

The vibrational energy per unit length along the string, or *wave energy density* [317] is given by the sum of potential and kinetic energy densities:

$$W(t,x) \triangleq \underbrace{\frac{1}{2}Ky'^2(t,x)}_{\text{potential}} + \underbrace{\frac{1}{2}\epsilon\dot{y}^2(t,x)}_{\text{kinetic}} \tag{C.50}$$

Sampling across time and space, and substituting traveling wave components, one can show in a few lines of algebra that the *sampled* wave energy density is given by

$$W(t_n, x_m) \triangleq W^+(n-m) + W^-(n+m), \tag{C.51}$$

where

$$\begin{aligned} W^+(n) &= \frac{\mathcal{P}^+(n)}{c} = \frac{f^+(n)v^+(n)}{c} = \epsilon\left[v^+(n)\right]^2 = \frac{\left[f^+(n)\right]^2}{K}, \\ W^-(n) &= \frac{\mathcal{P}^-(n)}{c} = -\frac{f^-(n)v^-(n)}{c} = \epsilon\left[v^-(n)\right]^2 = \frac{\left[f^-(n)\right]^2}{K}. \end{aligned}$$

Thus, traveling power waves (energy per unit time) can be converted to energy density waves (energy per unit length) by simply dividing by c, the speed of propagation. Quite naturally, the *total wave energy* in the string is given by the integral along the string of the energy density:

$$\mathcal{E}(t) = \int_{x=-\infty}^{\infty} W(t,x)dx \approx \sum_{m=-\infty}^{\infty} W(t,x_m)X \tag{C.52}$$

In practice, of course, the string length is finite, and the limits of integration are from the x coordinate of the left endpoint to that of the right endpoint, *e.g.*, 0 to L.

C.7.7 Root-Power Waves

It is sometimes helpful to *normalize* the wave variables so that signal power is uniformly distributed numerically. This can be especially helpful in fixed-point implementations.

From (C.49), it is clear that power normalization is given by

$$\begin{aligned} \tilde{f}^+ &\triangleq f^+/\sqrt{R} & \tilde{f}^- &\triangleq f^-/\sqrt{R} \\ \tilde{v}^+ &\triangleq v^+\sqrt{R} & \tilde{v}^- &\triangleq v^-\sqrt{R} \end{aligned} \tag{C.53}$$

where we have dropped the common time argument '(n)' for simplicity. As a result, we obtain

$$\begin{aligned} \mathcal{P}^+ &= f^+v^+ &&= \tilde{f}^+\tilde{v}^+ \\ &= R(v^+)^2 &&= (\tilde{v}^+)^2 \\ &= (f^+)^2/R &&= (\tilde{f}^+)^2 \end{aligned}$$

and

$$\begin{aligned} \mathcal{P}^- &= -f^-v^- &&= -\tilde{f}^+\tilde{v}^+ \\ &= R(v^-)^2 &&= (\tilde{v}^-)^2 \\ &= (f^-)^2/R &&= (\tilde{f}^-)^2. \end{aligned}$$

The normalized wave variables $\tilde{f}^{\pm}$ and $\tilde{v}^{\pm}$ behave physically like force and velocity waves, respectively, but they are scaled such that either can be squared to obtain instantaneous signal power. Waveguide networks built using normalized waves have many desirable properties [174, 172, 432]. One is the obvious numerical advantage of uniformly distributing signal power across available dynamic range in fixed-point implementations. Another is that only in the normalized case can the wave impedances be made *time varying* without modulating signal power. In other words, use of normalized waves eliminates "parametric amplification" effects; signal power is decoupled from parameter changes.

C.7.8 Total Energy in a Rigidly Terminated String

The total energy $\mathcal{E}$ in a length L, rigidly terminated, freely vibrating string can be computed as

$$\mathcal{E}(t) \triangleq \int_0^L W(t,x)dx = \int_{t_0}^{t_0+2L/c} \mathcal{P}(\tau,x)d\tau \tag{C.54}$$

$$\approx \sum_{m=0}^{N/2-1} W(t_n,x_m)X = \sum_{n=1}^{N} \mathcal{P}(t_0+t_n,x_m)T \tag{C.55}$$

for any $x \in [0, L]$. Since the energy never decays, t and t_0 are also arbitrary. Thus, because free vibrations of a doubly terminated string must be *periodic* in time, the total energy equals the integral of power over any period at any point along the string.

C.8 Scattering at Impedance Changes

When a traveling wave encounters a change in wave impedance, *scattering* occurs, *i.e.*, a traveling wave impinging on an impedance discontinuity will partially reflect and partially transmit at the junction in such a way that energy is conserved. This is a classical topic in transmission line theory [295], and it is well covered for acoustic tubes in a variety of references [297, 363]. However, for completeness, we derive the basic scattering relations below for plane waves in air, and for longitudinal stress waves in rods.

C.8.1 Plane-Wave Scattering

Consider a plane wave with peak pressure amplitude p_1^+ propagating from wave impedance R_1 into a new wave impedance R_2, as shown in Fig. C.15. (Assume R_1 and R_2 are real and positive.) The physical constraints on the wave are that

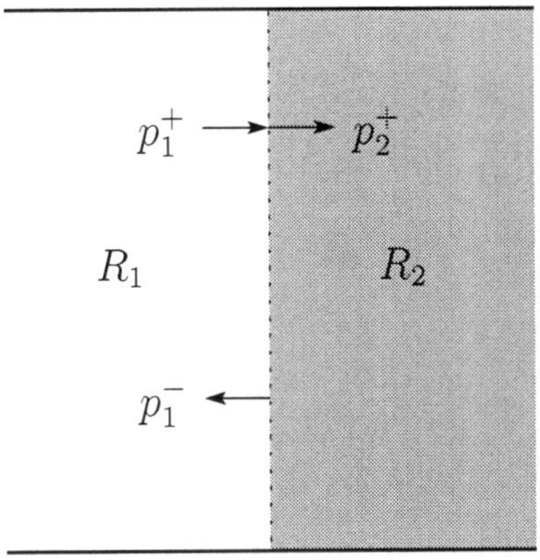

Figure C.15: Plane wave propagation in a medium changing from wave impedance R_1 to R_2.

- pressure must be continuous everywhere, and
- velocity in must equal velocity out (the junction has no state).

Since power is pressure times velocity, these constraints imply that signal power is conserved at the junction.[6] Expressed mathematically, the physical constraints at the junction can be written as follows:

$$\begin{aligned} p_1^+ + p_1^- &= p_2^+ \quad \text{(pressure continuous across junction)} \\ v_1^+ + v_1^- &= v_2^+ \quad \text{(velocity in = velocity out)} \end{aligned}$$

As derived in §C.7.3 on page 555, we also have the Ohm's law relations:

$$\begin{aligned} p_i^+ &= R_i v_i^+ \\ p_i^- &= - R_i v_i^- \end{aligned}$$

These equations determine what happens at the junction.

To obey the physical constraints at the impedance discontinuity, the incident plane-wave must split into a *reflected* plane wave p_1^- and a *transmitted* plane-wave p_2^+ such that pressure is continuous and signal power is conserved. The physical pressure on the left of the junction is $p_1 = p_1^+ + p_1^-$, and the physical pressure on the right of the junction is $p_2 = p_2^+ + p_2^- = p_2^+$, since $p_2^- = 0$ according to our set-up.

[6]When the wave impedance is complex, the junction effectively has state, so that energy, but not necessarily power, is conserved. See §C.18 for an example.

Scattering Solution

Define the junction pressure p_j and junction velocity v_j by

$$\begin{aligned}
p_j &\isdef p_1^+ + p_1^- = p_2^+ \quad \text{(pressure at junction)} \\
v_j &\isdef v_1^+ + v_1^- = v_2^+ \quad \text{(velocity at junction).}
\end{aligned}$$

Then we can write

$$\begin{aligned}
p_1^+ + p_1^- &= p_2^+ = p_j \\
\Rightarrow R_1 v_1^+ - R_1 v_1^- &= R_2 v_2^+ = R_2 v_j \\
\Rightarrow R_1 v_1^+ - R_1 (v_j - v_1^+) &= R_2 v_j \\
\Rightarrow 2\,R_1 v_1^+ - R_1 v_j &= R_2 v_j \\
\Rightarrow \boxed{v_j = \frac{2\,R_1}{R_1+R_2} v_1^+.}
\end{aligned}$$

Note that $v_j = v_2^+$, so we have found the velocity of the transmitted wave. Since $v_j = v_1^+ + v_1^-$, the velocity of the reflected wave is simply

$$v_1^- = v_j - v_1^+ = \left[\frac{2\,R_1}{R_1+R_2} - 1\right] v_1^+ = \frac{R_1 - R_2}{R_1+R_2} v_1^+.$$

We have solved for the transmitted and reflected velocity waves given the incident wave and the two impedances.

Using the Ohm's law relations, the pressure waves follow easily:

$$\begin{aligned}
p_2^+ &= R_2 v_2^+ = R_2 v_j = \frac{2\,R_2}{R_1+R_2} p_1^+ \\
p_1^- &= -R_1 v_1^- = \frac{R_2 - R_1}{R_1+R_2} p_1^+
\end{aligned}$$

Reflection Coefficient

Define the *reflection coefficient* of the scattering junction as

$$\boxed{\rho = \frac{R_2 - R_1}{R_1+R_2} = \frac{\text{Impedance Step}}{\text{Impedance Sum}}.}$$

Then we get the following scattering relations in terms of ρ for pressure waves:

$$\begin{aligned} p_2^+ &= (1+\rho)p_1^+ \\ p_1^- &= \rho p_1^+ \end{aligned}$$

Signal flow graphs for pressure and velocity are given in Fig. C.16.

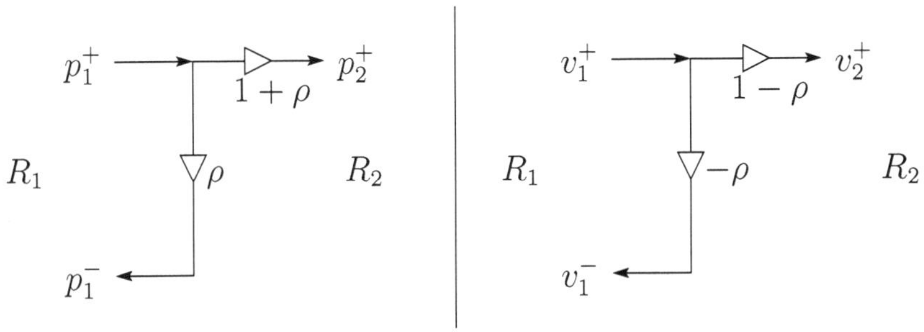

Figure C.16: Signal flow graph for the pressure and velocity components of a plane wave scattering at an impedance discontinuity R_1:R_2.

It is a simple exercise to verify that signal power is conserved by checking that $p_1^+v_1^+ = p_2^+v_2^+ + (-p_1^-v_1^-)$. (Left-going power is negated to account for its opposite direction-of-travel.)

So far we have only considered a plane wave incident on the left of the junction. Consider now a plane wave incident from the right. For that wave, the impedance steps from R_2 to R_1, so the reflection coefficient it "sees" is $-\rho$. By superposition, the signal flow graph for plane waves incident from either side is given by Fig. C.17. Note that the transmission coefficient is one plus the reflection coefficient in either direction. This signal flow graph is often called the "Kelly-Lochbaum" scattering junction [297].

There are some simple special cases:

- $R_2 = \infty \;\Rightarrow\; \rho = 1$ (*e.g.*, rigid wall reflection)
- $R_2 = 0 \;\Rightarrow\; \rho = -1$ (*e.g.*, open-ended tube)
- $R_2 = R_1 \;\Rightarrow\; \rho = 0$ (no reflection)

C.8.2 Plane-Wave Scattering at an Angle

Figure C.18 shows the more general situation (as compared to Fig. C.15 on page 561) of a sinusoidal traveling plane wave encountering an impedance discontinuity at some arbitrary angle of incidence, as indicated by the vector wavenumber $\underline{k}_1^+$. The mathematical details of general sinusoidal plane waves in air and vector wavenumber are reviewed in §B.8.1 on page 515.

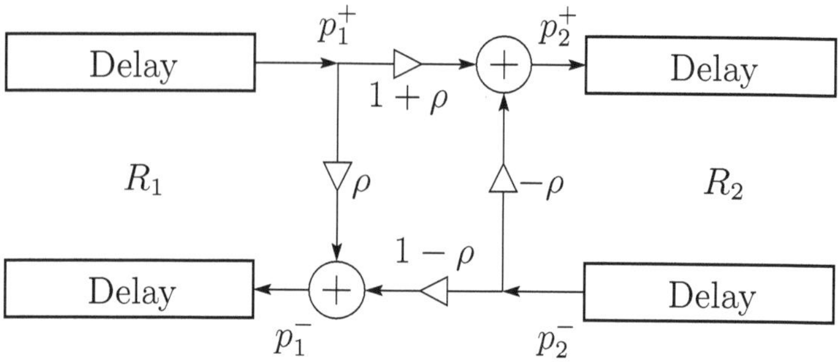

Figure C.17: Signal flow graph for plane waves incident on either the left or right of an impedance discontinuity. Also shown are delay lines corresponding to sampled traveling plane-wave components propagating on either side of the scattering junction.

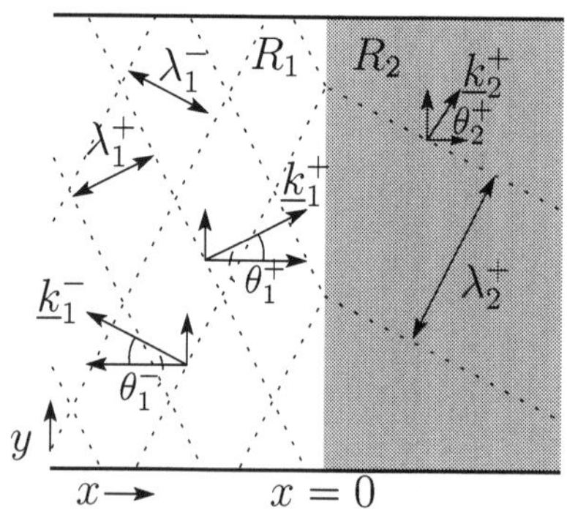

Figure C.18: Sinusoidal plane wave scattering at an impedance discontinuity—oblique angle of incidence θ_1^+.

At the boundary between impedance R_1 and R_2, we have, by continuity of pressure,

$$k_1 \sin(\theta_1^+) \quad = \quad k_1 \sin(\theta_1^-) \; = \; k_2 \sin(\theta_2^+)$$

as we will now derive.

Let the impedance change be in the $\underline{x} = (0, y, z)$ plane. Thus, the impedance is R_1 for $x \leq 0$ and R_2 for $x > 0$. There are three plane waves to consider:

- The incident plane wave with wave vector $\underline{k}_1^+$
- The reflected plane wave with wave vector $\underline{k}_1^-$
- The transmitted plane wave with wave vector $\underline{k}_2^+$

By continuity, the waves must agree on boundary plane:

$$\left\langle \underline{k}_1^+, \underline{r} \right\rangle = \left\langle \underline{k}_1^-, \underline{r} \right\rangle = \left\langle \underline{k}_2^+, \underline{r} \right\rangle$$

where $\underline{r} = (0, y, z)$ denotes any vector in the boundary plane. Thus, at $x = 0$ we have

$$k_{1y}^+ \, y + k_{1z}^+ \, z = k_{1y}^- \, y + k_{1z}^- \, z = k_{2y}^+ \, y + k_{2z}^+ \, z.$$

If the incident wave is constant along z, then $k_{1z}^+ = 0$, requiring $k_{1z}^- = k_{2z}^+ = 0$, leaving

$$k_{1y}^+ \, y = k_{1y}^- \, y = k_{2y}^+ \, y$$

or

$$\boxed{k_1 \sin(\theta_1^+) = k_1 \sin(\theta_1^-) = k_2 \sin(\theta_2^+)} \tag{C.56}$$

where θ is defined as zero when traveling in the direction of positive x for the incident ($\underline{k}_1^+$) and transmitted ($\underline{k}_2^+$) wave vector, and along *negative* x for the reflected ($\underline{k}_1^-$) wave vector (see Fig. C.18).

Reflection and Refraction

The first equality in Eq. (C.56) implies that the *angle of incidence equals angle of reflection:*

$$\boxed{\theta_1^+ = \theta_1^-}$$

We now wish to find the wavenumber in medium 2. Let c_i denote the phase velocity in wave impedance R_i:

$$c_i = \frac{\omega}{k_i}, \quad i = 1, 2$$

In impedance R_2, we have in particular

$$\omega^2 = c_2^2 k_2^2 = c_2^2\left[(k_{2x}^+)^2 + (k_{2y}^+)^2\right].$$

Solving for k_{2x}^+ gives

$$k_{2x}^+ = \sqrt{\frac{\omega^2}{c_2^2} - (k_{2y}^+)^2} = \sqrt{\frac{\omega^2}{c_2^2} - k_2^2\sin^2(\theta_2^+)}.$$

Since $k_1\sin(\theta_1^+) = k_2\sin(\theta_2^+)$ from above,

$$k_{2x}^+ = \sqrt{\frac{\omega^2}{c_2^2} - k_1^2\sin^2(\theta_1^+)} = \sqrt{\frac{\omega^2}{c_2^2} - \frac{\omega^2}{c_1^2}\sin^2(\theta_1^+)}.$$

We have thus derived

$$\boxed{k_{2x}^+ = \frac{\omega}{c_2}\sqrt{1 - \frac{c_2^2}{c_1^2}\sin^2(\theta_1^+)}.}$$

We earlier established $k_{2y}^+ = k_{1y}^+$ (for agreement along the boundary, by pressure continuity). This describes the *refraction* of the plane wave as it passes through the impedance-change boundary. The refraction angle depends on ratio of phase velocities c_2/c_1. This ratio is often called the *index of refraction*:

$$n \triangleq \frac{c_2}{c_1}$$

and the relation $k_1\sin(\theta_1^+) = k_2\sin(\theta_2^+)$ is called *Snell's Law* (of refraction).

Evanescent Wave due to Total Internal Reflection

Note that if $c_1 < c_2|\sin(\theta_1^+)|$, the horizontal component of the wavenumber in medium 2 becomes *imaginary*. In this case, the wave in medium 2 is said to be *evanescent*, and the wave in medium 1 undergoes *total internal reflection* (no power travels from medium 1 to medium 2). The evanescent-wave amplitude decays exponentially to the right and oscillates "in place" (like a standing wave). "Tunneling" is possible given a medium 3 beyond medium 2 in which wave propagation resumes.

To show explicitly the exponential decay and in-place oscillation in an evanescent wave, express the imaginary wavenumber as $k_x \triangleq -j\kappa_x$. Then we have

$$\begin{aligned} p(t,\underline{x}) &= \cos\left(\omega t - \underline{k}^T\underline{x}\right) = \text{re}\left\{e^{j(\omega t - k_x x - k_y y)}\right\} \\ &= \text{re}\left\{e^{j(\omega t - k_y y)}e^{-jk_x x}\right\}, \quad (k_x \stackrel{\Delta}{=} -j\kappa_x) \\ &\stackrel{\Delta}{=} \text{re}\left\{e^{j(\omega t - k_y y)}e^{-k_x x}\right\} \\ &= e^{-k_x x}\cos(\omega t - k_y y). \end{aligned}$$

Thus, an imaginary wavenumber corresponds to an exponentially decaying evanescent wave. Note that the time dependence (cosine term) applies to *all points* to the right of the boundary. Since evanescent waves do not really "propagate," it is perhaps better to speak of an "evanescent acoustic field" or "evanescent standing wave" instead of "evanescent waves".

For more on the physics of evanescent waves and tunneling, see [295].

C.8.3 Longitudinal Waves in Rods

In this section, elementary scattering relations will be derived for the case of *longitudinal* force and velocity waves in an ideal string or rod. In solids, force-density waves are referred to as *stress* waves [169, 261]. Longitudinal stress waves in strings and rods have units of (compressive) force per unit area and are analogous to longitudinal pressure waves in acoustic tubes.

A single waveguide section between two partial sections is shown in Fig. C.19. The sections are numbered 0 through 2 from left to right, and their wave impedances are R_0, R_1, and R_2, respectively. Such a rod might be constructed, for example, using three different materials having three different densities. In the ith section, there are two stress traveling waves: f_i^+ traveling to the right at speed c, and f_i^- traveling to the left at speed c. To minimize the numerical dynamic range, velocity waves may be chosen instead when $R_i > 1$.

As in the case of transverse waves (see the derivation of (C.46)), the traveling longitudinal plane waves in each section satisfy [169, 261]

$$\begin{aligned} f_i^+(t) &= \quad R_i v_i^+(t) \\ f_i^-(t) &= - \; R_i v_i^-(t) \end{aligned} \qquad \text{(C.57)}$$

where the wave impedance is now $R_i = \sqrt{E\rho}$, with ρ being the mass density, and E being the *Young's modulus* of the medium (defined as the stress over the strain, where *strain* means relative displacement—see §B.5.1 on page 499) [169, 261]. As before, velocity $v_i = v_i^+ + v_i^-$ is defined as positive to the right, and f_i^+ is the right-going traveling-wave component of the stress, and it is positive when the rod is locally *compressed*.

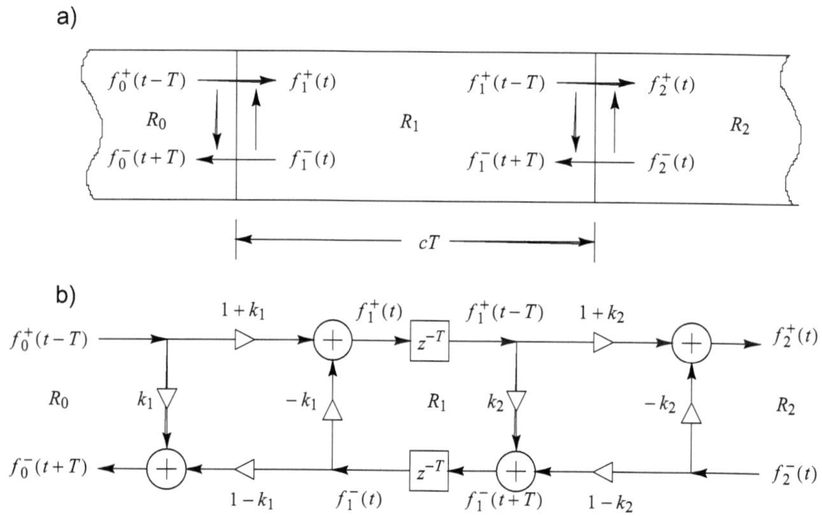

Figure C.19: A waveguide section between two partial sections. a) Physical picture indicating traveling waves in a continuous medium whose wave impedance changes from R_0 to R_1 to R_2. b) Digital simulation diagram for the same situation. The section propagation delay is denoted as z^{-T}. The behavior at an impedance discontinuity is characterized by a lossless splitting of an incoming wave into transmitted and reflected components.

If the wave impedance R_i is constant, the shape of a traveling wave is not altered as it propagates from one end of a rod-section to the other. In this case we need only consider f_i^+ and f_i^- at one end of each section as a function of time. As shown in Fig. C.19, we define $f_i^\pm(t)$ as the force-wave component at the *extreme left* of section i. Therefore, at the extreme right of section i, we have the traveling waves $f_i^+(t-T)$ and $f_i^-(t+T)$, where T is the travel time from one end of a section to the other.

For generality, we may allow the wave impedances R_i to vary with time. A number of possibilities exist which satisfy (C.57) in the time-varying case. For the moment, we will assume the traveling waves at the extreme right of section i are still given by $f_i^+(t-T)$ and $f_i^-(t+T)$. This definition, however, implies the velocity varies inversely with the wave impedance. As a result, signal energy, being the product of force times velocity, is "pumped" into or out of the waveguide by a changing wave impedance. Use of normalized waves $\tilde{f}_i^\pm$ avoids this. However, normalization increases the required number of multiplications, as we will see in §C.8.6 below.

As before, the physical force density (stress) and velocity at the left end of section i are obtained by summing the left- and right-going traveling wave components:

$$\begin{aligned} f_i &= f_i^+ + f_i^- \\ v_i &= v_i^+ + v_i^- \end{aligned} \tag{C.58}$$

Let $f_i(t, x_i)$ denote the force at position x_i and time t in section i, where $x_i \in [0, cT]$ is measured from the extreme left of section i along its axis. Then we have, for example, $f_i(t,0) \triangleq f_i^+(t) + f_i^-(t)$ and $f_i(t,cT) \triangleq f_i^+(t-T) + f_i^-(t+T)$ at the boundaries of section i. More generally, within section i, the physical stress may be expressed in terms of its traveling-wave components by

$$f_i(t,x_i) = f_i^+\left(t - \frac{x_i}{c}\right) + f_i^-\left(t + \frac{x_i}{c}\right), \quad 0 \le x_i \le cT.$$

C.8.4 Kelly-Lochbaum Scattering Junctions

Conservation of energy and mass dictate that, at the impedance discontinuity, force and velocity variables must be *continuous*

$$\begin{aligned} f_{i-1}(t,cT) &= f_i(t,0) \\ v_{i-1}(t,cT) &= v_i(t,0) \end{aligned} \tag{C.59}$$

where velocity is defined as positive to the right on both sides of the junction. Force (or stress or pressure) is a scalar while velocity is a vector with both a magnitude and direction (in this case only left or right). Equations (C.57), (C.58), and (C.59) imply the following *scattering equations* (a derivation is given in the next section for the more general case of N waveguides meeting at a junction):

$$\begin{aligned} f_i^+(t) &= [1+k_i(t)]\, f_{i-1}^+(t-T) - k_i(t) f_i^-(t) \\ f_{i-1}^-(t+T) &= k_i(t) f_{i-1}^+(t-T) + [1-k_i(t)]\, f_i^-(t) \end{aligned} \tag{C.60}$$

where

$$k_i(t) \triangleq \frac{R_i(t) - R_{i-1}(t)}{R_i(t) + R_{i-1}(t)} \tag{C.61}$$

is called the ith *reflection coefficient.* Since $R_i(t) \geq 0$, we have $k_i(t) \in [-1, 1]$. It can be shown that if $|k_i| > 1$, then either R_i or R_{i-1} is negative, and this implies an active (as opposed to passive) medium. Correspondingly, lattice and ladder recursive digital filters are *stable* if and only if all reflection coefficients are bounded by 1 in magnitude [297].

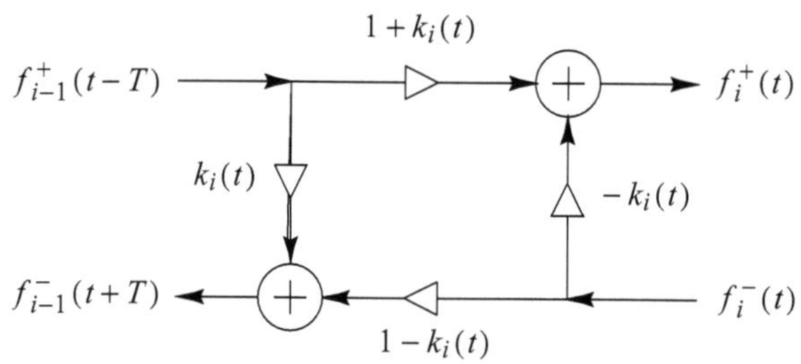

Figure C.20: The Kelly-Lochbaum scattering junction.

The scattering equations are illustrated in Figs. C.19b and C.20. In linear predictive coding of speech [483], this structure is called the *Kelly-Lochbaum scattering junction*, and it is one of several types of scattering junction used to implement *lattice* and *ladder* digital filter structures (§C.9.4, p. 579,[297]).

C.8.5 One-Multiply Scattering Junctions

By factoring out $k_i(t)$ in each equation of (C.60), we can write

$$\begin{aligned} f_i^+(t) &= f_{i-1}^+(t-T) + f_\Delta(t) \\ f_{i-1}^-(t+T) &= f_i^-(t) + f_\Delta(t) \end{aligned} \tag{C.62}$$

where

$$f_\Delta(t) \triangleq k_i(t)\left[f_{i-1}^+(t-T) - f_i^-(t)\right] \tag{C.63}$$

Thus, only *one multiplication* is actually necessary to compute the transmitted and reflected waves from the incoming waves in the Kelly-Lochbaum junction. This computation is shown in Fig. C.21, and it is known as the *one-multiply scattering junction* [297].

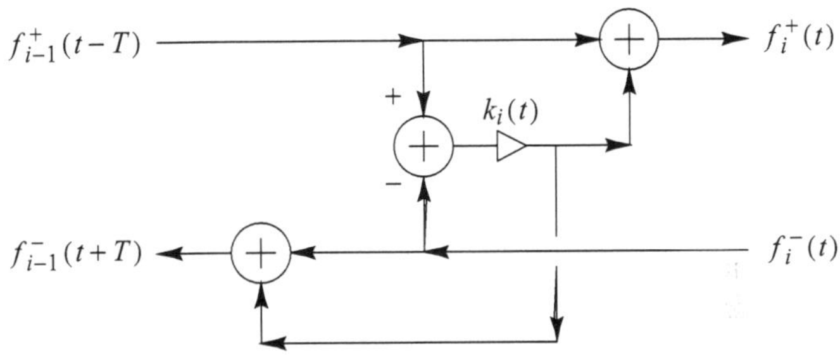

Figure C.21: The one-multiply scattering junction.

Another one-multiply form is obtained by organizing (C.60) as

$$\begin{aligned} f^+_i(t) &= f^-_i(t) + \alpha_i(t)\tilde{f}_d(t) \\ f^-_{i-1}(t+T) &= f^+_i(t) - \tilde{f}_d(t) \end{aligned} \tag{C.64}$$

where

$$\begin{aligned} \alpha_i(t) &\triangleq 1 + k_i(t) \\ \tilde{f}_d(t) &\triangleq f^+_{i-1}(t-T) - f^-_i(t). \end{aligned} \tag{C.65}$$

As in the previous case, only one multiplication and three additions are required per junction. This one-multiply form generalizes more readily to junctions of more than two waveguides, as we'll see in a later section.

A scattering junction well known in the LPC speech literature but not described here is the so-called *two-multiply* junction [297] (requiring also two additions). This omission is because the two-multiply junction is not valid as a general, *local*, physical modeling building block. Its derivation is tied to the reflectively terminated, cascade waveguide chain. In cases where it applies, however, it can be the implementation of choice; for example, in DSP chips having a fast multiply-add instruction, it may be possible to implement the inner loop of the two-multiply, two-add scattering junction using only two instructions.

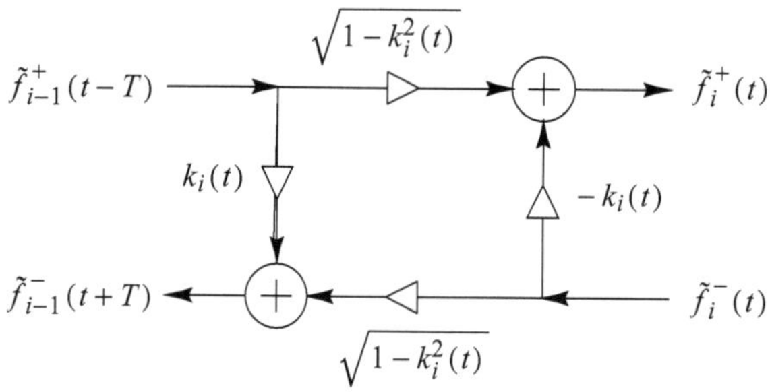

Figure C.22: The normalized scattering junction.

C.8.6 Normalized Scattering Junctions

Using (C.53) to convert to *normalized waves* $\tilde{f}^{\pm}$, the Kelly-Lochbaum junction (C.60) becomes

$$\begin{aligned}\tilde{f}^+_i(t) &= \sqrt{1-k_i^2(t)}\,\tilde{f}^+_{i-1}(t-T) - k_i(t)\,\tilde{f}^-_i(t)\\ \tilde{f}^-_{i-1}(t+T) &= k_i(t)\,\tilde{f}^+_{i-1}(t-T) + \sqrt{1-k_i^2(t)}\,\tilde{f}^-_i(t) \qquad \text{(C.66)}\end{aligned}$$

as diagrammed in Fig. C.22. This is called the *normalized scattering junction* [297], although a more precise term would be the "normalized-wave scattering junction."

It is interesting to define $\theta_i \triangleq \sin^{-1}(k_i)$, always possible for passive junctions since $-1 \leq k_i \leq 1$, and note that the normalized scattering junction is equivalent to a *2D rotation*:

$$\begin{aligned}\tilde{f}^+_i(t) &= \cos(\theta_i)\,\tilde{f}^+_{i-1}(t-T) - \sin(\theta_i)\,\tilde{f}^-_i(t)\\ \tilde{f}^-_{i-1}(t+T) &= \sin(\theta_i)\,\tilde{f}^+_{i-1}(t-T) + \cos(\theta_i)\,\tilde{f}^-_i(t) \qquad \text{(C.67)}\end{aligned}$$

where, for conciseness of notation, the time-invariant case is written.

While it appears that scattering of normalized waves at a two-port junction requires four multiplies and two additions, it is possible to convert this to three multiplies and three additions using a two-multiply "transformer" to power-normalize an ordinary one-multiply junction [432].

The *transformer* is a lossless two-port defined by [136]

$$\begin{aligned}f^+_i(t) &= g_i\,f^+_{i-1}(t-T)\\ f^-_{i-1}(t+T) &= \frac{1}{g_i}\,f^-_i(t). \qquad \text{(C.68)}\end{aligned}$$

The transformer can be thought of as a device which steps the wave impedance to a new value without scattering; instead, the traveling signal power is redistributed among the force and velocity wave variables to satisfy the fundamental relations $f^{\pm} = \pm R v^{\pm}$ (C.57) at the new impedance. An impedance change from R_{i-1} on the left to R_i on the right is accomplished using

$$g_i \triangleq \sqrt{\frac{R_i}{R_{i-1}}} = \sqrt{\frac{1+k_i(t)}{1-k_i(t)}} \tag{C.69}$$

as can be quickly derived by requiring $(f_{i-1}^{+})^2/R_{i-1} = (f_i^{+})^2/R_i$. The parameter g_i can be interpreted as the "turns ratio" since it is the factor by which force is stepped (and the inverse of the velocity step factor).

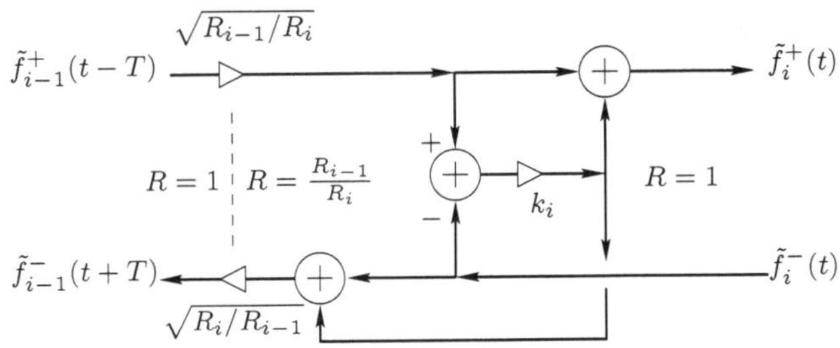

Figure C.23: Three-multiply normalized-wave scattering junction.

Figure C.23 illustrates a *three-multiply* normalized-wave scattering junction [432]. The impedance of all waveguides (bidirectional delay lines) may be taken to be $R = 1$. Scattering junctions may then be implemented as a denormalizing transformer $g = \sqrt{R_{i-1}}$, a one-multiply scattering junction k_i, and a renormalizing transformer $g = 1/\sqrt{R_i}$. Either transformer may be commuted with the junction and combined with the other transformer to give a three-multiply normalized-wave scattering junction. (The transformers are combined on the left in Fig. C.23).

In slightly more detail, a transformer $g = \sqrt{R_{i-1}}$ steps the wave impedance (left-to-right) from $R = 1$ to $R = R_{i-1}$. Equivalently, the normalized force-wave $\tilde{f}_{i-1}^{+}(t)$ is converted unnormalized form $f_{i-1}^{+}(t)$. Next there is a physical scattering from impedance R_{i-1} to R_i (reflection coefficient $k_i = (R_i - R_{i-1})/(R_i + R_{i-1})$). The outgoing wave to the right is then normalized by transformer $g = 1/\sqrt{R_i}$ to return the wave impedance back to $R = 1$ for wave propagation within a normalized-wave delay line to the right. Finally, the right transformer is commuted left and combined with the left transformer to reduce total computational complexity to one multiply and three adds.

It is important to notice that transformer-normalized junctions may have a large dynamic range in practice. For example, if $k_i \in [-1+\epsilon, 1-\epsilon]$, then Eq. (C.69) shows that the transformer coefficients may become as large as $\sqrt{2/\epsilon - 1}$. If ϵ is the "machine epsilon," *i.e.*, $\epsilon = 2^{-(n-1)}$ for typical n-bit two's complement arithmetic normalized to lie in $[-1, 1)$, then the dynamic range of the transformer coefficients is bounded by $\sqrt{2^n - 1} \approx 2^{n/2}$. Thus, while transformer-normalized junctions trade a multiply for an add, they require up to 50% more bits of dynamic range within the junction adders. On the other hand, it is very nice to have normalized waves (unit wave impedance) throughout the digital waveguide network, thereby limiting the required dynamic range to root physical power in all propagation paths.

C.8.7 Junction Passivity

In fixed-point implementations, the round-off error and other nonlinear operations should be confined when possible to physically meaningful wave variables. When this is done, it is easy to ensure that signal power is not increased by the nonlinear operations. In other words, nonlinear operations such as rounding can be made *passive*. Since signal power is proportional to the square of the wave variables, all we need to do is make sure amplitude is never increased by the nonlinearity. In the case of rounding, *magnitude truncation*, sometimes called "rounding toward zero," is one way to achieve passive rounding. However, magnitude truncation can attenuate the signal excessively in low-precision implementations and in scattering-intensive applications such as the digital waveguide mesh [519]. Another option is *error power feedback* in which case the cumulative round-off error power averages to zero over time.

A valuable byproduct of passive arithmetic is the suppression of *limit cycles* and *overflow oscillations* [432]. Formally, the signal power of a conceptually infinite-precision implementation can be taken as a Lyapunov function bounding the squared amplitude of the finite-precision implementation.

The Kelly-Lochbaum and one-multiply scattering junctions are *structurally lossless* [501, 461] (see also §C.17) because they have only one parameter k_i (or α_i), and all quantizations of the parameter within the allowed interval $[-1, 1]$ (or $[0, 2]$) correspond to lossless scattering.[7]

In the Kelly-Lochbaum and one-multiply scattering junctions, because they are structurally lossless, we need only double the number of bits at the output of each multiplier, and add one bit of extended dynamic range at

[7]Here it is assumed that $-k_i$ and $1 \pm k_i$ in the Kelly-Lochbaum junction can be computed exactly from k_i in the number system being used. This is the case in two's complement arithmetic as is typically used in practice.

the output of each two-input adder. The final outgoing waves are thereby *exactly* computed before they are finally rounded to the working precision and/or clipped to the maximum representable magnitude.

For the Kelly-Lochbaum scattering junction, given n-bit signal samples and m-bit reflection coefficients, the reflection and transmission multipliers produce $n+m$ and $n+m+1$ bits, respectively, and each of the two additions adds one more bit. Thus, the intermediate word length required is $n+m+2$ bits, and this must be rounded *without amplification* down to n bits for the final outgoing samples. A similar analysis gives also that the one-multiply scattering junction needs $n+m+2$ bits for the extended precision intermediate results before final rounding and/or clipping.

To formally show that magnitude truncation is sufficient to suppress overflow oscillations and limit cycles in waveguide networks built using structurally lossless scattering junctions, we can look at the signal power entering and leaving the junction. A junction is passive if the power flowing away from it does not exceed the power flowing into it. The total power flowing away from the ith junction is bounded by the incoming power if

$$\underbrace{\frac{[f_i^+(t)]^2}{R_i(t)} + \frac{[f_{i-1}^-(t+T)]^2}{R_{i-1}(t)}}_{\text{outgoing power}} \leq \underbrace{\frac{[f_{i-1}^+(t-T)]^2}{R_{i-1}(t)} + \frac{[f_i^-(t)]^2}{R_i(t)}}_{\text{incoming power}} \tag{C.70}$$

Let $\hat{f}$ denote the finite-precision version of f. Then a *sufficient* condition for junction passivity is

$$\left|\hat{f}_i^+(t)\right| \leq \left|f_i^+(t)\right| \tag{C.71}$$

$$\left|\hat{f}_{i-1}^-(t+T)\right| \leq \left|f_{i-1}^-(t+T)\right| \tag{C.72}$$

Thus, if the junction computations do not increase either of the output force amplitudes, no signal power is created. An analogous conclusion is reached for velocity scattering junctions.

Unlike the structurally lossless cases, the (four-multiply) normalized scattering junction has *two* parameters, $s_i \triangleq k_i$ and $c_i \triangleq \sqrt{1-k_i^2}$, and these can "get out of synch" in the presence of quantization. Specifically, let $\hat{s}_i \triangleq s_i - \epsilon_s$ denote the quantized value of s_i, and let $\hat{c}_i \triangleq c_i - \epsilon_c$ denote the quantized value of c_i. Then it is no longer the case in general that $\hat{s}_i^2 + \hat{c}_i^2 = 1$. As a result, the normalized scattering junction is not structurally lossless in the presence of coefficient quantization. A few lines of algebra shows that a passive rounding rule for the normalized junction must depend on the sign of the wave variable being computed, the sign of the coefficient quantization error, and the sign of at least one of the two incoming traveling waves. We can assume one of the coefficients is exact for passivity purposes, so assume $\epsilon_s = 0$ and define $\hat{c}_i = \left\lfloor \sqrt{1 - s_i^2} \right\rfloor$, where $\lfloor x \rfloor$ denotes largest quantized value less than or equal to x. In this case we have $\epsilon_c \geq 0$. Therefore,

$$\hat{f}_i^+ = \hat{c}_i f_{i-1}^+ - s_i f_i^- = f_i^+ - \epsilon_c f_{i-1}^+$$

and a passive rounding rule which guarantees $|\hat{f}_i^+| \leq |f_i^+|$ need only look at the sign bits of $\hat{f}_i^+$ and f_{i-1}^+.

The three-multiply normalized scattering junction is easier to "passify." While the transformer is not structurally lossless, its simplicity allows it to be made passive simply by using non-amplifying rounding on both of its coefficients as well as on its output wave variables. (The transformer is passive when the product of its coefficients has magnitude less than or equal to 1.) Since there are no additions following the transformer multiplies, double-precision adders are not needed. However, precision and a half is needed in the junction adders to accommodate the worst-case increased dynamic range. Since the one-multiply junction is structurally lossless, the overall junction is passive if non-amplifying rounding is applied to g_i, $1/g_i$, and the outgoing wave variables from the transformer and from the one-multiply junction.

In summary, a general means of obtaining passive waveguide junctions is to compute exact results internally, and apply *saturation* (clipping on overflow) and *magnitude truncation* (truncation toward zero) to the final outgoing wave variables. Because the Kelly-Lochbaum and one-multiply junctions are structurally lossless, exact intermediate results are obtainable using extended internal precision. For the (four-multiply) normalized scattering junction, a passive rounding rule can be developed based on two sign bits. For the three-multiply normalized scattering junction, it is sufficient to apply magnitude truncation to the transformer coefficients and all outgoing wave variables.

C.9 Digital Waveguide Filters

A *Digital Waveguide Filter* (DWF) can be defined as any digital filter built by stringing together digital waveguides.[8] In the case of a cascade combination of unit-length waveguide sections, they are essentially the same as *microwave filters* [372] and *unit element filters* [136] (Appendix F). This section shows how DWFs constructed as a cascade chain of waveguide sections, and reflectively terminated, can be transformed via elementary manipulations to well known *ladder* and *lattice* digital filters, such as those used in speech modeling [297, 363].

[8]This section is adapted from [434, 433].

DWFs may be derived conceptually by *sampling* the unidirectional traveling waves in a system of ideal, lossless waveguides. Sampling is across time and space. Thus, variables in a DWF structure are equal *exactly* (at the sampling times and positions, to within numerical precision) to variables propagating in the physical medium in an interconnection of uniform 1D waveguides.

C.9.1 Ladder Waveguide Filters

An important class of DWFs can be constructed as a linear cascade chain of digital waveguide sections, as shown in Fig. C.24 (a slightly abstracted version of Fig. C.19 on page 568). Each block labeled k_i stands for a stateless Kelly-Lochbaum scattering junction (§C.8.4, p. 569) interfacing between wave impedances R_{i-1} and R_i. We call this a *ladder waveguide filter* structure. It is an exact bandlimited discrete-time model of a sequence of wave impedances R_i, as used in the Kelly-Lochbaum piecewise-cylindrical acoustic-tube model for the vocal tract in the context of voice synthesis (Fig. 6.2, p. 209). The delays between scattering junctions along both the top and bottom signal paths make it possible to compute all of the scattering junctions in parallel—a property that is becoming increasingly valuable in signal-processing architectures.

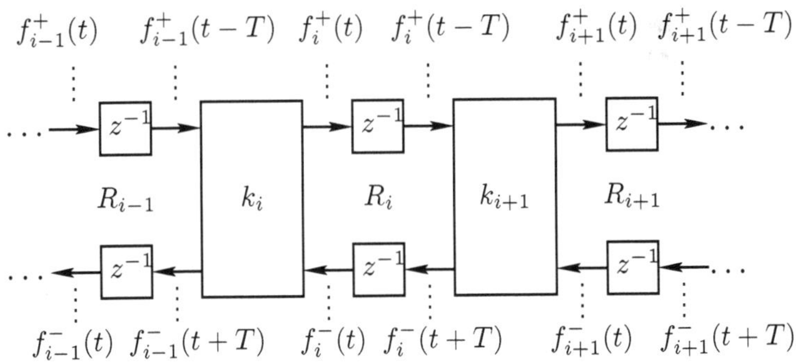

Figure C.24: Linear cascade, unit-delay, digital waveguide filter.

To transform the DWF of Fig. C.24 to a conventional *ladder digital filter* structure, as used in speech modeling [297, 363], we need to (1) terminate on the right with a pure reflection and (2) eliminate the delays along the top signal path. We will do this in stages so as to point out a valuable intermediate case.

C.9.2 Reflectively Terminated Waveguide Filters

A reflecting termination occurs when the wave impedance of the next section is either zero or infinity (§C.8.1, p. 562). Since Fig. C.24 is labeled for force waves, an infinite terminating impedance on the right results in

$$f_{i+1}^{-}(t+T) \;=\; f_{i+1}^{+}(t-T).$$

Similarly, a zero terminating impedance on the right gives

$$f_{i+1}^{-}(t+T) \;=\; -f_{i+1}^{+}(t-T).$$

For velocity waves, the signs are interchanged. Thus, a reflectively terminated ladder digital waveguide corresponds to a final feedback with gain ± 1 at the end of the ladder. Such a termination will be used below to derive conventional ladder/lattice filters below.

C.9.3 Half-Rate Ladder Waveguide Filters

The delays preceding the two inputs to each scattering junction can be "pushed" into the junction and then "pulled" out to the outputs and combine with the delays there. (This is easy to show using the Kelly-Lochbaum scattering junction derived in §C.8.4 on page 569.) By performing this operation on every other section in the DWF chain, the *half-rate ladder waveguide filter* shown in Fig. C.25 is obtained [432].

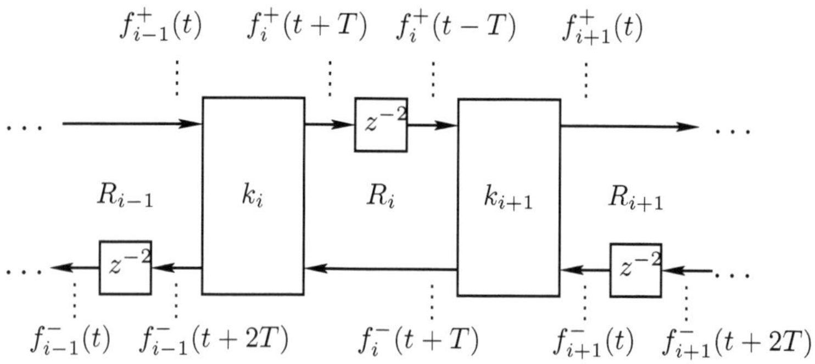

Figure C.25: Half-rate ladder waveguide filter (*cf.* Fig. C.24).

This "half-rate" ladder DWF is so-called because its sampling rate can be cut in half due to each delay being two-samples long. It has advantages worth considering, which will become clear after we have derived conventional ladder digital filters below. Note that now *pairs* of scattering junctions can be computed in parallel, and an exact physical interpretation remains. That is, it can still be used as a general purpose physical modeling building block in this more efficient (half-rate) form.

Note that when the sampling rate is halved, the physical wave variables (computed by summing two traveling-wave components) are at the same time only every other spatial sample. In particular, the physical transverse forces on the right side of scattering junctions i and $i+1$ in Fig. C.25 are

$$\begin{aligned} f_i(t+T) &= f_i^+(t+T) + f_i^-(t+T) \\ f_{i+1}(t) &= f_{i+1}^+(t) + f_{i+1}^-(t) \end{aligned}$$

respectively. In the half-rate case, adjacent spatial samples are separated in time by half a temporal sample $T/2$. If physical variables are needed only for even-numbered (or odd-numbered) spatial samples, then there is no relative time skew, but more generally, things like collision detection, such as for slap-bass string-models (§9.1.6), can be affected. In summary, the half-rate ladder waveguide filter has an alternating half-sample time skew from section to section when used as a physical modeling building block.

C.9.4 Conventional Ladder Filters

Given a reflecting termination on the right, the half-rate DWF chain of Fig. C.25 can be reduced further to the conventional ladder/lattice filter structure shown in Fig. C.26.

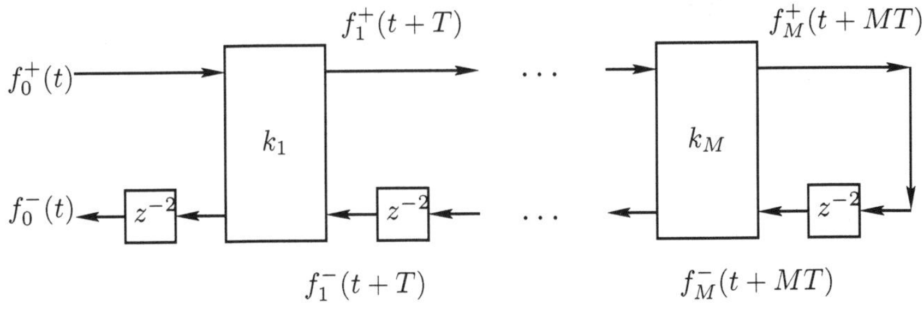

Figure C.26: Conventional ladder/lattice filter structure.

To make a standard ladder/lattice filter, the sampling rate is cut in half (*i.e.*, replace z^{-2} by z^{-1}), and the scattering junctions are typically implemented in one-multiply form (§C.8.5, p. 570) or normalized form (§C.8.6, p. 572), etc. Conventionally, if the graph of the scattering junction is nonplanar, as it is for the one-multiply junction, the filter is called a *lattice filter*; it is called a *ladder filter* when the graph is planar, as it is for normalized and Kelly-Lochbaum scattering junctions. For all-pole transfer functions $H(z) = F_0^-(z)/F_0^+(z)$, the *Durbin recursion* can be used to compute the reflection coefficients k_i from the desired transfer-function denominator polynomial coefficients [449]. To implement arbitrary transfer-function zeros, a linear combination of delay-element outputs is formed using weights that are called "tap parameters" [173, 297].

To create Fig. C.26 from Fig. C.24, all delays along the top rail are pushed to the right until they have all been worked around to the bottom rail. In the end, each bottom-rail delay becomes $2T$ seconds instead of T seconds. Such an operation is possible because of the termination at the right by an infinite (or zero) wave impedance. Note that there is a progressive one-sample time advance from section to section. The time skews for the right-going (or left-going) traveling waves can be determined simply by considering how many missing (or extra) delays there are between that signal and the unshifted signals at the far left.

Due to the reflecting termination, conventional lattice filters cannot be extended to the right in any physically meaningful way. Also, creating network topologies more complex than a simple linear cascade (or acyclic tree) of waveguide sections is not immediately possible because of the delay-free path along the top rail. In particular, the output f_M^+ cannot be fed back to the input f_0^+. Nevertheless, as we have derived, there is an exact physical interpretation (with time skew) for the conventional ladder/lattice digital filter.

C.9.5 Power-Normalized Waveguide Filters

Above, we adopted the convention that the time variation of the wave impedance did not alter the traveling force waves $f_i^\pm$. In this case, the power represented by a traveling force wave is modulated by the changing wave impedance as it propagates. The signal power becomes inversely proportional to wave impedance:

$$\mathcal{I}_i(t,x) = \mathcal{I}_i^+(t,x) + \mathcal{I}_i^-(t,x) = \frac{[f_i^+(t,x)]^2 - [f_i^-(t,x)]^2}{R_i(t)}$$

In some applications (such as time-varying waveguide reverberation [430]), it may be preferable to compensate for the power modulation so that changes in the wave impedances of the waveguides do not affect the power of the signals propagating within.

In [432, 433], three methods are discussed for making signal power *invariant* with respect to time-varying branch impedances:

1. The *normalized waveguide* scheme compensates for power modulation by scaling the signals leaving the delays so as to give them the same

power coming out as they had going in. It requires two additional scaling multipliers per waveguide junction.

2. The *normalized wave* approach [297] propagates *rms-normalized waves* in the waveguide. In this case, each delay-line contains $\tilde{f}_i^+(t,x) = f_i^+(t,x)/\sqrt{R_i(t)}$ and $\tilde{f}_i^-(t,x) = f_i^-(t,x)/\sqrt{R_i(t)}$. In this case, the power stored in the delays does not change when the wave impedance changes. This is the basis of the *normalized ladder filter* (NLF) [174, 297]. Unfortunately, four multiplications are obtained at each scattering junction.

3. The *transformer-normalized waveguide* approach changes the wave impedance at the output of the delay back to what it was at the time it entered the delay using a "transformer" (defined in §C.16, p. 616).

The transformer-normalized DWF junction is shown in Fig. C.27 [432]. As derived in §C.16, the transformer "turns ratio" g_i is given by

$$g_i = \sqrt{\frac{1-k_i}{1+k_i}}.$$

We can now modulate a single scattering junction, even in arbitrary network topologies, by inserting a transformer immediately to the left or right of the junction. Conceptually, the wave impedance is not changed over the delay-line portion of the waveguide section; instead, it is changed to the new time-varying value just before (or after) it meets the junction. When velocity is the wave variable, the coefficients g_i and g_i^{-1} in Fig. C.27 are swapped (or inverted).

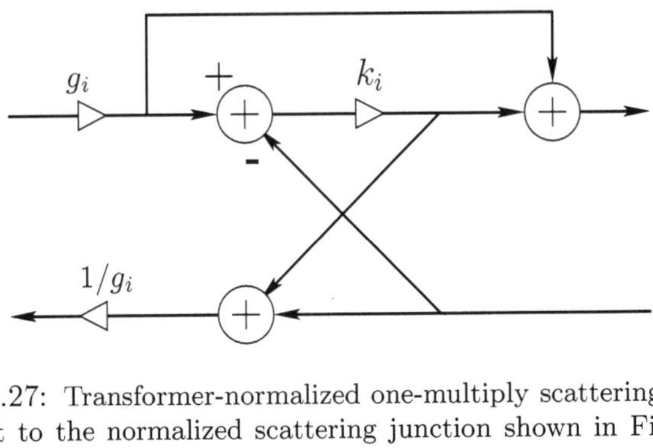

Figure C.27: Transformer-normalized one-multiply scattering junction, equivalent to the normalized scattering junction shown in Fig. C.22 on page 572

So, as in the normalized waveguide case, for the price of two extra multiplies per section, we can implement time-varying digital filters which do not modulate stored signal energy. Moreover, transformers enable the scattering junctions to be varied independently, without having to propagate time-varying impedance ratios throughout the waveguide network.

It can be shown [433] that cascade waveguide chains built using transformer-normalized waveguides are *equivalent* to those using normalized-wave junctions. Thus, the transformer-normalized DWF in Fig. C.27 and the wave-normalized DWF in Fig. C.22, p. 572 are equivalent. One simple proof is to start with a transformer (§C.16) and a Kelly-Lochbaum junction (§C.8.4), move the transformer scale factors inside the junction, combine terms, and arrive at Fig. C.22. One practical benefit of this equivalence is that the normalized ladder filter (NLF) can be implemented using only three multiplies and three additions instead of the usual four multiplies and two additions.

The transformer-normalized scattering junction is also the basis of the *digital waveguide oscillator* (§C.17).

C.10 "Traveling Waves" in Lumped Systems

One of the topics in classical network theory is the *reflection* and *transmission*, or *scattering* formulation for lumped networks [35]. Lumped scattering theory also serves as the starting point for deriving *wave digital filters* (the subject of Appendix F). In this formulation, forces (voltages) and velocities (currents) are replaced by so-called *wave variables*

$$\begin{aligned} f^+(t) &\isdef \frac{f(t)+R_0v(t)}{2} \\ f^-(t) &\isdef \frac{f(t)-R_0v(t)}{2} \end{aligned}$$

where R_0 is an arbitrary *reference impedance*. Since the above wave variables have dimensions of force, they are specifically *force waves*. The corresponding *velocity waves* are

$$\begin{aligned} v^+(t) &\isdef \frac{1}{2}[v(t)+f(t)/R_0], \\ v^-(t) &\isdef \frac{1}{2}[v(t)-f(t)/R_0]. \end{aligned}$$

Dropping the time argument, since it is always '(t)', we see that

$$\begin{aligned} f^+ &= R_0v^+ \\ f^- &= -R_0v^- \end{aligned} \quad \text{(C.73)}$$

and

$$\begin{aligned} f &= f^+ + f^- \\ v &= v^+ + v^- \end{aligned} \tag{C.74}$$

These are the basic relations for *traveling waves* in an ideal medium such as an ideal vibrating string or acoustic tube. Using voltage and current gives elementary *transmission line* theory.

C.10.1 Reflectance of an Impedance

Let $R(s)$ denote the driving-point impedance of an arbitrary continuous-time LTI system. Then, by definition, $F(s) = R(s)V(s)$ where $F(s)$ and $V(s)$ denote the Laplace transforms of the applied force and resulting velocity, respectively. The wave variable decomposition in Eq. (C.74) gives

$$\begin{aligned} F(s) &= R(s)V(s) \\ \Rightarrow F^+(s) + F^-(s) &= R(s)\left[V^+(s) + V^-(s)\right] \\ &= R(s)\left[\frac{F^+(s) - F^-(s)}{R_0}\right] \\ \Rightarrow F^-(s)\left[\frac{R(s)}{R_0} + 1\right] &= F^+(s)\left[\frac{R(s)}{R_0} - 1\right] \\ \Rightarrow F^-(s) &= F^+(s)\left[\frac{R(s) - R_0}{R(s) + R_0}\right] \\ &\triangleq F^+(s)\,\hat{\rho}_f(s) \end{aligned} \tag{C.75}$$

We may call $\hat{\rho}_f(s)$ the *reflectance* of impedance $R(s)$ relative to R_0. For example, if a transmission line with characteristic impedance R_0 were terminated in a lumped impedance $R(s)$, the reflection transfer function at the termination, as seen from the end of the transmission line, would be $\hat{\rho}_f(s)$.

We are working with reflectance for *force waves.* Using the elementary relations Eq. (C.73), *i.e.*, $F^+(s) = R_0V^+(s)$ and $F^-(s) = -R_0V^-(s)$, we immediately obtain the corresponding *velocity-wave reflectance:*

$$\hat{\rho}_v(s) \triangleq \frac{V^-(s)}{V^+(s)} = \frac{-F^-(s)/R_0}{F^+(s)/R_0} = -\frac{F^-(s)}{F^+(s)} = -\hat{\rho}_f(s)$$

C.11 Properties of Passive Impedances

It is well known that a real impedance R (in Ohms, for example) is *passive* so long as $R \geq 0$. A passive impedance cannot create energy. On the other hand, if $R < 0$, the impedance is *active* and has some energy source. The concept of passivity can be extended to complex frequency-dependent impedances $R(j\omega)$ as well: A complex impedance $R(j\omega)$ is *passive* if $R(s)$ is *positive real*, where s is the Laplace-transform variable. The positive-real property is discussed in §C.11.2 below.

This section explores some implications of the positive real condition for passive impedances. Specifically, §C.11.1 considers the nature of waves reflecting from a passive impedance in general, looking at the *reflection transfer function*, or *reflectance*, of a passive impedance. To provide further details, Section C.11.2 derives some mathematical properties of positive real functions, particularly for the discrete-time case. Application examples appear in §9.2.1 and §9.2.1.

C.11.1 Passive Reflectances

From Eq. (C.75), we have that the reflectance seen at a continuous-time impedance $R(s)$ is given for force waves by

$$\hat{\rho}_f(s) \triangleq \frac{F^-(s)}{F^+(s)} = \frac{R(s) - R_0}{R(s) + R_0} \tag{C.76}$$

where R_0 is the wave impedance connected to the impedance $R(s)$, and the corresponding velocity reflectance is $\hat{\rho}_v(s) = -\hat{\rho}_f(s)$. As mentioned above, all passive impedances are *positive real*. As shown in §C.11.2 on page 593, $R(s)$ is positive real if and only if $\hat{\rho}_f(s)$ is stable and has magnitude less than or equal to 1 on the $j\omega$ axis (and hence over the entire left-half plane, by the maximum modulus theorem), *i.e.*,

$$|\hat{\rho}_f(s)| \leq 1, \quad \text{re}\{s\} \leq 0. \tag{C.77}$$

In particular, $|\hat{\rho}_f(j\omega)| \leq 1$ for all radian frequencies $\omega \in (-\infty, \infty)$. Any stable $\hat{\rho}_f(s)$ satisfying Eq. (C.77) may be called a *passive reflectance.*

If the impedance $R(s)$ goes to infinity (becomes rigid), then $\hat{\rho}_f(s)$ approaches 1, a result which agrees with an analysis of rigid string terminations (p. 210). Similarly, when the impedance goes to zero, $\hat{\rho}_f(s)$ becomes -1, which agrees with the physics of a string with a free end. In acoustic stringed instruments, bridges are typically quite rigid, so that $\hat{\rho}_f(j\omega) \approx 1$ for all ω. If a body resonance is strongly coupled through the bridge, $|\hat{\rho}_f(j\omega_c)|$ can be significantly smaller than 1 at the resonant frequency ω_c.

Solving for $R(s)$ in Eq. (C.77), we can characterize every impedance in terms of its reflectance:

$$R(s) = R_0 \frac{1 + \hat{\rho}_f(s)}{1 - \hat{\rho}_f(s)}$$

Rewriting Eq. (C.76) in the form

$$\hat{\rho}_f(s) = \frac{\dfrac{R(s)}{R_0} - 1}{\dfrac{R(s)}{R_0} + 1},$$

we see that the reflectance is determined by the ratio of the "new impedance" $R(s)$ to the "old" impedance R_0 in which the incoming waves travel. In other words, the incoming waves see the wave impedance "step" from R_0 to $R(s)$, which results in a "scattering" of the incident wave into reflected and transmitted components, as discussed in §C.8 on page 560. The reflection and transmission coefficients depend on frequency when $R(j\omega)$ is not constant with respect to ω.

In the discrete-time case, which may be related to the continuous-time case by the bilinear transform (§7.3.2, p. 266), we have the same basic relations, but in the z plane:

$$\begin{aligned}
\hat{\rho}_f(z) &\triangleq \frac{F^-(z)}{F^+(z)} = \frac{R(z) - R_0}{R(z) + R_0} \\
R(z) &= R_0 \frac{1 + \hat{\rho}_f(z)}{1 - \hat{\rho}_f(z)} \\
\Gamma(z) &= \Gamma_0 \frac{1 - \hat{\rho}_f(z)}{1 + \hat{\rho}_f(z)}
\end{aligned} \tag{C.78}$$

where $\Gamma \triangleq 1/R$ denotes admittance, with

$$|\hat{\rho}_f(z)| \leq 1, \quad |z| \leq 1. \tag{C.79}$$

Mathematically, any stable transfer function having these properties may be called a *Schur function*. Thus, the discrete-time reflectance $\hat{\rho}_f(z)$ of an impedance $R(z)$ is a Schur function if and only if the impedance is passive (positive real).

Note that Eq. (C.79) may be obtained from the general formula for scattering at a loaded waveguide junction for the case of a single waveguide ($N = 1$) terminated by a lumped load (§C.12 on page 598).

In the limit as damping goes to zero (all poles of $R(z)$ converge to the unit circle), the reflectance $\hat{\rho}_f(z)$ becomes a digital *allpass filter*. Similarly, $\hat{\rho}_f(s)$ becomes a continuous-time allpass filter as the poles of $R(s)$ approach the $j\omega$ axis.

Recalling that a lossless impedance is called a *reactance* (§7.1, p. 252), we can say that every reactance gives rise to an *allpass reflectance*. Thus, for example, waves reflecting off a *mass* at the end of a vibrating string will be allpass filtered, because the driving-point impedance of a mass ($R(s) = ms$) is a pure reactance. In particular, the force-wave reflectance of a mass m terminating an ideal string having wave impedance R_0 is $\hat{\rho}_f(s) = (ms - R_0)/(ms + R_0)$, which is a continuous-time allpass filter having a pole at $s = -R_0/m$ and a zero at $s = R_0/m$.

It is intuitively reasonable that a passive reflection gain cannot exceed 1 at any frequency (*i.e.*, the reflectance is a Schur filter, as defined in Eq. (C.79)). It is also reasonable that lossless reflection would have a gain of 1 (*i.e.*, it is allpass).

Note that reflection filters always have an equal number of poles and zeros, as can be seen from Eq. (C.76) above. This property is preserved by the bilinear transform, so it holds in both the continuous- and discrete-time cases.

Reflectance and Transmittance of a Yielding String Termination

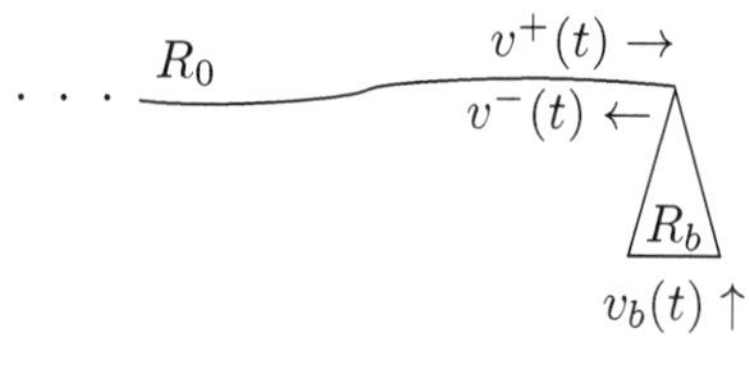

Figure C.28: Ideal vibrating string terminated on the right by a passive impedance $R_b(s)$.

Consider the special case of a reflection and transmission at a *yielding termination*, or "bridge", of an ideal vibrating string on its right end, as shown in Fig. C.28. Denote the incident and reflected velocity waves by $v^+(t)$ and $v^-(t)$, respectively, and similarly denote the force-wave components by $f^+(t)$ and $f^-(t)$. Finally, denote the velocity of the termination itself by $v_b(t) = v^+(t) + v^-(t)$, and its force-wave reflectance by

$$\hat{\rho}_f(s) \triangleq \frac{F^-(s)}{F^+(s)} = -\frac{V^-(s)}{V^+(s)} = \frac{R_b(s) - R_0}{R_b(s) + R_0},$$

where R_0 denotes the string wave impedance.

The bridge velocity is given by

$$V_b(s) \;=\; V^+(s) + V^-(s),$$

so that the *bridge velocity transmittance* is given by

$$\hat{\tau}_v(s) \;\isdef\; \frac{V_b(s)}{V^+(s)} \;=\; \frac{V^+(s) + V^-(s)}{V^+(s)} \;=\; 1 + \hat{\rho}_v(s) \;=\; 1 - \hat{\rho}_f(s),$$

and the bridge *force transmittance* is given by

$$\hat{\tau}_f(s) \;\isdef\; \frac{F_b(s)}{F^+(s)} \;=\; \frac{R_0[V^+(s) - V^-(s)]}{R_0 V^+(s)} \;=\; 1 + \hat{\rho}_f(s)$$

where the bridge force is defined as "up" so that it is given for small displacements by the string tension times minus the string slope at the bridge. (Recall from §C.7.2 on page 554 that force waves on the string are defined by $f = -Ky'$ where K and y' denote the string tension and slope, respectively.

Power-Complementary Reflection and Transmission

We can show that the reflectance and transmittance of the yielding termination are *power complementary*. That is, the reflected and transmitted signal-power sum to yield the incident signal-power.

The average power incident at the bridge at frequency ω can be expressed in the frequency domain as $F^+(e^{j\omega T})\overline{V^+(e^{j\omega T})}$. The reflected power is then $F^-\overline{V^-} = -\left|\hat{\rho}_f\right|^2 F^+\overline{V^+}$. Removing the minus sign, which can be associated with reversed direction of travel, we obtain that the *power reflection frequency response* is $\left|\hat{\rho}_f\right|^2$, which generalizes by analytic continuation to $\hat{\rho}_f(s)\hat{\rho}_f(-s)$. The power transmittance is given by

$$F_b\overline{V_b} \;=\; (\hat{\tau}_f F^+)\overline{(1-\hat{\rho}_f)V^+} \;=\; [(\hat{\rho}_f{+}1)F^+](1{-}\overline{\hat{\rho}_f V^+}) \;=\; (1{-}|\hat{\rho}_f|^2)F^+\overline{V^+}$$

which generalizes to the s plane as

$$F_b(s)V_b(-s) = \left[1 - \hat{\rho}_f(s)\hat{\rho}_f(-s)\right] F^+(s)V^+(-s)$$

Finally, we see that adding up the reflected and transmitted power yields the incident power:

$$-F^-(s)V^-(-s) + F_b(s)V_b(-s) \;=\; F^+(s)V^+(-s)$$

C.11.2 Positive Real Functions

Any *passive driving-point impedance*, such as the impedance of a violin bridge, is *positive real.* Positive real functions have been studied extensively in the continuous-time case in the context of *network synthesis* [68, 525]. Very little, however, seems to be available in the discrete time case. This section (reprinted from [428]) summarizes the main properties of positive real function in the z plane (*i.e.*, the discrete-time case).

Definition. A complex valued function of a complex variable $f(z)$ is said to be *positive real* (PR) if

1. z *real* $\implies f(z)$ *real*
2. $|z| \geq 1 \implies \text{re}\{f(z)\} \geq 0$

We now specialize to the subset of functions $f(z)$ representable as a ratio of finite-order polynomials in z. This class of "rational" functions is the set of all transfer functions of finite-order time-invariant linear systems, and we write $H(z)$ to denote a member of this class. We use the convention that stable, minimum phase systems are analytic and nonzero in the strict outer disk.[9] Condition (1) implies that for $H(z)$ to be PR, the polynomial coefficients must be real, and therefore complex poles and zeros must exist in conjugate pairs. We assume from this point on that $H(z) \neq 0$ satisfies (1). From (2) we derive the facts below.

Property 1. A real rational function $H(z)$ is PR iff $|z| \geq 1 \implies |\angle H(z)| \leq \frac{\pi}{2}$.

Proof. Expressing $H(z)$ in polar form gives

$$\begin{aligned}\text{re}\{H(z)\} &= \text{re}\left\{|H(z)|\, e^{j\angle H(z)}\right\} \\ &= |H(z)| \cos(\angle H(z)) \\ &\geq 0, \quad \text{whenever} \quad |\angle H(z)| \leq \frac{\pi}{2},\end{aligned}$$

since the zeros of $H(z)$ are isolated. □

Property 2. $H(z)$ is PR iff $1/H(z)$ is PR.

Proof. Assuming $H(z)$ is PR, we have by Property 1,

$$|\angle H^{-1}(z)| = |-\angle H(z)| = |\angle H(z)| \leq \frac{\pi}{2}, \quad |z| \geq 1.$$

□

Property 3. A PR function $H(z)$ is analytic and nonzero in the strict outer disk.

[9] The strict *outer disk* is defined as the region $|z| > 1$ in the extended complex plane.

Proof. (By contradiction)
Without loss of generality, we treat only n^{th} order polynomials

$$\alpha_0 z^n + \alpha_1 z^{n-1} + \cdots + \alpha_{n-1} z + \alpha_n$$

which are nondegenerate in the sense that $\alpha_0, \alpha_n \neq 0$. Since facts about $\alpha_0 H(z)$ are readily deduced from facts about $H(z)$, we set $\alpha_0 = 1$ at no great loss.

The general (normalized) causal, finite-order, linear, time-invariant transfer function may be written

$$\begin{aligned} H(z) &= z^{-\nu} \frac{b(z)}{a(z)} \\ &= z^{-\nu} \frac{1 + b_1 z^{-1} + \cdots + b_M z^{-M}}{1 + a_1 z^{-1} + \cdots + a_N z^{-N}} \\ &= z^{-\nu} \frac{\prod_{i=1}^{M} (1 - q_i z^{-1})}{\prod_{i=1}^{N} (1 - p_i z^{-1})} \\ &= z^{-\nu} \sum_{i=1}^{N_d} \sum_{j=1}^{\mu_i} \frac{z\, K_{i,j}}{(z - p_i)^j}, \qquad \nu \geq 0, \end{aligned} \tag{C.80}$$

where N_d is the number of distinct poles, each of multiplicity μ_i,and

$$\sum_{i=1}^{N_d} \mu_i = \max\{N, M\}.$$

Suppose there is a pole of multiplicity m outside the unit circle. Without loss of generality, we may set $\mu_1 = m$, and $p_1 = R\, e^{j\phi}$ with $R > 1$. Then for z near p_1, we have

$$\begin{aligned} z^{\nu} H(z) &= \frac{z\, K_{1,m}}{(z - R\, e^{j\phi})^m} + \frac{z\, K_{1,m-1}}{(z - R\, e^{j\phi})^{m-1}} + \cdots \\ &\approx \frac{z\, K_{1,m}}{(z - R\, e^{j\phi})^m}. \end{aligned}$$

Consider the circular neighborhood of radius ρ described by $z = R\, e^{j\phi} + \rho\, e^{j\psi}$, $-\pi \leq \psi < \pi$. Since $R > 1$ we may choose $\rho < R - 1$ so that all points z in this neighborhood lie outside the unit circle. If we write the residue of the factor $(z - R\, e^{j\phi})^m$ in polar form as $K_{1,m} = C\, e^{j\xi}$, then we have, for sufficiently small ρ,

$$z^{\nu} H(z) \approx \frac{K_{1,m} R\, e^{j\phi}}{(z - R\, e^{j\phi})^m} = \frac{K_{1,m} R\, e^{j\phi}}{\rho^m\, e^{jm\psi}} = \frac{C\, R}{\rho^m}\, e^{j(\phi + \xi - m\psi)}. \tag{C.81}$$

Therefore, approaching the pole $R\,e^{j\phi}$ at an angle ψ gives

$$\lim_{\rho\to 0}\left|\angle H(R\,e^{j\phi}+\rho\,e^{j\psi})\right| = \left|\phi(1-\nu)+\xi-m\psi\right|, \qquad -\pi\leq\psi<\pi$$

which cannot be confined to satisfy Property 1 regardless of the value of the residue angle ξ, or the pole angle ϕ (m cannot be zero by hypothesis). We thus conclude that a PR function $H(z)$ can have no poles in the outer disk. By Property 2, we conclude that positive real functions must be minimum phase. □

Corollary. In equation Eq. (C.80), $\nu = 0$.

Proof. If $\nu > 0$, then there are ν poles at infinity. As $|z| \to \infty$, $H(z) \to z^{-\nu} \implies |\angle H(z)| \to |\nu\angle z|$, we must have $\nu = 0$. □

Corollary. The log-magnitude of a PR function has zero mean on the unit circle.

This is a general property of stable, minimum-phase transfer functions which follows immediately from the *argument principle* [297, 326].

Corollary. A rational PR function has an equal number of poles and zeros all of which are in the unit disk.

This really a convention for numbering poles and zeros. In Eq. (C.80), we have $\nu = 0$, and all poles and zeros inside the unit disk. Now, if $M > N$ then we have $M - N$ extra poles at $z = 0$ induced by the numerator. If $M < N$, then $N - M$ zeros at the origin appear from the denominator.

Corollary. Every pole on the unit circle of a positive real function must be simple with a real and positive residue.

Proof. We repeat the previous argument using a semicircular neighborhood of radius ρ about the point $p_1 = e^{j\phi}$ to obtain

$$\lim_{\rho\to 0}\left|\angle H(e^{j\phi}+\rho e^{j\psi})\right| = \left|\phi+\xi-m\psi\right|, \qquad \phi-\frac{\pi}{2}\leq\psi\leq\phi+\frac{\pi}{2}. \tag{C.82}$$

In order to have $|\angle H(z)| \leq \pi/2$ near this pole, it is necessary that $m = 1$ and $\xi = 0$. □

Corollary. If $H(z)$ is PR with a zero at $z = q_1 = e^{j\phi}$, then

$$H'(z) \triangleq \frac{H(z)}{(1-q_1 z^{-1})}$$

must satisfy

$$\begin{aligned} H'(q_1) &\neq 0 \\ \angle H'(q_1) &= 0. \end{aligned}$$

Proof. We may repeat the above for $1/H(z)$.

Property. Every PR function $H(z)$ has a causal inverse z transform $h(n)$.

Proof. This follows immediately from analyticity in the outer disk [342, pp. 30-36] However, we may give a more concrete proof as follows. Suppose $h(n)$ is non-causal. Then there exists $k > 0$ such that $h(-k) \neq 0$. We have,

$$\begin{aligned} H(z) &\triangleq \sum_{n=-\infty}^{\infty} h(n)\, z^{-n} \\ &= h(-k)z^k + \sum_{n\neq -k} h(n)\, z^{-n}. \end{aligned}$$

Hence, $H(z)$ has at least one pole at infinity and cannot be PR by Property 3. Note that this pole at infinity cannot be cancelled since otherwise

$$\begin{aligned} & h(-k)z^k = \sum_{l\neq -k} \alpha(l)\, z^{-l} \\ \Longrightarrow\ & h(-k)\delta(n+k) = \sum_{m\neq -k} \alpha(m)\, \delta(m-n) \\ \Longrightarrow\ & h(-k) = 0 \end{aligned}$$

which contradicts the hypothesis that $h(n)$ is non-causal. □

Property. $H(z)$ is PR iff it is analytic for $|z| > 1$, poles on the unit circle are simple with real and positive residues, and re $\{\{\}\, H(e^{j\theta})\} \geq 0$ for $0 \leq \theta \leq \pi$.

Proof. If $H(z)$ is positive real, the conditions stated hold by virtue of Property 3 and the definition of positive real.

To prove the converse, we first show nonnegativity on the upper semicircle implies nonnegativity over the entire circle.

$$\begin{aligned} & \text{re}\left\{H(e^{j\theta})\right\} &\geq\ & 0, \qquad 0 \leq \theta \leq \pi \\ \Longrightarrow\ & \text{re}\left\{H(e^{-j\theta})\right\} &\triangleq\ & \text{re}\left\{\sum_{n=-\infty}^{\infty} h(n)e^{jn\theta}\right\} \\ & &=\ & \sum_{n=-\infty}^{\infty} h(n)\cos(n\theta) \\ & &=\ & \text{re}\left\{H(e^{j\theta})\right\} \\ & &\geq\ & 0, \qquad 0 \leq \theta \leq \pi \\ \Longrightarrow\ & \text{re}\left\{H(e^{j\theta})\right\} &\geq\ & 0, \qquad -\pi < \theta \leq \pi. \end{aligned}$$

Alternatively, we might simply state that $h(n)$ real implies re $\left\{H(e^{j\theta})\right\}$ is even in θ.

Next, since the function e^z is analytic everywhere except at $z = \infty$, it follows that $f(z) = e^{-H(z)}$ is analytic wherever $H(z)$ is finite. There are no poles of $H(z)$ outside the unit circle due to the analyticity assumption, and poles on the unit circle have real and positive residues. Referring again to the limiting form Eq. (C.81) of $H(z)$ near a pole on the unit circle at $e^{j\phi}$, we see that, as $\rho \to 0$, we have

$$\begin{aligned}
H(e^{j\phi} + \rho e^{j\psi}) &\to \frac{C}{\rho} e^{j(\phi-\psi)}, \qquad \phi - \frac{\pi}{2} \le \psi \le \phi + \frac{\pi}{2} \\
&\triangleq \frac{C}{\rho} e^{j\theta}, \qquad \theta \triangleq \phi - \psi, \quad -\frac{\pi}{2} \le \theta \le \frac{\pi}{2} \\
\Longrightarrow \quad f(z) &\to e^{-\frac{C}{\rho} e^{j\theta}} \\
&= e^{-\frac{C}{\rho}\cos\theta} e^{-j\frac{C}{\rho}\sin\theta} \\
&\to 0, \qquad\qquad\qquad\qquad (C.83)
\end{aligned}$$

since the residue C is positive, and the net angle θ does not exceed $\pm\pi/2$. From Eq. (C.83) we can state that for points z, z' with modulus ≥ 1, we have For all $\epsilon > 0$, there exists $\delta > 0$ such that $|z - z'| < \delta \implies |f(z) - f(z')| < \epsilon$. Thus $f(z)$ is analytic in the strict outer disk, and continuous up to the unit circle which forms its boundary. By the maximum modulus theorem [83],

$$\sup_{|z|\ge 1} |f(z)| \triangleq \sup_{|z|\ge 1} \left|e^{-H(z)}\right| = \sup_{|z|\ge 1} e^{-\text{re}\{H(z)\}} = \inf_{|z|\ge 1} \text{re}\left\{H(z)\right\}$$

occurs on the unit circle. Consequently,

$$\inf_{-\pi<\theta\le\pi} \text{re}\left\{H(e^{j\theta})\right\} \ge 0 \quad \Longrightarrow \quad \inf_{|z|\ge 1} \text{re}\left\{H(z)\right\} \ge 0 \quad \Longrightarrow \quad H(z)\ PR.$$

For example, if a transfer function is known to be asymptotically stable, then a frequency response with nonnegative real part implies that the transfer function is positive real.

Note that consideration of $1/H(z)$ leads to analogous necessary and sufficient conditions for $H(z)$ to be positive real in terms of its zeros instead of poles. □

Relation to Stochastic Processes

Property. If a stationary random process $\{x_n\}$ has a rational power spectral density $Re^{j\omega}$ corresponding to an autocorrelation function $r(k) = \mathcal{E}\left\{x_n x_{n+k}\right\}$, then

$$R_+(z) \triangleq \frac{r(0)}{2} + \sum_{n=1}^{\infty} r(n) z^{-n}$$

is positive real.

Proof.

By the representation theorem [19, pp. 98-103] there exists an asymptotically stable filter $H(z) = b(z)/a(z)$ which will produce a realization of $\{x_n\}$ when driven by white noise, and we have $Re^{j\omega} = H(e^{j\omega})H(e^{-j\omega})$. We define the analytic continuation of $Re^{j\omega}$ by $R(z) = H(z)H(z^{-1})$. Decomposing $R(z)$ into a sum of causal and anti-causal components gives

$$\begin{aligned} R(z) = \frac{b(z)b(z^{-1})}{a(z)a(z^{-1})} &= R_+(z) + R_-(z) \\ &= \frac{q(z)}{a(z)} + \frac{q(z^{-1})}{a(z^{-1})} \end{aligned}$$

where $q(z)$ is found by equating coefficients of like powers of z in

$$b(z)b(z^{-1}) = q(z)a(z^{-1}) + a(z)q(z^{-1}).$$

Since the poles of $H(z)$ and $R_+(z)$ are the same, it only remains to be shown that $\text{re}\left\{R_+(e^{j\omega})\right\} \geq 0,\ 0 \leq \omega \leq \pi$.

Since spectral power is nonnegative, $Re^{j\omega} \geq 0$ for all ω, and so

$$\begin{aligned} Re^{j\omega} &\triangleq \sum_{n=-\infty}^{\infty} r(n)\, e^{j\omega n} \\ &= r(0) + 2\sum_{n=1}^{\infty} r(n)\, \cos(\omega n) \\ &= 2\text{re}\left\{R_+(e^{j\omega})\right\} \\ &\geq 0. \end{aligned}$$

□

Relation to Schur Functions

Definition. A *Schur function* $S(z)$ is defined as a complex function analytic and of modulus not exceeding unity in $|z| \leq 1$.

Property. The function

$$S(z) \triangleq \frac{1 - R(z)}{1 + R(z)} \tag{C.84}$$

is a Schur function if and only if $R(z)$ is positive real.

Proof.

Suppose $R(z)$ is positive real. Then for $|z| \geq 1$, $\text{re}\{R(z)\} \geq 0 \Rightarrow 1 + \text{re}\{R(z)\} \geq 0 \Rightarrow 1 + R(z)$ is PR. Consequently, $1 + R(z)$ is minimum phase which implies all roots of $S(z)$ lie in the unit circle. Thus $S(z)$ is analytic in $|z| \leq 1$. Also,

$$\left|S(e^{j\omega})\right| = \frac{1 - 2\text{re}\left\{R(e^{j\omega})\right\} + \left|R(e^{j\omega}\right|^2}{1 + 2\text{re}\left\{R(e^{j\omega})\right\} + \left|R(e^{j\omega})\right|^2} \leq 1.$$

By the maximum modulus theorem, $S(z)$ takes on its maximum value in $|z| \geq 1$ on the boundary. Thus $S(z)$ is Schur.

Conversely, suppose $S(z)$ is Schur. Solving Eq. (C.84) for $R(z)$ and taking the real part on the unit circle yields

$$\begin{aligned}
R(z) &= \alpha \frac{1 - S(z)}{1 + S(z)} \\
\text{re}\left\{R(e^{j\omega})\right\} &= \alpha\,\text{re}\left\{\frac{1 - S(e^{j\omega})}{1 + S(e^{j\omega})} 1 + S(e^{-j\omega}) 1 + S(e^{-j\omega})\right\} \\
&= \alpha\,\text{re}\left\{\frac{1 - S(e^{j\omega}) + S(e^{-j\omega}) - \left|S(e^{j\omega})\right|^2}{\left|1 + S(e^{j\omega})\right|^2}\right\} \\
&= \alpha \frac{1 - \left|S(e^{j\omega})\right|^2}{\left|1 + S(e^{j\omega})\right|^2} \\
&\geq 0.
\end{aligned}$$

If $S(z) = \alpha$ is constant, then $R(z) = (1 - |\alpha|^2)/|1 + \alpha|^2$ is PR. If $S(z)$ is not constant, then by the maximum principle, $S(z) < 1$ for $|z| > 1$. By Rouche's theorem applied on a circle of radius $1 + \epsilon$, $\epsilon > 0$, on which $|S(z)| < 1$, the function $1 + S(z)$ has the same number of zeros as the function 1 in $|z| \geq 1 + \epsilon$. Hence, $1 + S(z)$ is minimum phase which implies $R(z)$ is analytic for $z \geq 1$. Thus $R(z)$ is PR.□

Relation to functions positive real in the right-half plane

Property. $\text{re}\{H(z)\} \geq 0$ for $|z| \geq 1$ whenever

$$\text{re}\left\{H\left(\frac{\alpha + s}{\alpha - s}\right)\right\} \geq 0$$

for $\text{re}\{s\} \geq 0$, where α is any positive real number.

Proof. We shall show that the change of variable $z \leftarrow (\alpha + s)/(\alpha - s)$, $\alpha > 0$, provides a conformal map from the z-plane to the s-plane that takes the region $|z| \geq 1$ to the region $\text{re}\{s\} \geq 0$. The general formula for a bilinear conformal mapping of functions of a complex variable is given by

$$\frac{(z-z_1)(z_2-z_3)}{(z-z_3)(z_2-z_1)} = \frac{(s-s_1)(s_2-s_3)}{(s-s_3)(s_2-s_1)}. \tag{C.85}$$

In general, a bilinear transformation maps circles and lines into circles and lines [83]. We see that the choice of three specific points and their images determines the mapping for all s and z. We must have that the imaginary axis in the s-plane maps to the unit circle in the z-plane. That is, we may determine the mapping by three points of the form $z_i = e^{j\theta_i}$ and $s_i = j\omega_i$, $i = 1, 2, 3$. If we predispose one such mapping by choosing the pairs $(s_1 = \pm\infty) \leftrightarrow (z_1 = -1)$ and $(s_3 = 0) \leftrightarrow (z_3 = 1)$, then we are left with transformations of the form

$$s = \left(s_2 \frac{z_2+1}{z_2-1}\right)\left(\frac{z-1}{z+1}\right) = \alpha\left(\frac{z-1}{z+1}\right)$$

or

$$z \leftarrow \frac{\alpha+s}{\alpha-s}, \tag{C.86}$$

Letting s_2 be some point $j\omega$ on the imaginary axis, and z_2 be some point $e^{j\theta}$ on the unit circle, we find that

$$\alpha = j\omega\frac{e^{j\theta}+1}{e^{j\theta}-1} = \omega\frac{\sin\theta}{1-\cos\theta} = \omega\cot(\theta/2)$$

which gives us that α is real. To avoid degeneracy, we require $s_2 \neq 0, \infty$, $z_2 \neq \pm 1$, and this translates to α finite and nonzero. Finally, to make the unit disk map to the *left*-half s-plane, ω and θ must have the same sign in which case $\alpha > 0$. □

There is a bonus associated with the restriction that α be real which is that

$$z = \frac{\alpha+s}{\alpha-s} \in \Re \quad \leftrightarrow \quad s = \alpha\frac{z-1}{z+1} \in \Re. \tag{C.87}$$

We have therefore proven

Property.

$$H(z)\ \text{PR} \Leftrightarrow H\left(\frac{\alpha+s}{\alpha-s}\right)\ \text{PR},$$

where α is any positive real number.

The class of mappings of the form Eq. (C.85) which take the exterior of the unit circle to the right-half plane is larger than the class Eq. (C.86). For example, we may precede the transformation Eq. (C.86) by any conformal map which takes the unit disk to the unit disk, and these mappings have the algebraic form of a first order complex allpass whose zero lies inside the unit circle.

$$z \leftarrow e^{j\theta} \frac{w - w_0}{\overline{w_0}\, w - 1}, \qquad |w_0| < 1 \tag{C.88}$$

where w_0 is the zero of the allpass and the image (also pre-image) of the origin, and θ is an angle of pure rotation. Note that Eq. (C.88) is equivalent to a pure rotation, followed by a *real* allpass substitution (w_0 real), followed by a pure rotation. The general preservation of condition (2) in Def. 2 forces the real axis to map to the real axis. Thus rotations by other than π are useless, except perhaps in some special cases. However, we may precede Eq. (C.86) by the first order *real* allpass substitution

$$z \leftarrow \frac{w - r}{r\, w - 1}, \qquad |r| < 1,\ r \text{ real},$$

which maps the real axis to the real axis. This leads only to the composite transformation,

$$z \leftarrow \frac{s + \left(\alpha \frac{1-r}{1+r}\right)}{s - \left(\alpha \frac{1-r}{1+r}\right)}$$

which is of the form Eq. (C.86) up to a minus sign (rotation by π). By inspection of Eq. (C.85), it is clear that sign negation corresponds to the swapping of points 1 and 2, or 2 and 3. Thus the only extension we have found by means of the general disk to disk pre-transform, is the ability to interchange two of the three points already tried. Consequently, we conclude that the largest class of bilinear transforms which convert functions positive real in the outer disk to functions positive real in the right-half plane is characterized by

$$z \leftarrow \pm \frac{\alpha + s}{\alpha - s}. \tag{C.89}$$

Riemann's theorem may be used to show that Eq. (C.89) is also the largest such class of conformal mappings. It is not essential, however, to restrict attention solely to conformal maps. The pre-transform $z \leftarrow \overline{z}$, for example, is not conformal and yet PR is preserved.

The bilinear transform is one which is used to map analog filters into digital filters. Another such mapping is called the *matched z transform* [362]. It also preserves the positive real property.

Property. $H(z)$ is PR if $H(e^{sT})$ is positive real in the analog sense, where $T > 0$ is interpreted as the sampling period.

Proof. The mapping $z \leftarrow e^{sT}$ takes the right-half s-plane to the outer disk in the z-plane. Also z is real if s is real. Hence $H(e^{sT})$ PR implies $H(z)$ PR. (Note, however, that rational functions do not in general map to rational functions.)□

These transformations allow application of the large battery of tests which exist for functions positive real in the right-half plane [525].

Special cases and examples

- The sum of positive real functions is positive real.
- The difference of positive real functions is conditionally positive real.
- The product or division of positive real functions is conditionally PR.
- $H(z)$ PR $\Longrightarrow$ $\alpha\, z^{\pm k} H(z)$ not PR for $\alpha > 0, k \geq 2$.

Minimum Phase (MP) polynomials in z

All properties of MP polynomials apply without modification to marginally stable allpole transfer functions (cf. Property 2):

- Every first-order MP polynomial is positive real.

- Every first-order MP polynomial $b(z) = 1 + b_1\, z^{-1}$ is such that $\frac{1}{b(z)} - \frac{1}{2}$ is positive real.

- A PR second-order MP polynomial with complex-conjugate zeros,

$$\begin{aligned} H(z) &= 1 + b_1 z^{-1} + b_2 z^{-2} \\ &= 1 - (2R\cos\phi) z^{-1} + R^2 z^{-2}, \quad R \leq 1 \end{aligned}$$

satisfies

$$R^2 + \frac{\cos^2\phi}{2} \leq 1.$$

If $2R^2 + \cos^2\phi = 2$, then $\text{re}\left\{H(e^{j\omega})\right\}$ has a double zero at

$$\begin{aligned} \omega &= \cos^{-1}\left(\pm\sqrt{\frac{1-R^2}{2R^2}}\right) = \cos^{-1}\left(\pm\frac{\cos\phi}{2R}\right) \\ &= \cos^{-1}\left(\pm\frac{1}{\sqrt{2}}\frac{\cos\phi}{\sqrt{1+\sin^2\phi}}\right). \end{aligned}$$

- All polynomials of the form

$$H(z) = 1 + R^n z^{-n}, \quad R \leq 1$$

are positive real. (These have zeros uniformly distributed on a circle of radius R.)

Miscellaneous Properties

- If all poles and zeros of a PR function are on the unit circle, then they alternate along the circle. Since this property is preserved by the bilinear transform, it is true in both the s and z planes. It can be viewed as a consequence of the $\pm 90°$ phase bounds for positive-real functions.

- If $B(s)/A(s)$ is PR, then so is $B'(s)/A'(s)$, where the prime denotes differentiation in s.

C.12 Loaded Waveguide Junctions

In this section, scattering relations will be derived for the general case of N waveguides meeting at a *load*. When a load is present, the scattering is no longer lossless, unless the load itself is lossless. (*i.e.*, its impedance has a zero real part). For $N > 2$, v_i^+ will denote a velocity wave traveling *into* the junction, and will be called an "incoming" velocity wave as opposed to "right-going."[10]

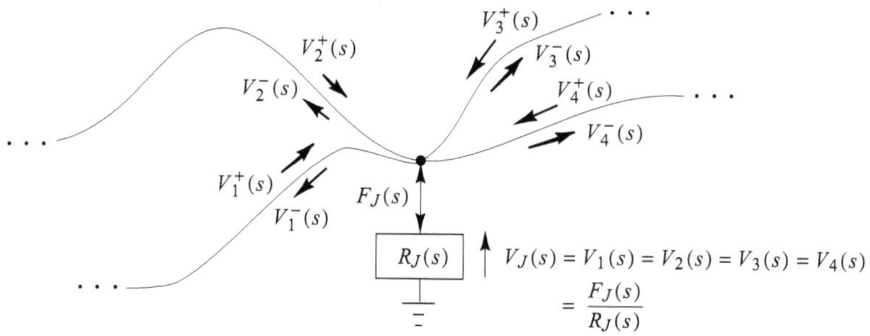

Figure C.29: Four ideal strings intersecting at a point to which a lumped impedance is attached. This is a series junction for transverse waves.

Consider first the *series* junction of N waveguides containing transverse force and velocity waves. At a series junction, there is a common velocity while the forces sum. For definiteness, we may think of N ideal strings intersecting at a single point, and the intersection point can be

[10]In the acoustic tube literature which involves only a cascade chain of acoustic waveguides, x^+ is taken to be traveling to the *right* along the axis of the tube [297]. In classical network theory [35] and in circuit theory, velocity (current) at the terminals of an N-port device is by convention taken to be positive when it flows *into* the device.

attached to a lumped load impedance $R_J(s)$, as depicted in Fig. C.29 for $N = 4$. The presence of the lumped load means we need to look at the wave variables in the frequency domain, *i.e.*, $V(s) = \mathcal{L}\{v\}$ for velocity waves and $F(s) = \mathcal{L}\{f\}$ for force waves, where $\mathcal{L}\{\cdot\}$ denotes the Laplace transform. In the discrete-time case, we use the z transform instead, but otherwise the story is identical. The physical constraints at the junction are

$$\begin{aligned} V_1(s) = V_2(s) = \cdots = V_N(s) &\stackrel{\Delta}{=} V_J(s) && \text{(C.90)} \\ F_1(s) + F_2(s) + \cdots + F_N(s) &= V_J(s)R_J(s) \stackrel{\Delta}{=} F_J(s) && \text{(C.91)} \end{aligned}$$

where the reference direction for the load force F_J is taken to be opposite that for the F_i. (It can be considered the "equal and opposite reaction" force at the junction.) For a wave traveling into the junction, force is positive pulling up, acting toward the junction. When the load impedance $R_J(s)$ is zero, giving a free intersection point, the junction reduces to the unloaded case, and signal scattering will be energy preserving. In general, the loaded junction is *lossless* if and only if $\text{re}\{R_J(j\omega)\} \equiv 0$, and it is *memoryless* if and only if $\text{im}\{R_J(j\omega)\} \equiv 0$.

The *parallel* junction is characterized by

$$\begin{aligned} F_1(s) = F_2(s) = \cdots = F_N(s) &\stackrel{\Delta}{=} F_J(s) && \text{(C.92)} \\ V_1(s) + V_2(s) + \cdots + V_N(s) &= F_J(s)/R_J(s) \stackrel{\Delta}{=} V_J(s) && \text{(C.93)} \end{aligned}$$

For example, $F_i(s)$ could be pressure in an acoustic tube and $V_i(s)$ the corresponding volume velocity. In the parallel case, the junction reduces to the unloaded case when the load impedance $R_J(s)$ goes to infinity.

The scattering relations for the series junction are derived as follows, dropping the common argument '(s)' for simplicity:

$$\begin{aligned} R_J V_J &= \sum_{i=1}^{N} F_i = \sum_{i=1}^{N} (F_i^+ + F_i^-) && \text{(C.94)} \\ &= \sum_{i=1}^{N} (R_i V_i^+ - R_i \underbrace{V_i^-}_{V_J - V_i^+}) && \text{(C.95)} \\ &= \sum_{i=1}^{N} (2R_i V_i^+ - R_i V_J) && \text{(C.96)} \end{aligned}$$

where R_i is the wave impedance in the ith waveguide, a real, positive constant. Bringing all terms containing V_J to the left-hand side, and solving for the junction velocity gives

$$V_J = 2\left(R_J + \sum_{i=1}^{N} R_i\right)^{-1} \sum_{i=1}^{N} R_i V_i^+ \tag{C.97}$$

$$\stackrel{\Delta}{=} \sum_{i=1}^{N} \mathcal{A}_i(s) V_i^+(s) \tag{C.98}$$

(written to be valid also in the multivariable case involving square impedance matrices R_i [433]), where

$$\mathcal{A}_i(s) \stackrel{\Delta}{=} \frac{2R_i}{R_J(s) + R_1 + \cdots + R_N} \tag{C.99}$$

Finally, from the basic relation $V_J = V_i = V_i^+ + V_i^-$, the outgoing velocity waves can be computed from the junction velocity and incoming velocity waves as

$$V_i^-(s) = V_J(s) - V_i^+(s) \tag{C.100}$$

Similarly, the scattering relations for the loaded *parallel* junction are given by

$$\begin{aligned} F_J(s) &= \sum_{i=1}^{N} \mathcal{A}_i(s) F_i^+(s), \quad \mathcal{A}_i(s) \stackrel{\Delta}{=} \frac{2\Gamma_i}{\Gamma_J(s) + \Gamma_1 + \cdots + \Gamma_N} \\ F_i^-(s) &= F_J(s) - F_i^+(s) \end{aligned} \tag{C.101}$$

where $F_J(s)$ is the Laplace transform of the force across all elements at the junction, $\Gamma_J(s)$ is the load admittance, and $\Gamma_i = 1/R_i$ are the branch admittances.

It is interesting to note that the junction load is equivalent to an $N+$ 1st waveguide having a (generalized) wave impedance given by the load impedance. This makes sense when one recalls that a transmission line can be "perfectly terminated" (*i.e.*, suppressing all reflections from the termination) using a lumped resistor equal in value to the wave impedance of the transmission line. Thus, as far as a traveling wave is concerned, there is no difference between a wave impedance and a lumped impedance of the same value.

In the unloaded case, $R_J(s) = 0$, and we can return to the time domain and define (for the series junction)

$$\alpha_i = \frac{2R_i}{R_1 + \cdots + R_N} \tag{C.102}$$

These we call the *alpha parameters*, and they are analogous to those used to characterize "adaptors" in wave digital filters (§F.2.2, p. 715). For unloaded junctions, the alpha parameters obey

$$0 \leq \alpha_i \leq 2 \tag{C.103}$$

and

$$\sum_{i=1}^{N} \alpha_i = 2 \tag{C.104}$$

In the unloaded case, the series junction scattering relations are given (in the time domain) by

$$\begin{aligned} v_J(t) &= \sum_{i=1}^{N} \alpha_i v_i^+(t) && \text{(C.105)} \\ v_i^-(t) &= v_J(t) - v_i^+(t) && \text{(C.106)} \end{aligned}$$

The alpha parameters provide an interesting and useful parametrization of waveguide junctions. They are explicitly the coefficients of the incoming traveling waves needed to compute junction velocity for a series junction (or junction force or pressure at a parallel junction), and losslessness is assured provided only that the alpha parameters be nonnegative and sum to 2. Having them sum to something less than 2 simulates a "resistive load" at the junction.

Note that in the lossless, equal-impedance case, in which all waveguide impedances have the same value $R_i = R$, (C.102) reduces to

$$\alpha_i = \frac{2}{N} \tag{C.107}$$

When, furthermore, N is a power of two, we have that there are *no multiplies* in the scattering relations (C.105). This fact has been used to build multiply-free reverberators and other structures using digital waveguide meshes [430, 519, 396, 521].

An elaborated discussion of $N = 2$ strings intersection at a load is given in in §9.2.1. Further discussion of the digital waveguide mesh appears in §C.14.

C.13 Two Coupled Strings

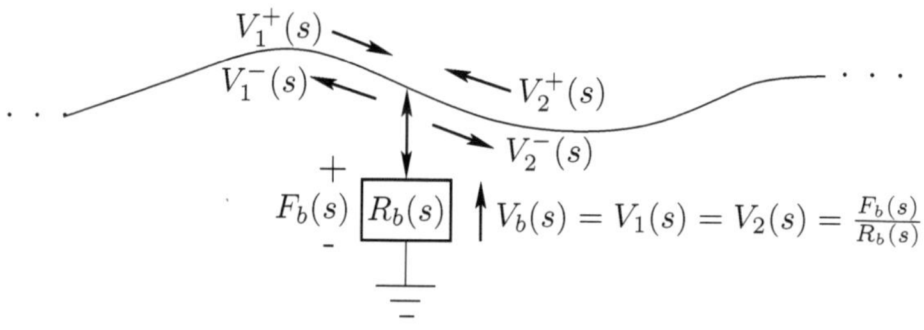

Figure C.30: Two strings terminated at a common bridge impedance.

C.13.1 Two Ideal Strings Coupled at an Impedance

A diagram for the two-string case is shown in Fig. C.30. This situation is a special case of the loaded waveguide junction, Eq. (C.98) on page 600, with the number of waveguides being $N = 2$, and the junction load being the transverse driving-point impedance $R_b(s)$ where the string drives the bridge. If the bridge is passive, then its impedance $R_b(s)$ is a positive real function (see §C.11.2 on page 588). For a direct derivation, we need only observe that (1) the string velocities of each string endpoint must each be equal to the velocity of the bridge, or $v_1 = v_2 = v_b$, and (2) the sum of forces of both strings equals the force applied to the bridge: $f_b = f_1 + f_2$. The bridge impedance relates the force and velocity of the bridge via $F_b(s) = R_b(s)V_b(s)$. Expanding into traveling wave components in the Laplace domain, we have

$$\begin{aligned} R_b(s)V_b(s) &= F_b(s) = F_1(s) + F_2(s) && \text{(C.108)} \\ &= [F_1^+(s) + F_1^-(s)] + [F_2^+(s) + F_2^-(s)] && \text{(C.109)} \\ &= R_1\{V_1^+(s) - [V_b(s) - V_1^+(s)]\} && \text{(C.110)} \\ &+ R_2\{V_2^+(s) - [V_b(s) - V_2^+(s)]\} && \text{(C.111)} \end{aligned}$$

or

$$V_b(s) = H_b(s)[R_1 V_1^+(s) + R_2 V_2^+(s)]$$

where R_i is the wave impedance of string i, and

$$H_b(s) \triangleq \frac{2}{R_b(s) + R_1 + R_2}$$

Thus, in the time domain, the incoming velocity waves are scaled by their respective wave impedances, summed together, and filtered according to the transfer function $H_b(s) = 2/[R_b(s) + R_1 + R_2]$ to obtain the velocity of the bridge $v_b(t)$.

Given the filter output $v_b(t)$, the outgoing traveling velocity waves are given by

$$v_1^-(t) = v_b(t) - v_1^+(t) \tag{C.112}$$
$$v_2^-(t) = v_b(t) - v_2^+(t) \tag{C.113}$$

Thus, the incoming waves are subtracted from the bridge velocity to get the outgoing waves.

Since $V_2^-(s) = H_b(s)R_1V_1^+(s) = H_b(s)F_1^+(s)$ when $V_2^+(s) = 0$, and vice versa exchanging strings 1 and 2, H_b may be interpreted as the *transmission admittance filter* associated with the bridge coupling. It can also be interpreted as the bridge admittance transfer function from every string, since its output is the bridge velocity resulting from the sum of incident traveling force waves.

A general coupling matrix contains a filter transfer function in each entry of the matrix. For N strings, each conveying a single type of wave (*e.g.*, horizontally polarized), the general linear coupling matrix would have N^2 transfer-function entries. In the present formulation, only one transmission filter is needed, and it is shared by all the strings meeting at the bridge. It is easy to show that the shared transmission filter for two coupled strings generalizes to N strings coupled at a common bridge impedance: From (C.98), we have

$$V_b(s) = H_b(s)\sum_{i=1}^{N} R_i V_i^+(s)$$

where

$$H_b(s) = \frac{2}{R_b(s) + \sum_{i=1}^{N} R_i}$$

Thus, $H_b(s)$ is the shared portion of the bridge filtering, leaving only a scaling according to relative impedance to be done in each branch.

The above sequence of operations is formally similar to the *one multiply scattering junction* frequently used in digital lattice filters [297]. In this context, it would be better termed the "one-filter scattering termination."

When the two strings are identical (as would be appropriate in a model for coupled piano strings), the computation of bridge velocity simplifies to

$$V_b(s) = H_b(s)[V_1^+(s) + V_2^+(s)] \tag{C.114}$$

where $H_b(s) \triangleq 2/[2 + R_b(s)/R]$ is the *velocity* transmission filter. In this case, the incoming velocities are simply summed and fed to the transmission filter which produces the bridge velocity at its output. A commuted simulation diagram appears in Fig. C.31.

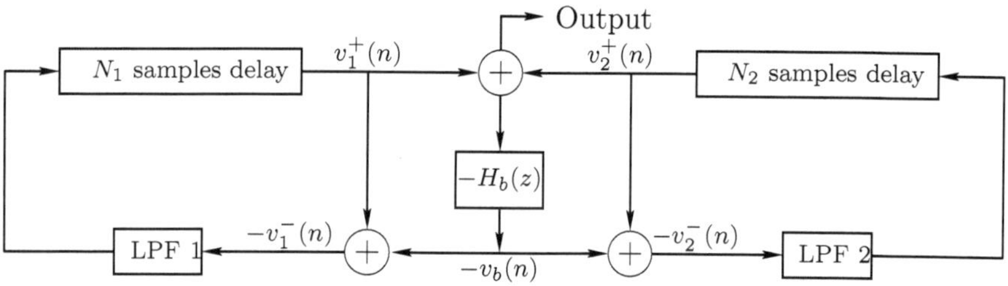

Figure C.31: General linear coupling of two equal-impedance strings using a common bridge filter.

Note that a yielding bridge introduces losses into all attached strings. Therefore, in a maximally simplified implementation, all string loop filters (labeled LPF_1 and LPF_2 in Fig. C.31) may be *eliminated*, resulting in only one filter—the transmission filter—serving to provide all losses in a coupled-string simulation. If that transmission filter has no multiplies, then neither does the entire multi-string simulation.

C.13.2 Coupled Strings Eigenanalysis

In §6.12.2 on page 239, general coupling of horizontal and vertical planes of vibration in an ideal string was considered. This eigenanalysis will now be applied here to obtain formulas for the *damping* and *mode tuning* caused by the coupling of two identical strings at a bridge. This is the case that arises in pianos [544].

The general formula for linear, time-invariant coupling of two strings can be written, in the frequency domain, as

$$\left[\begin{array}{c} V_1^-(s) \\ V_2^-(s) \end{array}\right] = \left[\begin{array}{cc} H_{11}(s) & H_{12}(s) \\ H_{21}(s) & H_{22}(s) \end{array}\right] \left[\begin{array}{c} V_1^+(s) \\ V_2^+(s) \end{array}\right]. \tag{C.115}$$

Filling in the elements of this coupling matrix $\mathbf{H}_c$ from the results of §C.13.1, we obtain

$$\mathbf{H}_c(s) = \left[\begin{array}{cc} 1 - H_b(s) & -H_b(s) \\ -H_b(s) & 1 - H_b(s) \end{array}\right]$$

where

$$H_b(s) = \frac{2}{2 + \tilde{R}_b}.$$

Here $\tilde{R}_b \triangleq R_b/R$ is the bridge impedance divided by the string impedance. Treating $\mathbf{H}_c(s)$ as a constant complex matrix for each fixed s, the eigenvectors are found[11] to be

$$\underline{e}_1 = \begin{bmatrix} 1 \\ 1 \end{bmatrix}, \qquad \underline{e}_2 = \begin{bmatrix} 1 \\ -1 \end{bmatrix},$$

respectively, and the eigenvalues are

$$\lambda_1(s) = 1 - 2H_b(s), \qquad \lambda_2 = 1.$$

Note that only one eigenvalue depends on $s = j\omega$, and neither eigenvector is a function of s.

We conclude that "in-phase vibrations" see a longer effective string length, lengthened by the phase delay of

$$1 - 2H_b = \frac{\tilde{R}_b(s) - 2}{\tilde{R}_b(s) + 2} = \frac{R_b(s) - 2R}{R_b(s) + 2R}$$

which is the reflectance seen from two in-phase strings each having impedance R. This makes physical sense because the in-phase vibrations will move the bridge in the vertical direction, causing more rapid decay of the in-phase mode.

We similarly conclude that the "anti-phase vibrations" see no length correction at all, because the bridge point does not move at all in this case. In other words, any bridge termination at a point is rigid with respect to anti-phase vibration of the two strings connected to that point.

The above analysis predicts that, in "stiffness controlled" frequency intervals (in which the bridge "looks like a damped spring"), the "initial fast decay" of a piano note should be a measurably flatter than the "aftersound" which should be exactly in tune as if the termination were rigid.

C.14 Digital Waveguide Mesh

In §C.12, the theory of multiport scattering was derived, *i.e.*, the reflections and transmissions that occur when N digital waveguides having wave impedances R_i are connected together. It was noted that when N is a power of two, there are *no multiplies* in the scattering relations Eq. (C.105) on page 601, and that this fact has been used to build multiply-free reverberators and other structures using digital waveguide meshes [430, 519, 146, 396, 521, 522, 398, 399, 401, 55, 202, 321, 320, 322, 422, 33].

C.14.1 The Rectilinear 2D Mesh

Figure C.32 shows the basic layout of the rectilinear 2D waveguide mesh. It can be thought of as simulating a plane using 1D digital waveguides in the

[11] `http://ccrma.stanford.edu/~jos/filters/Similarity_Transformations.html`

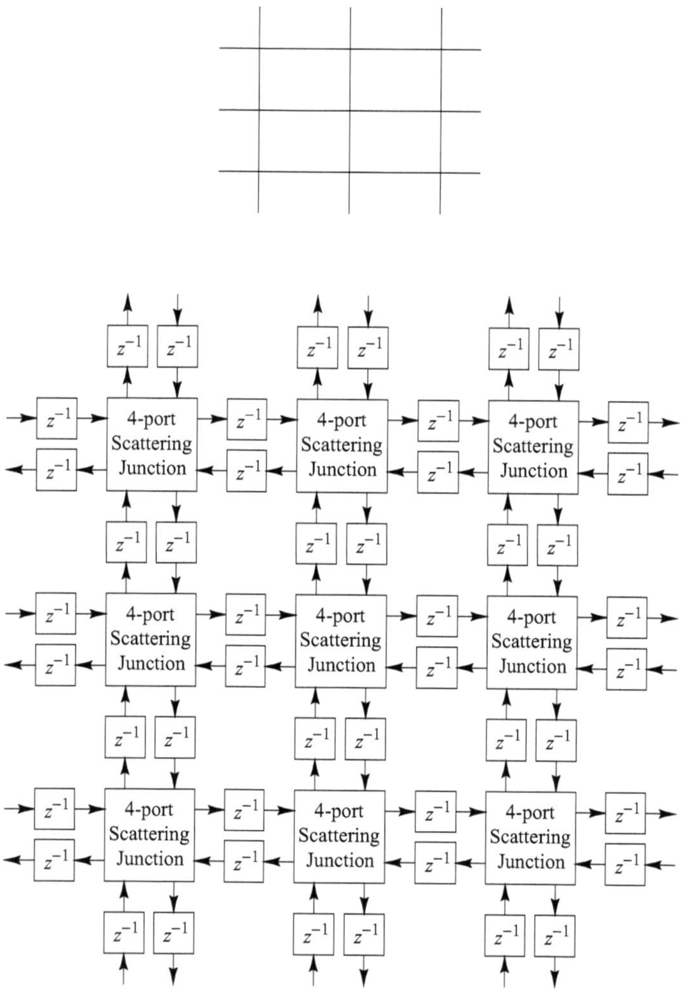

Figure C.32: The 2D rectilinear digital waveguide mesh.

same way that a tennis racket acts as a membrane composed of 1D strings.

At each node (string intersection), we have the following simple formula for the node velocity v in terms of the four incoming traveling-wave components:

$$v = \frac{\text{in}_1 + \text{in}_2 + \text{in}_3 + \text{in}_4}{2}$$

By continuity of velocity in a series connection (all string endpoints must move together at the node), the outgoing velocity-wave components must be given by

$$\text{out}_k = v - \text{in}_k, \qquad k = 1, 2, 3, 4.$$

C.14.2 Dispersion

Since the digital waveguide mesh is *lossless* by construction (when modeling lossless membranes and volumes), and since it is also linear and time-invariant by construction, being made of ordinary digital filtering computations, there is only one type of error exhibited by the mesh: *dispersion*. Dispersion can be quantified as an error in propagation speed as a function of frequency and direction along the mesh. The mesh geometry (rectilinear, triangular, hexagonal, tetrahedral, etc.) strongly influences the dispersion properties. Many cases are analyzed in [55] using von Neumann analysis (see also Appendix D).

The *triangular waveguide mesh* [146] turns out to be the simplest mesh geometry in 2D having the least dispersion *variation* as a function of direction of propagation on the mesh. In other terms, the triangular mesh is closer to *isotropic* than all other known elementary geometries. The *interpolated waveguide mesh* [398] can also be configured to optimize isotropy, but at a somewhat higher compuational cost.

C.14.3 Recent Developments

An interesting approach to dispersion compensation is based on *frequency-warping* the signals going into the mesh [399]. Frequency warping can be used to compensate frequency-dependent dispersion, but it does not address angle-dependent dispersion. Therefore, frequency-warping is used in conjunction with an isotropic mesh.

The 3D waveguide mesh [519, 522, 399] is seeing more use for efficient simulation of acoustic spaces [396, 182]. It has also been applied to statistical modeling of violin body resonators in [203, 202, 422, 428], in which the digital waveguide mesh was used to efficiently model only the "reverberant" aspects of a violin body's impulse response in statistically matched fashion (but close to perceptually equivalent). The "instantaneous" filtering by the violin body is therefore modeled using a separate equalizer capturing the important low-frequency body and air modes explicitly. A unified view of the digital waveguide mesh and *wave digital filters* (§F.1) as particular classes of energy invariant finite difference schemes (Appendix D) appears in [54]. The problem of modeling diffusion at a mesh boundary was addressed in [268], and maximally diffusing boundaries, using quadratic residue sequences, was investigated in [279]; an introduction to this topic is given in §C.14.6 below.

C.14.4 2D Mesh and the Wave Equation

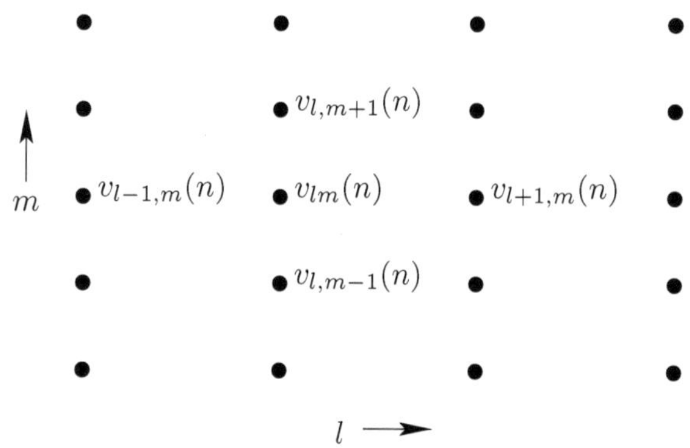

Figure C.33: Region of nodes in a rectilinear waveguide mesh.

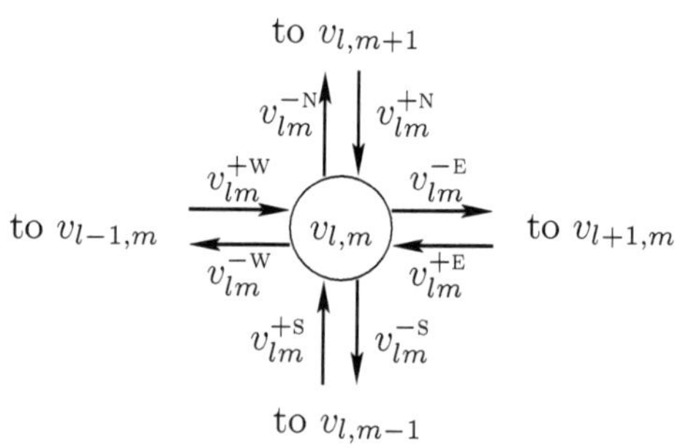

Figure C.34: Zoom-in about node (l, m) at time n in a rectilinear waveguide mesh, showing traveling-wave components entering and leaving the node. (All variables are at time n,)

Consider the 2D rectilinear mesh, with nodes at positions $x = lX$ and $y = mY$, where l and m are integers, and X and Y denote the spatial sampling intervals along x and y, respectively (see Fig. C.33). Then from Eq. (C.105) on page 601 the junction velocity v_{lm} at time n is given by

$$v_{lm}(n) = \frac{1}{2}\left[v_{lm}^{+\mathrm{N}}(n) + v_{lm}^{+\mathrm{E}}(n) + v_{lm}^{+\mathrm{S}}(n) + v_{lm}^{+\mathrm{W}}(n)\right]$$

where $v_{lm}^{+\mathrm{N}}(n)$ is the "incoming wave from the north" to node (l,m), and similarly for the waves coming from east, west, and south (see Fig. C.34).

These incoming traveling-wave components arrive from the four neighboring nodes after a one-sample propagation delay. For example, $v_{lm}^{+\mathrm{N}}(n)$, arriving from the north, departed from node $(l,m+1)$ at time $n-1$, as $v_{l,m+1}^{-\mathrm{S}}(n-1)$. Furthermore, the outgoing components at time n will arrive at the neighboring nodes one sample in the future at time $n+1$. For example, $v_{lm}^{-\mathrm{N}}(n)$ will become $v_{l,m+1}^{+\mathrm{S}}(n+1)$. Using these relations, we can write $v_{lm}(n+1)$ in terms of the four outgoing waves from its neighbors at time n:

$$v_{lm}(n+1) = \frac{1}{2}\left[v_{l,m+1}^{-\mathrm{S}}(n) + v_{l+1,m}^{-\mathrm{W}}(n) + v_{l,m-1}^{-\mathrm{N}}(n) + v_{l-1,m}^{-\mathrm{E}}(n)\right] \quad \text{(C.116)}$$

where, for instance, $v_{lm}^{-\mathrm{N}}(n)$ is the "outgoing wave to the north" from node (l,m). Similarly, the outgoing waves leaving $v_{lm}(n-1)$ become the incoming traveling-wave components of its neighbors at time n:

$$v_{lm}(n-1) = \frac{1}{2}\left[v_{l,m+1}^{+\mathrm{S}}(n) + v_{l+1,m}^{+\mathrm{W}}(n) + v_{l,m-1}^{+\mathrm{N}}(n) + v_{l-1,m}^{+\mathrm{E}}(n)\right] \quad \text{(C.117)}$$

This may be shown in detail by writing

$$\begin{aligned}
v_{lm}(n-1) &= \frac{1}{2}[v_{lm}^{+\mathrm{N}}(n-1) + \cdots + v_{lm}^{+\mathrm{W}}(n-1)]\\
&= \frac{1}{2}[v_{lm}(n-1) - v_{lm}^{-\mathrm{N}}(n-1) + \cdots + v_{lm}(n-1) - v_{lm}^{-\mathrm{W}}(n-1)]\\
&= 2v_{lm}(n-1) - \frac{1}{2}\left[v_{lm}^{-\mathrm{N}}(n-1) + \cdots + v_{lm}^{-\mathrm{W}}(n-1)\right]
\end{aligned}$$

so that

$$\begin{aligned}
v_{lm}(n-1) &= \frac{1}{2}[v_{lm}^{-\mathrm{N}}(n-1) + \cdots + v_{lm}^{-\mathrm{W}}(n-1)]\\
&= \frac{1}{2}\left[v_{l,m+1}^{+\mathrm{S}}(n) + v_{l+1,m}^{+\mathrm{W}}(n) + v_{l,m-1}^{+\mathrm{N}}(n) + v_{l-1,m}^{+\mathrm{E}}(n)\right].
\end{aligned}$$

Adding Equations (C.116–C.116), replacing terms such as $v_{l,m+1}^{+\mathrm{S}}(n)+v_{l,m+1}^{-\mathrm{S}}(n)$ with $v_{l,m+1}(n)$, yields a computation in terms of physical node velocities:

$$\boxed{\begin{aligned}
&v_{lm}(n+1) + v_{lm}(n-1) =\\
&\quad\frac{1}{2}\left[v_{l,m+1}(n) + v_{l+1,m}(n) + v_{l,m-1}(n) + v_{l-1,m}(n)\right]
\end{aligned}} \quad \text{(C.118)}$$

Thus, the rectangular waveguide mesh satisfies this equation giving a formula for the velocity at node (l, m), in terms of the velocity at its neighboring nodes one sample earlier, and itself two samples earlier. Subtracting $2v_{lm}(n)$ from both sides yields

$$\begin{aligned} & v_{lm}(n+1) - 2v_{lm}(n) + v_{lm}(n-1) \\ = \; & \frac{1}{2}\left\{\left[v_{l,m+1}(n) - 2v_{lm}(n) + v_{l,m-1}(n)\right]\right. \\ & \left. + \left[v_{l+1,m}(n) - 2v_{lm}(n) + v_{l-1,m}(n)\right]\right\} \end{aligned}$$

Dividing by the respective sampling intervals, and assuming $X = Y$ (square mesh-holes), we obtain

$$\begin{aligned} & \frac{v_{lm}(n+1) - 2v_{lm}(n) + v_{lm}(n-1)}{T^2} \\ = \; & \frac{X^2}{2T^2}\left[\frac{v_{l,m+1}(n) - 2v_{lm}(n) + v_{l,m-1}(n)}{Y^2}\right. \\ & \left. + \frac{v_{l+1,m}(n) - 2v_{lm}(n) + v_{l-1,m}(n)}{X^2}\right]. \end{aligned}$$

In the limit, as the sampling intervals X, Y, T approach zero such that $X/T = Y/T$ remains constant, we recognize these expressions as the definitions of the partial derivatives with respect to t, x, and y, respectively, yielding

$$\frac{\partial^2 v(t,x,y)}{\partial t^2} = \frac{X^2}{2T^2}\left[\frac{\partial^2 v(t,x,y)}{\partial x^2} + \frac{\partial^2 v(t,x,y)}{\partial y^2}\right].$$

This final result is the ideal 2D wave equation $\ddot{v} = c^2\nabla^2 v$, *i.e.*,

$$\frac{\partial^2 v}{\partial t^2} = c^2\left[\frac{\partial^2 v}{\partial x^2} + \frac{\partial^2 v}{\partial y^2}\right]$$

with

$$c = \frac{1}{\sqrt{2}}\frac{X}{T} = \frac{\sqrt{2}}{2}\frac{X}{T}. \tag{C.119}$$

Discussion regarding solving the 2D wave equation subject to boundary conditions appears in §B.8.3 on page 518. Interpreting this value for the wave propagation speed c, we see that every two time steps of $2T$ seconds corresponds to a spatial step of $\sqrt{2}X$ meters. This is the distance from one diagonal to the next in the square-hole mesh. We will show later that diagonal directions on the mesh support *exact* propagation (of plane waves traveling at 45-degree angles with respect to the x or y axes). In the x and y directions, propagation is highly *dispersive*, meaning that different frequencies travel at different speeds. The exactness of 45-degree angles can be appreciated by considering Huygens' principle on the mesh.

C.14.5 The Lossy 2D Mesh

Because the finite-difference form of the digital waveguide mesh is the more efficient computationally than explicitly computing scattering wave variables (too see this, count the multiplies required per node), it is of interest to consider the finite-difference form also in the case of frequency-dependent losses. The method of §C.5.5 on page 544 extends also to the waveguide mesh, which can be shown by generalizing the results of §C.14.4 above using the technique of §C.5.5.

The basic idea is once again that wave propagation during one sampling interval (in time) is associated with linear filtering by $G(z)$. That is, $G(z)$ is regarded as the per-sample wave propagation filter.

C.14.6 Diffuse Reflections in the Waveguide Mesh

In [416], Manfred Schroeder proposed the design of a diffuse reflector based on a *quadratic residue sequence*. A quadratic residue sequence $a_p(n)$ corresponding to a prime number p is the sequence n^2 mod p, for all integers n. The sequence is periodic with period p, so it is determined by $a_p(n)$ for $n = 0, 1, 2, \ldots, p-1$ (*i.e.*, one period of the infinite sequence).

For example, when $p = 7$, the first period of the quadratic residue sequence is given by

$$\begin{aligned} a_7 &= [0^2, 1^2, 2^2, 3^2, 4^2, 5^2, 6^2] \pmod 7 \\ &= [0, 1, 4, 2, 2, 4, 1] \end{aligned}$$

An amazing property of these sequences is that their Fourier transforms have precisely constant magnitudes. That is, the sequence

$$c_p(n) \triangleq e^{j\frac{2\pi}{p}a_p(n)}$$

has a DFT with exactly constant magnitude:[12]

$$|C_p(\omega_k)| \triangleq |\text{DFT}_k(c_p)| \triangleq \left|\sum_{n=0}^{p-1} c_p(n)e^{-j2\pi nk/p}\right| = \sqrt{p}, \quad \forall k \in [0, p-1]$$

This property can be used to give highly diffuse reflections for incident plane waves.

Figure C.35 presents a simple matlab script which demonstrates the constant-magnitude Fourier property for all odd integers from 1 to 99.

[12]While the literature seems to mention this property only for prime numbers p, it is straightforward to show that it holds in fact for all positive odd integers p. Prime values of p have advantages, however, when harmonics of the "design frequency" are considered.

```
function [c] = qrsfp(Ns)
%QRSFP Quadratic Residue Sequence Fourier Property demo
  if (nargin<1)
     Ns = 1:2:99; % Test all odd integers from 1 to 99
  end
  for N=Ns
    a = mod([0:N-1].^2,N);
    c = zeros(N-1,N);
    CM = zeros(N-1,N);
    c = exp(j*2*pi*a/N);
    CM = abs(fft(c))*sqrt(1/N);
    if (abs(max(CM)-1)>1E-10) || (abs(min(CM)-1)>1E-10)
       warn(sprintf("Failure for N=%d",N));
    end
  end
  r = exp(2i*pi*[0:100]/100); % a circle
  plot(real(r), imag(r),"k"); hold on;
  plot(c,"-*k"); % plot sequence in complex plane
end
```

Figure C.35: Matlab script for demonstrating the Fourier property of an odd-length quadratic residue sequence.

Quadratic residue diffusers have been applied as boundaries of a 2D digital waveguide mesh in [279]. An article reviewing the history of room acoustic diffusers may be found in [94].

C.15 FDNs as Digital Waveguide Networks

This section supplements §2.7 on page 65 on *Feedback Delay Networks* in the context of digital waveguide theory. Specifically, we review the interpretation of an FDN as a special case of a digital waveguide network, summarizing [464, 465, 385].

Figure C.36 illustrates an N-branch DWN. It consists of a single scattering junction, indicated by a white circle, to which N branches are connected. The far end of each branch is terminated by an ideal non-inverting reflection (black circle). The waves traveling into the junction are associated with the FDN delay line outputs $x_i(n - M_i)$, and the length of each waveguide is *half* the length of the corresponding FDN delay line M_i (since a traveling wave must traverse the branch twice to complete a round trip from the junction to the termination and back). When M_i is odd, we may replace the reflecting termination by a unit-sample delay.

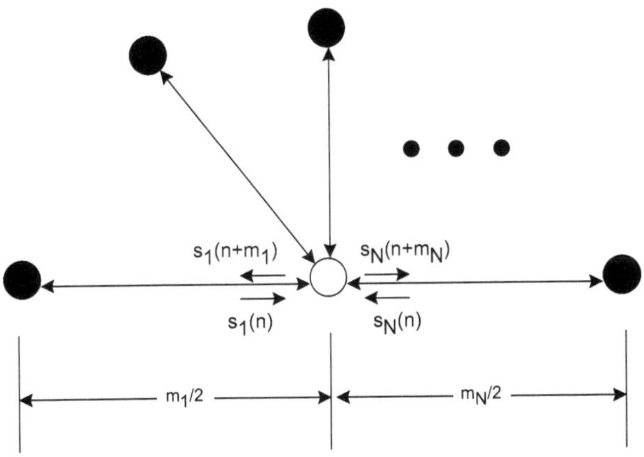

Figure C.36: Waveguide network consisting of a single scattering junction, indicated by an open circle, to which N branches are connected. The far end of each branch is terminated by an ideal, non-inverting reflection.

C.15.1 Lossless Scattering

The delay-line inputs (outgoing traveling waves) are computed by multiplying the delay-line outputs (incoming traveling waves) by the $N \times N$ feedback matrix (scattering matrix) $\mathbf{A} = [a_{i,j}]$. By defining $p_i^+ = x_i(n - M_i)$, $p_i^- = x_i(n)$, we obtain the more usual DWN notation

$$\mathbf{p}^- = \mathbf{A}\mathbf{p}^+ \tag{C.120}$$

where $\mathbf{p}^+$ is the vector of incoming traveling-wave samples arriving at the junction at time n, $\mathbf{p}^-$ is the vector of outgoing traveling-wave samples leaving the junction at time n, and $\mathbf{A}$ is the *scattering matrix* associated with the waveguide junction.

The junction of N physical waveguides determines the structure of the matrix $\mathbf{A}$ according to the basic principles of physics.

Considering the parallel junction of N lossless acoustic tubes, each having characteristic admittance $\Gamma_j = 1/R_j$, the continuity of pressure and conservation of volume velocity at the junction give us the following scattering matrix for the pressure waves [433]:

$$\mathbf{A} = \left[\begin{array}{cccc} \frac{2\Gamma_1}{\Gamma_J} - 1 & \frac{2\Gamma_2}{\Gamma_J} & \dots & \frac{2\Gamma_N}{\Gamma_J} \\ \frac{2\Gamma_1}{\Gamma_J} & \frac{2\Gamma_2}{\Gamma_J} - 1 & \dots & \frac{2\Gamma_N}{\Gamma_J} \\ \dots & & & \\ \frac{2\Gamma_1}{\Gamma_J} & \frac{2\Gamma_2}{\Gamma_J} & \dots & \frac{2\Gamma_N}{\Gamma_J} - 1 \end{array}\right] \tag{C.121}$$

where

$$\Gamma_J = \sum_{i=1}^{N} \Gamma_i. \tag{C.122}$$

Equation (C.121) can be derived by first writing the volume velocity at the j-th tube in terms of pressure waves as $v_j = (p_j^+ - p_j^-)\Gamma_j$. Applying the conservation of velocity we can find the expression

$$p = 2\sum_{i=1}^{N} \Gamma_i p_i^+ / \Gamma_J$$

for the junction pressure. Finally, if we express the junction pressure as the sum of incoming and outgoing pressure waves at any branch, we derive (C.121). See §C.12 for further details.

C.15.2 Normalized Scattering

For ideal numerical scaling in the L_2 sense, we may choose to propagate *normalized waves* which lead to normalized scattering junctions analogous to those encountered in normalized ladder filters [297]. Normalized waves may be either normalized pressure $\tilde{p}_j^+ = p_j^+\sqrt{\Gamma_i}$ or normalized velocity $\tilde{v}_j^+ = v_j^+/\sqrt{\Gamma_i}$. Since the signal power associated with a traveling wave is simply $\mathcal{P}_|^+ = (\tilde{p}_j^+)^2 = (\tilde{v}_j^+)^2$, they may also be called *root-power waves* [432]. Appendix C develops this topic in more detail.

The scattering matrix for normalized pressure waves is given by

$$\tilde{\mathbf{A}} = \begin{bmatrix} \frac{2\Gamma_1}{\Gamma_J} - 1 & \frac{2\sqrt{\Gamma_1\Gamma_2}}{\Gamma_J} & \dots & \frac{2\sqrt{\Gamma_1\Gamma_n}}{\Gamma_J} \\ \frac{2\sqrt{\Gamma_2\Gamma_1}}{\Gamma_J} & \frac{2\Gamma_2}{\Gamma_J} - 1 & \dots & \frac{2\sqrt{\Gamma_2\Gamma_n}}{\Gamma_J} \\ \dots & & \dots & \\ \dots & & \dots & \\ \frac{2\sqrt{\Gamma_n\Gamma_1}}{\Gamma_J} & \frac{2\sqrt{\Gamma_n\Gamma_2}}{\Gamma_J} & \dots & \frac{2\Gamma_n}{\Gamma_J} - 1 \end{bmatrix} \tag{C.123}$$

The normalized scattering matrix can be expressed as a negative Householder reflection

$$\tilde{\mathbf{A}} = \frac{2}{||\tilde{\mathbf{\Gamma}}||^2}\tilde{\mathbf{\Gamma}}\tilde{\mathbf{\Gamma}}^T - \mathbf{I} \tag{C.124}$$

where $\tilde{\mathbf{\Gamma}}^T = [\sqrt{\Gamma_1}, \dots, \sqrt{\Gamma_N}]$, and Γ_i is the wave admittance in the ith waveguide branch. To eliminate the sign inversion, the reflections at the far end of each waveguide can be chosen as -1 instead of 1. The geometric interpretation of (C.124) is that the incoming pressure waves are reflected about the vector $\tilde{\mathbf{\Gamma}}$. Unnormalized scattering junctions can be expressed in the form of an "oblique" Householder reflection $\mathbf{A} = 2\mathbf{1}\mathbf{\Gamma}^T / \langle \mathbf{1}, \mathbf{\Gamma} \rangle - \mathbf{I}$, where $\mathbf{1}^T = [1, \ldots, 1]$ and $\mathbf{\Gamma}^T = [\Gamma_1, \ldots, \Gamma_N]$.

C.15.3 General Conditions for Losslessness

The scattering matrices for lossless physical waveguide junctions give an apparently unexplored class of lossless FDN prototypes. However, this is just a subset of all possible lossless feedback matrices. We are therefore interested in the most general conditions for losslessness of an FDN feedback matrix. The results below are adapted from [464, 385].

Consider the general case in which $\mathbf{A}$ is allowed to be any scattering matrix, *i.e.*, it is associated with a not-necessarily-physical junction of N physical waveguides. Following the definition of losslessness in classical network theory, we may say that a waveguide scattering matrix $\mathbf{A}$ is said to be *lossless* if the *total complex power* [35] at the junction is scattering invariant, *i.e.*,

$$\begin{aligned} & \mathbf{p}^{+^*}\mathbf{\Gamma}\mathbf{p}^+ &=& \mathbf{p}^{-^*}\mathbf{\Gamma}\mathbf{p}^- \\ \Longrightarrow\quad & \mathbf{A}^*\mathbf{\Gamma}\mathbf{A} &=& \mathbf{\Gamma} \end{aligned} \tag{C.125}$$

where $\mathbf{\Gamma}$ is any Hermitian, positive-definite *matrix* (which has an interpretation as a generalized junction admittance). The form $x^*\mathbf{\Gamma}x$ is by definition the square of the *elliptic norm* of x induced by $\mathbf{\Gamma}$, or $\|\, x \,\|_{\mathbf{\Gamma}}^2 = x^*\mathbf{\Gamma}x$. Setting $\mathbf{\Gamma} = \mathbf{I}$, we obtain that $\mathbf{A}$ must be unitary. This is the case commonly used in current FDN practice.

The following theorem gives a general characterization of lossless scattering:

Theorem: A scattering matrix (FDN feedback matrix) $\mathbf{A}$ is lossless if and only if its eigenvalues lie on the unit circle and its eigenvectors are linearly independent.

Proof: Since $\mathbf{\Gamma}$ is positive definite, it can be factored (by the Cholesky factorization) into the form $\mathbf{\Gamma} = \mathbf{U}^*\mathbf{U}$, where $\mathbf{U}$ is an upper triangular matrix, and $\mathbf{U}^*$ denotes the Hermitian transpose of $\mathbf{U}$, *i.e.*, $\mathbf{U}^* \triangleq \overline{\mathbf{U}}^T$. Since $\mathbf{\Gamma}$ is positive definite, $\mathbf{U}$ is nonsingular and can be used as a similarity transformation matrix. Applying the Cholesky decomposition $\mathbf{\Gamma} = \mathbf{U}^*\mathbf{U}$ in Eq. (C.125) yields

$$\begin{aligned} & \mathbf{A}^*\mathbf{\Gamma}\mathbf{A} = \mathbf{\Gamma} \\ \Longrightarrow\quad & \mathbf{A}^*\mathbf{U}^*\mathbf{U}\mathbf{A} = \mathbf{U}^*\mathbf{U} \\ \Longrightarrow\quad & (\mathbf{U}^{-*}\mathbf{A}^*\mathbf{U}^*)(\mathbf{U}\mathbf{A}\mathbf{U}^{-1}) = \mathbf{I} \\ \Longrightarrow\quad & \tilde{\mathbf{A}}^*\tilde{\mathbf{A}} = \mathbf{I} \end{aligned}$$

where $\mathbf{U}^{-*} \triangleq (\mathbf{U}^{-1})^*$, and

$$\tilde{\mathbf{A}} \triangleq \mathbf{U}\mathbf{A}\mathbf{U}^{-1}$$

is similar to $\mathbf{A}$ using $\mathbf{U}^{-1}$ as the similarity transform matrix. Since $\tilde{\mathbf{A}}$ is unitary, its eigenvalues have modulus 1. Hence, the eigenvalues of every lossless scattering matrix lie on the unit circle in the z plane. It readily follows from similarity to $\tilde{\mathbf{A}}$ that $\mathbf{A}$ admits N linearly independent eigenvectors. In fact, $\tilde{\mathbf{A}}$ is a normal matrix ($\mathbf{A}\tilde{\mathbf{A}} = \tilde{\mathbf{A}}\mathbf{A}$), since every unitary matrix is normal, and normal matrices admit a basis of linearly independent eigenvectors [346].

Conversely, assume $|\lambda| = 1$ for each eigenvalue of $\mathbf{A}$, and that there exists a matrix $\mathbf{E}$ of linearly independent eigenvectors of $\mathbf{A}$. The matrix $\mathbf{E}$ diagonalizes $\mathbf{A}$ to give $\mathbf{E}^{-1}\mathbf{A}\mathbf{E} = \mathbf{D}$, where $\mathbf{D} = \text{diag}(\lambda_1, \dots, \lambda_N)$. Taking the Hermitian transform of this equation gives $\mathbf{E}^*\mathbf{A}^*\mathbf{E}^{-*} = \mathbf{D}^*$. Multiplying, we obtain $\mathbf{E}^*\mathbf{A}^*\mathbf{E}^{-*}\mathbf{E}^{-1}\mathbf{A}\mathbf{E} = \mathbf{D}^*\mathbf{D} = \mathbf{I} \implies \mathbf{A}^*\mathbf{E}^{-*}\mathbf{E}^{-1}\mathbf{A} = \mathbf{E}^{-*}\mathbf{E}^{-1}$. Thus, (C.125) is satisfied for $\mathbf{\Gamma} = \mathbf{E}^{-*}\mathbf{E}^{-1}$ which is Hermitian and positive definite. □

Thus, lossless scattering matrices may be fully parametrized as $\mathbf{A} = \mathbf{E}^{-1}\mathbf{D}\mathbf{E}$, where $\mathbf{D}$ is any unit-modulus diagonal matrix, and $\mathbf{E}$ is any invertible matrix. In the real case, we have $\mathbf{D} = \text{diag}(\pm 1)$ and $\mathbf{E} \in \Re^{N\times N}$.

Note that not all lossless scattering matrices have a simple *physical* interpretation as a scattering matrix for an intersection of N lossless reflectively terminated waveguides. In addition to these cases (generated by all non-negative branch impedances), there are additional cases corresponding to sign flips and branch permutations at the junction. In terms of classical network theory [35], such additional cases can be seen as arising from the use of "gyrators" and/or "circulators" at the scattering junction [433]).

C.16 Waveguide Transformers and Gyrators

The ideal transformer, depicted in Fig. C.37 a, is a lossless two-port electric circuit element which scales up voltage by a constant g [110, 35]. In other words, the voltage at port 2 is always g times the voltage at port 1. Since power is voltage times current, the current at port 2 must be $1/g$ times the current at port 1 in order for the transformer to be lossless. The scaling constant g is called the *turns ratio* because transformers are built by coiling wire around two sides of a magnetically permeable torus, and the number of winds around the port 2 side divided by the winding count on the port 1 side gives the voltage stepping constant g.

In the case of mechanical circuits, the two-port transformer relations

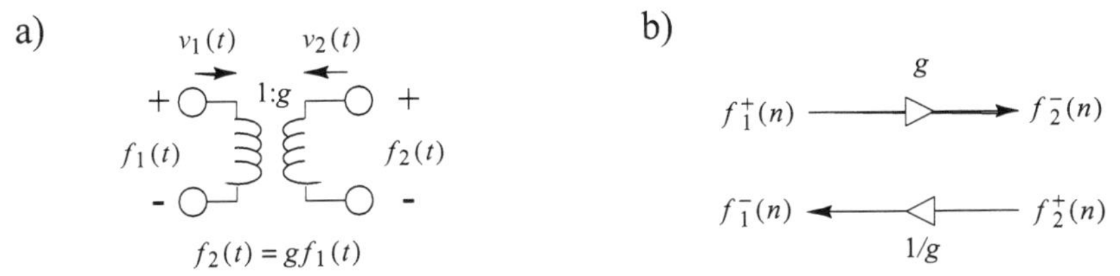

Figure C.37: a) Two-port description of the ideal transformer with "turns ratio" g. b) Corresponding wave digital transformer.

appear as

$$\begin{aligned} F_2(s) &= gF_1(s) \\ V_2(s) &= \frac{1}{g}V_1(s) \end{aligned}$$

where F and V denote force and velocity, respectively. We now convert these transformer describing equations to the wave variable formulation. Let R_1 and R_2 denote the wave impedances on the port 1 and port 2 sides, respectively, and define velocity as positive into the transformer. Then

$$\begin{aligned} f_1^+(t) &= \frac{f_1(t) + R_1 v_1(t)}{2} \\ f_1^-(t) &= \frac{f_1(t) - R_1 v_1(t)}{2} = \frac{\frac{1}{g}f_2(t) + R_1 g v_2(t)}{2} \\ &= \frac{1}{g}\frac{f_2(t) + R_1 g^2 v_2(t)}{2}. \end{aligned}$$

Similarly,

$$\begin{aligned} f_2^+(t) &= \frac{f_2(t) + R_2 v_2(t)}{2} \\ f_2^-(t) &= \frac{f_2(t) - R_2 v_2(t)}{2} = \frac{g f_1(t) + R_2 \frac{1}{g} v_1(t)}{2} \\ &= g\frac{f_1(t) + R_2 \frac{1}{g^2} v_1(t)}{2}. \end{aligned}$$

We see that choosing

$$g^2 = \frac{R_2}{R_1}$$

eliminates the scattering terms and gives the simple relations

$$\begin{aligned} f_2^-(t) &= g f_1^+(t) \\ f_1^-(t) &= \frac{1}{g} f_2^+(t). \end{aligned}$$

The corresponding wave flow diagram is shown in Fig. C.37 b.

Thus, a transformer with a voltage gain g corresponds to simply changing the wave impedance from R_1 to R_2, where $g = \sqrt{R_2/R_1}$. Note that the transformer implements a change in wave impedance *without scattering* as occurs in physical impedance steps (§C.8, p. 560).

C.16.1 Gyrators

Another way to define the ideal waveguide transformer is to ask for a two-port element that joins two waveguide sections of differing wave impedance in such a way that signal power is preserved and no scattering occurs. From Ohm's Law for traveling waves (Eq. (6.6), p. 207), and from the definition of power waves (§C.7.5, p. 557), we see that to bridge an impedance discontinuity between R_{i-1} and R_i with no power change and no scattering requires the relations

$$\frac{[f_i^+]^2}{R_i} = \frac{[f_{i-1}^+]^2}{R_{i-1}}, \qquad \frac{[f_i^-]^2}{R_i} = \frac{[f_{i-1}^-]^2}{R_{i-1}}.$$

Therefore, the junction equations for a transformer can be chosen as

$$f_i^+ = g_i\, f_{i-1}^+ \qquad f_{i-1}^- = g_i^{-1}\, f_i^- \tag{C.126}$$

where

$$g_i \isdef \pm\sqrt{\frac{R_i}{R_{i-1}}} \tag{C.127}$$

Choosing the *negative* square root for g_i^{-1} gives a *gyrator* [35]. Gyrators are often used in electronic circuits to replace inductors with capacitors. The gyrator can be interpreted as a transformer in cascade with a *dualizer* [433]. A dualizer converts one from wave variable type (such as force) to the other (such as velocity) in the waveguide.

The dualizer is readily derived from Ohm's Law for traveling waves:

$$\begin{aligned} f^+ &= Rv^+, & f^- &= -Rv^- \\ \Longleftrightarrow\quad f_i^+ &= Rv_{i-1}^+, & v_{i-1}^- &= -R^{-1} f_i^- \end{aligned}$$

In this case, velocity waves in section $i-1$ are converted to force waves in section i, and vice versa (all at wave impedance R). The wave impedance can be changed as well by cascading a transformer with the dualizer, which changes R to $R\sqrt{R_i/R} = \sqrt{RR_i}$ (where we assume $R = R_{i-1}$). Finally, the velocity waves in section $i-1$ can be scaled to equal their corresponding force waves by introducing a transformer $g = \sqrt{1/R}$ on the left, which then coincides Eq. (C.126) (but with a minus sign in the second equation).

C.17 The Digital Waveguide Oscillator

In this section, adapted from [461], a digital sinusoidal oscillator derived from digital waveguide theory is described which has good properties for VLSI implementation. Its main features are *no wavetable* and a computational complexity of only *one multiply* per sample when amplitude and frequency are constant. Three additions are required per sample. A piecewise exponential amplitude envelope is available for the cost of a second multiplication per sample, which need not be as expensive as the tuning multiply. In the presence of frequency modulation (FM), the amplitude coefficient can be varied to exactly cancel amplitude modulation (AM) caused by changing the frequency of oscillation.

C.17.1 Additive Synthesis

One of the very first computer music techniques introduced was *additive synthesis* [379]. It is based on Fourier's theorem which states that any sound can be constructed from elementary sinusoids, such as are approximately produced by carefully struck tuning forks. Additive synthesis attempts to apply this theorem to the synthesis of sound by employing large banks of sinusoidal oscillators, each having independent amplitude and frequency controls. Many analysis methods, *e.g.*, the phase vocoder, have been developed to support additive synthesis. A summary is given in [424].

While additive synthesis is very powerful and general, it has been held back from widespread usage due to its computational expense. For example, on a single DSP56001 digital signal-processing chip, clocked at 33 MHz, only about 60 sinusoidal partials can be synthesized in real time using non-interpolated, table-lookup oscillators. Interpolated table-lookup oscillators are much more expensive, and when all the bells and whistles are added, and system overhead is accounted for, only around 12 fully general, high-quality partials are sustainable at 44.1 KHz on a $33MHz$ DSP56001 (based on analysis of implementations provided by the NeXT Music Kit).

At CD-quality sampling rates, the note A1 on the piano requires $22050/55 \approx 400$ sinusoidal partials, and at least the low-frequency partials should use interpolated lookups. Assuming a worst-case average of 100 partials per voice, providing 32-voice polyphony requires 3200 partials, or around 64 DSP chips, assuming we can pack an average of 50 partials into each DSP. A more reasonable complement of 8 DSP chips would provide only 4-voice polyphony which is simply not enough for a piano synthesis. However, since DSP chips are getting faster and cheaper, DSP-based additive synthesis looks viable in the future.

The cost of additive synthesis can be greatly reduced by making special purpose VLSI optimized for sinusoidal synthesis. In a VLSI environment, major bottlenecks are *wavetables* and *multiplications*. Even if a single sinusoidal wavetable is shared, it must be accessed sequentially, inhibiting parallelism. The wavetable can be eliminated entirely if *recursive algorithms* are used to synthesize sinusoids directly.

C.17.2 Digital Sinusoid Generators

In [168], three techniques were examined for generating sinusoids digitally by means of recursive algorithms.[13] The recursions can be interpreted as implementations of second-order digital resonators in which the damping is set to zero. The three methods considered were (1) the *2D rotation* (2DR), or *complex multiply* (also called the "coupled form"), (2) the *modified coupled form* (MCF), or "magic circle" algorithm,[14] which is similar to (1) but with better numerical behavior, and (3) the direct-form, second-order, digital resonator (DFR) with its poles set to the unit circle.

These three recursions may be defined as follows:

$$\begin{array}{llcll}
(1) & x_1(n) & = & c_n x_1(n-1) + s_n x_2(n-1) & \\
 & x_2(n) & = & -s_n x_1(n-1) + c_n x_2(n-1) & \text{(2DR)} \\
(2) & x_1(n) & = & x_1(n-1) + \epsilon x_2(n-1) & \\
 & x_2(n) & = & -\epsilon x_1(n) + x_2(n-1) & \text{(MCF)} \\
(3) & x_1(n) & = & 2c_n x_1(n-1) - x_2(n-1) & \\
 & x_2(n) & = & x_1(n-1) & \text{(DFR)}
\end{array}$$

---

[13] See [105] for discussion of a wider variety of digital sine generation methods.

[14] The "magic circle" algorithm is so named because it generates closed curves in the (x_1, x_2) plane even when numerical precision is very low. For this reason, it has long been used as an algorithm for drawing "circles" (actually ellipses) in computer graphics. The algorithm is derived naturally by numerically integrating the derivative relationships between sine and cosine:

$$\begin{array}{rclcl}
d\cos(x)/dx & = & -\sin(x) & \Rightarrow & d\cos(x) = -dx \cdot \sin(x) \\
d\sin(x)/dx & = & \cos(x) & \Rightarrow & d\sin(x) = dx \cdot \cos(x).
\end{array}$$

The magic circle algorithm is easily shown to be equivalent to the lowpass output of the infinite-Q second-order *digital state variable filter* in "Chamberlin form" (defined in, *e.g.*, [478, 79]).

where $c_n \triangleq \cos(2\pi f_n T)$, $s_n \triangleq \sin(2\pi f_n T)$, f_n is the instantaneous frequency of oscillation (Hz) at time sample n, and T is the sampling period in seconds. The magic circle parameter is $\epsilon = 2\sin(\pi f_n T)$.

The digital waveguide oscillator appears to have the best overall properties yet seen for VLSI implementation. This structure, introduced in [461], may be derived from the theory of digital waveguides (see Appendix C, particularly §C.9 on page 576, and [433, 465]). Any second-order digital filter structure can be used as a starting point for developing a corresponding sinusoidal signal generator, so in this case we begin with the second-order waveguide filter.

C.17.3 The Second-Order Waveguide Filter

The first step is to make a second-order digital filter with zero damping by abutting two unit-sample sections of waveguide medium, and terminating on the left and right with perfect reflections, as shown in Fig. C.38. The wave impedance in section i is given by $R_i = \rho c/A_i$, where ρ is air density, A_i is the cross-sectional area of tube section i, and c is sound speed. The reflection coefficient is determined by the impedance discontinuity via $k = (R_1 - R_2)/(R_1 + R_2)$. It turns out that to obtain sinusoidal oscillation, one of the terminations must provide an inverting reflection while the other is non-inverting.

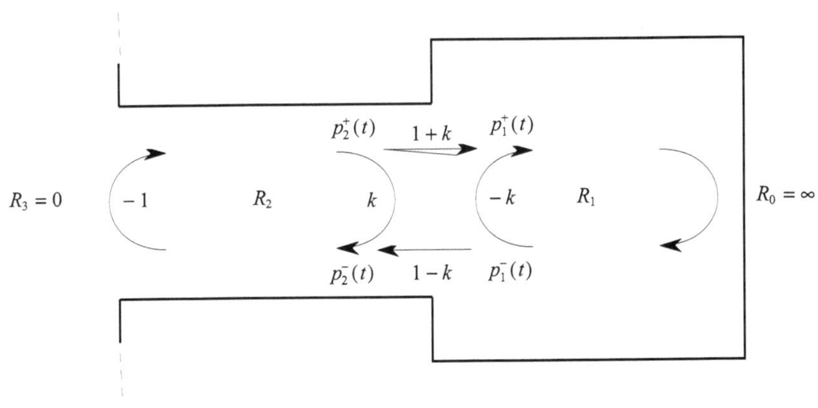

Figure C.38: The second-order, lossless, digital waveguide oscillator, built using two *acoustic tube* sections.

At the junction between sections 1 and 2, the signal is partially transmitted and partially reflected such that energy is conserved, *i.e.*, we have *lossless scattering*. The formula for the reflection coefficient k can be derived from the physical constraints that (1) pressure is continuous across the junction, and (2) there is no net flow into or out of the junction. For traveling pressure waves $p^{\pm}(t)$ and volume-velocity waves $u^{\pm}(t)$, we have $p^{+}(t) = Ru^{+}(t)$ and $p^{-}(t) = -Ru^{-}(t)$. The physical pressure and volume velocity are obtained by summing the traveling-wave components.

The discrete-time simulation for the physical system of Fig. C.38 is shown in Fig. C.39. The propagation time from the junction to a reflecting termination and back is one sample period. The half sample delay from the junction to the reflecting termination has been *commuted* with the termination and combined with the half sample delay *to* the termination. This is a special case of a "half-rate" waveguide filter [433].

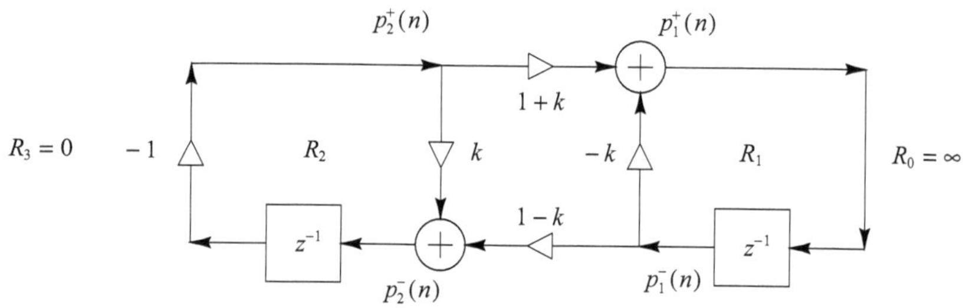

Figure C.39: The second-order, lossless, waveguide filter.

Since only two samples of delay are present, the digital system is at most second order, and since the coefficients are real, at most one frequency of oscillation is possible in $(0, \pi)$.

The scattering junction shown in the figure is called the Kelly-Lochbaum junction in the literature on lattice and ladder digital filters [173]. While it is the most natural from a physical point of view, it requires four multiplies and two additions for its implementation.

It is well known that lossless scattering junctions can be implemented in a variety of equivalent forms, such as the two-multiply and even one-multiply junctions. However, most have the disadvantage of not being *normalized* in the sense that changing the reflection coefficient k changes the amplitude of oscillation. This can be understood physically by noting that a change in k implies a change in R_2/R_1. Since the signal power contained in a waveguide variable, say $p_1^+(n)$, is $\left[p_1^+(n)\right]^2/R_1$, we find that modulating the reflection coefficient corresponds to modulating the signal energy represented by the signal sample in at least one of the two delay elements. Since energy is proportional to amplitude squared, energy modulation implies amplitude modulation.

The well-known normalization procedure is to replace the traveling pressure waves $p^{\pm}$ by "root-power" pressure waves $\tilde{p}^{\pm} = p^{\pm}/\sqrt{R}$ so that signal power is just the square of a signal sample $(\tilde{p}^{\pm})^2$. When this is done, the scattering junction transforms from the Kelly-Lochbaum or one-multiply form into the *normalized ladder* junction in which the reflection coefficients are again $\pm k$, but the forward and reverse transmission coefficients become $\sqrt{1-k^2}$. Defining $k = \sin(\theta)$, the transmission coefficients can be seen as $\cos(\theta)$, and we arrive essentially at the *coupled form*, or two-dimensional vector rotation considered in [168].

An alternative normalization technique is based on the digital waveguide *transformer* (§C.16, p. 616). The purpose of a "transformer" is to "step" the force variable (pressure in our example) by some factor g without scattering and without affecting signal energy. Since traveling signal power is proportional to pressure times velocity p^+u^+, it follows that velocity must be stepped by the inverse factor $1/g$ to keep power constant. This is the familiar behavior of transformers for analog electrical circuits: voltage is stepped up by the "turns ratio" and current is stepped down by the reciprocal factor. Now, since $p^+ = Ru^+$, traveling signal power is equal to $p^+u^+ = (p^+)^2/R$. Therefore, stepping up pressure through a transformer by the factor g corresponds to stepping up the wave impedance R by the factor g^2. In other words, the transformer raises pressure and decreases volume velocity by raising the wave impedance (narrowing the acoustic tube) like a converging cone.

If a transformer is inserted in a waveguide immediately to the left, say, of a scattering junction, it can be used to modulate the wave impedance "seen" to the left by the junction without having to use root-power waves in the simulation. As a result, the one-multiply junction can be used for the scattering junction, since the junction itself is not normalized. Since the transformer requires two multiplies, a total of three multiplies can effectively implement a normalized junction, where four were needed before. Finally, in just this special case, one of the transformer coefficients can be commuted with the delay element on the left and combined with the other transformer coefficient. For convenience, the -1 coefficient on the left is commuted into the junction so it merely toggles the signs of inputs to existing summers. These transformations lead to the final form shown in Fig. C.40.

The "tuning coefficient" is given by $C(n) = \cos(2\pi f_n T)$, where f_n is the desired oscillation frequency in Hz at sample n (in the undamped case), and T is the sampling period in seconds. The "amplitude coefficient" is $G(n) = r_n g_n/g_{n-1}$, where growth or decay factor per sample ($r_n \equiv 1$ for

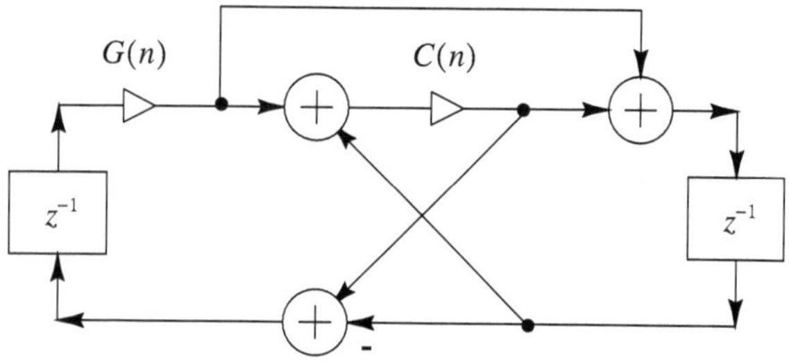

Figure C.40: The transformer-normalized, digital waveguide oscillator.

constant amplitude),[15] and g_n is the normalizing transformer "turns ratio" given by

$$g_n = \sqrt{\frac{1 - C(n)}{1 + C(n)}}.$$

When both amplitude and frequency are constant, we have $G(n) \equiv 1$, and only the tuning multiply is operational. When frequency changes, the amplitude coefficient deviates from unity for only one time sample to normalize the oscillation amplitude.

When amplitude and frequency are constant, there is no gradual exponential growth or decay due to round-off error. This happens because the only rounding is at the output of the tuning multiply, and all other computations are exact. Therefore, quantization in the tuning coefficient can only cause quantization in the frequency of oscillation. Note that any one-multiply digital oscillator should have this property. In contrast, the only other known normalized oscillator, the coupled form, *does* exhibit exponential amplitude drift because it has *two* coefficients $c = \cos(\theta)$ and $s = \sin(\theta)$ which, after quantization, no longer obey $c^2 + s^2 = 1$ for most tunings.

C.17.4 Application to FM Synthesis

The properties of the new oscillator appear well suited for FM applications in VLSI because of the minimized computational expense. However, in this application there are issues to be resolved regarding conversion from modulator output to carrier coefficients. Preliminary experiments indicate that FM indices less than 1 are well behaved when the output of a modulating oscillator simply adds to the coefficient of the carrier oscillator (bypassing

[15] When $r_n = r$ is constant, the time constant τ of the ensuing exponential growth or decay may be found by solving $r = e^{-T/\tau}$ for τ to obtain $\tau = -T/\ln(r)$, where T denotes the sampling interval, and $\ln(r)$ denotes the natural logarithm of r.

the exact FM formulas). Approximate amplitude normalizing coefficients have also been derived which provide a first-order approximation to the exact AM compensation at low cost. For music synthesis applications, we believe a distortion in the details of the FM instantaneous frequency trajectory and a moderate amount of incidental AM can be tolerated since they produce only second-order timbral effects in many situations.

C.17.5 Digital Waveguide Resonator

Converting a second-order oscillator into a second-order filter requires merely introducing damping and defining the input and output signals. In Fig. C.40, damping is provided by the coefficient $G(n)$, which we will take to be a constant

$$G(n) \equiv g.$$

When $g < 1$, the oscillator decays exponentially to zero from any initial conditions. The two delay elements constitute the *state* of the resonator. Let us denote by $x_1(n)$ the output of the delay element on the left in Fig. C.40 and let $x_2(n)$ be the delay-element output on the right. In general, an output signal $y(n)$ may be formed as any linear combination of the state variables:

$$y(n) = c_1 x_1(n) + c_2 x_2(n)$$

Similarly, input signals $u(n)$ may be summed into the state variables $x_i(n)$ scaled by arbitrary gain factors b_i.

The foregoing modifications to the digital waveguide oscillator result in the so-called *digital waveguide resonator* (DWR) [304]:

$$\begin{aligned} \tilde{x}_1(n) &= g x_1(n) && \text{(C.128)} \\ v(n) &= c[\tilde{x}_1(n) + x_2(n)] && \text{(C.129)} \\ x_1(n+1) &= v(n) - x_2(n) + \tilde{b}_1 u(n) && \text{(C.130)} \\ x_2(n+1) &= \tilde{x}_1(n) + v(n) + b_2 u(n) && \text{(C.131)} \\ \\ y(n) &= c_1 x_1(n) + c_2 x_2(n) && \text{(C.132)} \end{aligned}$$

where, as derived in the next section, the coefficients are given by

$$\begin{aligned} g &= r^2 && \text{(C.133)} \\ \tilde{b}_1 &= b_1 \sqrt{\frac{1-c}{1+c}} && \text{(C.134)} \\ c &= \sqrt{\frac{1}{1 + \frac{\tan^2(\theta)(1+g)^2 + (1-g)^2}{4g}}} && \text{(C.135)} \\ &\approx 1 - \frac{\tan^2(\theta)(1+g)^2 + (1-g)^2}{8g}. && \text{(C.136)} \end{aligned}$$

where $\lambda = re^{j\theta}$ denotes one desired pole (the other being at $\overline{\lambda}$). Note that $c = \cos(\theta)$ when $g = 1$ (undamped case). The DWR requires only two multiplies per sample. As seen earlier, when the decay time is set to ∞ ($g = 1$), one of the multiplies disappears, leaving only *one* multiply per sample for sinusoidal oscillation.

Figure C.41 shows an overlay of initial impulse responses for the three resonators discussed above. The decay factor was set to $r = 0.99$, and the output of each multiplication was quantized to 16 bits, as were all coefficients. The three waveforms sound and look identical. (There *are* small differences, however, which can be seen by plotting the differences of pairs of waveforms.)

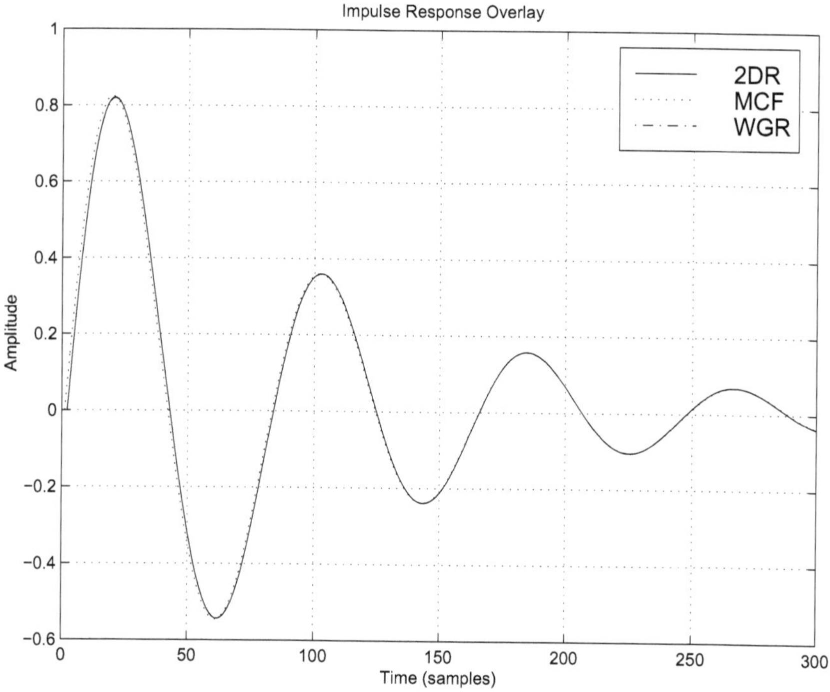

Figure C.41: Overlay of three resonator impulse responses, with $\mathbf{B}^T = [1, 0]$ and $\mathbf{C} = [0, 1]$, for the (1) complex-multiply resonator (labeled "2DR" for "2D rotation"), (2) modified coupled form (MCF), and (3) second-order digital waveguide resonator (DWR).

Figure C.42 shows the same impulse-response overlay but with $r = 1$ and only 4 significant bits in the coefficients and signals. The complex multiply oscillator can be seen to decay toward zero due to coefficient quantization ($x_1^2 + y_1^2 < 1$). The MCF and DWR remain steady at their initial amplitude. All three suffer some amount of tuning perturbation.

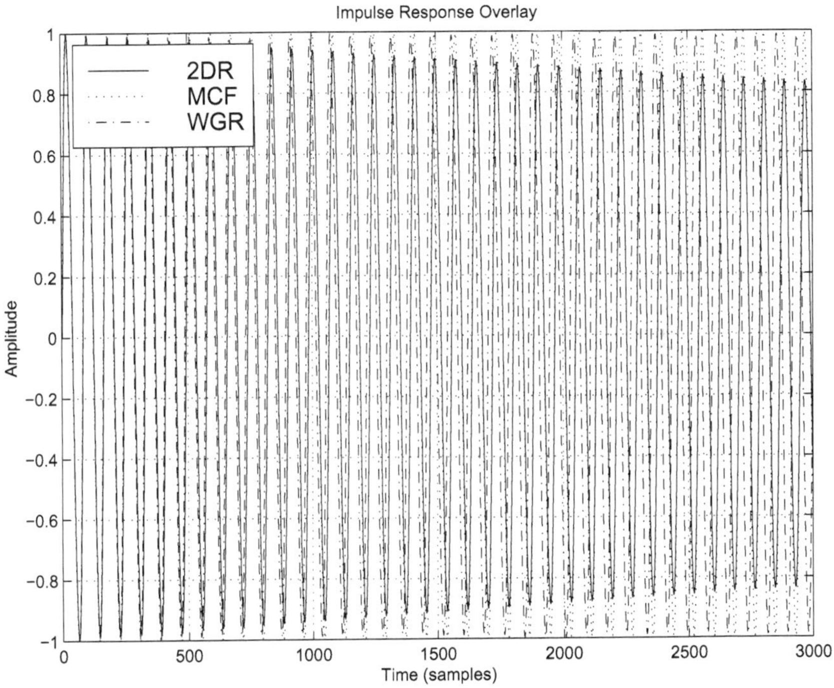

Figure C.42: Overlay of three resonator impulse responses, as in Fig. C.41, but with $r = 1$ and quantization of coefficients and signals to 4 significant bits.

C.17.6 State-Space Analysis

We will now use state-space analysis[16] [449] to determine Equations (C.133–C.136).

From Equations (C.128–C.132),

$$x_1(n+1) = c[gx_1(n) + x_2(n)] - x_2(n) = c\,gx_1(n) + (c-1)x_2(n)$$

and

$$x_2(n+1) = gx_1(n) + c[gx_1(n) + x_2(n)] = (1+c)gx_1(n) + c\,x_2(n)$$

[16] `http://ccrma.stanford.edu/~jos/filters/State_Space_Models.html`

In matrix form, the state time-update can be written

$$\left[\begin{array}{c} x_1(n+1) \\ x_2(n+1) \end{array}\right] = \underbrace{\left[\begin{array}{cc} gc & c-1 \\ g(c+1) & c \end{array}\right]}_{\mathbf{A}} \left[\begin{array}{c} x_1(n) \\ x_2(n) \end{array}\right] \tag{C.137}$$

or, in vector notation,

$$\begin{aligned} \underline{x}(n+1) &= \mathbf{A}\underline{x}(n) + \mathbf{B}u(n) & \text{(C.138)} \\ y(n) &= \mathbf{C}\underline{x}(n) & \text{(C.139)} \end{aligned}$$

where we have introduced an input signal $u(n)$, which sums into the state vector via the 2×1 (or $2 \times N_u$) vector $\mathbf{B}$. The output signal is defined as the 1×2 vector $\mathbf{C}$ times the state vector $\underline{x}(n)$. Multiple outputs may be defined by choosing $\mathbf{C}$ to be an $N_y \times 2$ matrix.

A basic fact from linear algebra is that the *determinant* of a matrix is equal to the *product of its eigenvalues*. As a quick check, we find that the determinant of A is

$$\det \mathbf{A} = c^2 g - g(c+1)(c-1) = c^2 g - g(c^2-1) = c^2 g - gc^2 + g = g. \tag{C.140}$$

When the eigenvalues λ_i of $\mathbf{A}$ (system poles) are complex, then they must form a complex conjugate pair (since $\mathbf{A}$ is real), and we have $\det \mathbf{A} = |\lambda_i|^2 = g$. Therefore, the system is *stable* if and only if $|g| < 1$. When making a digital sinusoidal oscillator from the system impulse response, we have $|g| = 1$, and the system can be said to be "marginally stable". Since an undriven sinusoidal oscillator must not lose energy, and since every lossless state-space system has unit-modulus eigenvalues (consider the modal representation), we expect $|\det A| = 1$, which occurs for $g = 1$.

Note that $\underline{x}(n) = \mathbf{A}^n \underline{x}(0)$. If we diagonalize this system to obtain $\tilde{\mathbf{A}} = \mathbf{E}^{-1}\mathbf{A}\mathbf{E}$, where $\tilde{\mathbf{A}} = \text{diag}[\lambda_1, \lambda_2]$, and $\mathbf{E}$ is the matrix of eigenvectors of $\mathbf{A}$, then we have

$$\underline{\tilde{x}}(n) = \tilde{A}^n \, \underline{\tilde{x}}(0) = \left[\begin{array}{cc} \lambda_1^n & 0 \\ 0 & \lambda_2^n \end{array}\right] \left[\begin{array}{c} \tilde{x}_1(0) \\ \tilde{x}_2(0) \end{array}\right]$$

where $\underline{\tilde{x}}(n) \triangleq \mathbf{E}^{-1}\underline{x}(n)$ denotes the state vector in these new "modal coordinates". Since $\tilde{A}$ is diagonal, the modes are decoupled, and we can write

$$\begin{aligned} \tilde{x}_1(n) &= \lambda_1^n \, \tilde{x}_1(0) \\ \tilde{x}_2(n) &= \lambda_2^n \, \tilde{x}_2(0) \end{aligned}$$

If this system is to generate a real sampled sinusoid at radian frequency ω, the eigenvalues λ_1 and λ_2 must be of the form

$$\begin{aligned} \lambda_1 &= e^{j\omega T} \\ \lambda_2 &= e^{-j\omega T}, \end{aligned}$$

(in either order) where ω is real, and T denotes the sampling interval in seconds.

Thus, we can determine the frequency of oscillation ω (and verify that the system actually oscillates) by determining the eigenvalues λ_i of A. Note that, as a prerequisite, it will also be necessary to find two linearly independent eigenvectors of A (columns of $\mathbf{E}$).

C.17.7 Eigenstructure

Starting with the defining equation for an eigenvector $\underline{e}$ and its corresponding eigenvalue λ,

$$\mathbf{A}\underline{e}_i = \lambda_i \underline{e}_i, \quad i = 1, 2$$

we get, using Eq. (C.137),

$$\begin{bmatrix} gc & c-1 \\ gc+g & c \end{bmatrix} \begin{bmatrix} 1 \\ \eta_i \end{bmatrix} = \begin{bmatrix} \lambda_i \\ \lambda_i \eta_i \end{bmatrix}. \tag{C.141}$$

We normalized the first element of $\underline{e}_i$ to 1 since $g\underline{e}_i$ is an eigenvector whenever $\underline{e}_i$ is. (If there is a missing solution because its first element happens to be zero, we can repeat the analysis normalizing the second element to 1 instead.)

Equation (C.141) gives us two equations in two unknowns:

$$gc + \eta_i(c-1) = \lambda_i \tag{C.142}$$

$$g(1+c) + c\eta_i = \lambda_i \eta_i \tag{C.143}$$

Substituting the first into the second to eliminate λ_i, we get

$$\begin{aligned} g + gc + c\eta_i &= [gc + \eta_i(c-1)]\eta_i = gc\eta_i + \eta_i^2(c-1) \\ \Rightarrow \eta_i &= \frac{c(g-1)}{2(1-c)} \pm \sqrt{\frac{c^2(1-g)^2 - 4g(1-c^2)}{(c-c)^2}} \\ &= \frac{(g-1)c}{2(1-c)} \pm j\sqrt{g\left(\frac{1+c}{1-c}\right) - \frac{c^2(1-g)^2}{4(1-c)^2}}. \end{aligned}$$

As g approaches 1 (no damping), we obtain

$$\eta_i = \pm j\sqrt{\frac{1+c}{1-c}} \qquad \text{(when } g = 1\text{)}.$$

Thus, we have found both eigenvectors:

$$\begin{aligned} \underline{e}_1 &= \begin{bmatrix} 1 \\ \eta \end{bmatrix}, \qquad \underline{e}_2 = \begin{bmatrix} 1 \\ -\eta \end{bmatrix}, \quad \text{where} \\ \eta &\stackrel{\Delta}{=} \frac{(g-1)c}{2(1-c)} + j\sqrt{g\left(\frac{1+c}{1-c}\right) - \frac{c^2(1-g)^2}{4(1-c)^2}} \end{aligned}$$

They are linearly independent provided $\eta \neq 0$. In the undamped case ($g = 1$), this holds whenever $c \neq -1$. The eigenvectors are finite when $c \neq 1$. Thus, the nominal range for c is the interval $c \in (-1, 1)$.

We can now use Eq. (C.142) to find the eigenvalues:

$$\begin{aligned}\lambda_i &= gc + \eta_i(c-1) \\ &= gc + \frac{(1-g)c}{2} \pm \sqrt{\left(\frac{c(1-g)}{2}\right)^2 - g(1-c^2)} \\ &= c\left(\frac{1+g}{2}\right) \pm j\sqrt{g(1-c^2) - \left[\frac{c(1-g)}{2}\right]^2}\end{aligned}$$

Damping and Tuning Parameters

The tuning and damping of the resonator impulse response are governed by the relation

$$\lambda_i = e^{\frac{T}{\tau}} e^{\pm j\omega T}$$

where T denotes the sampling interval, τ is the time constant of decay, and ω is the frequency of oscillation in radians per second. The eigenvalues are presumed to be complex, which requires, from Eq. (C.144),

$$g(1-c^2) \geq \frac{c^2(1-g)^2}{4} \;\Rightarrow\; c^2 \leq \frac{4g}{(1+g)^2}$$

To obtain a specific decay time-constant τ, we must have

$$\begin{aligned}e^{-2T/\tau} &= |\lambda_i|^2 = c^2\left(\frac{1+g}{2}\right)^2 + \left[g(1-c^2) - c^2\left(\frac{1-g}{2}\right)^2\right] \\ &= g\end{aligned}$$

Therefore, given a desired decay time-constant τ (and the sampling interval T), we may compute the damping parameter g for the digital waveguide resonator as

$$\boxed{g = e^{-2T/\tau}.}$$

Note that this conclusion follows directly from the determinant analysis of Eq. (C.140), provided it is known that the poles form a complex-conjugate pair.

To obtain a desired frequency of oscillation, we must solve

$$\begin{aligned}\theta = \omega T &= \tan^{-1}\left[\frac{\sqrt{g(1-c^2) - [c(1-g)/2]^2}}{c(1+g)/2}\right] \\ \Rightarrow \tan^2\theta &= \frac{g(1-c^2) - [c(1-g)/2]^2}{[c(1+g)/2]^2}\end{aligned}$$

for c, which yields

$$\boxed{c = \sqrt{\frac{1}{1+\frac{\tan^2(\theta)(1+g)^2+(1-g)^2}{4g}}} \approx 1 - \frac{\tan^2(\theta)(1+g)^2+(1-g)^2}{8g}}.$$

Note that this reduces to $c = \cos(\theta)$ when $g = 1$ (undamped case).

Eigenvalues in the Undamped Case

When $g = 1$, the eigenvalues reduce to

$$\lambda_i = c \pm j\sqrt{1-c^2}$$

Assuming $|c| < 1$, the eigenvalues can be expressed as

$$\lambda_i = c \pm j\sqrt{1-c^2} = \cos(\theta) \pm j\sin(\theta) = e^{\pm j\theta} \tag{C.144}$$

where $\theta = \omega T$ denotes the angular advance per sample of the oscillator. Since $c \in (-1, 1)$ corresponds to the range $\theta \in (-\pi, \pi)$, we see that c in this range can produce oscillation at any digital frequency.

For $|c| > 1$, the eigenvalues are real, corresponding to exponential growth and/or decay. (The values $c = \pm 1$ were excluded above in deriving Eq. (C.144).)

In summary, the coefficient c in the digital waveguide oscillator ($g = 1$) and the frequency of sinusoidal oscillation ω is simply

$$\boxed{c = \cos(\omega T)}.$$

Summary

This section introduced and analyzed the digital waveguide oscillator and resonator, as well as some related algorithms. As a recursive algorithm for digital sinusoid generation, it has excellent properties for VLSI implementation. It is like the 2D rotation (complex multiply) in that it offers instantaneous amplitude from its state and constant amplitude in the presence of frequency modulation. However, its implementation requires only two multiplies per sample instead of four. When used as a constant-frequency oscillator, it requires only one multiply per sample.

C.17.8 Matlab Sinusoidal Oscillator Implementations

This section provides Matlab/Octave program listings for the sinusoidal resonator/oscillator algorithms discussed above:

- Planar 2D Rotation (2DR) ("complex multiply")
- Modified Coupled Form (MCF) ("Magic Circle")
- Digital Waveguide Resonator (DWR)

The test program computes an impulse response of each resonator, and plots them overlaid for comparison. Next, it optionally plays each one as a sound and plots them individually.

```
% Filter test program in matlab

%N = 300;    % Number of samples to generate
N = 3000;    % Number of samples to generate
f = 100;    % Desired oscillation frequency (Hz)
fs = 8192; % Audio sampling rate (Hz)
%R = .99;    % Decay factor (1=never, 0=instant)
R = 1;   % Decay factor (1=never, 0=instant)
b1 = 1;     % Input gain to state variable x(n)
b2 = 0;     % Input gain to state variable y(n)

%nd = 16;    % Number of significant digits to use
nd = 4;    % Number of significant digits to use
base = 2;  % Mantissa base (two significant figures)
           % (see 'help chop' in Matlab)

u = [1,zeros(1,N-1)]; % driving input test signal
theta = 2*pi*f/fs; % resonance frequency, rad/sample

% ================================================
% 2D PLANAR ROTATION (COMPLEX MULTIPLY)

x1 = zeros(1,N); % Predeclare saved-state arrays
y1 = zeros(1,N);

x1(1) = 0;      % Initial condition
y1(1) = 0;      % Initial condition

c = chop(R*cos(theta),nd,base); % coefficient 1
s = chop(R*sin(theta),nd,base); % coefficient 2

for n=1:N-1,
  x1(n+1) = chop(   c*x1(n) - s*y1(n) + b1*u(n), nd,base);
  y1(n+1) = chop(   s*x1(n) + c*y1(n) + b2*u(n), nd,base);
end
```

```
% ================================================
% MODIFIED COUPLED FORM ("MAGIC CIRCLE")
% (ref: http://ccrma.stanford.edu/~jos/wgo/ )

x2 = zeros(1,N); % Predeclare saved-state arrays
y2 = zeros(1,N);

x2(1) = 0.0;      % Initial condition
y2(1) = 0.0;      % Initial condition

e = chop(2*sin(theta/2),nd,base); % tuning coefficient

for n=1:N-1,
  x2(n+1) = chop(R*(x2(n)-e*y2(n))+b1*u(n),nd,base);
  y2(n+1) = chop(R*(e*x2(n+1)+y2(n))+b2*u(n),nd,base);
end

% ================================================
% DIGITAL WAVEGUIDE RESONATOR (DWR)

x3 = zeros(1,N); % Predeclare saved-state arrays
y3 = zeros(1,N);

x3(1) = 0;       % Initial condition
y3(1) = 0;       % Initial condition

g = R*R;         % decay coefficient
t = tan(theta); % toward tuning coefficient
cp = sqrt(g/(g + t^2*(1+g)^2/4 + (1-g)^2/4)); % exact
%cp = 1 - (t^2*(1+g)^2 + (1-g)^2)/(8*g); % good approx
b1n = b1*sqrt((1-cp)/(1+cp)); % input scaling

% Quantize coefficients:
cp = chop(cp,nd,base);
g = chop(g,nd,base);
b1n = chop(b1n,nd,base);

for n=1:N-1,
  gx3 = chop(g*x3(n), nd,base); % mpy 1
  y3n = y3(n);
  temp = chop(cp*(gx3 + y3n), nd,base); % mpy 2
```

```
  x3(n+1) = temp - y3n + b1n*u(n);
  y3(n+1) = gx3 + temp + b2*u(n);
end

% ================================================
% playandplot.m

figure(4);
title('Impulse Response Overlay');
ylabel('Amplitude');
xlabel('Time (samples)');
alt=1;%alt = (-1).^(0:N-1); % for visibility
plot([y1',y2',(alt.*y3)']); grid('on');
legend('2DR','MCF','WGR');
title('Impulse Response Overlay');
ylabel('Amplitude');
xlabel('Time (samples)');

saveplot(sprintf('eps/tosc%d.ps',nd));

if 1
  playandplot(y1,u,fs,'2D rotation',1);
  playandplot(y2,u,fs,'Magic circle',2);
  playandplot(y3,u,fs,'WGR',3);
end

function playandplot(y,u,fs,ttl,fnum);
% sound(y,fs,16);
  figure(fnum);

  unwind_protect # protect graph state
  subplot(211);
  axis("labely");
  axis("autoy");
  xlabel("");
  title(ttl);
  ylabel('Amplitude');
  xlabel('Time (ms)');
  t = 1000*(0:length(u)-1)/fs;
  timeplot(t,u,'-');
  legend('input');
  subplot(212);
```

```
    ylabel('Amplitude');
    xlabel('Time (ms)');
    timeplot(t,y,'-');
    legend('output');
    unwind_protect_cleanup # restore graph state
      grid("off");
      axis("auto","label");
      oneplot();
    end_unwind_protect
  endfunction
```

Note that the chop utility only exists in Matlab. In Octave, the following "compatibility stub" can be used:

```
function [y] = chop(x,nd,base)
y = x;
```

C.17.9 Faust Implementation

The function oscw (file osc.lib) implements a digital waveguide sinusoidal oscillator in the Faust language [453, 154, 170]. There is also oscr implementing the 2D rotation case, and oscs implementing the modified coupled form (magic circle).

C.18 Non-Cylindrical Acoustic Tubes

In many situations, the wave impedance of the medium varies in a continuous manner rather than in discrete steps. This occurs, for example, in conical bores and flaring horns. In this section, based on [436], we will consider *non-cylindrical acoustic tubes.*

C.18.1 Horns as Waveguides

Waves in a horn can be analyzed as *one-parameter waves*, meaning that it is assumed that a constant-phase wavefront progresses uniformly along the horn. Each "surface of constant phase" composing the traveling wave has tangent planes normal to the horn axis and to the horn boundary. For cylindrical tubes, the surfaces of constant phase are planar, while for conical tubes, they are spherical [357, 317, 144]. The key property of a "horn" is that a traveling wave can propagate from one end to the other with negligible "backscattering" of the wave. Rather, it is smoothly "guided" from one end to the other. This is the meaning of saying that a horn is a "waveguide". The absence of backscattering means that the entire propagation path may be simulated using a pure delay line, which is very efficient computationally. Any losses, dispersion, or amplitude change due to horn radius variation ("spreading loss") can be implemented where the wave exits the delay line to interact with other components.

Overview of Methods

We will first consider the elementary case of a *conical* acoustic tube. All smooth horns reduce to the conical case over sufficiently short distances, and the use of many conical sections, connected via scattering junctions, is often used to model an arbitrary bore shape [71]. The conical case is also important because it is essentially the most general shape in which there are exact traveling-wave solutions (spherical waves) [357].

Beyond conical bore shapes, however, one can use a *Sturm-Liouville formulation* to solve for one-parameter-wave scattering parameters [50]. In this formulation, the *curvature* of the bore's cross-section (more precisely, the curvature of the one-parameter wave's constant-phase surface area) is treated as a *potential* function that varies along the horn axis, and the solution is an *eigenfunction* of this potential. Sturm-Liouville analysis is well known in *quantum mechanics* for solving *elastic scattering* problems and for finding the wave functions (at various energy levels) for an electron in a nonuniform potential well.

When the one-parameter-wave assumption breaks down, and multiple acoustic modes are excited, the *boundary element method* (BEM) is an effective tool [190]. The BEM computes the acoustic field from velocity data along any enclosing surface. There also exist numerical methods for simulating wave propagation in varying cross-sections that include "mode conversion" [336, 13, 117].

This section will be henceforth concerned with non-cylindrical tubes in which backscatter and mode-conversion can be neglected, as treated in [317]. The only exact case is the cone, but smoothly varying horn shapes can be modeled approximately in this way.

Back to the Cone

Note that the cylindrical tube is a limiting case of a cone with its apex at infinity. Correspondingly, a plane wave is a limiting case of a spherical wave having infinite radius.

On a fundamental level, all pressure waves in 3D space are composed of spherical waves [357]. You may have learned about the *Huygens-Fresnel principle* in a physics class when it covered waves [295]. The Huygens-Fresnel principle states that the propagation of any wavefront can be modeled as the superposition of spherical waves emanating from all points along the wavefront [122, page 344]. This principle is especially valuable for intuitively understanding *diffraction* and related phenomena such as *mode conversion* (which happens, for example, when a plane wave in a horn hits a sharp bend or obstruction and breaks up into other kinds of waves in the horn).

C.18.2 Conical Acoustic Tubes

The *conical acoustic tube* is a one-dimensional waveguide which propagates circular sections of *spherical pressure waves* in place of the plane wave which traverses a cylindrical acoustic tube [22, 349]. The wave equation in the spherically symmetric case is given by

$$c^2 p_x'' = \ddot{p}_x \tag{C.145}$$

where

$$\begin{aligned} c &\triangleq \text{sound speed} & p_x &\triangleq x\,p(t,x) \\ \dot{p}_x &\triangleq \frac{\partial}{\partial t}p_x(t,x) & p_x' &\triangleq \frac{\partial}{\partial x}p_x(t,x) \end{aligned}$$

and $p(t,x)$ is the *pressure* at time t and radial position x along the cone axis (or wall). In terms of p (rather than $p_x = xp$), Eq. (C.145) expands to *Webster's horn equation* [357]:

$$\frac{1}{c^2}\ddot{p} \;=\; \frac{1}{A}\left(Ap'\right)' \;=\; p'' + \frac{A'}{A}p'$$

where $A = \alpha x^2$ is the area of the spherical wavefront, so that $A'/A = 2/x$ (as discussed further in §C.18.4 below).

Spherical coordinates are appropriate because simple closed-form solutions to the wave equation are only possible when the forced boundary conditions lie along coordinate planes. In the case of a cone, the boundary conditions lie along a conical section of a sphere. It can be seen that the wave equation in a cone is identical to the wave equation in a cylinder, except that p is replaced by xp. Thus, the solution is a superposition of left- and right-going traveling wave components, scaled by $1/x$:

$$p(t,x) = \frac{f\left(t-\frac{x}{c}\right)}{x} + \frac{g\left(t+\frac{x}{c}\right)}{x} \tag{C.146}$$

where $f(\cdot)$ and $g(\cdot)$ are arbitrary twice-differentiable continuous functions that are made specific by the boundary conditions. A function of $(t - x/c)$ may be interpreted as a fixed waveshape traveling to the *right*, (*i.e.*, in the *positive* x direction), with speed c. Similarly, a function of $(t + x/c)$ may be seen as a wave traveling to the *left* (negative x direction) at c meters per second. The point $x = 0$ corresponds to the tip of the cone (center of the sphere), and $p(t, x)$ may be singular there.

In cylindrical tubes, the velocity wave is in phase with the pressure wave. This is not the case with conical or more general tubes. The velocity of a traveling may be computed from the corresponding traveling pressure wave by dividing by the wave impedance.

Digital Simulation

A discrete-time simulation of the above solution may be obtained by simply *sampling* the traveling-wave amplitude at intervals of T seconds, which implies a *spatial* sampling interval of $X \triangleq cT$ meters. Sampling is carried out mathematically by the change of variables

$$\begin{array}{rclcl} x & \to & x_m & = & x_0 + mX \\ t & \to & t_n & = & nT \end{array}$$

and this substitution in Eq. (C.146) gives

$$\begin{aligned} x_m p(t_n, x_m) &= f(t_n - x_m/c) + g(t_n + x_m/c) \\ &= f(nT - mX/c - x_0/c) + g(nT + mX/c + x_0/c) \\ &= f\left[(n-m)T - x_0/c\right] + g\left[(n+m)T + x_0/c\right]. \end{aligned}$$

Define

$$p^+(n) \triangleq f(nT - x_0/c) \qquad\qquad p^-(n) \triangleq g(nT + x_0/c)$$

where x_0 is arbitrarily chosen as the position along the cone closest to the tip. (We avoid a sample at the tip itself, where a pressure singularity may exist.) Then a section of the ideal cone can be simulated as shown in Figure C.43 (where pressure outputs are shown for $x = x_0$ and $x = x_0 + 3X$). A particle-velocity output is formed by dividing the traveling pressure waves by the wave impedance of air. Since the wave impedance is now a function of frequency and propagation direction (as derived below), a digital filter will replace what was a real number for cylindrical tubes.

A more compact simulation diagram which stands for either sampled or continuous simulation is shown in Figure C.44. The figure emphasizes that the ideal, lossless waveguide is simulated by a *bidirectional delay line.*

As in the case of uniform waveguides, the digital simulation of the traveling-wave solution to the lossless wave equation in spherical coordinates is exact at the sampling instants, to within numerical precision, provided that the traveling waveshapes are initially *bandlimited* to less than

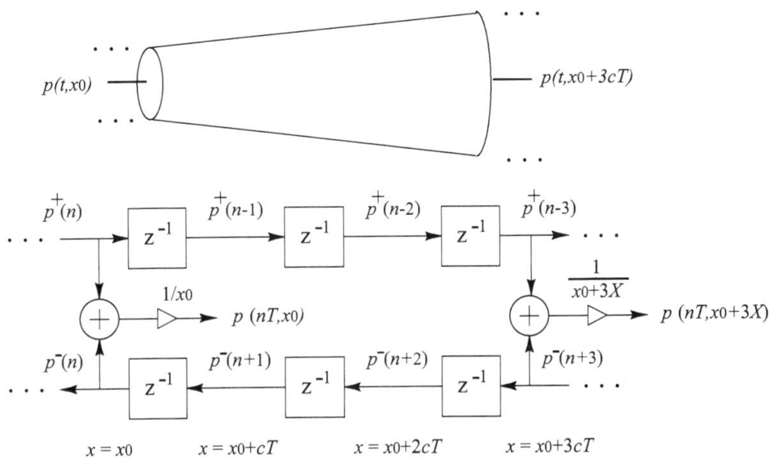

Figure C.43: Digital simulation of the ideal, lossless, conical waveguide with observation points at $x = x_0$ and $x = x_0 + 3X = x_0 + 3cT$. The symbol "z^{-1}" denotes a one-sample delay.

half the sampling frequency. Also as before, bandlimited interpolation can be used to provide time samples or position samples at points off the simulation grid. Extensions to include losses, such as air absorption and thermal conduction, or dispersion, can be carried out as described in §2.3 on page 45 and §C.5 on page 538 for plane-wave propagation (through a uniform wave impedance).

The simulation of Fig. C.44 suffices to simulate an isolated conical frustum, but what if we wish to interconnect two or more conical bores? Even more importantly, what driving-point impedance does a mouthpiece "see" when attached to the narrow end of a conical bore? The preceding only considered *pressure-wave* behavior. We must now also find the *velocity wave*, and form their ratio to obtain the driving-point impedance of a conical tube.

C.18.3 Momentum Conservation in Nonuniform Tubes

Newton's second law "force equals mass times acceleration" implies that the pressure gradient in a gas is proportional to the acceleration of a differential volume element in the gas. Let $A(x)$ denote the area of the surface of constant phase at radial coordinate x in the tube. Then the total force

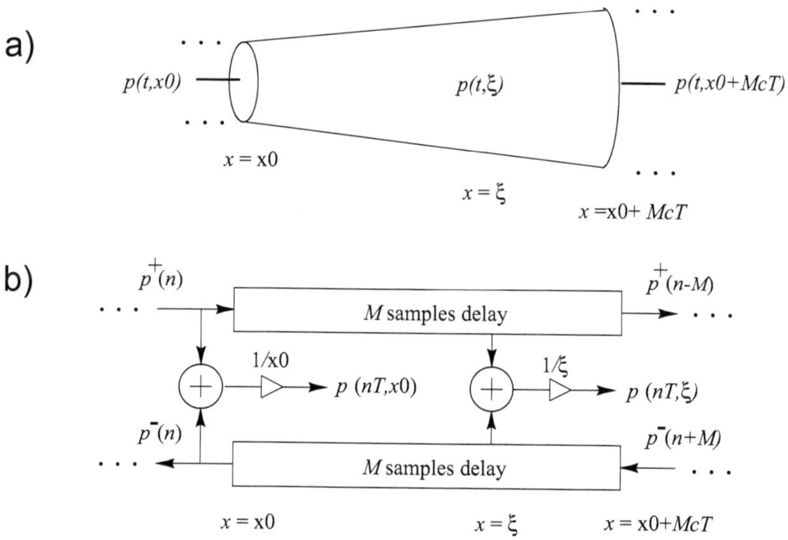

Figure C.44: Simplified picture of ideal waveguide simulation.

acting on the surface due to pressure is $f(t,x) = A(x)p(t,x)$, as shown in Fig. C.45.

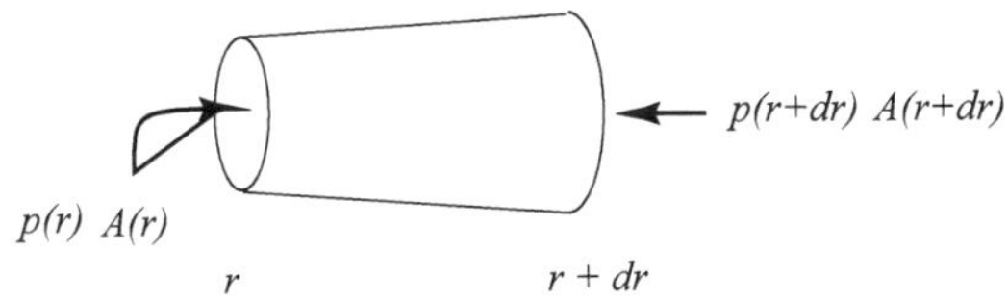

Figure C.45: Differential volume element for the conical acoustic tube.

The net force $df(t,x)$ *to the right* across the volume element between x and $x+dx$ is then

$$\begin{aligned} df(t,x) = f(t,x) - f(t,x+dx) &= f(t,x) - [f(t,x) + dx \cdot f'(x)] \\ &= -dx \cdot f'(x) \\ &= -dx \cdot [A(x)p(t,x)]' \\ &= -dx \cdot [A'p + A'p], \end{aligned}$$

where, when time and/or position arguments have been dropped, as in the last line above, they are all understood to be t and x, respectively. To apply Newton's second law equating net force to mass times acceleration, we need the mass of the volume element

$$\begin{aligned} dM(t,x) &= \int_x^{x+dx} \rho(t,\xi)A(\xi)\,d\xi \\ &\approx \int_x^{x+dx} [\rho(t,x) + \rho'(t,x)(\xi - x)][A(x) + A'(x)(\xi - x)]\,d\xi \\ &\approx \rho(t,x)A(x)\,dx + \int_x^{x+dx} [\rho'(t,x)A(x) + \rho(t,x)A'(x)]\,(\xi - x)\,d\xi \\ &= \rho\,A\,dx + (\rho' A + \rho A')\,\frac{(dx)^2}{2} \\ &\approx \rho\,A\,dx, \end{aligned}$$

where $\rho(t,x)$ denotes air density.

The center-of-mass acceleration of the volume element can be written as $\dot{u}(t,x)$ where $u(t,x)$ is particle velocity.[17] Applying Newton's second law

[17] More precisely, it could be expressed as $[\dot{u}(t,x) + \ddot{u}(t,x)\,dx]/2$, but this introduces a term that is second-order in dx that would ultimately be dropped.

$df = dM \cdot \dot{u}$, we obtain

$$-dx \cdot (A'p + Ap') = \rho\, A\, dx\, \dot{u} \tag{C.147}$$

or, dividing through by $-A\,dx$,

$$p' + p\,\frac{A'}{A} = -\rho\,\dot{u}. \tag{C.148}$$

In terms of the logarithmic derivative of A, this can be written

$$p' + p\,\ln' A = -\rho\,\dot{u}. \tag{C.149}$$

Note that p denotes *small-signal acoustic pressure*, while ρ denotes the full gas density (not just an acoustic perturbation in the density). We may therefore treat ρ as a constant.

Cylindrical Tubes

In the case of cylindrical tubes, the logarithmic derivative of the area variation, $\ln' A(x) = A'/A$, is zero, and Eq. (C.148) reduces to the usual momentum conservation equation $p' = -\rho\dot{u}$ encountered when deriving the wave equation for plane waves [318, 349, 47]. The present case reduces to the cylindrical case when

$$\frac{A'}{A} \ll \frac{p'}{p}$$

i.e., when the relative change in cross-sectional area is much less than the relative change in pressure along the tube. In other words, the tube area variation must be slower than the spatial variation of the wave itself. This assumption is also necessary for the "one-parameter-wave" approximation to hold in the first place.

If we look at sinusoidal spatial waves, $p = A_p e^{jk_p x}$ and $A = A_A e^{jk_A x}$, then $A'/A = k_A$ and $p'/p = k_p$, and the condition for cylindrical-wave behavior becomes $k_A \ll k_p$, *i.e.*, the spatial frequency of the wall variation must be much less than that of the wave. Another way to say this is that the wall must be approximately flat across a wavelength. This is true for smooth horns/bores at sufficiently high wave frequencies.

C.18.4 Wave Impedance in a Cone

From Eq. (C.146) we have that the traveling-wave solution of the wave equation in spherical coordinates can be expressed as

$$p(t,x) = \frac{f\left(t \mp \frac{x}{c}\right)}{x}$$

where the minus sign is associated with an expanding spherical wave, and the plus sign corresponds to a converging wave. The spatial derivative of this function is

$$p' = \frac{f'}{x} - \frac{f}{x^2} = \mp\frac{\dot{f}}{cx} - \frac{f}{x^2} = \mp\frac{\dot{p}}{c} - \frac{p}{x} \tag{C.150}$$

i.e., it can be expressed in terms of its own time derivative. This is a general property of any traveling wave.

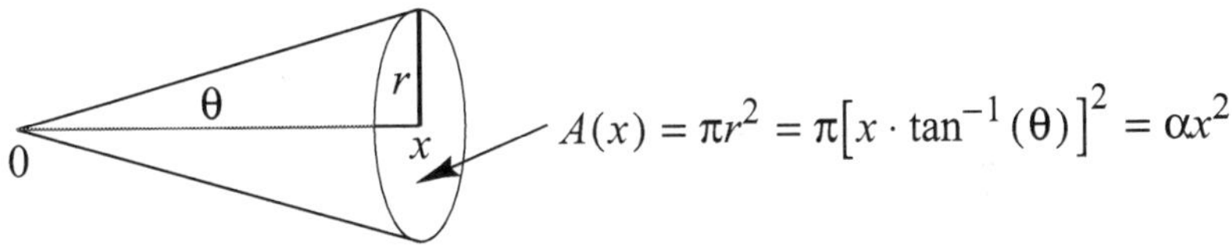

Figure C.46: Cone parameters.

Referring to Fig. C.46, the area function $A(x)$ can be written for any cone in terms of the distance from its apex as

$$A(x) = \alpha x^2$$

for some α. (We chose $x = 0$ at the tip of the cone in arriving at the basic traveling-wave solution to the wave equation.) This formula holds for both the planar cross-section indicated in Fig. C.46 and the spherical section having the same perimeter (*i.e.*, spherical surface area is proportional to radius squared). The logarithmic derivative of $A(x) = \alpha x^2$ is then

$$\ln' A = \frac{A'}{A} = \frac{2}{x},$$

which is independent of the conical taper angle $\theta = \tan\left(\sqrt{\alpha/\pi}\right)$. That is, one conical section of a spherical wave is like any other, as it must be due to spherical symmetry.

Substituting the logarithmic derivative of $A(x)$ and p' from Eq. (C.150) into the momentum-conservation equation Eq. (C.148) yields

$$\mp\frac{\dot{p}}{c} - \frac{p}{x} + p\frac{2}{x} = -\rho\dot{u}$$

or

$$\mp\frac{\dot{p}}{c} + \frac{p}{x} = -\rho\dot{u},$$

where the minus sign is for an expanding spherical wave, and the plus sign is for a converging spherical wave. Taking the Laplace transform of the above expression gives

$$\mp\frac{sP(s)}{c} + \frac{P(s)}{x} = -\rho s U(s)$$

in the case of zero initial conditions $p(0,x) = u(0,x) = 0$ for all x.

We can now solve for the *wave impedance* in each direction, where the wave impedance may be defined (§7.1) as the Laplace transform of the traveling pressure divided by the Laplace transform of the corresponding traveling velocity wave:

$$R^+(s) \triangleq \frac{P^+(s)}{U^+(s)}$$

$$R^-(s) \triangleq \frac{P^-(s)}{U^-(s)}$$

We introduce the shorthand

$$P^\pm(s) = \pm R^\pm(s)U^\pm(s)$$

where all of the upper signs or all of the lower signs are taken together, corresponding to expanding or converging spherical waves, respectively. In this notation, we may solve for the conical wave impedance as

$$R^\pm(s) \triangleq \frac{P^\pm(s)}{\pm U^\pm(s)} = \frac{\rho c}{1 \mp \frac{c}{sx}}.$$

Along the frequency axis $s = j\omega$ we get

$$R^\pm(j\omega) = \frac{\rho c}{1 \pm j\frac{c}{\omega x}} = \frac{\rho c}{1 \pm j\frac{1}{kx}}$$

where $\omega = 2\pi f$ is radian temporal frequency, and $k \triangleq \omega/c$ is the radian spatial frequency, or wavenumber.

Note that for a cylindrical tube, the wave impedance in both directions is $R^\pm(s) = \rho c$, and there is no frequency dependence. A wavelength or more away from the conical tip, *i.e.*, for $x \gg 1/k = \lambda/2\pi$, where λ is the spatial wavelength, the wave impedance again approaches that of a cylindrical bore. However, in conical musical instruments, the fundamental wavelength is typically twice the bore length, so the complex nature of the wave impedance is important throughout the bore and approaches being purely imaginary near the mouthpiece. This is especially relevant to conical-bore double-reeds, such as the bassoon.

Writing the wave impedance as

$$R^{\pm}(s) \stackrel{\Delta}{=} \frac{P^{\pm}(s)}{\pm U^{\pm}(s)} = \frac{1}{\frac{1}{\rho c} \mp \frac{1}{s\rho x}},$$

the reader familiar with circuit theory (§7.2.3) will recognize this as the parallel combination of the wave impedance ρc of a cylindrical bore and the lumped impedance of a mass formally equal to $\mp\rho x$. This equivalent circuit is explored in [37]. As x goes to 0 at the tip, the mass goes to zero and "shorts out" the traveling wave. Note that in the case of a converging cone, the mass is negative. This will give rise to an unstable one-pole filter in the waveguide model, and various solutions have been proposed for this problem [529, 50].

Up to now, we have been defining wave impedance as pressure divided by particle velocity. In acoustic tubes, *volume velocity* is what is conserved at a junction between two different acoustic tube types. Therefore, in acoustic tubes, we define the wave impedance as the ratio of pressure to *volume velocity*

$$R_A^{\pm}(s) \stackrel{\Delta}{=} \frac{P^{\pm}(s)}{\pm U^{\pm}(s)A(x)} = \frac{R^{\pm}(s)}{A(x)} = \frac{\rho c}{A(x)} \frac{1}{1 \pm \frac{c}{sx}}$$

or, defining $t_x \stackrel{\Delta}{=} x/c$ as the time to propagate over the distance x,

$$\boxed{R_A^{\pm}(s) = \frac{\rho c}{A(x)} \frac{s}{s \pm \frac{1}{t_x}}.} \tag{C.151}$$

This is the wave impedance we use to compute the generalized reflection and transmission coefficients at a change in cross-sectional area and/or taper angle in a conical acoustic tube. Note that it has a zero at $s = 0$ and a pole at $s = \pm 1/t_x$.

In this case, the equivalent mass is $\pm\rho x/A(x) = \pm\rho/(\alpha x)$. It would perhaps be more satisfying if the equivalent mass in the conical wave impedance were instead $\pm\rho x A(x)$ which is the mass of air contained in a cylinder of radius $A(x)$ projected back to the tip of the cone. However, the "acoustic mass" cannot be physically equivalent to mechanical mass. To see this, consider that the impedance of a mechanical mass is ms which is in physical units of mass per unit time, and by definition of mechanical impedance this equals force over velocity. The impedance in an acoustic tube, on the other hand, must be in units of pressure (force/area) divided by volume velocity (velocity × area) and this reduces to

$$\frac{\text{force/area}}{\text{velocity} \cdot \text{area}} = \frac{\text{mass}}{\text{time}} \cdot \frac{1}{\text{area}^2} = \frac{\text{mass-volume-density}}{\text{time} \cdot \text{distance}}$$

which is what we found.

The real part of the wave impedance corresponds to transportation of wave energy, the imaginary part is a so-called "reactance" and does not correspond to power transfer. Instead, it corresponds to a "standing wave" which is created by equal and opposite power flow, or an "evanescent wave" (§C.8.2, p. 566), which is a non-propagating, exponentially decaying, limiting form of a traveling wave in which the "propagation constant" is purely imaginary due to being at a frequency above or below a "cut off" frequency for the waveguide [295, 122]. Driving an ideal mass at the end of a waveguide results in total reflection of all incident wave energy along with a quarter-cycle phase shift. Another interpretation is that the traveling wave becomes a standing wave at the tip of the cone. This is one way to see how the resonances of a cone can be the same as those of a cylinder the same length which is open on *both* ends. (One might first expect the cone to behave like a cylinder which is open on one end and closed on the other.) Because the impedance approaches a purely imaginary zero at the tip, it looks like a mass (with impedance $\pm\rho x s$). The "piston of air" at the open end similarly looks like a mass [285].

C.18.5 More General One-Parameter Waves

The wave impedance derivation above made use of known properties of waves in cones to arrive at the wave impedances in the two directions of travel in cones. We now consider how this solution might be generalized to arbitrary bore shapes. The momentum conservation equation is already applicable to any wavefront area variation $A(x)$:

$$p' + p\,\ln' A \;=\; -\rho\dot{u}.$$

As we did for vibrating strings (§C.3.4, p. 529), suppose the pressure is sinusoidally driven so that we have

$$p(t,x) \;=\; P(x)e^{st}$$

where $s = j\omega$, $\omega = 2\pi f$, and f is the driving frequency. The partial derivatives become

$$p' \;=\; (\ln' P)p, \qquad\qquad \dot{p} \;=\; sp.$$

Substituting into the momentum equation gives

$$p(\ln' P + \ln' A) \;=\; -\rho\dot{u}.$$

Because the medium is linear and time-invariant, the velocity $u(t,x)$ must be of the form $U(x)e^{st}$, and we can define the spatially instantaneous wave impedance as

$$R(x) \stackrel{\Delta}{=} \frac{P(x)}{U(x)}.$$

The corresponding instantaneous wave admittance is then $\Gamma(x) \stackrel{\Delta}{=} U(x)/P(x)$. Then $u = \Gamma(x)p$, and the momentum equation becomes

$$\ln' P + \ln' A = -\rho s \Gamma$$

Solving for the wave impedance gives

$$R(x) = -\frac{s\rho}{\ln' P + \ln' A}.$$

Expressing $P(x)$ in exponential form as

$$P(x) \stackrel{\Delta}{=} e^{j\theta(x)}$$

where $\theta(x)$ may be complex, we may define the instantaneous spatial frequency (wavenumber) as

$$k(x) \stackrel{\Delta}{=} \theta'(x)$$

and since $\ln' P = jk(x)$, we have

$$R(x) = -\frac{s\rho}{jk(x) + \ln' A} = -\frac{\omega\rho}{k(x) + \ln' A}$$

Defining the spatially instantaneous phase velocity as

$$c(x) \stackrel{\Delta}{=} \frac{\omega}{k(x)}$$

we have

$$R(x) = -\frac{\rho c(x)}{1 + \frac{1}{jk(x)}\ln' A(x)} = -\frac{\rho c(x)}{1 + \frac{\ln' A(x)}{\ln' P(x)}} \tag{C.152}$$

This reduces to the simple case of the uniform waveguide when the logarithmic derivative of cross-sectional area $A(x)$ is small compared with the logarithmic derivative of the amplitude $P(x)$ which is proportional to the instantaneous spatial frequency. A traveling wave solution interpretation makes sense when the instantaneous wavenumber $k(x)$ is approximately real, and the phase velocity $c(x)$ is approximately constant over a number of wavelengths $\lambda(x) = 2\pi/k(x)$.

C.18.6 Generalized Wave Impedance

Figure C.47 depicts a section of a conical bore which widens to the right connected to a section which narrows to the right. In addition, the cross-sectional areas are not matched at the junction.

The horizontal x axis (taken along the boundary of the cone) is chosen so that $x = 0$ corresponds to the apex of the cone. Let $A(x) = \alpha x^2$ denote the cross-sectional area of the bore.

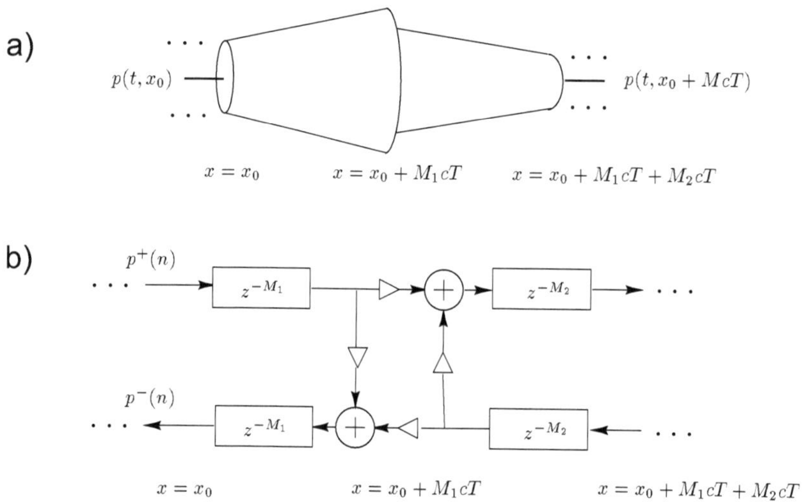

Figure C.47: a) Physical picture. b) Waveguide implementation.

Since a piecewise-cylindrical approximation to a general acoustic tube can be regarded as a "zeroth-order hold" approximation. A piecewise *conical* approximation then uses *first-order* (linear) segments. One might expect that quadratic, cubic, etc., would give better and better approximations. However, such a power series expansion has a problem: In zero-order sections (cylinders), plane waves propagate as traveling waves. In first-order sections (conical sections), spherical waves propagate as traveling waves. However, there are no traveling wave types for higher-order waveguide flare (*e.g.*, quadratic or higher) [357].

Since the digital waveguide model for a conic section is no more expensive to implement than that for a cylindrical section, (both are simply bidirectional delay lines), it would seem that modeling accuracy can be greatly improved for non-cylindrical bores (or parts of bores such as the bell) essentially for free. However, while the conic section itself costs nothing extra to implement, the *scattering junctions* between adjoining cone segments are more expensive computationally than those connecting cylindrical segments. However, the extra expense can be small. Instead of a single, real, reflection coefficient occurring at the interface between two cylinders of differing diameter, we obtain a *first-order reflection filter* at the interface between two cone sections of differing taper angle, as seen in the next section.

C.18.7 Generalized Scattering Coefficients

Generalizing the scattering coefficients at a multi-tube intersection (§C.12, p. 598) by replacing the usual real tube wave impedance $R_i = \rho c/A_i$ by the complex generalized wave impedance

$$R(x) = -\frac{\rho c(x)}{1 + \frac{\ln' A(x)}{\ln' P(x)}}$$

from Eq. (C.152) on page 647, or, as a special case, the conical-section wave impedance $R_A^{\pm}(s) = [\rho c/A(x)]/[s/(s \pm 1/t_x)]$ from Eq. (C.151) on page 645, we obtain the junction-pressure phasor [436]

$$P_J = \left(G_J + \sum_{i=1}^{N} G_i^- \right)^{-1} \sum_{i=1}^{N} \left(G_i^+ + G_i^- \right) P_i^+$$

where G_i^+ is the complex, frequency-dependent, incoming, acoustic admittance of the ith branch at the junction, G_i^- is the corresponding outgoing acoustic admittance, P_i^+ is the incoming traveling pressure-wave phasor in branch i, $P_i^- = P_J - P_i^+$ is the outgoing wave, and G_J is the admittance of a load at the junction, such as a coupling to another simulation. For generality, the formula is given as it appears in the multivariable case.

C.18.8 Cylinder with Conical Cap

Consider a cylindrical acoustic tube adjoined to a converging conical cap, as depicted in Figure C.48a. We may consider the cylinder to be either open or closed on the left side, but everywhere else it is closed. Since such a physical system is obviously passive, an interesting test of acoustic theory is to check whether theory predicts passivity in this case.

It is well known that a growing exponential appears at the junction of two conical waveguides when the waves in one conical taper angle reflect from a section with a smaller (or more negative) taper angle [7, 300, 8, 160, 9]. The most natural way to model a growing exponential in discrete time is to use an *unstable* one-pole filter [507]. Since unstable

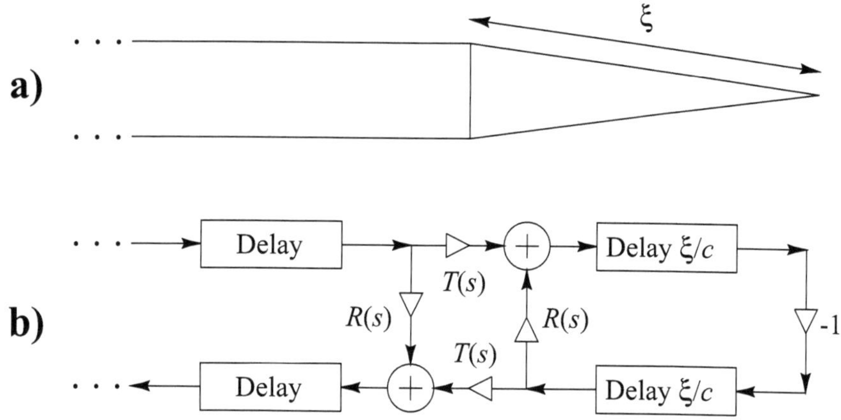

Figure C.48: Cylindrical acoustic tube with a conical cap. a) Physical outline. b) Continuous-time waveguide formulation.

filters do not normally correspond to passive systems, we might at first expect passivity to not be predicted. However, it turns out that all unstable poles are ultimately canceled, and the model is stable after all, as we will see. Unfortunately, as is well known in the field of automatic control, it is not practical to attempt to cancel an unstable pole in a real system, even when it is digital. This is because round-off errors will grow exponentially in the unstable feedback loop and eventually dominate the output.

The need for an unstable filter to model reflection and transmission at a converging conical junction has precluded the use of a straightforward recursive filter model [406]. Using special "truncated infinite impulse response" (TIIR) digital filters [541], an unstable recursive filter model can in fact be used in practice [529]. All that is then required is that the infinite-precision system be passive, and this is what we will show in the special case of Fig. C.48.

Scattering Filters at the Cylinder-Cone Junction

As derived in §C.18.4 on page 642, the wave impedance (for volume velocity) at frequency ω rad/sec in a converging cone is given by

$$Z_\xi(j\omega) = \frac{\rho c}{S(\xi)} \frac{j\omega}{j\omega - c/\xi} \tag{C.153}$$

where ξ is the distance to the apex of the cone, $S(\xi)$ is the cross-sectional area, and ρc is the wave impedance in open air. A cylindrical tube is the special case $\xi = \infty$, giving $Z_\infty(j\omega) = \rho c/S$, independent of position in the tube. Under normal assumptions such as pressure continuity and flow conservation at the cylinder-cone junction (see, *e.g.*, [300]), the junction reflection transfer function (reflectance) seen from the cylinder looking into the cone is derived to be

$$R(s) = -\frac{c/\xi}{c/\xi - 2s} \tag{C.154}$$

(where s is the Laplace transform variable which generalizes $s = j\omega$) while the junction transmission transfer function (transmittance) to the right is given by

$$T(s) = 1 + R(s) = -\frac{2s}{c/\xi - 2s} \tag{C.155}$$

The reflectance and transmittance from the right of the junction are the same when there is no wavefront area discontinuity at the junction [300]. Both $R(s)$ and $T(s)$ are first-order transfer functions: They each have a single real pole at $s = c/(2\xi)$. Since this pole is in the right-half plane, it corresponds to an unstable one-pole filter.

Reflectance of the Conical Cap

Let $t_\xi \triangleq \xi/c$ denote the time to propagate across the length of the cone in one direction. As is well known [22], the reflectance at the tip of an infinite cone is -1 for pressure waves. *I.e.*, it reflects like an open-ended cylinder. We ignore any absorption losses propagating in the cone, so that the transfer function from the entrance of the cone to the tip is e^{-st_ξ}. Similarly, the transfer function from the conical tip back to the entrance is also e^{-st_ξ}. The complete reflection transfer function from the entrance to the tip and back is then

$$R_{t_\xi}(s) = -e^{-2st_\xi} \tag{C.156}$$

Note that this is the reflectance a distance $\xi = ct_\xi$ from a conical tip *inside* the cone.

We now want to interface the conical cap reflectance $R_{t_\xi}(s)$ to the cylinder. Since this entails a change in taper angle, there will be reflection and transmission filtering at the cylinder-cone junction given by Eq. (C.154) and Eq. (C.155).

From inside the *cylinder*, immediately next to the cylinder-cone junction shown in Fig. C.48, the reflectance of the conical cap is readily derived from Fig. C.48b and Equations (C.154) and (C.155) to be

$$\begin{aligned} R_J(s) &= \frac{R(s) + 2R(s)R_{t_\xi}(s) + R_{t_\xi}(s)}{1 - R(s)R_{t_\xi}(s)} = \frac{1 + (1 + 2st_\xi)R_{t_\xi}(s)}{2st_\xi - 1 - R_{t_\xi}(s)} \\ &= \frac{1 - (1 + 2st_\xi)e^{-2st_\xi}}{2st_\xi - 1 + e^{-2st_\xi}} \triangleq \frac{N(s)}{D(s)} \end{aligned} \tag{C.157}$$

where

$$N(s) \triangleq 1 - e^{-2st_\xi} - 2st_\xi e^{-2st_\xi} \tag{C.158}$$

is the numerator of the conical cap reflectance, and

$$D(s) \triangleq 2st_\xi - 1 + e^{-2st_\xi} \tag{C.159}$$

is the denominator. Note that for very large t_ξ, the conical cap reflectance approaches $R_J = -e^{-2st_\xi}$ which coincides with the impedance of a length $\xi = ct_\xi$ open-end cylinder, as expected.

Stability Proof

A transfer function $R_J(s) = N(s)/D(s)$ is stable if there are no poles in the right-half s plane. That is, for each zero s_i of $D(s)$, we must have $\text{re}\{s_i\} \leq 0$. If this can be shown, along with $|R_J(j\omega)| \leq 1$, then the reflectance R_J is shown to be passive. We must also study the system zeros (roots of $N(s)$) in order to determine if there are any pole-zero cancellations (common factors in $D(s)$ and $N(s)$).

Since $\text{re}\{st_\xi\} \geq 0$ if and only if $\text{re}\{s\} \geq 0$, for $t_\xi > 0$, we may set $t_\xi = 1$ without loss of generality. Thus, we need only study the roots of

$$\begin{aligned} N(s) &= 1 - e^{-2s} - 2se^{-2s} \\ D(s) &= 2s - 1 + e^{-2s} \end{aligned}$$

If this system is stable, we have stability also for all $t_\xi > 0$. Since e^{-2s} is not a rational function of s, the reflectance $R_J(s)$ may have infinitely many poles and zeros.

Let's first consider the roots of the denominator

$$D(s) = 2s - 1 + e^{-2s}. \tag{C.160}$$

At any solution s of $D(s) = 0$, we must have

$$s = \frac{1 - e^{-2s}}{2} \tag{C.161}$$

To obtain separate equations for the real and imaginary parts, set $s = \sigma + j\omega$, where σ and ω are real, and take the real and imaginary parts of Eq. (C.161) to get

$$\begin{aligned} \text{re}\{D(s)\} &= (2\sigma - 1) + e^{-2\sigma}\cos(2\omega) = 0 \\ \text{im}\{D(s)\} &= 2\omega - e^{-2\sigma}\sin(2\omega) = 0 \end{aligned}$$

Both of these equations must hold at any pole of the reflectance. For stability, we further require $\sigma \leq 0$. Defining $\tau = 2\sigma$ and $\nu = 2\omega$, we obtain the somewhat simpler conditions

$$e^{\tau}(1-\tau) = \cos(\nu) \tag{C.162}$$

$$e^{\tau} = \frac{\sin(\nu)}{\nu} \tag{C.163}$$

For any poles of $R_J(s)$ on the $j\omega$ axis, we have $\tau = 0$, and Eq. (C.163) reduces to

$$\nu = \sin(\nu) \tag{C.164}$$

It is well known that the "sinc function" $\sin(\nu)/\nu$ is less than 1 in magnitude at all ν except $\nu = 0$. Therefore, Eq. (C.164) can hold only at $\omega = \nu = 0$.

We have so far proved that any poles on the $j\omega$ axis must be at $\omega = 0$.

The same argument can be extended to the entire right-half plane as follows. Going back to the more general case of Eq. (C.163), we have

$$\frac{\sin(\nu)}{\nu} = e^{\tau} \tag{C.165}$$

Since $|\sin(\nu)/\nu| \leq 1$ for all real ν, and since $|e^{\tau}| > 1$ for $\tau > 0$, this equation clearly has no solutions in the right-half plane. Therefore, the only possible poles in the right-half plane, including the $j\omega$ axis, are at $s = 0$.

In the left-half plane, there are many potential poles. Equation (C.162) has infinitely many solutions for each $\tau < 0$ since the elementary inequality $1 - \tau \leq e^{-\tau}$ implies $e^{\tau}(1-\tau) < e^{\tau}e^{-\tau} = 1$. Also, Eq. (C.163) has an increasing number of solutions as τ grows more and more negative; in the limit of $\tau = -\infty$, the number of solutions is infinite and given by the roots of $\sin(\nu)$ ($\nu = k\pi$ for any integer k). However, note that at $\tau = -\infty$, the solutions of Eq. (C.162) converge to the roots of $\cos(\nu)$ ($\nu = (2k+1)(\pi/2)$ for any integer k). The only issue is that the solutions of Eq. (C.162) and Eq. (C.163) must occur together.

Figure C.49 plots the locus of real-part zeros (solutions to Eq. (C.162)) and imaginary-part zeros (Eq. (C.163)) in a portion the left-half plane. The roots at $s = 0$ can be seen at the middle-right. Also, the asymptotic interlacing of these loci can be seen along the left edge of the plot. It is

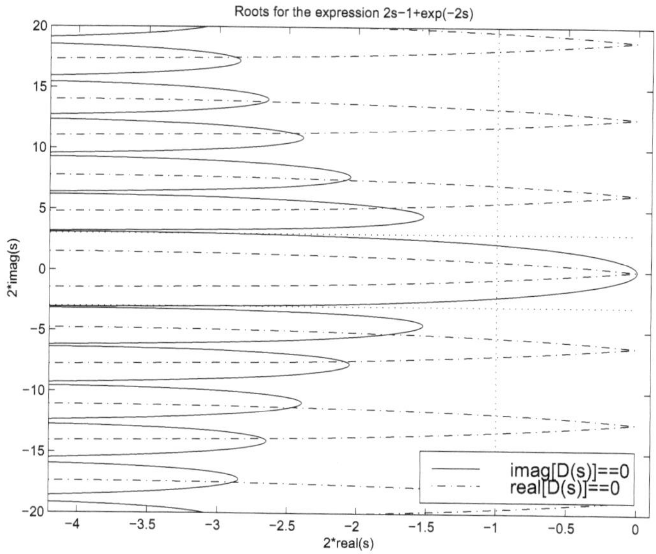

Figure C.49: Locus of solutions to Eq. (C.162) (dashed) (and Eq. (C.163) (solid). The dotted lines are lines of construction for an analytic proof that there are infinitely many roots in the left-half s-plane.

clear that the two loci must intersect at infinitely many points in the left-half plane near the intersections indicated in the graph. As $|\nu|$ becomes large, the intersections evidently converge to the peaks of the imaginary-part root locus (a log-sinc function rotated 90 degrees). At all frequencies ν, the roots occur near the log-sinc peaks, getting closer to the peaks as $|\nu|$ increases. The log-sinc peaks thus provide a reasonable estimate number and distribution in the left-half s-plane. An outline of an analytic proof is as follows:

- Rotate the loci in Fig. C.49 counterclockwise by 90 degrees.
- Prove that the two root loci are continuous, single-valued functions of ν (as the figure suggests).
- Prove that for $|\nu| > 3$, there are infinitely many extrema of the log-sinc function (imaginary-part root-locus) which have negative curvature and which lie below $\tau < -1$ (as the figure suggests). The $\tau = -1$ and $\nu = \pm 3$ lines are shown in the figure as dotted lines.
- Prove that the other root locus (for the real part) has infinitely many similar extrema which occur for $\tau > -1$ (again as the figure suggests).

- Prove that the two root-loci interlace at $\tau = -\infty$ (already done above).
- Then topologically, the continuous functions must cross at infinitely many points in order to achieve interlacing at $\tau = -\infty$.

The peaks of the log-sinc function not only indicate approximately where the left-half-plane roots occur

Reflectance Magnitude

We have shown that the conical cap reflectance has no poles in the strict right-half plane. For passivity, we also need to show that its magnitude is bounded by unity for all s on the $j\omega$ axis.

We have

$$R_J(j\omega) = \frac{1 - e^{-2j\omega} - 2j\omega e^{-2j\omega}}{2j\omega - 1 + e^{-2j\omega}} = e^{-2j\omega}\frac{e^{2j\omega} - 1 - 2j\omega}{e^{-2j\omega} - 1 + 2j\omega} = e^{-2j\omega}\frac{\overline{D(j\omega)}}{D(j\omega)}$$

so that $|R_J(j\omega)| = 1$, which is exactly lossless, as expected.

Poles at $s = 0$

We know from the above that the denominator of the cone reflectance has at least one root at $s = 0$. In this subsection we investigate this "dc behavior" of the cone more thoroughly.

A hasty analysis based on the reflection and transmission filters in Equations (C.154) and (C.155) might conclude that the reflectance of the conical cap converges to -1 at dc, since $R(0) = -1$ and $T(0) = 0$. However, this would be incorrect. Instead, it is necessary to take the limit as $\omega \to 0$ of the complete conical cap reflectance $R_J(s)$:

$$R_J(s) = \frac{1 - e^{-2s} - 2se^{-2s}}{2s - 1 + e^{-2s}} \tag{C.166}$$

We already discovered a root at $s = 0$ in the denominator in the context of the preceding stability proof. However, note that the numerator also goes to zero at $s = 0$. This indicates a pole-zero cancellation at dc. To find the reflectance at dc, we may use L'Hospital's rule to obtain

$$R_J(0) = \lim_{s\to 0}\frac{N'(s)}{D'(s)} = \lim_{s\to 0}\frac{4se^{-2s}}{2 - 2e^{-2s}} \tag{C.167}$$

and once again the limit is an indeterminate $0/0$ form. We therefore apply L'Hospital's rule again to obtain

$$R_J(0) = \lim_{s\to 0}\frac{N''(s)}{D''(s)} = \lim_{s\to 0}\frac{(4 - 8s)e^{-2s}}{4e^{-2s}} = +1 \tag{C.168}$$

Thus, two poles and zeros cancel at dc, and the dc reflectance is $+1$, not -1 as an analysis based only on the scattering filters would indicate. From a physical point of view, it makes more sense that the cone should "look like" a simple rigid termination of the cylinder at dc, since its length becomes small compared with the wavelength in the limit.

Another method of showing this result is to form a Taylor series expansion of the numerator and denominator:

$$N(s) = 2s^2 - \frac{8s^3}{3} + 2s^4 + \cdots \tag{C.169}$$

$$D(s) = 2s^2 - \frac{4s^3}{3} + \frac{2s^4}{3} + \cdots \tag{C.170}$$

Both series begin with the term $2s^2$ which means both the numerator and denominator have two roots at $s = 0$. Hence, again the conclusion is two pole-zero cancellations at dc. The series for the conical cap reflectance can be shown to be

$$R_J(s) = 1 - \frac{2s}{3} + \frac{2s^2}{9} - \frac{4s^3}{135} - \frac{2s^4}{405} + \cdots \tag{C.171}$$

which approaches $+1$ as $s \to 0$.

An alternative analysis of this issue is given by Benade in [37].

C.19 Exercises

1. Consider a semi-infinite ideal string with wave impedance R terminated on one end by an ideal mass m which slides vertically along a "frictionless vertical guide rod". The other end of the string stretches to infinity with uniform tension K.

   (a) Find a formula for the *mass density* of the string.

   (b) Find the *reflectance* (reflection transfer function)

   $$S_v(s) = \frac{V^-(s)}{V^+(s)}$$

   of the terminating mass, as seen from the string, for *velocity waves*.

   (c) Find the reflectance $S_f(s) = F^-(s)/F^+(s)$ of the terminating mass for *force waves*.

   (d) Find the reflectance of the terminating mass for

      i. *displacement waves*

ii. *acceleration waves*

iii. *slope waves*

(e) What is the magnitude of the force reflectance as a function of frequency, $|S_f(j\omega)|$?

(f) What is the phase of the force reflectance as a function of frequency, $\angle S_f(j\omega)$?

2. Suppose an infinitely long ideal string has wave impedance R and is sitting at rest with displacement $y(0, x) = 0$ for all $x \in (-\infty, \infty)$. Along comes a mass m which hits the string at time 0 with velocity v_0.

   (a) Find the mass velocity $v_m(t)$ for $t \geq 0$.

   (b) Find the mass position $y_m(t)$ for $t \geq 0$.

   (c) Assuming $R = m = c = 1$, sketch the shape of the string at times $t = 0, 1/4, 1/2, 1, \infty$, labeling all interesting points and asymptotes.

3. In §C.12 on page 598, we worked out the general reflectance of a load impedance $R_L(s)$ attached to an intersection of N ideal strings, each with wave impedance R_i, for the case of a *series* connection of the strings and load (all velocities equal at the junction, all forces summing to zero at the junction, appropriate for intersecting strings). Repeat this derivation for the case of a *parallel* connection at the junction (all forces equal at the junction, all velocities summing to zero at the junction, appropriate for intersecting acoustic tubes).

Appendix D

Finite-Difference Schemes

This appendix gives some simplified definitions and results from the subject of *finite-difference schemes* for numerically solving partial differential equations. Excellent references on this subject include Bilbao [53, 55] and Strikwerda [482].

The simplifications adopted here are that we will exclude nonlinear and time-varying partial differential equations (PDEs). We will furthermore assume constant step-sizes (sampling intervals) when converting PDEs to finite-difference schemes (FDSs), *i.e.*, sampling rates along time and space will be constant. Accordingly, we will assume that all initial conditions are *bandlimited* to less than half the *spatial* sampling rate, and that all excitations over time (such as summing input signals or "moving boundary conditions") will be bandlimited to less than half the *temporal* sampling rate. In short, the simplifications adopted here make the subject of partial differential equations isomorphic to that of linear systems theory [449]. For a more general and traditional treatment of PDEs and their associated finite-difference schemes, see, *e.g.*, [482].

D.1 Finite-Difference Schemes

Finite-Difference Schemes (FDSs) aim to solve differential equations by means of finite differences. For example, as discussed in §C.2 on page 524, if $y(t,x)$ denotes the displacement in meters of a vibrating string at time t seconds and position x meters, we may approximate the first- and second-order partial derivatives by

$$\begin{aligned}
\dot{y}(t,x) &\triangleq \frac{\partial}{\partial t}y(t,x) \approx \frac{y(t,x)-y(t-T,x)}{T} \\
y'(t,x) &\triangleq \frac{\partial}{\partial x}y(t,x) \approx \frac{y(t,x)-y(t,x-X)}{X} \\
\ddot{y}(t,x) &\triangleq \frac{\partial^2}{\partial t^2}y(t,x) \approx \frac{y(t+T,x)-2y(t,x)+y(t-T,x)}{T^2} \\
y''(t,x) &\triangleq \frac{\partial^2}{\partial x^2}y(t,x) \approx \frac{y(t,x+X)-2y(t,x)+y(t,x-X)}{X^2}
\end{aligned} \tag{D.1}$$

where T denotes the time sampling interval and X denotes the spatial sampling interval. Other types of finite-difference schemes were derived in Chapter 7 (§7.3.1 on page 265), including a look at frequency-domain properties. These finite-difference approximations to the partial derivatives may be used to compute solutions of differential equations on a discrete grid:

$$\begin{aligned}
x &\to x_m = mX \\
t &\to t_n = nT
\end{aligned}$$

Let us define an abbreviated notation for the grid variables

$$y_{n,m} \triangleq y(nT, mX)$$

and consider the ideal string wave equation (cf, §C.1):

$$y'' = \frac{1}{c^2}\ddot{y} \tag{D.2}$$

where c is a positive real constant (which turns out to be wave propagation speed). Then, as derived in §C.2 on page 524, setting $X = cT$ and substituting the finite-difference approximations into the ideal wave equation leads to the relation

$$y_{n+1,m} + y_{n-1,m} = y_{n,m+1} + y_{n,m-1}$$

everywhere on the time-space grid (*i.e.*, for all n and m). Solving for $y_{n+1,m}$ in terms of displacement samples at earlier times yields an *explicit finite-difference scheme* for string displacement:

$$y_{n+1,m} = y_{n,m+1} + y_{n,m-1} - y_{n-1,m} \tag{D.3}$$

The FDS is called *explicit* because it was possible to solve for the state at time n as a function of the state at earlier times (and any other positions m). This allows it to be implemented as a time recursion (or "digital filter") which computes a solution at time n from solution samples at earlier times (and any spatial positions). When an explicit FDS is not possible (*e.g.*, a non-causal filter is derived), the discretized differential equation is said to define an *implicit* FDS. An implicit FDS can often be converted to an explicit FDS by a rotation of coordinates [55, 482].

D.2 Convergence

A finite-difference scheme is said to be *convergent* if all of its solutions in response to initial conditions and excitations converge pointwise to the corresponding solutions of the original differential equation as the step size(s) approach zero.

In other words, as the step-size(s) shrink, the FDS solution must improve, ultimately converging to the corresponding solution of the original differential equation at every point of the domain.

In the vibrating string example, the limit is taken as the step sizes (sampling intervals) T and X approach zero. Since the finite-difference approximations in Eq. (D.1) converge in the limit to the very definitions of the corresponding partial derivatives, we expect the FDS in Eq. (D.3) based on these approximations to be convergent (and it is).

In establishing convergence, it is necessary to provide that any initial conditions and boundary conditions in the finite-difference scheme converge to those of the continuous differential equation, in the limit. See [482] for a more detailed discussion of this topic.

The *Lax-Richtmyer equivalence theorem* provides a means of showing convergence of a finite-difference scheme by showing it is both *consistent* and *stable* (and that the initial-value problem is *well posed*) [482]. The following subsections give basic definitions for these terms which applicable to our simplified scenario (linear, shift-invariant, fixed sampling rates).

D.2.1 Consistency

A finite-difference scheme is said to be *consistent* with the original partial differential equation if, given any sufficiently differentiable function $y(t, x)$, the differential equation operating on $y(t, x)$ approaches the value of the finite difference equation operating on $y(nT, mX)$, as T and X approach zero.

Thus, in the ideal string example, to show the consistency of Eq. (D.3) we must show that

$$\left(\frac{\partial^2}{\partial x^2} - \frac{1}{c^2}\frac{\partial^2}{\partial t^2}\right) y(t,x) = \lim_{T,X\to 0} \left[(\delta_x + \delta_x^{-1}) - (\delta_t + \delta_t^{-1})\right] y_{n,m}$$

for all $y(t,x)$ which are second-order differentiable with respect to t and x. On the right-hand side, we have introduced the following *shift operator notation*:

$$\begin{aligned} \delta_t^k y_{n,m} &\triangleq y_{n-k,m} \\ \delta_x^k y_{n,m} &\triangleq y_{n,m-k} \end{aligned} \tag{D.4}$$

In particular, we have

$$\begin{aligned} \delta_t y_{n,m} &\triangleq y_{n-1,m} \\ \delta_t^{-1} y_{n,m} &\triangleq y_{n+1,m} \\ \delta_x y_{n,m} &\triangleq y_{n,m-1} \\ \delta_x^{-1} y_{n,m} &\triangleq y_{n,m+1}. \end{aligned}$$

In taking the limit as T and X approach zero, we must maintain the relationship $X = cT$, and we must scale the FDS by $1/X^2$ in order to achieve an exact result:

$$\begin{aligned} &\lim_{T,X\to 0} \frac{1}{X^2}\left[(\delta_x + \delta_x^{-1}) - (\delta_t + \delta_t^{-1})\right] y_{n,m} \\ &\quad = \lim_{T,X\to 0} \left[\frac{\delta_x + 2 + \delta_x^{-1}}{X^2} - \frac{\delta_t + 2 + \delta_t^{-1}}{c^2T^2}\right] y_{n,m} \\ &\quad \triangleq \left(\frac{\partial^2}{\partial x^2} - \frac{1}{c^2}\frac{\partial^2}{\partial t^2}\right) y(t,x) \end{aligned}$$

as required. Thus, the FDS is consistent. See, *e.g.*, [482] for more examples.

In summary, consistency of a finite-difference scheme means that, in the limit as the sampling intervals approach zero, the original PDE is obtained from the FDS.

D.2.2 Well Posed Initial-Value Problem

For a proper authoritative definition of "well posed" in the field of finite-difference schemes, see, *e.g.*, [482]. The definition we will use here is less general in that it excludes amplitude growth from initial conditions which is faster than polynomial in time.

We will say that an initial-value problem is *well posed* if the linear system defined by the PDE, together with any bounded initial conditions is *marginally stable*.

As discussed in [449], a system is defined to be *stable* when its response to bounded initial conditions approaches zero as time goes to infinity. If the response fails to approach zero but does not exponentially grow over time (the *lossless* case), it is called *marginally stable*.

In the literature on finite-difference schemes, lossless systems are classified as stable [482]. However, in this book series, lossless systems are not considered stable, but only marginally stable.

When marginally stable systems are allowed, it is necessary to accommodate *polynomial growth* with respect to time. As is well known in linear systems theory, repeated poles can yield polynomial growth [449]. A very simple example is the ordinary differential equation (ODE)

$$\ddot{y} = 0$$

which, given the initial condition $y(0)$, has solutions

$$y(t) = y(0) + ct$$

for any constant c. Thus, the system is lossless and the initial condition is finite, yet solution is not bounded. Similarly, solutions to the ODE $\dddot{y} = 0$ can grow as t^2, and so on.

When all poles of the system are strictly in the left-half of the Laplace-transform s plane, the system is *stable*, even when the poles are repeated. This is because exponentials are faster than polynomials, so that any amount of exponential decay will eventually overtake polynomial growth and drag it to zero in the limit.

Marginally stable systems arise often in computational physical modeling. In particular, the ideal string is only marginally stable, since it is lossless. Even a simple unaccelerated mass, sliding on a frictionless surface, is described by a marginally stable PDE when the position of the mass is used as a state variable (see §7.1.2 on page 254). Given any nonzero initial velocity, the position of the mass approaches either + or − infinity, exactly as in the $\ddot{y} = 0$ example above. To avoid unbounded growth in practical systems, it is often preferable to avoid the use of displacement as a state variable. For ideal strings and freely sliding masses, force and velocity are usually good choices.

It should perhaps be emphasized that the term "well posed" normally allows for more general energy growth at a rate which can be bounded over all initial conditions [482]. In this book, however, the "marginally stable" case (at most polynomial growth) is what we need. The reason is simply that we wish to excluded unstable PDEs as a modeling target.

Note, however, that unstable systems can be used profitable over carefully limited time durations (see §9.7.2 on page 438 for an example).

In the ideal vibrating string, energy is conserved. Therefore, it is a marginally stable system. To show mathematically that the PDE Eq. (D.2) is marginally stable, we may show that

$$\| y(t,x) \|_2^2 (t) = \alpha \| y(0,x) \|_2^2 + \beta \| \dot{y}(0,x) \|_2^2 .$$

for some constants α and β. *I.e.*, we can show

$$\int_{-\infty}^{\infty} |y(t,x)|^2 \, dx = \alpha \int_{-\infty}^{\infty} |y(0,x)|^2 \, dx + \beta \int_{-\infty}^{\infty} |\dot{y}(0,x)|^2 \, dx$$

for all t.

Note that solutions on the ideal string are not bounded, since, for example, an infinitely long string (non-terminated) can be initialized with a constant positive velocity everywhere along its length. This corresponds physically to a nonzero transverse momentum, which is conserved. Therefore, the string will depart in the positive y direction, with an average displacement that grows linearly with t.

The well-posedness of a class of damped PDEs used in string modeling is analyzed in §D.2.2 on page 664.

A Class of Well Posed Damped PDEs

A large class of well posed PDEs is given by [45]

$$\ddot{y} + 2\sum_{k=0}^{M} q_k \frac{\partial^{2k+1} y}{\partial x^{2k} \partial t} + \sum_{k=0}^{N} r_k \frac{\partial^{2k} y}{\partial x^{2k}} \tag{D.5}$$

Thus, to the ideal string wave equation Eq. (C.1) on page 523 we may add any number of even-order partial derivatives in x, plus any number of mixed odd-order partial derivatives in x and t, where differentiation with respect to t occurs only once. Because every member of this class of PDEs is only second-order in time, it is guaranteed to be *well posed*, as we now show.

To show Eq. (D.5) is well posed [45], we must show that the roots of the characteristic polynomial equation (§D.3, p. 668) have negative real parts, *i.e.*, that they correspond to decaying exponentials instead of growing exponentials. To do this, we may insert the general eigensolution

$$y(t,x) = e^{st+vx}$$

into the PDE just like we did in §C.5 to obtain the so-called *characteristic polynomial equation*:

$$s^2 + 2q(v)s + r(v) = 0$$

where

$$q(v) \triangleq \sum_{k=0}^{M} q_k v^{2k}$$
$$r(v) \triangleq \sum_{k=0}^{N} r_k v^{2k}$$

Let's now set $v = jk$, where $k = 2\pi/\lambda$ is radian spatial frequency (called the "wavenumber" in acoustics) and of course $j = \sqrt{-1}$, thereby converting the implicit spatial Laplace transform to a spatial Fourier transform. Since there are only even powers of the spatial Laplace transform variable v, the polynomials $q(jk)$ and $r(jk)$ are *real.* Therefore, the roots of the characteristic polynomial equation (the natural frequencies of the time response of the system), are given by

$$s = -q \pm \sqrt{q^2 - r}.$$

Proof that the Third-Order Time Derivative is Ill Posed

For its tutorial value, let's also show that the PDE of Ruiz [392] is ill posed, *i.e.*, that at least one component of the solution is a growing exponential. In this case, setting $y(t, x) = e^{st+jkx}$ in Eq. (C.28), which we restate as

$$Ky'' = \epsilon\ddot{y} + \mu\dot{y} + \mu_3\dddot{y},$$

yields the characteristic polynomial equation

$$p(s, jk) = \mu_3 s^3 + \epsilon s^2 + \mu s + Kk^2 = 0.$$

By the Routh-Hurwitz theorem,[1] there is at least one root in the right-half s-plane.

It is interesting to note that Ruiz discovered the exponentially growing solution, but simply dropped it as being non-physical. In the work of Chaigne and Askenfelt [77], it is believed that the finite difference approximation itself provided the damping necessary to eliminate the unstable solution [45]. (See §7.3.2 on page 268 for a discussion of how finite difference approximations can introduce damping.) Since the damping effect is sampling-rate dependent, there is an upper bound to the sampling rate that can be used before an unstable mode appears.

---

[1] `http://mathworld.wolfram.com/Routh-HurwitzTheorem.html`

D.2.3 Stability of a Finite-Difference Scheme

A finite-difference scheme is said to be *stable* if it forms a *digital filter* which is at least *marginally stable* [449].

To distinguish between the stable and marginally stable cases, we may classify a finite-difference scheme as *strictly stable*, *marginally stable*, or *unstable*.

D.2.4 Lax-Richtmyer equivalence theorem

The Lax-Richtmyer equivalence theorem states that *"a consistent finite-difference scheme for a partial differential equation for which the initial-value problem is well posed is convergent if and only if it is stable."* For a proof, see [482, Ch. 10].

D.2.5 Passivity of a Finite-Difference Scheme

A condition stronger than stability as defined above is *passivity*. Passivity is not a traditional metric for finite-difference scheme analysis, but it arises naturally in special cases such as wave digital filters (§F.1 on page 701) and digital waveguide networks [55, 35]. In such modeling frameworks, all signals have a *physical interpretation* as wave variables, and therefore a physical energy can be associated with them. Moreover, each delay element can be associated with some real *wave impedance*. In such situations, passivity can be defined as the case in which all impedances are nonnegative. When complex, they must be *positive* real (see §C.11.2 on page 588).

To define passivity for all linear, shift-invariant finite difference schemes, irrespective of whether or not they are based on an impedance description, we will say that a finite-difference scheme is passive if all of its internal modes are stable. Thus, passivity is sufficient, but not necessary, for stability. In other words, there are finite difference schemes which are stable but not passive [55]. A stable FDS can have internal unstable modes which are not excited by initial conditions, or which always cancel out in pairs. A passive FDS cannot have such "hidden" unstable modes.

The absence of hidden modes can be ascertained by converting the FDS to a state-space model and checking that it is *controllable* (from initial conditions and/or excitations) and *observable* [449]. When the initial conditions can set the entire initial state of the FDS, it is then controllable from initial conditions, and only observability needs to be checked. A simple example of an unobservable mode is the second harmonic of an ideal string (and all even-numbered harmonics) when the only output observation is the midpoint of the string.

Summary

In summary, we have defined the following terms from the analysis of finite-difference schemes for the linear shift-invariant case with constant sampling rates:

- PDE *well posed* $\Leftrightarrow$ PDE at least marginally stable
- FDS *consistent* $\Leftrightarrow$ FDS shift operator $\rightarrow$ PDE operator as $T, X \rightarrow 0$
- FDS *stable* $\Leftrightarrow$ stable or marginally stable as a digital filter
- FDS *strictly stable* $\Leftrightarrow$ stable as a digital filter
- FDS *marginally stable* $\Leftrightarrow$ marginally stable as a digital filter

Finally, the Lax-Richtmyer equivalence theorem establishes that well posed + consistency + stability implies convergence, where, as defined in §D.2 above, convergence means that solutions of the FDS approach corresponding solutions of the PDE as $T, X \rightarrow 0$.

D.2.6 Convergence in Audio Applications

Because the range of human hearing is bounded (nominally between 20 and 20 kHz), spectral components of a signal outside this range are not audible. Therefore, when the solution to a differential equation is to be considered an audio signal, there are frequency regions over which convergence is not a requirement.

Instead of pointwise convergence, we may ask for the following two properties:

- *Superposition* holds.
- Convergence occurs *within the frequency band of human hearing.*

Superposition holds for all linear partial differential equations with constant coefficients (linear, shift-invariant systems [449]). We need this condition so that errors in the inaudible bands do not affect the audible bands. Inaudible errors are fine as long as they do not grow so large that they cause numerical overflow. An example in which this "bandlimited design" approach yields large practical dividends is in *bandlimited interpolator* design (see §4.4 on page 169).

In many cases, such as in digital waveguide modeling of vibrating strings, we can do better than convergence. We can construct finite difference schemes which agree with the corresponding continuous solutions *exactly* at the sample points. (See §C.4.1 on page 533.)

D.3 Characteristic Polynomial Equation

The *characteristic polynomial equation* for a linear PDE with constant coefficients is obtained by taking the 2D Laplace transform of the PDE with respect to x and t. A simple way of doing this is to substitute the general eigensolution

$$y(t,x) = e^{st+vx} \tag{D.6}$$

into the PDE, where $s = \sigma + j\omega$ denotes the complex variable associated with the Laplace-transform with respect to time, and $v = \beta + jk$ is the complex variable associated with the *spatial* Laplace transform.

As a simple example, the ideal-string wave equation (analyzed in §C.1 on page 523) is a simple second-order PDE given by

$$\frac{\partial^2 y(t,x)}{\partial x^2} = \frac{1}{c^2}\frac{\partial^2 y(t,x)}{\partial t^2} \tag{D.7}$$

where c is a positive constant (sound speed, as discussed in §C.1).

Substituting Eq. (D.6) into Eq. (D.7) results in the following *characteristic polynomial equation*:

$$v^2 - \frac{1}{c^2}s^2 = 0$$

Solving for s in terms of v gives the so-called *dispersion relation*:

$$s = \pm cv$$

or, looking only at the frequency axes (*i.e.*, using Fourier transforms in place of Laplace transforms),

$$\omega = \pm ck.$$

Since the *phase velocity* of a traveling wave is, by definition, the temporal frequency divided by spatial frequency, we have simply

$$v_p \triangleq \frac{\omega}{k} = \pm c.$$

This result can be interpreted as saying that all Fourier components of any solution of Eq. (D.7) must propagate along the string with speed c to either the left or the right along the string. In more general PDEs, propagation may be *dispersive*, in which case the phase velocity depends on frequency (see §C.6 on page 545 for an analysis of stiff vibrating strings, which are dispersive). Moreover, wave propagation may be *damped* in a frequency-dependent way, in which case one or more roots of the characteristic polynomial equation will have negative real parts; if any roots have positive real parts, we say the initial-value problem is *ill posed* since is has exponentially growing solutions in response to initial conditions.

D.4 Von Neumann Analysis

Von Neumann analysis is used to verify the *stability* of a finite-difference scheme. We will only consider one time dimension, but any number of spatial dimensions.

The procedure, in principle, is to perform a *spatial Fourier transform* along all spatial dimensions, thereby reducing the finite-difference scheme to a time recursion in terms of the spatial Fourier transform of the system. The system is then stable if this time recursion is at least marginally stable as a digital filter.

Let's apply von Neumann analysis to the finite-difference scheme for the ideal vibrating string Eq. (D.3):

$$y_{n+1,m} = y_{n,m+1} + y_{n,m-1} - y_{n-1,m}$$

There is only one spatial dimension, so we only need a single 1D Discrete Time Fourier Transform (DTFT) along m [451]. Using the *shift theorem* for the DTFT, we obtain

$$\begin{aligned} Y_{n+1}(k) &= (e^{jkX} + e^{-jkX})Y_n(k) - Y_{n-1}(k) \\ &= 2\cos(kX)Y_n(k) - Y_{n-1}(k) \\ &\triangleq 2c_k Y_n(k) - Y_{n-1}(k) \end{aligned} \tag{D.8}$$

where $k = 2\pi/\lambda$ denotes radian spatial frequency (wavenumber). (On a more elementary level, the DTFT along m can be carried out by substituting $Y_n(k)e^{jkX}$ for $y(n,m)$ in the finite-difference scheme.) This is now a second-order difference equation (digital filter) that needs its stability checked. This can be accomplished most easily using the Durbin recursion [449], or we can check that the poles of the recursion do not lie outside the unit circle in the z plane.

A method equivalent to checking the pole radii, and typically used when the time recursion is first order, is to compute the *amplification factor* as the complex gain $G(k)$ in the relation

$$Y_{n+1}(k) = G(k)Y_n(k).$$

The finite-difference scheme is then declared stable if $|G(k)| \leq 1$ for all spatial frequencies k.

Since the finite-difference scheme of the ideal vibrating string is so simple, let's find the two poles. Taking the z transform of Eq. (D.8) yields

$$zY(z,k) = 2c_k Y(z,k) - z^{-1}Y(z,k)$$

yielding the following characteristic polynomial:

$$z^2 - 2c_k z - 1 = 0$$

Applying the quadratic formula to find the roots yields

$$z = c_k \pm \sqrt{c_k^2 - 1}.$$

The squared pole moduli are then given by

$$|z|^2 = c_k^2 \pm (c_k^2 - 1) = \begin{cases} [2c_k^2 - 1, 1], & |c_k| \geq 1 \\ [1, 1], & |c_k| \leq 1 \end{cases}.$$

Thus, for marginal stability, we require $|c_k| \leq 1$, and the poles become

$$z = c_k \pm j\sqrt{1 - c_k^2} = \cos(kX) \pm j\sin(kX) = e^{\pm jkX}.$$

Since the range of spatial frequencies is $k \in [-\pi/X, \pi/X]$, we spontaneously have $|c_k| \leq 1$ for all k. Therefore, we have shown by von Neumann analysis that the finite-difference scheme Eq. (D.3) for the ideal vibrating string is *stable.*

In summary, von Neumann analysis verifies that no spatial Fourier components in the system are growing exponentially with respect to time.

Appendix E

Equivalence of Digital Waveguide and Finite Difference Schemes

It was shown in §C.4.3 on page 536 that the digital waveguide (DW) model for the ideal vibrating string performs the same "state transitions" as the more standard finite-difference time-domain (FDTD) scheme (also known as the "leapfrog" recursion). This appendix, initially published in [444], further establishes that the solution spaces of the two schemes are isomorphic. That is, a linear, one-to-one transformation is derived which converts any point in the state-space of one scheme to a unique point in the other scheme. Since boundary conditions and initial values are more intuitively transparent in the DW formulation, the simple means of converting back and forth can be useful in initializing and constructing boundaries for FDTD simulations, as we will see.

E.1 Introduction

The digital waveguide (DW) method has been used for many years to provide highly efficient algorithms for musical sound synthesis based on physical models [433, 447, 396]. For a much longer time, finite-difference time-domain (FDTD) schemes have been used to simulate more general situations, usually at higher cost [551, 392, 74, 77, 45, 397]. In recent years, there has been interest in relating these methods to each other [123] and in combining them for more general simulations. For example, modular hybrid methods have been devised which interconnect DW and FDTD simulations by means of a *KW converter* [223, 226]. The basic idea of the KW-converter adaptor is to convert the "Kirchoff variables" of the FDTD, such as string displacement, velocity, etc., to "wave variables" of the DW. The W variables are regarded as the traveling-wave components of the K variables.

In this appendix, we present an alternative to the KW converter. Instead of converting K variables to W variables, or vice versa, in the time domain, conversion formulas are derived with respect to the current state as a function of spatial coordinates. As a result, it becomes simple to convert any instantaneous state configuration from FDTD to DW form, or vice versa. Thus, instead of providing the necessary time-domain filter to implement a KW converter converting traveling-wave components to physical displacement of a vibrating string, say, one may alternatively set the displacement variables instantaneously to the values corresponding to a given set of traveling-wave components in the string model. Another benefit of the formulation is an exact physical interpretation of arbitrary initial conditions and excitations in the K-variable FDTD method. Since the DW formulation is exact in principle (though bandlimited), while the FDTD is approximate, even in principle, it can be argued that the true physical interpretation of the FDTD method is that given by the DW method. Since both methods generate the same evolution of state from a common starting point, they may only differ in computational expense, numerical sensitivity, and in the details of supplying initial conditions and boundary conditions.

The wave equation for the ideal vibrating string, reviewed in §C.1 on page 523, can be written as

$$Ky'' = \epsilon\ddot{y},$$

where the following notation is used:

$$\begin{array}{rclcrcl} K & \isdef & \text{string tension} & \qquad & y & \isdef & y(t,x) \\ \epsilon & \isdef & \text{linear mass density} & & \dot{y} & \isdef & \frac{\partial}{\partial t}y(t,x) \\ y & \isdef & \text{string displacement} & & y' & \isdef & \frac{\partial}{\partial x}y(t,x) \end{array}$$

In the following two subsections, we briefly recall *finite difference* and *digital waveguide* models for the ideal vibrating string.

E.1.1 Finite Difference Time Domain (FDTD) Scheme

As discussed in §C.2 on page 524, we may use *centered finite difference approximations* (FDA) for the second-order partial derivatives in the wave equation to obtain a *finite difference scheme* for numerically integrating the ideal wave equation [482, 311]:

$$\ddot{y}(t,x) \approx \frac{y(t+T,x)-2y(t,x)+y(t-T,x)}{T^2} \tag{E.1}$$

$$y''(t,x) \approx \frac{y(t,x+X)-2y(t,x)+y(t,x-X)}{X^2} \tag{E.2}$$

where T is the time sampling interval, and X is a spatial sampling interval.

Substituting the FDA into the wave equation, choosing $X = cT$, where $c \triangleq \sqrt{K/\epsilon}$ is sound speed (normalized to $c = 1$ below), and sampling at times $t = nT$ and positions $x = mX$, we obtain the following explicit finite difference scheme for the string displacement:

$$y(n+1,m) = y(n,m+1) + y(n,m-1) - y(n-1,m) \tag{E.3}$$

where the sampling intervals T and X have been normalized to 1. To initialize the recursion at time $n = 0$, past values are needed for all m (all points along the string) at time instants $n = -1$ and $n = -2$. Then the string position may be computed for all m by Eq. (E.3) for $n = 0, 1, 2, \ldots$. This has been called the *FDTD* or leapfrog finite difference scheme [127].

E.1.2 Digital Waveguide (DW) Scheme

We now derive the digital waveguide formulation by *sampling* the *traveling-wave* solution to the wave equation. It is easily checked that the lossless 1D wave equation $Ky'' = \epsilon\ddot{y}$ is solved by any string shape y which travels to the left or right with speed $c \triangleq \sqrt{K/\epsilon}$ [100]. Denote *right-going* traveling waves in general by $y_r(t-x/c)$ and *left-going* traveling waves by $y_l(t+x/c)$, where y_r and y_l are assumed twice-differentiable. Then, as is well known, the general class of solutions to the lossless, one-dimensional, second-order wave equation can be expressed as

$$y(t,x) = y_r\left(t - \frac{x}{c}\right) + y_l\left(t + \frac{x}{c}\right). \tag{E.4}$$

Sampling these traveling-wave solutions yields

$$\begin{aligned} y(nT,mX) &= y_r(nT - mX/c) + y_l(nT + mX/c) \\ &= y_r[(n-m)T] + y_l[(n+m)T] \\ &\triangleq y^+(n-m) + y^-(n+m) \end{aligned} \tag{E.5}$$

where a "+" superscript denotes a "right-going" traveling-wave component, and "−" denotes propagation to the "left". This notation is similar to that used for acoustic-tube modeling of speech [297].

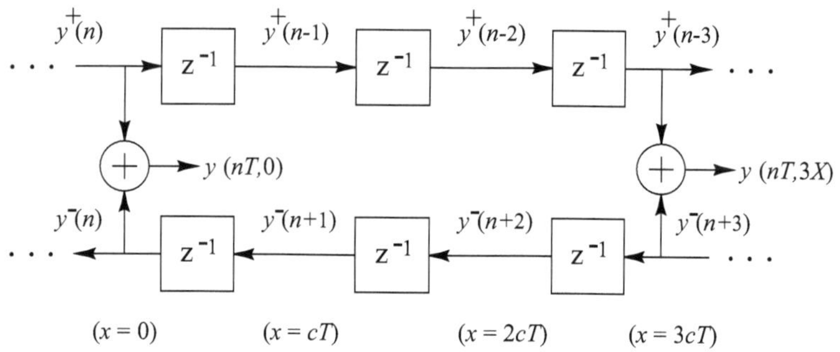

Figure E.1: Digital simulation of the ideal, lossless waveguide with observation points at $x = 0$ and $x = 3X = 3cT$.

Figure E.1 on page 674 (initially given as Fig. C.3 on page 534) shows a signal flow diagram for the computational model of Eq. (E.5), termed a *digital waveguide model* (developed in detail in Appendix C). Recall that, by the sampling theorem, it is an exact model so long as the initial conditions and any ongoing additive excitations are bandlimited to less than half the temporal sampling rate $f_s = 1/T$ [451, Appendix G]. Recall also that the position along the string, $x_m = mX = mcT$ meters, is laid out from left to right in the diagram, giving a physical interpretation to the horizontal direction in the diagram, even though spatial samples have been eliminated from explicit consideration. (The arguments of y^+ and y^- have physical units of time.)

The left- and right-going traveling wave components are summed to produce a physical output according to

$$y(nT, mX) = y^+(n-m) + y^-(n+m) \tag{E.6}$$

In Fig. E.1, "transverse displacement outputs" have been arbitrarily placed at $x = 0$ and $x = 3X$.

E.1.3 FDTD and DW Equivalence

The FDTD and DW recursions both compute time updates by forming fixed linear combinations of past state. As a result, each can be described in "state-space form" [449, Appendix G] by a constant matrix operator, the "state transition matrix", which multiplies the state vector at the current time to produce the state vector for the next time step. The FDTD operator propagates K variables while the DW operator propagates W variables. We may show equivalence by (1) defining a one-to-one transformation which will convert K variables to W variables or vice versa, and (2) showing that given any common initial state for both schemes, the state transition matrices compute the same next state in both cases.

The next section shows that the linear transformation from W to K variables,

$$y(n,m) = y^+(n-m) + y^-(n+m), \tag{E.7}$$

for all n and m, sets up a one-to-one linear transformation between the K and W variables. Assuming this holds, it only remains to be shown that the DW and FDTD schemes preserve mapping Eq. (E.7) after a state transition from one time to the next. While this has been shown previously [442], we repeat the derivation here for completeness. We also provide a state-space analysis reaching the same conclusion in §E.4.

From Fig. E.1, it is clear that the DW scheme preserves mapping Eq. (E.7) by definition. For the FDTD scheme, we expand the right-hand of Eq. (E.3) using Eq. (E.7) and verify that the left-hand side also satisfies the map, *i.e.*, that $y(n+1,m) = y^+(n+1-m) + y^-(n+1+m)$ holds:

$$\begin{aligned}
y(n+1,m) &= y(n,m+1) + y(n,m-1) - y(n-1,m) \\
&= y^+(n-m-1) + y^-(n+m+1) \\
&\quad +y^+(n-m+1) + y^-(n+m-1) \\
&\quad -y^+(n-m-1) - y^-(n+m-1) \\
&= y^-(n+m+1) + y^+(n-m+1) \\
&= y^+[(n+1)-m] + y^-[(n+1)+m] \\
&\triangleq y(n+1,m)
\end{aligned}$$

Since the DW method propagates sampled (bandlimited) solutions to the ideal wave equation without error, it follows that the FDTD method does the same thing, despite the relatively crude approximations made in Eq. (E.2). In particular, it is known that the FDA introduces artificial damping when applied to first order partial derivatives arising in lumped, mass-spring systems [447].

The equivalence of the DW and FDTD state transitions extends readily to the DW mesh [519, 447] which is essentially a lattice-work of DWs for simulating membranes and volumes. The equivalence is more important in higher dimensions because the FDTD formulation requires less computations per node than the DW approach in higher dimensions (see [33] for some quantitative comparisons).

Even in one dimension, the DW and finite-difference methods have unique advantages in particular situations [223], and as a result they are often combined together to form a hybrid traveling-wave/physical-variable simulation [351, 352, 222, 124, 123, 224, 263, 33].

E.2 State Transformations

In previous work, time-domain adaptors (digital filters) converting between K variables and W variables have been devised [223]. In this section, an alternative approach is proposed. Mapping Eq. (E.7) gives us an immediate conversion from W to K state variables, so all we need now is the inverse map for any time n. This is complicated by the fact that non-local spatial dependencies can go indefinitely in one direction along the string, as we will see below. We will proceed by first writing down the conversion from W to K variables in matrix form, which is easy to do, and then invert that matrix. For simplicity, we will consider the case of an infinitely long string.

To initialize a K variable simulation for starting at time $n+1$, we need initial spatial samples at all positions m for two successive times $n-1$ and n. From this state specification, the FDTD scheme Eq. (E.3) can compute $y(n+1,m)$ for all m, and so on for increasing n. In the DW model, all state variables are defined as belonging to the same time n, as shown in Fig. E.2.

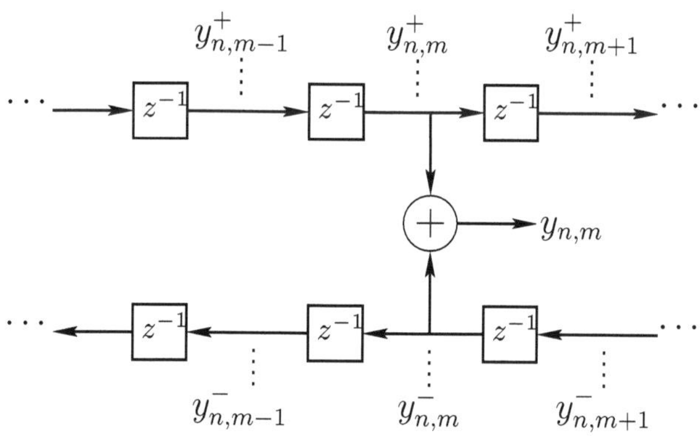

Figure E.2: DW flow diagram.

From Eq. (E.6), and referring to the notation defined in Fig. E.2, we may write the conversion from W to K variables as

$$\begin{aligned}
y_{n,m+1} &= y^+_{n,m+1} + y^-_{n,m+1} \\
y_{n,m-1} &= y^+_{n,m-1} + y^-_{n,m-1} \\
y_{n-1,m} &= y^+_{n-1,m} + y^-_{n-1,m} \\
&= y^+_{n,m+1} + y^-_{n,m-1}
\end{aligned} \tag{E.8}$$

where the last equality follows from the traveling-wave behavior (see Fig. E.2).

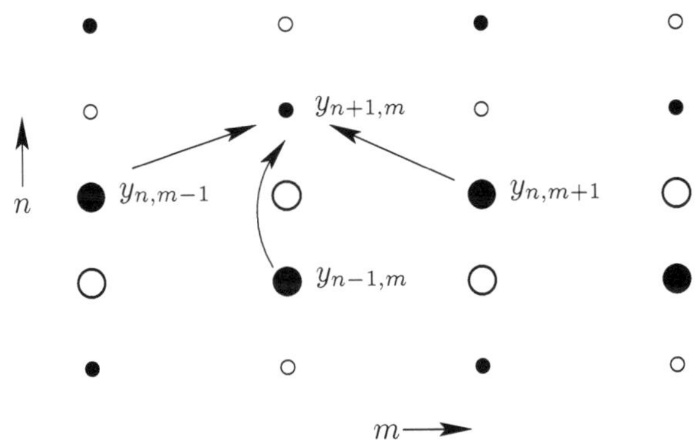

Figure E.3: Stencil of the FDTD scheme.

Figure E.3 shows the so-called "stencil" of the FDTD scheme. The larger circles indicate the state at time n which can be used to compute the state at time $n+1$. The filled and unfilled circles indicate membership in one of two interleaved grids [55]. To see why there are two interleaved grids, note that when m is even, the update for $y_{n+1,m}$ depends only on odd m from time n and even m from time $n-1$. Since the two W components of $y_{n-1,m}$ are converted to two W components at time n in Eq. (E.8), we have that the update for $y_{n+1,m}$ depends only on W components from time n and positions $m\pm 1$. Moving to the next position update, for $y_{n+1,m+1}$, the state used is independent of that used for $y_{n+1,m}$, and the W components used are from positions m and $m+2$. As a result of these observations, we see that we may write the state-variable transformation separately for even and odd m, *e.g.*,

$$\begin{bmatrix} \vdots \\ y_{n,m-1} \\ y_{n-1,m} \\ y_{n,m+1} \\ y_{n-1,m+2} \\ y_{n,m+3} \\ y_{n-1,m+4} \\ y_{n,m+5} \\ \vdots \end{bmatrix} = \begin{bmatrix} \ddots & & \vdots & \vdots & \vdots & \vdots & \vdots & \vdots & \vdots & 0 \\ \cdots & 1 & 1 & 0 & 0 & 0 & 0 & 0 & 0 & \cdots \\ \cdots & 0 & 1 & 1 & 0 & 0 & 0 & 0 & 0 & \cdots \\ \cdots & 0 & 0 & 1 & 1 & 0 & 0 & 0 & 0 & \cdots \\ \cdots & 0 & 0 & 0 & 1 & 1 & 0 & 0 & 0 & \cdots \\ \cdots & 0 & 0 & 0 & 0 & 1 & 1 & 0 & 0 & \cdots \\ \cdots & 0 & 0 & 0 & 0 & 0 & 1 & 1 & 0 & \cdots \\ \cdots & 0 & 0 & 0 & 0 & 0 & 0 & 1 & 1 & \cdots \\ 0 & \vdots & \vdots & \vdots & \vdots & \vdots & \vdots & \vdots & \ddots & \end{bmatrix} \begin{bmatrix} \vdots \\ y^{+}_{n,m-1} \\ y^{-}_{n,m-1} \\ y^{+}_{n,m+1} \\ y^{-}_{n,m+1} \\ y^{+}_{n,m+3} \\ y^{-}_{n,m+3} \\ y^{+}_{n,m+5} \\ y^{-}_{n,m+5} \\ \vdots \end{bmatrix}. \tag{E.9}$$

Denote the linear transformation operator by $\mathbf{T}$ and the K and W state vectors by $\underline{x}_K$ and $\underline{x}_W$, respectively. Then Eq. (E.9) can be restated as

$$\underline{x}_K = \mathbf{T}\underline{x}_W. \tag{E.10}$$

The operator $\mathbf{T}$ can be recognized as the Toeplitz operator associated with the linear, shift-invariant filter $H(z) = 1 + z^{-1}$. While the present context is not a simple convolution since $\underline{x}_W$ is not a simple time series, the inverse of $\mathbf{T}$ corresponds to the Toeplitz operator associated with

$$H(z) = \frac{1}{1+z^{-1}} = 1 - z^{-1} + z^{-2} - z^{-3} + \cdots .$$

Therefore, we may easily write down the inverted transformation:

$$\begin{bmatrix} \vdots \\ y^+_{n,m-1} \\ y^-_{n,m-1} \\ y^+_{n,m+1} \\ y^-_{n,m+1} \\ y^+_{n,m+3} \\ y^-_{n,m+3} \\ y^+_{n,m+5} \\ y^-_{n,m+5} \\ \vdots \end{bmatrix} = \begin{bmatrix} \ddots & & \vdots & \vdots & \vdots & \vdots & \vdots & \vdots & \vdots & \pm 1 \\ \cdots & 1 & -1 & 1 & -1 & 1 & -1 & 1 & -1 & \cdots \\ \cdots & 0 & 1 & -1 & 1 & -1 & 1 & -1 & 1 & \cdots \\ \cdots & 0 & 0 & 1 & -1 & 1 & -1 & 1 & -1 & \cdots \\ \cdots & 0 & 0 & 0 & 1 & -1 & 1 & -1 & 1 & \cdots \\ \cdots & 0 & 0 & 0 & 0 & 1 & -1 & 1 & -1 & \cdots \\ \cdots & 0 & 0 & 0 & 0 & 0 & 1 & -1 & 1 & \cdots \\ \cdots & 0 & 0 & 0 & 0 & 0 & 0 & 1 & -1 & \cdots \\ 0 & \vdots & \vdots & \vdots & \vdots & \vdots & \vdots & \vdots & \ddots & \end{bmatrix} \begin{bmatrix} \vdots \\ y_{n,m-1} \\ y_{n-1,m} \\ y_{n,m+1} \\ y_{n-1,m+2} \\ y_{n,m+3} \\ y_{n-1,m+4} \\ y_{n,m+5} \\ \vdots \end{bmatrix} \tag{E.11}$$

The case of the finite string is identical to that of the infinite string when the matrix $\mathbf{T}$ is simply "cropped" to a finite square size (leaving an isolated 1 in the lower right corner); in such cases, $\mathbf{T}^{-1}$ as given above is simply cropped to the same size, retaining its upper triangular ± 1 structure. Another interesting set of cases is obtained by inserting a 1 in the lower-left corner of the cropped $\mathbf{T}$ matrix to make it *circulant*; in these cases, the $M \times M$ matrix $\mathbf{T}^{-1}$ contains $\pm 1/2$ in every position for even M, and is singular for odd M (when there is one zero eigenvalue).

E.3 Excitation Examples

E.3.1 Localized Displacement Excitations

Whenever two adjacent components of $\underline{x}_K$ are initialized with equal amplitude, only a single W-variable will be affected. For example, the initial conditions (for time $n+1$)

$$\begin{aligned} y_{n,m-1} &= 1 \\ y_{n-1,m} &= 1 \end{aligned}$$

will initialize only $y^-_{n,m-1}$, a solitary left-going pulse of amplitude 1 at time $n = 0$, as can be seen from Eq. (E.11) by adding the leftmost columns explicitly written for $\mathbf{T}^{-1}$. Similarly, the initialization

$$\begin{aligned} y_{n-1,m-2} &= 1 \\ y_{n,m-1} &= 1 \end{aligned}$$

gives rise to an isolated right-going pulse $y^+_{n,m-1}$, corresponding to the leftmost column of $\mathbf{T}^{-1}$ plus the first column on the left not explicitly written in Eq. (E.11). The superposition of these two examples corresponds to a physical impulsive excitation at time 0 and position $m-1$:

$$\begin{aligned} y_{n-1,m-2} &= 1 \\ y_{n,m-1} &= 2 \\ y_{n-1,m} &= 1 \end{aligned} \tag{E.12}$$

Thus, the impulse starts out with amplitude 2 at time 0 and position $m-1$, and afterwards, impulses of amplitude 1 propagate away to the left and right along the string.

In summary, we see that to excite a single sample of displacement traveling in a single-direction, we must excite equally a pair of adjacent colums in $\mathbf{T}^{-1}$. This corresponds to equally weighted excitation of K-variable pairs the form $(y_{n,m}, y_{n-1,m\pm1})$.

Note that these examples involved only one of the two interleaved computational grids. Shifting over an odd number of spatial samples to the left or right would involve the other grid, as would shifting time forward or backward an odd number of samples.

E.3.2 Localized Velocity Excitations

Initial velocity excitations are straightforward in the DW paradigm, but can be less intuitive in the FDTD domain. It is well known that velocity in a displacement-wave DW simulation is determined by the *difference* of the right- and left-going waves [437]. Specifically, initial velocity waves $v^\pm$ can be computed from from initial displacement waves $y^\pm$ by spatially differentiating $y^\pm$ to obtain traveling *slope waves* $y'^\pm$, multiplying by minus the tension K to obtain *force waves*, and finally dividing by the wave impedance $R = \sqrt{K\epsilon}$ to obtain velocity waves:

$$\begin{aligned} v^+ &= -cy'^+ = \frac{f^+}{R} \\ v^- &= cy'^- = -\frac{f^-}{R}, \end{aligned} \tag{E.13}$$

where $c = \sqrt{K/\epsilon}$ denotes sound speed. The initial string velocity at each point is then $v(nT, mX) = v^+(n-m) + v^-(n+m)$. (A more direct derivation can be based on differentiating Eq. (E.4) with respect to x and solving for velocity traveling-wave components, considering left- and right-going cases separately at first, and arguing the general case by superposition.)

We can see from Eq. (E.11) that such asymmetry can be caused by unequal weighting of $y_{n,m}$ and $y_{n,m\pm 1}$. For example, the initialization

$$\begin{aligned} y_{n-1,m+1} &= +1 \\ y_{n-1,m} &= -1 \end{aligned}$$

corresponds to an impulse *velocity* excitation at position $m + 1/2$. In this case, both interleaved grids are excited.

E.3.3 More General Velocity Excitations

From Eq. (E.11), it is clear that initializing any single K variable $y_{n,m}$ corresponds to the initialization of an infinite number of W variables $y^+_{n,m}$ and $y^-_{n,m}$. That is, a single K variable $y_{n,m}$ corresponds to only a single column of $\mathbf{T}^{-1}$ for only one of the interleaved grids. For example, referring to Eq. (E.11), initializing the K variable $y_{n-1,m}$ to -1 at time n (with all other $y_{n,m}$ intialized to 0) corresponds to the W-variable initialization

$$\begin{aligned} y^+_{n,m-(2\mu+1)} &= +1, \quad \mu = 0,1,2,\cdots \\ y^-_{n,m-(2\mu+1)} &= -1, \quad \mu = 0,1,2,\cdots \end{aligned}$$

with all other W variables being initialized to zero. In view of earlier remarks, this corresponds to an impulsive velocity excitation on only one of the two subgrids. A schematic depiction from $\mu = m-5$ to $m+5$ of the W variables at time n is as follows:

$$\begin{array}{cccccc|ccccccc} \cdots & 1 & 0 & 1 & 0 & 1 & 0 & 0 & 0 & 0 & 0 & 0 & \cdots \\ \cdots & -1 & 0 & -1 & 0 & -1 & 0 & 0 & 0 & 0 & 0 & 0 & \cdots \\ \hline \cdots & 0 & 0 & 0 & 0 & 0 & 0 & 0 & 0 & 0 & 0 & 0 & \cdots \\ & & & & & & m & & & & & \mu & \rightarrow \end{array} \tag{E.14}$$

Below the solid line is the sum of the left- and right-going traveling-wave components, *i.e.*, the corresponding K variables at time n. The vertical lines divide positions $\mu = m-1$ and $\mu = m$. The initial displacement is zero everywhere at time n, consistent with an initial velocity excitation. At times $\nu = n+1, n+2, n+3, n+4$, the configuration evolves as follows:

$$\begin{array}{cccccc|ccccccc} \cdots & 0 & 1 & 0 & 1 & 0 & 1 & 0 & 0 & 0 & 0 & 0 & \cdots \\ \cdots & 0 & -1 & 0 & -1 & 0 & 0 & 0 & 0 & 0 & 0 & 0 & \cdots \\ \hline \cdots & 0 & 0 & 0 & 0 & 0 & 1 & 0 & 0 & 0 & 0 & 0 & \cdots \end{array} \tag{E.15}$$

$$\begin{array}{cccccc|cccccccc}
\cdots & 1 & 0 & 1 & 0 & 1 & 0 & 1 & 0 & 0 & 0 & 0 & \cdots \\
\cdots & -1 & 0 & -1 & 0 & 0 & 0 & 0 & 0 & 0 & 0 & 0 & \cdots \\
\hline
\cdots & 0 & 0 & 0 & 0 & 1 & 0 & 1 & 0 & 0 & 0 & 0 & \cdots
\end{array} \tag{E.16}$$

$$\begin{array}{cccccc|cccccccc}
\cdots & 0 & 1 & 0 & 1 & 0 & 1 & 0 & 1 & 0 & 0 & 0 & \cdots \\
\cdots & 0 & -1 & 0 & 0 & 0 & 0 & 0 & 0 & 0 & 0 & 0 & \cdots \\
\hline
\cdots & 0 & 0 & 0 & 1 & 0 & 1 & 0 & 1 & 0 & 0 & 0 & \cdots
\end{array} \tag{E.17}$$

$$\begin{array}{cccccc|cccccccc}
\cdots & 1 & 0 & 1 & 0 & 1 & 0 & 1 & 0 & 1 & 0 & 0 & \cdots \\
\cdots & -1 & 0 & 0 & 0 & 0 & 0 & 0 & 0 & 0 & 0 & 0 & \cdots \\
\hline
\cdots & 0 & 0 & 1 & 0 & 1 & 0 & 1 & 0 & 1 & 0 & 0 & \cdots
\end{array} \tag{E.18}$$

The sequence $[\ldots,1,0,1,0,1,\ldots]$ consists of a dc (zero-frequency) component with amplitude $1/2$, plus a sampled sinusoid of amplitude $1/2$ oscillating at half the sampling rate $f_s = 1/T$. The dc component is physically correct for an initial velocity point-excitation (a spreading square pulse on the string is expected). However, the component at $f_s/2$ is usually regarded as an artifact of the finite difference scheme. From the DW interpretation of the FDTD scheme, which is an exact, bandlimited physical interpretation, we see that physically the component at $f_s/2$ comes about from initializing velocity on only one of the two interleaved subgrids of the FDTD scheme. Any asymmetry in the excitation of the two interleaved grids will result in excitation of this frequency component.

Due to the independent interleaved subgrids in the FDTD algorithm, it is nearly always non-physical to excite only one of them, as the above example makes clear. It is analogous to illuminating only every other pixel in a digital image. However, joint excitation of both grids may be accomplished either by exciting adjacent spatial samples at the same time, or the same spatial sample at successive times instants.

In addition to the W components being non-local, they can demand a larger dynamic range than the K variables. For example, if the entire semi-infinite string for $m < 0$ is initialized with velocity $2X/T$, the initial displacement traveling-wave components look as follows:

$$\begin{array}{ccccccc|cccccc}
\cdots & 6 & 5 & 4 & 3 & 2 & 1 & 0 & 0 & 0 & 0 & 0 & \cdots \\
\cdots & -6 & -5 & -4 & -3 & -2 & -1 & 0 & 0 & 0 & 0 & 0 & \cdots \\
\hline
\cdots & 0 & 0 & 0 & 0 & 0 & 0 & 0 & 0 & 0 & 0 & 0 & \cdots
\end{array} \tag{E.19}$$

and the variables evolve forward in time as follows:

$$\begin{array}{ccccccc|cccccc}
\cdots & 7 & 6 & 5 & 4 & 3 & 2 & 1 & 0 & 0 & 0 & 0 & \cdots \\
\cdots & -5 & -4 & -3 & -2 & -1 & 0 & 0 & 0 & 0 & 0 & 0 & \cdots \\
\hline
\cdots & 2 & 2 & 2 & 2 & 2 & 2 & 1 & 0 & 0 & 0 & 0 & \cdots
\end{array} \tag{E.20}$$

$$\begin{array}{cccccccc|cccccc}
\cdots & 8 & 7 & 6 & 5 & 4 & 3 & 2 & 1 & 0 & 0 & 0 & \cdots \\
\cdots & -4 & -3 & -2 & -1 & 0 & 0 & 0 & 0 & 0 & 0 & 0 & \cdots \\
\hline
\cdots & 4 & 4 & 4 & 4 & 4 & 3 & 2 & 1 & 0 & 0 & 0 & \cdots
\end{array} \tag{E.21}$$

$$\begin{array}{ccccccc|cccccc}
\cdots & 9 & 8 & 7 & 6 & 5 & 4 & 3 & 2 & 1 & 0 & 0 & \cdots \\
\cdots & -3 & -2 & -1 & 0 & 0 & 0 & 0 & 0 & 0 & 0 & 0 & \cdots \\
\hline
\cdots & 6 & 6 & 6 & 6 & 5 & 4 & 3 & 2 & 1 & 0 & 0 & \cdots
\end{array} \tag{E.22}$$

Thus, the left semi-infinite string moves upward at a constant velocity of 2, while a ramp spreads out to the left and right of position $\mu = m$ at speed c, as expected physically. By Eq. (E.9), the corresponding initial FDTD state for this case is

$$\begin{aligned}
y_{n,\mu} &= 0, \quad \mu \in \mathbf{Z} \\
y_{n-1,m-1} &= -1, \\
y_{n-1,\mu} &= -2, \quad \mu < m-1,
\end{aligned}$$

where $\mathbf{Z}$ denotes the set of all integers. While the FDTD excitation is also not local, of course, it is bounded for all μ.

Since the traveling-wave components of initial velocity excitations are generally non-local in a displacement-based simulation, as illustrated in the preceding examples, it is often preferable to use velocity waves (or force waves) in the first place [447].

Another reason to prefer force or velocity waves is that displacement inputs are inherently impulsive. To see why this is so, consider that any physically correct driving input must effectively exert some finite force on the string, and this force is free to change arbitrarily over time. The "equivalent circuit" of the infinitely long string at the driving point is a "dashpot" having real, positive resistance $2R = 2\sqrt{K\epsilon}$. The applied force $f(t)$ can be divided by $2R$ to obtain the velocity $v(t)$ of the string driving point, and this velocity is free to vary arbitrarily over time, proportional to the applied force. However, this velocity must be *time-integrated* to obtain a displacement $y(t)$. Therefore, there can be *no instantaneous displacement response to a finite driving force.* In other words, any instantaneous effect of an input driving signal on an output displacement sample is non-physical except in the case of a massless system. Infinite force is required to move the string instantaneously. In sampled displacement simulations, we must interpret displacement changes as resulting from time-integration over a sampling period. As the sampling rate increases, any physically meaningful displacement driving signal must converge to zero.

E.3.4 Additive Inputs

Instead of initial conditions, ongoing input signals can be defined analogously. For example, feeding an input signal u_n into the FDTD via

$$\begin{aligned} y_{n,m-1} &= y_{n,m-1} + u_{n-1} \\ y_{n,m} &= y_{n,m} + 2u_n \\ y_{n,m+1} &= y_{n,m+1} + u_{n-1} \end{aligned} \tag{E.23}$$

corresponds to physically driving a single sample of string displacement at position m. This is the spatially distributed alternative to the temporally distributed solution of feeding an input to a single displacement sample via the filter $H(z) = 1 - z^{-2}$ as discussed in [223].

E.3.5 Interpretation of the Time-Domain KW Converter

As shown above, driving a single displacement sample $y_{n,m}$ in the FDTD corresponds to driving a velocity input at position m on two alternating subgrids over time. Therefore, the filter $H(z) = 1 - z^{-2}$ acts as the filter $H(z) = 1 - z^{-1}$ on either subgrid alone—a first-order difference. Since displacement is being simulated, velocity inputs must be numerically integrated. The first-order difference can be seen as canceling this integration, thereby converting a velocity input to a displacement input, as in Eq. (E.23).

E.4 State Space Formulation

In this section, we will summarize and extend the above discussion by means of a *state space analysis* [220].

E.4.1 FDTD State Space Model

Let $\underline{x}_K(n)$ denote the FDTD state for one of the two subgrids at time n, as defined by Eq. (E.10). The other subgrid is handled identically and will not be considered explicitly. In fact, the other subgrid can be dropped altogether to obtain a *half-rate, staggered grid* scheme [55, 147]. However, boundary conditions and input signals will couple the subgrids, in general. To land on the same subgrid after a state update, it is necessary to advance time by two samples instead of one. The state-space model for one subgrid of the FDTD model of the ideal string may then be written as

$$\begin{aligned} \underline{x}_K(n+2) &= \mathbf{A}_K\, \underline{x}_K(n) + \mathbf{B}_K\, \underline{u}(n+2) \\ \underline{y}(n) &= \mathbf{C}_K\, \underline{x}_K(n). \end{aligned} \tag{E.24}$$

To avoid the issue of boundary conditions for now, we will continue working with the infinitely long string. As a result, the state vector $\underline{x}_K(n)$ denotes a point in a space of countably infinite dimensionality. A proper treatment of this case would be in terms of operator theory [325]. However, matrix notation is also clear and will be used below. Boundary conditions are taken up in §E.4.3.

When there is a general input signal vector $\underline{u}(n)$, it is necessary to augment the input matrix $\mathbf{B}_K$ to accomodate contributions over both time steps. This is because inputs to positions $m \pm 1$ at time $n+1$ affect position m at time $n+2$. Henceforth, we assume $\mathbf{B}_K$ and $\underline{u}$ have been augmented in this way. Thus, if there are q input signals $\underline{v}(n) \triangleq [v_i(n)]$, $i = 1, \ldots, q$, driving the full string state through weights $\underline{\beta}_m \triangleq [\beta_{m,i}]$, $m \in \mathbf{Z}$, the vector $\underline{u}(n) =$ is of dimension $2q \times 1$:

$$\underline{u}(n+2) = \left[\begin{array}{c} \underline{v}(n+2) \\ \underline{v}(n+1) \end{array}\right]$$

When there is only one physical input, as is typically assumed for plucked, struck, and bowed strings, then $q = 1$ and $\underline{u}$ is 2×1. The matrix $\mathbf{B}_K$ weights these inputs before they are added to the state vector for time $n+2$, and its entries are derived in terms of the $\beta_{m,i}$ coefficients below.

$\mathbf{C}_K$ forms the output signal as an arbitrary linear combination of states. To obtain the usual displacement output for the subgrid, $\mathbf{C}_K$ is the matrix formed from the identity matrix by deleting every other row, thereby retaining all displacement samples at time n and discarding all displacement samples at time $n-1$ in the state vector $\underline{x}_K(n)$:

$$\underbrace{\left[\begin{array}{c} \vdots \\ y_{n,m-2} \\ y_{n,m} \\ y_{n,m+2} \\ y_{n,m+4} \\ \vdots \end{array}\right]}_{\underline{y}(n)} = \underbrace{\left[\begin{array}{cccccccccc} & \vdots & \vdots & \vdots & \vdots & \vdots & \vdots & \vdots & \vdots & \\ \cdots & 1 & 0 & 0 & 0 & 0 & 0 & 0 & 0 & \cdots \\ \cdots & 0 & 0 & 1 & 0 & 0 & 0 & 0 & 0 & \cdots \\ \cdots & 0 & 0 & 0 & 0 & 1 & 0 & 0 & 0 & \cdots \\ \cdots & 0 & 0 & 0 & 0 & 0 & 0 & 1 & 0 & \cdots \\ & \vdots & \vdots & \vdots & \vdots & \vdots & \vdots & \vdots & \vdots & \end{array}\right]}_{\mathbf{C}_K} \underbrace{\left[\begin{array}{c} \vdots \\ y_{n,m-2} \\ y_{n-1,m-1} \\ y_{n,m} \\ y_{n-1,m+1} \\ y_{n,m+2} \\ y_{n-1,m+3} \\ y_{n,m+4} \\ \vdots \end{array}\right]}_{\underline{x}_K(n)}$$

The state transition matrix $\mathbf{A}_K$ may be obtained by first performing a one-step time update,

$$y_{n+2,m} = y_{n+1,m-1} + y_{n+1,m+1} - y_{n,m} + \underline{\beta}_m^T \underline{v}(n+2),$$

and then expanding the two $n+1$ terms in terms of the state at time n:

$$
\begin{aligned}
y_{n+1,m-1} &= y_{n,m-2} + y_{n,m} - y_{n-1,m-1} + \underline{\beta}_{m-1}^T \underline{v}(n+1) \\
y_{n+1,m+1} &= y_{n,m} + y_{n,m+2} - y_{n-1,m+1} + \underline{\beta}_{m+1}^T \underline{v}(n+1)
\end{aligned}
\tag{E.25}
$$

The intra-grid state update for even m is then given by

$$
\begin{aligned}
& y_{n+2,m} \\
&= y_{n,m-2} - y_{n-1,m-1} + y_{n,m} - y_{n-1,m+1} + y_{n,m+2} \\
&\quad + \underline{\beta}_m^T \underline{v}(n+2) + (\underline{\beta}_{m-1} + \underline{\beta}_{m+1})^T \underline{v}(n+1) \\
&= [1,-1,1,-1,1] \begin{bmatrix} y_{n,m-2} \\ y_{n-1,m-1} \\ y_{n,m} \\ y_{n-1,m+1} \\ y_{n,m+2} \\ y_{n-1,m+3} \end{bmatrix} \\
&\quad + \left[\underline{\beta}_m^T \quad (\underline{\beta}_{m-1} + \underline{\beta}_{m+1})^T\right] \begin{bmatrix} \underline{v}(n+2) \\ \underline{v}(n+1) \end{bmatrix}.
\end{aligned}
\tag{E.26}
$$

For odd m, the update in Eq. (E.25) is used. Thus, every other row of $\mathbf{A}_K$, for time $n+2$, consists of the vector $[1,-1,1,-1,1]$ preceded and followed by zeros. Successive rows for time $n+2$ are shifted right two places. The rows for time $n+1$ consist of the vector $[1,-1,1]$ aligned similarly:

$$
\underbrace{\begin{bmatrix} \vdots \\ y_{n+1,m-1} \\ y_{n+2,m} \\ y_{n+1,m+1} \\ y_{n+2,m+2} \\ \vdots \end{bmatrix}}_{\underline{x}_K(n+2)}
\leftarrow
\underbrace{\begin{bmatrix}
 & \vdots & \vdots & \vdots & \vdots & \vdots & \vdots & \vdots & \\
\cdots & 1 & -1 & 1 & 0 & 0 & 0 & 0 & \cdots \\
\cdots & 1 & -1 & 1 & -1 & 1 & 0 & 0 & \cdots \\
\cdots & 0 & 0 & 1 & -1 & 1 & 0 & 0 & \cdots \\
\cdots & 0 & 0 & 1 & -1 & 1 & -1 & 1 & \cdots \\
 & \vdots & \vdots & \vdots & \vdots & \vdots & \vdots & \vdots &
\end{bmatrix}}_{\mathbf{A}_K}
\underbrace{\begin{bmatrix} \vdots \\ y_{n,m-2} \\ y_{n-1,m-1} \\ y_{n,m} \\ y_{n-1,m+1} \\ y_{n,m+2} \\ y_{n-1,m+3} \\ \vdots \end{bmatrix}}_{\underline{x}_K(n)}
$$

From Eq. (E.26) we also see that the input matrix $\mathbf{B}_K$ is given as defined in the following expression:

$$\underbrace{\begin{bmatrix} \vdots \\ y_{n+1,m-1} \\ y_{n+2,m} \\ y_{n+1,m+1} \\ y_{n+2,m+2} \\ y_{n+1,m+3} \\ y_{n+2,m+4} \\ y_{n+1,m+5} \\ \vdots \end{bmatrix}}_{\underline{x}_K(n+2)} \leftarrow \underbrace{\begin{bmatrix} \vdots & \vdots \\ 0 & \underline{\beta}^T_{m-1} \\ \underline{\beta}^T_m & \underline{\beta}^T_{m-1} + \underline{\beta}^T_{m+1} \\ 0 & \underline{\beta}^T_{m+1} \\ \underline{\beta}^T_{m+2} & \underline{\beta}^T_{m+1} + \underline{\beta}^T_{m+3} \\ 0 & \underline{\beta}^T_{m+3} \\ \underline{\beta}^T_{m+4} & \underline{\beta}^T_{m+3} + \underline{\beta}^T_{m+5} \\ 0 & \underline{\beta}^T_{m+5} \\ \vdots & \vdots \end{bmatrix}}_{\mathbf{B}_K} \underbrace{\begin{bmatrix} \underline{v}(n+2) \\ \underline{v}(n+1) \end{bmatrix}}_{\underline{u}(n+2)}.$$

E.4.2 DW State Space Model

As discussed in §E.2, the traveling-wave decomposition Eq. (E.4) defines a linear transformation Eq. (E.10) from the DW state to the FDTD state:

$$\underline{x}_K = \mathbf{T}\,\underline{x}_W \tag{E.27}$$

Since $\mathbf{T}$ is invertible, it qualifies as a linear transformation for performing a *change of coordinates* for the state space. Substituting Eq. (E.27) into the FDTD state space model Eq. (E.24) gives

$$\begin{aligned} \mathbf{T}\,\underline{x}_W(n+2) &= \mathbf{A}_K\,\mathbf{T}\,\underline{x}_W(n) + \mathbf{B}_K\,\underline{u}(n+2) && \text{(E.28)} \\ \underline{y}(n) &= \mathbf{C}_K\,\mathbf{T}\,\underline{x}_W(n). && \text{(E.29)} \end{aligned}$$

Multiplying through Eq. (E.28) by $\mathbf{T}^{-1}$ gives a new state-space representation of the same dynamic system which we will show is in fact the DW model of Fig. E.2:

$$\begin{aligned} \underline{x}_W(n+2) &= \mathbf{A}_W\,\underline{x}_W(n) + \mathbf{B}_W\,\underline{u}(n+2) \\ \underline{y}(n) &= \mathbf{C}_W\,\underline{x}_W(n) \end{aligned} \tag{E.30}$$

where

$$\begin{aligned} \mathbf{A}_W &\triangleq \mathbf{T}^{-1}\mathbf{A}_K\,\mathbf{T} \\ \mathbf{B}_W &\triangleq \mathbf{T}^{-1}\mathbf{B}_K \\ \mathbf{C}_W &\triangleq \mathbf{C}_K\,\mathbf{T} \end{aligned} \tag{E.31}$$

To verify that the DW model derived in this manner is the computation diagrammed in Fig. E.2, we may write down the state transition matrix for one subgrid from the figure to obtain the permutation matrix $\mathbf{A}_W$,

$$\underbrace{\begin{bmatrix} \vdots \\ y^+_{n+2,m-2} \\ y^-_{n+2,m-2} \\ y^+_{n+2,m} \\ y^-_{n+2,m} \\ y^+_{n+2,m+2} \\ y^-_{n+2,m+2} \\ \vdots \end{bmatrix}}_{\underline{x}_W(n+2)} \leftarrow \underbrace{\begin{bmatrix} \cdots & \vdots & \vdots & \vdots & \vdots & \vdots & \vdots & \vdots & \vdots & \vdots & \vdots & \\ \cdots & 1 & 0 & 0 & 0 & 0 & 0 & 0 & 0 & 0 & 0 & \cdots \\ \cdots & 0 & 0 & 0 & 0 & 0 & 1 & 0 & 0 & 0 & 0 & \cdots \\ \cdots & 0 & 0 & 1 & 0 & 0 & 0 & 0 & 0 & 0 & 0 & \cdots \\ \cdots & 0 & 0 & 0 & 0 & 0 & 0 & 0 & 1 & 0 & 0 & \cdots \\ \cdots & 0 & 0 & 0 & 0 & 1 & 0 & 0 & 0 & 0 & 0 & \cdots \\ \cdots & 0 & 0 & 0 & 0 & 0 & 0 & 0 & 0 & 0 & 1 & \cdots \\ \cdots & \vdots & \vdots & \vdots & \vdots & \vdots & \vdots & \vdots & \vdots & \vdots & \vdots & \end{bmatrix}}_{\mathbf{A}_W} \underbrace{\begin{bmatrix} \vdots \\ y^+_{n,m-4} \\ y^-_{n,m-4} \\ y^+_{n,m-2} \\ y^-_{n,m-2} \\ y^+_{n,m} \\ y^-_{n,m} \\ y^+_{n,m+2} \\ y^-_{n,m+2} \\ y^+_{n,m+4} \\ y^-_{n,m+4} \\ \vdots \end{bmatrix}}_{\underline{x}_W(n)} \tag{E.32}$$

and displacement output matrix $\mathbf{C}_W$:

$$\underbrace{\begin{bmatrix} \vdots \\ y_{n,m-2} \\ y_{n,m} \\ y_{n,m+2} \\ y_{n,m+4} \\ \vdots \end{bmatrix}}_{\underline{y}(n)} = \underbrace{\begin{bmatrix} & \vdots & \vdots & \vdots & \vdots & \vdots & \vdots & \vdots & \vdots & \\ \cdots & 1 & 1 & 0 & 0 & 0 & 0 & 0 & 0 & \cdots \\ \cdots & 0 & 0 & 1 & 1 & 0 & 0 & 0 & 0 & \cdots \\ \cdots & 0 & 0 & 0 & 0 & 1 & 1 & 0 & 0 & \cdots \\ \cdots & 0 & 0 & 0 & 0 & 0 & 0 & 1 & 1 & \cdots \\ & \vdots & \vdots & \vdots & \vdots & \vdots & \vdots & \vdots & \vdots & \end{bmatrix}}_{\mathbf{C}_W} \underbrace{\begin{bmatrix} \vdots \\ y^+_{n,m-2} \\ y^-_{n,m-2} \\ y^+_{n,m} \\ y^-_{n,m} \\ y^+_{n,m+2} \\ y^-_{n,m+2} \\ y^+_{n,m+4} \\ y^-_{n,m+4} \\ \vdots \end{bmatrix}}_{\underline{x}_W(n)}$$

DW Displacement Inputs

We define general DW inputs as follows:

$$y^+_{n,m} = y^+_{n-1,m-1} + (\underline{\gamma}^+_m)^T \underline{v}(n) \tag{E.33}$$

$$y^-_{n,m} = y^-_{n-1,m+1} + (\underline{\gamma}^-_m)^T \underline{v}(n) \tag{E.34}$$

The mth $2q \times 2$ block of the input matrix $\mathbf{B}_W$ driving state components $[y^+_{n+2,m}, y^-_{n+2,m}]^T$ and multiplying $[\underline{v}(n+2)^T, \underline{v}(n+1)^T]^T$ is then given by

$$(\mathbf{B}_W)_m = \begin{bmatrix} (\underline{\gamma}^+_m)^T & (\underline{\gamma}^+_{m-1})^T \\ (\underline{\gamma}^-_m)^T & (\underline{\gamma}^-_{m+1})^T \end{bmatrix}. \tag{E.35}$$

Typically, input signals are injected equally to the left and right along the string, in which case

$$\underline{\gamma}^+_m = \underline{\gamma}^-_m \stackrel{\Delta}{=} \underline{\gamma}_m.$$

Physically, this corresponds to applied forces at a single, non-moving, string position over time. The state update with this simplification appears as

$$\underbrace{\begin{bmatrix} \vdots \\ y^+_{n+2,m} \\ y^-_{n+2,m} \\ \vdots \end{bmatrix}}_{\underline{x}_W(n+2)} = \mathbf{A}_W \underline{x}_W(n) + \underbrace{\begin{bmatrix} \vdots & \vdots \\ \underline{\gamma}^T_m & \underline{\gamma}^T_{m-1} \\ \underline{\gamma}^T_m & \underline{\gamma}^T_{m+1} \\ \vdots & \vdots \end{bmatrix}}_{\mathbf{B}_W} \underbrace{\begin{bmatrix} \underline{v}(n+2) \\ \underline{v}(n+1) \end{bmatrix}}_{\underline{u}(n+2)}.$$

Note that if there are no inputs driving the adjacent subgrid ($\underline{\gamma}_{m-1} = \underline{\gamma}_{m+1} = 0$), such as in a half-rate staggered grid scheme, the input reduces to

$$\underline{x}_W(n+2) = \mathbf{A}_W \underline{x}_W(n) + \underbrace{\begin{bmatrix} \vdots \\ \underline{\gamma}^T_{m-2} \\ \underline{\gamma}^T_{m-2} \\ \underline{\gamma}^T_m \\ \underline{\gamma}^T_m \\ \underline{\gamma}^T_{m+2} \\ \underline{\gamma}^T_{m+2} \\ \vdots \end{bmatrix}}_{\mathbf{B}_W} \underline{v}(n+2).$$

To show that the directly obtained FDTD and DW state-space models correspond to the same dynamic system, it remains to verify that $\mathbf{A}_W =$

$\mathbf{T}^{-1}\mathbf{A}_K\mathbf{T}$. It is somewhat easier to show that

$$\begin{aligned}\mathbf{T}\,\mathbf{A}_W &= \mathbf{A}_K\,\mathbf{T}\\ &= \begin{bmatrix} & \vdots & \vdots & \vdots & \vdots & \vdots & \vdots & \vdots & \vdots & \vdots & \vdots & \\ \cdots & 1 & 0 & 0 & 0 & 0 & 1 & 0 & 0 & 0 & 0 & \cdots\\ \cdots & 0 & 0 & 1 & 0 & 0 & 1 & 0 & 0 & 0 & 0 & \cdots\\ \cdots & 0 & 0 & 1 & 0 & 0 & 0 & 0 & 1 & 0 & 0 & \cdots\\ \cdots & 0 & 0 & 0 & 0 & 1 & 0 & 0 & 1 & 0 & 0 & \cdots\\ & \vdots & \vdots & \vdots & \vdots & \vdots & \vdots & \vdots & \vdots & \vdots & \vdots & \end{bmatrix}.\end{aligned}$$

A straightforward calculation verifies that the above identity holds, as expected. One can similarly verify $\mathbf{C}_W = \mathbf{C}_K\,\mathbf{T}$, as expected. The relation $\mathbf{B}_W = \mathbf{T}^{-1}\,\mathbf{B}_K$ provides a recipe for translating any choice of input signals for the FDTD model to equivalent inputs for the DW model, or vice versa. For example, in the scalar input case ($q = 1$), the DW input-weights $\mathbf{B}_W$ become FDTD input-weights $\mathbf{B}_K$ according to

$$\begin{bmatrix}\vdots\\ y_{n+1,m-1}\\ y_{n+2,m}\\ y_{n+1,m+1}\\ y_{n+2,m+2}\\ \vdots\end{bmatrix} \leftarrow \underbrace{\begin{bmatrix}\vdots & \vdots\\ \gamma_m^+ + \gamma_{m-1}^- & \gamma_{m-1}^+ + \gamma_{m-1}^-\\ \gamma_m^+ + \gamma_m^- & \gamma_{m-1}^+ + \gamma_{m+1}^-\\ \gamma_m^- + \gamma_{m+1}^+ & \gamma_{m+1}^+\gamma_{m+1}^-\\ \gamma_{m+2}^+ + \gamma_{m+2}^- & \gamma_{m+1}^+ + \gamma_{m+3}^-\\ \vdots & \vdots\end{bmatrix}}_{\mathbf{B}_K}\begin{bmatrix}\underline{v}(n+2)\\ \underline{v}(n+1)\end{bmatrix}$$

The left- and right-going input-weight superscripts indicate the origin of each coefficient. Setting $\gamma_m^+ = \gamma_m^-$ results in

$$\mathbf{B}_K = \begin{bmatrix}\vdots & \vdots\\ \gamma_m + \gamma_{m-1} & 2\gamma_{m-1}\\ 2\gamma_m & \gamma_{m-1} + \gamma_{m+1}\\ \gamma_m + \gamma_{m+1} & 2\gamma_{m+1}\\ 2\gamma_{m+2} & \gamma_{m+1} + \gamma_{m+3}\\ \vdots & \vdots\end{bmatrix} \tag{E.36}$$

Finally, when $\gamma_m = 1$ and $\gamma_\mu = 0$ for all $\mu \neq m$, we obtain the result familiar from Eq. (E.23):

$$\mathbf{B}_K = \begin{bmatrix} \vdots & \vdots \\ 1 & 0 \\ 2 & 0 \\ 1 & 0 \\ \vdots & \vdots \end{bmatrix}$$

Similarly, setting $\gamma_\mu^\pm = 0$ for all $\mu \neq m+1$, the weighting pattern $(1, 2, 1)$ appears in the second column, shifted down one row. Thus, $\mathbf{B}_K$ in general (for physically stationary displacement inputs) can be seen as the superposition of weight patterns $(1, 2, 1)$ in the left column for even m, and the right column for odd m (the other subgrid), where the 2 is aligned with the driven sample. This is the general collection of displacement inputs.

DW Non-Displacement Inputs

Since a displacement input at position m corresponds to symmetrically exciting the right- and left-going traveling-wave components y_m^+ and y_m^-, it is of interest to understand what it means to excite these components *antisymmetrically*. As discussed in §E.3.3, an antisymmetric excitation of traveling-wave components can be interpreted as a *velocity* excitation. It was noted that localized velocity excitations in the FDTD generally correspond to non-localized velocity excitations in the DW, and that velocity in the DW is proportional to the *spatial derivative* of the difference between the left-going and right-going traveling displacement-wave components (see Eq. (E.13)). More generally, the antisymmetric component of displacement-wave excitation can be expressed in terms of any wave variable which is linearly independent relative to displacement, such as acceleration, slope, force, momentum, and so on. Since the state space of a vibrating string (and other mechanical systems) is traditionally taken to be position and velocity, it is perhaps most natural to relate the antisymmetric excitation component to velocity.

In practice, the simplest way to handle a velocity input $v_m(n)$ in a DW simulation is to first pass it through a first-order integrator of the form

$$H(z) = \frac{1}{1-z^{-1}} = 1 + z^{-1} + z^{-2} + \cdots \tag{E.37}$$

to convert it to a displacement input. By the equivalence of the DW and FDTD models, this works equally well for the FDTD model. However, in view of §E.3.3, this approach does not take full advantage of the ability of the FDTD scheme to provide localized velocity inputs for applications such as simulating a piano hammer strike. The FDTD provides such velocity inputs for "free" while the DW requires the external integrator Eq. (E.37).

Note, by the way, that these "integrals" (both that done internally by the FDTD and that done by Eq. (E.37)) are merely sums over discrete time—not true integrals. As a result, they are exact only at dc (and also trivially at $f_s/2$, where the output amplitude is zero). Discrete sums can also be considered exact integrators for impulse-train inputs—a point of view sometimes useful when interpreting simulation results. For normal bandlimited signals, discrete sums most accurately approximate integrals in a neighborhood of dc. The KW-converter filter $H(z) = 1 - z^{-2}$ has analogous properties.

Input Locality

The DW state-space model is given in terms of the FDTD state-space model by Eq. (E.31). The similarity transformation matrix $\mathbf{T}$ is bidiagonal, so that $\mathbf{C}_K$ and $\mathbf{C}_W = \mathbf{C}_K\mathbf{T}$ are both approximately diagonal when the output is string displacement for all m. However, since $\mathbf{T}^{-1}$ given in Eq. (E.11) is upper triangular, the input matrix $\mathbf{B}_W = \mathbf{T}^{-1}\mathbf{B}_K$ can replace sparse input matrices $\mathbf{B}_K$ with only half-sparse $\mathbf{B}_W$, unless successive columns of $\mathbf{T}^{-1}$ are equally weighted, as discussed in §E.3. We can say that local K-variable excitations may correspond to *non-local* W-variable excitations. From Eq. (E.35) and Eq. (E.36), we see that *displacement inputs are always local in both systems.* Therefore, local FDTD and non-local DW excitations can only occur when a variable dual to displacement is being excited, such as velocity. If the external integrator Eq. (E.37) is used, all inputs are ultimately displacement inputs, and the distinction disappears.

E.4.3 Boundary Conditions

The relations of the previous section do not hold exactly when the string length is finite. A finite-length string forces consideration of *boundary conditions.* In this section, we will introduce boundary conditions as perturbations of the state transition matrix. In addition, we will use the DW-FDTD equivalence to obtain physically well behaved boundary conditions for the FDTD method.

Consider an ideal vibrating string with $M = 8$ spatial samples. This is a sufficiently large number to make clear most of the repeating patterns in the general case. Introducing boundary conditions is most straightforward in the DW paradigm. We therefore begin with the order 8 DW model, for which the state vector (for the 0th subgrid) will be

$$\underline{x}_W(n) = \begin{bmatrix} y^+_{n,0} \\ y^-_{n,0} \\ y^+_{n,2} \\ y^-_{n,2} \\ y^+_{n,4} \\ y^-_{n,4} \\ y^+_{n,6} \\ y^-_{n,6} \end{bmatrix}.$$

The displacement output matrix is given by

$$\mathbf{C}_W = \begin{bmatrix} 1 & 1 & 0 & 0 & 0 & 0 & 0 & 0 \\ 0 & 0 & 1 & 1 & 0 & 0 & 0 & 0 \\ 0 & 0 & 0 & 0 & 1 & 1 & 0 & 0 \\ 0 & 0 & 0 & 0 & 0 & 0 & 1 & 1 \end{bmatrix}$$

and the input matrix $\mathbf{B}_W$ is an arbitrary $M \times 2q$ matrix. We will choose a scalar input signal $u(n)$ driving the displacement of the second spatial sample with unit gain:

$$\mathbf{B}_W = \begin{bmatrix} 0 & 0 \\ 0 & 0 \\ 1/2 & 1/2 \\ 1/2 & 1/2 \\ 0 & 0 \\ 0 & 0 \\ 0 & 0 \\ 0 & 0 \end{bmatrix}$$

The state transition matrix $\mathbf{A}_W$ is obtained by reducing Eq. (E.32) to finite order in some way, thereby introducing boundary conditions.

Resistive Terminations

Let's begin with simple "resistive" terminations at the string endpoints, resulting in the reflection coefficient g at each end of the string, where $|g| \leq 1$ corresponds to nonnegative (passive) termination resistances [447]. Inspection of Fig. E.2 makes it clear that terminating the left endpoint may be accomplished by setting

$$y^+_{n,0} = g_l y^-_{n,0},$$

and the right termination corresponds to

$$y^-_{n,6} = g_r y^+_{n,6}.$$

By allowing an additional two samples of round-trip delay in each endpoint reflectance (one sample in the chosen subgrid), we can implement these reflections within the state-transition matrix:

$$\tilde{\mathbf{A}}_W = \begin{bmatrix} 0 & g_l & 0 & 0 & 0 & 0 & 0 & 0 \\ 0 & 0 & 0 & 1 & 0 & 0 & 0 & 0 \\ 1 & 0 & 0 & 0 & 0 & 0 & 0 & 0 \\ 0 & 0 & 0 & 0 & 0 & 1 & 0 & 0 \\ 0 & 0 & 1 & 0 & 0 & 0 & 0 & 0 \\ 0 & 0 & 0 & 0 & 0 & 0 & 0 & 1 \\ 0 & 0 & 0 & 0 & 1 & 0 & 0 & 0 \\ 0 & 0 & 0 & 0 & 0 & 0 & g_r & 0 \end{bmatrix} \tag{E.38}$$

The simplest choice of state transformation matrix $\mathbf{T}$ is obtained by cropping it to size $M \times M$:

$$\mathbf{T} \triangleq \begin{bmatrix} 1 & 1 & 0 & 0 & 0 & 0 & 0 & 0 \\ 0 & 1 & 1 & 0 & 0 & 0 & 0 & 0 \\ 0 & 0 & 1 & 1 & 0 & 0 & 0 & 0 \\ 0 & 0 & 0 & 1 & 1 & 0 & 0 & 0 \\ 0 & 0 & 0 & 0 & 1 & 1 & 0 & 0 \\ 0 & 0 & 0 & 0 & 0 & 1 & 1 & 0 \\ 0 & 0 & 0 & 0 & 0 & 0 & 1 & 1 \\ 0 & 0 & 0 & 0 & 0 & 0 & 0 & 1 \end{bmatrix}$$

An advantage of this choice is that its inverse $\mathbf{T}^{-1}$ is similarly a simple cropping of the $M = \infty$ case. However, the corresponding FDTD system is not so elegant:

$$\begin{aligned} \tilde{\mathbf{A}}_K &\triangleq \mathbf{T}\tilde{\mathbf{A}}_W\mathbf{T}^{-1} \\ &= \begin{bmatrix} 0 & g_l & -g_l & h_l & -h_l & h_l & -h_l & h_l \\ 1 & -1 & 1 & 0 & 0 & 0 & 0 & 0 \\ 1 & -1 & 1 & -1 & 1 & 0 & 0 & 0 \\ 0 & 0 & 1 & -1 & 1 & 0 & 0 & 0 \\ 0 & 0 & 1 & -1 & 1 & -1 & 1 & 0 \\ 0 & 0 & 0 & 0 & 1 & -1 & 1 & 0 \\ 0 & 0 & 0 & 0 & 1 & -1 & h_r & -h_r \\ 0 & 0 & 0 & 0 & 0 & 0 & g_r & -g_r \end{bmatrix}, \end{aligned}$$

where $h_l \triangleq 1 + g_l$ and $h_r \triangleq 1 + g_r$. We see that the left FDTD termination is *non-local* for $g \neq -1$, while the right termination is local (to two adjacent spatial samples) for all g. This can be viewed as a consequence of having ordered the FDTD state variables as $[y_{n,m}, y_{n-1,m+1}, \ldots]$ instead of $[y_{n-1,m}, y_{n,m+1}, \ldots]$. Choosing the other ordering interchanges the end-point behavior. Call these orderings Type I and Type II, respectively. Then $\mathbf{T}_{II} = \mathbf{T}_I^T$; that is, the similarity transformation matrix $\mathbf{T}$ is transposed when converting from Type I to Type II or vice versa. By anechoically coupling a Type I FDTD simulation on the right with a Type II simulation on the left, general resistive terminations may be obtained on both ends which are localized to two spatial samples.

In nearly all musical sound synthesis applications, at least one of the string endpoints is modeled as rigidly clamped at the "nut". Therefore, since the FDTD, as defined here, most naturally provides a clamped end-point on the left, with more general localized terminations possible on the right, we will proceed with this case for simplicity in what follows. Thus, we set $g_l = -1$ and obtain

$$\overset{\vdash}{\mathbf{A}}_K \triangleq \begin{bmatrix} 0 & -1 & 1 & 0 & 0 & 0 & 0 & 0 \\ 1 & -1 & 1 & 0 & 0 & 0 & 0 & 0 \\ 1 & -1 & 1 & -1 & 1 & 0 & 0 & 0 \\ 0 & 0 & 1 & -1 & 1 & 0 & 0 & 0 \\ 0 & 0 & 1 & -1 & 1 & -1 & 1 & 0 \\ 0 & 0 & 0 & 0 & 1 & -1 & 1 & 0 \\ 0 & 0 & 0 & 0 & 1 & -1 & 1+g_r & -1-g_r \\ 0 & 0 & 0 & 0 & 0 & 0 & g_r & -g_r \end{bmatrix}$$

Boundary Conditions as Perturbations

To study the effect of boundary conditions on the state transition matrices $\mathbf{A}_W$ and $\mathbf{A}_K$, it is convenient to write the terminated transition matrix as the sum of of the "left-clamped" case $\overset{\vdash}{\mathbf{A}}_W$ (for which $g_l = -1$) plus a series of one or more rank-one perturbations. For example, introducing a right termination with reflectance g_r can be written

$$\overset{\vdash\!\dashv}{\mathbf{A}}_W = \overset{\vdash}{\mathbf{A}}_W + g_r \boldsymbol{\delta}_{8,7} = \mathbf{A}_W - \boldsymbol{\delta}_{1,2} + g_r \boldsymbol{\delta}_{8,7}, \tag{E.39}$$

where $\boldsymbol{\delta}_{ij}$ is the $M \times M$ matrix containing a 1 in its (i,j)th entry, and zero elsewhere. (Following established convention, rows and columns in matrices are numbered from 1.)

In general, when $i+j$ is odd, adding $\boldsymbol{\delta}_{ij}$ to $\overset{\vdash}{\mathbf{A}}_W$ corresponds to a *connection* from left-going waves to right-going waves, or vice versa (see Fig. E.2). When i is odd and j is even, the connection flows from the right-going to the left-going signal path, thus providing a termination (or partial termination) on the right. Left terminations flow from the bottom to the top rail in Fig. E.2, and in such connections i is even and j is odd. The spatial sample numbers involved in the connection are $2\lfloor(i-1)/2\rfloor$ and $2\lfloor(j-1)/2\rfloor$, where $\lfloor x\rfloor$ denotes the greatest integer less than or equal to x.

The rank-one perturbation of the DW transition matrix Eq. (E.39) corresponds to the following rank-one perturbation of the FDTD transition matrix $\overset{\vdash}{\mathbf{A}}_K$:

$$\overset{\vdash\!\dashv}{\mathbf{A}}_K \triangleq \overset{\vdash}{\mathbf{A}}_K + g\boldsymbol{\Delta}_{8,7}$$

where

$$\boldsymbol{\Delta}_{8,7} \triangleq \mathbf{T}\boldsymbol{\delta}_{8,7}\mathbf{T}^{-1} = \begin{bmatrix} 0 & 0 & 0 & 0 & 0 & 0 & 0 & 0 \\ \vdots & \vdots & \vdots & \vdots & \vdots & \vdots & \vdots & \vdots \\ 0 & 0 & 0 & 0 & 0 & 0 & 0 & 0 \\ 0 & 0 & 0 & 0 & 0 & 0 & 1 & -1 \\ 0 & 0 & 0 & 0 & 0 & 0 & 1 & -1 \end{bmatrix}. \tag{E.40}$$

In general, we have

$$\boldsymbol{\Delta}_{ij} = \sum_{\kappa=j}^{M}(-1)^{\kappa-j}\left(\boldsymbol{\delta}_{i\kappa} + \boldsymbol{\delta}_{i-1,\kappa}\right). \tag{E.41}$$

Thus, the general rule is that $\boldsymbol{\delta}_{ij}$ transforms to a matrix $\boldsymbol{\Delta}_{ij}$ which is zero in all but two rows (or all but one row when $i=1$). The nonzero rows are numbered i and $i-1$ (or just i when $i=1$), and they are identical, being zero in columns $1:j-1$, and containing $[1,-1,1,-1,\ldots]$ starting in column j.

Reactive Terminations

In typical string models for virtual musical instruments, the "nut end" of the string is rigidly clamped while the "bridge end" is terminated in a *passive reflectance* $S(z)$. The condition for passivity of the reflectance is simply that its gain be bounded by 1 at all frequencies [447]:

$$\left|S(e^{j\omega T})\right| \leq 1, \quad \forall\,\omega T \in [-\pi, \pi). \tag{E.42}$$

A very simple case, used, for example, in the Karplus-Strong plucked-string algorithm, is the two-point-average filter:

$$S(z) = -\frac{1+z^{-1}}{2}$$

To impose this lowpass-filtered reflectance on the right in the chosen subgrid, we may form

$$\overset{\vdash\!\dashv}{\mathbf{A}}_W = \overset{\vdash}{\mathbf{A}}_W - \frac{1}{2}\mathbf{\Delta}_{8,5} - \frac{1}{2}\mathbf{\Delta}_{8,7}$$

which results in the FDTD transition matrix

$$\overset{\vdash\!\dashv}{\mathbf{A}}_K \quad \triangleq \quad \begin{bmatrix} 0 & -1 & 1 & 0 & 0 & 0 & 0 & 0 \\ 1 & -1 & 1 & 0 & 0 & 0 & 0 & 0 \\ 1 & -1 & 1 & -1 & 1 & 0 & 0 & 0 \\ 0 & 0 & 1 & -1 & 1 & 0 & 0 & 0 \\ 0 & 0 & 1 & -1 & 1 & -1 & 1 & 0 \\ 0 & 0 & 0 & 0 & 1 & -1 & 1 & 0 \\ 0 & 0 & 0 & 0 & 1/2 & -1/2 & 0 & 0 \\ 0 & 0 & 0 & 0 & -1/2 & 1/2 & -1 & -1 \end{bmatrix}.$$

This gives the desired filter in a half-rate, staggered grid case. In the full-rate case, the termination filter is really

$$S(z) = -\frac{1+z^{-2}}{2}$$

which is still passive, since it obeys Eq. (E.42), but it does not have the desired amplitude response: Instead, it has a notch (gain of 0) at one-fourth the sampling rate, and the gain comes back up to 1 at half the sampling rate. In a full-rate scheme, the two-point-average filter must straddle both subgrids.

Another often-used string termination filter in digital waveguide models is specified by [447]

$$\begin{aligned} s(n) &= -g\left[\frac{h}{4}, \frac{1}{2}, \frac{h}{4}\right] \\ \longleftrightarrow \quad S(e^{j\omega T}) &= -e^{-j\omega T} g \frac{1 + h\cos(\omega T)}{2}, \end{aligned}$$

where $g \in (0,1)$ is an overall gain factor that affects the decay rate of all frequencies equally, while $h \in (0,1)$ controls the relative decay rate of low-frequencies and high frequencies. An advantage of this termination filter is that the delay is always one sample, for all frequencies and for all parameter settings; as a result, the tuning of the string is invariant with respect to termination filtering. In this case, the perturbation is

$$\overset{\vdash\!\dashv}{\mathbf{A}}_W = \overset{\vdash}{\mathbf{A}}_W - \frac{gh}{4}\delta(M-5,M) - \frac{g}{2}\delta(M-3,M) - \frac{gh}{4}\delta(M-1,M)$$

and, using Eq. (E.41), the order $M = 8$ FDTD state transition matrix is given by

$$\overset{\vdash\!\dashv}{\mathbf{A}}_K \triangleq \begin{bmatrix} 0 & -1 & 1 & 0 & 0 & 0 & 0 & 0 \\ 1 & -1 & 1 & 0 & 0 & 0 & 0 & 0 \\ 1 & -1 & 1 & -1 & 1 & 0 & 0 & 0 \\ 0 & 0 & 1 & -1 & 1 & 0 & 0 & 0 \\ 0 & 0 & 1 & -1 & 1 & -1 & 1 & 0 \\ 0 & 0 & 0 & 0 & 1 & -1 & 1 & 0 \\ 0 & 0 & g_1 & -g_1 & 1+g_2 & -1-g_2 & 1+g_3 & -1-g_3 \\ 0 & 0 & g_1 & -g_1 & g_2 & -g_2 & g_3 & -g_3 \end{bmatrix}$$

where

$$\begin{aligned} g_1 &\triangleq -\frac{gh}{4} \\ g_2 &\triangleq -\frac{g}{2} + g_1 \\ g_3 &\triangleq -\frac{gh}{4} + g_2. \end{aligned}$$

The filtered termination examples of this section generalize immediately to arbitrary finite-impulse response (FIR) termination filters $S(z)$. Denote the impulse response of the termination filter by

$$s(n) = [s_0, s_1, s_2, \ldots, s_N],$$

where the length N of the filter does not exceed $M/2$. Due to the DW-FDTD equivalence, the general stability condition is stated very simply as

$$\left|S(e^{j\omega T})\right| = \left|\sum_{n=0}^{N-1} s_n e^{-j\omega T}\right| \leq 1, \quad \forall\, \omega T \in [-\pi, \pi).$$

Interior Scattering Junctions

A so-called *Kelly-Lochbaum scattering junction* [297, 447] can be introduced into the string at the fourth sample by the following perturbation

$$\overset{\rightleftharpoons}{\mathbf{A}}_K = \overset{\vdash}{\mathbf{A}}_K + (1-k_l)\mathbf{\Delta}_{5,3} + k_r\mathbf{\Delta}_{5,8} + k_l\mathbf{\Delta}_{6,3} + (1-k_r)\mathbf{\Delta}_{6,8}.$$

Here, k_l denotes the reflection coefficient "seen" from left to right, and k_r is the reflectance of the junction from the right. When the scattering junction is caused by a change in string density or tension, we have $k_r = -k_l$. When it is caused by an externally imposed termination (such as a plectrum or piano-hammer touching the string), we have $k_r = k_l$, and the reflectances may become filters instead of real values in $[-1, 1]$. Energy conservation demands that the transmission coefficients be amplitude complementary with respect to the reflection coefficients [447].

A single time-varying scattering junction provides a reasonable model for plucking, striking, or bowing a string at a point. Several adjacent scattering junctions can model a distributed interaction, such as a piano hammer, finger, or finite-width bow spanning several string samples.

Note that scattering junctions separated by one spatial sample (as typical in "digital waveguide filters" [447]) will couple the formerly independent subgrids. If scattering junctions are confined to one subgrid, they are separated by two samples of delay instead of one, resulting in round-trip transfer functions of the form $H(z^2)$ (as occurs in the digital waveguide mesh). In the context of a half-rate staggered-grid scheme, they can provide general IIR filtering in the form of a ladder digital filter [297, 447].

E.4.4 Lossy Vibration

The DW and FDTD state-space models are equivalent with respect to lossy traveling-wave simulation. Figure E.4 shows the flow diagram for the case of simple attenuation by g per sample of wave propagation, where $g \in (0, 1]$ for a passive string.

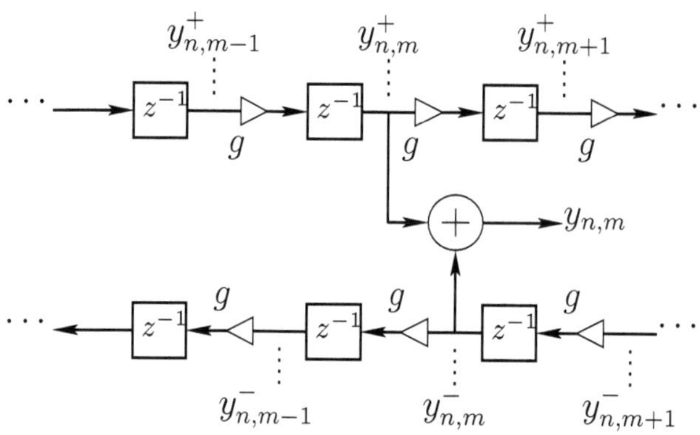

Figure E.4: DW flow diagram in the lossy case.

The DW state update can be written in this case as

$$\underline{x}_W(n+2) = g^2 \mathbf{A}_W \underline{x}_W(n) + \mathbf{B}_W \underline{u}(n+2).$$

where the loss associated with two time steps has been incorporated into the chosen subgrid for physical accuracy. (The neglected subgrid may now be considered lossless.) In changing coordinates to the FDTD scheme, the gain factor g^2 can remain factored out, yielding

$$\underline{x}_K(n+2) = g^2 \mathbf{A}_K \underline{x}_K(n) + \mathbf{B}_K \underline{u}(n+2).$$

When the input is zero after a particular time, such as in a plucked or struck string simulation, the losses can be implemented at the final output, and only when an output is required, *e.g.*,

$$y(n) = g^n y_0(n)$$

where $y_0(n)$ denotes the corresponding lossless simulation. When there is a general input signal, the state vector needs to be properly attenuated by losses. In the DW case, the losses can be lumped at two points per spatial input and output [447].

E.4.5 State Space Summary

We have seen that the DW and FDTD schemes correspond to state-space models which are related to each other by a simple change of coordinates (similarity transformation). It is well known that such systems exhibit the same transfer functions, have the same modes, and so on. In short, they are the same linear dynamic system. Differences may exist with respect to spatial locality of input signals, initial conditions, and boundary conditions.

State-space analysis was used to translate initial conditions and boundary conditions from one case to the other. Passive terminations in the DW paradigm were translated to passive terminations for the FDTD scheme, and FDTD excitations were translated to the DW case in order to interpret them physically.

E.5 Computational Complexity

The DW model is more efficient in one dimension because it can make use of delay lines to obtain an $\mathcal{O}(1)$ computation per time sample [437], whereas the FDTD scheme is $\mathcal{O}(M)$ per sample (M being the number of spatial samples along the string). There is apparently no known way to achieve $\mathcal{O}(1)$ complexity for the FDTD scheme. In higher dimensions, *i.e.*, when simulating membranes and volumes, the delay-line advantage disappears, and the FDTD scheme has the lower operation count (and memory storage requirements).

E.6 Summary

An explicit linear transformation was derived for converting state variables of the finite-difference time-domain (FDTD) scheme to those of the digital waveguide (DW) scheme. The equivalence of the FDTD and DW state transitions was reviewed, and the proof of state-space equivalence was completed. Since the DW scheme is exact within its bandwidth (being a sampled traveling-wave scheme instead of a finite difference scheme), it can be put forth as the proper physical interpretation of the FDTD scheme, and consequently be used to provide physically accurate initial conditions and excitations for the FDTD method. For its part, the FDTD method provides lower cost relative to the DW method in dimensions higher than one (for simulating membranes, volumes, and so on), and can be preferred in highly distributed nonlinear string simulation applications.

E.7 Future Work

The simple state translation formulas derived here for the one-dimensional case do not extend simply to higher dimensions. While straightforward extensions to higher dimensions are presumed to exist, a simple and intuitive result such as found here for the 1D case could be more useful for initializing and driving FDTD mesh simulations from a physical point of view. In particular, spatially localized initial conditions and boundary conditions in the DW framework should map to localized counterparts in the FDTD scheme. A generalization of the Toeplitz operator $\mathbf{T}$ having a known closed-form inverse could be useful in higher dimensions.

Acknowledgments

The author wishes to thank Stefan Bilbao, Georg Essl, and Patty Huang for fruitful discussions on topics addressed in this appendix.

Appendix F

Wave Digital Filters

A *Wave Digital Filter* (WDF) [136] is a particular kind of *digital filter* based on physical modeling principles. Unlike most digital filter types, every delay element in a WDF can be interpreted *physically* as holding the current state of a mass or spring (or capacitor or inductor). WDFs can also be viewed as a particular kind of *finite difference scheme* having unusually good numerical properties [55]. (See Appendix D for an introduction to finite difference schemes.) WDFs have been applied often in music signal processing [394, 338, 557, 361, 348, 555, 556, 56, 528, 524, 485].

Wave digital filters were developed initially by Alfred Fettweis [135] in the late 1960s for digitizing lumped *electrical* circuits composed of inductors, capacitors, resistors, transformers, gyrators, circulators, and other elements of classical network theory [136]. The WDF approach is based on the traveling-wave formulation of lumped electrical elements introduced by Belevitch [34].

A WDF is constructed by interconnecting simple discrete-time models of individual masses, springs, and dashpots (or inductors, capacitors, and resistors). The rules for interconnecting the elementary models are based on *scattering theory* (discussed in §C.8 on page 560). This is a direct result of the fact that all signals explicitly computed may be physically interpreted as *traveling wave components* of physical variables.

F.1 Wave Digital Elements

When modeling mechanical systems composed of masses, springs, and dashpots, it is best to begin with an *electrical equivalent circuit*. Equivalent circuits make clear the network-theoretic structure of the system, clearly indicating, for example, whether interacting elements should be connected in series or parallel. Each element of the equivalent circuit can then be replaced by a first-order wave digital element, and the elements are finally parallel or series connected by means of scattering-junction interfaces known as *adaptors*.

Wave digital elements may be derived from their describing differential equations (in continuous time) as follows:

- First express all physical quantities (such as force and velocity) in terms of traveling-wave components. The traveling wave components are called *wave variables*. For example, the force $f(n)$ on a mass is decomposed as $f(n) = f^+(n) + f^-(n)$, where $f^+(n)$ is regarded as a traveling wave propagating *toward* the mass, while $f^-(n)$ is seen as the traveling component propagating *away from* the mass. A "traveling wave" view of force mediation is actually much closer to physical reality than any instantaneous model.

- Second, digitize the resulting traveling-wave system using the *bilinear transform*. The bilinear transform is equivalent in the time domain to the *trapezoidal rule for numerical integration* (see §7.3.2).

- Connect N elementary units together by means of *N-port scattering junctions*. There are two basic types of scattering junction, one for parallel, and one for series connection. (See §C.8 on page 560 for the theory of scattering junctions.)

The next section will examine the above steps in greater detail.

An important benefit of introducing wave variables prior to bilinear transformation is the *elimination of delay-free loops* when connecting elementary building blocks. In other words, any number of elementary models can be interconnected, in series or in parallel, and the resulting finite-difference scheme remains *explicit* (free of delay-free loops).

F.1.1 A Physical Derivation of Wave Digital Elements

This section provides a "physical" derivation of Wave Digital Filters (WDF), which contrasts somewhat with the more formal derivation common in the literature. The derivation is presented as a numbered series of steps (some with rather long discussions):

1. To each element, such as a capacitor or inductor, attach a length of waveguide (electrical transmission line) having wave impedance R_0, and make it infinitesimally long. (Take the limit as its length goes to zero.) A schematic depiction of this is shown in Fig. F.1a. For consistency, all signals are Laplace transforms of their respective

time-domain signals. The length must approach zero in order not to introduce propagation delays into the signal path.

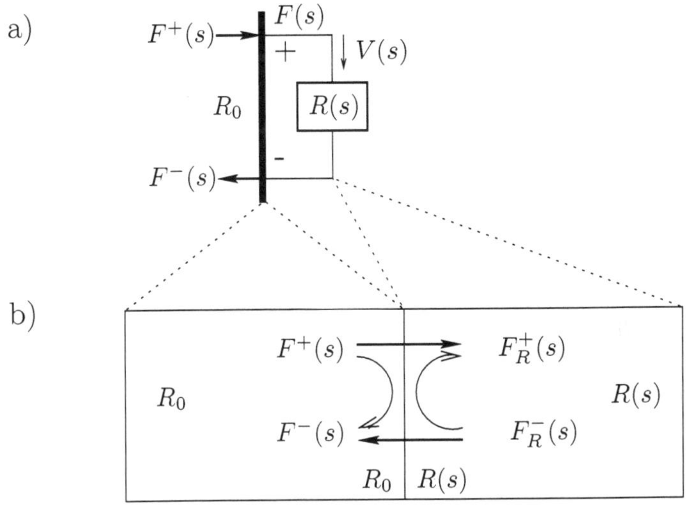

Figure F.1: a) Physical schematic for the derivation of a wave digital model of driving-point impedance $R(s)$. The inserted waveguide impedance R_0 is real and positive, but otherwise arbitrary. b) Expanded view of the interior of the infinitesimal waveguide section, also representing the termination impedance $R(s)$ as an impedance-step within the waveguide.

Points to note:

- The infinitesimal waveguide is *terminated* by the element. The element reflects waves as if it were a new waveguide section at impedance $R(s)$, as depicted in Fig. F.1b.
- The interface to the element is recast as traveling-wave *components* $F^+(s)$ and $F^-(s)$ at impedance R_0. In terms of these components, the physical force on the element is obtained by adding them together: $F(s) = F^+(s) + F^-(s)$.
- The waveguide impedance R_0 is *arbitrary* because it has been physically introduced. We will need to know it when we connect this element to other elements. The element's interface to other elements is now a waveguide (transmission line) at real impedance R_0.
- The junction is "parallel" (cf. §7.2 on page 257):
  - Force (voltage) must be continuous across the junction, since

otherwise there would be a finite force across a zero mass, producing infinite acceleration.

– The sum of velocities (currents) into the junction must be zero by conservation of mass (charge).

Reflectance of a General Lumped Waveguide Termination

2. Calculate the *reflectance* of the terminated waveguide. That is, find the Laplace transform of the return wave divided by the Laplace transform of the input wave going into the waveguide. In general, the reflectance of an impedance step for force waves (voltage waves in the electrical case) is

$$\boxed{\hat{\rho}_R(s) \isdef \frac{F^-(s)}{F^+(s)} = \frac{R(s)-R_0}{R(s)+R_0}} \tag{F.1}$$

This is easily derived from continuity constraints across the junction. Specifically, referring to Fig. F.1b, let $F_R(s) = F_R^+(s) + F_R^-(s)$ denote the physical force and its traveling-wave components within the "pseudo-infinitesimal-generalized-waveguide" defined by the element impedance $R(s)$, with the '+' superscript denoting a right-going wave.[1] Similarly, let $V(s) = V^+(s) + V^-(s)$ denote the velocity and its component wave variables on the side of the junction at impedance R_0, and let $V_R(s) = V_R^+(s) + V_R^-(s)$ denote the corresponding quantities on the element-side of the junction at impedance $R(s)$. Again, the '+' superscript denotes travel to the right. Then the physical continuity constraints imply

$$\begin{aligned} F(s) &= F_R(s) \\ 0 &= V(s) + V_R(s) \end{aligned}$$

By the definition of wave impedance in a waveguide, we have

$$\begin{aligned} F^+(s) &= R_0 V^+(s) \\ F^-(s) &= -R_0 V^-(s) \\ \\ F_R^+(s) &= R(s) V_R^+(s) \\ F_R^-(s) &= -R(s) V_R^-(s) \end{aligned}$$

[1] When there are more than two ports at a junction, the '+' superscript will denote a wave traveling *into* the junction.

Thus,

$$\begin{aligned}
0 &= V(s) + V_R(s) \\
&= \left[V^+(s) + V^-(s)\right] + \left[V_R^+(s) + V_R^-(s)\right] \\
&= \frac{F^+(s) - F^-(s)}{R_0} + \frac{F_R^+(s) - F_R^-(s)}{R(s)} \\
&= \frac{2F^+(s) - F(s)}{R_0} + \frac{2F_R^+(s) - F(s)}{R(s)} \\
\Longrightarrow \quad F(s)\left[\frac{1}{R_0} + \frac{1}{R(s)}\right] &= \frac{2}{R_0}F^+(s) + \frac{2}{R(s)}F_R^+(s)
\end{aligned}$$

Defining $\Gamma_0 \triangleq 1/R_0$ and $\Gamma(s) \triangleq 1/R(s)$, we have

$$F(s) = \frac{2\Gamma_0}{\Gamma_0 + \Gamma(s)}F^+(s) + \frac{2\Gamma(s)}{\Gamma_0 + \Gamma(s)}F_R^+(s) \triangleq \mathcal{A}(s)F^+(s) + \mathcal{A}_R(s)F_R^+(s) \tag{F.2}$$

Now that we've solved for the junction force $F(s)$, the outgoing waves are simply obtained from the force continuity constraint, $F(s) = F^+(s) + F^-(s) = F_R^+(s) + F_R^-(s)$:

$$F^-(s) = F(s) - F^+(s) \tag{F.3}$$

$$F_R^-(s) = F(s) - F_R^+(s) \tag{F.4}$$

Finally, the force-wave reflectance of an impedance step from R_0 to $R(s)$ can be found by solving Eq. (F.3) and (F.2) for $F^-(s)/F^+(s)$ with $F_R^+(s)$ set to zero:

$$\begin{aligned}
\hat{\rho}(s) &\triangleq \frac{F^-(s)}{F^+(s)} = \frac{F(s) - F^+(s)}{F^+(s)} = \mathcal{A}(s) - 1 \\
&= \frac{2\Gamma_0}{\Gamma_0 + \Gamma(s)} - 1 = \frac{\Gamma_0 - \Gamma(s)}{\Gamma_0 + \Gamma(s)} = \frac{R(s) - R_0}{R(s) + R_0}
\end{aligned}$$

as claimed.

Reflectances of Elementary Impedances

We now derive the reflectances of the elements used in LTI analog electric circuits, *viz.*, the capacitor, inductor, and resistor.

Capacitor Reflectance

For a capacitor of C Farads, the driving-point impedance is (see §7.1.3)

$$R_C(s) = \frac{1}{Cs}$$

(or k/s for a spring with constant k). Substituting into Eq. (F.1) gives the reflectance

$$\hat{\rho}_C(s) = \frac{R_C(s) - R_0}{R_C(s) + R_0} = \frac{1 - R_0 Cs}{1 + R_0 Cs} \tag{F.5}$$

Inductor Reflectance

For an inductor of L Henrys, we have

$$\begin{aligned} R_L(s) &= Ls \\ \Rightarrow\ \hat{\rho}_L(s) &= \frac{Ls - R_0}{Ls + R_0} = \frac{s - R_0/L}{s + R_0/L} \end{aligned} \tag{F.6}$$

Resistor Reflectance

Finally, for a resistor of R Ohms, we get

$$\hat{\rho}_R(s) = \frac{R - R_0}{R + R_0} = \frac{1 - R_0/R}{1 + R_0/R} \tag{F.7}$$

Note that both the capacitor and inductor reflectances are *stable allpass filters*, as they must be. Also, the resistor reflectance is always less than 1, no matter what waveguide impedance $R_0 > 0$ we choose.

Choosing Impedance to Simplify Element Reflectance

3. Observe that there is a natural choice for each waveguide impedance which will give us a normalized, "universal reflectance" for each element:

   - For the capacitor, setting $R_0 = 1/C$ gives

$$\boxed{\hat{\rho}_C(s) = \frac{1 - s}{1 + s}} \tag{F.8}$$

- For the inductor, setting $R_0 = L$ gives

$$\boxed{\hat{\rho}_L(s) = -\frac{1-s}{1+s}} \tag{F.9}$$

- And for the resistor, we set $R_0 = R$ to obtain

$$\boxed{\hat{\rho}_R(s) = 0} \tag{F.10}$$

Digitizing Elementary Reflectances by Bilinear Transform

4. Going to discrete time via the bilinear transform means making the substitution

$$s = c\frac{1-z^{-1}}{1+z^{-1}} \tag{F.11}$$

where $c > 0$ is an arbitrary real constant, usually taken to be $c = 2/T$. Solving for z^{-1} gives us the inverse bilinear transform:

$$z^{-1} = \frac{1-s/c}{1+s/c} \tag{F.12}$$

In this case, we see that setting $c = 1$ further simplifies our universal reflectances in the digital domain:

- For the "wave digital capacitor" (or spring), Eq. (F.8) becomes

$$\boxed{\hat{\rho}_C\left(\frac{z-1}{z+1}\right) = z^{-1}}$$

- For the "wave digital inductor" (or mass), Eq. (F.9) becomes

$$\boxed{\hat{\rho}_L\left(\frac{z-1}{z+1}\right) = -z^{-1}}$$

- And for the "wave digital resistor" (or dashpot), Eq. (F.10) becomes

$$\boxed{\hat{\rho}_R\left(\frac{z-1}{z+1}\right) = 0}$$

as before in the continuous-time case.

> Note that this choice of c is also the only one that eliminates *delay-free paths* in the fundamental elements. This allows them to be used as building blocks for *explicit* finite difference schemes.

We may still obtain the above results using the more typical value $c = 2/T$ (instead of $c = 1$) in the bilinear transform. From Eq. (F.12), it is clear that changing c amounts to a linear frequency scaling of $s = j\omega$. Such a scaling may be compensated by choosing the waveguide (port) impedances to be $R_L = Lc = 2L/T$ (instead of $R_L = L$) for the inductor, and $R_C = T/(2C)$ (instead of $1/C$) for the capacitor.

F.1.2 Summary of Wave Digital Elements

From Eq. (F.1), we have that the general reflectance of impedance $R(s)$ with respect to the reference impedance R_0 in the wave variable formulation is given by

$$\boxed{\hat{\rho}(s) \stackrel{\Delta}{=} \frac{R(s) - R_0}{R(s) + R_0}} \tag{F.13}$$

In WDF construction, the free constant in the bilinear transform is taken to be $c = 1$. Thus we obtain $\hat{\rho}_d(z) = \hat{\rho}[(1-z^{-1})/(1+z^{-1})]$. When $R(s)$ is first order, it is possible to choose the reference impedance R_0 so as to eliminate the delay-free path in the digital reflectance $\hat{\rho}_d(z)$, and so its value depends on the actual physical element being digitized.

F.1.3 Wave Digital Mass

In the case of a mass m, we have

$$R(s) = ms$$

which implies its reflectance is, from Eq. (F.13),

$$\hat{\rho}_m(s) = \frac{ms - R_0}{ms + R_0}$$

Setting $R_0 = m$ gives

$$\hat{\rho}_m(s) = \frac{s-1}{s+1}$$

and this choice also turns out to eliminate the delay-free path in the digital version. In view of the expression for the inverse bilinear transform in Eq. (F.12), *i.e.*, $z = (1+s)/(1-s)$, the bilinear transform of $\hat{\rho}(s)$ is immediately seen to be

$$\boxed{\hat{\hat{\rho}}_m(z) = -z^{-1}}$$

where we defined $\hat{\hat{\rho}}_m(z) \triangleq \hat{\rho}_m\left(\frac{z-1}{z+1}\right)$. The corresponding *difference equation* for the wave digital mass is

$$f^-(n) = -f^+(n-1)$$

and its wave flow diagram is drawn in Fig. F.2.

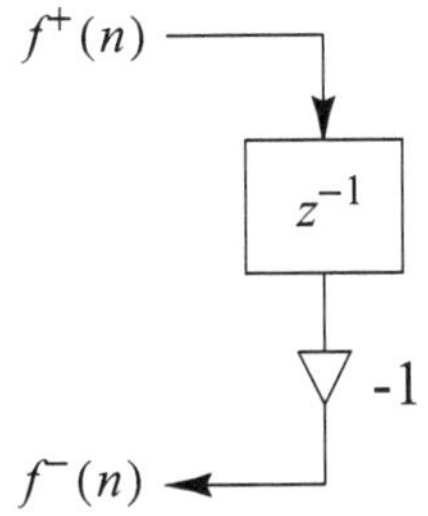

Figure F.2: Wave flow diagram for the wave digital mass. Note that the wave variables are written in the time domain as is customary in digital filter diagrams, while it would be more consistent (with the z^{-1} block) to keep them in the frequency domain as $F^+(z)$ and $F^-(z)$.

Thus, the wave digital mass is simply a unit-sample delay and a negation. The fact that the value of the mass has been canceled out will be addressed below in the subsection on "adaptors," *i.e.*, it only affects interconnection with other elements. For now, just remember that the reference impedance was chosen to be equal to the mass in order to get this simple wave flow diagram. Also note that the WDF mass simulator has no delay-free path from input to output.

F.1.4 Wave Digital Spring

In the case of a spring with stiffness k, we have the impedance

$$R(s) = k/s$$

which gives the reflectance

$$\hat{\rho}_k(s) = \frac{k/s - R_0}{k/s + R_0}$$

As before, we may eliminate k by choosing $R_0 = k$ to get

$$\hat{\rho}_k(s) = \frac{1-s}{1+s} = z^{-1}$$

under the bilinear transform. So we have the digital reflectance

$$\boxed{\hat{\hat{\rho}}_k(z) = z^{-1}} \qquad \text{(Wave Digital Spring)}$$

and corresponding difference equation

$$f^-(n) = f^+(n-1).$$

Again the delay-free path has been eliminated. The wave flow diagram is shown in Fig. F.3.

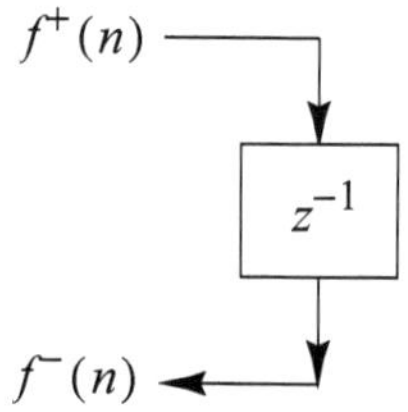

Figure F.3: Wave flow diagram for the Wave Digital Spring.

Thus, the WDF of a spring is simply a unit-sample delay, which is just the negative of the WDF mass. If we were to switch to velocity waves instead of force waves, both masses and springs would again correspond to unit-sample delays, but the spring would become inverting and the mass non-inverting.

F.1.5 Wave Digital Dashpot

Starting with a dashpot with coefficient μ, we have

$$R(s) = \mu$$

and reflectance

$$\hat{\rho}_\mu(s) = \frac{\mu - R_0}{\mu + R_0}$$

This time, choosing R_0 equal to the element value gives

$$\hat{\rho}_\mu(s) = 0$$

Conformally mapping the zero function yields the zero function so that

$$\boxed{\hat{\hat{\rho}}_\mu(z) = 0}$$

as well. Thus, the WDF of a dashpot is a "wave sink," as diagrammed in Fig. F.4.

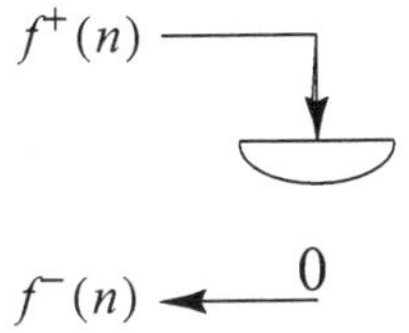

Figure F.4: Wave flow diagram for the Wave Digital Dashpot.

In the context of waveguide theory, a zero reflectance corresponds to a *matched impedance, i.e.*, the terminating transmission-line impedance equals the characteristic impedance of the line.

The difference equation for the wave digital dashpot is simply $f^-(n) = 0$. While this may appear overly degenerate at first, remember that the interface to the element is a port at impedance $R_0 = \mu$. Thus, in this particular case only, the infinitesimal waveguide interface is the element itself.

F.1.6 Limiting Cases

The force-wave reflectance of an infinite impedance (rigid wall or "open circuit") is

$$\hat{\rho}(s) = \frac{R(s) - R_0}{R(s) + R_0} = \frac{\infty - R_0}{\infty + R_0} = 1$$

Similarly, the force-wave reflectance of a zero impedance (free termination, frictionless surface, or "short circuit") is

$$\hat{\rho}(s) = \frac{0 - R_0}{0 + R_0} = -1$$

For velocity waves, we obtain the opposite results: rigid terminations are inverting, and free terminations are non-inverting.

F.1.7 Unit Elements

The unit element two-port is simply a bidirectional delay line with half a sample delay in each direction. As a result, it really belongs under the topic of distributed modeling. To avoid delay-free loops, Fettweis noted [135] that every pair of adaptors must be separated by at least one unit element. More recently, this objective is accomplished instead using "reflection-free ports" [136] (see also §F.2.2).

F.2 Adaptors for Wave Digital Elements

An *adaptor* is an N-port memoryless interface which interconnects wave digital elements. Since each element's "port" is a connection to an infinitesimal waveguide section at some real wave impedance R_i, and since the input/output signals are wave variables (traveling-waves within the waveguide), the adaptor must implement *signal scattering* appropriate for the connection of such waveguides. In other words, an N-port adaptor in a wave digital filter performs exactly the same computation as an N-port scattering junction in a digital waveguide network.[2]

This section first addresses the simpler two-port case, followed by a derivation of the general N-port adaptor, for both parallel and series connections of wave digital elements.

As discussed in §7.2, a physical connection of two or more ports can either be in *parallel* (forces are equal and the velocities sum to zero) or in *series* (velocities equal and forces sum to zero). Combinations of parallel and series connections are also of course possible.

F.2.1 Two-Port Parallel Adaptor for Force Waves

Figure F.5a illustrates a generic parallel two-port connection in terms of forces and velocities.

As discussed in §7.2, a *parallel connection* is characterized by a common force and velocities which sum to zero:

$$\begin{aligned} f_1(n) = f_2(n) &\stackrel{\Delta}{=} f_J(n) \\ v_1(n) + v_2(n) &= 0 \end{aligned}$$

Following the same derivation leading to Eq. (F.2), and defining $\Gamma_i = 1/R_i$

[2]Wave digital filters are not the same thing as digital waveguide networks, however, because the wave variables in a wave digital filter have a compressed frequency spectrum (from the bilinear transform), while the signals in a digital waveguide network have a bandlimited spectrum which is not warped.

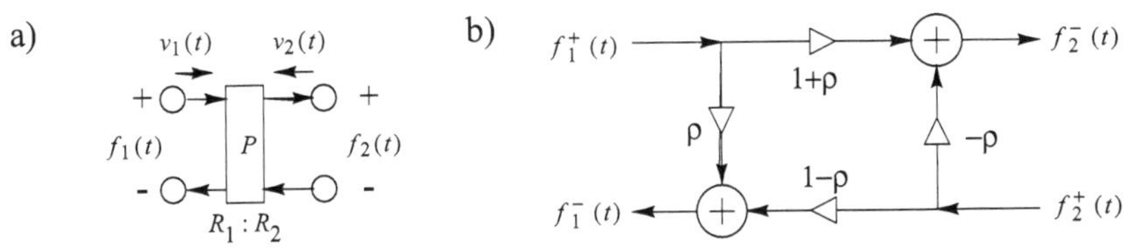

Figure F.5: a) Two-port description of the adaptor implementing a parallel connection between reference impedances R_1 and R_2. b) Corresponding parallel force scattering junction (adaptor wave flow diagram) in Kelly-Lochbaum form. Compare with Fig. F.7 on page 719.

for notational convenience, we obtain

$$\begin{aligned}
0 &= v_1 + v_2 \\
&= \frac{f_1^+ - f_1^-}{R_1} + \frac{f_2^+ - f_2^-}{R_2} \\
&= \frac{2f_1^+ - f_J}{R_1} + \frac{2f_2^+ - f_J}{R_2} \\
&\triangleq 2\Gamma_1 f_1^+ - \Gamma_1 f_J + 2\Gamma_2 f_2^+ - \Gamma_2 f_J \\
\Rightarrow \quad (\Gamma_1 + \Gamma_2) f_J &= 2\left[\Gamma_1 f_1^+ + \Gamma_2 f_2^+\right] \\
\Rightarrow \quad f_J &= 2\frac{\Gamma_1 f_1^+ + \Gamma_2 f_2^+}{\Gamma_1 + \Gamma_2}.
\end{aligned}$$

The outgoing wave variables are given by

$$\begin{aligned}
f_1^-(n) &= f_J(n) - f_1^+(n) \\
f_2^-(n) &= f_J(n) - f_2^+(n)
\end{aligned}$$

Defining the reflection coefficient as

$$\rho \triangleq \frac{R_2 - R_1}{R_2 + R_1}$$

we have that the scattering relations for the two-port parallel adaptor are

$$\begin{aligned}
f_1^- &= \rho f_1^+ + (1-\rho) f_2^+ \\
f_2^- &= (1+\rho) f_1^+ - \rho f_2^+
\end{aligned}$$

as diagrammed in Fig. F.5b. This can be called the *Kelly-Lochbaum* implementation of the two-port force-wave adaptor.

Now that we have a proper scattering interface between two reference impedances, we may connect two wave digital elements together, setting R_1 to the port impedance of element 1, and R_2 to the port impedance of element 2. An example is shown in Fig. F.35 on page 738.

The Kelly-Lochbaum adaptor in Fig. F.5b evidently requires four multiplies and two additions. Note that we can factor out the reflection coefficient in each equation to obtain

$$\begin{aligned} f_1^- &= f_2^+ + \rho(f_1^+ - f_2^+) \\ f_2^- &= f_1^+ + \rho(f_1^+ - f_2^+) \end{aligned}$$

which requires only one multiplication and three additions. This can be called the *one-multiply* form. The one-multiply form is most efficient in custom VLSI. The Kelly-Lochbaum form, on the other hand, may be more efficient in software, and slightly faster (by one addition) in parallel hardware.

Compatible Port Connections

Note carefully that to connect a wave digital element to port i of the adaptor, we route the signal $f^-(n)$ coming out of the element to become $f_i^+(n)$ on the adaptor port, and the signal $f_i^-(n)$ coming out of port i of the adaptor goes into the element as $f^+(n)$. Such a connection is said to be a *compatible port connection*. In other words, the connections must be made such that the arrows go in the same direction in the wave flow diagram.

F.2.2 General Parallel Adaptor for Force Waves

In the more general case of N wave digital element ports being connected in parallel, we have the physical constraints

$$f_1(n) = f_2(n) = \cdots = f_N(n) \triangleq f_J(n) \qquad \text{(F.14)}$$

$$v_1(n) + v_2(n) + \cdots + v_N(n) = 0 \qquad \text{(F.15)}$$

The derivation for the two-port case extends to the N-port case without modification:

$$\begin{aligned}
0 &= \sum_{i=1}^{N} v_i \\
&= \sum_{i=1}^{N} \frac{f_i^+ - f_i^-}{R_i} \\
&= \sum_{i=1}^{N} \frac{2f_i^+ - f_J}{R_i} \\
&\triangleq \sum_{i=1}^{N} (2\Gamma_i f_i^+ - \Gamma_i f_J) \\
\Rightarrow \sum_{j=1}^{N} \Gamma_j f_J &= \sum_{i=1}^{N} 2\Gamma_i f_i^+ \\
\Rightarrow f_J &= \frac{\sum_{i=1}^{N} 2\Gamma_i f_i^+}{\sum_{j=1}^{N} \Gamma_j}. \qquad \text{(F.16)}
\end{aligned}$$

The outgoing wave variables are given by

$$f_i^-(n) = f_J(n) - f_i^+(n)$$

Alpha Parameters

It is customary in the wave digital filter literature to define the *alpha parameters* as

$$\boxed{\alpha_i \triangleq \frac{2\Gamma_i}{\sum_{j=1}^{N} \Gamma_j}} \qquad \text{(F.17)}$$

where $\Gamma_i \triangleq 1/R_i$ are the admittances of the wave digital element interfaces (or "reference admittances," in WDF terminology). In terms of the alpha parameters, the force-wave parallel adaptor performs the following computations:

$$\begin{aligned}
f_J(n) &= \sum_{i=1}^{N} \alpha_i f_i^+(n) \qquad \text{(F.18)} \\
f_i^-(n) &= f_J(n) - f_i^+(n) \qquad \text{(F.19)}
\end{aligned}$$

We see that N multiplies and $2N-1$ additions are required. However, by observing from Eq. (F.17) that

$$\sum_{i=1}^{N} \alpha_i = 2,$$

we may implement α_1, say, as $2-\sum_{i=2}^{N}\alpha_i$ in order to eliminate one multiply. In WDF terminology, port 1 is then a *dependent port*.

Reflection Coefficient, Parallel Case

The *reflection coefficient* seen at port i is defined as

$$\rho_i \triangleq \left.\frac{f_i^-(n)}{f_i^+(n)}\right|_{f_j^+(n)=0, \forall j\neq i} \tag{F.20}$$

In other words, the reflection coefficient specifies what portion of the incoming wave $f_i^+(n)$ is *reflected* back to port i as part of the outgoing wave $f_i^-(n)$. The total outgoing wave on port i is the *superposition* of the reflected wave and the $N-1$ transmitted waves from the other ports:

$$f_i^-(n) = \rho_i f_i^+ + \sum_{j\neq i} \tau_{ji} f_j^+ \tag{F.21}$$

where τ_{ji} denotes the *transmission coefficient* from port j to port i. Starting with Eq. (F.19) and substituting Eq. (F.18) gives

$$\begin{aligned}
f_i^-(n) &= f_J(n) - f_i^+(n) \\
&= \left(\sum_{j=1}^{N} \alpha_j f_j^+(n)\right) - f_i^+(n) \\
&= (\alpha_i - 1) f_i^+(n) + \sum_{j\neq i} \alpha_j f_j^+(n)
\end{aligned}$$

Equating like terms with Eq. (F.21), we obtain

$$\rho_i = \alpha_i - 1 \tag{F.22}$$

$$\tau_{ji} = \alpha_j, \quad (i \neq j) \tag{F.23}$$

Thus, the jth alpha parameter is the force transmission coefficient from jth port to any other port (besides the ith). To convert the transmission coefficient from the ith port to the reflection coefficient for that port, we simply subtract 1. This general relationship is specific to force waves at a parallel junction, as we will soon see.

Physical Derivation of Reflection Coefficient

Physically, the reflection coefficient seen at port i is due to an impedance step from R_i, that of the port interface, to a new impedance consisting of the parallel combination of all *other* port impedances meeting at the junction. Let

$$\Gamma_J(i) \triangleq \sum_{i \neq j} \Gamma_i \tag{F.24}$$

denote this parallel combination, in admittance form. Then we must have

$$\rho_i = \frac{R_J(i) - R_i}{R_J(i) + R_i} = \frac{\Gamma_i - \Gamma_J(i)}{\Gamma_i + \Gamma_J(i)} \tag{F.25}$$

Let's check this "physical" derivation against the formal definition Eq. (F.20) leading to $\rho_i = \alpha_i - 1$ in Eq. (F.22). Toward this goal, let

$$\Gamma_J \triangleq \sum_{j=1}^{N} \Gamma_j$$

denote the parallel combination of *all* admittances connected to the junction. Then by Eq. (F.24), we have $\Gamma_J = \Gamma_i + \Gamma_J(i)$ for all i. Now, from Eq. (F.17),

$$\begin{aligned} \rho_i &\triangleq \alpha_i - 1 \triangleq \frac{2\Gamma_i}{\Gamma_J} - 1 \\ &\triangleq \frac{2\Gamma_i - \Gamma_J}{\Gamma_J} = \frac{2\Gamma_i - [\Gamma_i + \Gamma_J(i)]}{\Gamma_i + \Gamma_J(i)} \\ &= \frac{\Gamma_i - \Gamma_J(i)}{\Gamma_i + \Gamma_J(i)} = \frac{R_J(i) - R_i}{\Gamma_J(i) - R_i} \end{aligned}$$

and the result is verified.

Reflection Free Port

It is useful in practice, such as when connecting two adaptors together, to make one port *reflection free.* A *reflection-free port* is defined to have a zero reflection coefficient. For port i of a parallel adaptor to be reflection free, we must have, from Eq. (F.25),

$$R_i = R_J(i) \triangleq \frac{1}{\sum_{i \neq j} \Gamma_i}$$

Thus, the port's impedance must equal the parallel combination of the other port impedances at the junction. In this case, the junction as a whole "perfectly terminates" the reflection free port, so no reflections come back from it.

Connecting two adaptors at a reflection-free port prevents the formation of a *delay-free loop* which would otherwise occur [136]. As a result, multi-port junctions can be joined without having to insert unit elements (see §F.1.7) to avoid creating delay-free loops. Only one of the two ports participating in the connection needs to be reflection free.

We can always make a reflection-free port at the connection of two adaptors because the ports used for this connection (one on each adaptor) were created only for purposes of this connection. They can be set to any impedance, and only one of them needs to be reflection free.

To interconnect three adaptors, labeled A, B, and C, we may proceed as follows: Let A be augmented with *two* unconstrained ports, having impedances R_1 and R_2. Add a reflection-free port to B, and suppose its impedance has to be R_B. Add a reflection-free port to C, and suppose its impedance has to be R_C. Now set $R_1 = R_B$ and connect B to A via the corresponding ports. Similarly, set $R_2 = R_C$ and connect C to A accordingly. This adaptor-connection protocol clearly extends to any number of adaptors.

F.2.3 Two-Port Series Adaptor for Force Waves

Figure F.6a illustrates a generic two-port description of the *series* adaptor.

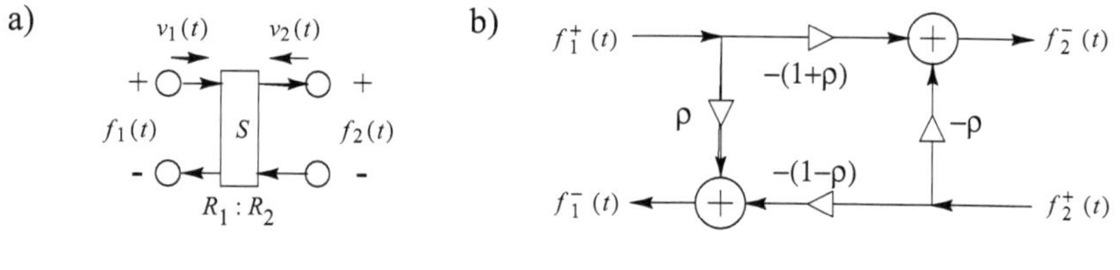

Figure F.6: a) Two-port description of the adaptor implementing a series connection between reference impedances R_1 and R_2. b) Corresponding series force scattering junction (adaptor wave flow diagram) in Kelly-Lochbaum form.

As discussed in §7.2, a *series connection* is characterized by a common velocity and forces which sum to zero at the junction:

$$f_1(n) + f_2(n) = 0$$
$$v_1(n) = v_2(n) \stackrel{\Delta}{=} v_J(n)$$

The derivation can proceed exactly as for the parallel junction in §F.2.1, but with force and velocity interchanged, *i.e.*, $f \leftrightarrow v$, and with impedance and admittance interchanged, *i.e.*, $R \leftrightarrow \Gamma$. In this way, we may take the *dual* of Eq. (F.14) to get

$$\begin{aligned} v_1^- &= -\rho v_1^+ + (1+\rho) v_2^+ \\ v_2^- &= (1-\rho) v_1^+ + \rho v_2^+ \end{aligned}$$

diagrammed in Fig. F.7. Converting back to force wave variables via $f_i^+ = R_i v_i^+$ and $f_i^- = -R_i v_i^-$, and noting that $(1+\rho)R_1/R_2 = 1-\rho$, we obtain, finally,

$$\begin{aligned} f_1^- &= \rho f_1^+ - (1-\rho) f_2^+ \\ f_2^- &= -(1+\rho) f_1^+ - \rho f_2^+ \end{aligned}$$

as diagrammed in Fig. F.6b. The one-multiply form is now

$$\begin{aligned} f_1^- &= -f_2^+ + \rho(f_1^+ + f_2^+) \\ f_2^- &= -f_1^+ - \rho(f_1^+ + f_2^+). \end{aligned}$$

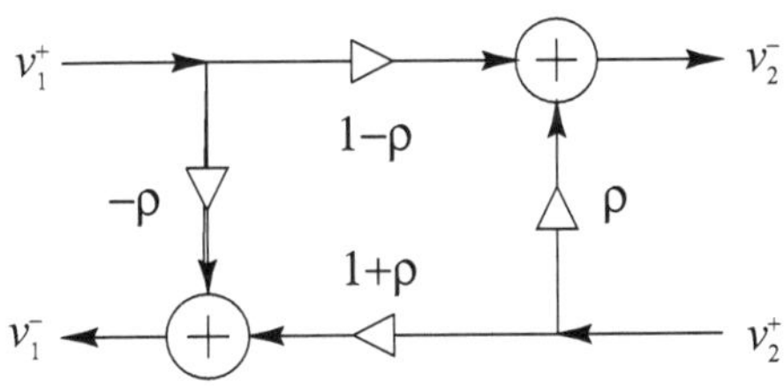

Figure F.7: Series velocity scattering junction in Kelly-Lochbaum form.

F.2.4 General Series Adaptor for Force Waves

In the more general case of N ports being connected in *series*, we have the physical constraints

$$\begin{aligned} v_1(n) = v_2(n) = \cdots = v_N(n) \stackrel{\Delta}{=} v_J(n) \\ f_1(n) + f_2(n) + \cdots + f_N(n) = 0 \end{aligned}$$

The derivation is the dual of that in the parallel case (cf. Eq. (F.16)), *i.e.*, force and velocity are interchanged, and impedance and admittance are interchanged:

$$\begin{aligned}
0 &= \sum_{i=1}^{N} f_i \\
&= \sum_{i=1}^{N} R_i\left(v_i^+ - v_i^-\right) \\
&= \sum_{i=1}^{N} R_i\left(2v_i^+ - v_J\right) \\
\Rightarrow \quad \sum_{j=1}^{N} R_j v_J &= \sum_{i=1}^{N} 2R_i v_i^+ \\
\Rightarrow \quad v_J &= \frac{\sum_{i=1}^{N} 2R_i v_i^+}{\sum_{j=1}^{N} R_j}.
\end{aligned}$$

The outgoing wave variables are given by

$$v_i^-(n) = v_J(n) - v_i^+(n)$$

Beta Parameters

It is customary in the wave digital filter literature to define the *beta parameters* as

$$\boxed{\beta_i \isdef \frac{2R_i}{\sum_{j=1}^{N} R_j}} \tag{F.26}$$

where R_i are the port impedances (attached element reference impedances). In terms of the beta parameters, the force-wave series adaptor performs the following computations:

$$\begin{aligned}
v_J(n) &= \sum_{i=1}^{N} \beta_i v_i^+(n) && \text{(F.27)}\\
v_i^-(n) &= v_J(n) - v_i^+(n) && \text{(F.28)}
\end{aligned}$$

However, we normally employ a mixture of parallel and series adaptors, while keeping a force-wave simulation. Since $f_i^+(n) = R_i v_i^+(n)$, we obtain, after a small amount of algebra, the following recipe for the series force-wave adaptor:

$$\begin{aligned}
f_J^+(n) &= \sum_{i=1}^{N} f_i^+(n) && \text{(F.29)}\\
f_i^-(n) &= f_i^+(n) - \beta_i f_J^+(n) && \text{(F.30)}
\end{aligned}$$

We see that we have N multiplies and $2N-1$ additions as in the parallel-adaptor case. However, we again have from Eq. (F.26) that

$$\sum_{i=1}^{N} \beta_i = 2,$$

so that we may implement one beta parameter as 2 minus the sum of the rest, thus eliminating a multiplication by creating a *dependent port.*

Reflection Coefficient, Series Case

The *velocity reflection coefficient* seen at port i is defined as

$$\rho_i^v \triangleq \left.\frac{v_i^-(n)}{v_i^+(n)}\right|_{v_j^+(n)=0,\forall j\neq i} \tag{F.31}$$

Representing the outgoing velocity wave $v_i^-(n)$ as the superposition of the reflected wave $\rho_i^v v_i^+(n)$ plus the $N-1$ transmitted waves from the other ports, we have

$$v_i^-(n) = \rho_i^v v_i^+ + \sum_{j\neq i} \tau_{ji}^v v_j^+ \tag{F.32}$$

where τ_{ji}^v denotes the *velocity transmission coefficientvelocity!transmission coefficient* from port j to port i. Substituting Eq. (F.29) into Eq. (F.30) yields

$$\begin{aligned}
v_i^-(n) &= v_J(n) - v_i^+(n)\\
&= \left(\sum_{j=1}^{N} \beta_j v_j^+(n)\right) - v_i^+(n)\\
&= (\beta_i - 1)v_i^+(n) + \sum_{j\neq i} \beta_j v_j^+(n)
\end{aligned}$$

Equating like terms with Eq. (F.32) gives

$$\rho_i^v = \beta_i - 1 \tag{F.33}$$

$$\tau_{ji}^v = \beta_j, \quad (i \neq j) \tag{F.34}$$

Thus, the jth beta parameter is the velocity transmission coefficient from jth port to any other port (besides the ith). To convert the transmission coefficient from the ith port to the reflection coefficient for that port, we simply subtract 1. These relationships are specific to velocity waves at a series junction (cf. Eq. (F.22)). They are exactly the dual of Equations (F.22–F.23) for force waves at a parallel junction.

Physical Derivation of Series Reflection Coefficient

Physically, the *force-wave* reflection coefficient seen at port i of a series adaptor is due to an impedance step from R_i, that of the port interface, to a new impedance consisting of the series combination of all *other* port impedances meeting at the junction. Let

$$R_J(i) \triangleq \sum_{i \neq j} R_i \tag{F.35}$$

denote this series combination. Then we must have, as in Eq. (F.25),

$$\rho_i = \frac{R_J(i) - R_i}{R_J(i) + R_i} \tag{F.36}$$

Let's check this "physical" derivation against the formal definition Eq. (F.31) leading to $\rho_i^v = \beta_i - 1$ in Eq. (F.33). Define the total junction impedance as

$$R_J \triangleq \sum_{j=1}^{N} R_j$$

This is the series combination of *all* impedances connected to the junction. Then by Eq. (F.35), $R_J = R_i + R_J(i)$ for all i. From Eq. (F.26), the *velocity* reflection coefficient is given by

$$\begin{aligned}
\rho_i^v &\triangleq \beta_i - 1 \triangleq \frac{2R_i}{R_J} - 1 \\
&\triangleq \frac{2R_i - R_J}{R_J} = \frac{2R_i - [R_i + R_J(i)]}{R_i + R_J(i)} \\
&= \frac{R_i - R_J(i)}{R_i + R_J(i)} \\
&= -\rho_i
\end{aligned}$$

Since

$$\rho_i^v \triangleq \frac{v_i^-(n)}{v_i^+(n)} = \frac{-f_i^-(n)/R_i}{f_i^+(n)/R_i} = -\frac{f_i^-(n)}{f_i^+(n)} \triangleq -\rho_i$$

the result follows.

Series Reflection Free Port

For port i to be reflection free in a series adaptor, we require

$$R_i = R_J(i) \triangleq \sum_{i \neq j} R_i \tag{F.37}$$

That is, the port's impedance must equal the series combination of the other port impedances at the junction. This result can be compared with that for the parallel junction in §F.2.2.

The series adaptor has now been derived in a way which emphasizes its duality with respect to the parallel adaptor.

F.3 Wave Digital Modeling Examples

This section presents a series of examples, working up from very simple to more complicated situations.

F.3.1 "Piano hammer in flight"

Suppose we wish to model a situation in which a mass of size m kilograms is traveling with a constant velocity. This is an appropriate model for a piano hammer after its key has been pressed and before the hammer has reached the string.

Figure F.2 shows the "wave digital mass" derived previously. The derivation consisted of inserting an infinitesimal waveguide[3] having (real) impedance m, solving for the force-wave reflectance of the mass as seen from the waveguide, and then mapping it to the discrete time domain using the bilinear transform.

We now need to attach the other end of the transmission line to a "force source" which applies a force of zero newtons to the mass. In other words, we need to terminate the line in a way that corresponds to zero force.

Let the force-wave components entering and leaving the mass be denoted f^+ and f^-, respectively (*i.e.*, we are dropping the subscript 'd' in Fig. F.2). The physical force associated with the mass is

$$f(n) = f^+(n) + f^-(n) = f^+(n) - f^+(n-1)$$

The zero-force case is therefore obtained when $f^+(n) = -f^-(n) = f^+(n-1)$. This is illustrated in Fig. F.8.

Figure F.8a (left portion) illustrates what we derived by physical reasoning, and as such, it is most appropriate as a physical model of the constant-velocity mass. However, for actual *implementation*, Fig. F.8b would be more typical in practice. This is because we can always negate the state variable $x(n)$ if needed to convert it from $f^+(n-1)$ to $f^-(n)$. It is very common to see final simplifications like this to maximize efficiency.

[3]Here, it is perhaps most concrete to think in terms of electrical equivalent circuits, so that the mass is an inductor and a "waveguide" is a "transmission line."

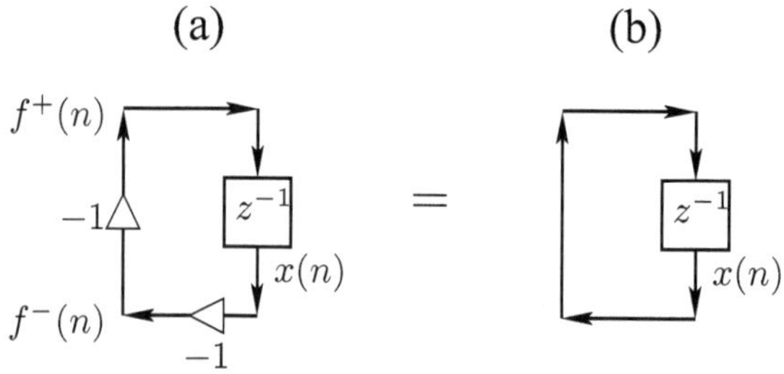

Figure F.8: Wave digital mass in flight at a constant velocity.

Note that Fig. F.8b can be interpreted physically as a wave digital *spring* displaced by a constant force $f(n) = 2x(n)$.

Extracting Physical Quantities

Since we are using a force-wave simulation, the state variable $x(n)$ (delay element output) is in units of physical force (newtons). Specifically, $x(n) = f^+(n-1)$. (The physical force is, of course, 0, while its traveling-wave components are not 0 unless the mass is at rest.) Using the fundamental relations relating traveling force and velocity waves

$$\begin{aligned} f^+(n) &\triangleq R_0 v^+(n) \\ f^-(n) &\triangleq -R_0 v^-(n) \end{aligned}$$

where $R_0 = m$ here, it is easy to convert the state variable $x(n)$ to other physical units, as we now demonstrate.

The velocity of the mass, for example, is given by

$$v(n) = v^+(n) + v^-(n) = \frac{f^+(n)}{m} - \frac{f^-(n)}{m} = \frac{2f^+(n)}{m} = \frac{2}{m}x(n)$$

Thus, the state variable $x(n)$ can be scaled by $2/m$ to "read out" the mass velocity.

The kinetic energy of the mass is given by

$$\mathcal{E}_m = \frac{1}{2}mv^2(n) = \frac{2}{m}x^2(n)$$

I.e., the square of the state variable $x(n)$ can be scaled by $2/m$ to produce the physical kinetic energy associated with the mass.

F.3.2 Force Driving a Mass

Suppose now that we wish to *drive* the mass along a frictionless surface using a variable force $f(n)$. This is similar to the previous example, except that we now want the traveling-wave components of the force on the mass to sum to $f(n)$ instead of 0:

$$f(n) = f^+(n) + f^-(n)$$

Since $f(n)$ and $f^-(n)$ are given, $f^+(n)$ must be computed as $f^+(n) = f(n) - f^-(n)$. This is shown in Fig. F.9.

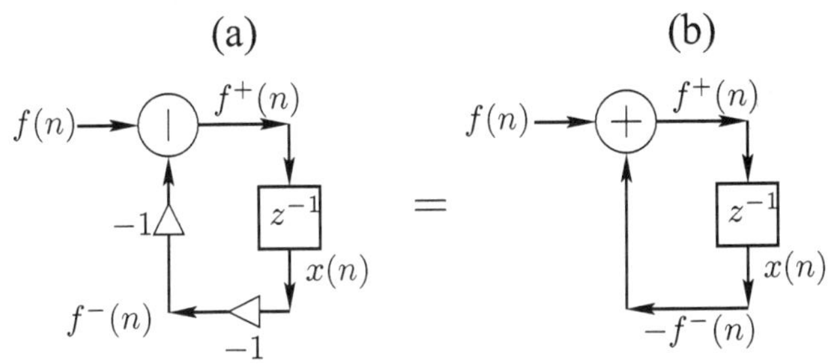

Figure F.9: Wave digital mass driven by external force $f(n)$.

The simplified form in Fig. F.9b can be interpreted as a wave digital spring with applied force $f(n)$ delivered from an infinite source impedance. That is, when the applied force goes to zero, the termination remains rigid at the current displacement.

A More Formal Derivation of the Wave Digital Force-Driven Mass

Above we derived how to handle the external force by direct physical reasoning. In this section, we'll derive it using a more general step-by-step procedure which can be applied systematically to more complicated situations.

Figure F.10 gives the physical picture of a free mass driven by an external force in one dimension. Figure F.11 shows the electrical equivalent circuit for this scenario in which the external force is represented by a voltage source emitting $f(t)$ *volts*, and the mass is modeled by an inductor having the value $L = m$ *Henrys*.

The next step is to convert the voltages and currents in the electrical equivalent circuit to *wave variables*. Figure F.12 gives an intermediate equivalent circuit in which an infinitesimal transmission line section with

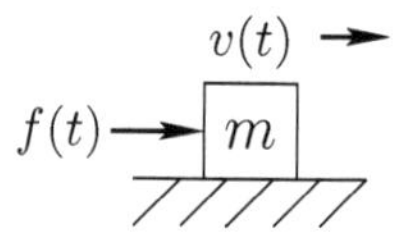

Figure F.10: Physical diagram of an external force driving a mass sliding on a frictionless surface.

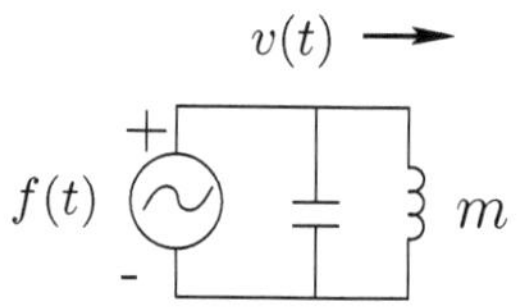

Figure F.11: Electrical equivalent circuit of the force-driven mass in Fig. F.10.

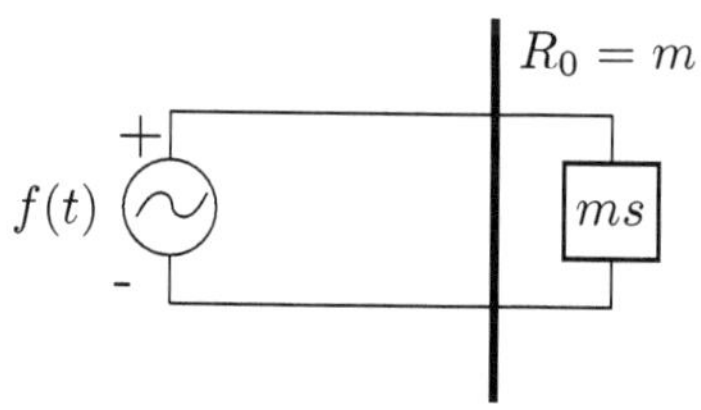

Figure F.12: Intermediate equivalent circuit for the force-driven mass in which an infinitesimal transmission line section has been inserted to facilitate conversion of the mass impedance ms into a wave-variable reflectance.

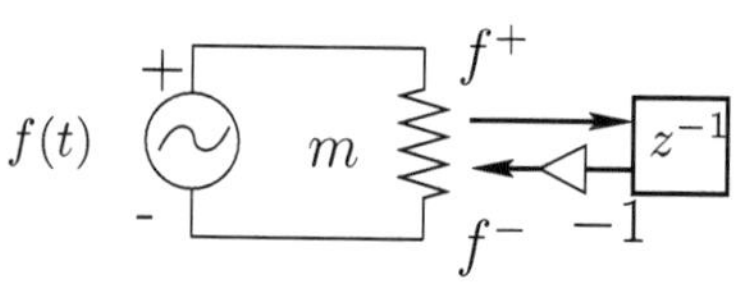

Figure F.13: Intermediate wave-variable model of the force-driven mass of Fig. F.11.

real impedance $R_0 = m$ has been inserted to facilitate the computation of the wave-variable reflectance, as we did in §F.1.1 to derive Eq. (F.1).

Figure F.13 depicts a next intermediate equivalent circuit in which the mass has been replaced by its reflectance (using "z^{-1}" to denote the continuous-time reflectance $(1-s)/(1+s)$, as derived in §F.1.1). The infinitesimal transmission-line section is now represented by a "resistor" since, when the voltage source is initially "switched on", it only "sees" a real resistance having the value m Ohms (the waveguide interface). After a short period of time determined by the reflectance of the mass,[4] "return waves" from the mass result in an ultimately *reactive* impedance. This of course must be the case because the mass does not dissipate energy. Therefore, the "resistor" of m Ohms is not a resistor in the usual sense since it does not convert potential energy (the voltage drop across it) into heat. Instead, it converts potential energy into propagating waves with 100% efficiency. Since all of this wave energy is ultimately reflected by the terminating element (mass, spring, or any combination of masses and springs), the net effect is a purely reactive impedance, as we know it must be.

To complete the wave digital model, we need to connect our wave digital mass to an ideal force source which asserts the value $f(n)$ each sample time. Since an ideal force source has a zero internal impedance, we desire a parallel two-port junction which connects the impedances $R_1 = 0$ ($\Gamma_1 = \infty$) and $R_2 = m$ ($\Gamma_2 = 1/m$), as shown in Fig. F.14. From Eq. (F.18)

---

[4]The phase delay, for example, of the reflectance $(1-s)/(1+s)$ of an ideal spring is given by

$$P(\omega) = -\frac{\angle\frac{1-j\omega}{1+j\omega}}{\omega}.$$

At $\omega = 1$ rad/sec, in particular, the phase delay is $P(1) = -\angle[(1-j)/(1+j)] = -\angle(-j) = \pi/2$. Since the period is 2π when $\omega = 1$, this is a quarter-cycle of delay.

After the bilinear transform $s = (z-1)/(z+1)$, the phase delay becomes

$$P_d(\omega_d) = -\frac{\angle e^{-j\omega_d T}}{\omega_d} = -\frac{-\omega_d T}{\omega_d} = T,$$

where ω_d is defined by $\omega = \tan(\omega_d T/2)$ (the relationship between continuous-time frequency and discrete-time frequency imposed by the bilinear transform). Thus, the phase delay comes out to exactly one sample at each frequency. Checking the point $\omega = 1 \leftrightarrow \omega_d T = \pi/2 \Rightarrow f_d = f_s/4$, *i.e.*, four samples per period, we find that one sample of delay does in fact correspond to a quarter-cycle of delay.

The same holds true for the group delay of the spring reflectance:

$$G_d(\omega_d) \triangleq -\frac{d}{d\omega_d}\angle e^{-j\omega_d T} = T$$

We see that the frequency-warping under the bilinear transform serves to move each frequency to the unique point along the unit circle at which the reflectance of a mass or spring happens to be exactly one sample. The mass reflectance additionally inverts.

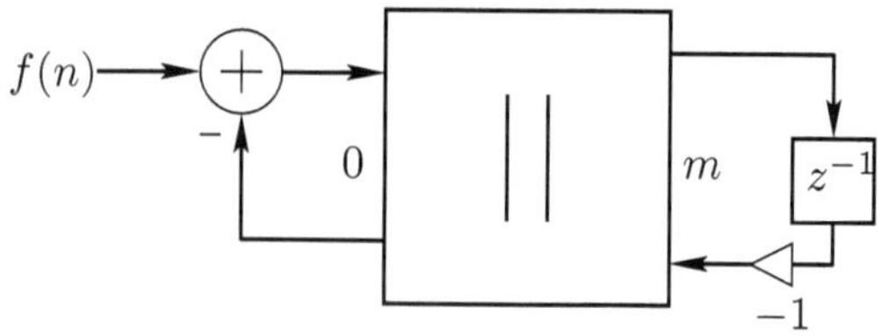

Figure F.14: Interconnection of the wave digital mass with an ideal force source by means of a two-port parallel adaptor. The symbol "||" is used in the WDF literature to signify a parallel adaptor.

we have that the common junction force is equal to

$$\begin{aligned}
f(n) &= \alpha_1 f_1^+(n) + \alpha_2 f_2^+(n) \\
&= \frac{2\Gamma_1}{\Gamma_1+\Gamma_2} f_1^+(n) + \frac{2\Gamma_2}{\Gamma_1+\Gamma_2} f_2^+(n) \\
&= \frac{2\cdot\infty}{\infty+\frac{1}{m}} f_1^+(n) + \frac{\frac{2}{m}}{\infty+\frac{1}{m}} f_2^+(n) \\
&= 2f_1^+(n)
\end{aligned}$$

from which we conclude that

$$\boxed{f_1^+(n) = \frac{f(n)}{2}}$$

The outgoing waves are, by Eq. (F.19),

$$\begin{aligned}
f_1^-(n) &= f(n) - f_1^+(n) = \frac{f(n)}{2} \\
f_2^-(n) &= f(n) - f_2^+(n) = f(n) - f_m^-(n)
\end{aligned}$$

Since $\alpha_1 = 1$ and $\alpha_2 = 0$ for this model, the reflection coefficient seen on port 1 is $\rho = \alpha_1 - 1 = 1$. The transmission coefficient from port 1 is $1+\rho = 2$. In the opposite direction, the reflection coefficient on port 2 is $-\rho = -1$, and the transmission coefficient from port 2 is $1-\rho = 0$. The final result, drawn in Kelly-Lochbaum form (see §F.2.1), is diagrammed in Fig. F.15, as well as the result of some elementary simplifications. The final model is the same as in Fig. F.9, as it should be.

F.3.3 Force Driving a Spring against a Wall

For this example, we have an external force $f(n)$ driving a spring k which is terminated on the other end at a rigid wall. Figure F.16 shows the physical diagram and the electrical equivalent circuit is given in Fig. F.17.

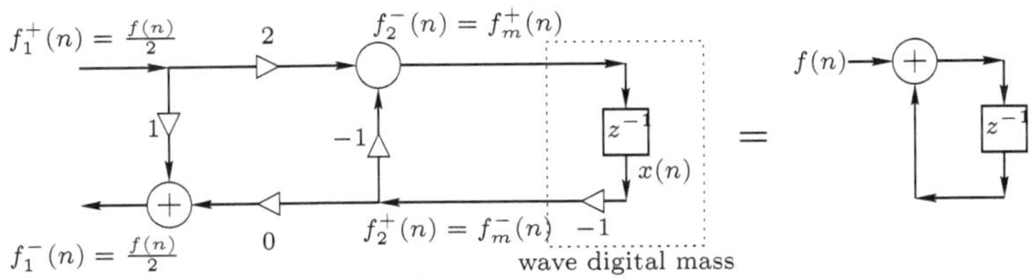

Figure F.15: Wave digital mass driven by external force $f(n)$.

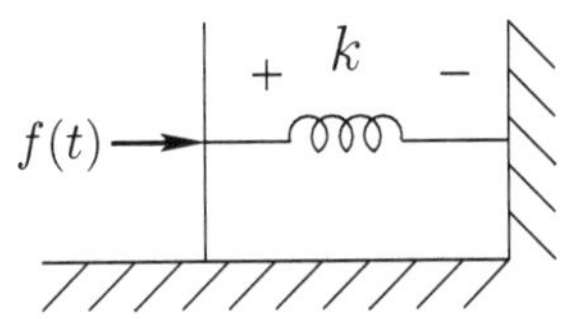

Figure F.16: External force driving a spring terminated by a rigid wall.

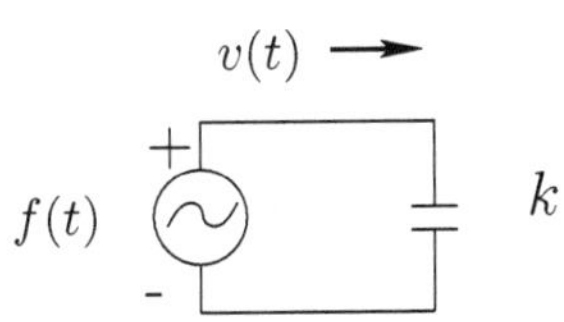

Figure F.17: Electrical equivalent circuit of the compressed spring of Fig. F.16.

Figure F.18 depicts the insertion of an infinitesimal transmission line, and Fig. F.19 shows the result of converting the spring impedance to wave variable form.

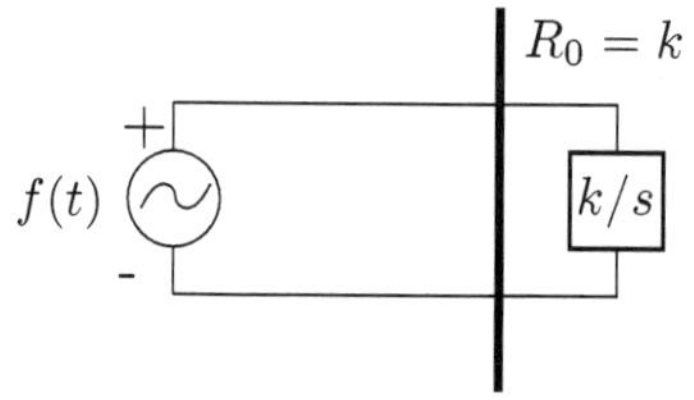

Figure F.18: Intermediate equivalent circuit for the force-driven spring in which an infinitesimal transmission line section has been inserted to facilitate conversion of the spring impedance k/s into a wave-variable reflectance.

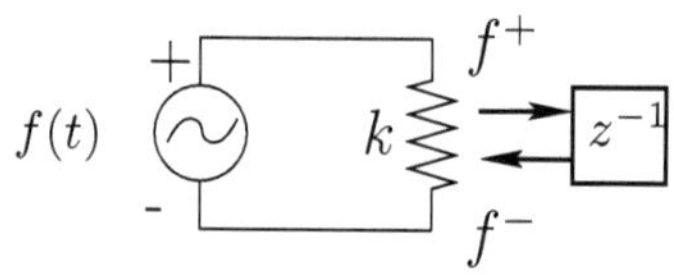

Figure F.19: Intermediate wave-variable model of Fig. F.17.

The two-port adaptor needed for this problem is the same as that for the force-driven mass, and the final result is shown in Fig. F.20.

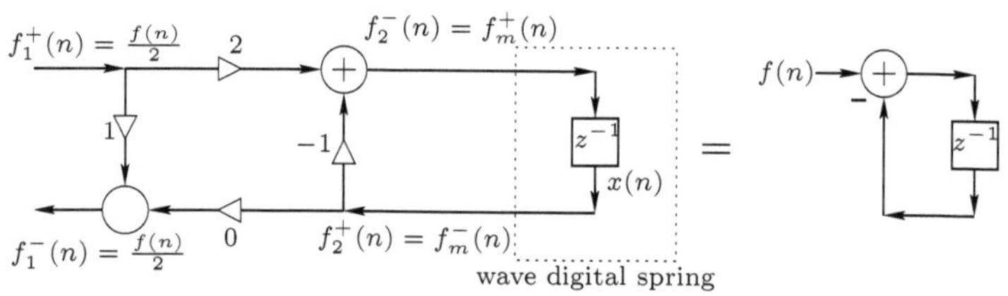

Figure F.20: Wave digital spring driven by external force $f(n)$.

Note that the spring model is being driven by a force from a zero source impedance, in contrast with the infinite source impedance interpretation of Fig. F.8b as a compressed spring. In this case, if the driving force goes to zero, the spring force goes immediately to zero ("free termination") rather than remaining fixed.

F.3.4 Spring and Free Mass

For this example, we have an external force $f(n)$ driving a spring k which in turn drives a free mass m. Since the force on the spring and the mass are always the same, they are formally "parallel" impedances.

This problem is easier than it may first appear since an ideal "force source" (*i.e.*, one with a zero source impedance) driving impedances in parallel can be analyzed separately for each parallel branch. That is, the system is equivalent to one in which the mass and spring are not connected at all, and each has its own copy of the force source. With this insight in mind, one can immediately write down the final wave-digital model shown in Fig. F.25. However, we will go ahead and analyze this case more formally since it has some interesting features.

Figure F.21 shows the physical diagram of the spring-mass system driven by an external force. The electrical equivalent circuit appears in Fig. F.22, and the first stage of a wave-variable conversion is shown in Fig. F.23.

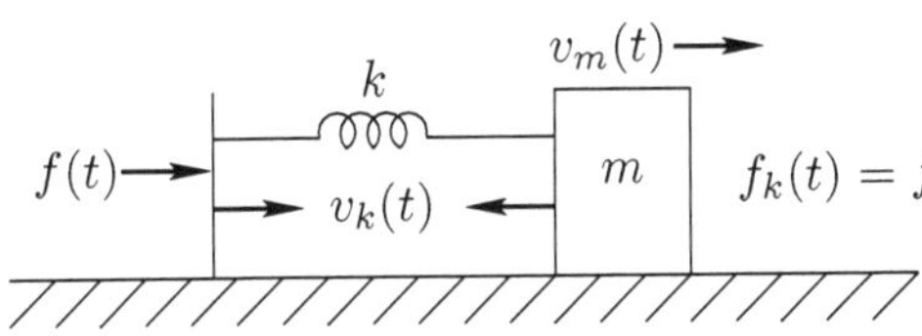

Figure F.21: External force driving a spring which in turn drives a free mass sliding on a frictionless surface.

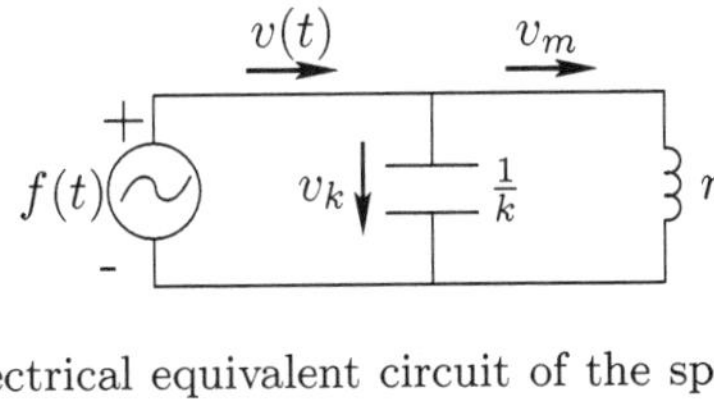

Figure F.22: Electrical equivalent circuit of the spring/mass system of Fig. F.21.

For this example we need a three-port parallel adaptor, as shown in Fig. F.24 (along with its attached mass and spring). The port impedances are 0, k, and m, yielding alpha parameters $\alpha_1 = 2$ and $\alpha_2 = \alpha_3 = 0$. The final result, after the same sorts of elementary simplifications as in the previous example, is shown in Fig. F.25. As predicted, a force source driving elements in parallel is equivalent to a set of independent force-driven elements.

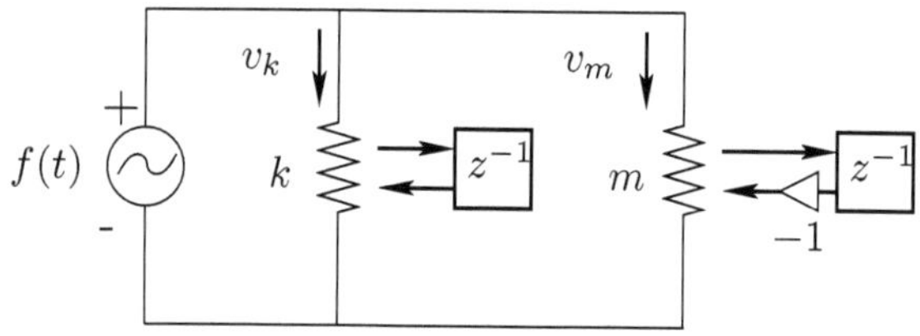

Figure F.23: Intermediate wave-variable model of the mass and dashpot of Fig. F.22.

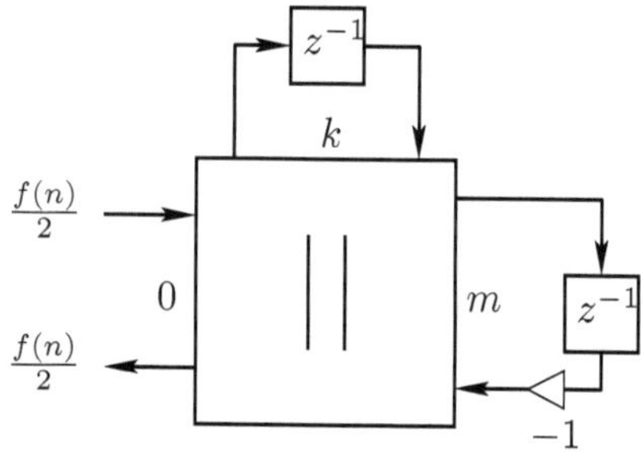

Figure F.24: Wave digital model for the parallel combination of a wave digital mass m, wave digital spring k, and driving force $f(n)$. The corresponding port impedances are m, k, and 0, respectively.

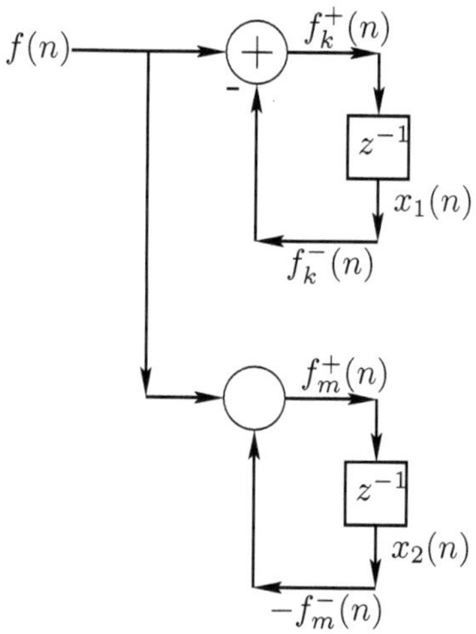

Figure F.25: Wave digital filter model of an external force driving a mass through a spring. The mass force-wave components are denoted $f_m^{\pm}$, while those for the spring are denoted $f_k^{\pm}$.

From this and the preceding example, we can see a general pattern: Whenever there is an ideal force source driving a parallel junction, then $\Gamma_1 = \infty$ and all other port admittances are finite. In this case, we always obtain $\alpha_1 = 2$ and $\alpha_i = 0$, $i \neq 1$.

F.3.5 Mass and Dashpot in Series

This is our first example illustrating a *series* connection of wave digital elements. Figure F.26 gives the physical scenario of a simple mass-dashpot system, and Fig. F.27 shows the equivalent circuit. Replacing element voltages and currents in the equivalent circuit by wave variables in an infinitesimal waveguides produces Fig. F.28.

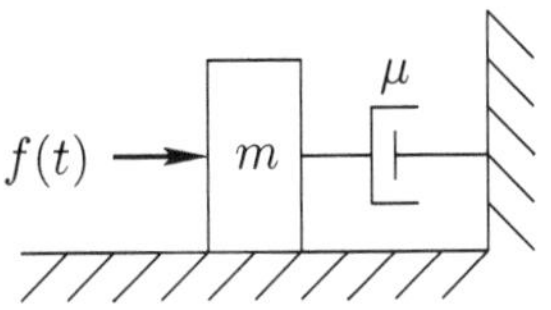

Figure F.26: External force driving a mass which in turn drives a dashpot terminated on the other end by a rigid wall.

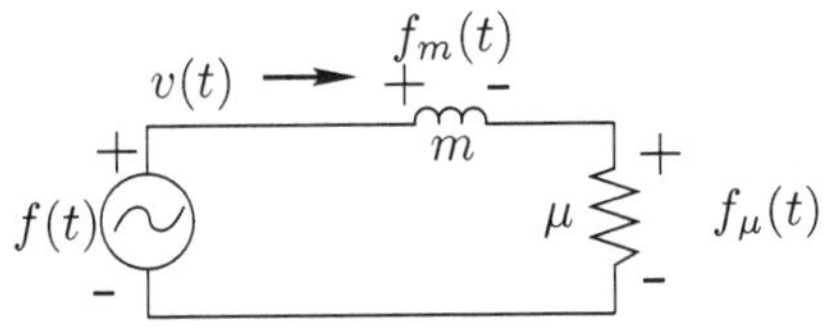

Figure F.27: Electrical equivalent circuit of the mass and dashpot system of Fig. F.26.

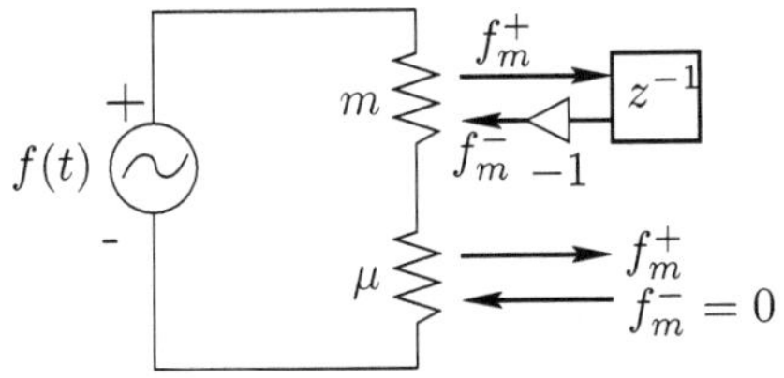

Figure F.28: Intermediate wave-variable model of the mass and dashpot of Fig. F.27.

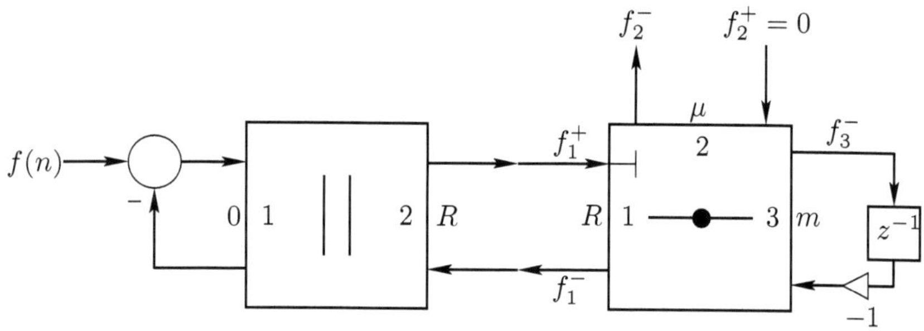

Figure F.29: Wave digital filter for an ideal force source in parallel with the series combination of a mass m and dashpot μ. The parallel and series adaptors are joined at an impedance R which is calculated to suppress reflection from port 1 of the series adaptor.

The system can be described as an ideal force source $f(t)$ connected in *parallel* with the *series* connection of mass m and dashpot μ. Figure F.29 illustrates the resulting wave digital filter. Note that the ports are now numbered for reference. Two more symbols are introduced in this figure: (1) the horizontal line with a dot in the middle indicating a series adaptor, and (2) the indication of a *reflection-free port* on input 1 of the series adaptor (signal $f_1^+(n)$). Recall that a reflection-free port is always necessary when connecting two adaptors together, to avoid creating a delay-free loop.

Let's first calculate the impedance R necessary to make input 1 of the series adaptor reflection free. From Eq. (F.37), we require

$$R = m + \mu$$

That is, the impedance of the reflection-free port must equal the series combination of all other port impedances meeting at the junction.

The parallel adaptor, viewed alone, is equivalent to a force source driving impedance $R = m + \mu$. It is therefore realizable as in Fig. F.20 with the wave digital spring replaced by the mass-dashpot assembly in Fig. F.29. However, we can also carry out a quick analysis to verify this: The alpha parameters are

$$\begin{aligned}
\alpha_1 &\isdef \frac{2\Gamma_1}{\Gamma_1+\Gamma_2} = \frac{2\cdot\infty}{\infty+\frac{1}{m}+\frac{1}{\mu}} = 2 \\
\alpha_2 &\isdef \frac{2\Gamma_2}{\Gamma_1+\Gamma_2} = \frac{2\left(\frac{1}{m}+\frac{1}{\mu}\right)}{\infty+\left(\frac{1}{m}+\frac{1}{\mu}\right)} = 0
\end{aligned}$$

Therefore, the reflection coefficient seen at port 1 of the parallel adaptor is $\rho = \alpha_1 - 1 = 1$, and the Kelly-Lochbaum scattering junction depicted in Fig. F.20 is verified.

Let's now calculate the internals of the series adaptor in Fig. F.29. From Eq. (F.26), the beta parameters are

$$\begin{aligned}
\beta_1 &\triangleq \frac{2R_1}{R_1+R_2+R_3} = \frac{2(m+\mu)}{(m+\mu)+m+\mu} = 1 \\
\beta_2 &\triangleq \frac{2R_2}{R_1+R_2+R_3} = \frac{2\mu}{(m+\mu)+m+\mu} = \frac{\mu}{m+\mu} \\
\beta_3 &\triangleq \frac{2R_3}{R_1+R_2+R_3} = \frac{2m}{(m+\mu)+m+\mu} = \frac{m}{m+\mu}
\end{aligned}$$

Following Eq. (F.30), the series adaptor computes

$$\begin{aligned}
f_J^+(n) &= f_1^+(n) + f_2^+(n) + f_3^+(n) = f(n) + f_3^+(n) \\
f_1^-(n) &= f_1^+(n) - \beta_1 f_J^+(n) = f_3^+(n) \\
f_3^-(n) &= f_3^+(n) - \beta_3 f_J^+(n) \\
&= (1-\beta_3) f_3^+(n) - \beta_3 f(n) \\
&= \frac{\mu}{m+\mu} f_3^+(n) - \frac{m}{m+\mu} f(n)
\end{aligned}$$

We do not need to explicitly compute $f_2^-(n)$ because it goes into a purely resistive impedance μ and produces no return wave. For the same reason, $f_2^+(n) \equiv 0$. Figure F.30 shows a wave flow diagram of the computations derived, together with the result of elementary simplifications.

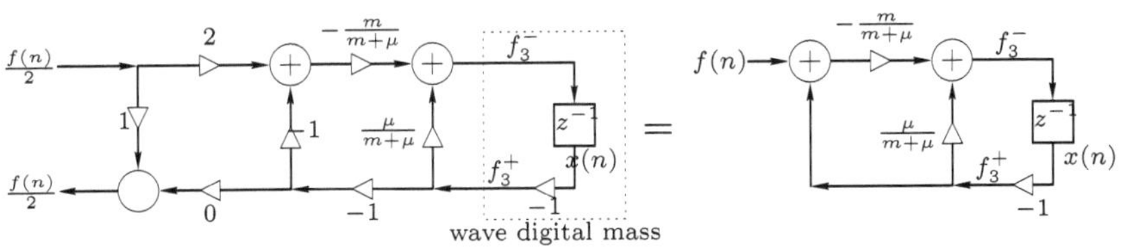

Figure F.30: Wave flow diagram for the WDF modeling an ideal force source in parallel with the series combination of a mass m and dashpot μ.

Because the difference of the two coefficients in Fig. F.30 is 1, we can easily derive the one-multiply form in Fig. F.31.

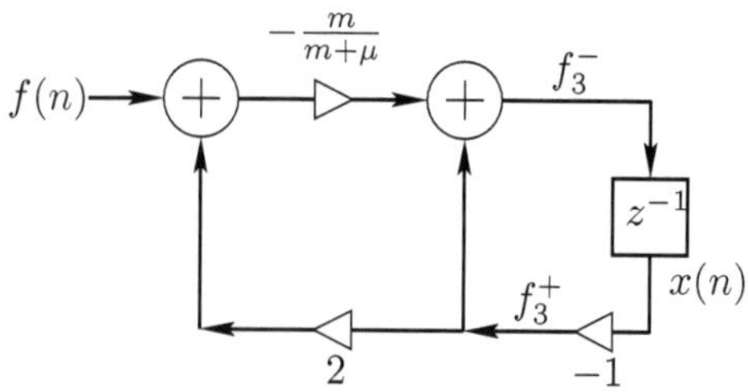

Figure F.31: One-multiply form of the WDF in Fig. F.30.

Checking the WDF against the Analog Equivalent Circuit

Let's check our result by comparing the transfer function from the input force to the force on the mass in both the discrete- and continuous-time cases.

For the discrete-time case, we have

$$H_m(z) \triangleq \frac{F_3(z)}{F(z)} = \frac{F_3^+(z) + F_3^-(z)}{F(z)} = (1 - z^{-1})\frac{F_3^-(z)}{F(z)}$$

where the last simplification comes from the mass reflectance relation $F_3^+(z) = -z^{-1}F_3^-(z)$. (Note that we are using the standard traveling-wave notation for the *adaptor*, so that the ± signs are swapped relative to element-centric notation.)

We now need $F_3^-(z)/F(z)$. To simplify notation, define the two coefficients as

$$\begin{aligned} a &= \frac{m}{m+\mu} \\ b &= \frac{\mu}{m+\mu} \end{aligned}$$

From Figure F.30, we can write

$$\begin{aligned} F_3^-(z) &= -a\left[F(z) - z^{-1}F_3^-(z)\right] + b\left[-z^{-1}F_3^-(z)\right] \\ &= -aF(z) + (a-b)z^{-1}F_3^-(z) \\ \Rightarrow \quad F_3^-(z) &= -a\frac{F(z)}{1-(a-b)z^{-1}F_3^-(z)} \end{aligned}$$

Thus, the desired transfer function is

$$H_m(z) = -a\frac{1-z^{-1}}{1-(a-b)z^{-1}} = -\frac{m}{m+\mu}\frac{1-z^{-1}}{1-\left(\frac{m-\mu}{m+\mu}\right)z^{-1}}$$

We now wish to compare this result to the bilinear transform of the corresponding analog transfer function. From Figure F.27, we can recognize the mass and dashpot as *voltage divider*:

$$H_m^a(s) = \frac{ms}{ms+\mu}$$

Applying the bilinear transform yields

$$\begin{aligned}
H_m^a\left(\frac{1-z^{-1}}{1+z^{-1}}\right) &= \frac{m\left(\frac{1-z^{-1}}{1+z^{-1}}\right)}{m\left(\frac{1-z^{-1}}{1+z^{-1}}\right)+\mu} = \frac{m\left(1-z^{-1}\right)}{m\left(1-z^{-1}\right)+\left(1+z^{-1}\right)\mu} \\
&= m\frac{1-z^{-1}}{(m+\mu)+(\mu-m)z^{-1}} = \left(\frac{m}{m+\mu}\right)\frac{1-z^{-1}}{1-\left(\frac{m-\mu}{m+\mu}\right)z^{-1}} \\
&= H_m(z)
\end{aligned}$$

Thus, we have verified that the force transfer-function from the driving force to the mass is identical in the discrete- and continuous-time models, except for the bilinear transform frequency warping in the discrete-time case.

F.3.6 Wave Digital Mass-Spring Oscillator

Let's look again at the mass-spring oscillator of §F.3.4, but this time without the driving force (which effectively decouples the mass and spring into separate first-order systems). The physical diagram and equivalent circuit are shown in Fig. F.32 and Fig. F.33, respectively.

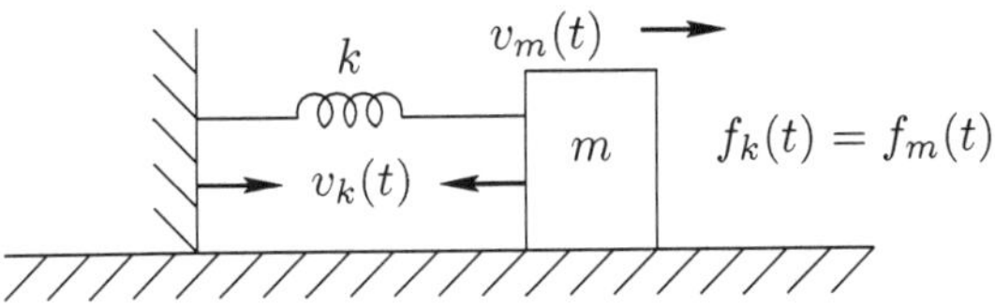

Figure F.32: Elementary mass-spring oscillator.

Note that the mass and spring can be regarded as being in parallel or in series. Under the parallel interpretation, we have the WDF shown in Fig. F.34 and Fig. F.35.[5] The reflection coefficient ρ can be computed, as usual, from the first alpha parameter:

$$\rho = \alpha_1 - 1 = \frac{2\Gamma_1}{\Gamma_1+\Gamma_2} - 1 = \frac{\Gamma_1-\Gamma_2}{\Gamma_1+\Gamma_2} = \frac{R_2-R_1}{R_2+R_1} = \frac{m-k}{m+k}$$

[5]Note that the wave variables are now labeled in element-centric notation as opposed

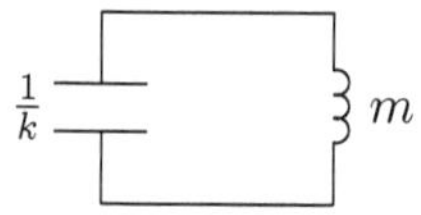

Figure F.33: Equivalent circuit for the mass-spring oscillator.

This result, $\rho = (m-k)/(m+k)$, is just the "impedance step over impedance sum", so no calculation was really necessary.

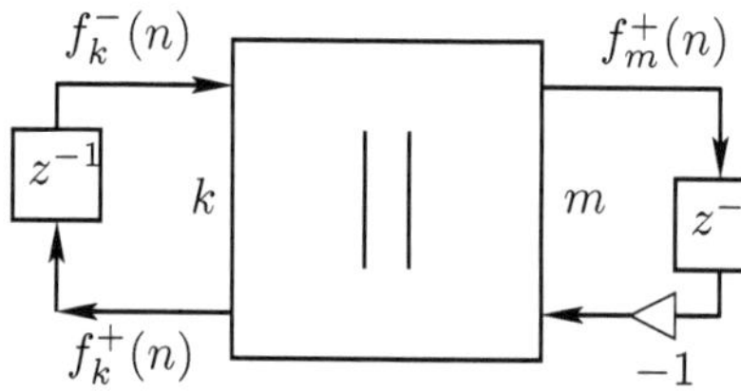

Figure F.34: Wave digital mass-spring oscillator.

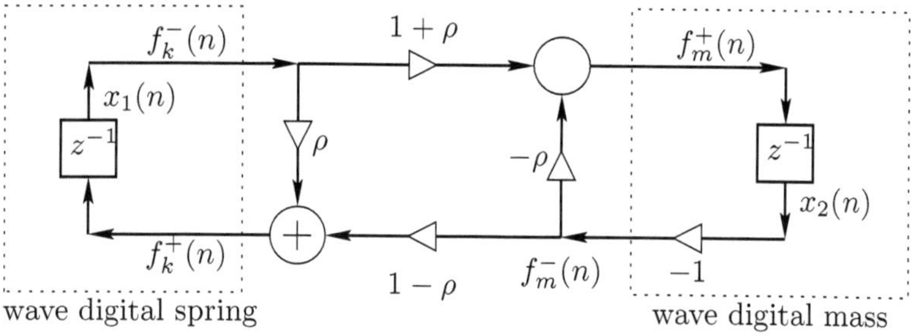

Figure F.35: Detailed wave-flow diagram for the wave digital mass-spring oscillator.

---

to adaptor-centric notation:

$$\begin{aligned} f_k(n) &= f_k^+(n) + f_k^-(n) \quad \text{(Spring Force)} \\ f_m(n) &= f_m^+(n) + f_m^-(n) \quad \text{(Mass Force)} \end{aligned}$$

Oscillation Frequency

From Fig. F.33, we can see that the impedance of the parallel combination of the mass and spring is given by

$$R_{m||k}(s) \triangleq \frac{k}{s} \Bigg|\Bigg| ms = \frac{\frac{k}{s} ms}{\frac{k}{s} + ms} = \frac{ks}{s^2 + \frac{k}{m}} \tag{F.38}$$

(using the product-over-sum rule for combining impedances in parallel). The poles of this impedance are given by the roots of the denominator polynomial in s:

$$s = \pm j\sqrt{\frac{k}{m}} \tag{F.39}$$

The resonance frequency of the mass-spring oscillator is therefore

$$\omega_0 = \sqrt{\frac{k}{m}} \tag{F.40}$$

Since the poles $s = \pm j\omega_0$ are on the $j\omega$ axis, there is no damping, as we expect.

We can now write reflection coefficient ρ (see Fig. F.35) as

$$\rho = \frac{m-k}{m+k} = \frac{1-\frac{k}{m}}{1+\frac{k}{m}} = \frac{1-\omega_0^2}{1+\omega_0^2}$$

We see that dc ($\omega_0 = 0$) corresponds to $\rho = 1$, and $\omega_0 = \infty$ corresponds to $\rho = -1$.

DC Analysis of the WD Mass-Spring Oscillator

Considering the dc case first ($\rho = 1$), we see from Fig. F.35 that the state variable $x_1(n)$ will circulate unchanged in the isolated loop on the left. Let's call this value $x_1(n) \equiv x_0$. Then the physical force on the spring is always equal to

$$f_k(n) = f_k^+(n) + f_k^-(n) = 2x_1(n) = 2x_0. \qquad \text{(spring force, dc case)} \tag{F.41}$$

The loop on the right in Fig. F.35 receives $2x_0$ and adds $x_2(n)$ to that. Since $x_2(n+1) = 2x_1(n) + x_2(n)$, we see it is *linearly growing* in amplitude. For example, if $x_2(0) = 0$ (with $x_1(0) = x_0$), we obtain $x_2 = [0, 2x_0, 4x_0, 6x_0, \ldots]$, or

$$x_2(n) = 2nx_0, \quad n = 0, 1, 2, 3, \ldots. \tag{F.42}$$

At first, this result might appear to contradict conservation of energy, since the state amplitude seems to be growing without bound. However, the *physical* force is fortunately better behaved:

$$f_m(n) = f_m^+(n) + f_m^-(n) = x_2(n+1) - x_2(n) = 2x_0. \tag{F.43}$$

Since the spring and mass are connected in parallel, it must be the true that they are subjected to the same physical force at all times. Comparing Equations (F.41–F.43) verifies this to be the case.

WD Mass-Spring Oscillator at Half the Sampling Rate

Under the bilinear transform, the $s = \infty$ maps to $z = -1$ (half the sampling rate). It is therefore no surprise that given $\omega_0 = \infty$ ($\rho = -1$), inspection of Fig. F.35 reveals that any alternating sequence (sinusoid sampled at half the sampling rate) will circulate unchanged in the loop on the right, which is now isolated. Let $x_2(n) = (-1)^n x_0$ denote this alternating sequence. The loop on the left receives $-2x_2(n)$ and adds $-x_1(n-1)$ to it, *i.e.*, $x_1(n+1) = -x_1(n) - 2x_2(n) = -x_1(n) - 2x_2(0)(-1)^n$. If we start out with $x_1(0) = 0$ and $x_2(0) = x_0$, we obtain $x_1 = [0, -2x_0, 4x_0, -6x_0, \ldots]$, or

$$x_1(n) = (-1)^n 2nx_0, \quad n = 0, 1, 2, 3, \ldots .$$

However, the physical spring force is well behaved, since

$$f_k(n) = f_k^+(n) + f_k^-(n) = x_1(n+1) + x_1(n) = (-1)^{n+1} 2x_0$$

As a check, the mass force is found to be

$$\begin{aligned} f_m(n) &= x_2(n+1) - x_2(n) \\ &= (-1)^{n+1} x_0 - (-1)^n x_0 \\ &= (-1)^{n+1} x_0 + (-1)^{n+1} x_0 \\ &= (-1)^{n+1} 2x_0, \end{aligned}$$

which agrees with the spring, as it must.

Linearly Growing State Variables in WD Mass-Spring Oscillator

It may seem disturbing that such a simple, passive, physically rigorous simulation of a mass-spring oscillator should have to make use of state variables which grow without bound for the limiting cases of simple harmonic motion at frequencies zero and half the sampling rate. This is obviously a valid concern in practice as well. However, it is easy to show that this only happens at dc and $f_s/2$, and that there is a true degeneracy at these frequencies, even in the physics. For all frequencies in the audio range (*e.g.*, for typical sampling rates), such state variable growth cannot occur. Let's take closer look at this phenomenon, first from a signal processing point of view, and second from a physical point of view.

A Signal Processing Perspective on Repeated Mass-Spring Poles

Going back to the poles of the mass-spring system in Eq. (F.39), we see that, as the imaginary part of the two poles, $\pm\omega_0 = \pm\sqrt{k/m}$, approach zero, they come together at $s = 0$ to create a *repeated pole.* The same thing happens at $\omega_0 = \infty$ since both poles go to "the point at infinity".

It is a well known fact from linear systems theory that two poles at the same point $s = s_0 = \sigma_0$ in the s plane can correspond to an impulse-response component of the form $te^{\sigma_0 t}$, in addition to the component $e^{\sigma_0 t}$ produced by a single pole at $s = \sigma_0$. In the discrete-time case, a double pole at $z = r_0$ can give rise to an impulse-response component of the form nr_0^n. This is the fundamental source of the linearly growing internal states of the wave digital sine oscillator at dc and $f_s/2$. It is interesting to note, however, that such modes are always *unobservable* at any *physical* output such as the mass force or spring force that is not actually linearly growing.

Physical Perspective on Repeated Poles in the Mass-Spring System

In the physical system, dc and infinite frequency are in fact strange cases. In the case of dc, for example, a nonzero constant force implies that the mass m is under constant acceleration. It is therefore the case that its *velocity is linearly growing.* Our simulation predicts this, since, using Eq. (F.43) and Eq. (F.42),

$$\begin{aligned} v_m(n) &= \frac{f_m^+(n)}{m} - \frac{f_m^-(n)}{m} = \frac{1}{m}\left[x_2(n+1) + x_2(n)\right] \\ &= \frac{1}{m}\left[2(n+1) + 2n\right]x_0 = \frac{1}{m}(4nx_0 + 2x_0). \end{aligned}$$

The dc term $2x_0/m$ is therefore accompanied by a linearly growing term $2nx_0/m$ in the *physical mass velocity.* It is therefore unavoidable that we have some means of producing an unbounded, linearly growing output variable.

Mass-Spring Boundedness in Reality

To approach the limit of $\omega_0 = \sqrt{k/m} = 0$, we must either take the spring constant k to zero, or the mass m to infinity, or both.

In the case of $k \to 0$, the constant force must approach zero, and we are left with at most a constant mass velocity in the limit (not a linearly growing one, since there can be no dc force at the limit). When the spring force reaches zero, $x_1(n) = 0$, so that only zeros will feed into the loop on the right in Fig. F.35, thus avoiding a linearly growing velocity, as demanded by the physics. (A constant velocity is free to circulate in the loop on the right, but the loop on the left must be zeroed out in the limit.)

In the case of $m \to \infty$, the mass becomes unaffected by the spring force, so its final velocity must be zero. Otherwise, the attached spring would keep compressing or stretching forever, and this would take infinite energy. (Another way to arrive at this conclusion is to note that the final kinetic energy of the mass would be $mv^2/2 = \infty$.) Since the total energy in an undriven mass-spring oscillator is always constant, the infinite-mass limit must be accompanied by a zero-velocity limit.[6] This means the mass's state variable $x_2(n)$ in Fig. F.35 must be forced to zero in the limit so that there will be no linearly growing solution at dc.

In summary, when two or more system poles approach each other to form a repeated pole, care must be taken to ensure that the limit is approached in a physically meaningful way. In the case of the mass-spring oscillator, for example, any change in the spring constant k or mass m must be accompanied by the physically appropriate change in the state variables $x_1(n)$ and/or $x_2(n)$. It is obviously incorrect, for example, to suddenly set $k = 0$ in the simulation without simultaneously clearing the spring's state variable $x_1(n)$, since the force across an infinitely compliant spring can only be zero.

Similar remarks apply to repeated poles corresponding to $\omega_0 = \infty$. In this case, the mass and spring basically change places.

Energy-Preserving Parameter Changes (Mass-Spring Oscillator)

If the change in k or m is deemed to be "internal", that is, involving no external interactions, the appropriate accompanying change in the internal state variables is that which *conserves energy*. For the mass and its velocity, for example, we must have

$$\frac{1}{2}m_1 v_1^2 = \frac{1}{2}m_2 v_2^2$$

where m_1, m_2 denote the mass values before and after the change, respectively, and v_1, v_2 denote the corresponding velocities. The velocity must

[6]Note that the mass and velocity limits are tied together such that $mv^2/2 =$ constant. This information is lost in the final limits because the expression $\infty \cdot 0^2/2$ is "indeterminate".

therefore be scaled according to

$$v_2 = v_1\sqrt{\frac{m_1}{m_2}},$$

since this holds the kinetic energy of the mass constant. Note that the *momentum* of the mass *is* changed, however, since

$$m_2 v_2 = m_2 v_1\sqrt{\frac{m_1}{m_2}} = v_1\sqrt{m_1 m_2} = m_1 v_1\sqrt{\frac{m_2}{m_1}}$$

If the spring constant k is to change from k_1 to k_2, the instantaneous spring displacement x must satisfy

$$\frac{1}{2}k_1 x_1^2 = \frac{1}{2}k_2 x_2^2$$

In a velocity-wave simulation, displacement is the integral of velocity. Therefore, the energy-conserving velocity correction is impulsive in this case.

Exercises in Wave Digital Modeling

1. **Comparing digital and analog frequency formulas.** This first exercise verifies that the elementary "tank circuit" always resonates at exactly the frequency it should, according to the bilinear transform frequency mapping $\omega_a = \tan(\omega_d T/2)$, where ω_a denotes "analog frequency" and ω_d denotes "digital frequency".

   (a) Find the poles of Fig. F.35 in terms of ρ.

   (b) Show that the resonance frequency is given by

   $$f_s \arccos(\rho)$$

   where f_s denotes the sampling rate.

   (c) Recall that the mass-spring oscillator resonates at $\omega_0 = \sqrt{k/m}$. Relate these two resonance frequency formulas via the analog-digital frequency map $\omega_a = \tan(\omega_d T/2)$.

   (d) Show that the trig identity you discovered in this way is true. *I.e.*, show that

   $$f_s \arccos\left[\frac{k-m}{k+m}\right] = 2 f_s \arctan\left[\sqrt{\frac{m}{k}}\right].$$

Appendix G

Resources on the Internet

- STK Home Page: `https://ccrma.stanford.edu/software/stk/`
- HUT Acoustics Lab Home Page: `http://www.acoustics.hut.fi/`

  Sound examples: `http://www.acoustics.hut.fi/demos/`
- CCRMA course "Signal Processing Models in Musical Acoustics":
  `https://ccrma.stanford.edu/courses/420/`

  Lab exercises: `https://ccrma.stanford.edu/realsimple/`

  Overview presentation:
  `https://ccrma.stanford.edu/~jos/pdf/AES-Masterclass.pdf`
- Digital Waveguide Modeling of Musical Instruments (Home Page):
  `https://ccrma.stanford.edu/~jos/wg.html`
- JOS Home Page: `https://ccrma.stanford.edu/~jos/`
- Perry Cook Home Page:
  `http://www.cs.princeton.edu/~prc/NewWork.html`
- Gary Scavone Home Page: `http://www.music.mcgill.ca/~gary/`
- Digital Audio FX (DAFX) Web Site: `http://www.dafx.de/`
- SoftSynth: `http://www.softsynth.com/links/dsp_theory.html`
- Music-DSP Source Code Archive: `http://www.musicdsp.org/`
- Articles, online books, blogs, and discussions at *dsprelated.com:*
  `http://www.dsprelated.com`
- Numerical Recipes:
  `http://www.numerical-recipes.com/nronline_switcher.php`
- C++: `http://www.cpp4u.com/`

Appendix H

Sound Examples

In the Web version of this document at

`https://ccrma.stanford.edu/~jos/pasp/Sound_Examples.html`,

the sound examples below are links:

- Plucked and Struck Acoustic Strings
  - Plucked/Struck Strings (NeXT Computer, 1989): (WAV) (MP3)
  - Classical Guitar: (WAV) (MP3)
  - Harpsichord: (WAV) (MP3)
  - Piano: (WAV) (MP3)
- Bowed Strings
  - Early Cello (1982): (WAV) (MP3)
  - Double Bass: (WAV) (MP3)
  - Cello: (WAV) (MP3)
  - Viola 1: (WAV) (MP3)
  - Viola 2: (WAV) (MP3)
  - Violin 1: (WAV) (MP3)
  - Violin 2: (WAV) (MP3)
  - Violin Ensemble: (WAV) (MP3)
- Electric Guitars
  - Electric Guitar, Bass Pickup: (WAV) (MP3)
  - Electric Guitar, Single-Coil Pickup: (WAV) (MP3)
  - Hollow-Body Electric Guitar: (WAV) (MP3)
  - Distorted Electric Guitar: (WAV) (MP3)

- Distorted Electric Guitar + Feedback: (WAV) (MP3)
- Distorted Electric Guitar + Feedback: (WAV) (MP3)
- Electric Guitar + Wah-Wah Effect: (WAV) (MP3)
- Nonlinear String: (WAV) (MP3)
- Sitar: (WAV) (MP3)

- Yamaha "Virtual Lead" (VL) Synthesizer (1994)
  - Oboe and Bassoon: (WAV) (MP3)
  - Shakuhachi: (WAV) (MP3)
  - Trumpet: (WAV) (MP3)
  - Tenor Saxophone: (WAV) (MP3)

- Staccato Systems
  - Distorted Electric Guitar + Feedback: (WAV) (MP3)
  - Distorted Electric Guitar + Feedback: (WAV) (MP3)
  - "Slide Flute" (Perry Cook 1992): (WAV) (MP3)
  - Finger Bass: (WAV) (MP3)
  - Harpsichord: (WAV) (MP3)
  - Jazz Trio: (WAV) (MP3)
  - Latin Ensemble: (WAV) (MP3)
  - Percussion
    - Gongs: (WAV) (MP3)
    - Marimba: (WAV) (MP3)

C++ software that generates relevant sound examples may be found in the Synthesis Tool Kit[1] (STK) distribution [86]. There are also relevant examples in the Faust[2] language [454].[3]

[1] http://ccrma.stanford.edu/CCRMA/Software/STK/
[2] http://faust.grame.fr/
[3] http://ccrma.stanford.edu/realsimple/faust_strings

Bibliography

[1] J. Abel, V. Välimäki, and J. O. Smith, "Robust, efficient design of allpass filters for dispersive string sound synthesis," *IEEE Signal Processing Letters*, vol. 17, no. 4, pp. 406–409, 2010.

[2] J. Abel and J. O. Smith, "Robust design of very high-order allpass dispersion filters," *Proceedings of the Conference on Digital Audio Effects (DAFx-06), Montreal, Canada*, Sept. 2006, `http://www.dafx.de/`.

[3] J. S. Abel and D. P. Berners, *Music 424 Handouts: Signal Processing Techniques for Digital Audio Effects*, `http://ccrma.stanford.edu/courses/424/`, Spring 2004.

[4] S. Adachi and M. Sato, "Trumpet sound simulation using a two-dimensional lip vibration model," *Journal of the Acoustical Society of America*, vol. 99, pp. 1200–1209, 1996.

[5] J.-M. Adrien, "The missing link: Modal synthesis," in *Representations of Musical Signals* (G. De Poli, A. Picialli, and C. Roads, eds.), pp. 269–297, Cambridge, MA: MIT Press, 1991.

[6] J.-M. Adrien and J. Morrison, "Mosaïc: A modular program for synthesis by modal superposition," in *Proceedings of the Colloquium on Physical Modeling*, (Grenoble, France), ACROE, 1990.

[7] J. Agullo, A. Barjau, and J. Martinez, "Alternatives to the impulse response h(t) to describe the acoustical behavior of conical ducts," *Journal of the Acoustical Society of America*, vol. 84, pp. 1606–1612, Nov. 1988.

[8] J. Agullo, A. Barjau, and J. Martinez, "Conical bores. Part II: Multiconvolution," *Journal of the Acoustical Society of America*, vol. 84, pp. 1620–1627, Nov. 1988.

[9] J. Agullo, A. Barjau, and J. Martinez, "On the time-domain description of conical bores," *Journal of the Acoustical Society of America*, vol. 91, pp. 1099–1105, Feb. 1992.

[10] J. B. Allen, "private communication," 1983.

[11] J. B. Allen and D. A. Berkley, "Image method for efficiently simulating small-room acoustics," *Journal of the Acoustical Society of America*, vol. 65, pp. 943–950, 1979.

[12] J. B. Allen and L. R. Rabiner, "A unified approach to short-time Fourier analysis and synthesis," *Proc. IEEE*, vol. 65, pp. 1558–1564, Nov. 1977.

[13] N. Amir, V. Pagneux, and J. Kergomard, "A study of wave propagation in varying cross-section waveguides by modal decomposition. part 2. results," *Journal of the Acoustical Society of America*, vol. 101, no. (5), Part 1, pp. 2504–2517, 1996.

[14] N. Amir, G. Rosenhouse, and U. Shimony, "Reconstructing the bore of brass instruments: Theory and experiment," in *Proceedings of the Stockholm Musical Acoustics Conference (SMAC-93)*, (Stockholm), pp. 470–475, Royal Swedish Academy of Music, July 1993.

[15] N. Amir, G. Rosenhouse, and U. Shimony, "Discrete model for tubular acoustic systems with varying cross section—the direct and inverse problems. Part 1: Theory," *Acta Acustica*, vol. 81, pp. 450–462, 1995.

[16] N. Amir, G. Rosenhouse, and U. Shimony, "Discrete model for tubular acoustic systems with varying cross section - the direct and inverse problems. Part 1: Theory, and Part 2: Experiment," *Acta Acustica*, vol. 81, pp. 450–474, 1995.

[17] C. Anderton, *Digital Delay Handbook*, New York: Amsco Publications, 1985.

[18] A. Askenfelt, ed., *Five Lectures on the Acoustics of the Piano*, Stockholm: Royal Swedish Academy of Music, 1990, lectures by H. A. Conklin, Anders Askenfelt and E. Jansson, D. E. Hall, G. Weinreich, and K. Wogram. Sound example CD included. Publication number 64. `http://www.speech.kth.se/music/5_lectures/`.

[19] K. J. Åstrom, *Introduction to Stochastic Control Theory*, New York: Academic Press, 1970.

[20] B. S. Atal and L. S. Hanauer, "Speech analysis and synthesis by linear prediction of the speech wave," *Journal of the Acoustical Society of America*, vol. 50, pp. 637–655, 1971.

[21] F. Avanzini, S. Serafin, and D. Rocchesso, "Modeling interactions between rubbed dry surfaces using an elasto-plastic friction model," in *Proceedings of the COST-G6 Conference on Digital Audio Effects (DAFx-02), Hamburg, Germany*, pp. 111–116, September 2002, `http://www.dafx.de/`.

[22] R. D. Ayers, L. J. Eliason, and D. Mahgerefteh, "The conical bore in musical acoustics," *American Journal of Physics*, vol. 53, pp. 528–537, June 1985.

[23] B. Bank and L. Sujbert, "Modeling the longitudinal vibration of piano strings," in *Proceedings of the Stockholm Musical Acoustics Conference (SMAC-03), http://www.speech.kth.se/smac03/*, (Stockholm), pp. 143–146, Royal Swedish Academy of Music, Aug. 2003.

[24] B. Bank, *Physics-Based Sound Synthesis of String Instruments including Geometric Nonlinearities*, PhD thesis, Budapest University of Technology

and Economics, Budapest, Department of Measurement and Information Systems, Hungary, Feb. 2006, `http://www.mit.bme.hu/~bank/phd`.

[25] B. Bank and M. Karjalainen, "Passive admittance matrix modeling for guitar synthesis," in *Proceedings of the 13th International Conference on Digital Audio Effects (DAFx-10), Graz, Austria, September 6–10*, 2010.

[26] B. Bank and H.-M. Lehtonen, "Perception of longitudinal components in piano string vibration," *JASA Express Letters*, vol. 128, Aug. 2010.

[27] B. Bank and L. Sujbert, "Efficient modeling strategies for the geometric nonlinearities of musical instrument strings," in *Proceedings of the Forum Acusticum Meeting, Budapest*, pp. 2268–2278, Apr. 2005.

[28] B. Bank and L. Sujbert, "Generation of longitudinal vibrations in piano strings: From physics to sound synthesis," *Journal of the Acoustical Society of America*, vol. 117, pp. 2268–2278, Apr. 2005.

[29] B. Bank, F. Avanzini, G. Borin, G. D. Poli, F. Fontana, , and D. Rocchesso, "Physically informed signal processing methods for piano sound synthesis: A research overview," *EURASIP Journal on Applied Signal Processing*, vol. 2003, Sept. 2003, `http://www.mit.bme.hu/~bank/publist/jasp03.pdf`.

[30] B. Bank, S. Zambon, and F. Fontana, "A modal-based real-time piano synthesizer," *IEEE Transactions on Audio, Speech, and Language Processing*, vol. 18, no. 4, pp. 809–821, 2010.

[31] E. Barbour, "The cool sound of tubes," *IEEE Spectrum*, pp. 24–35, Aug. 1998.

[32] B. Bartlett, "A scientific explanation of phasing (flanging)," *Journal of the Audio Engineering Society*, vol. 18, pp. 674–675, Jun 1970.

[33] M. J. Beeson and D. T. Murphy, "RoomWeaver: A digital waveguide mesh based room acoustics research tool," in *Proceedings of the Conference on Digital Audio Effects (DAFx-04), Naples, Italy*, Oct. 2004, `http://www.dafx.de/`.

[34] V. Belevitch, "Summary of the history of circuit theory," *Proceedings of the IRE*, vol. 50, pp. 848–855, May 1962.

[35] V. Belevitch, *Classical Network Theory*, San Francisco: Holden Day, 1968.

[36] B. Beliczynski, I. Kale, and G. D. Cain, "Approximation of FIR by IIR digital filters: An algorithm based on balanced model reduction," *IEEE Transactions on Acoustics, Speech, Signal Processing*, vol. 40, pp. 532–542, Mar. 1992.

[37] A. H. Benade, "Equivalent circuits for conical waveguides," *Journal of the Acoustical Society of America*, vol. 83, pp. 1764–1769, May 1988.

[38] A. H. Benade, *Fundamentals of Musical Acoustics*, New York: Dover, 1990.

[39] A. H. Benade and E. V. Jansson, "On plane and spherical waves in horns with nonuniform flare. I. theory of radiation, resonance frequencies, and mode conversion," *Acustica*, vol. 31, no. 2, pp. 80–98, 1974.

[40] G. Bennett and X. Rodet, "Synthesis of the singing voice," in *Current Directions in Computer Music Research* (M. V. Mathews and J. R. Pierce, eds.), pp. 19–44, Cambridge, MA: MIT Press, 1989.

[41] J. Bensa, "Stiff piano string modeling: Computational comparison between finite differences and digital waveguide," in *Acoustical Society of America, Program of the 148th Meeting (abstract and presentation), San Diego, California, Nov. 15–19*, (New York), Acoustical Society of America, Nov. 2004.

[42] J. Bensa, *Analysis and Synthesis of Piano Sounds using Physical and Signal Models*, PhD thesis, Université de la Méditérranée, Marseille, France, 2003, `http://www.lma.cnrs-mrs.fr/ bensa/`.

[43] J. Bensa, S. Bilbao, R. Kronland-Martinet, and J. O. Smith, "From the physics of piano strings to digital waveguides," in *Proceedings of the 2002 International Computer Music Conference, Sweden*, 2002.

[44] J. Bensa, R. Kronland-Martinet, S. Bilbao, and J. O. Smith, "A power-normalized non-linear lossy piano hammer," in *Proceedings of the Stockholm Musical Acoustics Conference (SMAC-03), http://www.speech.kth.se/smac03/*, (Stockholm), pp. 365–368, Royal Swedish Academy of Music, Aug. 2003.

[45] J. Bensa, S. Bilbao, R. Kronland-Martinet, and J. O. Smith, "The simulation of piano string vibration: from physical models to finite difference schemes and digital waveguides," *Journal of the Acoustical Society of America*, vol. 114(2), pp. 1095–1107, 2003.

[46] J. Bensa, S. Bilbao, R. Kronland-Martinet, J. O. Smith, and T. Voinier, "Computational modeling of stiff piano strings using digital waveguides and finite differences," *Acta Acustica united with Acustica*, vol. 91(2), pp. 289–298, 2005.

[47] L. L. Beranek, *Acoustics*, `http://asa.aip.org/publications.html`: American Institute of Physics, for the Acoustical Society of America, 1986, 1st ed. 1954.

[48] L. Beranek, *Concert Halls and Opera Houses: Music, Acoustics, and Architecture*, Berlin: Springer-Verlag, 2004.

[49] A. Berkhout, D. de Vries, and P. Vogel, "Acoustic control by wave field synthesis," *Journal of the Acoustical Society of America*, vol. 93, pp. 2764–2778, 1993.

[50] D. P. Berners, *Acoustics and Signal Processing Techniques for Physical Modeling of Brass Instruments*, PhD thesis, Elec. Engineering Dept., Stanford University (CCRMA), June 1999, `http://ccrma.stanford.edu/~dpberner/`.

[51] D. P. Berners and J. O. Smith, "On the use of Schrodinger's equation in the analytic determination of horn reflectance," in *Proceedings of the 1994 International Computer Music Conference, Århus*, pp. 419–422, Computer Music Association, 1994.

[52] R. T. Beyer, *Nonlinear Acoustics*, American Institute of Physics, for the Acoustical Society of America, 1997 (orig. 1974), `http:`.

[53] S. Bilbao, "Numerical sound synthesis: Finite difference schemes and simulation in musical acoustics," 2009.

[54] S. Bilbao, *Wave and Scattering Methods for the Numerical Integration of Partial Differential Equations*, PhD thesis, Stanford University, June 2001, `http://ccrma.stanford.edu/~bilbao/`.

[55] S. Bilbao, *Wave and Scattering Methods for Numerical Simulation*, New York: John Wiley and Sons, Inc., July 2004.

[56] S. Bilbao, J. O. Smith, J. Bensa, and R. Kronland-Martinet, "A power-normalized wave digital piano hammer," *Journal of the Acoustical Society of America, Program of the 144th Meeting, Cancun, Mexico, Dec. 2–6*, Dec. 2002, see paper on the conference CD-ROM.

[57] J. Blauert, *Spatial Hearing. The Psychophysics of Human Sound Localization*, Cambridge, MA, USA: MIT Press, 1997.

[58] B. Blesser, "Interdisciplinary synthesis of reverberation viewpoint," *Journal of the Audio Engineering Society*, vol. 49, pp. 867–903, Oct. 2001.

[59] H. Bode, "History of electronic sound modification," *Journal of the Audio Engineering Society*, vol. 32, pp. 730–739, Oct. 1984.

[60] G. Borin, G. De Poli, and A. Sarti, "A modular approach to excitator-resonator interaction in physical models synthesis," in *Proceedings of the 1989 International Computer Music Conference, Ohio*, pp. 46–50, Computer Music Association, 1989.

[61] J. Borish, *Electronic Simulation of Auditorium Acoustics*, PhD thesis, Elec. Engineering Dept., Stanford University (CCRMA), 1984, CCRMA Technical Report STAN-M-18, `http://ccrma.stanford.edu/STANM/stanms/stanm18/`.

[62] M. Bosi and R. E. Goldberg, *Introduction to Digital Audio Coding and Standards*, Boston: Kluwer Academic Publishers, 2003.

[63] X. Boutillon, "Model for piano hammers: Experimental determination and digital simulation," *Journal of the Acoustical Society of America*, vol. 83, pp. 746–754, Feb. 1988.

[64] S. Boyd and L. Vandenberghe, *Convex Optimization*, Cambridge University Press, Feb. 2004, `http://www.stanford.edu/~boyd/cvxbook/`.

[65] C. B. Boyer, *A History of Mathematics*, Princeton University Press, 1968.

[66] K. Bradley, M.-H. Cheng, and V. L. Stonick, "Automated analysis and computationally efficient synthesis of acoustic guitar strings and body," in

Proceedings of the IEEE Workshop on Applications of Signal Processing to Audio and Acoustics, New Paltz, NY, (New York), pp. 238–241, IEEE Press, Oct. 1995, Session 9a, paper 7, 4 pages.

[67] E. Brandt, "Hard sync without aliasing," in *Proceedings of the 2001 International Computer Music Conference, Havana*, 2001, http://www-2.cs.cmu.edu/~eli/papers/icmc01-hardsync.pdf.

[68] O. Brune, "Synthesis of a finite two terminal network whose driving point impedance is a prescribed function of frequency," in *J. Math. and Phys.*, vol. 10, pp. 191–236, 1931.

[69] C. Cadoz, A. Luciani, and J. L. Florens, "CORDIS-ANIMA: A modeling and simulation system for sound and image synthesis — the general formalism," *Computer Music Journal*, vol. 17, no. 4, pp. 19–29, 1993.

[70] D. M. Campbell, "Brass instruments es we know them today," in *Proceedings of the Stockholm Musical Acoustics Conference (SMAC-03), http://www.speech.kth.se/smac03/*, (Stockholm), pp. 185–188, Royal Swedish Academy of Music, Aug. 2003.

[71] R. Caussé, J. Kergomard, and X. Lurton, "Input impedance of brass musical instruments—comparison between experiment and numerical models," *Journal of the Acoustical Society of America*, vol. 75, pp. 241–254, Jan. 1984.

[72] C. Chafe, "Pulsed noise in self-sustained oscillations of musical instruments," in *Proceedings of the International Conference on Acoustics, Speech, and Signal Processing, Albuquerque*, (New York), IEEE Press, 1990, available as CCRMA Technical Report STAN-M-65, Music Dept., Stanford University.

[73] A. Chaigne, "Viscoelastic properties of nylon guitar strings," *Catgut Acoustical Society Journal, Series II*, vol. 1, pp. 21–27, May 1991.

[74] A. Chaigne, "On the use of finite differences for musical synthesis. application to plucked stringed instruments," *Journal d'Acoustique*, vol. 5, no. 2, pp. 181–211, 1992.

[75] A. Chaigne and A. Askenfelt, "Numerical simulations of piano strings. I. A physical model for a struck string using finite difference methods," *Journal of the Acoustical Society of America*, vol. 95, no. 2, pp. 1112–1118, 1994.

[76] A. Chaigne and A. Askenfelt, "Numerical simulations of piano strings. II. comparisons with measurements and systematic exploration of some hammer-string parameters," *Journal of the Acoustical Society of America*, vol. 95, pp. 1631–1640, Mar. 1994.

[77] A. Chaigne and A. Askenfelt, "Numerical simulations of piano strings, parts I and II," *Journal of the Acoustical Society of America*, vol. 95, pp. 1112–1118, 1631–1640, Feb.–March 1994.

[78] H. Chamberlin, *Musical Applications of Microprocessors*, New Jersey: Hayden Book Co., Inc., 1980.

[79] H. Chamberlin, *Musical Applications of Microprocessors, Second Edition*, New Jersey: Hayden Books, 1987, second printing.

[80] J. M. Chowning, "The simulation of moving sound sources," *Journal of the Audio Engineering Society*, vol. 19, pp. 2–6, 1971.

[81] J. M. Chowning, "Frequency modulation synthesis of the singing voice," in *Current Directions in Computer Music Research* (M. V. Mathews and J. R. Pierce, eds.), pp. 57–63, Cambridge, MA: MIT Press, 1989.

[82] J. M. Chowning and D. Bristow, *FM Theory and Applications*, Tokyo: Yamaha Corporation, 1986.

[83] R. V. Churchill, *Complex Variables and Applications*, New York: McGraw-Hill, 1960.

[84] I. Cohen and T. Hélie, "Real-time simulation of a guitar power amplifier," in *Proceedings of the 13th International Conference on Digital Audio Effects (DAFx-10), Graz, Austria, September 6–10*, 2010.

[85] H. A. Conklin, "Generation of partials due to nonlinear mixing in a stringed instrument," *Journal of the Acoustical Society of America*, vol. 105, pp. 536–545, 1999.

[86] P. Cook and G. Scavone, *Synthesis ToolKit in C++, Version 4*, `http://ccrma.stanford.edu/CCRMA/Software/STK/`, 2007, see also `http://ccrma.stanford.edu/~jos/stkintro/`.

[87] P. R. Cook, *Identification of Control Parameters in an Articulatory Vocal Tract Model, with Applications to the Synthesis of Singing*, PhD thesis, Elec. Engineering Dept., Stanford University (CCRMA), Dec. 1990, `http://www.cs.princeton.edu/~prc/`.

[88] P. R. Cook, "Tbone: An interactive waveguide brass instrument synthesis workbench for the NeXT machine," in *Proceedings of the 1991 International Computer Music Conference, Montreal*, pp. 297–299, Computer Music Association, 1991.

[89] P. R. Cook, "A meta-wind-instrument physical model, and a meta-controller for real time performance control," in *Proceedings of the 1992 International Computer Music Conference, San Jose*, pp. 273–276, Computer Music Association, 1992.

[90] P. R. Cook, "Singing voice synthesis: History, current work, and future directions," *Computer Music Journal*, vol. 20, no. 3, pp. 38–46, 1996.

[91] P. R. Cook, "Synthesis toolkit in C++, version 1.0," in *SIGGRAPH Proceedings*, Assoc. Comp. Mach., May 1996.

[92] P. R. Cook, *Real Sound Synthesis for Interactive Applications*, A. K. Peters, L.T.D., 2002.

[93] C. Cooper and J. Abel, "Digital simulation of 'brassiness' and amplitude-dependent propagation speed in wind instruments," in *Proceedings of the 13th International Conference on Digital Audio Effects (DAFx-10), Graz, Austria, September 6–10*, 2010.

[94] T. J. Cox and P. D'Antonio, "A brief history of room acoustic diffusers," *Acoustics Today*, pp. 18–25, July 2006.

[95] L. Cremer, *The Physics of the Violin*, Cambridge, MA: MIT Press, 1984.

[96] R. Crochiere, L. Rabiner, and R. Shively, "A novel implementation of digital phase shifters," *Bell System Technical Journal*, vol. 65, pp. 1497–1502, Oct. 1975.

[97] R. Crochiere and L. R. Rabiner, *Multirate Digital Signal Processing*, Englewood Cliffs, NJ: Prentice-Hall, 1983.

[98] J. Cullen, J. Gilbert, and D. M. Campbell, "Brass instruments: Linear stability analysis and experiments with an artificial mouth," *Acta Acustica*, vol. 86, pp. 704–724, 2000.

[99] G. Dahlquist and Å. Björck, *Numerical Methods*, Englewood Cliffs, NJ: Prentice-Hall, 1974.

[100] J. l. R. d'Alembert, "Investigation of the curve formed by a vibrating string, 1747," in *Acoustics: Historical and Philosophical Development* (R. B. Lindsay, ed.), pp. 119–123, Stroudsburg: Dowden, Hutchinson & Ross, 1973.

[101] J. P. Dalmont, C. J. Nederveen, V. Dubos, S. Ollivier, V.Mserette, and E. te Sligte, "Experimental determination of the equivalent circuit of an open side hole: Linear and nonlinear behaviour," *Acta Acustica united with Acustica*, vol. 88, pp. 567–575, 2002.

[102] J.-P. Dalmont, J. Gilbert, and S. Ollivier, "Nonlinear characteristics of single-reed instruments: Quasistatic volume flow and reed opening measurements," *Journal of the Acoustical Society of America*, vol. 114, pp. 2253–2262, Oct. 2003.

[103] O. Darrigol, "The acoustic origins of harmonic analysis," *Archive for History of the Exact Sciences*, vol. 61, July 2007.

[104] J. Dattorro, "Effect design," *Journal of the Audio Engineering Society*, vol. 45, pp. 660–684, September 1997.

[105] J. Dattorro, "Effect design: Part 3 oscillators: Sinusoidal and pseudonoise," *Journal of the Audio Engineering Society*, vol. 50, no. 3, pp. 115–146, 2002.

[106] J. R. Deller Jr., J. H. L. Hansen, and J. G. Proakis, *Discrete-Time Processing of Speech Signals*, New York: John Wiley and Sons, Inc., Mar. 2001.

[107] P. Depalle and S. Tassart, "State space sound synthesis and a state space synthesiser builder," in *Proceedings of the 1995 International Computer Music Conference, Banff*, pp. 88–95, Computer Music Association, 1995, `http://mediatheque.ircam.fr/articles/textes/Depalle95a/`.

[108] S. Dequand, J. F. H. Willems, M. Leroux, R. Vullings, M. van Weert, C. Thieulot, and A. Hirschberg, "Experimental study of the influence of

mouth geometry on sound production in a recorder-like instrument: Windway length and chamfers," *Journal of the Acoustical Society of America*, vol. 113, pp. 1724–1735, 2003.

[109] G. Derveaux, A. Chaigne, P. Joly, and E. Bécache, "Time-domain simulation of a guitar: Model and method," *Journal of the Acoustical Society of America*, vol. 114, pp. 3368–3383, Dec. 2003.

[110] C. A. Desoer and E. S. Kuh, *Basic Circuit Theory*, New York: McGraw-Hill, 1969.

[111] P. H. Dietz, "Simulation of trumpet tones via physical modeling," Master's thesis, Elec. and Comp. Engineering Dept., Carnegie Mellon University, Dec 1988, available as CMU Report no. LASIP-89-07, Aug. 1989.

[112] P. A. M. Dirac, *The Principles of Quantum Mechanics, Fourth Edition*, New York: Oxford University Press, 1958–2001.

[113] S. Dostrovsky, "Early vibration theory: Physics and music in the seventeenth century," *Archive for History of the Exact Sciences*, vol. 14, no. 3, 1975.

[114] B. Doval and X. Rodet, "Fundamental frequency estimation using a new harmonic matching method," in *Proceedings of the 1991 International Computer Music Conference, Montreal*, pp. 555–558, Computer Music Association, 1991.

[115] DSP Committee, ed., *Programs for Digital Signal Processing*, New York: IEEE Press, 1979.

[116] DSP Committee, ed., *Programs for Digital Signal Processing II*, New York: IEEE Press, 1979.

[117] V. Dubos, J. Kergomard, A. Khettabi, D. H. K. J. P. Dalmont, and C. J. Nederveen, "Theory of sound propagation in a duct with a branched tube using modal decomposition," *Acta Acustica*, vol. 85, pp. 153–169, 1999.

[118] E. Ducasse, "An alternative to the traveling-wave approach for use in two-port descriptions of acoustic bores," *Journal of the Acoustical Society of America*, vol. 112, pp. 3031–3041, Dec. 2002.

[119] H. W. Dudley, "The vocoder," *Bell Labs Rec.*, vol. 18, pp. 122–126, 1939, Reprinted in [408, pp. 347–351].

[120] H. K. Dunn, "The calculation of vowel resonances, and an electrical vocal tract," *Journal of the Acoustical Society of America*, vol. 22, pp. 740–753, Nov. 1950, Reprinted in [408, pp. 31–44].

[121] G. E. F. Eckerholm, "Player-instrument interaction models for digital waveguide synthesis of guitar: Touch and collisions," *IEEE Transactions on Audio, Speech, and Language Processing*, vol. 18, no. 4, pp. 822–832, 2010.

[122] W. C. Elmore and M. A. Heald, *Physics of Waves*, New York: Dover, 1969.

[123] C. Erkut and M. Karjalainen, "Finite difference method vs. digital waveguide method in string instrument modeling and synthesis," in *Proceedings of the International Symposium on Musical Acoustics (ISMA-02), Mexico City*, 2002, `http://www.acoustics.hut.fi/~cerkut/publications.html`.

[124] C. Erkut and M. Karjalainen, "Virtual strings based on a 1-D FDTD waveguide model," *Proceedings of the Audio Engineering Society 22nd International Conference, Espoo, Finland*, pp. 317–323, June 15–17, 2002, `http://www.acoustics.hut.fi/~cerkut/publications.html`.

[125] P. A. A. Esquef, V. Välimäki, and M. Karjalainen, "Restoration and enhancement of solo guitar recordings based on sound source modeling," *Journal of the Audio Engineering Society*, vol. 2002, no. 4, pp. 227–236, 2002.

[126] G. Essl, *Physical Wave Propagation Modeling for Real-Time Synthesis of Natural Sounds*, PhD thesis, Computer Science Dept., Princeton University, Nov. 2002, `http://ncstrl.cs.princeton.edu/expand.php?id=TR-659-02`.

[127] G. Essl, Jan 13, 2004, `http://arxiv.org/abs/physics/0401065v1`.

[128] G. Evangelista and J. O. Smith, "Structurally passive scattering element for modeling guitar pluck action," in *Proceedings of the 13th International Conference on Digital Audio Effects (DAFx-10), Graz, Austria, September 6–10*, 2010.

[129] B. Fabre, "From sound synthesis to instrument making: An overview of recent researches on woodwinds and organs," in *Proceedings of the Stockholm Musical Acoustics Conference (SMAC-03), http://www.speech.kth.se/smac03/*, (Stockholm), pp. 239–242, Royal Swedish Academy of Music, Aug. 2003.

[130] B. Fabre and A. Hirschberg, "Physical modeling of flue instruments: a review of lumped models," *Acta Acustica*, vol. 86, pp. 599–610, 2000.

[131] G. Fant, "The acoustics of speech," in *Report of the Third International Congress on Acoustics*, pp. 188–201, 1959, Reprinted in [408, pp. 17–30].

[132] G. Fant, *Acoustic Theory of Speech Production*, The Hague: Mouton, 1960.

[133] G. Fant, "The LF-model revisited. transformations and frequency domain analysis," *STL-QPSR*, vol. 2-3, pp. 119–156, 1995.

[134] C. W. Farrow, "A continuously variable digital delay element," in *Proceedings of the IEEE Conference on Circuits and Systems, Espoo, Finland*, vol. 3, pp. 2641–2645, June 7–9 1988.

[135] A. Fettweis, "Digital filters related to classical structures," *AEU: Archive für Elektronik und Übertragungstechnik*, vol. 25, pp. 79–89, Feb 1971, see also U.S. Patent 3,967,099, 1976, now expired.

[136] A. Fettweis, "Wave digital filters: Theory and practice," *Proceedings of the IEEE*, vol. 74, pp. 270–327, Feb. 1986.

[137] R. P. Feynman, *Feynman Lectures On Physics (3 Volume Set)*, Reading MA: Addison-Wesley, 1970.

[138] R. Fitzpatrick, *Thermodynamics and Statistical Mechanics: An intermediate level course*, Lulu Marketplace, 2006.

[139] J. L. Flanagan, *Speech Analysis, Synthesis, and Perception*, New York: Springer Verlag, 1972.

[140] J. L. Flanagan, "Voices of men and machines," *Journal of the Acoustical Society of America*, vol. 51, pp. 1375–1387, Mar. 1972, Reprinted in [408, pp. 4–16].

[141] J. L. Flanagan and K. Ishizaka, "Automatic generation of voiceless excitation in a vocal cord-vocal tract speech synthesizer," *IEEE Transactions on Acoustics, Speech, Signal Processing*, vol. 24, pp. 163–170, Apr. 1976.

[142] J. L. Flanagan and L. R. Rabiner, eds., *Speech Synthesis*, Stroudsburg, Penn.: Dowden, Hutchinson, and Ross, Inc., 1973.

[143] N. H. Fletcher and T. D. Rossing, *The Physics of Musical Instruments*, New York: Springer Verlag, 1991, See [144] for a later edition.

[144] N. H. Fletcher and T. D. Rossing, *The Physics of Musical Instruments, 2nd Edition*, New York: Springer Verlag, 1998.

[145] J.-L. Florens and C. Cadoz, "The physical model: Modeling and simulating the instrumental universe," in *Representations of Musical Signals* (G. De Poli, A. Picialli, and C. Roads, eds.), pp. 227–268, Cambridge, MA: MIT Press, 1991.

[146] F. Fontana and D. Rocchesso, "A new formulation of the 2D-waveguide mesh for percussion instruments," in *Proceedings of the XI Colloquiun on Musical Informatics*, (Bologna, Italy), pp. 27–30, Nov. 1995.

[147] B. Fornberg, *A Practical Guide to Pseudo-Spectral Methods*, Cambridge University Press, 1998.

[148] A. Franck, "Efficient algorithms and structures for fractional delay filtering based on lagrange interpolation," *Journal of the Audio Engineering Society*, vol. 56, pp. 1036–1056, Dec. 2008.

[149] W. Frank and A. Lacroix, "Improved vocal tract models for speech synthesis," *Proceedings of the International Conference on Acoustics, Speech, and Signal Processing, Tokyo*, vol. 3, pp. 2011–2014, 1986.

[150] G. F. Franklin, J. D. Powell, and M. L. Workman, *Digital Control of Dynamic Systems, Third Edition*, Englewood Cliffs, NJ: Prentice-Hall, 1998.

[151] F. G. Friedlander, "On the oscillations of the bowed string," *Proceedings of the Cambridge Philosophy Society*, vol. 49, pp. 516–530, 1953.

[152] W. G. Gardner, "The virtual acoustic room," Master's thesis, MIT, 1992, `http://www.harmony-central.com/Computer/Programming/-virtual-acoustic-room.ps.gz`.

[153] W. G. Gardner, "Reverberation algorithms," in *Applications of Digital Signal Processing to Audio and Acoustics* (M. Kahrs and K. Brandenburg, eds.), pp. 85–131, Boston/Dordrecht/London: Kluwer Academic Publishers, 1998, see http://www.wkap.nl/book.htm/0-7923-8130-0.

[154] E. Gaudrain and Y. Orlarey, "A Faust tutorial," Sept. 2007, `http://www.grame.fr/pub/faust_tutorial.pdf`.

[155] Y. Genin, "An introduction to the model reduction problem with hankel norm criterion," in *Proc. European Conf. Circuit Theory and Design, The Hague*, Aug 1981.

[156] M. A. Gerzon, "Synthetic stereo reverberation, parts i and ii," *Studio Sound*, vol. 13(I), 14(II), pp. 632–635(I), 24–28(II), 1971(I),1972(II).

[157] M. A. Gerzon, "Unitary (energy preserving) multichannel networks with feedbacks," *Electronics Letters*, vol. 12, no. 11, pp. 278–279, 1976.

[158] M. A. Gerzon, "The design of precisely coincident microphone arrays for stereo and surround sound," *50th AES Convention*, Mar. 1975.

[159] J. Gilbert and J. F. Petiot, "Brass instruments: Some theoretical and experimental results," in *Proceedings of the International Symposium on Musical Acoustics (ISMA-97), Edinburgh, Scotland*, pp. 391–400, 1997.

[160] J. Gilbert, J. Kergomard, and J. D. Polack, "On the reflection functions associated with discontinuities in conical bores," *Journal of the Acoustical Society of America*, vol. 87, pp. 1773–1780, Apr. 1990.

[161] J. Gilbert, S. Ponthus, and J. F. Petiot, "Artificial buzzing lips and brass instruments: Experimental results," *Journal of the Acoustical Society of America*, vol. 104, pp. 1627–1632, 1998.

[162] P. E. Gill, W. Murray, and M. H. Wright, *Practical Optimization*, New York: Academic Press, 1981.

[163] N. Giordano and A. J. Korty, "Motion of a piano string: Longitudinal vibrations and the role of the bridge," *Journal of the Acoustical Society of America*, vol. 100, pp. 3899–3908, 1996.

[164] N. Giordano and J. P. Millis, "Hysteretic behavior of piano hammers," in *Proceedings of the International Symposium on Musical Acoustics (ISMA-01), Perugia, Umbria, Italy*, pp. 237–240, 2001.

[165] B. R. Glasberg and B. C. J. Moore, "A model of loudness applicable to time-varying sounds," *Journal of the Audio Engineering Society*, vol. 50, pp. 331–342, May 2002.

[166] A. A. Goldstein, *Constructive Real Analysis*, New York: Harper and Row, 1967.

[167] G. H. Golub and C. F. Van Loan, *Matrix Computations, 2nd Edition*, Baltimore: The Johns Hopkins University Press, 1989.

[168] J. W. Gordon and J. O. Smith, "A sine generation algorithm for VLSI applications," in *Proceedings of the 1985 International Computer Music Conference, Vancouver*, Computer Music Association, 1985.

[169] K. F. Graff, *Wave Motion in Elastic Solids*, New York: Dover, 1991 (1975 version).

[170] Grame, "Faust quick reference," Aug. 2010, `http://faudiostream.cvs.sourceforge.net/viewvc/faudiostream/-faust/documentation/faust-quick-reference.pdf`.

[171] R. A. Granger, *Fluid Mechanics*, New York: Dover, 1995.

[172] A. H. Gray, "Passive cascaded lattice digital filters," *IEEE Transactions on Acoustics, Speech, Signal Processing*, vol. ASSP-27, pp. 337–344, May 1980.

[173] A. H. Gray and J. D. Markel, "Digital lattice and ladder filter synthesis," in *Selected Papers in Digital Signal Processing* (Digital Signal Processing Committee, ed.), IEEE Press, 1975.

[174] A. H. Gray and J. D. Markel, "A normalized digital filter structure," *IEEE Transactions on Acoustics, Speech, Signal Processing*, vol. ASSP-23, pp. 268–277, June 1975.

[175] R. M. Gray and L. D. Davisson, *An Introduction to Statistical Signal Processing*, Cambridge University Press, 2003, `http://www-ee.stanford.edu/~gray/sp.pdf`.

[176] K. Guettler and A. Askenfelt, "On the kinematics of spiccato and ricochet bowing," *Catgut Acoustical Society Journal, Series II*, vol. 3, pp. 9–15, Nov. 1998.

[177] M. Gutknecht, J. O. Smith, and L. N. Trefethen, "The Caratheodory-Fejer (CF) method for recursive digital filter design," *IEEE Transactions on Acoustics, Speech, Signal Processing*, vol. 31, no. 6, pp. 1417–1426, 1983.

[178] D. E. Hall and A. Askenfelt, "Piano string excitation v: Spectra for real hammers and strings," *Journal of the Acoustical Society of America*, vol. 83, pp. 1627–1638, Apr. 1988.

[179] D. Halliday and R. Resnick, *Fundamentals of Physics*, New York: John Wiley and Sons, Inc., 1970.

[180] D. Halliday, R. Resnick, and J. Walker, *Extended, Fundamentals of Physics, 6th Edition*, New York: John Wiley and Sons, Inc., 2000.

[181] R. Hanson, "Measurements of nonlinear effects in a driven vibrating wire," *Journal of the Acoustical Society of America*, vol. 96, pp. 1549–1556, Sept. 1994.

[182] A. Härmä, M. Karjalainen, L. Savioja, V. Välimäki, U. K. Laine, and J. Huopaniemi, "Frequency-warped signal processing for audio applications," *Journal of the Audio Engineering Society*, vol. 48, pp. 1011–1031, Nov. 2000.

[183] C. M. Harris, "Absorption of sound in air versus humidity and temperature," *Journal of the Acoustical Society of America*, vol. 40, no. 1, pp. 148–159, 1966.

[184] F. J. Harris, "Performance and design considerations of the Farrow filter when used for arbitrary resampling of sampled time series," in *Proc. 31st Asilomar Conference on Signals, Systems, and Computers*, vol. 2, pp. 1745–1749, Nov. 1997.

[185] J. Hawkins, *On Intelligence*, Times Books, 2004.

[186] T. Hélie and M. Hasler, "Representation of the weakly nonlinear propagation in air-filled pipes with Volterra series," in *11th InternationalIEEE Workshop on Nonlinear Dynamics of Electronic Systems*, 2003.

[187] T. Hélie and D. Matignon, "Numerical simulation of acoustic waveguides for Webster-Lokshin model using diffusive representations," in *Sixth InternationalConference onMath. and Numer. Aspects of Wave Prop. Phenomena, Jyväskylä, Finland*, 2003.

[188] T. Hélie and X. Rodet, "Radiation of a pulsating portion of a sphere: Application to radiation of horns," *Acta Acustica united with Acustica*, vol. 89, no. 4, pp. 565–577, 2003.

[189] C. A. Henricksen, "Unearthing the mysteries of the Leslie cabinet," *Recording Engineer/Producer magazine*, Apr. 1981, `http://www.theatreorgans.com/hammond/faq/mystery/mystery.html`.

[190] D. J. Henwood, "The boundary-element method and horn design," *Journal of the Audio Engineering Society*, vol. 41, no. 6, pp. 485–496, 1993.

[191] J. Herrera, C. Hanson, and J. S. Abel, "Discrete time emulation of the leslie speaker," *Audio Engineering Society Convention*, Oct. 2009, Paper 7925.

[192] W. J. Hess, *Algorithms and Devices for Pitch Determination of Speech-Signals*, Berlin: Springer-Verlag, 1983.

[193] F. B. Hildebrand, *Introduction to Numerical Analysis*, New York: Dover, 1974, Second edition.

[194] L. Hiller and P. Ruiz, "Synthesizing musical sounds by solving the wave equation for vibrating objects. Part I.," *Journal of the Audio Engineering Society*, vol. 19, pp. 462–470, June 1971, Part II: vol. 19, no. 7, pp. 542–551, July/Aug. 1971.

[195] S. Hirschman, *Digital Waveguide Modelling and Simulation of Reed Woodwind Instruments*, Engineer's thesis, Elec. Engineering Dept., Stanford University (CCRMA), May 1991, available as CCRMA Technical Report Stan-M-72, Music Dept., Stanford University, July 1991.

[196] A. Hirschberg, "Aero-acoustics of wind instruments," in Hirschberg *et al.* [197], pp. 291–361.

[197] A. Hirschberg, J. Kergomard, and G. Weinreich, eds., *Mechanics of Musical Instruments*, Berlin: Springer-Verlag, 1995.

[198] A. Hirschberg, J. G. R. Msallam, , and A. P. Wijnands, "Shock waves in trombones," *Journal of the Acoustical Society of America*, vol. 99, pp. 1754–1758, Mar. 1996.

[199] A. Horner, J. Beauchamp, and L. Haken, "Methods for multiple wavetable synthesis of musical instrument tones," *Journal of the Audio Engineering Society*, vol. 41, pp. 336–356, May 1993.

[200] N. G. Horton and T. R. Moore, "Modeling the magnetic pickup of an electric guitar," *American Journal of Physics*, vol. 77, Feb. 2009.

[201] P. Huang and J. O. Smith, "Loudness-based display and analysis applied to artificial reverberation," in *Proceedings of the 2004 International Computer Music Conference, Miami, Florida*, Computer Music Association, Nov. 2004.

[202] P. Huang, S. Serafin, and J. O. Smith, "Modeling high-frequency modes of complex resonators using a waveguide mesh," in *Proceedings of the International Conference on Digital Audio Effects (DAFx-00), Verona, Italy*, Dec. 2000, `http://www.dafx.de/`.

[203] P. Huang, S. Serafin, and J. O. Smith, "A waveguide mesh model of high-frequency violin body resonances," in *Proceedings of the 2000 International Computer Music Conference, Berlin*, Aug. 2000.

[204] B. Hutchins, *Musical Engineer's Handbook*, Ithaca, New York: Electronotes, 1975.

[205] C. M. Hutchins, "The air and wood modes of the violin," *Journal of the Audio Engineering Society*, vol. 46, pp. 751–765, Sept. 1998.

[206] K. Ishizaka and J. L. Flanagan, "Synthesis of voiced sounds from a two-mass model of the vocal cords," *Bell System Technical Journal*, vol. 51, pp. 1233–1268, 1972.

[207] D. A. Jaffe and J. O. Smith, "Extensions of the Karplus-Strong plucked string algorithm," *Computer Music Journal*, vol. 7, no. 2, pp. 56–69, 1983, Reprinted in [380, pp. 481–494].

[208] D. A. Jaffe and J. O. Smith, "Performance expression in commuted waveguide synthesis of bowed strings," in *Proceedings of the 1995 International Computer Music Conference, Banff*, pp. 343–346, Computer Music Association, 1995.

[209] E. V. Jansson, N.-E. Molin, and H. O. Saldner, "On eigenmodes of the violin — electronic holography and admittance measurements," *Journal of the Acoustical Society of America*, vol. 95, pp. 1100–1105, Feb. 1994.

[210] H. Järveläinen, *Perception of Attributes in Real and Synthetic String Instrument Sounds*, Doctoral thesis, Helsinki University of Technology, Espoo, Finland, January 2003, `http://lib.hut.fi/Diss/2003/isbn9512263149/`.

[211] H. Järveläinen, V. Välimäki, and M. Karjalainen, "Audibility of inharmonicity in string instrument sounds, and implications to digital

sound synthesis," in *Proceedings of the 1999 International Computer Music Conference, Beijing*, pp. 359–362, Oct. 22-27, 1999, http://www.acoustics.hut.fi/~hjarvela/publications/.

[212] P. Jinachitra, *Robust Structured Voice Extraction for Flexible Expressive Resynthesis*, PhD thesis, Elec. Engineering Dept., Stanford University (CCRMA), 2007, downloadable PDF and sound examples at http://ccrma.stanford.edu/~pj97/Thesis/index.html.

[213] P. Jinachitra and J. O. Smith, "Joint estimation of glottal source and vocal tract for vocal synthesis using Kalman smoothing and EM algorithm," in *Proceedings of the IEEE Workshop on Applications of Signal Processing to Audio and Acoustics, New Paltz, NY*, (New York), IEEE Press, Oct. 2005.

[214] P. Jinachitra and J. O. Smith, "Generative model of voice in noise for structured coding applications," in *Proceedings of the International Conference on Acoustics, Speech, and Signal Processing, Honolulu, USA*, (New York), IEEE Press, 2007.

[215] J.-M. Jot, "An analysis/synthesis approach to real-time artificial reverberation," in *Proceedings of the International Conference on Acoustics, Speech, and Signal Processing, San Francisco*, (New York), pp. II.221–II.224, IEEE Press, 1992.

[216] J.-M. Jot, *Etude et Réalization d'un Spatialisatuer de Sons par Modèles Physiques et Perceptifs*, PhD thesis, French Telecom, Paris, Paris 92 E 019, 1992, written in French.

[217] J.-M. Jot and A. Chaigne, "Digital delay networks for designing artificial reverberators," *Audio Engineering Society Convention*, Feb. 1991.

[218] J.-P. Jullien, E. Kahle, M. Marin, O. Warusfel, G. Bloch, and J.-M. Jot, "Spatializer: A perceptual approach," *AES 94th Convention*, vol. Preprint 3465 (B1-5), pp. 1–13, Mar. 16–19 1993.

[219] Kaegi, W., and S. Tempelaars, "Vosim—a new sound synthesis system," *Journal of the Audio Engineering Society*, vol. 26, no. 6, pp. 418–24, 1978.

[220] T. Kailath, *Linear Systems*, Englewood Cliffs, NJ: Prentice-Hall, 1980.

[221] J. F. Kaiser, "Using the $I_0 - \sinh$ window function," *IEEE Transactions on Circuits and Systems—I: Fundamental Theory and Applications*, pp. 20–23, April 22–25 1974, Reprinted in [116], pp. 123–126.

[222] M. Karjalainen, "1-D digital waveguide modeling for improved sound synthesis," in *Proceedings of the International Conference on Acoustics, Speech, and Signal Processing, Orlando, Florida, USA*, vol. 2, (New York), pp. 1869–1872, IEEE Press, May 2002.

[223] M. Karjalainen, "Mixed physical modeling: DWG + FDTD + WDF," in *Proceedings of the IEEE Workshop on Applications of Signal Processing to Audio and Acoustics, New Paltz, NY*, (New York), pp. 225–228, IEEE Press, Oct. 2003.

[224] M. Karjalainen, "Time-domain physical modeling and real-time synthesis using mixed modeling paradigms," in *Proceedings of the Stockholm Musical Acoustics Conference (SMAC-03)*, *http://www.speech.kth.se/smac03/*, (Stockholm), pp. 393–396, Royal Swedish Academy of Music, Aug. 2003.

[225] M. Karjalainen, "private communication," 2005.

[226] M. Karjalainen and C. Erkut, "Digital waveguides vs. finite difference schemes: Equivalance and mixed modeling," *EURASIP Journal on Applied Signal Processing*, vol. 2004, pp. 978–989, June 15, 2004.

[227] M. Karjalainen and U. K. Laine, "A model for real-time sound synthesis of guitar on a floating-point signal processor," in *Proceedings of the International Conference on Acoustics, Speech, and Signal Processing, Toronto*, vol. 5, (New York), pp. 3653–3656, IEEE Press, May 1991.

[228] M. Karjalainen and J. Pakarinen, "Wave digital simulation of a vacuum-tube amplifier," in *Proceedings of the International Conference on Acoustics, Speech, and Signal Processing, Dallas, TX, USA*, vol. V, pp. 153–, 2006.

[229] M. Karjalainen and J. O. Smith, "Body modeling techniques for string instrument synthesis," in *Proceedings of the 1996 International Computer Music Conference, Hong Kong*, pp. 232–239, Computer Music Association, Aug. 1996.

[230] M. Karjalainen, U. K. Laine, T. I. Laakso, and V. Välimäki, "Transmission-line modeling and real-time synthesis of string and wind instruments," in *Proceedings of the 1991 International Computer Music Conference, Montreal*, pp. 293–296, Computer Music Association, 1991.

[231] M. Karjalainen, J. Backman, and J. Pölkki, "Analysis, modeling, and real-time sound synthesis of the kantele, a traditional Finnish string instrument," in *Proceedings of the International Conference on Acoustics, Speech, and Signal Processing, Minneapolis*, (New York), pp. 229–232, IEEE Press, 1993.

[232] M. Karjalainen, V. Välimäki, and Z. Jánosy, "Towards high-quality sound synthesis of the guitar and string instruments," in *Proceedings of the 1993 International Computer Music Conference, Tokyo*, pp. 56–63, Computer Music Association, Sept. 10–15 1993, `http://www.acoustics.hut.fi/~vpv/publications/icmc93-guitar.htm`.

[233] M. Karjalainen, V. Välimäki, and T. Tolonen, "Plucked string models: From the Karplus-Strong algorithm to digital waveguides and beyond," *Computer Music Journal*, vol. 22, no. 3, pp. 17–32, 1998, available online at `http://www.acoustics.hut.fi/~vpv/publications/cmj98.htm`.

[234] M. Karjalainen, P. Antsalo, A. Mäkivirta, T. Peltonen, and V. Välimäki, "Estimation of modal decay parameters from noisy response measurements," *Journal of the Audio Engineering Society*, vol. 2002, no. 11, pp. 867–878, 2002.

[235] M. Karjalainen, P. A. A. Esquef, P. Antsalo, A. Mäkivirta, and V. Välimäki, "Frequency-zooming ARMA modeling of resonant and reverberant systems," *Journal of the Audio Engineering Society*, vol. 2002, no. 12, pp. 1012–1029, 2002.

[236] K. Karplus and A. Strong, "Digital synthesis of plucked string and drum timbres," *Computer Music Journal*, vol. 7, no. 2, pp. 43–55, 1983, Reprinted in [380, pp. 467–479].

[237] S. M. Kay, *Modern Spectral Estimation*, Englewood Cliffs, NJ: Prentice-Hall, 1988.

[238] D. H. Keefe, "Theory of the single woodwind tone hole. experiments on the single woodwind tone hole," *Journal of the Acoustical Society of America*, vol. 72, pp. 676–699, Sept. 1982.

[239] D. H. Keefe, "Acoustical wave propagation in cylindrical ducts: Transmission line approximations for isothermal and nonisothermal boundary conditions," *Journal of the Acoustical Society of America*, vol. 75, pp. 58–62, Jan. 1984.

[240] D. H. Keefe, "Woodwind air column models," *Journal of the Acoustical Society of America*, vol. 88, pp. 35–51, Jan. 1990.

[241] R. G. Keen, *The Technology of the Univibe*, `http://www.geofex.com/Article_Folders/univibe/uvfrindx.htm`, 1998.

[242] R. Keen, *The Technology of Phase Shifters and Flangers*, `http://www.geofex.com/Article_Folders/phasers/phase.html`, 1999.

[243] E. Keller, *Fundamentals of Speech Synthesis*, New York: John Wiley and Sons, Inc., 1994.

[244] J. B. Keller, "Bowing of violin strings," *Comm. Pure Applied Math.*, vol. 6, pp. 483–495, 1953.

[245] J. L. Kelly and C. C. Lochbaum, "Speech synthesis," *Proceedings of the Fourth International Congress on Acoustics, Copenhagen*, pp. 1–4, September 1962, Paper G42. Reprinted in [142, pp. 127–130].

[246] J. L. Kelly and C. C. Lochbaum, "Speech synthesis," in *Proceedings of the Speech Communications Seminar*, (Stockholm), p. Paper F7, Speech Transmission Lab., Royal Institute of Technology, 1962.

[247] J. Kemp, *Theoretical and Experimental Study of Wave Propagation in Brass Musical Instruments*, PhD thesis, University of Edinburgh, 2002, `http://www.ph.ed.ac.uk/~jonathan/thesis/`.

[248] G. S. Kendall and W. L. Martens, "Simulating the cues of spatial hearing in natural environments," *Proceedings of the 1984 International Computer Music Conference, Paris*, pp. 111–125, 1984.

[249] J. Kergomard, "Elementary considerations on reed-instrument oscillations," in Hirschberg *et al.* [197], pp. 229–114.

[250] Y. E. Kim, "A framework for parametric singing voice analysis/synthesis," in *Proceedings of the IEEE Workshop on Applications of Signal Processing to Audio and Acoustics, New Paltz, NY*, (New York), IEEE Press, Oct. 2003.

[251] Y. E. Kim, "Singing voice analysis, synthesis, and modeling," in *Handbook on Signal Processing for Acoustics* (D. Havelock, ed.), Berlin: Springer-Verlag, 2006, in press.

[252] A. Klapuri, "Multipitch estimation and sound separation by the spectral smoothness principle," in *IEEE International Conference on Acoustics, Speech, and Signal Processing*, (Salt Lake City, Utah, USA), May 2001.

[253] A. Klapuri, "Multiple fundamental frequency estimation by harmonicity and spectral smoothness," *IEEE Transactions on Speech and Audio Processing*, vol. 11, no. 6, pp. 804–816, 2003.

[254] A. Klapuri, "A perceptually motivated multiple-F0 estimation method," in *Proceedings of the IEEE Workshop on Applications of Signal Processing to Audio and Acoustics, New Paltz, NY*, (New York), IEEE Press, Oct. 2005, `http://www.cs.tut.fi/sgn/arg/klap/waspaa2005.pdf`.

[255] D. Klatt, "Software for a cascade/parallel formant synthesizer," *Journal of the Acoustical Society of America*, vol. 67, pp. 13–33, 1980.

[256] D. Klatt, "Review of text-to-speech conversion for english," *Journal of the Acoustical Society of America*, vol. 82, pp. 737–793, Sept. 1987.

[257] D. H. Klatt and L. C. Klatt, "Analysis, synthesis, and perception of voice quality variations among female and make talkers," *Journal of the Acoustical Society of America*, vol. 87, no. 2, pp. 820–857, 1990.

[258] D. Kleppner and R. J. Kolenkow, *An Introduction to Mechanics*, New York: McGraw-Hill, 1973.

[259] M. Kline, *Mathematics and the Physical World*, New York: Dover, 1991, 1st ed. 1894.

[260] A. v. Knesebeck and U. Zölzer, "Comparison of pitch trackers for real-time guitar effects," in *Proceedings of the 13th International Conference on Digital Audio Effects (DAFx-10), Graz, Austria, September 6–10*, 2010.

[261] H. Kolsky, *Stress Waves in Solids*, New York: Dover, 1963.

[262] P. J. Kootsookos and R. C. Williamson, "FIR approximation of fractional sample delay systems," *IEEE Transactions on Circuits and Systems—II: Analog and Digital Signal Processing*, vol. 43, pp. 40–48, Feb 1996.

[263] A. Krishnaswamy and J. O. Smith, "Methods for simulating string collisions with rigid spatial objects," in *Proceedings of the IEEE Workshop on Applications of Signal Processing to Audio and Acoustics, New Paltz, NY*, (New York), IEEE Press, Oct. 2003.

[264] U. Z. Kristjan Dempwolf, Martin Holters, "Discretization of parametric analog circuits for real-time simulations," in *Proceedings of the 13th International Conference on Digital Audio Effects (DAFx-10), Graz, Austria, September 6–10*, 2010.

[265] J. Kuroda, J. Smith, and J. Perng, "Modeling the piano string as a mass-spring chain," *Acoustical Society of America, Program of the 2nd Pan-American/Iberian Meeting on Acoustics (abstract and presentation), Cancun, Mexico, Nov. 15–19*, 2010, (abstract only).

[266] J. Kuroda, J. Smith, K. V. Heusen, and J. Perng, "A stiff mass-spring-chain model for piano strings," *Acoustical Society of America, Program of the 2nd Pan-American/Iberian Meeting on Acoustics (abstract and presentation), Cancun, Mexico, Nov. 15–19*, 2010, (abstract only).

[267] T. I. Laakso, V. Välimäki, M. Karjalainen, and U. K. Laine, "Splitting the Unit Delay—Tools for Fractional Delay Filter Design," *IEEE Signal Processing Magazine*, vol. 13, pp. 30–60, Jan. 1996.

[268] J. Laird, P. Masri, and N. Canagarajah, "Modelling diffusion at the boundary of a digital waveguide mesh," in *Proceedings of the International Computer Music Conference*, (Beijing, China), Oct. 1999.

[269] C. Lambourg and A. Chaigne, "Measurements and modeling of the admittance matrix at the bridge in guitars," *Proceedings of the Stockholm Musical Acoustics Conference (SMAC-93)*, pp. 448–453, July 1993.

[270] L. D. Landau and E. Lifshitz, *Mechanics, Third Edition: Volume 1 (Course of Theoretical Physics)*, Oxford, UK: Butterworth-Heinemann, Jan. 1976.

[271] M. Lang, "Allpass filter design and applications," *IEEE Transactions on Signal Processing*, vol. 46, no. 9, pp. 2505–2514, 1998.

[272] M. Lang and T. I. Laakso, "Simple and robust method for the design of allpass filters using least-squares phase error criterion," *IEEE Transactions on Circuits and Systems—I: Fundamental Theory and Applications*, vol. 41, no. 1, pp. 40–48, 1994.

[273] J. Laroche, "A new analysis/synthesis system of musical signals using prony's method. Application to heavily damped percussive sounds," *Proc. IEEE ICASSP-89, Glasgow*, pp. 2053–2056, May 1989.

[274] J. Laroche, "The use of the matrix pencil method for the spectrum analysis of musical signals," *Journal of the Acoustical Society of America*, vol. 94, pp. 1958–1965, Oct. 1993.

[275] J. Laroche, "Time and pitch scale modification of audio signals," in *Applications of DSP to Audio & Acoustics* (M. Kahrs and K. Brandenburg, eds.), pp. 279–309, Boston/Dordrecht/London: Kluwer Academic Publishers, 1998.

[276] P. Larson, "Creating a virtual cello," tech. rep., Elec. Engineering Dept., Stanford University (CCRMA), June 2003, Music 421 Project Report, `http://ccrma.stanford.edu/~jos/plarson/`.

[277] M. Laurson and M. Kuuskankare, "Aspects on time modification in score-based performance control," in *Proceedings of the Stockholm Musical Acoustics Conference (SMAC-03), http://www.speech.kth.se/smac03/*, (Stockholm), pp. 545–548, Royal Swedish Academy of Music, Aug. 2003.

[278] M. Laurson, C. Erkut, V. Välimäki, and M. Kuuskankare, "Methods for Modeling Realistic Playing in Acoustic Guitar Synthesis," *Computer Music Journal*, vol. 25, no. 3, pp. 38–49, 2001, `http://lib.hut.fi/Diss/2002/isbn9512261901/article6.pdf`.

[279] K. Lee and J. O. Smith, "Implementation of a highly diffusing 2-D digital waveguide mesh with a quadratic residue diffuser," in *Proceedings of the 2004 International Computer Music Conference, Miami, Florida*, Computer Music Association, Nov. 2004.

[280] N. Lee and J. O. Smith, "Use of the energy decay relief (EDR) to estimate partial-overtone decay-times in a freely vibrating string," *Acoustical Society of America, Program of the 152nd Meeting (abstract and presentation), Honolulu, Hawaii*, Nov. 28–Dec. 2 2006, (abstract) — associated laboratory exercises available online: `http://ccrma.stanford.edu/realsimple/vguitar/`.

[281] N. Lee and J. O. Smith, "Virtual stringed instruments," Dec. 2007, `http://ccrma.stanford.edu/realsimple/phys_mod_overview/`.

[282] N. Lee, Z. Duan, and J. O. Smith III, "Excitation signal extraction for guitar tones," in *Proceedings of the 2007 International Computer Music Conference, Aug. 27–31, Copenhagen, Denmark*, Aug. 27–31 2007.

[283] N. Lee, J. O. Smith, J. Abel, and D. Berners, "Pitch glide analysis and synthesis from recorded tones," in *Proceedings of the 12th International Conference on Digital Audio Effects (DAFx-09), Como, Italy, September 1–4*, 2009.

[284] W. R. LePage, *Complex Variables and the Laplace Transform for Engineers*, New York: Dover, 1961.

[285] H. Levine and J. Schwinger, "On the radiation of sound from an unflanged circular pipe," *Physical Review*, vol. 73, no. 4, pp. 383–406, 1948.

[286] M. Lewisohn, *The Beatles: Recording Sessions*, New York: Harmony Books, 1988.

[287] L. Ljung and T. L.Soderstrom, "The Steiglitz-McBride algorithm revisited—convergence analysis and accuracy aspects," *IEEE Transactions on Automatic Control*, vol. 26, pp. 712–717, June 1981, See also the function `stmcb()` in the Matlab Signal Processing Toolbox.

[288] L. Ljung and T. L. Soderstrom, *Theory and Practice of Recursive Identification*, Cambridge, MA: MIT Press, 1983.

[289] N. J. Loy, *An Engineer's Guide to FIR Digital Filters*, Englewood Cliffs, NJ: Prentice-Hall, 1988.

[290] H.-L. Lu, *Toward a High-Quality Singing Synthesizer with Vocal Texture Control*, PhD thesis, Elec. Engineering Dept., Stanford University (CCRMA), July 2002, `http://ccrma.stanford.edu/~vickylu/thesis/`.

[291] H.-L. Lu and J. O. Smith, "Glottal source modeling for singing voice synthesis," in *Proceedings of the 2000 International Computer Music*

Conference, Berlin, Aug. 2000, http://ccrma.stanford.edu/~vickylu/-research/glottalSource/glottal.htm.

[292] H.-L. Lu and J. O. Smith, "Estimating glottal aspiration noise via wavelet thresholding and best basis thresholding," in *Proceedings of the IEEE Workshop on Applications of Signal Processing to Audio and Acoustics, New Paltz, NY*, (New York), IEEE Press, Oct. 2001.

[293] J. Macak and J. Schimmel, "Real-time guitar tube amplifier simulation using an approximation of differential equations," in *Proceedings of the 13th International Conference on Digital Audio Effects (DAFx-10), Graz, Austria, September 6–10*, 2010.

[294] E. Maestre, J. Bonada, M. B. A. Pérez, and E. Guaus, "Acquisition of violin instrumental gestures using a commercial EMF tracking device," in *Proceedings of the 2007 International Computer Music Conference, Aug. 27–31, Copenhagen, Denmark*, 2007.

[295] I. G. Main, *Vibrations and Waves in Physics, Third Edition*, Cambridge University Press, 1993.

[296] J. Makhoul, "Linear prediction: A tutorial review," *Proceedings of the IEEE*, vol. 63, pp. 561–580, Apr. 1975.

[297] J. D. Markel and A. H. Gray, *Linear Prediction of Speech*, New York: Springer Verlag, 1976.

[298] R. Marogna, F. Avanzin, and B. Bank, "Energy based synthesis of tension modulation in membranes," in *Proceedings of the 13th International Conference on Digital Audio Effects (DAFx-10), Graz, Austria, September 6–10*, 2010.

[299] K. D. Marshall, "Modal analysis: A primer on theory and practice," *Catgut Acoustical Society Journal*, pp. 7–17, Nov. 1986.

[300] J. Martinez and J. Agullo, "Conical bores. Part I: Reflection functions associated with discontinuities," *Journal of the Acoustical Society of America*, vol. 84, pp. 1613–1619, Nov. 1988.

[301] S. J. Mason, "Feedback theory—some properties of signal flow graphs," *Proceedings of the IRE*, vol. 41, pp. 1144–1156, Sept. 1953.

[302] S. J. Mason, "Feedback theory—further properties of signal flow graphs," *Proceedings of the IRE*, vol. 44, pp. 920–926, July 1956.

[303] D. C. Massie, "Wavetable sampling synthesis," in *Applications of DSP to Audio & Acoustics* (M. Kahrs and K. Brandenburg, eds.), pp. 311–341, Boston/Dordrecht/London: Kluwer Academic Publishers, 1998.

[304] M. Mathews and J. O. Smith, "Methods for synthesizing very high Q parametrically well behaved two pole filters," in *Proceedings of the Stockholm Musical Acoustics Conference (SMAC-03), http://www.speech.kth.se/smac03/*, (Stockholm), Royal Swedish Academy of Music, Aug. 2003, available online, with sound examples, at http://ccrma.stanford.edu/~jos/smac03maxjos/.

[305] M. V. Mathews, "The digital computer as a musical instrument," *Science*, vol. 142, no. 11, pp. 553–557, 1963.

[306] I. G. Mattingly, "Speech synthesis for phonetic and phonological models," in *Current Trends in Linguistics* (T. A. Sebeok, ed.), vol. 12, pp. 2451–2487, The Hague: Mouton, 1974, `http://www.mindspring.com/~ssshp/ssshp_cd/im_home.htm`.

[307] M. E. McIntyre and J. Woodhouse, "On the fundamentals of bowed string dynamics," *Acustica*, vol. 43, pp. 93–108, Sept. 1979.

[308] M. E. McIntyre, R. T. Schumacher, and J. Woodhouse, "On the oscillations of musical instruments," *Journal of the Acoustical Society of America*, vol. 74, pp. 1325–1345, Nov. 1983.

[309] E. Meijering, "A chronology of interpolation: From ancient astronomy to modern signal and image processing," *Proceedings of the IEEE*, vol. 90, pp. 319–342, Mar. 2002.

[310] S. K. Mitra, *Digital Signal Processing: A Computer-Based Approach, Fourth Edition*, New York: McGraw-Hill, 2010.

[311] P. Moin, *Engineering Numerical Analysis*, Cambridge University Press, 2001.

[312] B. C. J. Moore, *An Introduction to the Psychology of Hearing*, New York: Academic Press, 1997.

[313] F. R. Moore, *Elements of Computer Music*, Englewood Cliffs, NJ: Prentice Hall, 1990.

[314] J. A. Moorer, "About this reverberation business," *Computer Music Journal*, vol. 3, no. 2, pp. 13–18, 1979, reprinted in [382].

[315] J. A. Moorer, "The manifold joys of conformal mapping: Applications to digital filtering in the studio," *Journal of the Audio Engineering Society*, vol. 31, pp. 826–841, Nov 1983.

[316] A. Morgan, *Solving Polynomial Systems Using Continuation for Engineering and Scientific Computations*, Englewood Cliffs, NJ: Prentice-Hall, 1987.

[317] P. M. Morse, *Vibration and Sound*, `http://asa.aip.org/publications.html`: American Institute of Physics, for the Acoustical Society of America, 1948, 1st edition 1936, last author's edition 1948, ASA edition 1981.

[318] P. M. Morse and K. U. Ingard, *Theoretical Acoustics*, New York: McGraw-Hill, 1968.

[319] R. Msallam, S. Dequidt, R. Caussé, and S. Tassart, "Physical model of the trombone including nonlinear effects. application to the sound synthesis of loud tones," *Acta Acustica*, vol. 86, pp. 725–736, 2000.

[320] D. T. Murphy and D. M. Howard, "2-D digital waveguide mesh topologies in room acoustics modelling," in *Proceedings of the COST G-6 Conference on Digital Audio Effects (DAFx)*, (Verona, Italy), pp. 211–216, Dec. 2000, `http://www.dafx.de/`.

[321] D. T. Murphy, D. M. Howard, and A. M. Tyrrell, "Multi-channel reverberation for computer music applications," in *Proceedings of the IEEE Workshop on Signal Processing Systems*, pp. 210–219, 1998.

[322] D. T. Murphy, C. J. C. Newton, and D. M. Howard, "Digital waveguide mesh modelling of room acoustics: Surround-sound, boundaries and plugin implementation," in *Proceedings of the COST G-6 Conference on Digital Audio Effects (DAFx)*, (Limerick, Ireland), Dec. 2001, `http:`

[323] J. Nam, V. Välimäki, J. S. Abel, and J. O. Smith, "Alias-free oscillators using a feedback delay loop," in *Proceedings of the 12th International Conference on Digital Audio Effects (DAFx-09), Como, Italy, September 1–4*, 2009.

[324] J. Nam, V. Välimäki, J. S. Abel, and J. O. Smith, "Efficient antialiasing oscillator algorithms using low-order fractional delay filters," *IEEE Transactions on Audio, Speech, and Language Processing*, vol. 18, no. 4, pp. 773–785, 2010.

[325] A. W. Nayor and G. R. Sell, *Linear Operator Theory in Engineering and Science*, New York: Springer Verlag, 1982.

[326] Z. Nehari, *Conformal Mapping*, New York: Dover, 1952.

[327] A. Ng and A. Horner, "Iterative combinatorial basis spectra in wavetable matching," *Journal of the Audio Engineering Society*, vol. 50, no. 12, pp. 1054–1063, 2002.

[328] I. Nikolić, "Improvements of artificial reverberation by use of subband feedback delay networks," *Proc. 112th Convention of the Audio Engineering Society, Munich*, 2002, Paper 5630.

[329] B. Noble, *Applied Linear Algebra*, Englewood Cliffs, NJ: Prentice-Hall, 1969.

[330] D. D. Nolte, "The tangled tale of phase space," *Physics Today*, vol. 63, pp. 33–38, Apr. 2010.

[331] D. Noreland, *Numerical Techniques for Acoustic Modelling and Design of Brass Wind Instruments*, PhD thesis, Faculty of Science and Engineering, Uppsala Universitet, 2002.

[332] H. F. Olson, *Music, Physics and Engineering (formerly titled Musical Engineering)*, New York: Dover, 1967, 2nd ed., 1st ed. 1952.

[333] A. V. Oppenheim and R. W. Schafer, *Digital Signal Processing*, Englewood Cliffs, NJ: Prentice-Hall, 1975.

[334] A. V. Oppenheim, D. H. Johnson, and K. Steiglitz, "Computation of spectra with unequal resolution using the fast Fourier transform," *Proceedings of the IEEE*, vol. 59, pp. 299–301, 1971.

[335] D. O'Shaughnessy, *Speech Communication*, Reading MA: Addison-Wesley, 1987.

[336] V. Pagneux, N. Amir, and J. Kergomard, "A study of wave propagation in varying cross-section waveguides by modal decomposition. part 1. theory and validation," *Journal of the Acoustical Society of America*, vol. 100, no. 4, pp. 2034–2048, 1996.

[337] T. Painter and A. Spanias, "Perceptual coding of digital audio," *Proceedings of the IEEE*, vol. 88, pp. 451–513, Apr. 2000.

[338] J. Pakarinen and M. Karjalainen, "Enhanced wave digital triode model for real-time tube amplifier emulation," *IEEE Transactions on Audio, Speech, and Language Processing*, vol. 18, pp. 738–746, May 2010.

[339] J. Pakarinen and D. T. Yeh, "A review on digital guitar tube amplifier modeling techniques," *Computer Music Journal*, vol. 33, pp. 85–, 2009.

[340] A. Paladin and D. Rocchesso, "A dispersive resonator in real-time on mars workstation," in *Proceedings of the 1992 International Computer Music Conference, San Jose*, pp. 146–149, Computer Music Association, 1992.

[341] B. H. Pandy, G. S. Settles, and J. D. Miller, "Schlieren imaging of shock waves from a trumpet," *Journal of the Acoustical Society of America*, vol. 114, pp. 3363–3367, Dec. 2003.

[342] A. Papoulis, *Signal Analysis*, New York: McGraw-Hill, 1977.

[343] T. W. Parks and C. S. Burrus, *Digital Filter Design*, New York: John Wiley and Sons, Inc., June 1987, contains FORTRAN software listings.

[344] J. Pekonen and V. Välimäki, "Filter-based alias reduction in classical waveform synthesis," *Proceedings of the International Conference on Acoustics, Speech, and Signal Processing, Las Vegas, Nevada, USA*, pp. 133–136, March 30-April 4 2008.

[345] H. Penttinen and V. Välimäki, "Time-domain approach to estimating the plucking point of guitar tones obtained with an under-saddle pickup," *Applied Acoustics*, vol. 65, pp. 1207–1220, Dec. 2004.

[346] S. Perlis, *Theory of Matrices*, New York: Dover, 1991, 1st ed. 1952.

[347] C.-Y. J. Perng, J. Smith, and T. Rossing, "Physical modeling of the harpsichord plectrum-string interaction," in *Proceedings of the 13th International Conference on Digital Audio Effects (DAFx-10), Graz, Austria, September 6–10*, 2010.

[348] S. Petrausch and R. Rabenstein, "Wave digital filters with multiple nonlinearities," in *in Proc. 12th European Signal Processing Conf. (EUSIPCO-2004*, pp. 77–80, 2004.

[349] A. D. Pierce, *Acoustics*, American Institute of Physics, for the Acoustical Society of America, 1989, `http://asa.aip.org/publications.html`.

[350] J. R. Pierce and S. A. Van Duyne, "A passive nonlinear digital filter design which facilitates physics-based sound synthesis of highly nonlinear musical instruments," *Journal of the Acoustical Society of America*, vol. 101, pp. 1120–1126, Feb. 1997.

[351] R. Pitteroff and J. Woodhouse, "Mechanics of the contact area between a violin bow and a string, part ii: Simulating the bowed string," *Acta Acustica*, vol. 84, pp. 744–757, 1998.

[352] R. Pitteroff and J. Woodhouse, "Mechanics of the contact area between a violin bow and a string, part iii: Parameter dependence," *Acta Acustica*, vol. 84, pp. 929–946, 1998.

[353] N. Porcaro, P. Scandalis, J. O. Smith, D. A. Jaffe, and T. Stilson, "SynthBuilder—a graphical real-time synthesis, processing and performance system," in *Proceedings of the 1995 International Computer Music Conference, Banff*, pp. 61–62, Computer Music Association, 1995.

[354] W. H. Press, S. A. Teukolsky, W. T. Vetterling, and B. P. Flannery, *Numerical Recipes in C: The Art of Scientific Computing*, Cambridge University Press, 2 ed., 1992.

[355] M. Puckette, *Pure Data (PD)*, `http://www.puredata.org`, July 2004.

[356] E. M. Purcell, *Electricity and Magnetism (Berkeley Physics Course, Volume 2), Second Edition*, New York: McGraw-Hill, Aug. 1984.

[357] G. Putland, "Every one-parameter acoustic field obeys webster's horn equation," *Journal of the Audio Engineering Society*, vol. 41, pp. 435–451, June 1993.

[358] W. Putnam and J. O. Smith, "Design of fractional delay filters using convex optimization," in *Proceedings of the IEEE Workshop on Applications of Signal Processing to Audio and Acoustics, New Paltz, NY*, (New York), IEEE Press, Oct. 1997, `http://ccrma.stanford.edu/~jos/resample/optfir.pdf`.

[359] T. F. Quatieri and R. J. McAulay, "Audio signal processing based on sinusoidal analysis/synthesis," in *Applications of DSP to Audio & Acoustics* (M. Kahrs and K. Brandenburg, eds.), pp. 343–416, Boston/Dordrecht/London: Kluwer Academic Publishers, 1998.

[360] R. Rabenstein and L. Trautmann, "Digital sound synthesis of string instruments with the functional transformation method," *Signal Processing*, vol. 83, pp. 1673–1688, Aug. 2003.

[361] R. Rabenstein, S. Petrausch, A. Sarti, G. D. Sanctis, C. Erkut, and M. Karjalainen, "Block-based physical modeling for digital sound synthesis," *IEEE Signal Processing Magazine*, pp. 42–54, Mar. 2007.

[362] L. R. Rabiner and B. Gold, *Theory and Application of Digital Signal Processing*, Prentice-Hall, 1975.

[363] L. R. Rabiner and R. W. Schafer, *Theory and Applications of Digital Speech Processing*, Prentice-Hall, 2010.

[364] C. Raffel and J. Smith, "Practical modeling of bucket-brigade device circuits," in *Proceedings of the 13th International Conference on Digital Audio Effects (DAFx-10), Graz, Austria, September 6–10*, 2010.

[365] C. V. Raman, "On the mechanical theory of vibrations of bowed strings, etc," *Indian Assoc. Cult. Sci. Bull.*, vol. 15, pp. 1–158, 1918.

[366] E. Rank and G. Kubin, "A waveguide model for slapbass synthesis," in *Proceedings of the International Conference on Acoustics, Speech, and Signal Processing*, vol. 1, (New York), pp. 443–446, IEEE Press, 1997.

[367] J. Rauhala and V. Välimäki, "Dispersion modeling in waveguide piano synthesis using tunable allpass filters," *Proceedings of the Conference on Digital Audio Effects (DAFx-06), Montreal, Canada*, pp. 71–76, Sept. 2006, `http://www.dafx.ca/proceedings/papers/p_071.pdf`, `http://www.acoustics.hut.fi/research/asp/piano/`.

[368] J. Rauhala and V. Välimäki, "Tunable dispersion filter design for piano synthesis," *IEEE Signal Processing Letters*, vol. 13, pp. 253–256, May 2006.

[369] G. R. Reddy and M. N. S. Swamy, "Digital all-pass filter design through discrete hilbert transform," in *Proceedings of the International Conference on Acoustics, Speech, and Signal Processing, Albuquerque*, 1998.

[370] P. A. Regalia, S. K. Mitra, P. P. Vaidyanathan, M. K. Renfors, and Y. Neuvo, "Tree-structured complementary filter banks using all-pass sections," *IEEE Transactions on Circuits and Systems*, vol. CAS-34, pp. 1470–1484, Dec. 1987.

[371] P. A. Regalia, S. K. Mitra, and P. P. Vaidyanathan, "The digital all-pass filter: A versatile signal processing building block," *Proceedings of the IEEE*, vol. PROC-76, pp. 19–37, Jan. 1988.

[372] J. D. Rhodes, P. C. Marston, and D. C. Youla, "Explicit solution for the synthesis of two-variable transmission-line networks," *IEEE Transactions on Circuit Theory*, vol. CT-20, pp. 504–567, Sept. 1973.

[373] J. R. Rice, *The Approximation of Functions, Volume I*, Reading MA: Addison-Wesley, 1964.

[374] J. R. Rice, *The Approximation of Functions, Volume II*, Reading MA: Addison-Wesley, 1969.

[375] R. Rideout, "Yamaha VL1 virtual acoustic synthesizer," *Keyboard Magazine*, vol. 20, pp. 104–118, June 1994.

[376] D. C. Rife and R. R. Boorstyn, "Single-tone parameter estimation from discrete-time observations," *IEEE Transactions on Information Theory*, vol. IT-20, pp. 591–598, Sep. 1974.

[377] D. C. Rife and R. R. Boorstyn, "Multiple tone parameter estimation from discrete-time observations," *Bell System Technical Journal*, vol. 55, pp. 1389–1410, Nov. 1976.

[378] J. Riionheimo and V. Välimäki, "Parameter estimation of a plucked string synthesis model using genetic algorithm with perceptual fitness calculation," *EURASIP Journal on Applied Signal Processing*, vol. 3, pp. 791–805, July 2003, sound examples available at `http://www.acoustics.hut.fi/publications/papers/jasp-ga/`.

[379] J.-C. Risset, "Computer music experiments, 1964—...," *Computer Music Journal*, vol. 9, no. 1, pp. 11–18, 1985, Reprinted in [380].

[380] C. Roads, ed., *The Music Machine*, Cambridge, MA: MIT Press, 1989.

[381] C. Roads, *The Computer Music Tutorial*, Cambridge, MA: MIT Press, 1996.

[382] C. Roads and J. Strawn, eds., *Foundations of Computer Music*, Cambridge, MA: MIT Press, 1985.

[383] D. Rocchesso, "Fractionally addressed delay lines," *IEEE Transactions on Speech and Audio Processing*, vol. 8, pp. 717–727, Nov. 2000.

[384] D. Rocchesso and F. Scalcon, "Accurate dispersion simulation for piano strings," in *Proc. Nordic Acoustical Meeting (NAM'96)*, (Helsinki, Finland), June 12-14 1996, 9 pages.

[385] D. Rocchesso and J. O. Smith, "Circulant and elliptic feedback delay networks for artificial reverberation," *IEEE Transactions on Speech and Audio Processing*, vol. 5, pp. 51–63, Jan. 1997, preprint `http://ccrma.stanford.edu/~jos/cfdn/`.

[386] X. Rodet, "Time-domain formant wave-function synthesis," *Computer Music J.*, pp. 6–14, 1984.

[387] X. Rodet, "Flexible yet controllable physical models: A nonlinear dynamics approach," in *Proceedings of the 1993 International Computer Music Conference, Tokyo*, pp. 10–15, Computer Music Association, 1993.

[388] X. Rodet, "One and two mass model oscillations for voice and instruments," in *Proceedings of the 1995 International Computer Music Conference, Banff*, pp. 207–214, Computer Music Association, 1995.

[389] X. Rodet, Y. Potard, and J. Barrière, "The chant project: From the synthesis of the singing voice to synthesis in general," in *The Music Machine* (C. Roads, ed.), pp. 449–465, Cambridge, MA: MIT Press, 1989.

[390] T. D. Rossing, F. R. Moore, and P. A. Wheeler, *The Science of Sound (3rd Edition)*, Reading MA: Addison-Wesley, 2003.

[391] D. R. Rowland and C. Pask, "The missing wave momentum mystery," *American Journal of Physics*, vol. 67, pp. 378–388, May 1999.

[392] P. M. Ruiz, *A Technique for Simulating the Vibrations of Strings with a Digital Computer*, PhD thesis, Music Master Diss., Univ. Ill., Urbana, 1969.

[393] P. Samson, "A general-purpose synthesizer," *Journal of the Audio Engineering Society*, vol. 28, no. 3, pp. 106–113, 1980.

[394] G. D. Sanctis and A. Sarti, "Virtual analog modeling in the wave-digital domain," *IEEE Transactions on Audio, Speech, and Language Processing*, vol. 18, pp. 715–727, May 2010.

[395] B. Santo, "Volume cranked up in amp debate," *Electronic Engineering Times*, pp. 24–35, October 3, 1994, `http://www.trueaudio.com/at_eetjlm.htm`.

[396] L. Savioja, J. Backman, A. Järvinen, and T. Takala, "Waveguide mesh method for low-frequency simulation of room acoustics," *Proceedings of the 15th International Conference on Acoustics (ICA-95), Trondheim, Norway, April 4–9*, pp. 637–640, June 1995.

[397] L. Savioja, "Real-time 3d finite-difference time-domain simulation of low- and mid-frequency room acoustics," in *Proceedings of the 13th International Conference on Digital Audio Effects (DAFx-10), Graz, Austria, September 6–10*, 2010.

[398] L. Savioja and V. Välimäki, "Reduction of the dispersion error in the interpolated digital waveguide mesh using frequency warping," in *Proceedings of the International Conference on Acoustics, Speech, and Signal Processing, Phoenix*, vol. 2, (New York), pp. 973–976, IEEE Press, Mar. 15–19 1999, `http://www.acoustics.hut.fi/~vpv/publications/icassp99-warp.htm`.

[399] L. Savioja and V. Välimäki, "Reducing the dispersion error in the digital waveguide mesh using interpolation and frequency-warping techniques," *IEEE Transactions on Speech and Audio Processing*, pp. 184–194, Mar. 2000.

[400] L. Savioja and V. Välimäki, "Interpolated 3-D digital waveguide mesh with frequency warping," *Proceedings of the IEEE International Conference on Acoustics, Speech, and Signal Processing*, vol. 5, pp. 3345–3348, May 2001.

[401] L. Savioja and V. Välimäki, "Interpolated rectangular 3-d digital waveguide mesh algorithms with frequency warping," *IEEE Transactions on Speech and Audio Processing*, vol. 11, pp. 783–790, Nov. 2003.

[402] G. Scavone and J. O. Smith, "Digital waveguide modeling of woodwind toneholes," *132nd meeting of the Acoustical Society of America, Honolulu*, Dec. 1996.

[403] G. Scavone and J. O. Smith, "Digital waveguide modeling of woodwind toneholes," in *Proceedings of the 1997 International Computer Music Conference, Greece*, Computer Music Association, 1997.

[404] G. Scavone and J. O. Smith, "Scattering parameters for the Keefe clarinet tonehole model," in *Proceedings of the International Symposium on Musical Acoustics (ISMA-97), Edinburgh, Scotland*, pp. 433–438, Aug. 1997.

[405] G. P. Scavone and P. R. Cook, "Real-time computer modeling of woodwind instruments," in *Proceedings of the International Symposium on Musical Acoustics (ISMA-98), Leavenworth, Washington*, pp. 197–202, July 1998, `http://ccrma.stanford.edu/~jos/tonehole/`.

[406] G. P. Scavone, *An Acoustic Analysis of Single-Reed Woodwind Instruments with an Emphasis on Design and Performance Issues and Digital Waveguide Modeling Techniques*, PhD thesis, CCRMA, Music Dept., Stanford University, Mar. 1997, `http://ccrma.stanford.edu/~gary/`.

[407] R. W. Schafer and L. R. Rabiner, "A digital signal processing approach to interpolation," *Proceedings of the IEEE*, vol. 61, pp. 692–702, June 1973.

[408] R. W. Schafer and J. D. Markel, eds., *Speech Analysis*, New York: IEEE Press, 1979.

[409] J. C. Schelleng, "The violin as a circuit," *Journal of the Acoustical Society of America*, vol. 35, pp. 326–338, 1963.

[410] J. C. Schelleng, "The bowed string and the player," *Journal of the Acoustical Society of America*, vol. 53, pp. 26–41, Jan. 1973.

[411] K. M. Schmidt, *Prelude for Strings: The Origins of Spectral Theory*, Cardiff University: Cardiff School of Mathematics, 2004, `http://www.cf.ac.uk/maths/diffeq/strings.html`.

[412] M. R. Schroeder, "Natural-sounding artificial reverberation," *Journal of the Audio Engineering Society*, vol. 10, no. 3, pp. 219–223, 1962.

[413] M. R. Schroeder, "New method for measuring reverberation time," *Journal of the Acoustical Society of America*, vol. 37, no. 5, pp. 232–235, 1965.

[414] M. R. Schroeder, "Vocoders: Analysis and synthesis of speech (a review of 30 years of applied speech research)," *Proceedings of the IEEE*, vol. 54, pp. 720–734, May 1966, Reprinted in [408, pp. 352–366].

[415] M. R. Schroeder, "Digital simulation of sound transmission in reverberant spaces (part i)," *Journal of the Acoustical Society of America*, vol. 47, no. 2, pp. 424–431, 1970.

[416] M. R. Schroeder, "Diffuse sound reflection by maximum-length sequence," *Journal of the Acoustical Society of America*, vol. 57, no. 1, pp. 149–150, 1975.

[417] M. R. Schroeder and B. F. Logan, "Colorless artificial reverberation," *IRE Transactions*, vol. AU-9, pp. 209–214, Nov–Dec 1961.

[418] C. Segoufin, B. Fabre, M. P. Verge, A. Hirschberg, and A. P. J. Wijnands, "Experimental study of the influence of mouth geometry on sound production in a recorder-like instrument: Windway length and chamfers," *Acta Acustica united with Acustica*, vol. 86, pp. 649–661, 2000.

[419] S. Serafin, *The sound of friction: real-time models, playability and musical applications*, PhD thesis, Music Department, Stanford University, 2004.

[420] S. Serafin and D. Young, "Bowed string physical model validation through use of a bow controller and examination of bow strokes," in *Proceedings of the Stockholm Musical Acoustics Conference (SMAC-03), http://www.speech.kth.se/smac03/*, 2003.

[421] S. Serafin, J. O. Smith, and J. Woodhouse, "An investigation of the impact of torsion waves and friction characteristics on the playability of virtual bowed strings," in *Proceedings of the IEEE Workshop on Applications of Signal Processing to Audio and Acoustics, New Paltz, NY*, (New York), IEEE Press, Oct. 1999.

[422] S. Serafin, P. Huang, and J. O. Smith, "The banded digital waveguide mesh," in *Workshop on Current Research Directions in Computer Music*, (Audiovisual Institute, Pompeu Fabra University), Nov. 15–17, 2001.

[423] S. Serafin, P. Huang, S. Ystad, C. Chafe, and J. O. Smith, "Analysis and synthesis of unusual friction-driven musical instruments," in *Proceedings of the 2002 International Computer Music Conference, Sweden*, 2002.

[424] X. Serra and J. O. Smith, "Spectral modeling synthesis: A sound analysis/synthesis system based on a deterministic plus stochastic decomposition," *Computer Music Journal*, vol. 14, no. 4, pp. 12–24, 1990, Software and recent papers URL: `//www.iua.upf.es/~xserra/`.

[425] D. B. Sharp, *Acoustic Pulse Reflectometry for the Measurement of Musical Wind Instruments*, PhD thesis, Dept. of Physics and Astronomy, University of Edinburgh, 1996, `http://acoustics.open.ac.uk/pdf/thesis.pdf`.

[426] J. H. Smith and J. Woodhouse, "The tribology of rosin," *Journal of Mechanics and Physics of Solids*, vol. 48, pp. 1633–1681, Aug. 2000.

[427] J. Smith, J. Kuroda, J. Perng, K. V. Heusen, and J. Abel, "Efficient computational modeling of piano strings for real-time synthesis using mass-spring chains, coupled finite differences, and digital waveguide sections," *Acoustical Society of America, Program of the 2nd Pan-American/Iberian Meeting on Acoustics (abstract and presentation), Cancun, Mexico, Nov. 15–19*, 2010, invited presentation. Presentation overheads: `https://ccrma.stanford.edu/~jos/pdf/ASA-2010-jos.pdf`.

[428] J. O. Smith, *Techniques for Digital Filter Design and System Identification with Application to the Violin*, PhD thesis, Elec. Engineering Dept., Stanford University (CCRMA), June 1983, CCRMA Technical Report STAN-M-14, `http://ccrma.stanford.edu/STANM/stanms/stanm14/`.

[429] J. O. Smith, "An allpass approach to digital phasing and flanging," in *Proceedings of the 1984 International Computer Music Conference, Paris*, pp. 103–109, Computer Music Association, 1984, CCRMA Technical Report STAN-M-21, `http://ccrma.stanford.edu/STANM/stanms/stanm14/`.

[430] J. O. Smith, "A new approach to digital reverberation using closed waveguide networks," in *Proceedings of the 1985 International Computer Music Conference, Vancouver*, pp. 47–53, Computer Music Association, 1985, also available in [433].

[431] J. O. Smith, "Efficient simulation of the reed-bore and bow-string mechanisms," in *Proceedings of the 1986 International Computer Music Conference, The Hague*, pp. 275–280, Computer Music Association, 1986, also available in [433].

[432] J. O. Smith, "Elimination of limit cycles and overflow oscillations in time-varying lattice and ladder digital filters," in *Proceedings of the IEEE Conference on Circuits and Systems, San Jose*, pp. 197–299, May 1986, conference version; full version available in [433].

[433] J. O. Smith, "Music applications of digital waveguides," Tech. Rep. STAN-M-39, CCRMA, Music Department, Stanford University, 1987, CCRMA Technical Report STAN-M-39, `http://ccrma.stanford.edu/STANM/stanm39/`.

[434] J. O. Smith, "Waveguide filter tutorial," in *Proceedings of the 1987 International Computer Music Conference, Champaign-Urbana*, pp. 9–16, Computer Music Association, 1987.

[435] J. O. Smith, "Viewpoints on the history of digital synthesis," in *Proceedings of the 1991 International Computer Music Conference, Montreal*, pp. 1–10, Computer Music Association, 1991, `http://ccrma.stanford.edu/~jos/kna/`.

[436] J. O. Smith, "Waveguide simulation of non-cylindrical acoustic tubes," in *Proceedings of the 1991 International Computer Music Conference, Montreal*, pp. 304–307, Computer Music Association, 1991.

[437] J. O. Smith, "Physical modeling using digital waveguides," *Computer Music Journal*, vol. 16, no. 4, pp. 74–91, 1992, special issue: Physical Modeling of Musical Instruments, Part I. `http://ccrma.stanford.edu/~jos/pmudw/`.

[438] J. O. Smith, "The window method for digital filter design," Winter 1992, Mathematica notebook for Music 420 (EE 367A), URL:`ftp://ccrma-ftp.stanford.edu/pub/DSP/Tutorials/Kaiser.ma.Z`.

[439] J. O. Smith, "Efficient synthesis of stringed musical instruments," in *Proceedings of the 1993 International Computer Music Conference, Tokyo*, pp. 64–71, Computer Music Association, 1993, incorporated into this book.

[440] J. O. Smith, "Efficient synthesis of stringed musical instruments," in *Proceedings of the 1993 International Computer Music Conference, Tokyo*, pp. 64–71, Computer Music Association, 1993, incorporated into [456].

[441] J. O. Smith, "Physical modeling synthesis update," *Computer Music Journal*, vol. 20, no. 2, pp. 44–56, 1996, `http://ccrma.stanford.edu/~jos/pmupd/`.

[442] J. O. Smith, "Principles of digital waveguide models of musical instruments," in *Applications of Digital Signal Processing to Audio and Acoustics* (M. Kahrs and K. Brandenburg, eds.), pp. 417–466, Boston/Dordrecht/London: Kluwer Academic Publishers, 1998.

[443] J. O. Smith, *Introduction to Matlab and Octave*, `http://ccrma.stanford.edu/~jos/matlab/`, 2003.

[444] J. O. Smith, "On the equivalence of digital waveguide and finite difference time domain schemes," July 21, 2004, `http://arxiv.org/abs/physics/0407032/`.

[445] J. O. Smith, "Virtual acoustic musical instruments: Review and update," *Journal of New Music Research*, vol. 33, no. 3, pp. 283–304, 2004.

[446] J. O. Smith, "A history of ideas leading to virtual acoustic musical instruments," in *Proceedings of the IEEE Workshop on Applications of Signal Processing to Audio and Acoustics, New Paltz, NY*, (New York), IEEE Press, Oct. 2005, keynote paper.

[447] J. O. Smith, *Physical Audio Signal Processing: Digital Waveguide Modeling of Musical Instruments and Audio Effects*, `http://ccrma.stanford.edu/~jos/pasp/`, Dec. 2005.

[448] J. O. Smith, *Virtual Acoustic Musical Instruments: Review of Models and Selected Research*, `http://ccrma.stanford.edu/~jos/Mohonk05/`, 2005.

[449] J. O. Smith, *Introduction to Digital Filters with Audio Applications*, `http://ccrma.stanford.edu/~jos/filters/`, Sept. 2007, online book.

[450] J. O. Smith, "Making virtual electric guitars and associated effects using Faust [draft]," Dec. 2007, `http://ccrma.stanford.edu/realsimple/faust_strings/`.

[451] J. O. Smith, *Mathematics of the Discrete Fourier Transform (DFT), with Audio Applications, Second Edition*, `http://ccrma.stanford.edu/~jos/mdft/`, Apr. 2007, online book.

[452] J. O. Smith, *Physical Audio Signal Processing*, `http://ccrma.stanford.edu/~jos/pasp/`, Aug. 2007, online book.

[453] J. O. Smith, "Signal processing in Faust and Pd," Nov. 2007, `http://ccrma.stanford.edu/realsimple/faust/`.

[454] J. O. Smith, "Virtual electric guitars and effects using Faust and Octave," in *Proceedings of the 6th International Linux Audio Conference (LAC2008), http://lac.linuxaudio.org/*, 2008, paper: `http://-ccrma.stanford.edu/realsimple/faust_strings/faust_strings.pdf`, presentation overheads: `http://ccrma.stanford.edu/~jos/pdf/LAC2008-jos.pdf`, supporting website: `http://ccrma.stanford.edu/realsimple/faust_strings/`.

[455] J. O. Smith, "Audio FFT filter banks," in *Proceedings of the 12th International Conference on Digital Audio Effects (DAFx-09), Como, Italy, September 1–4*, Sept. 2009.

[456] J. O. Smith, *Physical Audio Signal Processing*, `http://ccrma.stanford.edu/~jos/pasp/`, 2010, online book.

[457] J. O. Smith, *Spectral Audio Signal Processing*, `http://ccrma.stanford.edu/~jos/sasp/`, Mar. 2010, online book.

[458] J. O. Smith, "`prime_power_delays()`," Nov. 2010, function checked into Faust git repository, file `effect.lib`.

[459] J. O. Smith and J. S. Abel, "The Bark bilinear transform," in *Proceedings of the IEEE Workshop on Applications of Signal Processing to Audio and Acoustics, New Paltz, NY*, (New York), IEEE Press, Oct. 1995, Session 8, paper 6, 4 pages. `http://ccrma.stanford.edu/~jos/gz/bbtmh.tgz`.

[460] J. O. Smith and J. S. Abel, "Bark and ERB bilinear transforms," *IEEE Transactions on Speech and Audio Processing*, pp. 697–708, Nov. 1999.

[461] J. O. Smith and P. R. Cook, "The second-order digital waveguide oscillator," in *Proceedings of the 1992 International Computer Music Conference, San Jose*, pp. 150–153, Computer Music Association, 1992, `http://ccrma.stanford.edu/~jos/wgo/`.

[462] J. O. Smith and B. Friedlander, "Adaptive interpolated time-delay estimation," *IEEE Transactions on Aerospace and Electronic Systems*, vol. 21, pp. 180–199, Mar. 1985.

[463] J. O. Smith and P. Gossett, "A flexible sampling-rate conversion method," in *Proceedings of the International Conference on Acoustics, Speech, and Signal Processing, San Diego*, vol. 2, (New York), pp. 19.4.1–19.4.2, IEEE Press, Mar. 1984, expanded tutorial and associated free software available at the Digital Audio Resampling Home Page: `http://ccrma.stanford.edu/~jos/resample/`.

[464] J. O. Smith and D. Rocchesso, "Connections between feedback delay networks and waveguide networks for digital reverberation," in *Proceedings of the 1994 International Computer Music Conference, Århus*, pp. 376–377, Computer Music Association, 1995.

[465] J. O. Smith and D. Rocchesso, "Aspects of digital waveguide networks for acoustic modeling applications," *http://ccrma.stanford.edu/~jos/wgj/*, December 19, 1997.

[466] J. O. Smith and G. Scavone, "The one-filter Keefe clarinet tonehole," in *Proceedings of the IEEE Workshop on Applications of Signal Processing to Audio and Acoustics, New Paltz, NY*, (New York), IEEE Press, Oct. 1997.

[467] J. O. Smith and X. Serra, "PARSHL: A program for the analysis/synthesis of inharmonic sounds based on a sinusoidal representation," in *Proceedings of the 1987 International Computer Music Conference, Champaign-Urbana*, Computer Music Association, 1987, Also available as Stanford Music Department Technical Report STAN-M-43. Expanded version available on-line at `http://ccrma.stanford.edu/~jos/parshl/`.

[468] J. O. Smith and S. A. Van Duyne, "Commuted piano synthesis," in *Proceedings of the 1995 International Computer Music Conference, Banff*, pp. 319–326, Computer Music Association, 1995, `http://ccrma.stanford.edu/~jos/pdf/svd95.pdf`.

[469] J. O. Smith, S. Serafin, J. Abel, and D. Berners, "Doppler simulation and the leslie," in *Proceedings of the COST-G6 Conference on Digital Audio Effects (DAFx-02), Hamburg, Germany*, pp. 13–20, September 26 2002, `https://ccrma.stanford.edu/ jos/doppler/`.

[470] T. Smyth, J. S. Abel, and J. O. Smith, "The estimation of birdsong control parameters using maximum likelihood and minimum action," in *Proceedings of the Stockholm Musical Acoustics Conference (SMAC-03), http://www.speech.kth.se/smac03/*, (Stockholm), Royal Swedish Academy of Music, Aug. 2003.

[471] T. Smyth, J. S. Abel, and J. O. Smith, "The feathered clarinet reed," in *Proceedings of the Conference on Digital Audio Effects (DAFx-04), Naples, Italy*, Oct. 2004.

[472] Sondius Development Team, "Sondius sound examples," 1993–1996, `http://ccrma.stanford.edu/~jos/pasp/Sound_Examples.html`.

[473] A. Spanias, T. Painter, and V. Atti, *Audio Signal Processing and Coding*, New York: John Wiley and Sons, Inc., 2007.

[474] J. Stautner and M. Puckette, "Designing multichannel reverberators," *Computer Music Journal*, vol. 6, no. 1, pp. 52–65, 1982.

[475] K. Steiglitz, *A Digital Signal Processing Primer with Applications to Audio and Computer Music*, Reading MA: Addison-Wesley, 1996.

[476] S. S. Stevens and H. Davis, *Hearing: Its Psychology and Physiology*, American Inst. of Physics, for the Acoustical Society of America, 1983, copy of original 1938 edition, `http://asa.aip.org/publications.html`.

[477] T. Stilson, "Forward-going wave extraction in acoustic tubes," in *Proceedings of the 1995 International Computer Music Conference, Banff*, pp. 517–520, Computer Music Association, 1995.

[478] T. Stilson, *Efficiently Variable Algorithms in Virtual-Analog Music Synthesis—A Root-Locus Perspective*, PhD thesis, Elec. Engineering Dept., Stanford University (CCRMA), June 2006, `http://ccrma.stanford.edu/~stilti/`.

[479] T. Stilson and J. O. Smith, "Alias-free synthesis of classic analog waveforms," in *Proceedings of the 1996 International Computer Music Conference, Hong Kong*, pp. 332–335, Computer Music Association, 1996, `http://ccrma.stanford.edu/~stilti/`.

[480] T. Stilson and J. O. Smith, "Analyzing the Moog VCF with considerations for digital implementation," in *Proceedings of the 1996 International Computer Music Conference, Hong Kong*, Computer Music Association, 1996, `http://ccrma.stanford.edu/~stilti/`.

[481] G. Strang and G. J. Fix, *An Analysis of the Finite Element Method*, Englewood Cliffs, NJ: Prentice-Hall, 1973.

[482] J. C. Strikwerda, *Finite Difference Schemes and Partial Differential Equations*, Pacific Grove, CA: Wadsworth and Brooks, 1989.

[483] H. W. Strube, "Linear prediction on a warped frequency scale," *Journal of the Acoustical Society of America*, vol. 68, no. 4, pp. 1071–1076, 1980.

[484] H. W. Strube, "How to make an all-pass filter with a desired impulse response," *IEEE Transactions on Acoustics, Speech, Signal Processing*, vol. 30, no. 2, pp. 326–327, 1982.

[485] H. W. Strube, "Time-varying wave digital filters for modeling analog systems," *IEEE Transactions on Acoustics, Speech, Signal Processing*, vol. 30, no. 6, pp. 864–868, 1982.

[486] R. D. Strum and D. E. Kirk, *First Principles of Discrete Systems and Digital Signal Processing*, Reading MA: Addison-Wesley, 1988.

[487] A. Stulov, "Hysteretic model of the grand piano hammer felt," *Journal of the Acoustical Society of America*, vol. 97, pp. 2577–2585, Apr. 1995.

[488] A. Stulov, "Experimental and theoretical studies of piano hammer," in *Proceedings of the Stockholm Musical Acoustics Conference (SMAC-03), http://www.speech.kth.se/smac03/*, vol. I, pp. 175–178, 2003.

[489] A. Stulov, "Experimental and computational studies of piano hammers," *Acta Acustica united with Acustica*, vol. 91(6), pp. 1086–1097, 2005.

[490] C. R. Sullivan, "Extending the Karplus-Strong algorithm to synthesize electric guitar timbres with distortion and feedback," *Computer Music Journal*, vol. 14, no. 3, pp. 26–37, 1990.

[491] J. Sundberg, "Synthesis of singing by rule," in *Current Directions in Computer Music Research* (M. V. Mathews and J. R. Pierce, eds.), pp. 45–55, Cambridge, MA: MIT Press, 1989.

[492] J. Sundberg, "Synthesizing singing," in *Representations of Musical Signals* (G. De Poli, A. Picialli, and C. Roads, eds.), pp. 299–320, Cambridge, MA: MIT Press, 1991.

[493] J. Sundberg, *The Science of the Singing Voice*, Dekalb, Illinois: Northern Illinois University Press, 1987.

[494] H. Suzuki, "Model analysis of a hammer-string interaction," *Journal of the Acoustical Society of America*, vol. 82, pp. 1145–1151, Oct. 1987.

[495] T. Tolonen, "Model-based analysis and resynthesis of acoustic guitar tones," Master's thesis, Helsinki University of Technology, Espoo, Finland, Jan. 1998, Report 46, Laboratory of Acoustics and Audio Signal Processing.

[496] T. Tolonen, V. Välimäki, and M. Karjalainen, "Modeling of tension modulation nonlinearity in plucked strings," *IEEE Transactions on Speech and Audio Processing*, vol. SAP-8, pp. 300–310, May 2000.

[497] C. Traube and J. O. Smith, "Estimating the plucking point on a guitar string," in *Proceedings of the International Conference on Digital Audio Effects (DAFx-00), Verona, Italy*, Dec. 2000, `http://www.dafx.de/`.

[498] C. Traube and J. O. Smith, "Estimating the fingering and the plucking points on a guitar string from a recording," in *Proceedings of the IEEE Workshop on Applications of Signal Processing to Audio and Acoustics, New Paltz, NY*, (New York), IEEE Press, Oct. 2001.

[499] L. Trautmann, B. Bank, V. Välimäki, and R. Rabenstein, "Combining Digital Waveguide and Functional Transformation Methods for Physical Modeling of Musical Instruments," in *the AES 22nd International Conference on Virtual, Synthetic and Entertainment Audio*, (Espoo, Finland), pp. 307–316, June 15-17 2002.

[500] L. Trautmann and R. Rabenstein, "Stable systems for nonlinear discrete sound synthesis with the functional transformation method," in *Proceedings of the International Conference on Acoustics, Speech, and Signal Processing, Orlando, Florida, USA*, (New York), IEEE Press, May 2002.

[501] P. P. Vaidyanathan, *Multirate Systems and Filter Banks*, Prentice-Hall, 1993.

[502] V. Välimäki, M. Laurson, and C. Erkut, "Commuted waveguide synthesis of the clavichord," *Computer Music Journal*, vol. 27, pp. 71–82, Spring 2003.

[503] V. Välimäki, *Discrete-Time Modeling of Acoustic Tubes Using Fractional Delay Filters*, PhD thesis, Report no. 37, Helsinki University of Technology, Faculty of Electrical Engineering, Laboratory of Acoustic and Audio Signal Processing, Espoo, Finland, Dec. 1995, `http://www.acoustics.hut.fi/~vpv/publications/vesa_phd.html`.

[504] V. Välimäki, "Discrete-time synthesis of the sawtooth waveform with reduced aliasing," *IEEE Signal Processing Letters*, vol. 12, no. 3, pp. 214–217, 2005.

[505] V. Välimäki and A. Huovilainen, "Antialiasing oscillators in subtractive synthesis," *IEEE Signal Processing Magazine*, vol. 24, pp. 116–125, Mar. 2007.

[506] V. Välimäki and A. Huovilainen, "Oscillator and filter algorithms for virtual analog synthesis," *Computer Music Journal*, vol. 30, pp. 19–31, summer 2007.

[507] V. Välimäki and M. Karjalainen, "Digital waveguide modeling of wind instrument bores constructed of truncated cones," in *Proceedings of the 1994 International Computer Music Conference, Århus*, pp. 423–430, Computer Music Association, 1994.

[508] V. Välimäki and M. Karjalainen, "Improving the Kelly-Lochbaum vocal tract model using conical tube sections and fractional delay filtering techniques," in *Proc. 1994 International Conference on Spoken Language Processing (ICSLP-94)*, vol. 2, (Yokohama, Japan), pp. 615–618, IEEE Press, Sept. 18–22 1994.

[509] V. Välimäki and T. Tolonen, "Development and calibration of a guitar synthesizer," *Journal of the Audio Engineering Society*, vol. 46, Sept. 1998, Proceedings of the 103rd Convention of the Audio Engineering Society, New York, Sept. 1997.

[510] V. Välimäki, M. Karjalainen, and T. I. Laakso, "Modeling of woodwind bores with finger holes," in *Proceedings of the 1993 International Computer Music Conference, Tokyo*, pp. 32–39, Computer Music Association, 1993.

[511] V. Välimäki, T. I. Laakso, and J. Mackenzie, "Elimination of transients in time-varying allpass fractional delay filters with application to digital

waveguide modeling," in *Proceedings of the 1995 International Computer Music Conference, Banff*, pp. 327–334, Computer Music Association, 1995.

[512] V. Välimäki, J. Huopaniemi, M. Karjalainen, and Z. Jánosy, "Physical modeling of plucked string instruments with application to real-time sound synthesis," *Journal of the Audio Engineering Society*, vol. 44, pp. 331–353, May 1996.

[513] V. Välimäki, T. Tolonen, and M. Karjalainen, "Signal-dependent nonlinearities for physical models using time-varying fractional delay filters," in *Proceedings of the 1998 International Computer Music Conference, Michigan*, pp. 264–267, Computer Music Association, 1998.

[514] V. Välimäki, M. Karjalainen, T. Tolonen, and C. Erkhut, "Nonlinear modeling and synthesis of the kantele—a traditional finnish string instrument," in *Proceedings of the 1999 International Computer Music Conference, Beijing*, 1999.

[515] V. Välimäki, H. Penttinen, J. Knif, M. Laurson, and C. Erkut, "Sound synthesis of the harpsichord using a computationally efficient physical model," *EURASIP Journal on Applied Signal Processing*, vol. 7, pp. 934–948, 2004, special issue on Model-Based Sound Synthesis.

[516] V. Välimäki, J. Pakarinen, C. Erkhut, and M. Karjalainen, "Discrete-time modelling of musical instruments," *Reports on Progress in Physics*, vol. 69, no. 1, 2005.

[517] V. Välimäki, J. Nam, J. O. Smith, and J. S. Abel, "Alias-suppressed oscillators based on differentiated polynomial waveforms," *IEEE Transactions on Audio, Speech, and Language Processing*, vol. 18, May 2010.

[518] C. Vallette, "The mechanics of vibrating strings," in Hirschberg *et al.* [197], pp. 115–183.

[519] S. A. Van Duyne and J. O. Smith, "Physical modeling with the 2-D digital waveguide mesh," in *Proceedings of the 1993 International Computer Music Conference, Tokyo*, pp. 40–47, Computer Music Association, 1993, `http://ccrma.stanford.edu/~jos/pdf/mesh.pdf`.

[520] S. A. Van Duyne and J. O. Smith, "Developments for the commuted piano," in *Proceedings of the 1995 International Computer Music Conference, Banff*, pp. 335–343, Computer Music Association, 1995, `http://ccrma.stanford.edu/~jos/pdf/vds95.pdf`.

[521] S. A. Van Duyne and J. O. Smith, "The tetrahedral waveguide mesh: Multiply-free computation of wave propagation in free space," in *Proceedings of the IEEE Workshop on Applications of Signal Processing to Audio and Acoustics, New Paltz, NY*, (New York), pp. 9a.6.1–4, IEEE Press, Oct. 1995.

[522] S. A. Van Duyne and J. O. Smith, "The 3D tetrahedral digital waveguide mesh with musical applications," in *Proceedings of the 1996 International Computer Music Conference, Hong Kong*, Computer Music Association, 1996.

[523] S. A. Van Duyne and J. O. Smith, "The 3D tetrahedral digital waveguide mesh with musical applications," in *Proceedings of the International Computer Music Conference*, (Hong Kong), pp. 9–16, Aug. 1996.

[524] S. A. Van Duyne, J. R. Pierce, and J. O. Smith, "Traveling-wave implementation of a lossless mode-coupling filter and the wave digital hammer," in *Proceedings of the 1994 International Computer Music Conference, Århus*, pp. 411–418, Computer Music Association, 1994, Also presented at the conference of the Acoustical Society of America, Nov., 1994.

[525] M. E. Van Valkenburg, *Introduction to Modern Network Synthesis*, New York: John Wiley and Sons, Inc., 1960.

[526] M. O. Van Walstijn and D. Campbell, "Discrete-time modelling of woodwind instrument bores using wave variables," *Journal of the Acoustical Society of America*, vol. 113, pp. 575–585, 2003.

[527] M. van Walstijn and G. de Bruin, "Conical waveguide filters," in *Proceedings of the International Conference on Acoustics and Musical Research*, (Ferrara, Italy), pp. 47–54, CIARM, May 1995.

[528] M. van Walstijn and G. Scavone, "The wave digital tonehole," in *Proceedings of the 2000 International Computer Music Conference, Berlin*, pp. 465–468, Computer Music Association, 2000, `http://ccrma.stanford.edu/~jos/wdth/`.

[529] M. van Walstijn and J. O. Smith, "Use of truncated infinite impulse response (TIIR) filters in implementing efficient digital waveguide models of flared horns and piecewise conical bores with unstable one-pole filter elements," in *Proc. Int. Symp. Musical Acoustics (ISMA-98), Leavenworth, Washington*, pp. 309–314, Acoustical Society of America, June 28 1998, `http://ccrma.stanford.edu/~jos/tiirts/`.

[530] M. O. van Walstijn, *Discrete-Time Modelling of Brass and Reed woodwind Instruments with Application to Musical Sound Synthesis*, PhD thesis, Department of Physics and Astronomy, University of Edinburgh, 2002, `http://www.ph.ed.ac.uk/ maarten/`.

[531] M. P. Verge, *Aeroacoustics of Confined Jets with Applications to the Physical Modeling of Recorder-Like Instruments*, PhD thesis, Eindhoven University, 1995.

[532] M. P. Verge and A. Hirschberg, "Turbulence noise in flue instruments," in *Proceedings of the International Symposium on Musical Acoustics (ISMA-95), Dourdan, France*, (France), pp. 93–99, Société Français d'Acoustique, July 1995.

[533] M. P. Verge, R. Caussé, and A. Hirschberg, "A physical model of recorder-like instruments," in *Proceedings of the 1995 International Computer Music Conference, Banff*, pp. 37–44, Computer Music Association, 1995.

[534] C. Vergez and X. Rodet, "Model of the trumpet functioning: Real time simulation and experiments with an artificial mouth," in *Proceedings of the International Symposium on Musical Acoustics (ISMA-97), Edinburgh, Scotland*, pp. 425–432, Aug. 1997.

[535] C. Vergez and X. Rodet, "Experiments with an artificial mouth for trumpet," *Proceedings of the International Symposium on Musical Acoustics (ISMA-98), Leavenworth, Washington*, 1998.

[536] C. Vergez and X. Rodet, "Dynamic systems and physical models of trumpet-like instruments analytical study and asymptotical properties," *Acta Acustica*, 2000.

[537] C. Vergez and X. Rodet, "New algorithm for nonlinear propagation of a sound wave application to a physical model of a trumpet," *Journal of Signal Processing*, 2000.

[538] H. L. F. von Helmholtz, *Die Lehre von den Tonempfindungen als Physiologische Grundlage für die Theorie der Musik*, Braunschweig: F. Vieweg und Sohn, 1863.

[539] H. L. F. von Helmholtz, *On the Sensations of Tone as a Physiological Basis for the Theory of Music*, New York: Dover, 1954, English translation of 1863 (German) edition.

[540] J. von Neumann, *Mathematical Foundations of Quantum Mechanics*, Princeton University Press, 1932.

[541] A. Wang and J. O. Smith, "On fast FIR filters implemented as tail-canceling IIR filters," *IEEE Transactions on Signal Processing*, vol. 45, pp. 1415–1427, June 1997.

[542] G. N. Watson, *A Treatise on the Theory of Bessel Functions*, Cambridge University Press, 2 ed., 1944.

[543] J. Wawrzynek, "VLSI models for sound synthesis," in *Current Directions in Computer Music Research* (M. V. Mathews and J. R. Pierce, eds.), pp. 113–148, Cambridge, MA: MIT Press, 1989.

[544] G. Weinreich, "Coupled piano strings," *Journal of the Acoustical Society of America*, vol. 62, pp. 1474–1484, Dec 1977, see also [18] and *Scientific American*, vol. 240, p. 94, 1979.

[545] D. A. Wells, *Lagrangian Dynamics*, New York: McGraw-Hill, 1967, Schaum's Outline series.

[546] E. M. Wenzel, "Localization in virtual acoustic displays," *Presence*, vol. 1, pp. 80–107, Summer 1992.

[547] R. E. Williamson, R. H. Crowell, and H. F. Trotter, *Calculus of Vector Functions*, Englewood Cliffs, NJ: Prentice-Hall, 1972.

[548] J. D. Wise, J. R. Caprio, and T. W. Parks, "Maximum likelihood pitch estimation," *IEEE Transactions on Acoustics, Speech, Signal Processing*, vol. ASSP-24, pp. 418–423, Oct. 1976.

[549] J. Woodhouse, "On the playability of violins. Part I: Reflection functions. Part II: Minimum bow force and transients," *Acustica*, vol. 78, pp. 125–136,137–153, 1993.

[550] J. Woodhouse, "Bowed string simulation using a thermal friction model," *Acta Acustica united with Acustica*, vol. 89, pp. 355–368, 2003.

[551] K. S. Yee, "Numerical solution of initial boundary value problems involving Maxwell's equations," *IEEE Trans. Antennas and Propagation*, vol. 14, pp. 302–307, 1966.

[552] B. Yegnanarayana, "Design of recursive group-delay filters by autoregressive modeling," *IEEE Transactions on Acoustics, Speech, Signal Processing*, vol. 30, pp. 632–637, Aug. 1982.

[553] D. Yeh and J. O. Smith III, "Discretization of the '59 Fender Bassman tone stack," *Proceedings of the Conference on Digital Audio Effects (DAFx-06), Montreal, Canada*, Sept. 2006, `http://www.dafx.de/`.

[554] D. Yeh, J. S. Abel, A. Vladimirescu, and J. O. Smith, "Numerical methods for simulation of guitar distortion circuits," *Computer Music Journal*, 2008, `http://www.mitpressjournals.org/toc/comj/32/2`.

[555] D. Yeh, J. Abel, and J. O. Smith, "Automated physical modeling of nonlinear audio circuits for real-time audio effects—part i: Theoretical development," *IEEE Transactions on Audio, Speech, and Language Processing*, vol. 18, no. 4, pp. 728–737, 2010.

[556] D. T. Yeh, *Digital Implementation of Musical Distortion Circuits by Analysis and Simulation*, PhD thesis, CCRMA, Elec. Engineering Dept., Stanford University (CCRMA), June 2009, `http://ccrma.stanford.edu/~dtyeh/`.

[557] D. T. Yeh and J. O. Smith, "Simulating guitar distortion circuits using wave digital and nonlinear state-space formulations," *Proceedings of the Conference on Digital Audio Effects (DAFx-08), Helsinki University of Technology, Espoo, Finland*, Sept. 1–4 2008, `http://www.dafx.de/`.

[558] M. M. J. Yekta, "Equivalence of the Lagrange interpolator for uniformly sampled signals and the scaled binomially windowed shifted sinc function," *Digital Signal Processing*, vol. 19, pp. 838–842, Sept. 2009.

[559] L. A. Zadeh and C. A. Desoer, *Linear System Theory: The State Space Approach*, New York: McGraw-Hill, 1963, reprinted by Krieger, 1979.

[560] U. Zölzer, *Digital Audio Signal Processing*, New York: John Wiley and Sons, Inc., 1999.

Index

8200630R0

Made in the USA
Lexington, KY
14 January 2011